KU-330-542

# Crustal evolution in northwestern Britain and adjacent regions

*Geological Journal Special Issue No. 10*

GEOLOGICAL JOURNAL SPECIAL ISSUES

No. 1. Controls of Metamorphism
Edited by *W. S. Pitcher and G. W. Flinn*
(out of print)

No. 2. Mechanism of Igneous Intrusion
Edited by *G. Newall and N. Rast*

No. 3. Trace fossils
Edited by *T. P. Crimes and J. C. Harper*

No. 4. Faunal Provinces in Space and Time
Edited by *F. A. Middlemiss, P. F. Rawson and G. Newall*

No. 5. The Boreal Lower Cretaceous
Edited by *R. Casey and P. F. Rawson*

No. 6. Ice Ages: Ancient and Modern
Edited by *A. E. Wright and F. Moseley*

No. 7. Quaternary History of the Irish Sea
Edited by *C. Kidson and M. J. Tooley*

No. 8. A Stratigraphical Index of British Ostracoda
Edited by *R. H. Bate and J. E. Robinson*

No. 9. Trace fossils 2
Editea by *T. P. Crimes and J. C. Harper*

No. 10. Crustal evolution in northwestern Britain and adjacent regions
Edited by *D. R. Bowes and B. E. Leake*

ISSN 0435–3951
ISBN 0 902354 10 8

*Dedicated*

*to*

PROFESSOR T. NEVILLE GEORGE

D.Sc., Ph.D., Sc.D., D.ès-Sc., LL.D., F.R.S., F.R.S.E.

*Views of the new Geology Building, University of Glasgow.*

# Crustal evolution in northwestern Britain and adjacent regions

Proceedings of an International Conference held in Glasgow University,
5-8th April 1977 to mark the opening of the
new Geology Building in the University

Edited by
D. R. BOWES
and
B. E. LEAKE
Department of Geology
University of Glasgow

Series Editor
G. NEWALL

Seel House Press, Liverpool

THE SEEL HOUSE PRESS
Seel Street, Liverpool, L1 4AY

First Edition, August 1978



Printed in Great Britain by
SEEL HOUSE PRESS
Seel Street, Liverpool L1 4AY

# Contents

# Preface

This volume contains papers read at a symposium held in the Department of Geology, University of Glasgow, Scotland, in early April 1977 to mark the opening of the new Geology Building in the University. The formal opening on 4th April 1977, was by Sir Peter Kent, F.R.S. who also delivered the 1977 Celebrity Lecture of the Geological Societies of Glasgow and Edinburgh which was given as part of the symposium.

The subject of the meeting, "Crustal processes and evolution in northwestern Britain and adjacent regions", was chosen because of Glasgow's situation and the involvement by members of the Department in the study of this area, and because northwestern Britain is a classic geological area. The crustal segment considered is of great significance not only in the understanding of the geological evolution of the British Isles and indeed northwestern Europe, but also over an even larger area as a critical part of the jigsaw that was created by the opening of the Atlantic Ocean in Mesozoic times and the closing of the proto-Atlantic Ocean, between Scotland and the Anglo-Welsh area, in early Palaeozoic times. With the new geochronological and other data on the nature of the pre-Caledonian basement in Britain to assist in correlations across the Atlantic Ocean to northeastern America, Newfoundland and to the Grenville basement, and with the volume of new data emerging from geochemical, geophysical, igneous, sedimentary and related studies in the North Atlantic area, by an international group of investigators, the symposium naturally comprised an international group of speakers and attenders, totalling about 200 geologists.

It is not solely because northwestern Britain is in a key position that it is an important geological area, but also because it has been studied in great detail compared with most other regions of the Earth's crust; much of it is superbly exposed and most of it has been mapped and remapped in detail on the 1:10,000 scale being, of course, the cradle of James Hutton's geology. Even so, there are substantial tracts that are little understood and geochronological, geophysical and structural investigations are revealing complexities unthought of until recently. As illustrative of this, research at the associated Isotope Geology Unit of the Scottish Universities Research and Reactor Centre, East Kilbride, was reported by Drs R. T. Pidgeon and M. Aftalion, who showed that some of the 400–500 m.y. old Caledonian granites preserved zircons with histories going back 1500 m.y., conveying cryptic messages about the parent rocks from which the granitic magmas were derived. Such information will lead to further understanding of the nature of the basement beneath the Caledonides and to the origin of the granite magmas themselves. Both of these subjects were further dealt with by other contributors.

Dr. H. Williams of the Memorial University of Newfoundland gave a stimulating keynote address in which he examined the available information from Newfoundland and the northeastern Appalachians and made some comparisons with the British Isles.

Plate tectonic models proposed for the British Isles are conflicting. Nevertheless, all of them suppose that a plate margin originally crossed part of Scotland, and the Girvan district features in many of the hypotheses. This area was considered in detail, mostly by contributors from the Glasgow Department, much new factual geological information being reported and illustrated in the course of a field excursion which was part of the symposium.

Off-shore, the geology of the seas around northwestern Britain, Ireland, and Norway is the subject of intensive exploration in the search for oil and gas. The off-shore geology, it must be admitted, has greatly surprised geologists because the thickness of Mesozoic and Tertiary rocks is far in excess of that found on the adjacent land, a situation that reveals the fallacy of supposing that the coastline of Britain is an ephemeral erosive feature of no deep-seated significance.

This book presents the accounts received for publication on all these aspects, from the formation of the Lewisian basement to present-day sedimentation in the North Atlantic area and volcanic activity in Iceland. The book is not intended to be either a definitive history of the region or a balanced account of each part of the long and complex history of the crust in this region. Rather the volume is designed to highlight particular aspects in which modern work has led to significant advances in our knowledge both of the history of this well-studied region and the processes involved in crustal formation and modification, in the belief that some of this knowledge will have wide application elsewhere.

It was appropriate that at a symposium that marked the opening of the first specifically designed Geology Building in the University, the subject chosen should have been one in which Glasgow geologists have made notable contributions over a period of many years. Many of the scientists concerned were internationally famous for their advancement of geology either while in Glasgow or later, and at a time when the Department of Geology looks forward to a new period of development, a brief mention of the past researches of geologists in the University, particularly those concerned with northwestern Britain, and of the development of the study of geology in the University, is appropriate.

In the first 400 years of the University's existence (the foundation was in 1451) geological knowledge had not advanced sufficiently to justify its inclusion as a formal part of the University curriculum. With the establishment of the Regius Chair of Natural History by King George III in 1807, geology began to make its appearance in the University with the appointment to that Chair of, first Lockhart Muirhead in 1807, then William Couper (1829) and particularly in 1857 when Henry Darwin Rogers was appointed. Rogers had been a geologist in the State of Virginia (1835), Director of the Geological Survey of New Jersey and also State Geologist of Pennsylvania. These early occupants of the Chair of Natural History, although possessing as much expertise in geology and mineralogy as in zoology or botany, published little. Nevertheless, the University continued to employ geologically minded Professors of Natural History with the appointment of John Young (who had been mapping in the Southern Uplands) from the Geological Survey in 1866.

During this period the University's main contributions to geological advancement came from Sir William Thomson, F.R.S. (Lord Kelvin from 1872), Professor of Natural Philosophy, whose attempts to calculate the age of the Earth sparked off vigorous controversy and whose interest and involvement in geology are well summarised by the fact that he was elected President of the Geological Society of Glasgow for the unprecedented period of just over 21 years (1872–1893) being succeeded as President by Sir Archibald Geikie, F.R.S. (1893–1899) and then Professor Charles Lapworth, F.R.S. (1899–1902).

It was not until 1903 that a separate Chair and Department of Geology were founded, partly with funds from the Carnegie Trust, and the dynamic J. W. Gregory held the first appointment in 1904. Professor Gregory, F.R.S. (1904–1929), was the first to recognise the origin of the great East African Rift Valley, which modern work, like that on the Midland Valley of Scotland, has shown to be far more complex than simplistic models originally suggested. His habitual travelling—East Africa, Libya, Angola, Australia, India, Spitzbergen, Tibet, the Rocky Mountains and finally the Andes—gave him an encyclopaedic view of geology that was expressed in a profuse outpouring of published work, ranging from local glacial geology to global tectonics. His views, set out in over 300 papers and 20 books, cannot, in fairness, be said to have been always correct or generally accepted; thus his trenchant views on the permanency of the oceans and continents have not proved enduring. Nevertheless, his global thinking, his love of field work, his persuasive lecturing, his stimulating interest and his sheer vitality served to establish the new Department in a grand way and to inspire his students and colleagues to vigorous investigations and travels well exemplified by J. V. Harrison whose work after leaving Glasgow in Persia, Borneo, Colombia, Honduras, Mexico, Venezuela, Trinidad and above all Peru, has all the stamp of Gregory's exploratory mind.

Gregory's assistant, G. W. Tyrrell, who served the Department from 1906 to 1954, was the foil to Gregory. While his investigations were on rocks from all over the world, including Spitzbergen, East Greenland and Antarctica, Tyrrell laid a foundation of detailed information about the igneous rocks of the region around Glasgow that pointed towards many of the concepts of petrogenesis that have emerged in recent years. His work on the igneous rocks of the Midland Valley, the Lugar Sill, his memorable work on the Island of Arran, his best seller *The Principles of Petrology* (1926: still in print 1977) and his inspired teaching of petrology were among the substantial contributions to petrology that made the name of G. W. Tyrrell world famous.

Perhaps the greatest advancement of our understanding of Dalradian tectonics came from Professor E. B. Bailey, F.R.S. (1929–1937) who succeeded Gregory. Bailey brought a vast experience of mapping in the Geological Survey, the ready ability to synthesise the results of detailed mapping over wide areas and, like Gregory, a love of field work. Of course, field work had to be conducted in short trousers and a walk through a river at the beginning of a day meant that any subsequent rain was hardly noticed! His interests were concentrated around the regional structure of the Caledonides extending from Scandinavia into Ireland, but he also worked on igneous rocks, the concept of primary magmas, and sedimentary and geomorphological aspects of geology. His period in Glasgow, before he returned to the Geological Survey of Great Britain as Director, enabled him to complete his synthesis of the structure of the Tay nappe. Contributions to this work came from W. J. McCallien, for 20 years (1922–1942) a member of the Department, and A. Allison whose investigations in Inishtrahull, Antrim, Inishowen and the south-western Highlands, allowed the firm correlation of the Scottish and Irish Precambrian and Dalradian rocks, while subsequently J. G. C. Anderson continued in the same field. This period of structural studies on a grand scale, that became possible following the comprehensive detailed mapping of the Highlands, continued in a less sweeping manner with the work of B. C. King, N. Rast and T. C. R. Pulvertaft (in Donegal) during their times in Glasgow. More recently, although work has continued on the Dalradian rocks of the Highlands, especially under D. R. Bowes, there has been a marked shift towards the petrological and structural study of the Lewisian basement, mainly due to the work of D. R. Bowes, but involving various workers at one time in the Department, such as R. G. Park and A. E. Wright. Joint studies with the Isotope Geology Unit of the Scottish Universities Research and Reactor Centre at East Kilbride have contributed to the elucidation of the chronological history of the Lewisian complex and its place in crustal evolution. Meantime, work on the Dalradian moved to Ireland, particularly to Connemara where A. Rothstein pursued his studies of the Dawros peridotite, B. W. Evans examined partial melting near ultrabasic intrusions and B. E. Leake, the present occupant of the Chair of Geology, has worked for over 25 years.

The major field of continuing study in the Department has been that of stratigraphy and palaeontology, investigated by a long succession of staff and students including Ethel D. Currie (1922–1962), John Weir (1925–1959), R. M. C. Eagar, W. E. Swinton and particularly Professor A. E. Trueman, F.R.S. (1937–1946) who followed Bailey into the Chair, Trueman's successor, Professor T. N. George, F.R.S. (1947–1974), and A. Williams, F.R.S. (1950–1954; now Principal of the University). Both Trueman and George pursued long-continued palaeontological, structural and stratigraphical studies into the Carboniferous period, both possessed a deep interest in the geomorphological evolution of Britain and were able to paint a very broad geomorphological canvas and communicate these results intelligibly to the general public. They encouraged and inspired numerous students and members of the staff in studies of Carboniferous stratigraphy and palaeontology, including D. Leitch, L. R. Moore, R. H. Cummings, K. A. G. Shiells, R. B. Wilson, O. Dixon, G. M. Dunlop and G. Y. Craig, the last of whom branched out with pioneer palaeoecological studies.

The rocks of the Girvan–Ballantrae area, of which detailed accounts are included in this volume, and the Southern Uplands, have puzzled and occupied the attention of many geologists at one time in the Department, including E. B. Bailey and W. J. McCallien, T. W. Bloxam, A. Williams, S. M. K. Henderson and E. K. Walton. From this ground came early turbidite and Steinmann Trinity studies and the complexity of the rocks still defies a convincing synthesis.

Accordingly, from the far northwest, through the present off-shore geological investigations involving both geophysics and sedimentology that are currently underway in the Department, to the geology of the Highlands, the Midland Valley and the ground to the south, the Department has been deeply involved in varied studies over many years.

Closely associated with the Department, but regrettably now geographically separated, is the Hunterian Museum, the basis of which was bequeathed to the University by William Hunter in 1783, with a munificent gift of books and manuscripts, coins, paintings, archaeological and ethnographical material, geological, zoological and botanical specimens and an extensive series of anatomical preparations. The staff of the geological section are also members of the Geology Department and some of their researches have appeared in the Monographs of the Hunterian Museum.

At one time temporary junior appointments to the teaching staff ensured constant staff change. This system has now disappeared, but a Glasgow feature that continues is that of very large (for Britain) first year classes in geology, commonly 200–300, most of the members of which study geology as an ancillary subject for wider educational purposes rather than for professional training. Nevertheless, over the last 30 years at least 15 former members of the Department have become university professors, six others have become directors of national geological surveys, and the fields of oil and gas exploration, coal geology, mineral exploration and mining have all received significant contributions to their staff from Glasgow graduates.

The new Geology Building contains five floors devoted to research laboratories, workshops, staff rooms and offices with an attached two storey teaching block (see frontispiece). It is equipped for geochemical, mineralogical, palaeontological, sedimentological and geophysical work with normal supporting facilities.

The main credit for negotiating the construction of this new building must undoubtedly be given to the indefatigable Professor Emeritus Thomas Neville George, D.Sc., Ph.D., Sc.D., D.-ès-Sc., LL.D., F.R.S., F.R.S.E., who was Head of the Department for 27 years (1947–1974), and whom it is a privilege to have as a continuing member of the Department and to whom this book is dedicated. Much of the work reported to this symposium by members of the Department was carried out under his Headship. His particular fields of research in no way adequately represent the breadth of his interests. This meant that he was able to lead the Department through the transitional period from the now traditional views of geology into the exciting new concepts and developments which this symposium seeks to consider in this superb geological segment of the Earth's crust.

The editors would like to thank the Series Editor, Dr. G. Newall, and Mr. G. Wilding of Seel House Press, for their continued help in the production of this volume. The cover design is by Dr. G. Newall and Mr. J. Lynch.

Bernard E. Leake

Head of Department,
Department of Geology,
University of Glasgow,
Glasgow G12 8QQ, Scotland

# *Part 1*

# Introduction

# Geological development of the northern Appalachians: its bearing on the evolution of the British Isles

*H. Williams*

The northern Appalachians and the British Caledonides represent the now widely separated central parts of a Palaeozoic orogen that was continuous before the opening of the present North Atlantic Ocean. During its evolution, the northern extremity of the Appalachians in Newfoundland and the western extremity of the Caledonides in Ireland were separated only by the combined width of the modern continental shelves, approximately 700 km.

The western margin of the Appalachians, embracing rocks equivalent to those of the Hebridean foreland and Dalradian Supergroup, is interpreted in terms of an evolving stable continental margin. Destruction of this margin in the Canadian crustal segment led to the emplacement of classic ophiolite suites that now occupy a structural position analogous to ophiolite complexes of Oman, Zagros and the Himalayas. Equivalents in Newfoundland of the Dalradian Supergroup overlie reworked basement gneisses and include ophiolitic mélanges and dismembered ophiolite complexes in easterly exposures that are affected by the full range of local deformation and metamorphism. The relationships indicate that this orthotectonic zone formed during or after earliest ophiolite transport. Mafic-ultramafic complexes of the Shetland Islands and elsewhere in Scotland may represent local analogues to Appalachian transported ophiolite suites.

Central parts of the orogen in Canada include lower Ordovician rocks interpreted as *in situ* oceanic crust overlain by thick volcanic arc sequences that are equated with rocks in the Girvan–Ballantrae district of Scotland. Intrusion of granite, metamorphism and penetrative deformation accompanied the early Ordovician evolution of the volcanic centres that were then far removed from adjacent continental margins. Everywhere these central parts of the orogen were intruded and deformed during later events, but rarely to the same intensity as rocks at the destroyed continental margin (equivalent of the Dalradian Supergroup). The Dunnage mélange in central Newfoundland possibly represents the vestige of an ancient oceanic trench to the east of an island arc. The Gander Group and its underlying basement to the east appear to be equivalents of the Bray Series and Rosslare complex of Ireland. Locally, the central volcanic terrane of the Appalachian orogen is entirely missing in narrow constrictions within the system where complete suturing has occurred.

The eastern margin of the Appalachians has extensive, thick upper Precambrian volcanic and sedimentary accumulations (analogues of the Uriconian and Longmyndian rocks of England) overlain by a full complement of Cambrian strata. Late Precambrian fossils, similar to forms in the Charnian assemblage of England, are common in eastern Newfoundland and the Cambrian succession with its Atlantic realm trilobites is similar in detail to the Cambrian succession of Wales. An extensive thick tillite unit occurs toward the base of the Precambrian succession in

Newfoundland and it was contemporaneous, in part, with subaerial volcanism. The evolution of the eastern margin of the Appalachians and the significance of its rocks and structures are presently less well-known compared with central and western parts of the orogen. The late Precambrian development of the eastern margin may belong to an earlier orogenic cycle unrelated to that which involved the evolution of western parts of the system.

## 1. Introduction

All the main structural elements of the British Caledonides, from the Hebridean foreland southwards across the deformed zone of Dalradian rocks and the Midland Valley to the Precambrian and Cambrian rocks of the Welsh Borderlands, can be matched with similar elements in the northern Appalachians. In fact, the geology of Newfoundland at the northern extremity of the Appalachian orogen, is more akin to that of the British Isles than it is to parts of the southern Appalachians. Obviously, the interpretation of the geological developments of the Appalachians on one side of the present North Atlantic Ocean bears heavily upon the interpretation of crustal processes in the Caledonides on its opposite side.

The Appalachian orogen extends 3500 km southwestwards from Newfoundland along the Atlantic seaboard to Alabama in the southeast United States. It is continuous in subsurface with the Ouachita System of Arkansas, and it extends to southern Mexico. The British Caledonides have natural continuations northward through East Greenland and Scandinavia and beyond to Spitzbergen. Before the opening of the present oceans, the Appalachian–Caledonian orogen had a combined length, therefore, of about 10,000 km, with extremities undefined. The extent of this orogen is comparable with that of the present North American Cordillera, the Andes of South America, or the Alpine fold belt of the Mediterranean region.

Insular Newfoundland at the northeastern extremity of the Appalachian system displays a virtually complete, well-exposed cross-section through the orogen along its highly indented northeastern coastline. The superb exposures of lower Palaeozoic rocks there permit a searching scrutiny of major tectonic elements and fundamental junctions among contrasting rock groups. Unfortunately no comparable cross-section exists for the British Isles, and the important early Palaeozoic elements of the orogen are hidden in large part by post-orogenic cover sequences.

Apart from past geological contiguity and present maritime setting, the understanding of the geology of the northern Appalachians and British Caledonides has benefited because of historic cultural ties between eastern Canada and the British Isles. The first geological survey of Newfoundland was conducted by a young Irishman, J. B. Jukes (1842) during the year 1839 and 1840. The father of Canadian geology, Sir William Logan, while born in Montreal was educated in Britain. And in more recent years there has been a steady exchange of geologists between the British Isles and North America. Not unexpectedly therefore, some geological interpretations and conceptual models for the Appalachians draw heavily upon experience gained in the British Caledonides. The geological experience of the author is strictly North American so that slight differences in outlook and shifts in emphasis may be discernible in some of the comparisons and interpretations that follow.

The Appalachian–Caledonian orogen is composed mainly of upper Precambrian and Palaeozoic rocks that are sharply contrasted in thickness, facies, and structural style in different parts of the system. The most recent geological syntheses are based upon the recognition of contrasting tectonic–stratigraphic zones across the

Table 1. Zonal subdivisions of Appalachians and Caledonides

| NORTHERN APPALACHIANS | | BRITISH CALEDONIDES | | |
|---|---|---|---|---|
| Present zones | Williams *et al.* 1972, 1974 | Present zones | Dewey 1974 | Phillips *et al.* 1976 |
| HUMBER | A (Lomond)<br>B (Hampden)<br>C (Fleur de Lys–west) | HEBRIDES | A<br>B | A<br>C |
| DUNNAGE | C (Fleur dy Lys–east)<br>D (Notre Dame)<br>E (Exploits)<br>F (Botwood) | DUNDEE | C<br>D | D<br>E<br>F |
| GANDER | G (Gander) | GREENORE | E, F | F, G |
| AVALON | H (Avalon) | ANGLESEY | F, G, H, I, J, K | H |
| | I (Meguma) | | | |

system. Nine zones are defined in the Canadian Appalachians (designated alphabetically A to I; Williams *et al.* 1972, 1974), and eleven comparable zones are defined in the British Caledonides (designated A to K; Dewey 1974). Although a sophisticated subdivision into a multitude of zones may serve to elucidate local contrasts, long range comparisons are possible only if the orogen is divided into a few broad zones. For this reason the northern Appalachians has been divided by Williams (1976) into four major divisions and each has been extrapolated over the full length of the system from Newfoundland to Alabama. Local names are applied to these broad divisions in Newfoundland, which from west to east are Humber, Dunnage, Gander and Avalon (Table 1). A similar four-fold subdivision is recognizable in the British Caledonides, and for the purposes of descriptions and comparisons that follow, each British zone is assigned a local name chosen purposely with phonetics similar to that of the name of the comparable zone in Newfoundland, viz: Hebrides (Humber), Dundee (Dunnage), Greenore (Gander) and Anglesey (Avalon). These names are introduced merely as convenient geographic references, for experience has shown that it is easier to mentally locate or remember a local geographical name than a letter or number. An additional zone, the Meguma zone of Nova Scotia, lies to the east of the Avalon zone in the northern Appalachians. It has no apparent analogue in the British Caledonides unless its Ordovician and older greywackes and slates are equivalent to some rocks of the Welsh basin. If so, they may represent cover sequences upon upper Precambrian rocks of the Avalon zone.

The major divisions, which imply gross similarities between the northern Appalachians and British Caledonides are outlined in Figure 1 (fold out). The geological development of these major divisions and comparisons between the northern Appalachians and British Caledonides form the basis of this contribution. No attempt is made to summarize and synthesize all of the pertinent geological data that bear upon the geological evolution of the Appalachians and British Caledonides.

Instead, a brief description of the tectonic elements that constitute each zone is followed by broad comparisons between the Appalachians and Caledonides and then discussion of some specific problems that bear upon conceptual evolutionary models. Emphasis is placed on contrasts in the records of Cambrian and Ordovician times within the zonal subdivisions of the orogen, rather than on development in Silurian and later times.

Only a few recent references are cited in the text, especially with respect to the geology of the British Isles. For more complete bibliographies on the geology of the British Caledonides the reader is referred to Johnson and Stewart (1963), Craig (1965), Rast and Crimes (1969) and Dewey (1974); and for more complete bibliographies on Appalachian geology to Zen *et al.* (1968), Kay (1969), Poole *et al.* (1970), Rodgers (1970) and Williams *et al.* (1972, 1974).

## 2. Definition and broad comparison of Appalachian zones in Newfoundland with those of the British Caledonides

### 2a. Humber zone

The Humber zone (as summarized in Stevens 1970; Bird and Dewey 1970; Williams *et al.* 1972, 1974; Williams and Stevens 1974), consists of a crystalline Grenvillian basement overlain by clastic sedimentary rocks (e.g. Bradore Formation, Bateau Formation) and a prominent Cambrian–Ordovician carbonate sequence (e.g. St George Group, Table Head Formation) in the west, and polydeformed clastic sedimentary rocks with local volcanic rocks (Fleur de Lys Supergroup) upon a reworked Grenvillian basement in the east. The relatively undeformed carbonate sequence is overlain by easterly derived clastic rocks (e.g. Goose Tickle Formation), mélange and a variety of transported sequences (e.g. Humber Arm Supergroup, Cow Head Group, Maiden Point Formation, Skinner Cove Volcanics, Little Port complex) capped by an ophiolite suite (e.g. Bay of Islands complex, White Hills peridotite). Towards the east, the carbonate sequence is imbricated by thrust faults and progressively deformed and metamorphosed (e.g. western White Bay) and it is followed further east by the polydeformed and metamorphosed Fleur de Lys Supergroup. The eastern margin of the Humber zone is drawn at the Baie Verte lineament (Williams *et al.* 1977), a steep structural belt marked by deformed ophiolites (e.g. Advocate complex, Point Rousse complex), and mafic volcanic rocks (Flatwater Group). The main structures in the Humber zone are of early to middle Ordovician (Taconic) generation as indicated by local sub-Silurian (Sops Arm Group) and sub-Caradocian (Long Point Formation) unconformities. Easterly parts of the Humber zone were also involved in Devonian (Acadian) and local Carboniferous (Alleghanian) deformation, and Acadian granites cut the deformed Fleur de Lys Supergroup.

The Humber zone of Newfoundland extends the full length of the Appalachian system to Alabama of the southeastern United States (Fig. 1). Grenvillian inliers can be traced all the way along the western margin of the Appalachians through the Virginia Blue Ridge to Georgia, where they are overlain everywhere by thick clastic sequences similar to those of the Fleur de Lys Supergroup and less deformed equivalents found in the north. Likewise the carbonate sequence can be traced from western Newfoundland to the Valley and Ridge Province of the southern Appalachians with local correlations extending to details. As well, transported sequences with associated mélange occupy much of the Quebec-Gaspé segment of the Appalachians and continue through the Taconic Ranges of New York to the Hamburg klippe of Pennsylvania. Middle Ordovician (Taconic) orogenesis is recognized as the first major event throughout most of this zone wherever Silurian cover rocks or isotopic data allow its definition. Later events are of less regional significance. Ophiolite suites along the eastern margin of the Humber zone (Baie Verte lineament) can be traced from Newfoundland to the eastern tip of Gaspé Peninsula. Further to the southwest the Ordovician structures are hidden by Silurian and Devonian cover rocks, but the ophiolite suites reappear in the Eastern Townships of Quebec, where with Newfoundland examples, they form the world's richest asbestos belt. From there a continuous string of small ultramafic plutons extend through West Vermont to Staten Island, New York and to the transported Baltimore gabbro complex of Maryland. Further south, ultramafic rocks occur in the Shores mélange (Brown 1976) of Virginia and throughout the metamorphic rocks of the eastern Blue Ridge. The Brevard zone of the Carolinas, Georgia and Alabama possibly represents a suture zone and a continuation of the eastern limit of the Humber zone in the extreme southern Appalachians.

## 2b. Hebrides zone

In Scotland, the basement to the Hebrides zone (as summarized in Johnson and Stewart 1963; Craig 1965 and Dewey 1974) is represented by the Lewisian complex (Bowes 1978), except in West Ireland where gneisses and granites dated at 1100–1000 m.y. are related to Grenville rocks of Canada (van Breemen *et al.* 1978) and between the Moine thrust and Great Glen fault of Scotland where some rocks within the outcrop of the Moine assemblage have been dated at *c.* 1000 m.y. (Brook *et al.* 1976) and others at 800–700 m.y. (van Breemen *et al.* 1978). In the Hebrides zone, some metamorphic rocks of the Moine assemblage and the rocks of the Dalradian Supergroup (Harris *et al.* 1978) compare with the Fleur de Lys terrane of the Humber zone (Church 1969; Kennedy *et al.* 1972) and the Durness sequence in Scotland is a direct analogue of the carbonate terrane of western Newfoundland (Swett and Smit 1972). The Highland Boundary fault, which delineates the Hebrides zone to the south and is marked by a discontinuous belt of metavolcanic rocks, cherts and serpentinites, appears to be a natural continuation of the Baie Verte lineament. Conspicuously absent in the Hebrides zone are transported sequences and ophiolotes like those so common in western Newfoundland. Thrust imbricated mafic–ultramafic complexes of the Shetland Islands (Flinn 1958) may afford a local example. If so, they possibly root to the south of the Hebrides zone in the vicinity of the Highland Boundary fault.

## 2c. Dunnage zone

The Dunnage zone (as summarized in Williams *et al.* 1972, 1974) is recognized above all by its preponderance of mainly mafic volcanic rocks with associated cherts, slates, greywackes and minor limestone. These rocks are of late Cambrian to middle Ordovician age and they contrast sharply with rocks of the same age in the Humber zone to the west. Stratigraphic sections exceed 7 km in thickness with rapid facies transitions that make correlation difficult. The zone includes an extensive mélange in northeastern Newfoundland consisting of a variety of volcanic and sedimentary blocks, some up to 1 km in width, set in a black and green shale matrix (Dunnage mélange—Kay 1976). Nearby to the east, a similar mélange at the eastern margin of the zone contains local ultramafic blocks (Kennedy and McGonigal 1972).

East of the Baie Verte lineament at the western margin of the Dunnage zone, volcanic sequences conformably overlie the ophiolite suite of rock units (Point Rousse complex, Betts Cove complex). The volcanic sequences pass upwards into Caradocian black shales that are common throughout this zone. Toward the east, similar black shales overlie a Dunnage mélange basement and they are succeeded by greywackes and conglomerates that constitute an upward shoaling sequence.

Rocks of the Dunnage zone in Newfoundland are much less deformed compared with nearby parts of the Humber and Gander zones, except locally where the Twillingate granite and older mafic volcanic rocks are in amphibolite facies and surrounded by relatively undeformed and unmetamorphosed mafic volcanic rocks. Along the western margin of the Dunnage zone, ophiolite suites (e.g. at Nippers Harbour), nearby mafic volcanic assemblages (Snooks Arm Group) and granitic intrusions (Burlington granodiorite) are all overlain unconformably by Silurian–Devonian rocks of the Cape St John and Mic Mac Lake groups. The eastern margin of the Dunnage zone includes a linear belt of discontinuous mafic-ultramafic plutons, and locally (Gander Lake), these are overlain unconformably by conglomerates at the base of a sequence of Caradocian black shales (Davidsville Group).

The Dunnage zone of northeastern Newfoundland is absent at the southwestern corner of the island where opposing gneissic terranes of the Humber and Gander zones are separated only by a narrow belt of mylonite (Cape Ray suture). In mainland North America, Ordovician volcanic correlatives of rocks of the Dunnage zone are represented in New Brunswick (Tetagouche Group), Maine (Shin Pond Formation) and from New Hampshire to Vermont (Ammonoosuc and Partridge formations), extending to Long Island Sound. Possible correlatives still further south include the James Run Formation of Maryland, the Chapawamsic–Quantico formations of northern Virginia, and volcanic rocks of the Evington Group in central Virginia. The Tetagouche Group of New Brunswick and southward correlatives in Maine and Vermont all overlie sedimentary rocks that are underlain in turn by metamorphic rocks (Miramichi anticlinorium in New Brunswick, Oliverian gneiss domes of Vermont). It would appear therefore that while the volcanic rocks of the Dunnage zone are locally underlain by the ophiolite suite in westerly exposures, easterly examples overlie a continental basement.

## 2d. Dundee zone

In the Dundee zone of Scotland and Ireland, the Ballantrae complex (Bluck 1978), Highland Border Series and Ordovician rocks of the Lough Nafooey Group are possible correlatives of ophiolite complexes and volcanic rocks of the Dunnage zone. The northwestern boundary of the Dundee zone in Scotland is drawn at the Highland Boundary fault, but this boundary is more difficult to define in western Ireland where the possibly equivalent Clew Bay fault is bordered to the south by the Dalradian rocks of Connemara, a metamorphosed sedimentary assemblage that would fit more comfortably within the Hebrides zone. The southeastern boundary of the Dundee zone is also nebulous and is largely hidden by Devonian and younger cover rocks where tentatively located at the Solway Firth.

## 2e. Gander zone

The Gander zone of Newfoundland consists mainly of pre-middle Ordovician arenaceous rocks (Gander Group) that are in most places polydeformed thus resembling, in structural pattern, the Fleur de Lys Supergroup on the opposite side of the Dunnage zone (Kennedy and McGonigal 1972). Towards the east, migmatities, granitic gneisses and some foliated megacrystic granites are all thought to represent basement to the polydeformed arenaceous cover sequence. Palaeozoic megacrystic biotite granites and garnetiferous muscovite leucogranites and associated pegmatites are everywhere common throughout this zone. The western boundary of the Gander zone is marked by a black shale mélange in northeastern Newfoundland and elsewhere it is drawn at the contact between Caradocian black shales of the Dunnage zone and the Gander Group. Arenites and basement gneisses of the Gander zone may therefore underlie extensive eastern areas of the Dunnage zone in Newfoundland. The eastern boundary of the Gander zone is drawn at the Dover fault (Blackwood and Kennedy 1975) in northeastern Newfoundland that separates metamorphic rocks to the west from little deformed and essentially unmetamorphosed upper Precambrian sedimentary and volcanic rocks of the Avalon zone to the east.

The Gander zone, though well-defined in Newfoundland, is difficult to recognise on existing geological maps over large parts of the Appalachian orogen to the south. Rocks possibly equivalent to the Gander Group occur in the Miramichi region of New Brunswick where they are possibly underlain by basement metamorphic complexes. The Cookson Group (New Brunswick) and the Grand Pitch and Ellsworth groups (Maine) may be further correlatives of the Gander Group. Basement gneisses in Maine (Cushing, Chain Lakes complex) may be equivalent to Gander zone basement in Newfoundland and some gneisses of the Oliverian domes may also equate to Gander zone basement. Other gneissic complexes with associated foliated granites are common in a comparable position at Fredericksburg in Virginia and further south through the Charlotte belt of the Appalachian Piedmont in North Carolina.

## 2f. Greenore zone

In the Greenore zone of southeastern Ireland, parts of the Cullenstown Group and the Bray Series may be Gander Group equivalents. Nearby gneisses and metamorphosed intrusions of the Rosslare complex possibly correlate with basement rocks of the Gander zone in Newfoundland. Ordovician volcanic rocks above the Cullenstown Group may represent a similar situation to that of the Tetagouche Group in New Brunswick. In northern England, the Manx and Skiddaw slates may also correlate with parts of the Gander Group.

## 2g. Avalon zone

The Avalon zone is composed mainly of upper Precambrian volcanic and sedimentary rocks that are relatively unmetamorphosed and in places they pass upwards into Cambrian and lower Ordovician shales and sandstones. Locally the Precambrian rocks are cut by granite (Holyrood) that is nonconformably overlain by gently dipping Cambrian shales. Underwater sampling of the Virgin Rocks and Eastern Shoals in the eastern part of the Grand Banks (Lilly 1965) and further east at Flemish Cap (Pelletier 1971) has revealed rocks of late Precambrian age like those of the Avalon zone. If these rocks are continuous in subsurface beneath the Grand Banks, then the Avalon zone is far wider than the combined width of other zones in the Appalachian orogen.

The upper Precambrian rocks of the Avalon zone in Newfoundland can be broadly subdivided into three groups: (1) a basal assemblage of predominantly subaerial volcanic rocks but including some sedimentary units (Harbour Main and Love Cove groups), (2) an intermediate assemblage of marine siliceous slates and greywackes (Conception Group) with local volcanic interlayers and including an important tillite unit near its base (Gaskiers Formation) and beds containing abundant Precambrian fossils at its top (Mistaken Point Formation), and (3) an upper assemblage of sedimentary rocks that in places includes a thick volcanic unit near its base and everywhere includes large amounts of arkose, red sandstone and conglomerate of shallow water continental facies.

The upper Precambrian beds are overlain by a white quartzite unit (Random Formation), in turn followed by lower Cambrian fossiliferous shales with Atlantic trilobite faunas. The Cambrian shales are remarkably uniform throughout the zone and represent a fairly complete history of deposition. Locally, these are overlain by Ordovician and younger rocks.

No crystalline basement rocks are known in the Avalon zone of Newfoundland, except possibly for local exposures at the northern tip of the island of Miquelon. Elsewhere at Cape Breton Island and southeastern New Brunswick, marbles and quartzites with associated gneisses (George River and Green Head groups) underlie upper Precambrian assemblages equivalent to those in Newfoundland (Rast *et al.* 1975).

Upper Precambrian and Cambrian rocks like those of the Avalon zone in Newfoundland can be traced with confidence to eastern Massachusetts in the United States. Similar rocks reappear further south in the Slate Belt of Virginia, the Carolinas and Georgia. Marbles and quartzites, possibly equivalent to basement rocks in the Avalon zone further north, occur in the Kings Mountain Belt of North Carolina and further south in the Pine Mountain Belt of Georgia and Alabama.

### 2h. Anglesey zone

The Anglesey zone of the British Caledonides contains upper Precambrian rocks like those of the Avalon zone in Newfoundland. The Arvonian volcanic rocks of North Wales and the Uriconian volcanic rocks and Longymyndian assemblages of England appear to be direct correlatives. In addition, some of the Anglesey zone upper Precambrian rocks at Charnwood Forest contain frond-like organisms almost identical to those of the Mistaken Point Formation in eastern Newfoundland. The Cambrian succession of Wales is a further direct analogue of Cambrian rocks of the Avalon zone in Newfoundland, and both are characterized by the same European realm trilobite faunas. Small basement inliers in the Anglesey zone (Rushton Schists, Malvernian gneisses, Mona complex) may be equivalent to inliers elsewhere in the Avalon zone of the Appalachian orogen. More distant correlatives occur in the Pentevrian and Brioverian rock units of Brittany.

## 3. Possible relationship of zonal subdivision of the Appalachian–Caledonian orogen to the evolution of a late Precambrian–early Palaeozoic (Iapetus) ocean

Since the advent of the concept of plate tectonics, models for the development of the Appalachians and Caledonides follow the suggestion of Wilson (1966) and involve the generation and destruction of a late Precambrian–early Palaeozoic Iapetus ocean (Dewey 1969; Bird and Dewey 1970; Stevens 1970; McKerrow and Ziegler

1971; Williams and Stephens 1974; Phillips *et al.* 1976). Early stages of development relate to rifting of continental crust, the build-up of thick clastic sedimentary accumulations at the juvenile continental margins, the initiation of spreading centres, and the generation of oceanic crust. Intermediate development involved continued spreading and the evolution of its stable continental margins. Final development embraced plate convergence, the disappearance of large tracts of oceanic crust, island-arc volcanism, and the destruction of its continental margins.

The tectonic evolution of the northern Appalachians is viewed in a simplistic way as related to the opening and then closing of a late Precambrian–early Palaeozoic Iapetus ocean. The Humber zone records the development and destruction of an Atlantic-type continental margin, the ancient continental margin of eastern North America, along the western margin of Iapetus. Ordovician rocks of the Dunnage zone are vestiges of oceanic crust, island-arc sequences and subduction related mélanges that originated in the oceanic domain. The Gander and Avalon zones developed upon continental crust that clearly lay to the east and southeast of the Iapetus ocean. The Gander zone may have represented an Atlantic-type margin in some places and at some stages of its development (Kennedy 1976). In other places, and at later stages of development, it appears to have evolved as an Andean margin with volcanic arc deposits, similar to those of the Dunnage zone, resting directly upon sedimentary rocks and continental basement. The evolution of the Avalon zone is even less well understood. It was largely platformal during early Palaeozoic times and its late Precambrian history of volcanism, magmatism and local deformation may relate to a cycle that was earlier than that involving the birth and destruction of Iapetus.

Faunal contrasts across the orogen support this model and provide an additional parameter that strengthens the proposed zonal subdivision. Cambrian (Cowie 1971) and lower Ordovician (Williams 1972) shelly faunas of the Humber–Hebrides zone at the northwestern margin of Iapetus have Pacific affinities and contrast with those of the Avalon–Anglesey zone that have Atlantic or European affinities. A similar contrast is displayed among graptolites (Skevington 1974) of the Gander-Greenore zone (Atlantic) and the Dunnage-Dundee zone (Pacific). Local provincialism is also displayed within the Dunnage-Dundee zone (Neuman 1971; McKerrow and Cocks 1977) and the Humber-Hebrides zone (Kindle and Whittington 1958; Epstein *et al.* 1972). Palaeomagnetic data relating to deposits of early Ordovician times provide further support for a wide Iapetus ocean during this period (Briden *et al.* 1973).

## 4. Tectonic evolution of the northern Appalachians and implications for the development of the British Caledonides

Some of the following comparisons between the northern Appalachians and British Caledonides are ideas and suggestions presented more as a basis for discussion and a target for criticism rather than as real solutions to local problems. The ultimate solutions will emerge only when due regard is given to all the rocks and relationships throughout the entire length of the Appalachian–Caledonian system.

### 4a. Humber zone

The evolution of the Humber zone involved rifting of Grenvillian basement, mafic dyke intrusion, mafic volcanism and the accumulation of thick clastic sequences. All these events were initiated during late Precambrian times, as

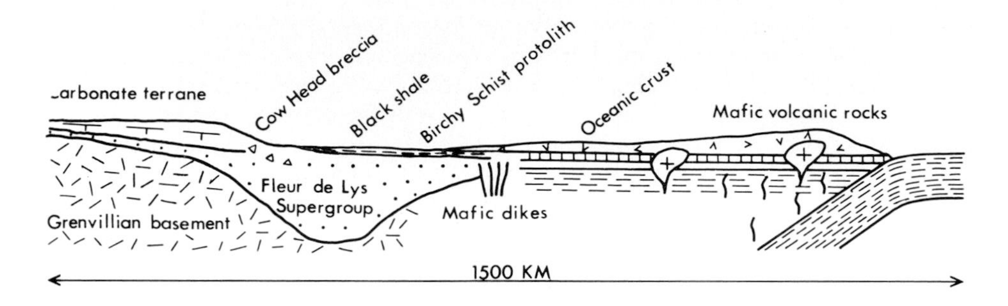


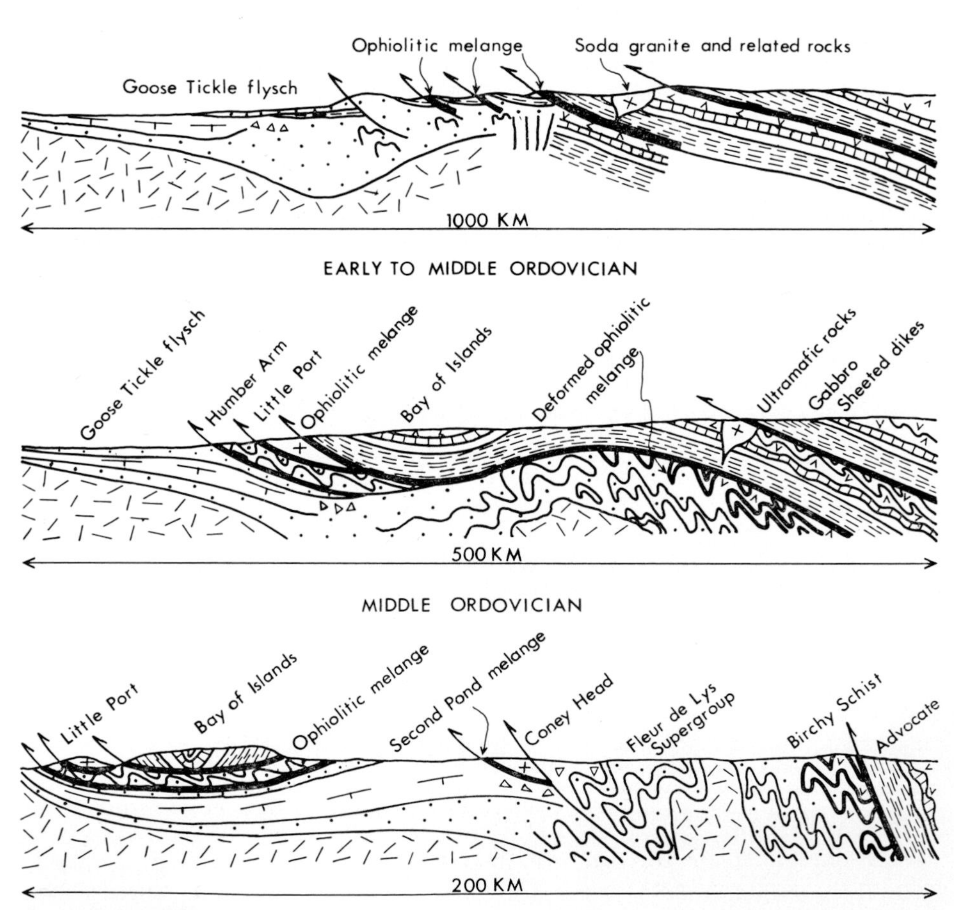


Fig. 2. A model for the structural development of the Humber zone in western Newfoundland.

indicated by isotopic studies of the rift facies volcanic and dyke rocks that yield ages from 800 m.y. (Pringle *et al.* 1971) to 600 m.y. (Stukas and Reynolds 1974). Stratigraphic relationships suggest rifting in latest Precambrian or earliest Palaeozoic times with the 800 m.y. radiometric dates difficult to reconcile with the fact that a stable morphological Atlantic-type margin did not evolve until early Cambrian times, i.e. about 200 m.y. after the initial rifting. Clastic rocks that accumulated during this stage of development (Fleur de Lys Supergroup and equivalents) are possibly as much as 10 km thick and they are represented all along the western flank of the Appalachians.

A thinner sequence of mainly carbonate rocks, locally supratidal, formed at the continental shelf as lower Cambrian to uppermost lower Ordovician deposits. These sediments thicken eastward and record an upward transition from immature arkosic sandstones to mature quartzites, and then limestones and dolomites.

Rocks deposited at the morphological margin, between the carbonate shelf sequence and thicker parts of the rise prism (Fleur de Lys Supergroup) are now preserved as allochthonous sequences structurally above the shelf sequence. These consist of coarse calc-breccias (Cow Head Group) and turbidites and shales (Humber Arm Supergroup) with distinctive shelf-edge faunas.

Uppermost slices of the allochthonous sequences consist of ophiolite suites up to 10 km thick that are interpreted as oceanic crust and mantle. As such they provide direct evidence that the Appalachian system evolved during a cycle of oceanic growth and destruction, rather than one of ensialic rifting. Similar ophiolites (Point Rousse and Betts Cove complexes), along the Baie Verte lineament to the east of the Fleur de Lys continental rise prism, may be remnants of the same ophiolite sheet. If so, then it was obducted across the ancient continental margin in the same way as the Semail nappe of Oman or the Papuan nappe of New Guinea. Transport of the ophiolites across an initially undeformed Fleur de Lys continental rise prism is suggested by the occurrence of ophiolitic mélange within the Birchy Schist complex of the Fleur de Lys Supergroup that shows the effects of the full range of deformational events that affected the Fleur de Lys rocks. This simple model (Fig. 2) is contrasted with an alternative interpretation that relates each ophiolite belt to an equal number of small ocean basins that formed, at least in part, after deformation and metamorphism of the Fleur de Lys prism (Dewey and Bird 1971; Kennedy 1975).

Destruction of the ancient stable margin began toward the end of early Ordovician times. The first indication of instability is indicated by the development of an ancient karst topography across an upwarped carbonate shelf. Later subsidence is recorded, first by the deposition of deeper water carbonates (Table Head Formation) across the disturbed bank, then by a flood of clastic rocks bearing ophiolite detritus from the east (Goose Tickle Formation). These were, in turn, structurally overridden by a sequence of contrasting rock assemblages in separate slices. The structurally lowest slices are composed of sedimentary rocks from the nearby continental margin, with the highest slice of ophiolites representing even further travelled oceanic crust and mantle. The geometry of the slices and facies relationships indicate that the structural pile was assembled from the east, possibly through peeling of successively landward sections by the overriding lithosphere. A contact between the stratigraphic base of the ophiolite sequence and a metamorphic aureole of supracrustal rocks, now frozen into the folded ophiolite slice, represents the actual zone of obduction where the hot oceanic plate moved across the continental margin. The contacts of latest emplacements of the structural slices are now marked by thin zones of shale mélange with exotic blocks. These are the result of mass wastage and tectonic mixing associated with gravity sliding.

Deformation during Ordovician times and prograde metamorphism are recorded in eastern exposures of the carbonate shelf terrane in western White Bay. Further east polyphase deformation and metamorphism in the continental rise prism probably accompanied ophiolite obduction and an attempt to submerge the leading edge of the North American continent along an easterly dipping subduction zone. Subsequent rebound and later imbrication explains the present position of the orthotectonic Fleur de Lys zone with respect to transported ophiolites in western Newfoundland and less deformed ophiolites at the Baie Verte lineament along the eastern boundary of the Humber zone (Fig. 2).

The modern margin of eastern North America parallels the ancient margin as recorded in the evolution of the Humber zone. In a similar fashion, the Humber zone parallels the Grenville structural province and it may therefore mimic a still older continental edge ancestral to this structural province.

## 4b. Hebrides zone

Similarities of certain tectonic elements of the Hebrides zone with those of the Humber zone of the Appalachians suggest parallels in structural evolution. However there are major differences also.

The basement to the Hebrides zone in the north is the Lewisian complex which shows evidence of the effects of the Scourian, Inverian and Laxfordian orogenic episodes that pre-date the Grenville cycle (Bowes 1978). The Lower Torridonian arkose of the Stoer Group, which immediately overlies the crystalline rocks in parts of northwestern Scotland, post-dates the Grenville orogenic episode (Moorbath 1969; van Breemen *et al.* 1978) and may represent molasse associated with the epeirogenic uplift stage of the Grenville cycle, i.e. be unrelated to the Caledonian cycle. Other upper Precambrian sedimentary accumulations (Upper Torridonian assemblage), which resemble sequences in fault-bounded basins and whose accumulation could be connected with an early opening of Iapetus (Stewart 1975), have been correlated with metamorphosed Moine rocks to the southeast of the Moine thrust. However recent work now suggests that rock assemblages for which the term "Moine" has been used may not all be of one age. The *c.* 1000 m.y. age of the Ardgour granitic gneiss (Brook *et al.* 1976) which occurs in the Glenfinnan division of the Moine rocks (Johnstone *et al.* 1969) is indicative of activity during the Grenville cycle. For Moine rocks in the Morar division there is isotopic evidence of a *c* 800–700 m.y. tectonothermal episode (van Breemen *et al.* 1978) which includes the development of the discordant Knoydart and related pegmatites (Long and Lambert 1963). This is the Knoydartian orogeny of Bowes (1968)—subsequently referred to by Lambert (1969) as the 'Morarian' orogeny—which is related by van Breemen *et al.* (1978) to an early stage of activity in the plate tectonic regime that subsequently produced the British Caledonides. The rock assemblage between the Great Glen and Highland Boundary faults, that has been correlated with Moine rocks (Central Highland Granulites), is overlain conformably by the Dalradian Supergroup (late Precambrian – early Palaeozoic in age) and must have only been involved in Caledonian movements. Harris *et al.* (1978) refer to this assemblage as the Grampian Group of the Dalradian Supergroup.

Proximity of Lewisian rocks, unaffected by Caledonian movements, and of rocks, both in western Ireland and northwestern Scotland, that are the products of the Grenville cycle but with a thermal and tectonic overprint due to the Caledonian orogeny, indicates that the Appalachian-Caledonian orogenic front acutely transgresses the Grenville structural province along its course from the northern Appalachians to the Caledonides of Scotland. The reappearance of products of the Grenville cycle on the eastern side of the Caledonides in Scandinavia indicates that the Appalachian–Caledonian deformed zone successfully crossed the Grenville orogen.

The Dalradian Supergroup of Scotland and Ireland has been interpreted as a prism of sediments deposited at a continental margin (Dewey 1969). Other interpretations suggest an ensialic crustal situation (Phillips *et al.* 1976; Harris *et al.* 1978) with a source area to the northwest during the deposition of the lower part of the assemblage and the emergence of a continental area to the south that supplied detritus to a northward basin during the deposition of the upper part of the assemblage. However, if the Dalradian Supergroup was entirely deposited in an ensialic crustal situation, there is a problem concerning the control of the accumulation of such a thick sedimentary sequence. Furthermore, if there was an early Palaeozoic Iapetus in Britain, where are the upper Precambrian–lower Palaeozoic deposits that accumulated at its northwestern margin? Southerly derived clastic assemblages in the upper part of the Dalradian Supergroup suggest a parallel with southerly derived clastic assemblages above the carbonate terrane in the northern Appalachians. In both areas these rocks may reflect the destruction of the northwestern margin of Iapetus. Recycling of easterly parts of the stable margin during uplift and imbrication could possibly provide a source for the clastic rocks, and according to this model the clastic assemblages in the upper parts of the Dalradian Supergroup should contain volcanic rock fragments and chromite and serpentinite clasts like Newfoundland analogues.

Relationships that define the age of earliest deformation of the Dalradian Supergroup in the British Isles are conflicting. In Ireland, the Grampian event is interpreted as Arenigian or earlier where mafic volcanic rocks, cherts and slates of the Lough Nafooey Group presumably overlie deformed Dalradian rocks. The Lough Nafooey Group, an unlikely post-orogenic cover sequence, resembles the Arenigian Highland Border Series of Scotland. There, the Highland Border Series was affected during the Grampian event, suggesting an Arenigian or later age for this event. The Scottish situation compares more favourably with interpretations for the time of earliest deformation in the Humber zone. Diachronism of the Grampian event may reconcile the relationships, but it does seem incongruous that oceanic rocks are involved in the deformation in one place (Highland Boundary) whereas correlatives are interpreted to unconformably overlie deformed rocks in another (South Mayo trough). The occurrence of Dalradian rocks to the south of Clew Bay and the South Mayo trough (Connemara) lend difficulties in defining a local Hebrides–Dundee zone boundary, unless early Palaeozoic transcurrent movements are envisaged.

The Durness succession of Scotland with its basal pipe rock and Cambrian fucoid beds compares well with similar rocks of the Humber zone of the Appalachians. In North America, the equivalent rocks are interpreted as a carbonate bank that evolves at the border of an Iapetus ocean (Rodgers 1968; Williams and Stevens 1974). The absence of southerly derived flysch above the Scottish carbonate sequence, and the absence of transported rocks like those in western Newfoundland, vastly hinders geological interpretations of the Hebrides zone according to a model developed for the northern Appalachians. Perhaps significant in this regard, is the realization that less than 1000 m of erosion in westerly parts of the Humber zone would entirely remove all sign of transported rocks, leaving an essentially undeformed carbonate terrane. If the ultramafic complexes of the Shetland Islands are transported oceanic crust and mantle, they will provide an important adjunct to the development of the Hebrides zone.

In the Humber zone structures verge westward and toward the continental foreland. This contrasts with the southward facing Loch Tay Nappe which affects Dalradian rocks in Scotland. A similar structure in the Appalachians in Quebec has been interpreted as essentially due to slumping, where a continental prism of sediments approached a trench at an eastward dipping subduction zone (St Julien and Hubert 1975).

As in the case with the Fleur de Lys Supergroup of the Humber zone, the highest grades of metamorphism are localized in central parts of the Moine–Dalradian terrane. Burial to effect metamorphism was most likely accomplished by structural imbrication rather than stratigraphic burial. Early granites emplaced into Dalradian rocks may be the result of melting due to extreme structural thickening of the crust (cf. Pidgeon and Aftalion 1978).

## 4c. Dunnage zone

The dominant volcanic assemblages of the Dunnage zone in Newfoundland are interpreted as ancient analogues of modern island-arcs. These sequences commonly show a lithological evolution from lowermost mafic dykes, gabbros and pillow lavas upward through marine cherts and turbidites into pyroclastic rocks and volcaniclastic sedimentary rocks capped by limestone and subaerial tuffs. The overall deep– to shallow-water lithic change is accompanied by geochemical changes in the volcanic stratigraphy from low-potassium tholeiites at the base to calc-alkaline low-silica andesites toward the top that show progressive enrichment in $Al_2O_3$ and $K_2O$ and a decrease in CaO and MgO (Kean and Strong 1975). Many of these sequences are much too thick to correspond to oceanic crustal layer 2, although some of them are clearly built upon an ophiolite suite.

Near the western margin of the Dunnage zone, the Betts Cove ophiolite complex (Upadhyay *et al.* 1971) consists of a basal ultramafic member, transitionally overlain by a poorly developed gabbroic member, in turn overlain by a sheeted dyke complex that consists of practically 100% mafic dykes. Overlying lower Ordovician volcanic rocks, cherts and argillites are nearly 4 km thick. A lowermost pillow lava unit constitutes the upper part of the Betts Cove ophiolite complex so that a completely conformable transition exists from the ultramafic member of the ophiolite complex to the top of the thick overlying mainly volcanic succession. There is little doubt therefore that rocks of the Dunnage zone, especially in westerly exposures, directly overlie oceanic crust.

Tonalites or trondjhemites such as those at Twillingate (Strong and Payne 1973; Williams and Payne 1975) are dated at 510 m.y. and are thought to represent anatectic extracts related to subduction. Local metamorphism and intense mylonitization in the granitic and surrounding rocks are also interpreted as effects of subduction within the oceanic domain.

The Dunnage mélange occurs in a wide belt near the eastern margin of the Dunnage zone. The chaotic terrane is the locus of a set of distinctive intrusions that include quartz–feldspar porphyries, diorites and dolerite dykes. Like nearby volcanic suites, the mélange is overlain by Caradocian black shale and its matrix is dated locally as Tremadocian (Hibbard *et al.* 1977). The mélange is interpreted as a trench-slope deposit implying subduction in this easterly area.

The age of volcanism in the Dunnage zone indicates that island-arcs were growing east of the Humber zone continental margin during emplacement of the west Newfoundland allochthons and furthermore, that the evolution of the volcanic islands ceased at the time of final emplacement of the allochthons (Caradocian).

Mixed volcanic rocks and dark shales along the eastern margin of the Dunnage zone overlie siliceous sediments of the Gander zone, implying a continental basement to easterly parts of the zone. Nearby small mafic–ultramafic complexes have been interpreted either as disrupted ophiolites or mantle diapirs that rose through the crust above an easterly dipping subduction zone (Stevens *et al.* 1974).

The most plausible method for ophiolite obduction in western Newfoundland, namely eastward subduction, would imply that the volcanic arc sequences of the Dunnage zone developed above an eastward dipping subduction zone (Strong *et al.* 1974). However, the position of the Dunnage mélange, located to the east of the volcanic arc sequences, implies subduction in the opposite direction. Possibly more than one subduction zone was active during the early Ordovician evolution of the Dunnage zone. However, subduction appears to have ceased everywhere by Caradocian time.

Silurian rocks of the Dunnage zone are sharply contrasted in facies with underlying Ordovician rocks. Caradocian black shales are overlain conformably and transitionally by Silurian greywackes that pass upwards into coarse plutonic boulder conglomerates with local coralline shale interbeds. These are in turn overlain by Llandovery and Wenlock terrestrial volcanic rocks and fluviatile, mainly red, micaceous sandstones (Williams 1967). Devonian rocks are everywhere terrestrial. These changes indicate that the Ordovician Iapetus ocean was essentially closed by Silurian time.

## 4d. Dundee zone

The Ballantrae complex of Ayrshire that includes serpentinites, gabbros, trondhjemites, pillow lavas and cherts is interpreted by Church and Gayer (1973) as a dismembered ophiolite suite (see Bluck 1978). Its presence implies oceanic crust and mantle in this central area of the Caledonides. Blueschist metamorphism displayed locally by some of these rocks suggests conditions like those characteristic of subduction zones. Nearby unconformable Caradocian strata suggest that this phase of activity at Ballantrae was contemporaneous with similar events in the Dunnage zone of Newfoundland.

Serpentinites, gabbros, pillow lavas and Arenigian sediments along the Highland Boundary fault (Highland Border Series) present a situation similar to that of the Baie Verte lineament in Newfoundland. Here, the lineament is interpreted as the ancient continent–ocean interface (Williams 1975; Miller and Deutch 1977) and the most westerly possible root zone for transported ophiolites such as the Bay of Islands complex in the Humber zone.

A similar enigma to the timing of the Grampian event concerns the age of earliest deformations in the Fleur de Lys Supergroup with respect to the time of generation and emplacement of ophiolites in the Newfoundland Appalachians (Church 1969; Dewey and Bird 1971; Kennedy 1975; Bursnall and De Wit 1975; Williams *et al.* 1977). The present consensus is that the Fleur de Lys Supergroup (equated with the Dalradian Supergroup) was undeformed before the generation and transport of nearby ophiolite complexes, and this is supported by the recent discovery of deformed ophiolitic mélanges along the eastern margin of the Fleur de Lys terrane (Williams *et al.* 1977).

## 4e. Gander zone

The Gander zone evolved on a continental gneissic basement that must have lain along the eastern margin of Iapetus. A polydeformed thick arenite sequence that overlies reworked basement has been interpreted as a prism of sediment built up along the eastern margin of Iapetus (Kennedy 1976). Deformation of the cover sequence is further interpreted to pre-date nearby Caradocian marine black shales of the Dunnage zone (Davidsville Group). Ultramafic rocks along the Dunnage–Gander zone contact are unconformably overlain by Caradocian conglomerates that contain clasts of penetratively deformed ultramafic rocks. The relationships suggest emplacement of the ultramafic rocks during Ganderian orogeny (Kennedy 1975).

Megacrystic biotite granites, foliated biotite granites and garnetiferous muscovite leucogranites all seem to be most common in the Gander zone. Increasing $K_2O$ content from west to east across this area of Newfoundland has been interpreted to favour eastward subduction (Strong *et al.* 1974). The isotopic ages of the granitic rocks vary from early Ordovician to Carboniferous (Bell *et al.* 1977).

## 4f. Greenore zone

In the Greenore zone of Ireland, the Rosslare complex appears to be a direct correlative of basement gneisses in the Gander zone, thus indicating continental crust on this eastern side of Iapetus. Other rocks of the Cullenstown Group and Bray Series may be equivalent to the Gander Group, i.e. cover sequences to the Rosslare complex. A nearby thick pile of Caradocian mixed volcanic rocks could represent the development of a volcanic arc sequence upon continental crust, a situation similar to that presented by the Tetagouche Group in the New Brunswick Gander zone.

In northern England the Manx and Skiddaw slates are possible further correlatives of the Newfoundland Gander Group. The overlying Borrowdale Volcanic Group in this area would further suggest the development of an on-land arc sequence.

## 4g. Avalon zone

The position of the Avalon zone with respect to the Dunnage zone indicates that the former did not directly face or abut the Iapetus ocean. Its mainly Precambrian development is also difficult to reconcile with an essentially early Palaeozoic development of Iapetus. Furthermore, the Avalon zone was essentially a stable platform during Cambrian and Ordovician times when the generation and destruction of Iapetus was most active.

Possibly the Gander–Avalon zones were linked during their Precambrian development. Increasing difficulty in separating the Gander–Avalon zones by comparing their oldest basement rocks would support this suggestion. Furthermore, the Dover fault that defines the zone boundary in Newfoundland is interpreted as having developed during Precambrian times (Blackwood and Kennedy 1975).

The controlling feature of the Precambrian development of the Avalon zone was either an archipelago of volcanic islands related to subduction and a cycle that pre-dated the opening of Iapetus (Rast *et al.* 1976), or else its upper Precambrian volcanic rocks relate to rifting and the initiation of Iapetus (Papezik 1972). If the Avalon zone evolved independently then its present proximity to the Humber zone and the ancient North American margin is due to the random interactions of wandering continents. If the Avalon zone is related to the initiation of the continental rise prism of the Humber zone, then the Iapetus ocean formed behind a migrating Avalon microcontinent (Schenk 1971).

### 4h. Anglesey zone

The Uriconian, Arvonian, Charnian and Longmyndian successions of the Anglesey zone can all be matched with upper Precambrian successions of the Avalon zone in Newfoundland. Overlying Cambrian rocks are identical in North America and the British Isles. Small Precambrian inliers, e.g. Primrose Hill Gneiss and Schists, Ruston Schists and Malvernian gneisses (cf. Bennison and Wright 1969), indicate development upon a continental basement, as is indicated for the Avalon zone. Similarly, the proposed correlation of the Mona complex and Cullenstown Group (Dewey 1974) implies a basement linkage between the Greenore–Anglesey zones.

The Brioverian rocks of Brittany are probable correlatives of Avalonian upper Precambrian rocks and other possible correlatives are to be sought in the Iberian Peninsula, Northwest Africa, Czechoslovakia and possibly Turkey. The Avalon–Anglesey zone, although parallel to the Humber–Hebrides zone throughout the entire Appalachians and British Caledonides, clearly follows a separate course beyond the British Isles, passing to the southeast, rather than to the northwest of the Baltic Shield.

Locally, the Anglesey zone displays Precambrian ophiolitic mélange and blueschist metamorphism in Wales and a well-preserved ophiolite complex occurs in possible correlative rocks at Bou Azzer of Northwest Africa. A true understanding of the Avalon–Anglesey zone must therefore account for these diverse tectonic elements, and proper synthesis can only be attempted when all are located and duly considered.

## 5. Development in Silurian and later times

The history of Silurian–Devonian development of the Appalachian–Caledonides orogen presents a picture entirely different from that which prevailed during its Cambrian–Ordovician development. A sub-Silurian or sub-middle Ordovician (Caradocian) unconformity across the entire system, except for local areas of the Dunnage zone, indicates destruction of earlier Ordovician depositional environments. And even in local areas of the Dunnage zone, where the record is complete, marine Ordovician rocks pass upward into Silurian conglomerates and continental red beds.

The common view that a Silurian or Devonian Iapetus eventually closed in late Silurian or Devonian times to produce the Cymrian and Acadian orogenic events (Dewey 1969; McKerrow and Ziegler 1971; Schenk 1971; McKerrow and Cocks 1977) is based more on the premise that orogeny is the result of moving plates and continental collision, rather than the stratigraphic record. The existence of a Silurian or later Iapetus cannot be demonstrated convincingly on stratigraphic grounds. Furthermore, Silurian or Devonian ophiolite complexes are unknown anywhere within the Applachian–Caledonian orogen. Silurian and Devonian magmatism and deformation that affect the Appalachian–Caledonian orogen indicate, more than anything else, that these processes persist long after the main phases of horizontal movements and ocean destruction.

In the Canadian Appalachians, granitic and related intrusions show a broad preference for particular zones. Ordovician intrusions within the Dunnage zone clearly relate to its early development (Twillingate granite, Coaker porphyry; Kay 1976). Later Devonian intrusions that cut Silurian red beds in the same zone are mainly composite batholiths with earlier mafic phases and later granitic phases (e.g. Mount Peyton batholith). In contrast, the common Gander zone granites and the sparse Humber zone granites are potassic biotite granites, commonly megacrystic, and garnetiferous muscovite leucogranites. Some of these (e.g. Ackley batholith) clearly intrude the Gander–Avalon zone boundary, implying little crustal distinction between these zones at the time of intrusion. The occurrence of coarse-grained potassic biotite granites in the southeastern part of the Dunnage zone that resemble intrusions of the Gander zone may mean that this part of the Dunnage zone is underlain by Gander-type continental basement.

Local belts within the northern Appalachians have been affected by Carboniferous deformation. Most of these deformed belts are coincident with earlier troughs of especially thick and coarse Carboniferous sandstones and conglomerates, so that deformation was localized in areas of prior crustal instability. There is no stratigraphic evidence in the Canadian Appalachians to suggest the presence of oceans or moving plates during this late phase of development.

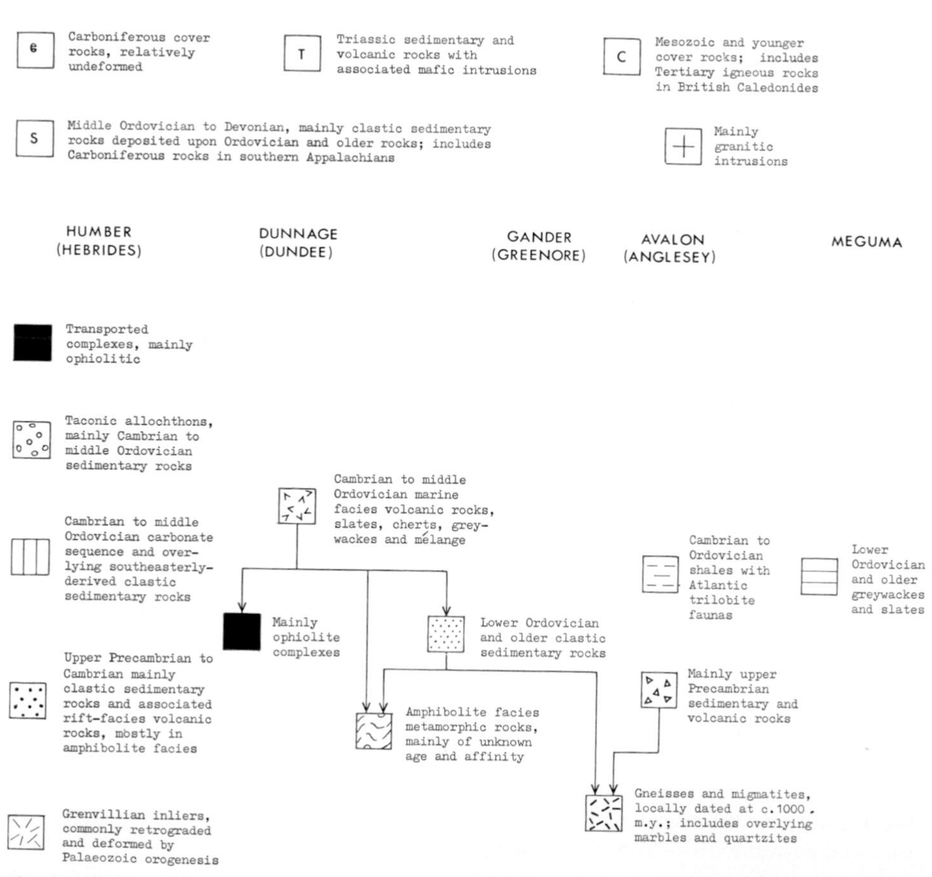


Fig. 1. Major lithological units and zonal subdivision of the Appalachians and British Caledonides.

## 6. Concluding remarks

Since the suggestion of Wilson (1966) that the Appalachian–Caledonian orogen was the result of generation, then destruction, of an old ocean, a number of workers have attempted to explain various facets of Appalachian–Caledonian evolution in terms of lithospheric plate tectonics (Dewey 1969; Bird and Dewey 1970; Fitton and Hughes 1970; Stevens 1970; Church and Stevens 1971; McKerrow and Ziegler 1971; Williams and Stevens 1974; Kennedy 1975; Phillips *et al.* 1976). Major differences among some models arise mainly because of the emphasis placed upon local relationships and disregard for broad regional generalities. Large scale correlations like the type attempted here, must necessarily precede the development of models that explain relationships along the full length of the orogen.

Probably the surest path toward regional correlations will involve the compilation of new and innovative tectonic maps for the entire orogen such as the new tectonic map of the Appalachians (Williams 1978); similar maps are in preparation for most of the Caledonides. Continued cooperation, future comparisons, and further compilations will undoubtly greatly advance our knowledge of this most intriguing mountain belt.

*Acknowledgments.* I wish to thank W. R. Church, J. F. Dewey, M. J. Kennedy, A. F. King, M. D. Max, E. R. W. Neale, P. D. Ryan and R. K. Stevens who through discussions helped extrapolate the zonal subdivision of the northern Appalachians into the British Isles. I also wish to acknowledge financial support for fieldwork in the Appalachians from the National Research Council of Canada, the Canadian Department of Energy, Mines and Resources, and an Izaac Walton Killam Special Senior Research Scholarship.

## References

Bell, K., Blenkinsop, J. and Strong, D. F. 1977. The geochemistry of some granitic bodies from eastern Newfoundland and its bearing on Appalachian evolution. *Can. J. Earth Sci.* **14,** 456–476.

Bennison, G. M. and Wright, A. E. 1969. *The Geological History of the British Isles.* Arnold London.

Bird, J. M. and Dewey, J. F. 1970. Lithosphere Plate–Continental Margin Tectonics and the Evolution of the Appalachian Orogen. *Geol. Soc. Am. Bull.* **81,** 1031–1060.

Blackwood, R. F. and Kennedy, M. J. 1975. The Dover Fault: Western Boundary of the Avalon Zone in Northeastern Newfoundland. *Cam. J. Earth Sci.* **12,** 320–325.

Bluck, B. J. 1978. Geology of a continental margin 1: the Ballantrae Complex. *In* Bowes, D. R. and Leake, B. E. (Editors). *Crustal evolution in northwestern Britain and adjacent regions,* 151–162. *Geol. J. Spec. Issue* No. 10.

Bowes, D. R. 1968. The absolute time scale and the subdivision of Precambrian rocks in Scotland. *Geol. Fören. Stockholm Förh.* **90,** 175–188.

—— 1978. Shield formation in early Precambrian times: the Lewisian complex. *In* Bowes, D. R. and Leake, B. E. (Editors). *Crustal evolution in northwestern Britain and adjacent regions,* 39–80. *Geol. J. Spec. Issue* No. 10.

Briden, J. C., Morris, W. A. and Piper, J. D. A. 1973. Palaeomagnetic studies in the British Caledonides—VI Regional and Global Implications. *Geophys. J. R. astron. Soc.* **34,** 107–134.

Brook, M., Brewer, M. S. and Powell, D. 1976. Grenville age for rocks in the Moine of northwestern Scotland. *Nature, Lond.* **260,** 515–517.

Brown, W. R. 1976. Tectonic Mélange in the Arvonia Slate District of Virginia. *Geol. Soc. Am. Abstracts with Programs* **8** (2), 142.

Bursnall, J. T. and De Wit, M. J. 1975. Timing and development of the orthotectonic zone in the Appalachian Orogen of Northwest Newfoundland. *Can. J. Earth. Sci.* **12,** 1712–1722.

CHURCH, W. R. 1969. Metamorphic Rocks of Burlington Peninsula and adjoining areas of Newfoundland and their bearing on continental drift in North Atlantic. *In* Kay, M. (Editor). *North Atlantic—geology and continental drift*, a symposium, 212–233. *Mem. Am. Assoc. Petrol. Geol.* **12.**

CHURCH, W. R. and GAYER, R. A. 1973. The Ballantrae Ophiolite. *Geol. Mag.* **110,** 497–510.

—— and STEVENS, R. K. 1971. Early Paleozoic ophiolite complexes of the Newfoundland Appalachians as mantle-oceanic crust sequences. *J. Geophys. Res.* **76,** 1460–1466.

COWIE, J. W. 1971. Lower Cambrian faunal provinces. *In* Middlesmiss, F. A., Rawson, P. F. and Newall, G. (Editors). *Faunal Provinces in Space and Time,* 31–46. *Geol. J. Spec. Issue* No. 4.

CRAIG, G. Y. (Editor) 1965. *The Geology of Scotland.* Oliver and Boyd, Edinburgh.

DEWEY, J. F. 1969. Evolution of the Appalachian/Caledonian Orogen. *Nature, Lond.* **222,** 124–129.

—— 1974. The Geology of the Southern Termination of the Caledonides. *In* Nairn, A.E.M. and Stehli, F. G. (Editors). *The Ocean Basins and Margins* 2 *The North Atlantic,* 205–231. Plenum Press, New York.

—— and BIRD, J. M. 1971. Origin and emplacement of the ophiolite suite: Appalachian ophiolites in Newfoundland. *J. Geophys. Res.* **76,** 3179–3206.

EPSTEIN, J. B., EPSTEIN, A. G. and BERGSTROM, S. M. 1972. Significance of Lower Ordovician Exotic blocks in the Hamburg Klippe, eastern Pennsylvania. *U.S. geol. Surv. Prof. Pap.* **800**–D, 29–36.

FITTON, J. G. and HUGHES, D. J. 1970. Volcanism and plate tectonics in the British Ordovician. *Earth Planet. Sci. Lett.* **8,** 223–228.

FLINN, D. 1958. On the nappe structure of North-east Shetland. *Q. J. geol. Soc. Lond.* **114,** 107–136.

HARRIS, A. L., BALDWIN, C. T., BRADBURY, H. J., JOHNSON, H. D. and SMITH, R. A. 1978. Ensialic basin sedimentation : the Dalradian Supergroup. *In* Bowes, D. R. and Leake, B. E. (Editors). *Crustal evolution in northwestern Britain and adjacent regions,* 115–138. *Geol. J. Spec. Issue* No. 10.

HIBBARD, J. P., STOUGE, S. and SKEVINGTON, D. 1977. Fossils from the Dunnage Mélange, north-central Newfoundland. *Can. J. Earth Sci.* **14,** 1176–1178.

JOHNSON, M. R. W. and STEWART, F. H. (Editors). 1963. *The British Caledonides.* Oliver and Boyd, Edinburgh.

JOHNSTONE, G. S., SMITH, D. I. HARRIS, A. L. 1969. Moinian Assemblage of Scotland. *In* Kay, M. (Editor). North Atlantic-geology and continental drift a symposium, 1959–180. *Mem. Am. Ass. Petrol. Geol.* **12.**

JUKES, J. B. 1842. *Excursions in and about Newfoundland during the years 1839 and 1840* **1,** 2. John Murray, London.

KAY, M. (Editor). 1969. North Atlantic—geology and continental drift, a symposium, *Mem. Am. Ass. Petrol. Geol.* **12.**

—— 1976. Dunnage Mélange and Subduction of the Protacadic Ocean, Northeast Newfoundland. *Geol. Soc. Am. Spec. Pap.* **175,** 49 pp.

KEAN, B. F. and STRONG, D. F. 1975. Geochemical Evolution of an Ordovician Island arc of the central Newfoundland Appalachians. *Am. J. Sci.* **275,** 97–118.

KENNEDY, M. J. 1975. Repetitive Orogeny in the northeastern Appalachians—new plate models based upon Appalachian examples. *Tectonophysics* **28,** 39–87.

—— 1976. Southeastern Margin of the Northeastern Appalachians: Late Precambrian Orogeny on a continental margin. *Geol. Soc. Am. Bull.* **87,** 1317–1325.

—— and McGONIGAL, M. 1972. The Gander Lake and Davidsville groups of northeastern Newfoundland: new data and geotectonic implications. *Can. J. Earth Sci.* **9,** 452–459.

——, NEALE, E. R. W. and PHILLIPS, W. E. A. 1972. Similarities in the early structural development of the Northwestern margin of the Newfoundland Appalachians and Irish Caledonides. *Proc. Int. geol. Congr.* **24**(3), 516–531.

KINDLE, C. H. and WHITTINGTON, H. B. 1958. Stratigraphy of the Cow Head Region, Western Newfoundland. *Geol. Soc. Am. Bull.* **69,** 315–342.

LAMBERT, R. ST J. 1969. Isotopic studies relating to the Pre-Cambrian history of the Moinian of Scotland. *Proc. geol. Soc. Lond.* **1652,** 243–245.

LILLY, H. D. 1965. Submarine exploration of the Virgin Rocks Area, Grand Banks, Newfoundland: a preliminary note. *Geol. Soc. Am. Bull.* **76,** 131–132.

LONG, L. E. and LAMBERT, R. ST J. 1963. Rb-Sr isotopic ages from the Moine Series. *In* Johnson, M. W. R. and Stewart, F. H. (Editors). *The British Caledonides,* 217–247. Oliver & Boyd, Edinburgh.

MCKERROW, W. S. and COCKS, L. R. M. 1977. The location of the Iapetus Ocean suture in Newfoundland. *Can. J. Earth Sci.* **14,** 488–495.

—— and ZEIGLER, A. M. 1971. The Lower Silurian Palaeogeography of New Brunswick and adjacent areas. *J. Geol.* **79,** 635–646.

MILLER, H. G. and DEUTCH, E. R. 1976. New gravitational evidence for the subsurface extent of oceanic crust in north-central Newfoundland. *Can. J. Earth Sci.* **13,** 459–469.
MOORBATH, S. 1969. Evidence for the age of deposition of the Torridonian sediments of north-west Scotland. *Scott. J. Geol.* **5,** 154–170.
NEUMAN, R. B. 1971. An Early Middle Ordovician Brachiopod Assemblage from Maine, New Brunswick, and Northern Newfoundland. *In* Dutro, J. T. Jr. (Editor). *Paleozoic perspectives: A paleontologic tribute to G. Arthur Cooper,* 113–124. *Smithsonian Contr. Paleobiology,* No. 3.
PAPEZIK, V. S. 1972. Late Precambrian ignimbrites in eastern Newfoundland and their tectonic significance. *Proc. Int. geol. Congr.* **24**(1), 147–152.
PELLETIER, B. R. 1971. A granodiorite drill core from the Flemish Cap, eastern Canadian Continental shelf. *Can. J. Earth Sci.* **8,** 1499–1503.
PHILLIPS, W. E. A., STILLMAN, C. J. and MURPHY, T. 1976. A Caledonian plate tectonic model. *J. geol. Soc. Lond.* **132,** 579–609.
PIDGEON, R. T. and AFTALION, M. 1978. Cogenetic and inherited zircon U–Pb systems in granites: Palaeozoic granites of Scotland and England. *In* Bowes, D. R. and Leake, B. E. (Editors). *Crustal evolution in northwestern Britain and adjacent regions,* 183–220. *Geol. J. Spec. Issue* No. 10.
POOLE, W. H., SANDFORD, B. V., WILLIAMS, H. and KELLEY, D. G. 1970. Geology of South-eastern Canada. *In* Douglas, R. J. W. (Editor). *Geology and Economic Minerals of Canada,* 227–304. *Geol. Surv. Can. Econ. Geol. Rep.* No. 1.
PRINGLE, I. R., MILLER, J. A. and WARRELL, D. M. 1971. Radiometric age determinations from the Long Range Mountains, Newfoundland. *Can. J. Earth Sci.* **8,** 1325–1330.
RAST, N. and CRIMES, T. P. 1969. Caledonian orogenic episodes in the British Isles and North-western France and their tectonic and geochronologic interpretation. *Tectonophysics* **7,** 277–307.
——, O'BRIEN, B. H. and WARDLE, R. J. 1976. Relationships between Precambrian and Lower Paleozoic rocks of the 'Avalon Platform' in New Brunswick, the northeast Appalachians and the British Isles. *Tectonophysics* **30,** 315–338.
RODGERS, J. 1968. The Eastern edge of the North American continent during the Cambrian and Early Ordovician. *In* Zen, E–an, White, W. S., Hadley, J. B., and Thompson, J. B. Jr. (Editors). *Studies of Appalachian Geology: Northern and Maritime,* 141–149. Interscience, New York.
—— 1970. The tectonics of the Appalachians. Wiley-Interscience, New York.
SCHENK, P. E. 1971. Southeastern Atlantic Canada, Northwestern Africa, and Continental Drift. *Can. J. Earth Sci.* **8,** 1218–1251.
SKEVINGTON, D. 1974. Controls influencing the composition and distribution of Ordovician graptolite faunal provinces. *In* Rickards, R. B., Jackson, D. E. and Hughes, C. P. (Editors). *Graptolite studies in honour of O. M. B. Bulman,* 59–73. *Spec. Pap. Palaeont.* No. 13.
STEVENS, R. K. 1970. Cambro-Ordovician flysch sedimentation and tectonics in west Newfoundland and their possible bearing on a Proto-Atlantic Ocean. *In* Lajoie, J. (Editor). *Flysch sedimentology in North America,* 165–177. *Geol. Assoc. Can. Spec. Pap.* No. 7.
——, STRONG, D. F. and KEAN, B. F. 1974. Do some eastern Appalachian Ultramafic rocks represent Mantle Diapirs produced above a subduction zone? *Geology* **2,** 175–178.
STEWART, A. D. 1975. 'Torridonian' rocks of western Scotland. *In* Harris, A. L., Shackleton, R. M., Watson, J., Downie, C., Harland, W. B. and Moorbath, S. (Editors). *A Correlation of the Precambrian rocks of the British Isles,* 43–51. *Geol. Soc. Lond. Spec. Rep.* No. 6.
ST JULIEN, P. and HUBERT, C. 1975. Evolution of the Taconian Orogen in the Quebec Appalachians. *Am. J. Sci.* **275,** 337–362.
STRONG, D. F., DICKSON, W. L., O'DRISCOLL, C. F., KEAN, B. F. and STEVENS, R. K. 1974. Geochemical Evidence for an east-dipping Appalachian subduction zone in Newfoundland. *Nature, Lond.* **248,** 37–39.
—— and PAYNE, J. G. 1973. Early Paleozoic volcanism and metamorphism of the Mortons Harbour-Twillingate Area, Newfoundland. *Can. J. Earth Sci.* **10,** 1363–1379.
STUKAS, V. and REYNOLDS, P. H. 1974. $^{40}Ar/^{39}Ar$ dating of the Long Range dikes, Newfoundland. *Earth Planet. Sci. Lett.* **22,** 256–266.
SWETT, K. and SMIT, D. E. 1972. Cambro–Ordovician shelf sedimentation of Western Newfoundland, Northwest Scotland and Central East Greenland. *Proc. Int. geol. Congr.* **24**(6), 33–41.
UPADHYAY, H. D., DEWEY, J. F. and NEALE, E. R. W. 1971. The Betts Cove Ophiolite Complex, Newfoundland: Appalachian Oceanic crust and mantle. *Proc. geol. Assoc. Can.* **24,** 27–34.
VAN BREEMEN, O., HALLIDAY, A. N., JOHNSON, M. R. W. and BOWES, D. R. 1978. Crustal additions in late Precambrian times. *In* Bowes, D. R. and Leake, B. E. (Editors). *Crustal evolution in northwestern Britain and adjacent regions,* 81–106. *Geol. J. Spec. Issue* No. 10.
WILLIAMS, A. 1972. Distribution of brachiopod assemblages in relation to Ordovician palaeogeography. *In* Hughes, N. F. (Editor). *Organisms and Continents through time,* 241–269. *Spec. Pap. Palaeont.* No. 12.

WILLIAMS, H. 1967. Silurian Rocks of Newfoundland. *Geol. Ass. Can. Spec. Pap.* No. 4. 93–138.
—— 1975. Structural Succession, Nomenclature, and Interpretation of Transported rocks in Western Newfoundland. *Can. J. Earth Sci.* **12,** 1874–1894.
WILLIAMS, H. 1976. Tectonic-Stratigraphic Subdivision of the Appalachian Orogen. *Geol. Soc. Am. Abstracts with Programs* **8,** 300.
—— 1978. *Tectonic–lithofacies map of the Appalachian orogen (1:1 000 000)*. Dept. of Geol., Memorial Univ. of Newfoundland, St. Johns. Map No. 1.
——, HIBBARD, J. P. and BURSNALL, J. T. 1977. Geologic Setting of asbestos-bearing ultramafic rocks along the Baie Verte Lineament, Newfoundland. *Geol. Surv. Can. Pap.* 77–1A, 351–360.
——, KENNEDY, M. J. and NEALE, E. R. W. 1972. The Appalachian Structural Province. *In* Price, R. A. and Douglas, R. J. W. (Editors). *Variations in Tectonic Styles in Canada,* 181–261 *Geol. Assoc. Can. Spec. Pap.* No. 11.
——, ——and —— 1974. The northeastward termination of the Appalachian Orogen. *In* Nairn, A. E. M. and Stehli, F. G. (Editors). *Ocean Basins and Margins* 2, 79–123. Plenum Press, New York.
—— and PAYNE, J. G. 1975. The Twillingate Granite and nearby Volcanic Groups: An Island Arc Complex in Northeast Newfoundland. *Can. J. Earth Sci.* **12,** 982–995.
—— and STEVENS, R. K. 1974. The Ancient Continental Margin of Eastern North America. *In* Burk, C. A. and Drake, C. L. (Editors). *The Geology of Continental Margins,* 781–796. Springer Verlag, New York.
WILSON, J. T. 1966. Did the Atlantic close and then reopen? *Nature. Lond.* **211,** 676–681.
ZEN, E-AN, WHITE, W. S., HADLEY, J. B. and THOMPSON, J. B. JR. (Editors) 1968. *Studies of Appalachian Geology: Northern and Maritime.* Interscience, New York.

Harold Williams,
Department of Geology,
Memorial University,
St. John's,
Newfoundland, A1B 3X5, Canada.

# Crustal structure of the eastern North Atlantic seaboard

*J. Hall*

The arrangement of Precambrian crustal units, the evolution of the Caledonides, and the formation of the Atlantic Ocean are three significant events in the geological history of northwest Britain. The extent to which present knowledge of local deep crustal structure helps to understand these events is discussed and proposals made concerning the definition of continental crust, the evolution of continental crust, discrimination between alternative models of lower crustal layering and the relationship between North Sea subsidence and Atlantic evolution.

## 1. Introduction

Knowledge of the crustal structure in the vicinity of northwest Britain should assist in understanding the arrangement of the Precambrian crustal units, the evolution of the Caledonides and the fragmentation of Laurasia by the formation of the Atlantic Ocean. These three aspects of the geological evolution of the region are first briefly described working backwards through time. Then the main crustal units of the area are defined and discussed in relation to some questions with which later papers in this volume are concerned.

## 2. Major geological events

### 2a. Formation of the Atlantic and contemporaneous continental structures

The matching of geological features and the fit of continental margins across the Atlantic (e.g. Bullard *et al.* 1965; Bott and Watts 1971; Vann 1974), together with the pattern of oceanic magnetic strip anomalies and the location of the Mid-

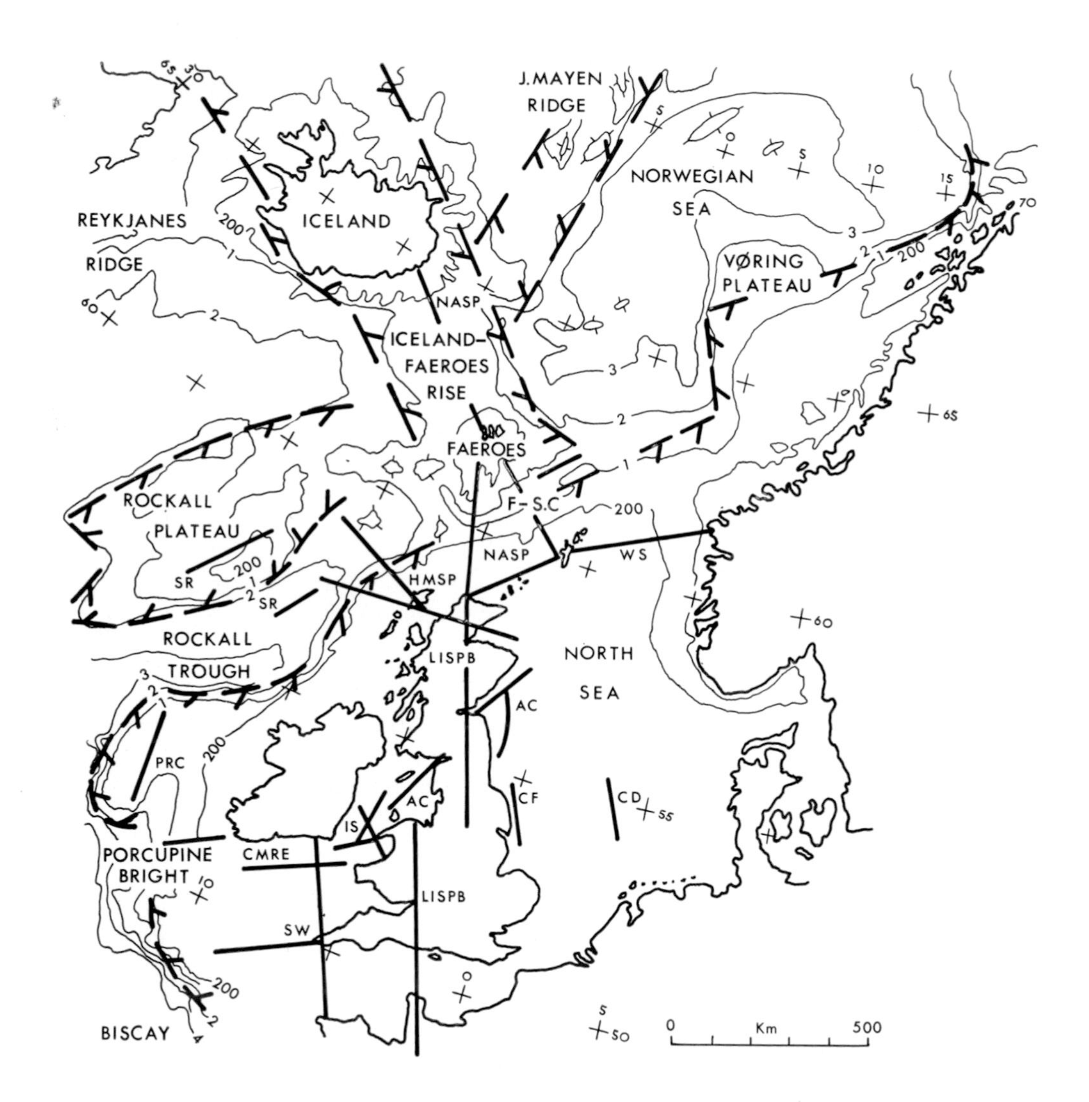


Fig. 1. Map of the Atlantic seaboard of Britain, showing the main physiographical units and locations of seismic surveys penetrating to the Moho. Bathymetric contours are at 200 m, 1 km, 2 km, 3 km, 4 km. F–SC = Faeroe–Shetland channel. Seismic surveys are referenced as follows: AC = Agger and Carpenter (1964); CD, CF = Collette *et al.* (1970); CMRE = Bamford (1971, 1972), Bamford and Blundell (1970); HMSP = University of Durham (1975, pers. comm.); IS = Blundell and Parks (1969); LISPB = Bamford *et al.* (1976); NASP = Bott *et al.* (1974, 1976), Smith and Bott (1975); PRC = Whitmarsh *et al.* (1974); SR = Scrutton (1970, 1972), Scrutton and Roberts (1971); SW = Bott *et al.* (1970); WS = Willmore (1973), see also Sellevoll (1973). Results of many surveys of the adjacent oceanic areas are summarised in Bott and Watts (1971).

Atlantic ridge and its transform faults (Vogt and Avery 1974; Laughton 1971; Le Pichon and Fox 1971), has lead via the theory of plate tectonics, to the concept of a multi-stage opening of the North Atlantic generally as follows:

(i) From approximately 200 m.y. to 80 m.y. ago, separation of Africa from North America and the rotation of Iberia, opening up the Atlantic and the Bay of Biscay. This was terminated northwards against a transform fault from Biscay to Labrador–Greenland.

(ii) In a phase that ended 60 m.y. ago, Greenland, Europe and Africa moved away together from North America; the opening extended up the Labrador Sea.

(iii) In a phase that lasted until 47 m.y. ago, Greenland moved away from North America, Europe from Greenland and Africa from North America. There was triple point spreading with new opening in the northeast Atlantic creating the Reykjanes ridge.

(iv) Subsequently the Labrador Sea stabilised while Europe continued to move away from Greenland and North America until the present. During part of the spreading about it, the Reykjanes ridge was crossed by fracture zones (approximately 38–20 m.y. ago).

Such a four-stage history explains the sequence of magnetic "unconformities" (Williams and McKenzie 1971), the variable width of the ocean and the corresponding maximum depth, the history of igneous activity in early Mesozoic times and subsequent graben development on the western flank of the North Atlantic. There are obvious spatial and temporal correlations of the beginning of sea-floor spreading between Greenland and Europe with the early Tertiary volcanicity of East Greenland and northwest Britain.

That this is by no means a satisfactory explanation of the development of the continental margin of northwest Britain is clear from Figure 1. The continental margin off western Britain is fairly sharply defined at the Biscay edge and may be followed northwards around (or across?) the Porcupine seabight and up the western edge of the Porcupine ridge (Bailey *et al.* 1971). Immediately to the west lies the Rockall trough, probably floored by oceanic crust of unknown age (Scrutton and Roberts 1971), which is separated from the main area of the northeast Atlantic by the Rockall and Hatton banks on which continental basement occurs (Roberts *et al.* 1972). To the north, the simple pattern of magnetic strip anomalies about the Reykjanes ridge is terminated by a cross-cutting aseismic ridge forming the Iceland–Faeroes rise and the Iceland–Greenland ridge, and crossing the active ridge on Iceland (Bott *et al.* 1971; Fleischer 1971). This aseismic ridge appears to coincide with a transverse fracture zone which has allowed the northeast Atlantic to develop somewhat independently to the north, where the nearness of the active ridge to Greenland, the depth of the Norwegian Sea and the presence of aseismic ridges that may be continental fragments (the Jan Mayen ridge) suggests a multi-phase sea-floor spreading during the last 60 m.y. with the axis jumping eastwards from the Norwegian Sea to the present location in two steps (Talwani and Eldholm 1972; Talwani and Udintsev 1976). The Norwegian Sea is linked to the Rockall trough via the Faeroe–Shetland channel across a saddle, the Wyville–Thomson ridge, with the Faeroes as the northern end of the Rockall microcontinent. The Faeroe–Shetland channel contains thick sediments, as do Rockall trough and the Porcupine seabight: all might be pre-Tertiary aborted oceans.

Knowledge of the deep structure would clarify many of the problems of our understanding of this region. Matters of particular importance are (i) the criteria for the recognition of continental crust, (ii) the position of the continental margins,

(iii) structure of the Iceland–Faeroes rise, (iv) the age of the Rockall trough and (v) whether the Vøring plateau is oceanic, continental or, in parts, both.

Formation of new ocean by sea-floor spreading is usually accompanied by the development of sympathetic structures on the newly-created continental edges. An initial period of uplift is accompanied by volcanism and graben formation, followed by oceanward subsidence as the lithosphere cools and lateral support is removed. These effects are not universally observed and the time constants related to them are not well defined. However, the thermal time constant (which dictates the post-splitting subsidence of oceanic crust as it migrates away from the ridge) is of the order of 50 m.y. (McKenzie 1967). Because of this it is not always easy to ascribe specific events in the sequence of changes in the pattern of spreading in the North Atlantic to the structures observed. The distance away from a newly-forming margin to which the related stresses are transmitted effectively is thought to be dependent on the inelasticity of the lithosphere (Bott 1973), but is estimated empirically to be of the order of 100 km which is about the width of the zone of Mesozoic graben on the continental margin of eastern U.S.A. This would suggest that the North Sea basins are not directly related to the formation of the Atlantic Ocean, unless the hot creep mechanism (Bott 1971) is operative at ranges of up to 1000 km from the marginal sink.

Another difficulty is the difference between the Tertiary coastal flexure of Greenland with its parallel dyke swarm (parallel to the Reykjanes ridge) and the perpendicularly oriented Tertiary dyke swarms of northwest Britain. Perhaps the difference is explained by the differing distances from the new spreading axis. The lateral relative tension applied to the base of the lithosphere by divergence of a convective upwelling is largely dissipated by splitting, and thereafter lateral transport is predominantly by basal shearing. However, gravitationally-caused sliding of lithosphere away from the ocean ridge may be sufficient to produce a compressive stress on the flanks of the thermal uplift that leads to dyke swarm intrusion perpendicular to the ridge. By contrast the ridge-parallel swarm of Greenland must be oriented by the shelf-marginal collapse implicit in the flexure.

Another difficulty related to the correlation of shelf structure and ocean evolution lies with the Rockall–Faeroe microcontinent. As this moved off the thermal high and away from Greenland 60 m.y. ago, it should have subsided at a progressively decreasing rate (Detrick *et al.* 1977). However, the timing of subsidence was not in phase with this (Laughton *et al.* 1972) and overall subsidence of about 1 km is much less than the 3 km of subsidence of the Rockall–Hatton basin with its large accumulation of Miocene sediments. The internal deformation implicit in the relative subsidence casts some doubt on the applicability of some models of plate subsidence away from accreting margins.

Reconstruction of the continental positions prior to opening cannot be entirely satisfactory until the questions posed here are answered. These answers are especially needed to reduce the permissible number of degrees of freedom offered by plate tectonic interpretations of the Caledonides.

## 2b. Evolution of the Caledonides

The Caledonides include those rocks of early Palaeozoic age and older, confined between the Moine thrust zone in northwest Scotland (and its equivalent in East Greenland) and a less well determined line joining the southeastern thrust front in Norway with the Welsh Borders. The rocks vary greatly in age, structure and metamorphic grade but were all deformed and welded into a Caledonian mountain belt

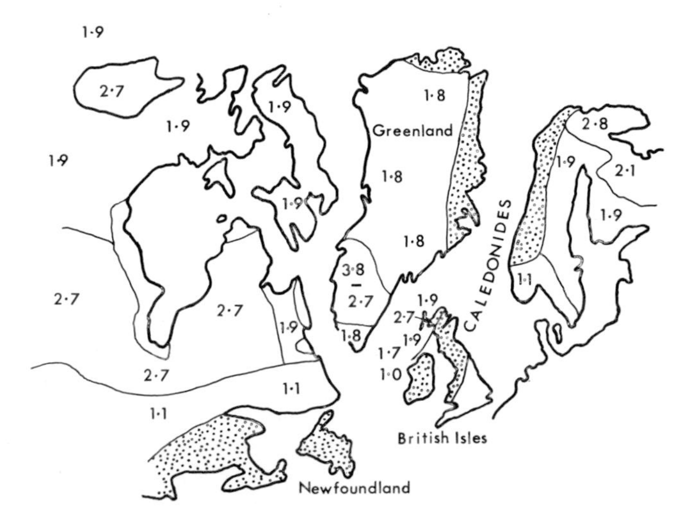


Fig. 2. Map showing ages in billions of years of the peaks of the last isotopically significant thermal events in different units of pre-Caledonian crustal blocks, which are shown as they were distributed in North America, Greenland and northwest Europe 400 m.y. ago.

by 400 m.y. ago and intruded by masses of igneous rock. The belt has variable width as does any element, e.g. the southwesterly convergence of the Moine thrust and the Highland Boundary fault zones effects a narrowing of from 250 km to 100 km in the exposed metamorphic Caledonides. The lateral discontinuities of the belt are matched by temporal variations which suggest that the Scandinavian nappes may be much younger than those in the Scottish Highlands. The age range of events included in the Caledonides is from 800–750 m.y. (pegmatites in the Moine rocks cf. van Breemen *et al.* 1978) to 390 m.y. (latest granites). The gross contrasts of the high grade metamorphic character of the Highlands with the low grade of the Caledonides to the south of the Midland Valley, together with the Cambrian and Ordovician faunal contrasts of the Girvan and Lake District successions (Cowie 1971; Williams 1972; Skevington 1974), have lead to the erection of evolutionary models that postulate the existence of a Proto-Atlantic ocean (Harland and Gayer 1972) whose closure gave rise to subduction of ocean below continent and continental collision. There have been many proposed plate tectonic models of the Caledonides (e.g. Dewey 1969; Fitton and Hughes 1970; Phillips *et al.* 1976; Wright 1976; Lambert and McKerrow 1976), later ones invoking transform movements induced by oblique convergence of plates, and subduction of ocean ridges. All models break the history down into distinct events, so that, for instance, the early evolution of the Scottish Highlands is postulated as occurring in one or more orogenic phases (e.g. Wright 1976) prior to the final continental (or continent–arc–continent) collision that was responsible for the deformation of the Southern Uplands.

Suturing of continental plates and the obduction of oceanic crust might be discernible from the evidence of variation of crustal structure across the Caledonides offered by deep crustal seismic traverses.

### 2c. Reconstruction of the Laurasian Shield

Palaeomagnetic evidence suggests that the Proto-Atlantic may have had a width of 1000±800 km (Briden *et al.* 1973), and that displacement of the continental blocks on either side by movement parallel to the orogen may have occurred during the lifetime of the ocean. Figure 2 gives an indication of the stabilisation dates of the main Archaean and Proterozoic provinces on either side of the Caledonides. In most provinces there is evidence for earlier events and the dates given are only a general indication (±100 m.y.) of the age of the last regionally significant thermal event. Several phases can be matched in time across the Caledonides, but the spatial matching, allowing for some modest movement across, and greater movement along, the strike is more difficult. There do not appear to be other events to match the 1200 m.y. Gardar intrusions of South Greenland or the 1400 m.y. Elsonian event of the Nain province of Canada.

One possible link is the continuation of the (1100–1000 m.y. old) Grenville province across the southern part of the Rockall microcontinent, through western Ireland and Scotland (van Breemen *et al.* 1978), to join with the Gothides of southern Norway and southwest Sweden. It is not clear how to remove any Caledonide strike-slip effect to bring the Grenville province into the correct position opposite the Gothides, but the Grenville Front would have been crossed by the Caledonides and might be located by isotopic dating of older basement within them. The Front may have been slid out along the Moine thrust.

## 3. Main crustal units

The most useful guide to deep crustal structure comes from seismic surveys that have detected the Mohorovičić Discontinuity. The location of such surveys around the eastern North Atlantic seaboard is shown in Figure 1. The surveys must be interpreted with due regard for gravitational variation, magnetic characteristics of the rocks of the sea-bed, and the form of shallow geological features revealed by sampling and high resolution seismic reflection profiling.

The problems of interpretation of seismic surveys and the lack of sufficient areal coverage means that it would be misleading to consider a map of Moho depth variation as a guide to discussion. However, certain generalizations can be drawn from the main conclusions of the surveys of the region (Fig. 3).

Considering the variety of tectonic environments encountered in the surveys, crustal thickness is surprisingly uniform, between 25 km and 35 km, with these extremes found either side of the Faeroe–Shetland channel. The crust is nowhere thicker than average for continental areas and is rather thinner than elsewhere in the foreland of the Caledonides, below the northern North Sea and below the Moray Firth. Smith and Bott (1975) suggested that the crust below the Caledonides on land is about 20% thicker than below the foreland. The L.I.S.P.B. data (Bamford *et al.* 1976) confirm this and the rise in middle crustal boundaries inferred by McLean and Qureshi (1966). The L.I.S.P.B. data suggest continuity of deeper crustal layering from the foreland across the main belt and into the Midland Valley, but the Southern Uplands crustal layering is not so simply correlated (cf. Powell 1970). All

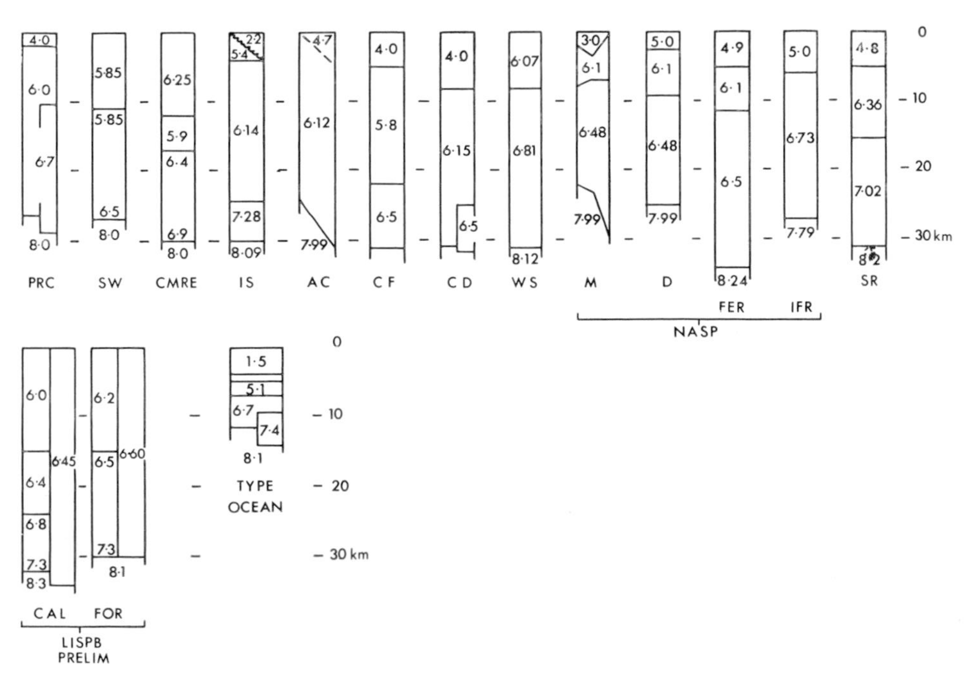


Fig. 3. Summary of P–wave velocity structure of the crust from the surveys shown in Figure 1. Layer velocities are in km/s; alternative interpretations are blocked together. Section WS pertains to the Shetlands; SR to the Rockall plateau; in N.A.S.P., M refers to the Moray Firth, D to the Cape Wrath–Shetlands, FER to the Faeroes and IFR to the Iceland–Faeroes rise; in L.I.S.P.B., FOR refers to the Caledonian foreland north of Cape Wrath, CAL to the Grampian Highlands.

these surveys penetrate thick crust. "Standard" oceanic crust has been revealed off the continental margin of the Western Approaches (cf. Bott and Watts 1971) and between Hatton bank and the Reykjanes ridge, but no surveys have defined the Moho satisfactorily in the Rockall trough: Scrutton and Roberts (1971) obtained an unreversed 4·7 km/s refractor at a depth of 3 km below sea level in the trough. This refractor might be Layer 2 of the oceanic crust.

Regional gravity profiles suggest that the crust is in approximate isostatic equilibrium, so that thin crust is to be expected below major sedimentary basins. This appears to be the case in the Moray Firth district and Sellevoll (1973) reports the crust below the Viking graben to be 4 km thinner than that below the Shetlands and 16 km thinner than that below the Scandinavian Caledonides. However, the amount of thinning is probably underestimated by not taking into account the existence of a thick sediment layer. Compensation is not total for narrow basins like those west of the Orkneys. Here the Moho depth is rather constant, suggesting that graben formation at shallow crustal levels is accompanied by limited compensation by creep in the lower crust.

The Faeroes and the Rockall–Hatton block are confirmed as being composed of continental crust. Gravity profiles across the Rockall–Hatton basin (Scrutton 1972) show crustal thinning comparable with that in the North Sea.

The Iceland–Faeroes rise has a crustal thickness only a little less than the average for the British region and the interpretation of this rise as being anomalously thick oceanic crust is as much the result of the difficulty of retaining it in continental reconstructions as of unambiguous geophysical evidence. It does have a denser crust than the adjacent Faeroes block (Bott *et al.* 1971; Fleischer 1971) and it lacks material of velocity 6·0±0·3 km/s which is typical of continental upper crust.

It is becoming increasingly clear both from recent surveys in and around Britain and elsewhere that the lower parts of the continental crust are complicated. The determination of the deeper structure of the crust depends on the recognition of late-arriving phases such as P* refractions and $P_MP$ reflections and on their time and amplitude relationships to the first arrivals. The data then is hidden comparatively and requires more sophisticated interpretational techniques. As these are being brought into play, it appears that under large areas of Britain a three (or four) layer crust can be discerned. The layers are:

(1) sediments and lavas above crystalline basement ($V_P$ = 2–5 km/s);

(2) upper crustal crystalline basement ($V_P$ = 5·8–6·3 km/s);

(3) mid-crustal rocks ($V_P$ = 6·3–6·5 km/s);

(4) a deep crustal layer in which velocity varies between 6·8 km/s and 7·3 km/s often with strong vertical velocity gradients (up to 0·1 km/s/km) and occasionally increasing gradually to sub-Moho velocities (e.g. 8·0 km/s) so that the Moho is transitional over distances of the order of several kilometres (it usually requires a sharp change over no more than 100–500 m in order to give a clear $P_MP$ reflection).

The deep structure of the crust is thus defined in skeletal form only. Though distinct units can be discerned, the positions of the lateral boundaries are usually undefined (but cf. Bott *et al.* 1976) by the seismic surveys and are only located by shallower features, for example, bathymetry, gravitational edge effects, and magnetic contrasts.

In the following discussion, certain aspects of the deep structure are examined in more detail in relation to their potential uses in forming models of crustal evolution.

## 4. Discussion

### 4a. What is continental crust?

The juxtaposition of the Iceland–Faeroes rise and the Faeroes with their different deep structures prompts the question of what is continental crust. A thickness criterion is scarcely good enough since it would not admit of the possibility of continental crust being substantially thinned to oceanic crustal thicknesses. Inspection of the seismic layerings of Figure 3 indicates the lack of velocities around 6 km/s in oceanic crusts and the significant contrast between velocities of oceanic and continental crust. The former is likely to reflect changes from Layer 2 basalts and dykes (with intense jointing developed because of rapid cooling, and greenschist facies mineral assemblages) to Layer 3 gabbros (comparatively crack-free, and with amphibolite facies assemblages) with an increase in the P wave velocity over a small depth range from 4·5–5·5 km/s to over 6·5 km/s. By contrast, continental crusts contain great thicknesses of acidic-intermediate igneous rocks, low grade meta-sediments and quartzofeldspathic gneisses with P wave velocities in the range 5·8–6·5 km/s. All these rocks are produced in the process of continental manufacture from oceanic crust in island-arc evolution.

In view of these considerations continental crust is here defined, seismologically, as supra-Moho layering containing at least 5 km thickness of material with a P wave velocity in the range 5·8–6·5 km/s. Usually the thickness of that layer will be in the range 10–30 km.

## 4b. A model of continental crustal evolution

Substantial thicknesses of relatively acidic crystalline rocks characterise the upper continental crust. The deeper parts contain higher velocity rocks up to mantle velocities of 8 km/s. The continental crust is laterally as well as vertically inhomogeneous, but this should not detract from generalised models of evolution that account for:

(1) a fairly shallow discontinuity in which velocity jumps to 6·4–6·6 km/s;

(2) a deeper discontinuity across which the velocity increases to 6·7–7·3 km/s and below which velocity may increase steadily towards 8·0 km/s.

The evolutionary model is now described, following Ringwood (summarised in 1975) and then applied to northwest Britain. Oceanic crust is taken as the starting point. On subduction, frictional heating on the shearing surface causes dehydration of amphibolites and the uprising water causes partial melting in the overlying mantle wedge. Fractionation of these hydrous magmas gives rise to the early tholeiitic stage of island-arc development. Residual eclogite partially melts at greater depth to produce acidic magmas which react with overlying mantle to give a calc-alkaline suite of igneous rocks which will come to overlie the earlier tholeiitic rocks as the subduction zone migrates back from the position of the early tholeiitic magmatism.

The tholeiitic phase may be missing if the temperatures are so high that the amphibolite-ecolgite transition produces an acidic silicate melt directly. By this means a thick igneous crust is created, a crust that is heterogeneous but with an overall acid-intermediate composition, except possibly for a basal, basic, layer. Increasing thickness of the volcanic arc then brings about metamorphism of the pile that converts it into a continental crust.

Alternatively, if oceanic crust is being subducted below a continental margin, the andesitic igneous pile will be incorporated within pre-existing continental crustal rocks, and a lower crustal layer of basic composition may be produced by tholeiitic intrusion or earlier basal underplating in the genesis of the margin at a mid-ocean ridge. This may be a more likely way of generating the 50 km thick crust that existed in the metamorphic Caledonides.

If there is a temperature gradient of 30°C/km (Barrovian) in the upper crust reducing slightly with depth to give temperatures of not more than 1200°C at 50 km, then rock of granodioritic composition with high $P_{H_2O}$ above 650°C and at about 24 km depth would be metamorphosed under upper amphibolite facies conditions with the production of granitic magmas that migrate upwards to intrude shallower amphibolite and greenschist facies metavolcanic rocks and derived metasediments. At 700°C and 25–28 km depth the rocks enter granulite facies and the rocks having already passed through amphibolite facies anatexis would be further depleted in incompatible elements and acquired an intermediate, andesitic composition. Metamorphism of the basal tholeiitic layer at say 40–50 km depth (1000–1200°C) would produce garnet granulite, and at greater than 50 km in areas of rather lower temperature eclogite would form. The resultant gabbro-garnet granulite-ecologite transition would be gradual. In traversing the amphibolite facies the basal layer composition would have stayed constant even in the presence of excess water since the production of quartz+clinopyroxene (hydrated to amphibole) from plagioclase

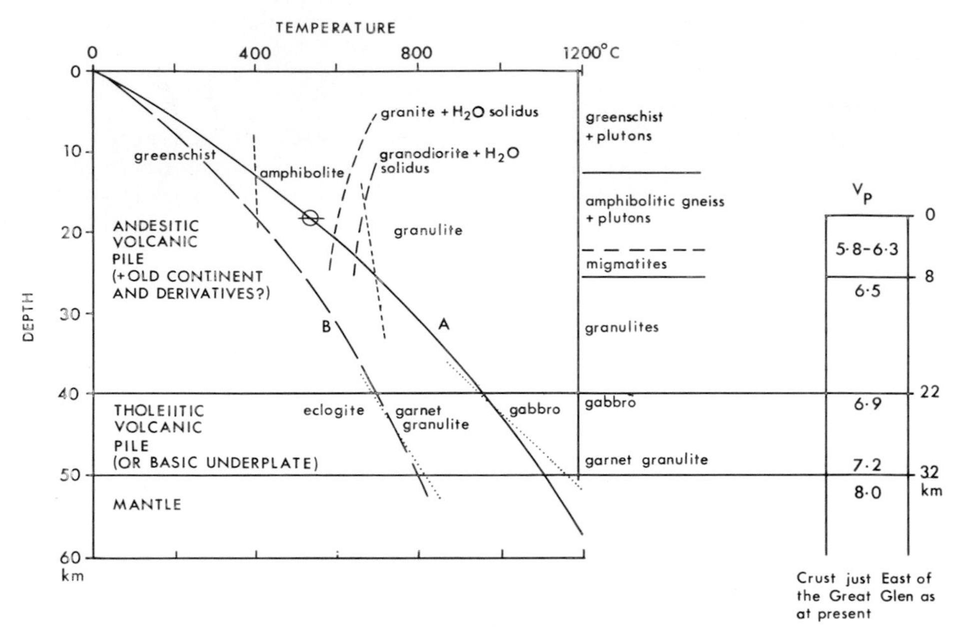


Fig. 4. A model of the evolution of a 3-layer crust applicable to the present structure below the Scottish Highlands. Temperature–depth curve A passes through the fixed point (18·5 km, 535°C) of Richardson and Powell (1976), and its intersections with compositional, phase, facies and melting boundaries signifies the transformation of the original 50 km pile of new igneous rocks (plus old crust) to the layered sequence shown to the right. The rapid uplift and erosion of this sequence preserves them in a metastable state to give the crustal section (P–wave velocity in km/s) on the right hand edge. Temperature–depth curve B suggests that in an environment of lower temperature gradient there might be a more complete transition of gabbro to eclogite which could result in a Moho which, while still a compositional boundary, might be relatively indistinct seismically (like that below the south of Scotland according to Bamford *et al.* 1976).

and the solution of the quartz to give an acidic liquid occurs at pressures too high to occur with this temperature gradient (see e.g. Wyllie 1971 fig. 8–18). Rapid uplift and cooling would preserve a layered continental crust, like that shown in Figure 4, with a heterogeneous upper crust containing a mixture of metamorphosed volcanic rocks and volcanic detritus invaded by granites passing downwards into amphibolitic quartzofeldspathic gneisses of acidic-intermediate composition. Underneath would be granulites of intermediate composition, which in turn could overlie a basic layer progressing from gabbro towards eclogitic rocks.

This is a "dry" model of the continental crust favoured for northwest Britain because of the apparent continuity of Lewisian-like granulites below the Highlands. In Figure 4 the model is applied to the metamorphic Caledonides using the Spean Bridge calculation of Richardson and Powell (1976) that adds, in Caledonian time, 18·5 km to the L.I.S.P.B. crustal thickness of approximately 32 km, so giving a total crustal thickness of 50 km and an upper crustal temperature gradient of 29°C/km. The three-layer crust so formed would have the right characteristics to match the L.I.S.P.B. interpretation. Additionally it is interesting to note that a lower tempera-

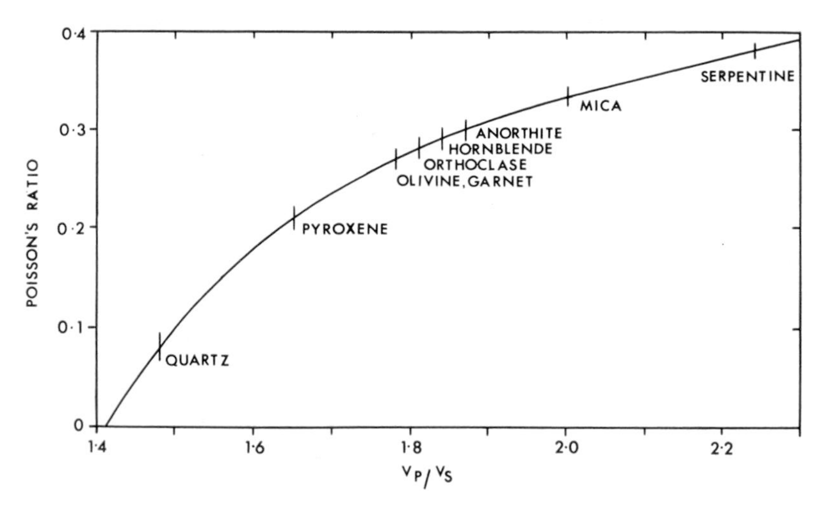


Fig. 5. A graph of Poisson's Ratio (P.R.) against the ratio of velocities of P– and S–waves, with approximate values of isotropic aggregates of minerals estimated from Anderson and Lieberman (1968); anorthite = $An_{50}$.

## Table 1. Some estimates of Poisson's Ratio

| *Survey (See Figs 1, 3)* | $V_p$ *km/s* | *Poisson's Ratio* ($\pm 0{\cdot}02$) | *Comment* |
|---|---|---|---|
| S.W. | 5·85 | 0·18 | Quartz-rich granites |
| L.I.S.P.B. | 6·2 | 0·24 | Gneisses |
| | 5·7 | 0·20 | Low grade quartz schists |
| | 6·26 | 0·23 | Gneisses |
| N.A.S.P. | 6·4 | 0·24 | Granulites |
| | 7·84 | 0·30 | Mantle |

ture gradient (say 15–18°C/km) below the Southern Uplands would push the gabbro-eclogite transition of a deep basal layer much further towards the eclogite composition giving velocities that, in approaching sub-Moho levels, may explain the interpreted "transitional" Moho there.

The model defined above applies to island arc-continent transformation. Andean type orogeny may be viewed similarly but with added ingredients, such as the ingestion of pre-existing continental material and possibly a greater contribution to crustal thickening by direct intrusion of acidic plutons from the mantle. Such a model would not differ radically from that outlined. However, it is necessary to point out that such a model is over simplified, that a rock may describe a complex path in P–T–$P_{H_2O}$ space and that the curve defined merely indicates the facies stabilised by rapid uplift and cooling.

## 4c. The value of Poisson's Ratio in discriminating between alternative interpretations of deep crustal layering

Various suggestions have been made from time to time about the likely constituents of deep crustal layers, both oceanic and continental. With several mineral groups incorporated in any one rock type, it is impossible to differentiate models on P wave velocity ($V_P$) alone. Since S wave velocities ($V_S$) do not depend solely on $V_P$ and there is quite a marked variation in $V_P/V_S$ ratios among the mineral groups (Fig. 5), a knowledge of $V_S$ can be a useful indicator of rock type. Poisson's Ratio is directly calculable from $V_P/V_S$ and is unusually low in quartz and exceptionally high in some of the hydrous minerals, particularly serpentine.

Some examples of Poisson's Ratio in Table 1 (using very rough estimates of $V_S$ from published refraction surveys) show the usefulness of the parameter. However, because of limited accuracy of the measurements, imprecise knowledge of mineral velocities, and the fact that the simple S waves envisaged are not generated in anisotropic materials, it might appear that the usefulness is somewhat limited. If $V_P$ increases with Poisson's Ratio within the limits of our resolution, then nothing new is learnt from the latter parameter. To illustrate the potential two radically different rock types of similar $V_P$ that might both exist in the lower continental crust are quoted by Ringwood (1975). Below are tabulated the densities, $V_P$ and Poisson's Ratio.

| | *Density* g/cm$^3$ | *Velocity* *km/s* | *Poisson's* *Ratio* |
|---|---|---|---|
| Diopside hornblendite | 3·20 | 7·08 | 0·26 |
| High-pressure diorite | 3·13 | 7·3 | 0·21 |

Accordingly consideration of Poisson's Ratio could be a rewarding exercise in investigations of the deeper crust (e.g. Christensen 1972), but better knowledge of velocities of minerals in aggregate form may be required to make effective use of this tool.

## 4d. North Sea subsidence and Atlantic evolution

The North Sea basin has existed since Permian times and is linked northwards to equivalents in Greenland and the Barents Sea shelf. Subsidence in the northern North Sea to accumulate 8 km of Mesozoic and Tertiary sediment, if achieved by thinning of continental crust and maintaining isostatic equilibrium, requires a thinning of about 50 %, equivalent to a stretching of 100 % or a movement of Britain away from Norway of 50–100 km. If that is incompatible with the extension implicit in normal faulting of North Sea sediments, the obvious alternative is to envisage a pre-Mesozoic aborted ocean opening, later filled with sediment. However, this is not consistent with the shallow water facies encountered, particularly of the Zechstein evaporites.

The regional significance of the North Sea subsidence may be seen in the pattern of asymmetrical, tilted block, basins with which it appears to be associated. Both in the Viking graben and, to the west, in the Moray Firth (Chesher and Bacon 1975), Mesozoic sediments dip and thicken (especially lower to middle Mesozoic rocks) towards the west or northwest and are truncated by contemporaneously developed normal faults that uplifted older, often basement, rocks on their western side. The pattern appears again strikingly in the Minch (Hall and Smythe 1972), west of Scotland, and in northeast Greenland (Callomon *et al.* 1972). The similarity with the Red Sea (Lowell and Genik 1972) is marked and a similar cause, namely incipient ocean-floor spreading, might be invoked, perhaps in this instance as a southerly-extending arm of an Arctic Ocean created as an adjustment to the building of the Uralides. The increase in crustal extension northwards suggests that the thermal uplift may have increased to sufficiency for successful sea-floor spreading in the Arctic. The North Sea would thus be remote from the thermal maximum, so explaining the limited volcanicity. This leaves major difficulties with hypotheses of the evolution of the Rockall trough (to which the Minch basin is so close) in either early or late Mesozoic times (Roberts 1975; Russell 1976), but accounts for the Greenland–Europe split in Tertiary times as a geometrical solution to the independent spreading of the North Atlantic southwest of Britain and the slow stretching of Greenland and Britain from Norway, initiated by a massive injection of heat below Iceland. This would also provide an obvious reason for the westward jumps in the spreading axes from the Norwegian Sea to the present Iceland–Jan Mayen ridge (Talwani and Udintsev 1976). However, again there is a difficulty in reconciling the easterly dip of ?Mesozoic sediments on the continental fragment of the Jan Mayen ridge with the asymmetry noted above, for between the Orkneys and the Faeroe–Shetland channel both senses of asymmetry occur in sedimentary basins disposed about gravity high "A" (Watts 1971).

There is as yet no entirely satisfactory relationship of the North Sea evolution to that of the Rockall trough and the other parts of the northeast Atlantic.

*Acknowledgments.* I am grateful to A. C. McLean, D. W. Powell, M. J. Russell, D. K. Smythe and I. R. Vann for countless stimulating discussions, and to N.E.R.C. for a Research Grant (GR3/2491) to study the seismic velocities of metamorphic rocks.

## References

AGGER, H. E. and CARPENTER, E. W. 1964. A crustal study in the vicinity of the Eskdalemuir Seismological Array Station. *Geophys. J. R. astron. Soc.* **9,** 69–83.
ANDERSON, O. L. and LIEBERMANN, R. C. 1968. Sound velocities in rocks and minerals: experimental methods, extrapolations to very high pressures and results. *In* Mason, W. P. (Editor). *Physical Acoustics* IV B, 329–472. Academic Press, New York.
BAILEY, R. J., BUCKLEY, J. S. and CLARKE, R. H. 1971. A model for the early evolution of the Irish continental margin. *Earth Planet. Sci. Lett.* **13,** 79–84.
BAMFORD, S. A. D. and BLUNDELL, D. J. 1970. The South-West Britain Continental Margin Experiment. *Rep. Inst. geol. Sci.* No. 70/14, 143–156.
—— 1971. An interpretation of first-arrival data from the Continental Margin Refraction Experiment. *Geophys. J. R. astron. Soc.* **24,** 213–229.
—— 1972. Evidence for a low velocity zone in the crust beneath the Western British Isles. *Geophys. J. R. astron. Soc.* **30,** 101–105.
——, FABER, S., JACOB, B., KAMINSKI, W., NUNN, K., PRODEHL, C., FUCHS, K., KING, R. and WILLMORE, P. 1976. A Lithospheric Seismic Profile in Britain – I. Preliminary Results. *Geophys. J. R. astron. Soc.* **44,** 145–160.
BLUNDELL, D. J. and PARKS, R. 1969. A study of the crustal structure beneath the Irish Sea. *Geophys. J. R. astron. Soc.* **17,** 45–62.
BOTT, M. H. P. 1971. Evolution of young continental margins and formation of shelf basins. *Tectonophysics* **11,** 319–327.
—— 1973. Shelf subsidence in relation to the evolution of young continental margins. *In* Tarling, D. H. and Runcorn, S. K. (Editors). *Implications of Continental Drift to the Earth Sciences* **2,** 675–683. Academic Press, London.
——, BROWITT, C. W. A. and STACEY. A. P. 1971. The deep structure of The Iceland–Faeroe Ridge. *Mar. geophys. Res.* **1,** 328–351.
—— and WATTS, A. B. 1971. Deep structure of the continental margin adjacent to the British Isles. *Rep. Inst. geol. Sci.* No. 70/14, 89–109.
——, HOLDER, A. P., LONG, R. E. and LUCAS, A. L. 1970. Crustal structure beneath the granites of southwest England. *In* Newall, G. and Rast, N. (Editors). *Mechanism of igneous intrusion,* 93–102. *Geological Journal Special Issue* No. 2.
——, NIELSEN, P. H. and SUNDERLAND, J. 1976. Converted P–waves at the continental margin between the Iceland–Faeroe Ridge and the Faeroe Block. *Geophys. J. R. astron. Soc.* **44,** 229–238.
——, SUNDERLAND, J., SMITH, P. J., CASTEN, U. and SAXOV, S. 1974. Evidence for continental crust beneath the Faeroe Islands. *Nature, Lond.* **248,** 202–204.
BRIDEN, J. C., MORRIS, W. A. and PIPER, J. D. A. 1973. Palaeomagnetic studies in the British Caledonides—VI. Regional and Global implications. *Geophys. J. R. astron. Soc.* **34,** 107–134.
BULLARD, E., EVERETT, J. E. and SMITH, A. G. 1965. The fit of the continents around the Atlantic. In *A symposium on Continental Drift. Philos. Trans. R. Soc. Lond.* **258,** 41–51.
CALLOMON, J. H., DONOVAN, D. T. and TRÜMPY, R. 1972. An annotated map of the Permian and Mesozoic formations of East Greenland. *Medd. Grønland* **168,** No. 3, 1–35.
CHESHER, J. A. and BACON, M. 1975. A deep seismic survey in the Moray Firth. *Rep. Inst. geol. Sci.* No. 75/11, 13 pp.
CHRISTENSEN, N. I. 1972. The abundance of serpentinites in the ocean crust. *J. Geol.* **80,** 709–719.
COLLETTE, B. J., LAGAAY, R. A., RITSEMA, A. R. and SCHOUTEN, J. A. 1970. Seismic investigations in the North Sea, 3–7. *Geophys. J. R. astron. Soc.* **19,** 183–199.
COWIE, J. W. 1971. Lower Cambrian faunal provinces. *In* Middlemiss, F. A., Rawson, P. F. and Newall, G. (Editors). *Faunal Provinces in Space and Time,* 31–46. *Geological J. Special Issue* No. 4.
DETRICK, R. S., SCLATER, J. G. and THIEDE, J. 1977. The subsidence of aseismic ridges. *Earth Planet. Sci. Lett.* **34,** 185–196.

DEWEY, J. F. 1969. Evolution of the Appalachian–Caledonian Orogen. *Nature, Lond.* **222,** 124–129.
FITTON, J. G. and HUGHES, D. J. 1970. Volcanism and plate tectonics in the British Ordovician. *Earth Planet. Sci. Lett.* **8,** 223–228.
FLEISCHER, U. 1971. Gravity surveys over the Reykjanes Ridge and between Iceland and the Faeroe Islands. *Mar. geophys. Res.* **1,** 314–327.
HALL, J. and SMYTHE, D. K. 1972. Discussion of the relation of Palaeogene ridge and basin structures of Britain to the North Atlantic. *Earth Planet. Sci. Lett.* **17,** 54–60.
HARLAND, W. B. and GAYER, R. A. 1972. The Arctic Caledonides and earlier oceans. *Geol. Mag.* **109,** 289–314.
LAMBERT, R. ST J. and MCKERROW, W. S. 1976. The Grampian Orogeny. *Scott. J. Geol.* **12,** 271–292.
LAUGHTON, A. S. 1971. South Labrador Sea and the evolution of the North Atlantic. *Nature, Lond.* **232,** 612–617.
——, BERGGREN, W. A. *et al.* 1972. Sites 116 and 117, In *Initial reports of the Deep Sea Drilling Project* **12,** 395–671.
LE PICHON, X. and FOX, P. J. 1971. Marginal offsets, fracture zones, and the early opening of the North Atlantic. *J. geophys. Res.* **76,** 6294–6308.
LOWELL, J. D. and GENIK, G. J. 1972. Sea-floor spreading and structural evolution of the Red Sea. *Bull. Am. Ass. Petrol. Geol.* **56,** 247–259.
MCKENZIE, D. 1967. Some remarks on heat flow and gravity anomalies. *J. geophys. Res.* **72,** 6261–6273.
MCLEAN, A. C. and QURESHI, I. R. 1966. Regional gravity anomalies in the western Midland Valley of Scotland. *Trans. R. Soc. Edinb.* **66,** 267–283.
PHILLIPS, W. E. A., STILLMAN, C. J. and MURPHY, T. 1976. A Caledonian plate tectonic model. *J. geol. Soc. Lond.* **132,** 579–609.
POWELL, D. W. 1970. Magnetised rocks within the Lewisian of Western Scotland and under the Southern Uplands. *Scott. J. Geol.* **6,** 353–369.
RICHARDSON, S. W. and POWELL, R. 1976. Thermal causes of Dalradian metamorphism in the central Highlands of Scotland. *Scott. J. Geol.* **12,** 237–268.
RINGWOOD, A. E. 1975. *Composition and petrology of the Earth's Mantle.* McGraw-Hill, New York.
ROBERTS, D. G., MATTHEWS, D. H. and EDEN, R. A. 1972. Metamorphic rocks from the southern end of Rockall Bank. *J. geol. Soc. Lond.* **128,** 501–506.
—— 1975. Marine geology of the Rockall Plateau and Trough. *Philos. Trans. R. Soc.* **278A,** 447–509.
RUSSELL, M. J. 1976. A possible Lower Permian age for the onset of ocean floor spreading in the northern North Atlantic. *Scott. J. Geol.* **12,** 315–323.
SCRUTTON, R. A. 1970. Results of a seismic refraction experiment on Rockall Bank. *Nature, Lond.* **227,** 826–827.
—— and ROBERTS, D. G. 1971. Structure of the Rockall Plateau and Trough, N.E. Atlantic. *Rep. Inst. geol. Sci.* No. 70/14, 77–87.
—— 1972. The crustal structure of Rockall Plateau micro-continent. *Geophys. J. R. astron. Soc.* **27,** 259–275.
SELLEVOLL, M. A. 1973. Mohorovicic Discontinuity beneath Fennoscandia and adjacent parts of the Norwegian Sea and the North Sea. *Tectonophysics* **20,** 359–366.
SKEVINGTON, D. 1974. Controls influencing the composition and distribution of Ordovician graptolite faunal provinces. *In* Rickards, R. B., Jackson, D. E. and Hughes, C. P. (Editors). *Graptolite studies in honour of O. M. B. Bulman,* 359–373. *Spec. Pap. Palaeont.* **13.**
SMITH, P. J. and BOTT, M. H. P. 1975. Structure of the crust beneath the Caledonian foreland and Caledonian belt of the North Scottish Shelf region. *Geophys. J. R. astron. Soc.* **40,** 187–205.
TALWANI, M. and ELDHOLM, O. 1972. Continental margin off Norway: a geophysical study. *Geol. Soc. Am. Bull.* **83,** 3575–3606.
—— and UDINTSEV, G. 1976. Tectonic synthesis. In *Initial Reports of the Deep Sea Drilling Project* **38,** 1213–1242.
VAN BREEMEN, HALLIDAY, A. N., JOHNSON, M. R. W. and BOWES, D. R. 1978. Crustal additions in late Precambrian times. *In* Bowes, D. R. and Leake, B. E. (Editors). *Crustal evolution in northwestern Britain and adjacent regions,* 81–106. *Geol. J. Spec. Issue* No. 10.
VANN, I. R. 1974. A modified predrift fit of Greenland and western Europe. *Nature, Lond.* **251,** 209–211.
VOGT, P. R. and AVERY, O. E. 1974. Detailed magnetic surveys in the northeast Atlantic and Labrador Sea. *J. geophys. Res.* **79,** 363–389.
WATTS, A. B. 1971. Geophysical investigations on the continental shelf and slope north of Scotland. *Scott. J. Geol.* **7,** 189–218.
WHITMARSH, R. B., LANGFORD, J. J., BUCKLEY, J. S., BAILEY, R. J. and BLUNDELL, D. J. 1974. The crustal structure beneath Porcupine Ridge as determined by explosion seismology. *Earth Planet. Sci. Lett.* **22,** 197–204.

WILLIAMS, A. 1972. Distribution of brachiopod assemblages in relation to Ordovician palaeography. *In* Hughes, N. F. (Editor). *Organisms and continents through time,* 241–269. *Spec. Pap. Palaeontol.* **12.**

WILLIAMS, C. A. and MCKENZIE, D. 1971. The evolution of the northeast Atlantic. *Nature, Lond.* **232,** 168–173.

WILLMORE, P. L. 1973. Crustal structure in the region of the British Isles. *Tectonophysics* **20,** 341–357.

WRIGHT, A. E. 1976. Alternating subduction direction and the evolution of the Atlantic Caledonides. *Nature, Lond.* **264,** 156–160.

WYLLIE, P. J. 1971. *The dynamic Earth: textbook in geosciences.* Wiley, New York.

Jeremy Hall,
Department of Geology,
University of Glasgow,
Glasgow G12 8QQ, Scotland.

# *Part 2*

# Basement formation and distribution

# Shield formation in early Precambrian times: the Lewisian complex

*D. R. Bowes*

The Lewisian complex of northwestern Britain, which has a history from 2·8 to 1·5–1·3 b.y., is largely made up of mantle-derived products of a major crust-forming episode in late Archaean times. Isotopic data rule out derivation both by 'reactivation' of crust formed before this time and by subsequent large scale 'reactivation' in Proterozoic times. The varying effects and expressions of both tectonic overprinting and reheating are assessed on the basis of isotopic data related to specific events in structural, metamorphic and igneous sequences determined for the various regions. These and the histories of sedimentary and igneous crustal additions are interpreted in terms of three major cycles of crustal history whose time spans of 2·8–2·5 (2·4) b.y. (Scourian), 2·4–2·2 b.y. (Inverian) and 2·2–1·5 (or younger) b.y. (Laxfordian) and separate parts have been established on the basis of integrated field and isotopic studies.

## 1. Introduction

The Precambrian crystalline rocks of northwestern Britain, referred to as the Lewisian complex, have probably been subjected to more detailed investigation than any other shield area rocks in the world. The earlier studies of Peach *et al.* (1907, 1910), Peach and Horne (1930), Richey and Thomas (1930) and Sutton and Watson (1951a) in the Northwest Highlands of Scotland, of Jehu and Craig (1923, 1925, 1926, 1927, 1934), Dougal (1928) and Davidson (1943) in the Outer Hebrides, Hull *et al.* (1890) and McCallien (1930) in Eire, and many others (references in Bailey 1951; Phemister 1960; Anderson 1965; Bowes 1969), have been followed by intense field and laboratory activity over the past fifteen years. Marine geophysical work, particularly, has demonstrated the extent of the complex (Fig. 1) and its possible link with Precambrian rocks in Northwest Norway (Talwani and Eldholm 1972) while the nature of the rocks of the complex have been significant in interpreting the crustal structure of Europe adjacent to the North Atlantic Ocean (Smith and Bott

1975; Hall and Al-Haddad 1976; Bamford *et al.* 1977; Hall 1978). Much of the complex has been remapped at 1:10,000 and considerable parts at 1:2,500. This, together with related structural and petrographical investigations, have revealed extensive polyphase deformational, polymetamorphic and igneous histories for the various regimes which provide bases for interpretation of crustal history.

The corresponding period has also seen a marked increase in knowledge of the behaviour of isotopes in rock and mineral systems that has led to a clearer understanding of the relationships of isotopic data to geological events. With progressively more of the isotopic data being linked to specific igneous and metamorphic events (cf. Fig. 2), whose positions in the overall sequence of events has been established on the basis of field evidence, various phases or cognate groups of phases of structural, metamorphic and igneous activity can be assigned to, or correlated with major episodes of earth history with far more confidence than previously. In addition, both the geographical distribution and time span of the Lewisian complex have been generally established. The isotopic evidence of van Breemen *et al.* (1976, 1978) precludes the interpretation of the Annagh Gneiss Complex of West Mayo, Eire, as being an extension of the Lewisian complex (Max 1970; Dunning and Max 1975). Likewise suggestions linking the Moine rocks and the Lewisian rocks (cf. Read 1934) are not supported by isotopic evidence (Brook *et al.* 1977; van Breemen *et al.* 1978) or field evidence (Ramsay 1958, 1963). However, the presence of 'inliers' of the Lewisian complex within the Moine assemblage (Peach *et al.* 1913) is confirmed by the isotopic data of Moorbath and Taylor (1974)—they are now generally regarded as thrust slices or cores of recumbent folds and have a wide distribution in the Northern Highlands (Figs 1, 2; Johnstone 1975 fig. 6). This, together with geophysical evidence and the U-Pb zircon systems of some of the Caledonian granites (Pidgeon and Aftalion 1978), is indicative of Lewisian rocks underlying at least a considerable part of the Caledonides.

The time span of the complex is *c.* 2900–2800 m.y.—*c.* 1300–1500 m.y. The isotopic evidence precludes the long Archaean history (back to *c.* 3800 m.y.) proposed by Davies (1975a) and inferred by Bridgewater *et al.* (1973), with no early Archaean ages demonstrated for any part of the complex except possibly for some detrital grains (Bowes 1976a; Bowes *et al.* 1976; Moorbath 1976; Chapman and Moorbath 1977). Until the recent work of Brook *et al.* (1976) and van Breemen *et al.* (1978), that showed there was isotopic evidence for a tectonothermal episode *c.* 1100–1000 m.y. ago in other parts of Scotland and in Ireland, the time of deposition of the base of the unconformably overlying Torridonian sediments was generally taken as limiting the time span of the Lewisian complex (cf. Sutton 1967, pp. 504–5); this time is *c.* 975 m.y. ago (Moorbath 1969—all ages calculated or recalculated with $\lambda^{87}Rb = 1.42 \times 10^{-11}\ yr^{-1}$). However, the demonstration of strong influence of the Grenville cycle in part of this crustal region accounts for some effects shown by isotopic systems of minerals in the Lewisian complex (p. 60) and puts its lower limit at the initiation of the new cycle, probably about 1300–1500 m.y. ago (Fig. 3). This means that a *c.* 1500 m.y. time span, from late Archaean to middle Proterozoic times, is represented in rocks of the complex. Such a time span does not cover the development of very early crustal material, as seen in Greenland and elsewhere. However, it does cover a period during which there were major crustal developments that are shown not only in the Baltic Shield of northwest Europe, but also in other shield areas such as Canada, U.S.A. and Western Australia. These include the development of granite–greenstone terrain in the

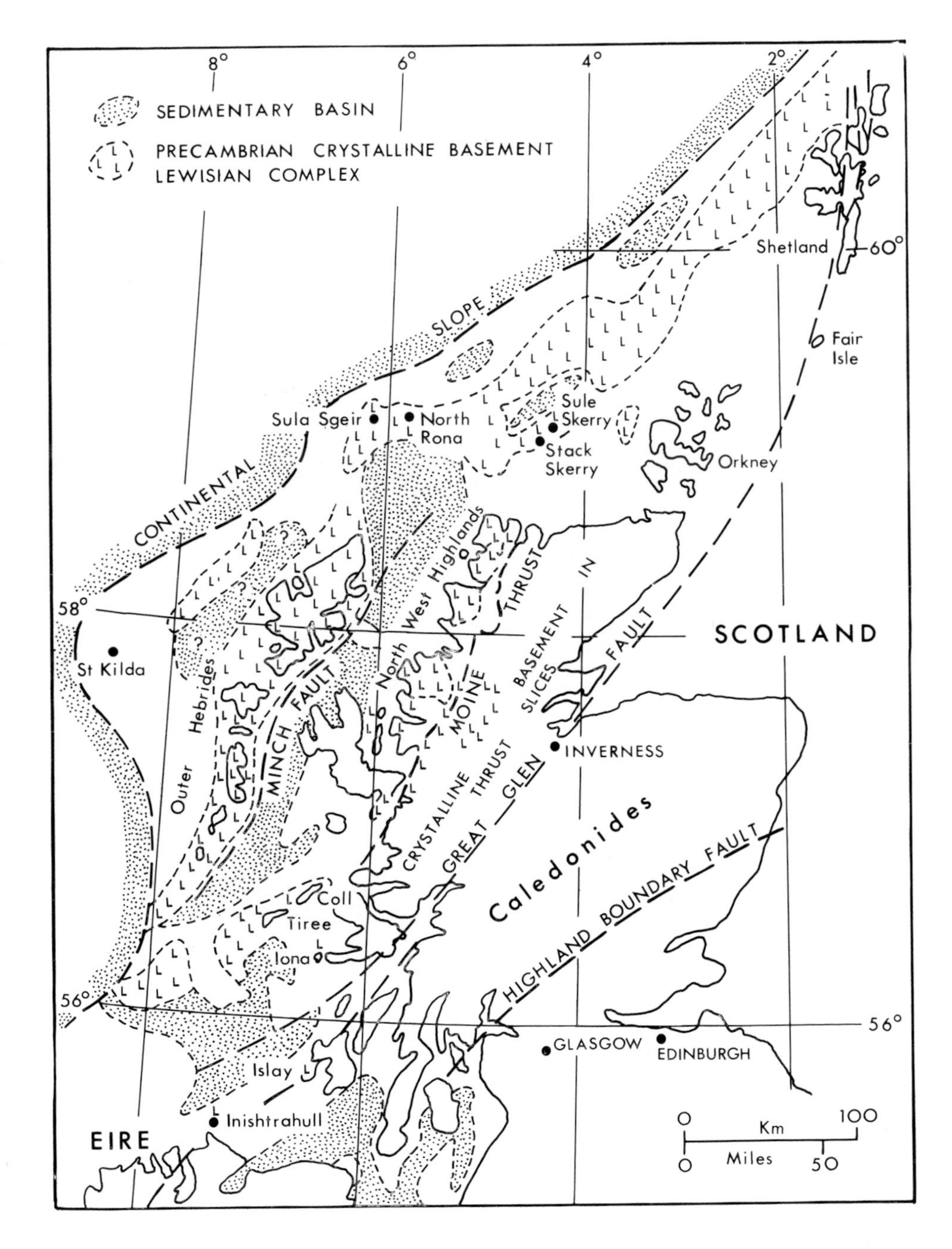


Fig. 1. Distribution of the Lewisian complex in northwestern and northern Britain (mainly based on 1:2,500,000 map—The sub-Pleistocene Geology of the British Isles and Adjacent Continental Shelf; Institute of Geological Sciences, 1972).

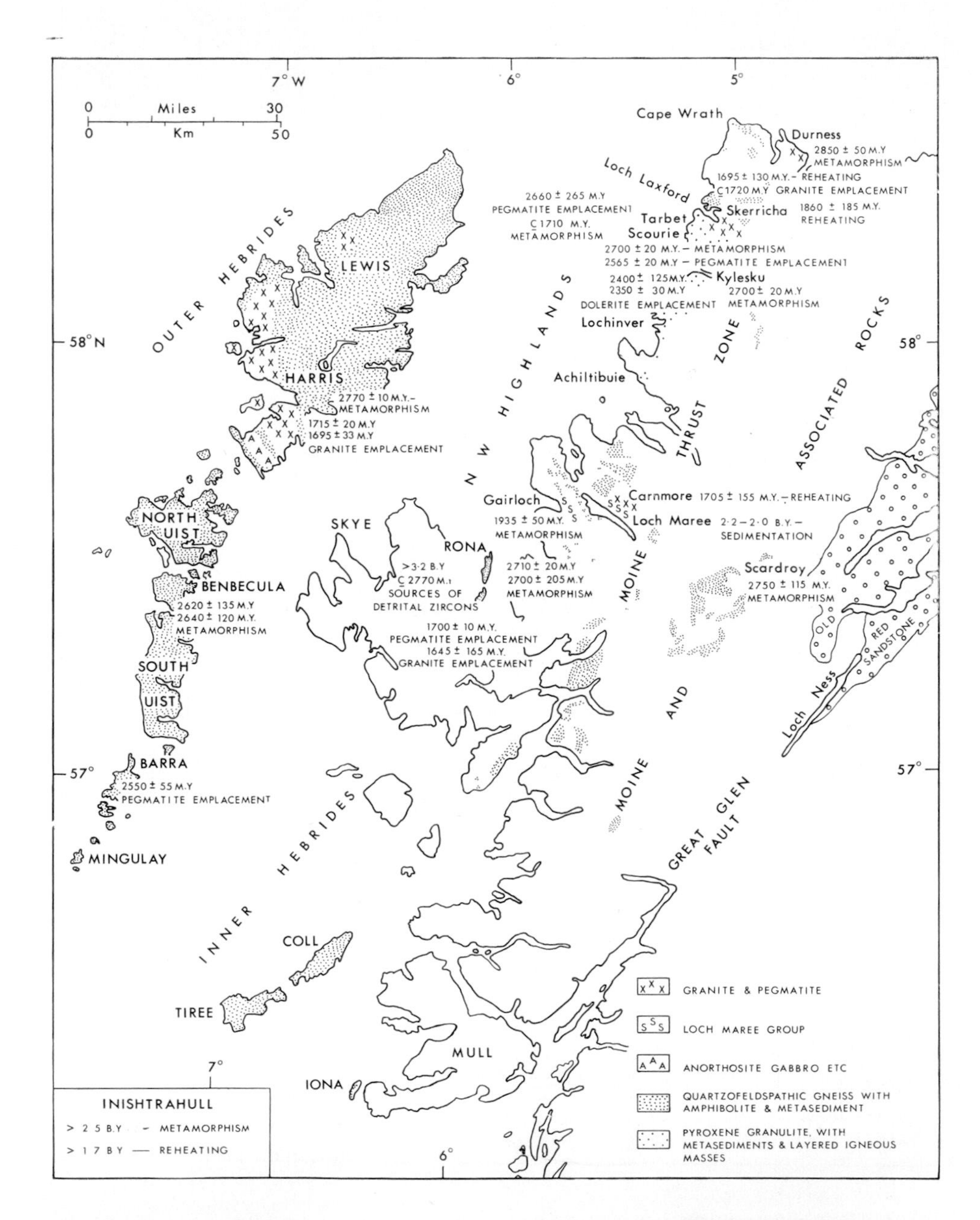


Fig. 2. Major lithological units of the Lewisian complex and a summary of isotopic data obtained from Rb–Sr whole-rock, Rb–Sr mineral—whole-rock and U–Pb zircon studies.

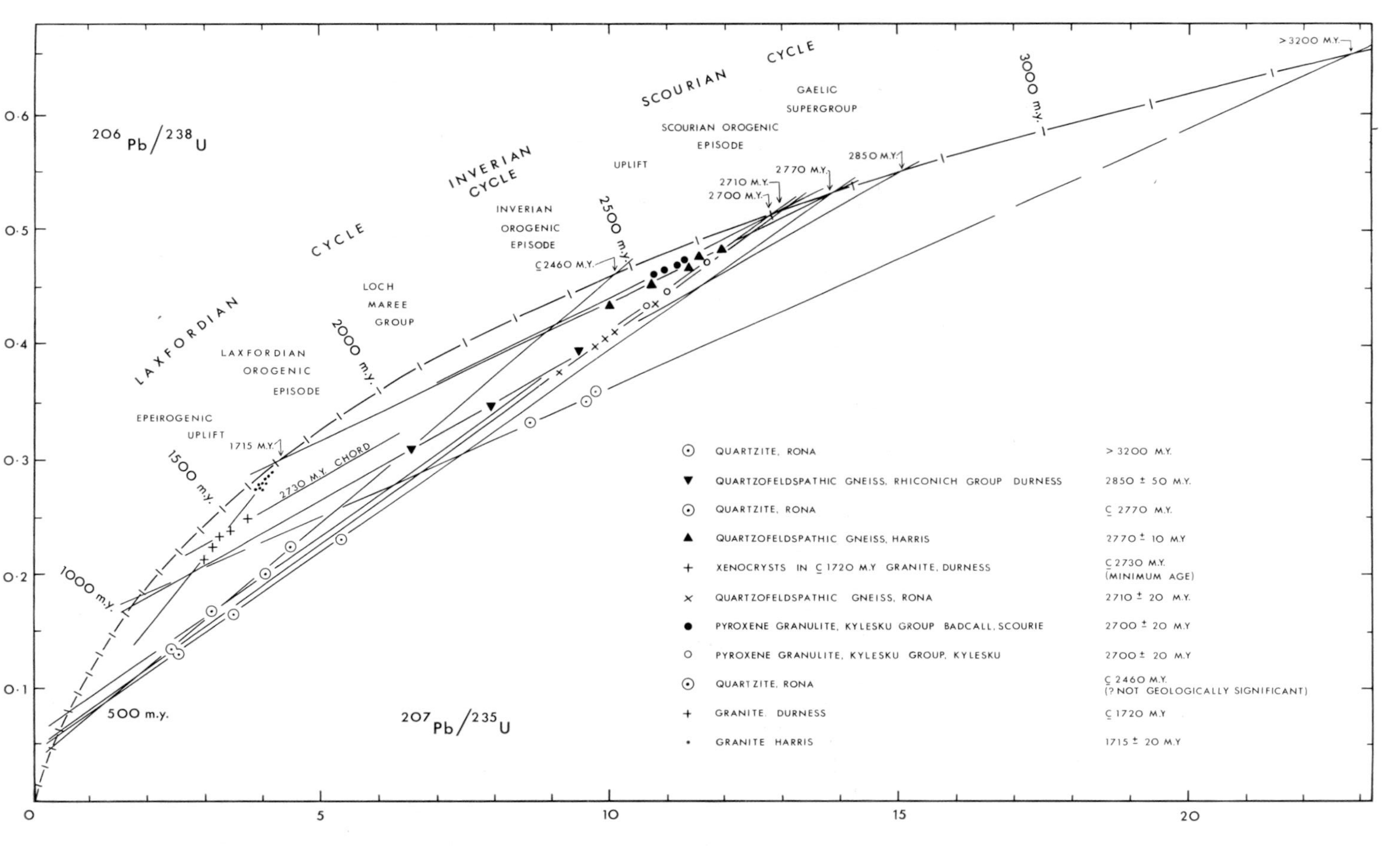


Fig. 3. U–Pb zircon data and time spans of major episodes of earth history for the Lewisian complex.

2800–2600 m.y. period, with vast new additions of crustal material (e.g. Mutanen 1976; Bowes 1976b—Finland; Krogh and Davis 1971 and Arth and Hanson 1975—North America; Roddick *et al.* 1976—Western Australia). Also included is the period in early Proterozoic times when thick sedimentary piles were accumulated and the period in early–middle Proterozoic times during which there was widespread orogenic activity, e.g. in Fennoscandia (Bowes 1976c).

Not only have the isotopic data permitted the geographical distribution and time span of the Lewisian complex to be established with some confidence, but they provide a means of assessing the nature and relative importance of the processes operative in its development. In particular they have been used to assess the extent to which the complex represents, on the one hand, the products of massive 'reworking' or 'reactivation' of existing crustal material or, on the other hand, of crustal additions from the upper mantle that have largely retained their characters despite subsequent tectonic and thermal overprinting. The results of this assessment bear not only on the origin and evolution of the Precambrian basement complex of Scotland, but also on the interpretation of Precambrian shield areas in general. This is because the concept that much of the Lewisian complex represents the products of intense structural and metamorphic modification of earlier-formed rocks, with 'reactivation' producing virtually new rocks over areas of crustal dimensions (Sutton and Watson 1951a pp. 293–5), has been applied generally to other shield areas (e.g. Watson 1967, 1973; Read and Watson 1975), including Greenland (cf. Pulvertaft 1968; Chadwick *et al.* 1974). The crustal model presented for the Lewisian complex, based on this concept, is that of relict kernels of pyroxene–bearing rocks (cf. Read and Watson 1975 pp. 44–6) which retain structural patterns and textures developed during the Scourian orogeny (*c.* 2800–2600 m.y.), amongst a 'reactivated', 'reworked' or 'regenerated' complex mainly of quartzofeldspathic gneiss and amphibolite (the 'Laxfordian complex'). These amphibolite facies rocks, which constitute the main bulk of the Lewisian complex (Fig. 2), have been interpreted by Sutton and Watson (1951a fig. 13; 1962 fig. 6; 1969 fig. 1) and Watson (1965 fig. 2.2; 1975) as being the result of massive 'reworking' during the Laxfordian orogeny (*c.* 1950–1700 m.y.—see Bikerman *et al.* 1975) and as showing rapid variations in tectonic styles. A key factor in this interpretation is the relationship of structural and metamorphic features of the banded and foliated rocks to basic minor intrusions which were assumed to belong to one single related suite ('Scourie dykes') and to be a stratigraphic time-marker (Sutton and Watson 1951a p. 292).

While this interpretation has been considered by some workers to be inconsistent both with field evidence (e.g. Bowes 1962, 1968a,b, 1969, 1976c; Bowes and Ghaly 1964; Dash *et al.* 1974; Bowes and Hopgood 1975a) and also with theoretical considerations (Whitten 1951; Hopgood 1966), the results of Rb–Sr and K–Ar isotopic studies of minerals have been cited as confirmation of such an interpretation (Sutton and Watson 1969 p. 119, 121; Dalziel 1969 p. 9; Owen 1976 p. 43). However, as set out later, Rb–Sr and Pb–Pb whole-rock, U–Pb zircon and initial $^{87}Sr/^{86}Sr$ ratio data place severe restraints on the extent to which 'reactivation' has been operative during the development of the Lewisian complex and point to a crystalline basement whose major chemical, mineralogical and fabric features are indicative of the nature of materials and processes *c.* 2800–2700 m.y. ago, i.e. of late Archaean times (Bowes 1978a).

These isotopic studies, like corresponding studies of Precambrian rocks in Greenland (Moorbath 1977 p. 166–71), bear on the interpretation of crustal evolution of shield areas bordering the eastern part of the North Atlantic Ocean. They also bear on attempts to determine continental configurations at various stages of Precambrian times, not least because the much quoted correlation of Precambrian rocks between Northwest Scotland and West Greenland (e.g. Sabine and Watson 1965 p. 484; Windley 1977 fig. 1. 2), relies heavily on an interpretation of the Lewisian complex that, to a large extent, is not consistent with the available isotopic data (cf. Figs 3, 9 and Bridgewater *et al.* 1973, table 1).

## 2. Rock units

In outline, though not in detail, the gross distribution of lithological units in the Lewisian complex is simple (cf. Peach *et al.* 1907, folding map). Throughout large parts it is generally similar to that between Loch Laxford and Durness (Figs 2, 7) where the Rhiconich group (Dash 1969 p. 349; Rhiconich group is preferred to 'Laxford assemblage'—Holland and Lambert 1973—on grounds of priority) consists predominantly of grossly banded amphibolite facies quartzofeldspathic and related gneiss ($SiO_2$ 64–86%) together with amphibolite and subordinate sedimentary units including quartzites and calc-silicate rocks (Chowdhary 1971; Chowdhary *et al.* 1971). The average composition of the gneiss (major elements as standard cell, trace elements as parts per million) is: $K_{2\cdot1}$ $Na_{7\cdot5}$ $Ca_{2\cdot7}$ $Mg_{2\cdot0}$ $Mn_{0\cdot03}$ $Fe_{1\cdot9}$ $Al_{15\cdot0}$ $Si_{60\cdot9}$ $Ti_{0\cdot3}$ $P_{0\cdot1}$ $(O_{153\cdot6}\ [OH]_{6\cdot4})_{160}$ Ba 592, Ce 71, Co 39, Cr 32, Cu 26, Ga 17·5, La 58, Nb 6, Ni 30, Rb 101, Sr 588, Zn 57, Zr 144 (Bowes 1976a cf. Bowes 1972; Holland and Lambert 1972c). The overall composition of the gneiss varies somewhat, and possibly significantly in other regions (cf. Nisbet 1961; Bowes 1972; Tarney *et al.* 1972; Holland and Lambert 1972a, 1973) and within relatively small areas a wide variety of gneisses can occur (Myers 1970). 'Grey gneiss' mainly of tonalitic to granodioritic composition is common in considerable parts of the Outer Hebrides (Moorbath *et* al. 1975 fig. 1). There, also, metasediments in the gneiss complex are more varied and more widespread than on the mainland and include semipelitic and pelitic schists and gneisses (Coward *et al.* 1969; Watson and Lisle 1973). Metasediments make up a marked proportion of the complex in the Inner Hebrides, particularly on the islands of Tiree (Sinclair 1964; Westbrook 1972) and Iona (Jehu 1922), with the metasedimentary—igneous rock association on Iona like that in some greenstone belts, but metamorphosed.

In some places associated with metasediments (e.g. Watson 1969 fig. 2), and in other places amongst the gneisses, are ultrabasic to basic, including anorthositic meta-igneous masses. Many are small; they are commonly banded and some show features indicative of rhythmic layering. Primary olivine and pyroxene are generally partially or completely replaced by tremolite–actinolite, talc, serpentine and related products of hydrous retrogressive metamorphism (Bowes *et al.* 1964). Banded or streaky amphibolites with hornblendite layers are thought by Davies and Watson (1977) to form the predominant part of this ultramafic–mafic association. If this is so, then the proportion of these metamorphosed igneous masses in the complex is correspondingly increased. Other amphibolites, which locally are clearly discordant to the dominant gneissic foliation (Figs. 4d, 7e; Dash *et al.* 1974; cf. Park and Cresswell, 1973 p. 126–7), are generally concordant or nearly so; comparable relations between metadolerite and gneiss have been described by Ghaly (1966) and Park and Cresswell (1972). The result is a leucocratic/melanocratic complex (Fig. 4a) similar in general appearance to those parts where gneiss and amphibolite sharing a common metamorphic and structural history are interbanded (Fig. 4b). In places the mafic bands in the predominantly gneissic complex are of more than one generation (Ghaly 1966 fig. 10).

Granulite facies rocks are common in the segment between Scourie and Achiltibuie (Figs. 2, 8) with banded two-pyroxene and hornblende granulites, showing variable effects of retrogression (Ghose 1958; Howie 1964; Beach 1974a), together with very subordinate proportions of metasediment, making up the Kylesku group (Khoury 1968a, p. 44; Muecke 1969; Barooah 1970; Kylesku group is preferred to 'Scourie assemblage'—Holland and Lambert 1973, 1975—on grounds of priority). Only locally do metasediments make up a considerable proportion of the rock assemblage (Davies 1974). Lithological layering is on various scales, 10–20 m, 1–2 m and a few cm–mm, the fine layering possibly resulting from metamorphic segregation before granulite facies metamorphism. However, there is little dimension alignment of minerals and granulose texture predominates in both mafic and felsic (plagioclase $An_{33-44}$ $\pm$ quartz $\pm$ scapolite) bands. Hence, at least for considerable parts of the Kylesku group, 'banded granulite' is a more accurate description of the nature of the rocks than 'gneiss', as used by some authors. It is garnet-bearing, represents metamorphic activity at considerable crustal depth (O'Hara 1975, 1976, 1977) and its average composition between Kylesku and Scourie is $K_{0\cdot9}$ $Na_{6\cdot7}$ $Ca_{6\cdot6}$ $Mg_{4\cdot8}$ $Mn_{0\cdot1}$ $Fe_{4\cdot6}$ $Al_{16\cdot7}$ $Ti_{0\cdot4}$ $Si_{55\cdot0}$ $(O_{152\cdot9}\ [OH]_{7\cdot1})_{160}$ Ba 440, Ce 50, Co 48, Cr 150, Cu 48, Ga 18, La 19, Nb 5·5, Ni 88, Rb 3, Sr 460, Zn 83, Zr 115 (Bowes 1976a cf. Bowes 1972; Holland and Lambert 1972c). This generally andesitic composition (quartz-dioritic—Holland and Lambert 1975) corresponds to that for granulites north of Lochinver (Sheraton 1970). However, there is a wide compositional range of rocks showing a calc-alkaline trend (Bowes, Barooah and Khoury 1971), from basaltic ($SiO_2$ 47%, MgO 9%) to rhyolitic ($SiO_2$ 76%, MgO 0·5%), but from which there has been conspicuous depletion of K, Rb, Th and U as well as water. The K/Rb ratio of 700–1200 is markedly higher than that for the Rhiconich group gneisses (240) with the K/Sr, K/Ba and Rb/Sr ratios very much lower (15 v 40, 10 v 25–50, 0·02 v 0·2, respectively) (Tarney *et al.* 1972; Sheraton, Skinner and Tarney 1973). These geochemical characteristics largely survived retrogressive metamorphism in localised zones of tectonic activity in which new rock fabrics were formed in Proterozoic times.

Amongst the rocks of the Kylesku group are lens-shaped masses of ultrabasic and basic rock of igneous derivation (Fig. 8d; O'Hara 1961a, 1965; Bowdidge 1969; Vernon 1970; Kerr 1974). This includes metaperidotite, metapyroxenite, garnet-pyroxene-plagioclase rocks of gabbroic-troctolitic composition (garnet pyriclasite) and meta-anorthosites whose layering and structures resemble wedge and current bedding. They pre-date the granulite facies metamorphism which also profoundly affected their host rocks (Bowes *et al.* 1964, 1966). Forsteritic olivine-chrome spinel and calcic andesine bands and lenses occur locally with some anorthositic material of pegmatitic character showing textures reminiscent of plagioclase cumulate textures (Davies 1974). Many of the features of this rock association correspond to those of very much larger layered igneous masses in other high grade Archaean terrain, e.g. Fiskenaesset, West Greenland and Sittampundi, South India (Windley 1973; Escher and Watt 1976), while petrochemical fields and trends, including the iron enrichment in the pyriclasites (Bowes *et al.* 1966, 1970; Bowes, Barooah and Khoury 1971) correspond to those for layered sills in low grade Archaean terrain (cf. Williams and Hallberg 1973). The extent to which rocks composed predominantly of hornblende and plagioclase, included by Barooah (1967), Khoury (1968a) and others as part of the Kylesku group, represent tectonically detached parts of layered igneous complexes, as proposed by Davies (1974) for an igneous complex south of Loch Laxford, awaits determination of criteria indicative of their petrogenesis.

Granulite facies rocks, which are strongly magnetised with properties like those of the mainland (Powell 1970 fig. 6) occur as strips on the eastern side of both Barra (Hopgood 1971a) and South Uist (Kursten 1957; Coward 1972) above the Hebridean (Outer Isles) thrust (Francis and Sibson, 1973; Moorbath *et al.* 1975 fig. 1). Their structural position above amphibolite facies tonalitic and related gneiss is the result of movements that post-date the Lewisian complex. The granulite facies rocks of South Harris include spinel– and garnet–amphibole lherzolites (Livingstone 1976). They are much affected by dynamic metamorphism (Dearnley 1963), also appear to have been tectonically emplaced, but during the Lewisian history of the rocks. (cf. Wood 1975).

A considerable variety of igneous rocks cut the gneiss-granulite complex. These include dykes of dolerite, picrite and related rocks, now variously affected by metamorphism, which are common in the Scourie–Kylesku–Assynt region (Fig. 8e; Peach *et al.* 1907; Sutton and Watson 1951a, b; Burns 1966; Tarney 1973) and include the Scourie dyke (Teall 1885). Basic minor intrusions with varying structural and age relations and chemical characteristics are common in the Gairloch–Loch Maree region (e.g. Peach *et al.* 1907; Bowes and Park 1966; Ghaly 1966; Park 1966; Power and Park 1969), but generally of less widespread occurrence elsewhere (eg. Hopgood 1971a; Park and Cresswell 1972, 1973; Dash *et al.* 1974; Standley 1975a; Taft 1978). The anorthosite-gabbro-tonalite complex of South Harris appears to be unique in the region (Davidson 1943; Dearnley 1963; Palmer 1971). Abundant granitic and pegmatitic intrusions occur particularly in the Loch Laxford and Durness districts of the Northwest Highlands (Fig. 7d; Peach *et al.* 1907; Inglis 1966; Bowes 1967; Findlay 1970; Gillen 1975) and in South Harris (von Knorring and Dearnley 1960; Myers 1971a).

The metasediments of the Loch Maree Group (including the Gairloch metasediments) whose deposition took place in early Proterozoic times (Bikerman *et al.* 1975) include banded ironstones, graphitic schists, marbles and manganese-rich units in addition to the siliceous mica schists of greywacke derivation that predominate (Peach *et al.* 1907, 1913; Bowes and Bhattacharjee 1967; Keppie 1969). Intense penetrative deformation under amphibolite facies conditions has destroyed primary sedimentary structures but the distinctive nature of lithological units and their chemical compositions are indicative of a sedimentary derivation. Associated amphibole- and chlorite-bearing lithologies could contain a fraction of basic igneous material (Lambert and Holland 1974) possibly of pyroclastic derivation.

The nature of the lithological types making up the tectonic units of the Lewisian complex within the Caledonides (Figs. 1, 2; cf. Sutton and Watson 1959; Johnson 1960; Ramsay 1963; Johnstone 1975; Barber and May 1976) is generally comparable with that to the west of the Moine thrust. In addition to various kinds of gneiss, some of which are migmatitic in character, and amphibolites with varying structural relations, there are ultramafic bodies and metasediments. Pelitic and semipelitic rocks are represented by kyanite- and garnet-bearing rocks that are interbanded with calc-silicate rocks and marbles (Peach *et al.* 1910; Ramsay 1958); eulysites also occur (Tilley 1936). Locally there are lenticles or bands of eclogite (Alderman 1936; Mercy and O'Hara 1968). These show variable amounts of retrogression, are possibly genetically associated with ultrabasic rocks (Mercy and O'Hara 1965; Essene *et al.* 1966) and a *c.* 1500 m.y. K–Ar pyroxene age has been recorded (Miller *et al.* 1963; Miller and Brown 1965).

The distinctive geochemical features of the Lewisian rocks (cf. Lambert and Holland 1973, 1974) have permitted the existence of small inliers within the Moine assemblage to be confirmed (e.g. Winchester and Lambert 1970; Winchester 1971; Moorhouse 1976).

## 3. Sequence of events

### 3a. In the amphibolite facies gneissic terrain

In the amphibolite facies gneissic terrain, the sequence of events is, in general pattern, similar to that shown in many intensely polyphase deformed mobile zones. Tight to isoclinal folds (Fig. 7a), commonly associated with intense penetrative foliation are representative of early phases, large asymmetrical folds (Fig. 4h) with quartzofeldspathic material in hinge zones and localized new schistosity are representative of middle phases, while open, upright folds (Fig. 4j), together with much granitic injection in parts, are representative of later phases. Consequently refolded folds are common (Fig. 4e), successively emplaced igneous masses may show different relationships to the same structural element, and the different expressions from place to place of successively formed structures results in variations in structural trend from place to place (cf. Bowes and Hopgood 1976).

Such a sequence of events is particularly well shown on the island of *Rona* (Fig. 2) where metasedimentary quartzites (Fig. 4f) amongst the banded gneissic complex (Fig. 4a) include some heavy mineral bands. In summary the sequence of geological events is:

(1) Formation of a lithologically layered sequence, at least in part of sedimentary derivation. The hornblende-rich rocks (Fig. 4b) may represent basic igneous masses either contemporaneous with or later than the sediments or may represent sediments (possibly including tuffs) of basic composition.

(2) Development of an intense penetrative planar mineral alignment which parallels bedding in metasediments and is strong in both gneiss and amphibolite; no associated folds have been identified.

(3) Formation of intrafolial folds ($F_1$)*, mainly isoclinal, and axial planar foliation which reinforces the earlier-formed foliation to make the dominant foliation, which is composite. Intrafolial lenses and dissected parts of folds, particularly hinges, are abundant and there is a strong rodding parallel to the fold hinges and a corresponding mineral lineation. Patchy pegmatite, largely autochthonous and probably the result of anatexis, hornblende lenses and pods resulting from metamorphic differentiation and possibly some agmatite formed at this stage in the sequence.

*For convenience and ease of reference, the various structural elements are here numbered sequentially—folds $F_1$, $F_2$ . . . , foliations/schistosities $S_1$, $S_2$ . . . , lineations $L_1$, $L_2$ . . . Here the commonly occurring foliation that pre-dates the first recognized set of folds is not numbered and the first fold set labelled $F_1$ with its genetically associated features $S_1$ and $L_1$ (cf. Hopgood and Bowes 1972a). In some other publications (e.g. Bowes 1976a), the early foliation is labelled $S_1$ and subsequently formed structures labelled $F_2$, $S_2$, $L_2$, $F_3$, $S_3$, $L_3$ etc. with corresponding labels for deformational phases ($D_2$, $D_3$ etc.).

(4) Emplacement of basic minor intrusions, one set affected by agmatite formation and a later set cross-cutting the agmatites: some of the intrusions show banding which is possibly flow banding that has been subjected to mimetic recrystallization.
(5) Formation of folds ($F_2$), some large and some near isoclinal, which have round and often teat-shaped hinges. They fold the intrafolial folds (Fig. 4c), the dominant foliation (Fig. 4e) and the banding of earlier-formed basic intrusions.
(6) Emplacement of abundant basic sheets, now amphibolites, most of which show near-concordant relationships; clear discordance (Fig. 4d) and emplacement in hinge zones of $F_2$ folds shown in parts. Where banding is shown, it parallels the margins and the rock may superficially resemble amphibolites that have undergone penetrative dynamothermal metamorphism.
(7) Formation of large and smaller asymmetrical folds ($F_3$) with shorter vertical (Fig. 4a, b) or overturned limbs and longer shallowly-dipping limbs (Figs. 4f, h). Axial planes, like new axial planar mineral growth which is prominent in some hinge zones (Fig. 4g), are flat-lying except where they have been re-oriented as the result of superposed deformation (Fig. 4f). This generation of structures is a key for correlation of the structural sequence, both within Rona and outside. The axial trend of the folds is ESE–SE except in the hinge zones of tight, later-formed folds, which is not common.
(8) Emplacement of muscovite-bearing pegmatites, some in the hinge zones of $F_3$ folds.
(9) Development of open to moderately tight upright folds ($F_4$) with rodding and some parallel mineral lineation, and a SE axial trend.
(10) Formation of more or less symmetrical, open, upright folds ($F_5$), with local development of parallel linear fabric elements (Fig. 4j) and an easterly axial trend.
(11)–(12) Emplacement of pegmatite and granite, with $F_5$ axial planes as a structural control; emplacement of basic minor intrusions (Fig. 4k). Mutual time relations have yet to be established.
(13) Formation of very open, upright folds ($F_6$) with a NNE–NE axial trend.
(14) Development of high angle faults with shatter belts and low angled faults with pseudotachylite.

The presence of a few minor intrusions of metadolerite, which show relict ophitic texture and resemble metadolerites *c.* 25 km to the northeast, near Gairloch (Park 1964; Bowes and Ghaly 1964), suggest an additional phase of basic igneous activity. These metadolerites show evidence of deformation during event (9), but no obvious effects of deformation in event (7). However, their very limited occurernce has not permitted definite confirmation of emplacement after event (7).

From the isotopic data for various rocks and minerals, the crustal significance of at least some of the events in the extensive sequence on Rona can be elucidated. Holland and Lambert (1972b) interpreted *c.* 1750–1650 m.y. K–Ar hornblende ages as indicative of *c.* 1750 m.y. metamorphic activity during the Laxfordian orogeny associated with the development of the gneissic complex which accordingly was interpreted as representing a supacrustal assemblage showing the effects of the Laxfordian orogeny. Such an interpretation corresponds to that suggested by Bowes (1968a, 1969), not just for Rona, but for the gneissic part of the Lewisian complex generally: this interpretation was based on *c.* 1600 m.y. mineral ages (Giletti 1959; Giletti *et al.* 1961a, b; Evans 1965) representing the time of metamorphism in a gneiss complex that was not considered to show features of a pre-existing crystalline complex that had been subjected to 'regeneration'. For those workers who considered the complex to show the effects of intense 'reworking',

Fig. 4. The Lewisian complex of Rona: (a) gross banding in steep limb of $F_3$ fold due to near-concordant basic minor intrusions in quartzofeldspathic gneiss; (b) lithological layering, foliation and $F_1$ fold in gneiss/amphibolite—*c.* 2.7 b.y.; (c) $F_2$ fold refolding $F_1$ fold; (d) post-$F_2$ amphibolite discordant to quartzofeldspathic gneiss; (e) $F_1$, $F_2$, $F_3$ relations.

Fig. 4. (cont.). The Lewisian complex of Rona: (f) $F_3/S_3$ in quartzite/amphibolite; (g) $F_3/S_3$ in post-$F_2$ amphibolite; (h) $F_3$ asymmetrical fold deformed by steeply plunging $F_4$ fold; (j) open $F_5$ fold; (k) post-$F_5$ basic minor intrusion.

the mineral isotopic data were interpreted as supporting the hypothesis that the major part of the Lewisian complex represented 'virtually new rocks' formed during the Laxfordian orogeny at the expense of much older rocks formed during the Scourian orogeny in late Archaean times (Sutton and Watson 1969).

The elucidation of the Rona sequence, with its many phases of basic igneous emplacement, showed that the concept of subdividing the Lewisian complex with reference to a single suite of basic minor intrusions, that could be used as a stratigraphic time-marker (Sutton and Watson 1951a), was not applicable to the Lewisian complex generally. But the 1710 m.y. muscovite age for a pegmatite in the middle of the sequence (event 8) and a *c.* 1650 m.y. age for a granite late in the sequence (event 11 or 12; Fig. 6h; Lyon *et al.* 1973) indicated that at least a considerable part of the sequence represented activity during the Laxfordian orogeny. Demonstration that the sequence was almost certainly not that of one orogenic episode, but a combination of parts of sequences widely separated in time, was shown by both Rb–Sr whole-rock and U–Pb zircon data for the gneisses indicating a *c.* 2700 m.y. age (Figs. 3, 5c). Had the Rb–Sr whole-rock data not been available in addition to the U–Pb zircon data, there still could have existed the possibility that the zircon age was the age of source rocks of a sedimentary pile. However, together the U–Pb and Rb–Sr data, not only demonstrate the existence on Rona of a gneiss complex formed during the Scourian orogenic episode but also permit a precise age for the associated major metamorphic event (2710 $\pm$ 20 m.y.) to be established. In addition the U–Pb zircon data points, like the Rb–Sr data, give no indication of disturbance during the Laxfordian episode (Fig. 3) indicating that the 'reactivation' did not play a significant part and that those parts of the sequence on Rona formed during this episode were the result of tectonic overprinting and igneous emplacement, i.e. penetrative effects associated with dynamothermal metamorphism were not widespread during the Laxfordian episode.

The low initial $^{87}Sr/^{86}Sr$ ratio for the gneisses was interpreted by Lyon *et al.* (1973 p. 398) "as reflecting mantle Sr rather than Sr from regenerated crustal material" (cf. Moorbath 1975). This conclusion corresponds to that of Moorbath *et al.* (1969), based on Pb isotopic studies, for the Lewisian complex in general: the formation of the complex in late Archaean times is related to intense metamorphic modification of material very soon after its separation from the upper mantle, not of pre-existing crustal material, with later metamorphic activity causing only "minor redistribution of uranium and/or lead leading to some scatter about the $^{206}Pb/^{204}Pb$ versus $^{207}Pb/^{204}Pb$ lead line" (p. 245).

Isotopic evidence for the existence of material that does pre-date the 2710 $\pm$ 20 m.y. old gneiss complex has come from a study of the zircon population of quartzite (Bowes *et al.* 1976). Data points for some grains fall on a chord with *c.* 2770 and *c.* 400 m.y. intersections (Fig. 3). This, together with the near euhedral grain shape, indicates the possibility of the existence of material derived from an igneous pile itself derived from a mantle source very shortly before the Scourian orogenic episode. Other almost round grains are older than the initiation of the Lewisian complex and their source region may have a history going back to 3250 m.y., or even earlier.

At the other end of the time span represented by the sequence shown by the Lewisian complex on Rona are *c.* 1550 m.y. K–Ar ages for the muscovite dated at 1710 m.y. by Rb–Sr methods (Lyon *et al.* 1973) and *c.* 1400 m.y. K–Ar biotite ages (Holland and Lambert 1972b). These are interpreted as indicative of minor opening of minerals to argon loss, possibly associated with the development of

## Table 1. Sequence of events in Lewisian complex, Outer Hebrides

| Cycle | Events | | Age |
|---|---|---|---|
| SCOURIAN CYCLE | Lithological layering — tonalitic-granodioritic masses amongst sedimentary and volcanogenic units and disrupted basic minor intrusions; strong flat-lying foliation | | |
| | Intrafolial folds; gneiss-formation mainly under amphibolite facies conditions | | 2.8-2.7 b.y. |
| | Basic minor intrusions Pegmatites Boudins Agmatites | --- localized | |
| | No open folds formed in the middle and later phases of the Scourian orogenic episode recognized because of tectonic overprinting in Laxfordian orogenic episode | | |
| | Uplift | | |
| | No effects of deformation and metamorphism in the Inverian episode recognized | | |
| | No conclusive evidence for basic minor intrusions co-eval with the Scourie dyke and related intrusions of NW Highlands | | |
| | No representatives of Loch Maree Group | | |
| ? | Basement deformation of 2.7 b.y. old rocks; tight to isoclinal folds; ? related to anorthosite | | |
| | Pegmatites Boudins Agmatites Basic minor intrusions --- localized | time relations uncertain | |
| | Anorthosite - gabbro - tonalite | | |
| LAXFORDIAN CYCLE | No recognized effects of deformation or metamorphism in the early phases of the Laxfordian orogenic episode equivalent to those seen in Loch Maree Group of NW Highlands | | |
| | Asymmetrical folds with shallowly-dipping axial planes; middle phase of Laxfordian orogenic episode | | |
| | Pegmatites Basic minor intrusions Granulite facies metamorphism | --- localized | |
| | Open, upright folds with SE axial trend | Pegmatites in parts | |
| | Open, upright folds with E axial trend | Pegmatites and granites with localised development of injection complexes; Basic minor intrusions in parts | 1.7 b.y. |
| | Very open, upright folds with NNE axial trend | Quartzofeldspathic veins (minor) | |
| | Very open, upright folds with ENE axial trend; kink bands | | |
| | Epeirogenic uplift | | 1.5 b.y. |

shear belts. Whether this was at an end stage of the Laxfordian cycle, of which at least events 8 to 11 or 12 are a part, or of a later cycle, is not known. However, the ages for the biotites correspond to some from the area between Rona and Gairloch (Fig. 2; Giletti *et al.* 1961b; Moorbath and Park 1972) where chloritization and effects of shearing have been described.

Throughout most of the *Outer Hebrides* the sequence of events shown in the Lewisian complex generally corresponds to that in Rona (Table 1) with metamorphic mineral assemblages indicating progressive changes in P/T conditions from 8–11 Kb and 650°–800°C to *c.* 3 Kb and 300°C (Dickinson and Watson 1976). The time span also corresponds to the gneiss-forming event *c.* 2·7 b.y. ago (Figs 3, 5a, b; Pidgeon and Aftalion 1972; Moorbath *et al.* 1975), granite emplacement between $F_5$ and $F_6$ *c.* 1·7 b.y. ago (Figs 3, 6g; van Breemen *et al.* 1971; Bowes, Hopgood and Taft 1971; cf. Smales *et al.* 1958; Myers 1971a), K–Ar biotite ages down to *c.* 1550 m.y. (Moorbath *et al.* 1975) and a uraninite from a pegmatite at *c.* 1500 m.y. ago (Bowie 1962, 1964—the even younger K–Ar feldspar ages given by Holmes *et al.* (1955) may not be geologically significant). This means that, as

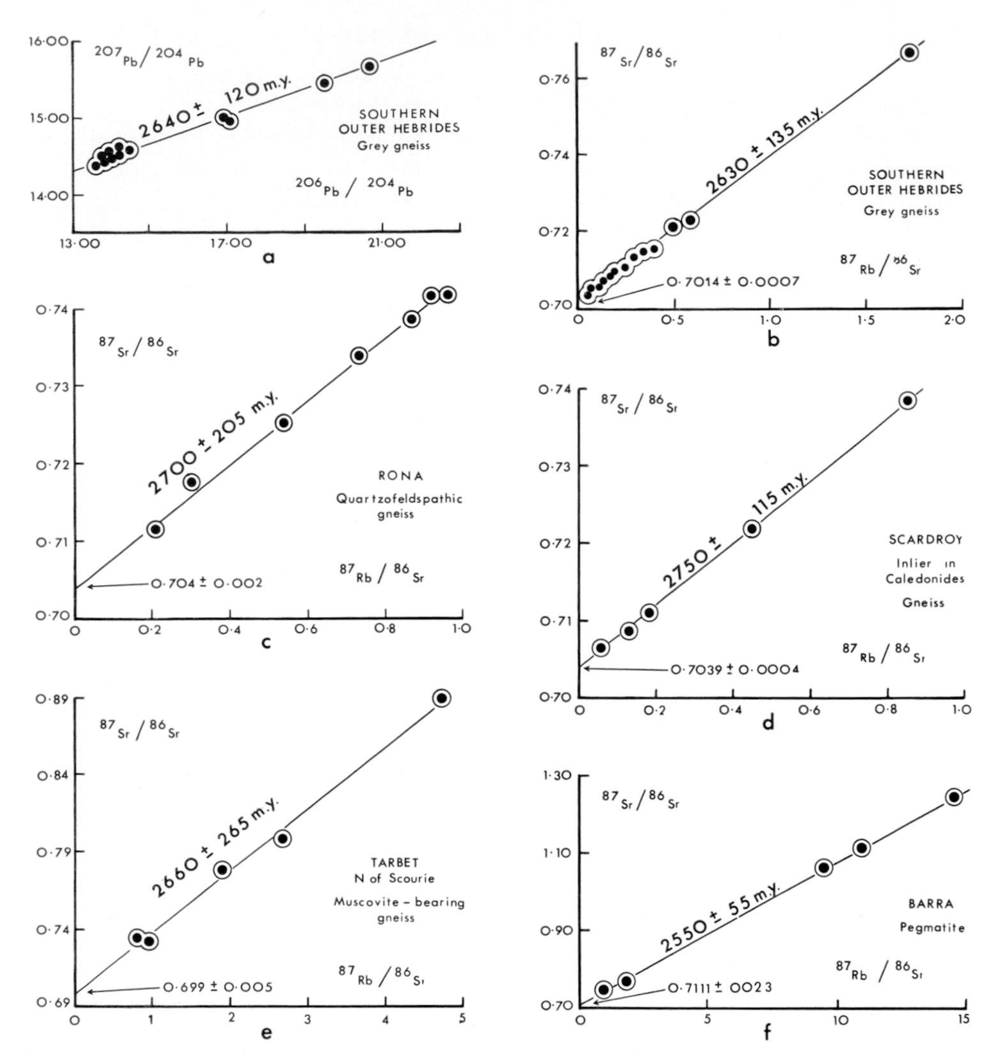


Fig. 5. Rb–Sr and Pb–Pb whole-rock data for products of the Scourian episode (references in text).

for Rona, the early part of the extensive structural, metamorphic and igneous sequence is representative of the Scourian episode and the *c.* 1000 m.y. younger latter part of the sequence representative of the Laxfordian episode. Available isotopic data does not permit definite allocation of events of the middle part of the sequence to one or other episode, but the large asymmetrical $F_3$ structures, which are key structures for correlation from island to island, are correlated with the $F_3$ folds of Rona whose hinge zones contain the 1710 m.y. old pegmatites. However, evidence is available to show that a number of the events early in the sequence, in addition to the gneiss-forming event, are part of the Scourian episode. A *c.* 2550 m.y. Rb–Sr whole-rock age of pegmatites in Barra (Fig. 5f; Francis *et al.* 1971) is similar to that for pegmatitic gneiss in South Uist (Lambert, Evans and Dearnley

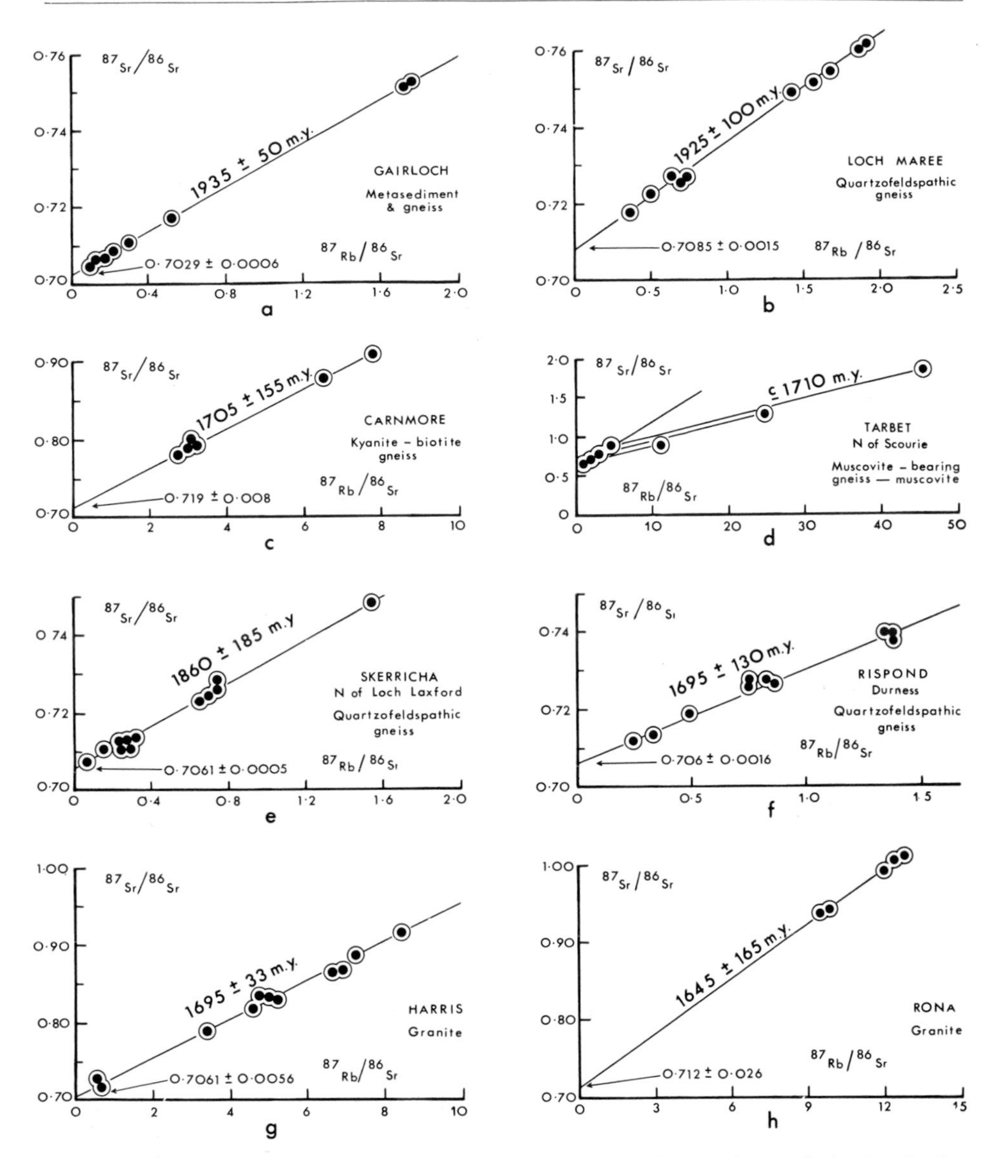


Fig. 6. Rb–Sr whole-rock and mineral—whole-rock data for products of the Laxfordian episode (references in text).

1970; cf. Dearnley and Dunning 1968). The initial $^{87}Sr/^{86}Sr$ ratio of 0·711 ± 0·0023 strongly suggests derivation by partial melting from the gneissic country rock which the Barra pegmatites cut, as do intervening basic–ultrabasic intrusions dated by K–Ar hornblende studies at *c.* 2600 m.y. (Francis *et al.* 1971).

Suggestions that the time span of the Lewisian complex in the Outer Hebrides includes a "pre-Scourian orogenic cycle of possible Katarchaean (>3000 m.y.) age" (Dearnley and Dunning 1968 p. 353; Dearnley 1966; Salop 1977) with the 'grey gneiss' constituting a 'Pre-Scourian complex' (Sutton and Dearnley 1964

p. 189; Sutton 1973 p. 10; Garson and Livingstone 1973 p. 74) is not in accord with either the determined initial $^{87}Sr/^{86}Sr$ ratio of 0·7014 $\pm$ 0·0007 (Fig. 5b; cf. Moorbath 1975 fig. 1a) or the Pb isotope ratios (Fig. 9). Taken together these isotopic data strongly suggest that "the precursors of the gneiss complex were derived from upper mantle source regions not more than 100–200 m.y. prior to the Scourian metamorphism, and rule out the possibility that the complex represents the reworking of a significantly older gneissic basement complex with anything like normal crustal Rb/Sr ratios" (Moorbath *et al.* 1975 p. 213). Such arguments, based on isotopic data, also lend no support to the proposed existence of a cover–basement relationship in the Barra Isles that pre-dates the Scourian orogeny (Francis 1973). Likewise there is no isotopic evidence to support the proposal of Garson and Livingstone (1973 p. 76) that the anorthositic complex of South Harris (Fig. 2) was in existence, as an upper layer of oceanic crust, before the metamorphic and igneous activity of the Scourian episode; zircons in the tonalitic part of the complex give a U–Pb age consistent with formation in an early part of the Laxfordian episode, with no memory of the much earlier Scourian episode (Pidgeon and Aftalion 1978). This is consistent with evidence of the anorthositic complex post-dating the metamorphic fabric associated with the $F_1$ event and being deformed by $F_3$ folds.

That there are a series of phases of basic magmatism in the Outer Hebrides has been shown by Hopgood (1971a) for Barra, Taft (1978) for southern Lewis and Hopgood and Bowes (1972a) for various places throughout the islands. In addition to amphibolites that form part of the gneissic complex and the basic and ultrabasic intrusions that pre-date the *c.* 2550 m.y. pegmatites, emplacement of basic minor intrusions has been recorded between $F_2$ and $F_3$, $F_3$ and $F_4$, and $F_5$ and $F_6$ (cf. events 1, 4, 6, possibly between 7 and 9 and 11–12 of the Rona sequence). The youngest set is only sparsely distributed. However, the basic sheets emplaced between $F_2$ and $F_3$ are prominent in southern Lewis (Taft 1978; cf. Myers 1971b), as they are in Rona (event 6), but in parts of Barra the prominent intrusions are those emplaced between $F_3$ and $F_4$, with the effects of co-eval granulite facies metamorphism localized in the particular region (Hopgood 1971a).

With the known problems of interpreting K–Ar data in regions of complex metamorphic activity (cf. Pidgeon 1970), the K–Ar whole-rock ages given by Lambert, Myers and Watson (1970) and Lambert, Evans and Dearnley (1970) for basic masses in Lewis (*c.* 2500–2400 m.y.) and Benbecula and South Uist (1700–1500 m.y.) cannot be assigned with confidence to any of the events in the sequence. Certainly there is no clear isotopic evidence to support or refute the correlation of basic minor intrusions in the Outer Hebrides with the *c.* 2·4 b.y. tholeiitic Scourie dyke and other dykes in the Northwest Highlands made by Dearnley (1962, 1963, 1973), Watson (1965, 1968) and others. However, if correlation of basic masses in the Outer Hebrides with basic masses on the mainland can be made, it can only be in relation to *one* of the sets of minor intrusions if the unique stratigraphic time significance given by Sutton and Watson (1951a p. 292) to the Scourie dykes and other basic minor intrusions is to be maintained.

The fold sets $F_{1-6}$ and related fabric features, which show generally comparable development throughout the various islands of the Outer Hebrides (Bowes and Hopgood 1971, 1975a, table 1), show marked correspondence with the structures developed during events 3, 5, 7, 9, 10 and 13 for Rona. This includes correspondence in the axial directions of the respective $F_4$, $F_5$ and $F_6$ folds (Bowes and Hopgood 1973). However, the fold set most prominently developed can be different from one place to another. Where open $F_6$ folds are present on the flat limbs of previously formed structures, as on the island of Mingulay (Fig. 2), these may form the major mappable structures of the district (Robertson *et al.* 1964; Bowes and Hopgood 1969 fig. 2—Skipisdale fold phase, referred to as $F_5$). In parts of Barra $F_5$ folds are the most prominent structural feature (Hopgood 1971a p. 45 fig. 28) while $F_4$ folds are prominent in southern Lewis (Taft 1971). The coaxial but not coaxial-planar nature of $F_4$ and $F_3$ folds results in the prominent structural trend shown in parts. However, the successive development, separated by the emplacement of pegmatite, of these two fold sets has not been consistently recognized and regional correlations have been made assuming co-eval development (Dearnley 1962; Coward *et al.* 1970; Bowes and Hopgood 1976).

The effects of heterogeneous deformation (Ramsay and Graham 1970), with competence variations acting as major controls on all scales (Coward 1973a), are shown in the products of the second to seventh deformational phases, with the position on previously formed structures being important for the attitudes of superposed folds (cf. Elliott 1964, 1965, 1968). Likewise there is evidence of variation in strain rates (Coward 1973b; Graham and Coward 1973). These factors cause variations in structural and metamorphic patterns for features that can be shown to have developed co-evally. However, co-eval development has not consistently been demonstrated (eg. Coward *et al.* 1970), particularly where basic minor intrusions have been used as strain markers on the assumption that they represent a single set whose deformational histories correspond, something that is not the case in the Outer Hebrides (Lisle 1977). In such cases the resultant patterns may be significant in terms of localizing areas in which the time of emplacement of basic masses corresponds, rather than establishing variations in deformational intensity and expression (Davies *et al.* 1975 fig. 3).

Neither the developmental sequence of structural elements, nor the general direction of axial *trend* of successively-formed folds (Table 1) are affected by the variable expression of structural and fabric elements, and the sequence in the Outer Hebrides is sufficiently distinctive to permit it to be distinguished from that of other polyphase deformed basement complexes (Hopgood 1973). This, together with the isotopic evidence for the existence of parts of two widely separated orogenic episodes, lend no support to the suggestion of Holland and Lambert (1969, 1975 p. 162) that the polyphase structural sequence corresponds with the "well-documented sequential and repetitious character of folded structures within many orogenic and basement terrains" that is "a function of the standard rheological properties of rock, not of time".

Correlation of the Lewisian complex in the Outer Hebrides can be made with the gneiss complex of *Inishtrahull*, Eire (Fig. 1). There the overall sequence of events corresponds to that for the Outer Hebrides, and for Rona, as do the axial trends of folds developed in the latter half of the deformational sequence (Bowes *et al.* 1968; Bowes and Hopgood 1975b; cf. Hopgood 1971b; Hopgood and Bowes 1972b). The scatter of the isotopic Rb–Sr whole-rock data points prevents an unambiguous subdivision of the complex, but suggests the presence of rocks formed in the Scourian episode and reflects superimposed effects of the Laxfordian episode. The *c.* 370 m.y. K–Ar mineral ages are related to uplift associated with movement on the Great Glen fault system, and not to the history of the Lewisian complex (Macintyre *et al.* 1975).

For the region between *Loch Laxford* and *Durness* in the Northwest Highlands, which includes the type area for the Laxfordian orogeny/complex/metamorphism

(Sutton and Watson 1951a p. 294–5), an extensive polyphase deformational, metamorphic and igneous history has been demonstrated (Gillen 1975). Like Rona, the Outer Hebrides and Inishtrahull, penetrative foliation that can be shown to be largely composite (Findlay 1970; Chowdhary and Bowes 1972), together with tight to isoclinal folds (Fig. 7a) and amphibolite, gneiss and migmatite formation characterize the early (Durness and Gualin) deformational phases. Large asymmetrical and symmetrical folds with SE axial trends (Fig. 7b; Bowes 1969 fig.6) characterize the middle (Dionard) deformational phase (Dash 1969), with Coward (1974) suggesting considerable movement between the gneisses and the granulites below. In the late (Leacach) phases, there is the successive development of upright, generally open folds with SE, E and NNE axial trends ($F_{4a, 4b, 4c}$ of Gillen 1975; $F_{4, 5, 6}$ of Bowes 1975). These structures are generally weakly expressed. In places where there has been abundant granitic injection in addition to intense boudinage with curvature associated with boudin separation (Fig. 7c), as at Durness (Findlay 1970), only weak cleavages have been recognized. Like the other parts of the Lewisian complex of the mainland, the sequence differs from that of the Outer Hebrides in the absence of anorthosite–gabbro emplacement. Other differences with much of the Lewisian complex are (1) the apparent absence of folds equivalent to $F_2$ of Rona (event 5) and the Outer Hebrides, (2) the great abundance of granitic and pegmatitic material emplaced in the middle of the sequence, particularly in the vicinity of Loch Laxford (Peach and Horne 1930 fig. 2; Inglis 1966; Bowes 1967) and Durness (Findlay 1970) where injection complexes were formed (Fig. 7d) and (3) the recognition of only one phase of basic minor igneous intrusion after the formation of the gneiss complex, this being after the development of the injection complexes (Fig. 7e; Dash *et al.* 1974; cf. Standley 1975b).

The *c.* 1550 m.y. (and younger) Rb–Sr ages of minerals from gneisses of the Rhiconich group (Giletti *et al.* 1961b) have been interpreted as indicative of a metamorphic event during the Laxfordian orogeny. Likewise the Rb–Sr whole-rock age of 1860 $\pm$ 185 m.y. (recomputed uncertainly using McIntyre *et al.* 1966) for gneisses (Fig. 6e) was interpreted by Lambert and Holland (1972 p. 3) as the "minimum age for the climax of the Laxfordian metamorphic episode". Corresponding ages, within errors, have been determined by Lyon and Bowes (1977) for Rhiconich group gneisses from near Durness (1695 $\pm$ 130 m.y.—Rb–Sr whole-rock; Fig. 6f) and for granite forming parts of the injection complex near Durness (1720 $\pm$ 30 m.y.—U–Pb zircon; Fig. 3). In addition, Lyon and Bowes (1977) showed the presence of *c.* 2730 m.y. old zircon xenocrysts in this granite (Fig. 3) and that the gneiss, which on the evidence of Moorbath *et al.* (1969 had no crustal history before *c.* 2900 m.y. ago, had a U–Pb zircon age of 2850 $\pm$ 50 m.y. (Fig. 3). The zircon population in the gneiss is homogeneous and has morphological characteristics consistent with growth under conditions of high grade metamorphism. This evidence indicates that the dominant rock type of the region, the quartzofeldspathic gneiss, was formed during the Scourian orogenic episode together with the amphibolite which shares with it a community of structural elements. The abundant emplacement of granitic and pegmatitic masses after the Dionard deformational phase led to the rehomogenization of rubidium and strontium isotopic systems under thermal, not dynamothermal, conditions while the U–Pb systems in zircons remained closed due to annealing of the lattices. A 1700 $\pm$ 50 m.y. $^{40}K/^{40}Ar$ isochron age for hornblendes from amphibolites (Lyon and Bowes 1977), which is in agreement with K–Ar hornblende ages of Lambert and Holland (1972), represents the time at which hornblende cooled to

Fig. 7. The Lewisian complex of the Durness (a–d)–Loch Laxford (e) region: (a) Gualin phase (*c*. 2.8 b.y.) fold deforming Durness phase foliation, in amphibolite; (b) Dionard phase folds; (c) boudins on steep limb of a Dionard phase fold; (d) part of injection complex (*c*. 1.7 b.y.) with granite and pegmatite injected into steep limb of a Dionard phase fold; (e) basic minor intrusion cutting Dionard phase fold and *c*. 1.7 b.y. pegmatite.

its 'blocking temperature'. This must have been near, or at, the close of the Laxfordian orogenic episode while the spread of K–Ar biotite ages from *c.* 1750–1300 m.y. (Lambert and Holland 1972; Lyon and Bowes 1977) appears to reflect the complex interplay of reheating associated with granite intrusion, cooling history, uneven uplift and partial argon loss during a younger mild event. The existence of such an event in this region *c.* 1000 m.y. ago (ie. part of the Grenville cycle) is suggested by the U–Pb zircon data for the gneiss (Fig. 3) and by radiation damage ages (Lyon 1973).

The combination of the sequence of events and isotopic data related to particular events indicates that, like the major part of the Lewisian complex, the rocks of the region from Loch Laxford to Durness show products of parts of two episodes separated by *c.* 1000 m.y.; the gneisses and related rocks formed in the early phases are products of the Scourian orogeny while the tectonic overprinting, igneous activity and thermal effects of the middle and late phases took place in the Laxfordian orogeny. Not only does the isotopic data place severe restraints on the extent to which the concept of 'reworking' or 'reactivation' can be applied to the rock complex, but it means that in its type area, the 'Laxfordian complex' is largely composed of the products of gneiss-forming and related processes operative at least 700 m.y. before the Laxfordian orogeny. Hence this orogeny, in its type area, exhibits the effects of *basement* tectonics together with igneous emplacement and associated reheating. From isotopic and structural evidence these can be shown to have taken place in the middle and later parts of the orogenic episode.

In the early part of the recent remapping activity, a considerable amount of evidence concerning sequences of events came from the region with *Gairloch–Loch Maree* at its centre (Fig. 2) in which metasediments of the Loch Maree Group occur in addition to the gneisses and amphibolites. In both metasedimentary and gneissic terrain early deformational phases are represented by tight to isoclinal folds and related dimensionally oriented mineral growths, the middle phase by large asymmetrical folds with pegmatite and granite in some hinge zones, and later phases by upright and generally open folds with localized intense development of pseudotachylite (Park 1961, 1964, 1965, 1969a; Bhattacharjee 1963, 1968, 1969; Elliott 1964; Ghaly 1966). Isotopic studies were limited to a small number of K–Ar determinations (Giletti *et al.* 1961b; Evans and Park 1965) and the ages generally interpreted as related to metamorphic events in the early phases. Large asymmetrical folds, including the Tollie antiform, and related parasitic folds figured prominently in correlations while the nature and successive development of structures (cf. Elliott 1965, 1968) as well as the metamorphic state of the rocks played significant roles in interpretations of crustal history, including separation of 'basement' and 'cover' assemblages (Bhattacharjee 1964; Park 1969b, 1970a). Also used for subdivision were the relations of structures and fabrics to basic minor intrusions: this followed the proposal of Sutton and Watson (1951a) that there was one set of basic minor intrusions in the Lewisian complex that represented a unique stratigraphic marker. The chronologies based on the assumption of a unique set of basic igneous masses (e.g. Park 1970b, 1971) differ from those using field relations as a basis for establishing the existence of more than one set of basic minor intrusions and fitting their emplacement into the overall structural sequence (e.g. Bowes 1969, 1971).

The genesis and age of the commonly occurring quartzofeldspathic and related gneisses were also subject to various interpretations (cf. Peach and Horne 1930 p. 40–1). The presence of small masses of metasediment amongst these rocks was used as evidence of gneiss formation from the sediments of the Loch Maree Group during early–middle Proterozoic times in the Gairloch–Loch Maree district (Bowes and Bhattacharjee 1967; Bowes 1969), while K–Ar isotopic data were used as evidence of gneiss formation during the Scourian episode in the Gruinard Bay district (Evans and Lambert 1974). Relations with basic minor intrusions were interpreted as indicative of gneiss formation as the result of intense 'reactivation' of an earlier-formed metamorphic complex (Sutton and Watson 1962, 1969; Sutton 1969) while development during a succession of events, including during an episode (Inverian) between the Scourian and Laxfordian events, has also been proposed (Cresswell 1972; Cresswell and Park 1973; Park 1973). This is partly based on correlations of structures and rock fabrics, but partly on the interpretation of K–Ar isotopic data by Moorbath and Park (1972) as recording a succession of metamorphic events including three metamorphic phases in the Laxfordian episode at *c.* 1700, 1500 and 1400 m.y. ago. Cataclasis and reheating after the Laxfordian episode *c.* 1150 m.y. ago was also suggested for rocks of the Torridon area.

Rb–Sr whole-rock isotopic data and initial $^{87}Sr/^{86}Sr$ considerations (Bikerman *et al.* 1974, 1975) have clarified a number of these controversial matters. They indicate that (1) gneisses north of Loch Maree, towards Gruinard Bay (cf. Crane 1973), have crustal histories going back to 2·7–2·8 b.y., (2) there is no isotopic evidence for the proposed activity during the Inverian episode, (3) the Loch Maree Group was deposited, presumably on an older metamorphic complex, after 2·2 b.y. ago, (4) metamorphism of the Loch Maree Group, including gneiss formation took place 1935 $\pm$ 50 m.y. ago (Fig. 6a), (5) there was some development of gneisses from the pre-existing metamorphic complex 1925 $\pm$ 100 m.y. ago (Fig. 6b), (6) reheating took place 1705 $\pm$ 155 m.y. ago (Fig. 6c)—this is related to abundant pegmatite emplacement in the hinge zone of a large asymmetrical fold—and (7) final closure of isotopic systems during epeirogenic uplift took place *c.* 1500 m.y. ago. On this evidence the *c.* 1700, 1500 and 1400 m.y. K–Ar ages of Moorbath and Park (1972) fall either at the end of the orogenic episode or during the epeirogenic uplift stage and do not relate to metamorphic events. The evidence also points to the presence of two groups of sediments (pre-2·7 b.y. and between 2·2 and 2·0 b.y.) and to two gneiss-forming events (2·7 and 1·9 b.y.). It also permits the Laxfordian orogenic cycle to be defined as consisting of a depositional episode 2·2–2·0 b.y. ago, an orogenic episode 2·0–1·7 b.y. ago and an epeirogenic episode 1·7–1·5 (or 1·4) b.y. ago. Thus while rocks of the Loch Maree Group, and igneous rocks emplaced into them, show structures, rock fabrics and mineral assemblages resulting from tectonism and metamorphism solely in the Laxfordian episode (Keppie 1969), some of the gneisses and amphibolites of the district show features produced in the Scourian episode that were modified by tectonic overprinting and thermal activity in the Laxfordian episode. For the gneisses derived from pre-existing crystalline material and dated at 1·9 b.y. (Fig. 6b), the effects of early as well as middle and late deformational phases of the Laxfordian orogeny are shown. However, for others, only the effects of the middle and later phases are recognized as being superimposed on the *c.* 1000 m.y. older material: the 1·9 b.y. dynamo-thermal metamorphic event, which had such a powerful effect on the Loch Maree Group has not been recognized in these rocks from the isotopic studies. This is apparently the situation for the gneisses throughout much of the Lewisian complex.

Correlation of the middle and later phases of the Laxfordian episode from the Gairloch–Loch Maree district to other parts of the Lewisian complex is made on the basis of (1) the large asymmetrical folds with SE axial trends (cf. event 7 ($F_3$) of Rona), (2) the correspondence of the 1·7 b.y. reheating event associated with pegmatite emplacement into asymmetrical folds, with the 1710 m.y. emplacement of pegmatite on Rona (event 8), and the 1720 m.y. emplacement of granite near Durness, both into the hinge zones of large asymmetrical ($F_3$) folds, and (3) the successive development of generally open and upright folds with SE, E and NNE axial trends in the late phases (Bhattacharjee 1963), which show correspondence respectively with those developed on Rona, the Outer Hebrides, Inishtrahull and north of Loch Laxford. The reheating at Torridon, for which the *c.* 1150 m.y. age is given, could be an expression of the Grenville cycle.

## 3b. In the granulite facies terrain of Northwest Highlands

An extensive sequence of events has been demonstrated for the Kylesku group and related rocks, including those in the area around Scourie, the type area for the Scourian orogeny/complex/metamorphism (Sutton and Watson 1951a pp. 294–5). A polyphase deformational, and a polymetamorphic sequence that includes a phase of granulite facies metamorphism, has superimposed on it two sets of structures that are largely limited to discrete linear belts and represent the products of successive episodes of basement block movement (Barooah 1967; Khoury 1968a; Bowes 1969; Sheraton, Tarney *et al.* 1973).

The phase of dominantly static granulite facies metamorphism that follows the successive development of flat-lying foliation and much dissected and apressed isoclinal folds ($F_1$; cf. Khoury 1968a Fig. 8a) in the polyphase sequence has been dated at 2700 $\pm$ 20 m.y. (U–Pb zircon—Fig. 3; Pidgeon and Bowes 1972). This corresponds to a Pb–Pb whole-rock age of 2680 $\pm$ 60 m.y. for gneisses and granulites (Fig. 9; Chapman and Moorbath 1977). These ages, when taken together with the *c.* 2·9–2·8 b.y. estimates for the time of separation of the Kylesku group material from the upper mantle (Moorbath *et al.* 1969; Moorbath 1976), and the initial $^{87}Sr/^{86}Sr$ ratios of the granulites that reflect mantle Sr rather than Sr from regenerated crustal material (Spooner and Fairbairn 1970), show that the early phases of the polyphase deformational sequence, which included tectonic fragmentation and agmatization (Fig. 8b), must have affected a dominantly igneous assemblage not long after it began its crustal history. And for these granulites, including the fragments of metasediments, this deformation must have been at a very considerable depth in the crust (O'Hara 1975, 1976, 1977). Deformational phases after the granulite facies event resulted in successive development of four sets of generally upright folds. The first of these are commonly asymmetrical or monoclinal with NNE to NE axial trends (Fig. 8c; $F_2$ of Khoury 1968a) and granitic or pegmatitic masses in some hinge zones. However, due to the refractory nature of the host granulites, these quartzofeldspathic masses are very limited in extent. The Rb–Sr isotopic data, which show considerable scatter either due to diachronous development or widespread open system conditions, indicate ages of *c.* 2·6–2·5 b.y. (Fig. 5e; Evans and Lambert 1974; Lyon *et al.* 1975), i.e. generally co-eval with pegmatites of Barra. Particular ages have not been or cannot be determined for successively formed fold sets with NW (Kylesku fold—$F_3$; Fig. 8d), E and NNE axial trends, for very small granitic intrusions that post-date the last set, or for basic minor intrusions whose position in the sequence has yet to be determined. However, these events all pre-date the formation of the NW-trending zones of complex structures which control the uprise of a suite of basic dykes,

Fig. 8. The Lewisian complex of the Scourie (d)–Kylesku (a, c)–Lochinver (b, e) region: (a) $F_1$ folds in *c.* 2.7 b.y. granulites; (b) tectonically fragmented and agmatized product of an early phase of the Scourian episode; (c) asymmetrical $F_2$ fold in granulite; (d) strong joints axial planar to Kylesku fold in layered ultrabasic mass; (e) member of Assynt dyke suite emplaced in steep limb of monoclinal fold formed in the Inverian episode.

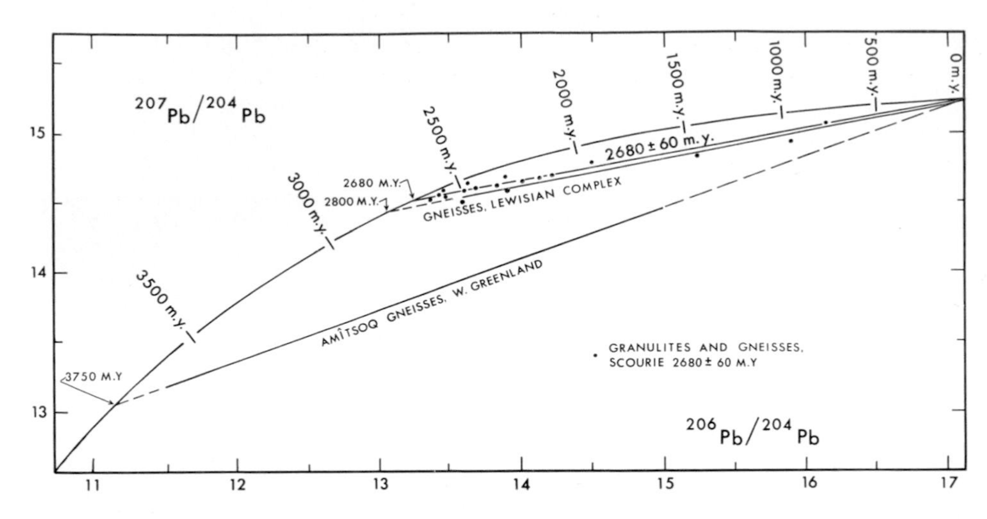


Fig. 9. Pb–Pb whole-rock data for the Lewisian complex formed in late Archaean times compared with that for Amîtsoq gneisses of W. Greenland formed in early Archaean times (*after* Moorbath 1976 and Chapman and Moorbath 1977).

representatives of which have been dated (Rb–Sr whole-rock) at *c.* 2400 m.y. (Chapman 1976) and whose initial $^{87}Sr/^{86}Sr$ ratios are *c.* 0·702.

The cognate deformational, metamorphic and igneous sequence of events which includes the *c.* 2700 m.y. granulite facies phase of metamorphism and ends with (minor) granite emplacement after the formation of the NNE-trending and other upright open folds with axial planar joints (Fig. 8d) is referred to as the Scourian orogeny (cf. Sutton and Watson 1951a p. 294; 'Scourian' is preferred to 'Early Scourian'—Sutton and Watson 1969—and 'Badcallian'—Park 1970b—on grounds of priority; cf. Sheraton, Tarney *et al.* 1973 p. 43). The retrogression of pyroxene granulites to amphibolite facies gneisses in the NW-trending zones of complex structures and the evidence that the *c.* 2400 m.y. dolerite dykes intruded into these zones cooled under amphibolite facies conditions (O'Hara 1961b) together are indicative of epeirogenic uplift towards or after the end of the Scourian orogeny and *c.* 2400 m.y. ago (cf. O'Hara 1977 fig. 6). Accordingly it is possible to establish the existence of a Scourian cycle (Fig. 3; cf. Bennison and Wright 1969) with the development of the Kylesku group and related rocks taking place *c.* 2·8 (2·9)–2·7 b.y. ago, the orogenic stage *c.* 2·7–2·5 b.y. ago and the uplift stage *c.* 2·5–2·4 b.y. ago. There are clearly many differences between the nature and development of this cycle and orogenic cycles of Phanerozoic times, but the gross overall pattern is generally similar.

The first set of linear belts and associated folds to affect the products of the Scourian episode trend WNW to NW, parallel to the axial trend of the earlier-formed isoclinal folds and Kylesku and related folds (Khoury 1968a; Bennison and Wright 1969 fig. 3.3). Many of the belts have the gross form of monoclinal to asymmetrical folds, some of which deform complex flow folds with axes perpendicular to the trend of the belts (Khoury 1968b; cf. King 1955). Retrogression to amphibolite facies mineral assemblages, which involved the addition of large amounts of water and the localized production of vertically disposed gneissic foliation, was associated with the development of agmatites and pegmatites that cross-cut some of the structures of the belts. These pegmatites appear to be those dated at *c.* 2300 m.y. (Rb–Sr whole-rock) by Evans and Lambert (1974) and at *c.* 2200 m.y. by Evans (1964, 1965).

The emplacement of the Scourie dyke suite, that includes the NW- to WNW-trending abundant dykes of the Assynt district (Fig. 8e; Peach *et al.* 1907 p. 89–96, 138–42, 161–5; O'Hara 1962; Tarney 1963, 1973), followed this sequence of tectonothermal and igneous events with many exploiting the steep limbs of the monoclinal folds as a control of emplacement. Partial or complete amphibolitization of the igneous masses and development of open folds followed in turn (Bowes 1969 table 1). The dolerites ('epidiorites') are cut by picritic and gabbroic dykes whose very coarse grained margins of pyroxenite resulted from magmatic intrusion into hot country rock (Tarney 1963). Rb–Sr whole-rock and K–Ar mineral and whole-rock dates of emplacement are *c.* 2400 and *c.* 2200 m.y., respectively (Chapman 1976; Evans and Tarney 1964). The older K–Ar ages recorded for some dykes result from the presence of excess argon, either due to mantle-derived argon brought up by the magma or to clinopyroxenes accepting radiogenic argon from host gneisses undergoing metamorphism. Successively emplaced dolerites, giving K–Ar ages down to *c.* 1900 m.y., came into country rock as its temperature was progressively decreasing.

The cognate tectonothermal and igneous products of the *c.* 2·4–2·2 b.y. time span, together with evidence for the addition of water from below in large quantities along master joints (cf. Fig. 8d; Khoury 1968b) as part of mantle degassing (Tarney 1973), are indicative of the basement part of an orogenic regime whose cover parts (if they existed) would resemble the folded Huronian Supergroup of East Canada. On these grounds the earth movements, with their related magmatism, have been referred to as the Inverian orogeny (Fig. 3; Bowes 1968a, b) and having taken place in the Inverian episode (Bowes 1965). This follows on from the conclusion of Evans and Lambert (1964), based on studies in the Loch Inver district (Fig. 2), that the metamorphic activity for which Evans (1965) gave an age of *c.* 2200 m.y. and referred to as the 'Inverian metamorphism' took place in an episode entirely separate from the Scourian and Laxfordian episodes (Evans and Lambert 1974). In its original usage the term 'Inverian metamorphism' referred to activity before the emplacement of basic dykes (cf. Lambert 1965). The 'Inverian orogeny' includes the tectonothermal and related activity but also includes the amphibolitization and deformation that followed dyke emplacement; hence it includes igneous activity when host rocks were still hot at the crustal level now exposed. It does not include the emplacement of the dolerites, with *c.* 1950 m.y. K–Ar ages, that Tarney (1973 p. 117) considers to have been intruded under more brittle and probably cooler conditions, although recognition of criteria indicative of the cessation of the orogenic activity is difficult and the cessation would probably have been diachronous with depth ('Inverian', when used to designate a distinct episode or orogeny, is preferred to 'late Scourian'—Sutton 1974, Watson 1975—on the grounds of priority, while its use avoids the confusion associated with the many uses of the term 'Scourian'—cf. Lambert 1974).

The second set of linear belts, many of which can be followed for many km in NW to WNW directions, formed in response to what Peach *et al.* (1907) referred to as "pre-Torridonian" movements. They are ductile shear zones which cut across the products of both the Scourian and Inverian episodes (Khoury 1968a; Beach 1974b). Tight to isoclinal folds of variable plunge occur between strongly refoliated bands in which localized retrogression, at upper greenschist facies, is associated with the formation of axial planar foliation. Locally, upper amphibolite facies assemblages occur whose formation was associated with the adiabatic transport of water upwards through the crust (Beach 1973, 1976). Basic, picritic and pegmatitic dykes cut the newly formed foliation and are themselves affected by subsequent deformation (Bowes and Khoury 1965; Barooah 1967; Khoury 1968a; Bowes 1969 table 1). A Rb–Sr mineral age for one of the pegmatites is *c.* 1600 m.y. (Giletti *et al.* 1961b). However, a *c.* 1710 m.y. Rb–Sr mineral—whole-rock age (Fig. 6d) for dimensionally aligned muscovite growth related to the Dionard phase of the Laxford orogeny in the region immediately to the north (Lyon *et al.* 1975) suggest that the pegmatites are likely to form part of the *c.* 1720 m.y. suite of granites and pegmatites commonly emplaced in the hinge zones of Dionard phase folds. Accordingly the ductile shear zones, including the wide one containing granitic sheets that forms the boundary between the granulite facies rocks around Scourie and the quartzofeldspathic gneisses and related rocks around Loch Laxford, are considered to represent the products of the middle and late phases of the Laxfordian orogeny although initiation in the Inverian episode has been suggested (e.g. Holland and Lambert 1975). The interpretation of this boundary (variously referred to as the Ben Stack line and the Laxford front) as a metamorphic-migmatitic front (Sutton and Watson 1951a, 1962) is not supported by other workers who assign a major role to tectonic activity in its development (Bowes 1962; Beach *et al.* 1974; Park and Cresswell 1975; Davies 1975b, 1976). The model for the district produced by Bott *et al.* (1972) from gravity and magnetic studies is consistent with high angle movement parallel to the *c.* 60° SW dipping axial planes of Dionard phase folds, bringing granulite facies rocks in to juxtaposition with the gneisses (cf. Bowes 1969 fig. 1). The seismic studies of Hall (1978) indicate that if this was the operative process, then a vertical movement of many km would be required. The granitic sheets around Loch Laxford are generally inclined at *c.* 60° SW and show evidence of considerable upward migration of material, including $K_2O$, $H_2O$ and $CO_2$ (Beach and Fyfe 1962), along a zone that may be of significance on a crustal scale (Davies and Windley 1976).

The ductile shear zones, the granites and the earlier-formed features are all displaced by two prominent sets of near-vertical faults, one of which has almost straight and continuous NNW-trending outcrop traces; for the other, the outcrop traces are arcuate, discontinuous and NE- to NNE-trending (Bowes 1969 fig. 2). These faults, which are related to the epeirogenic stage of the Laxfordian episode, represent the first set of major faults developed in the crustal segment and appear to have exerted a strong influence on the subsequent development of the region (Bowes 1976c; cf. Fig. 1 and Watson 1977, fig. 1 for the arcuate NNE- to NE-trending Minch fault and margins of sedimentary basins and basement blocks). This is reflected in the sub-surface geology (e.g. McQuillin and Watson 1973).

## 4. Nature of crustal history

The model of shield development that has gradually taken shape as the isotopic data, particularly from U–Pb zircon and Rb–Sr whole-rock studies related to the geologically determined sequence of events, have become available, is that of the development of the Lewisian complex as a new crustal addition in late Archaean times and its essential permanence throughout subsequent geological time. There were some further crustal additions during early and middle Proterozoic times while reheating and tectonic overprinting variously affected the late Archaean assemblage. Locally the younger features may be prominent, but most of the rocks have characters that essentially represent the nature of products and processes 2·7 ± 0·1 b.y. ago. This means that the results obtained from the intense study of the Lewisian complex over the past fifteen years provide a basis for elucidating the nature and operation of processes involved in shield formation in late Archaean times where there were large additions of crustal material from upper mantle sources. The studies also provide case histories of (1) the testing and rejecting of hypotheses of shield formation involving massive 'reactivation' or 'reworking', either of early Archaean crust in late Archaean times or late Archaean crust in early to middle Proterozoic times, and (2) the use of isotopic data without sufficient understanding of their meaning in relationship to geological processes.

## 4a. Orogenic cycles

The availability of data from a variety of isotopic systems whose relationships to specific geological events can be assessed has meant that time spans can be given, not only to the three cycles of crustal activity previously recognized, at least in part, from geological evidence, but also to their constituent parts (Fig. 3). Both the Scourian and Laxfordian cycles show evidence of rock assemblages that were affected by extensive polyphase deformational activity and then subject to crustal uplift. For the Inverian cycle, only the effects of activity in the basement are shown, there being no supracrustal assemblage in the region whose development immediately preceded the rise of geoisotherms in the localized belts. However, the progressively lower temperatures of the rocks into which the dykes were injected, following the last deformational phase, may be indicative of a stage of uplift before the deposition of the Loch Maree Group.

## 4b. Expression of orogenic cycles

In no part of the complex is the whole sequence of events relating to the three cycles expressed. What is interpreted as being a continuous sequence through the Scourian cycle is seen in the granulite facies terrain of the Northwest Highlands. Likewise, what is taken to be a continuous sequence for the Laxfordian cycle is expressed by the rocks of the Loch Maree Group in the Gairloch–Loch Maree district. In the gneissose terrain that makes up most of the exposed part of the complex, only those parts of the Scourian cycle up to early phases in the orogenic stage are recognized generally, while the Laxfordian cycle is generally represented by the middle to late phases of the orogenic stage together with the epeirogenic uplift stage. Only locally amongst the gneisses have products of the Scourian orogenic episode developed after the *c.* 2·7 b.y. gneiss-forming event been recognized, e.g. the *c.* 2550 m.y. pegmatites of Barra, and no Rb–Sr whole-rock isotopic systems in gneisses been recorded that are indicative of homogenization during the Inverian episode. In addition only locally do gneisses have isotopic systems and structural features indicative of development during an early phase of the Laxfordian episode. Hence, apart from igneous masses which intrude into the gneisses and for which there is generally little good isotopic data, the representatives of other features known to have formed elsewhere in the complex in the 2·7–1·7 b.y. period are largely unrecognized. This could largely be the result of (1) tectonic overprinting during the Laxfordian orogenic episode, particularly in its middle phase, with the development of large NW-trending folds that tightened pre-existing NW-trending folds and made other very open folds unrecognizable, and (2) recumbent structures in the early phases of deformation of the Loch Maree Group having little effect on basement crystalline rock except near to major dislocation zones, such as thrusts. However, if the $F_2$ folds of Rona (event 5), the Outer Hebrides and Inishtrahull express changes in ductility associated with the emplacement of high temperature anorthositic magma, as seen in South Harris, they could represent a stage in the initiation of the Laxfordian episode.

## 4c. Mobile belt formation in late Archaean times

On the basis of isotopic and geochemical considerations, the original nature of much of the material making up the new crustal additions *c.* 2·8 b.y. ago can be shown to have been igneous and derived from the mantle (Moorbath *et al.* 1969; Tarney 1976). A calc-alkaline volcanic suite of average andesitic composition, with derived pyroclastic and sedimentary material was suggested by Sheraton (1970) and Bowes, Barooah and Khoury (1971) for the Kylesku group rocks and a crustal model, involving a thick supracrustal pile (Gaelic Supergroup—Bowes 1976a), erected for these and the Rhiconich group and related rocks (Bowes 1975, fig. 2). This pile was compared in composition with the pile of basalts, andesites, dacites and rhyodacites (and derived rocks) present in greenstone belts formed in late Archaean times, except for some constituents that had been removed from parts of the pile during granulite facies metamorphism. However, the rocks show some features of chemical composition that Holland and Lambert (1975) consider to be more consistent with a plutonic rather than a volcanic derivation. In addition rare earth patterns for gneisses formed in Archaean times generally do not correspond with what would be expected for rocks of volcanic derivation (Tarney 1976) while rare earth data for high grade rocks in the Lewisian complex in the Inner Hebrides (Drury 1972a, b; 1973a, b; 1974) suggest derivation from the products of plutonic crystallization of dacitic melts derived from hydrous garnet-bearing basaltic rocks at mantle depths (Drury 1978a). Accordingly the present evidence points to much of the Lewisian complex being derived from tonalitic-granodioritic material (cf. 'grey gneiss' of the Outer Hebrides; Windley 1977) that began its crustal history *c.* 2·8 b.y. ago, with incorporated sedimentary masses in part derived from igneous material and in part derived from pre-existing crystalline material (Bowes *et al.* 1976).

Plutonic masses of tonalite-granodiorite would have been emplaced at some depth in the crust, possibly as a plexus of subhorizontal sheets that were strongly deformed by subhorizontal movements with the production of flat-lying foliation and recumbent isoclinal folds and affected by amphibolite to granulite facies metamorphism. These are associated with disrupted relics of metasediment and layered peridotite–gabbro–anorthosite complexes, showing variable effects of deformation that relate to competence contrasts. The concordant amphibolite masses now seen in the gneissose terrain would represent successive emplacement of basic igneous sheets that were disrupted and then brought into tectonic conformity with lithological layering that parallels foliation while the open folds of the late phases would express successive weak deformatiom of the generally flat-lying layering. Major faulting would not be expected within the belt, as distinct from the margins of the belt.

Such a high grade gneiss-granulite 'mobile belt' situation is compared by Tarney and Windley (1978) to that of the deeply eroded parts of the batholiths of the southern Andes and the North American cordillera. There the results of crystallization of dacitic magma, under plutonic conditions, contain metasedimentary relics and tectonically disrupted layered ultrabasic–basic complexes. In parts the plutonic masses show evidence of strong deformation at high grade and they are associated with calc-alkaline volcanicity. The geochemical characters of these modern day igneous rocks are the result of a multi-stage fractionation process that has been operative over a time span that is well within the time span which the isotopic data for the Lewisian rocks places on the time of separation of igneous products from the mantle in late Archaean times and their subsequent involvement in the Scourian orogenic episode. The modern day fractionation process involves melting of the mantle below mid-ocean ridges and then differentiation in magma chambers below the ridge. This already differentiated material is taken down a subduction zone and further differentiated during emplacement in the mobile zone. A comparable multi-stage fractionation process, associated in the mobile zone with the upward movement of large volumes of gas due to mantle degassing, could account for very many of the characters of the *c.* 2·8–2·7 b.y. old parts of the Lewisian complex. The whole process could be related to the existence of 'rolls', immediately below the crust, related to very large convection cells in an undifferentiated mantle that would lead to the development of features somewhat comparable with ridge and subduction zones—the hot line tectonic regime of Sun and Hanson (1975; cf. Bowes 1978b). The recurrence of northwest-trending structural features in the products of the Scourian episode suggest a NW–SE direction for such 'rolls', 'ridges' and 'subduction zones' with any zone of downgoing slabs being at a much shallower angle than in the present day analogues and any zones of magmatism much wider (Drury 1978b). Steeper thermal gradients in Archaean times would probably have resulted in higher rates of generation of 'sialic' crust than is known from the present day, as the result of greater 'ridge' and 'subduction' activity. In addition, the high proportion of olivine in the products of basic and ultrabasic magmatism in Archaean times means that the hydrated material would take with it down 'subduction zones' a greater proportion of water than is the case in present day, and so assist in providing conditions under which melting occurs.

The foregoing model for late Archaean mobile belts is consistent with the present state of knowledge of the Lewisian complex. In addition, the existence, adjacent to the plutonic zone of southern Chile, of an ensialic back-arc marginal basin containing the association of volcanic, sedimentary and plutonic rock types that is very similar to that of many late Archaean greenstone belts (Tarney *et al.* 1976), could provide a basis for explaining any greenstone belt association that may be demonstrated for a part of the Lewisian complex. Likewise it could form a basis for explaining the presence of greenstone belts in and on a dominantly tonalitic-granodioritic gneiss complex developed in late Archaean times in the Baltic Shield (North and East Finland and northwestern USSR) which Bowes (1975; 1976b) correlates with the Lewisian complex.

## 4d. Crustal history in early–middle Proterozoic times

The crustal significance of events in early–middle Proterozoic times, expressed in the Lewisian complex, seems more likely to emerge when they are considered along with events affecting other Precambrian shield areas, than when considered alone. Such an approach is made possible by the demonstration of the existence of super-continental masses at that time, on the basis of palaeomagnetic studies (e.g. Piper 1976).

The great similarities between the events of parts of North America (e.g. Van Schmus 1976) and parts of the Baltic Shield (e.g. Bowes 1976c) in early–middle Proterozoic times, and the strong evidence for correlation of Finland and Scotland in Precambrian times (Bowes 1975) means that the products of the Inverian episode could be related to deformation of basement below the equivalents of the Huronian Supergroup of Canada. Likewise the Loch Maree Group, which appears to be a continuation of the Karelian schists of Finland (Bowes 1976c), could be related to the Animikie Group of northern USA (cf. Bowes 1968a). All these sedimentary units were affected by generally co-eval orogenic episodes, these being the Laxfordian, Svecokarelian and Penokean episodes, respectively. The concept of the initiation of the Svecokarelian episode being associated with the emplacement of large masses of anorthositic and related high temperature magma into the middle levels of the crust (Bowes *et al.* in press) could be applied with equal validity to the Laxfordian episode with the emplacement of anorthositic magma early in Proterozoic times known for the Scottish region. However, the subsequent development of rapakivi granites in Fennoscandia (cf. Bridgewater *et al.* 1974) is not matched in NW Britain, at least not at the levels now seen. Such magmatic emplacement and continued pulses of basic igneous magmatism throughout the Laxfordian episode (cf. Rona sequence), could be related to the occurrence of the type of crustal conditions in the basement, as seen in the Lewisian complex, where there was extensive tectonic overprinting and the development and emplacement of large masses of granitic material with associated thermal effects.

## 5. Stratigraphical nomenclature and subdivision

The stratigraphical nomenclature and subdivisions used in relation to the Lewisian complex over more than a quarter of a century have been so varied and so confusing that all but those involved in research must have found terminology a great hindrance to comprehension of fundamental principles illustrated by the development of the complex. Much of the confusion in relation to nomenclature stems from the original definition of the terms 'Scourian' and 'Laxfordian' (Sutton and Watson 1951a pp. 294–5) that covered their use not only in both lithostratigraphic and chronostratigraphic senses, but also in terms of processess (metamorphism and orogeny) and, by implication, in terms of geographical distribution (trend of orogenic belt). This has led to the use of just 'the Scourian' and 'the Laxfordian' to cover whichever usage the particular author wished, which was not always clear. It also meant that it has not been possible to use correctly one aspect of the original definition without misusing another, e.g. the 'Scourian complex' shows evidence of two major separate episodes of metamorphism (*c.* 2·7 and *c.* 2·4–2·2 b.y.) for which Evans (1965) used the terms 'Scourian' and 'Inverian'. This usage was generally followed by Bowes (1968a, b, 1969) who considered that both the 'Scourian episode' and the 'Inverian episode' were orogenic episodes. Subsequently, in order to retain the uses of 'Scourian' with reference to both a rock complex and a long period of time, 'Early Scourian' (Sutton and Watson 1969), 'Badcallian' (Park 1970b) or 'early Scourian (Badcallian)' (Watson 1975) and 'late Scourian' (Sutton 1974; Watson 1975) were introduced to generally correspond with 'Scourian' and 'Inverian' of Evans (1965). However, these new terms were used to include deformation as well as metamorphism and Sutton and Watson (1969) used 'Inverian' rather than 'late Scourian'.

Much of the confusion relating to subdivision of the Lewisian complex has arisen from the inherent weakness of using basic dykes as a means of separating a pre-dyke 'Scourian' complex from a post-dyke 'Laxfordian' complex on the assumption that the dykes represented a unique and recognizable time plane. Apart from the numerous demonstrations of more than one phase of basic minor intrusion in many parts of the Lewisian complex, with different phases separated by dated products, the original field data presented by Sutton and Watson (1951a, pl. xx) are not consistent with the proposed subdivision, e.g. a dolerite dyke cuts the hinge zone of an isoclinal fold formed of 'Laxfordian gneisses'. The unsoundness of the basis of this subdivision is emphasized by isotopic studies in the type area for the 'Laxfordian complex' (p. 57–60). There (1) amphibolites, previously interpreted as being part of the *c.* 2·4 b.y. basic minor intrusions taken to separate the Scourian and Laxfordian complexes, have been shown to be younger than *c.* 1720 m.y. old granites and (2) the 'Laxfordian complex' is largely composed of the products of the Scourian episode, i.e. the formation of the 'Laxfordian' and 'Scourian' complexes was co-eval (cf. Davies 1977).

The difficulties inherent in, and the inadequacies of, these original definitions and subdivisions have been pointed out by many workers from a whole range of considerations, beginning with Whitten (1951). With Davies and Watson (1977) finding it necessary to radically change the use of the term 'Laxfordian complex' to cover the products of "intermittent tectonothermal activity . . . . . from 2800–1700 Ma" without the apparent need to change the original premises, it clearly desirable to end the confusion by discarding its causes and with them the concept that "no terminology defined by reference to purely stratigraphic criteria can be consistently applied" (Watson 1975 p. 15).

The combination of field and isotopic data now provide adequate criteria for subdivision of the 'Lewisian Gneiss' of Peach *et al.* (1907) into cognate units of earth history. The synoptic Figure 3 sets out such a subdivision with details of its basis in the text. It uses 'Scourian' and 'Laxfordian' in one of the ways they were originaly defined, together with 'Inverian'; these terms are used in preference to 'Badcallian', 'early Scourian' and 'late Scourian' on the basis of priority of usage. The Loch Maree Group (previously 'Series') is assigned to its stratigraphic position on the basis of isotopic and geochemical considerations and the crustal pile formed in the initial stage of the Scourian cycle is referred to as the Gaelic super-group rather than 'Supergroup' (Bowes 1976a) in view of the model of development proposed (p. 68); it consists of the Kylesku group (not 'Group' as used by some authors), the Rhiconich group and related rocks. With the information now available linking structural/metamorphic/igneous sequences and isotopic data, this subdivision of earth history is considered to be soundly and unambiguously based. It should form a firm foundation for the many aspects of the development and crustal significance of the Lewisian complex that await clarification from detailed mineralogical, geochemical, geophysical, fabric and structural studies.

*Acknowledgements* It is a pleasure to acknowledge the stimulus of B. C. Barooah, C. C. Bhattacharjee, P. K. Chowdhary, B. Dash, D. W. Elliott, D. Findlay, C. Gillen, T. S. Ghaly, A. M. Hopgood, J. D. Keppie, S. G. Khoury, R. G. Park, R. T. Pidgeon, O. van Breemen and A. E. Wright whose contributions to the elucidation of the development of the Lewisian complex go far beyond the content of their published work. The help of T. N. George, B. E. Leake and other members of staff of the Department of Geology, University of Glasgow and the financial support of the University Court of the University of Glasgow are gratefully acknowledged.

## References


ALDERMAN, A. R. 1936. Eclogites from the neighbourhood of Glenelg, Inverness-shire. *Q. J. geol. Soc. Lond.* **92,** 488–533.

ANDERSON, J. G. C. 1965. The Precambrian of the British Isles. *In* Rankama, K. (Editor). *The Precambrian, Vol. 2,* 25-111. Wiley, New York.

ARTH, J. G. and HANSON, G. N. 1975. Geochemistry and origin of the Early Precambrian crust of northeastern Minnesota. *Geochim. Cosmochim. Acta* **39,** 325–362.

BAILEY, E. B. 1951. Scourie dykes and Laxfordian metamorphism. *Geol. Mag.* **88,** 153–165.

BAMFORD, D., NUNN, K., PRODEHL, C. and JACOB, B. 1977. LISPB–III. Upper crustal structure of Northern Britain. *J. geol. Soc. Lond.* **133,** 481–488.

BARBER, A. J. and MAY, F. 1976. The history of the Western Lewisian in the Glenelg Inlier, Lochalsh, Northern Highlands. *Scott. J. Geol.* **12,** 35–50.

BAROOAH, B. C. 1967. *The geology of the Lewisian rocks south-east of Scourie, Sutherland.* Univ. Glasgow Ph.D. thesis (unpubl.).

—— 1970. Significance of calc-silicate rocks and meta-arkose in the Lewisian complex south-east of Scourie. *Scott. J. Geol.* **6,** 221–225.

BEACH, A. 1973. The mineralogy of high temperature shear zones at Scourie, N.W. Scotland. *J. Petrol.* **14,** 231–248.

—— 1974a. Amphibolitization of Scourian granulites. *Scott. J. Geol.* **10,** 35–44.

—— 1974b. The measurement and significance of displacements on Laxfordian shear-zones, north-west Scotland. *Proc. Geol. Assoc.* **85,** 13–22.

—— 1976. The interrelations of fluid transport, deformation, geochemistry and heat flow in early Proterozoic shear zones in the Lewisian complex. *Phil. Trans. R. Soc. Lond.* A, **280,** 569–604.

——, COWARD, M. P. and GRAHAM, R. H. 1974. An interpretation of the structural evolution of the Laxford front. *Scott. J. Geol.* **9,** 297–308.

—— and FYFE, W. S. 1972. Fluid transport and shear zones at Scourie, Sutherland: evidence of overthrusting? *Contrib. Mineral. and Petrol.* **36,** 175–180.

BENNISON, G. M. and WRIGHT, A. E. 1969. *The Geological History of the British Isles.* Arnold, London.

BHATTACHARJEE, C. C. 1963. The late structural and petrological history of the Lewisian rocks of the Meall Deise area, north of Gairloch, Ross-shire. *Trans. geol. Soc. Glasg.* **25,** 31–60.

—— 1964. The Lewisian geology of the Little Sand dioritic mass, Gairloch. *Geol. Mag.* **101,** 48–62.

—— 1968. The structural history of the Lewisian rocks north-west of Loch Tollie, Ross-shire, Scotland. *Scott. J. Geol.* **4,** 235–264.

—— 1969. Structural history of Lewisian rocks of Tollie, Ross-shire. *Scott. J. Geol.* **5,** 296–301.

BIKERMAN, M., BOWES, D. R. and VAN BREEMEN, O. 1974. Geochronology of some Lewisian (Precambrian) metamorphic rocks in north-western Scotland by the Rb–Sr whole rock method. *Geol. Soc. Am. Abstr.* **6,** 654–655.

——, —— and —— 1975. Rb–Sr whole rock isotopic studies of Lewisian metasediments and gneisses in the Loch Maree region, Ross-shire. *J. geol. Soc. Lond.* **131,** 237–254.

BOTT, M. H. P., HOLLAND, J. G., STORRY, P. G. and WATTS, A. B. 1972. Geophysical evidence concerning the structure of the Lewisian of Sutherland, N.W. Scotland. *J. geol. Soc. Lond.* **128,** 599–612.

BOWDIDGE, C. R. 1969. *Petrological studies of Lewisian basic and ultrabasic rocks near Scourie, Sutherland.* Univ. Edinburgh Ph.D. thesis (unpubl.).

BOWES, D. R. 1962. Discussion of Depth and tectonics as factors in regional metamorphism. *Proc. geol. Soc. Lond.* **1594,** 28–30.

—— 1965. Discussion *of* Sabine, P. A. and Watson, J. Isotopic age-determination of rocks from the British Isles, 1955–64. *Q. J. geol. Soc. Lond.* **121,** 523–525.

—— 1967. Petrochemistry of some Lewisian granitic rocks. *Mineral. Mag.* **36,** 342–362.

—— 1968a. An orogenic interpretation of the Lewisian of Scotland. *Rep. XXIII Int. geol. Congr.* **4,** 225–236.

—— 1968b. The absolute time scale and the subdivision of Precambrian rocks in Scotland. *Geol. Forën. Stockholm Förh.* **90,** 175–188.

—— 1969. The Lewisian of Northwest Highlands of Scotland. *In* Kay, M. (Editor). *North Atlantic—geology and continental drift—a symposium,* 575–594. *Mem. Am. Assoc. Petrol. Geol.* **12.**

—— 1971. Lewisian chronology. *Scott. J. Geol.* **7,** 179–182.

—— 1972. Geochemistry of Precambrian crystalline basement rocks, North-West Highlands of Scotland. *Rep. XXIV Int. geol. Congr.* **1,** 97–103.

—— 1975. Scotland–Finland Precambrian correlations. *Bull. geol. Soc. Finl.* **47,** 1–12.

—— 1976a. Archaean crustal history in North-western Britain. *In* Windley, B. F. (Editor). *The Early History of the Earth,* 469–479. Wiley, London.

—— 1976b. Archaean crustal history in the Baltic Shield. *In* Windley, B. F. (Editor). *The Early History of the Earth,* 481–488. Wiley, London.

BOWES, D. R. 1976c. Tectonics in the Baltic Shield in the period 2000–1500 m.y. ago. *Acta geol. Pol.* **26,** 355–376.
—— 1978a. Application of U–Pb zircon and other isotopic studies to the identification of Archaean rocks in thermally and tectonically overprinted terrain: Lewisian complex of Scotland. *In* Windley, B. F. and Naqui, S. M. (Editors). *Archaean Geochemistry.* Elsevier, Amsterdam.
—— 1978b. Characterization of regimes of polyphase deformed metamorphic rocks in the Baltic Shield. *In* Verwoerd ,W. J. (Editor). *Mineralization in Metamorphic Terranes. Spec. Publ. geol. Soc. S. Afr.* **4** (in press).
——, BAROOAH, B. C. and KHOURY, S. G. 1971. Original nature of Archaean rocks of north-west Scotland. *Spec. Publ. geol. Soc. Aust.* **3,** 77–92.
—— and BHATTACHARJEE, C. C. 1967. The metamorphic and migmatitic history of the Lewisian rocks north-west of Loch Tollie, Ross-shire, Scotland. *Krystalinikum* **5,** 7–60.
——, CAMPBELL, D. S., GAÁL, G., HOPGOOD, A. M. and KOISTINEN, T. In press. A model for the evolution of the Svecokarelides.
—— and GHALY, T. S. 1964. Age relations of Lewisian basic rocks, south of Gairloch, Ross-shire. *Geol. Mag.* **101,** 150–160.
—— and HOPGOOD, A. M. 1969. The Lewisian gneiss complex of Mingulay, Outer Hebrides, Scotland. *In* Larsen, L. H. (Editor). *Igneous and Metamorphic Geology,* 317–356. *Mem. geol. Soc. Am.* **115.**
—— and —— 1971. Correlation by structural sequence in polyphase deformed terrain. *Geol. Soc. Am. Abstr.* **3,** 510–511.
—— and —— 1973. Framework of the Precambrian crystalline complex of northwestern Scotland. *In* Pidgeon, R. T., Macintyre, R. M., Sheppard, S. M. F. and van Breemen, O. (Editors). *Geochronology-isotope geology of Scotland Field guide and reference,* A1–14, *3rd Europ. Congr. of Geochronologists.* Scottish Universities Research and Reactor Centre, East Kilbride.
—— and —— 1975a. Framework of the Precambrian crystalline complex of the Outer Hebrides, Scotland. *Krystalinikum* **11,** 7–23.
—— and —— 1975b. Structure of the gneiss complex of Inishtrahull, Co. Donegal. *Proc. R. Irish Acad.* **75B,** 369–390.
—— and —— 1976. Significance of structural trend in Precambrian terrain. *Acta geol. Pol.* **26,** 57–82.
——, —— and PIDGEON, R. T. 1976. Source ages of zircons in an Archaean quartzite, Rona, Inner Hebrides, Scotland. *Geol. Mag.* **113,** 545–552.
——, —— and SMART, J. 1968. Glasgow University Exploration Society Expedition to Inishtrahull. *Nature, Lond.* **217,** 344–345.
——, —— and TAFT, M. B. 1971. Granitic injection complex of Harris, Outer Hebrides. *Scott. J. Geol.* **7,** 289–291.
—— and KHOURY, S. G. 1965. Successive periods of basic dyke emplacement in the Lewisian complex south of Scourie, Sutherland. *Scott. J. Geol.* **1,** 295–299.
—— and PARK, R. G. 1966. Metamorphic segregation banding in the Loch Kerry basite sheet from the Lewisian of Gairloch, Ross-shire, Scotland. *J. Petrol.* **7,** 306–330.
——, SKINNER, W. R. and WRIGHT, A. E. 1970. Petrochemical comparison of the Bushveld Igneous Complex with some other mafic complexes. *In* Visser, D. J. L. and von Gruenewaldt, G. (Editors). *Symposium on the Bushveld Igneous Complex and other layered intrusions,* 425–440. *Spec. Publ. geol. Soc. S. Afr.* **1.**
——, WRIGHT, A. E. and PARK, R. G. 1964. Layered intrusive rocks in the Lewisian of the North-West Highlands of Scotland. *Q. J. geol. Soc. Lond.* **120,** 153–192.
——, —— and —— 1966. Origin of ultrabasic and basic masses in the Lewisian. *Geol. Mag.* **103,** 280–284.
BOWIE, S. H. U. 1962. Report of Atomic Energy Division. *Summ. Prog. geol. Surv. Gt Br.* for 1961, 68.
—— 1964. Report of Atomic Energy Division. *Summ. Prog. geol. Surv. Gt Br.* for 1963, 76.
BRIDGEWATER, D., SUTTON, J. and WATTERSON, J. 1974. Crustal downfolding associated with igneous activity. *Tectonophysics* **21,** 57–77.
——, WATSON, J. and WINDLEY, B. F. 1973. The Archaean craton of the North Atlantic region. *Phil. Trans. R. Soc. Lond.* A **273,** 493–512.
BROOK, M., BREWER, M. S. and POWELL, D. 1976. Grenville age for rocks in the Moine of north-western Scotland. *Nature, Lond.* **260,** 515–517.
——, POWELL, D. and BREWER, M. S. 1977. Grenville events in Moine rocks of the Northern Highlands, Scotland. *J. geol. Soc. Lond.* **133,** 489–496.
BURNS, D. J. 1966. Chemical and mineralogical changes associated with the Laxfordian metamorphism of dolerite dykes in the Scourie-Loch Laxford area, Sutherland, Scotland. *Geol. Mag.* **103,** 19–35.
CHADWICK, B., COE, K., GIBBS, A. D., SHARP, M. R. and WELLS, P. R. A. 1974. Field evidence relating to the origin of ~ 3000/Myr gneisses in southern West Greenland. *Nature, Lond.* **248,** 136–137.
CHAPMAN, H. 1976. Strontium isotope geochemistry of Precambrian basic dykes from north-west Scotland. *Abstr. Eur. Colloq. of Geochronologists,* Amsterdam.

CHAPMAN, H. and MOORBATH, S. 1977. Lead isotope measurements from the oldest recognized Lewisian gneisses of north-west Scotland. *Nature, Lond.* **268,** 41–42.
CHOWDHARY, P. K. 1971. Zircon populations in Lewisian quartzite, gneiss and granite north of Loch Laxford, Sutherland. *Geol. Mag.* **108,** 255–262.
—— and BOWES, D. R. 1972. Structure of Lewisian rocks between Loch Inchard and Loch Laxford, Sutherland, Scotland. *Krystalinikum* **9,** 21–51.
——, DASH, B. and FINDLAY, D. 1971. Metasediments in the Rhiconich group of the Lewisian between Loch Laxford and Durness, Sutherland. *Scott. J. Geol.* **7,** 1–9.
COWARD, M. P. 1972. The Eastern Gneisses of South Uist. *Scott. J. Geol.* **8,** 1–12.
—— 1973a. Heterogeneous deformation in the development of the Laxfordian complex of South Uist, Outer Hebrides. *J. geol. Soc. Lond.* **129,** 139–158.
—— 1973b. The structure and origin of areas of anomalously low–intensity finite deformation in the Outer Hebrides. *Tectonophysics* **16,** 117–140.
—— 1974. Flat-lying structures within the Lewisian Basement Gneiss Complex of N.W. Scotland. *Proc. Geol. Assoc.* **85,** 459–472.
——, FRANCIS, P. W., GRAHAM, R. H., MYERS, J. S. and WATSON, J. 1969. Remnants of an early metasedimentary assemblage in the Lewisian complex of the Outer Hebrides. *Proc. Geol. Assoc.* **80,** 387–408.
——, ——, —— and WATSON, J. 1970. Large-scale Laxfordian structures of the Outer Hebrides in relation to those of the Scottish mainland. *Tectonophysics* **10,** 425–435.
CRANE, A. 1973. *The geology of the Lewisian complex near Poolewe, Ross-shire.* Univ. Keele Ph.D. thesis (unpubl.).
CRESSWELL, D. 1972. The structural development of the Lewisian rocks of the north shore of Loch Torridon, Ross-shire. *Scott. J. Geol.* **8,** 293–308.
—— and PARK, R. G. 1973. The metamorphic history of the Lewisian rocks of the Torridon area in relation to that of the remainder of the southern Laxfordian belt. *In* Park, R. G. and Tarney, J. (Editors). *The Early Precambrian of Scotland and Related Rocks of Greenland,* 77–84. Univ. Keele.
DALZIEL, I. W. D. 1969. Pre-Permian history of the British Isles. *In* Kay, M. (Editor). *North Atlantic—geology and continental drift—a symposium,* 5–31. *Mem. Am. Assoc. Petrol. Geol.* **12.**
DASH, B. 1969. Structure of the Lewisian rocks between Strath Dionard and Rhiconich, Sutherland, Scotland. *Scott. J. Geol.* **5,** 347–374.
——, CHOWDHARY, P. K. and BOWES, D. R. 1974. Basic minor intrusions north of Loch Laxford, Sutherland and their significance in Lewisian chronology. *Scott. J. Geol.* **10,** 45–52.
DAVIDSON, C. F. 1943. The Archaean rocks of the Rodil district, South Harris, Outer Hebrides. *Trans. R. Soc. Edinb.* **61,** 71–112.
DAVIES, F. B. 1974. A layered basic complex in the Lewisian, south of Loch Laxford, Sutherland. *J. geol. Soc. Lond.* **130,** 279–284.
—— 1975a. Origin and ancient history of gneisses older than 2800 m.y. in the Lewisian complex. *Nature, Lond.* **258,** 589–591.
—— 1975b. Evolution of Proterozoic basement patterns in the Lewisian complex. *Nature, Lond.* **256,** 568–570.
—— 1976. Early Scourian structures in the Scourie–Laxford region and their bearing on the evolution of the Laxford Front. *J. geol. Soc. Lond.* **132,** 543–554.
—— 1977. Archaean evolution of the Lewisian complex of Gruinard Bay, Ross-shire. *Scott. J. Geol.* **13,** 189–196.
——, LISLE, R. J. and WATSON, J. 1975. The tectonic evolution of the Lewisian complex in northern Lewis, Outer Hebrides. *Proc. Geol. Assoc.* **86,** 45–61.
—— and WATSON, J. V. 1977. Early basic bodies in the type Laxfordian complex, NW Scotland and their bearing on its origin. *J. geol. Soc. Lond.* **133,** 123–131.
—— and WINDLEY, B. F. 1976. The significance of major Proterozoic high-grade linear belts in continental evolution. *Nature, Lond.* **263,** 383–385.
DEARNLEY, R. 1962. An outline of the Lewisian complex of the Outer Hebrides in relation to that of the Scottish mainland. *Q. J. geol. Soc. Lond.* **118,** 143–176.
—— 1963. The Lewisian complex of South Harris; with some observations on the metamorphosed basic intrusions of the Outer Hebrides. *Q. J. geol. Soc. Lond.* **119,** 243–312.
—— 1966. Orogenic fold–belts and a hypothesis of earth evolution. *Physics Chem. Earth* **7,** 1–114.
—— 1973. Scourie dykes of the Outer Hebrides. *In* Park, R. G. and Tarney, J. (Editors). *The Early Precambrian of Scotland and Related Rocks of Greenland,* 131–135. Univ. Keele.
—— and DUNNING, F. W. 1968. Metamorphosed and deformed pegmatites and basic dykes in the Lewisian complex of the Outer Hebrides and their geological significance. *Q. J. geol. Lond.* **123** (for 1967), 353–378.
DICKINSON, B. B. and WATSON, J. 1976. Variations in crustal level and geothermal gradient during the evolution of the Lewisian complex of north-west Scotland. *Precambrian Res.* **3,** 363–374.
DOUGAL, J. W. 1928. Observations in the geology of Lewis. *Trans. Edinb. geol. Soc.* **12,** 12–18.
DRURY, S. A. 1972a. The tectonic evolution of a Lewisian complex on Coll, Inner Hebrides. *Scott. J. Geol.* **8,** 309–333.
—— 1972b. The chemistry of some granitic veins from the Lewisian of Coll and Tiree, Argyllshire, Scotland. *Chem. Geol.* **9,** 175–193.

DRURY, S. A. 1973a. The geochemistry of Precambrian granulite facies rocks from the Lewisian complex of Tiree, Inner Hebrides, Scotland. *Chem. Geol.* **11,** 167–188.
—— 1973b. Structure and metamorphism of the Lewisian of East Tiree, Inner Hebrides. *Scott. J. Geol.* **9,** 89–94.
—— 1974. Chemical changes during retrogressive metamorphism of Lewisian granulite facies rocks from Coll and Tiree. *Scott. J. Geol.* **10,** 237–256.
—— 1977a. REE distributions in a high grade Archaean gneiss complex in Scotland: implications for the genesis of ancient salic crust. *IUGS–UNESCO–IGCP Archaean geochemistry. The origin and evolution of Archaean continental crust,* 122–123. Symposium Abstracts, Hyderabad.
—— 1977b. Basic factors in Archaean geotectonics. *IUGS–UNESCO–IGCP Archaean geochemistry. The origin and evolution of Archaean continental crust,* 83–84. Symposium Abstracts, Hyderabad.
DUNNING, F. W. and MAX, M. D. 1975. Explanatory notes to the geological map of the exposed and concealed Precambrian basement of the British Isles. *In* Harris, A. L., Shackleton, R. M., Watson, J., Downie, C., Harland, W. B. and Moorbath, S. (Editors). *A correlation of Precambrian rocks in the British Isles,* 11–14. *Spec. Rep. geol. Soc. Lond.* No. 6.
ELLIOTT, D. W. 1964. *Geology of the Lewisian Complex of the Slattadale area, south of Loch Maree, Ross-shire.* Univ. Glasgow Ph.D. thesis (unpubl.).
—— 1965. The quantitative mapping of directional minor structures. *J. Geol.* **73,** 865–880.
—— 1968. Interpretation of fold geometry from lineation isogonic maps. *J. Geol.* **76,** 171–190.
ESCHER, A. and WATT, W. S. (Editors). 1976. *Geology of Greenland.* Geological Survey of Greenland, Copenhagen.
ESSENE, E. J., EVANS, B. W. and CARMICHAEL, I. S. E. 1966. Websterite from Glenelg, Inverness-shire. *Scott. J. Geol.* **2,** 224–225.
EVANS, C. R. 1964. Geochronology of the Lewisian basement complex near Lochinver, Sutherland. *Advancement Sci. Lond.* **20,** 446.
—— 1965. Geochronology of the Lewisian basement near Lochinver, Sutherland. *Nature, Lond.* **207.** 54–56.
—— and LAMBERT, R. ST J. 1964. Discussion *of* Park, R. G. The structural history of the Lewisian rocks of Gairloch, Wester Ross. *Q. J. geol. Soc. Lond.* **120,** 429–430.
—— and —— 1974. The Lewisian of Lochinver, Sutherland; the type area for the Inverian metamorphism. *J. geol. Soc. Lond.* **130,** 125–150.
—— and PARK, R. G. 1965. Potassium-argon age determinations from the Lewisian of Gairloch, Ross-shire, Scotland. *Nature, Lond.* **205,** 350–352.
—— and TARNEY, J. 1964. Isotopic ages of Assynt dykes. *Nature, Lond.* **207,** 54–56.
FINDLAY, D. 1970. *Studies of fold tectonics in the Lewisian of the Durness region, Sutherland.* Univ. Glasgow Ph.D. thesis (unpubl.).
FRANCIS, P. W. 1973. Scourian-Laxfordian relationships in the Barra Isles. *J. geol. Soc. Lond.* **129,** 161–182.
——, MOORBATH, S. and WELKE, H. J. 1971. Isotopic age data for Scourian intrusive rocks on the Isle of Barra, Outer Hebrides, northwest Scotland. *Geol. Mag.* **108,** 13–22.
—— and SIBSON, R. H. 1973. The Outer Hebrides Thrust. *In* Park, R. G. and Tarney, J. (Editors). *The Early Precambrian of Scotland and Related Rocks of Greenland,* 95–104. Univ. Keele.
GARSON, M. S. and LIVINGSTONE, A. 1973. Is the South Harris complex in North Scotland a Precambrian overthrust slice of oceanic crust and island arc? *Nature Phys. Sci., Lond.* **243,** 74–76.
GHALY, T. S. 1966. The Lewisian geology of the area between Loch Shieldaig and Loch Bràigh Horrisdale, Gairloch, Ross-shire. *Scott. J. Geol.* **2,** 282–305.
GHOSE, C. 1958. *The structural petrology of the Lewisian rocks around Scourie, North-west Highlands of Scotland.* Univ. Bristol Ph.D. thesis (unpubl.).
GILETTI, B. J. 1959. Rubidium-strontium ages of Lewisian rocks from north-west Scotland. *Nature, Lond.* **184,** 1793–1794.
——, LAMBERT, R. ST J. and MOORBATH, S. 1961a. The basement rocks of Scotland and Ireland. *Ann. N.Y. Acad. Sci.* **91,** 464–468.
——, MOORBATH, S. and LAMBERT, R. ST J. 1961b. A geochronological study of the metamorphic complexes of the Scottish Highlands. *Q. J. geol. Soc. Lond.* **117,** 233–272.
GILLEN, C. 1975. Structural and metamorphic history of the Lewisian gneiss around Loch Laxford, Sutherland, Scotland. *Krystalinikum* **11,** 63–85.
GRAHAM, R. H. and COWARD, M. P. 1973. The Laxfordian of the Outer Hebrides. *In* Park, R. G. and Tarney, J. (Editors). *The Early Precambrian of Scotland and Related Rocks of Greenland,* 85–93, Univ. Keele.
HALL, J. 1978. 'LUST'—a seismic refraction survey of the Lewisian basement complex in NW Scotland. *J. geol. Soc. Lond.* **135** (in press).
—— and AL-HADDAD, F. M. 1976. Seismic velocities in the Lewisian metamorphic complex, north-west Britain—*in situ* measurements. *Scott. J. Geol.* **12,** 305–314.
HOLLAND, J. G. and LAMBERT, R. ST J. 1969. Structural regimes and metamorphic facies. *Tectonophysics* **7,** 197–217.
—— and —— 1972a. The geochemistry of lithium in the mainland Lewisian of Scotland. *Rep. XXIV Int. geol. Congr.* **10,** 169–178.

HOLLAND, J. G. and LAMBERT. R ST J. 1972b. Chemical petrology and K–Ar ages of the Lewisian gneisses of Rona, near Skye, Scotland. *Geol. Mag.* **109,** 339–347.
—— and —— 1972c. Major element composition of shields and the continental crust. *Geochim. Cosmochin. Acta* **36,** 673–683.
—— and —— 1973. Comparative major element geochemistry of the Lewisian of the mainland of Scotland. *In* Park, R. G. and Tarney, J. (Editors). *The Early Precambrian of Scotland and Related Rocks of Greenland,* 51–62. Univ. Keele.
—— and —— 1975. The chemistry and origin of the Lewisian gneisses of the Scottish mainland: the Scourie and Inver assemblages and sub-crustal accretion. *Precambrian Res.* **2,** 161–188.
HOLMES, A., SHILLIBEER, H. A. and WILSON, J. T. 1955. Potassium-argon ages of some Lewisian and Fennoscandian pegmatites. *Nature, Lond.* **176,** 390–392.
HOPGOOD, A. M. 1966. Theoretical consideration of the mechanics of tectonic re-orientation of dykes. *Tectonophysics* **3,** 17–28.
—— 1971a. Structure and tectonic history of Lewisian Gneiss, Isle of Barra, Scotland. *Krystalinikum* **7,** 27–60.
—— 1971b. Correlation by tectonic sequence in Precambrian gneiss terrains. *Spec. Publ. geol. Soc. Aust.* **3,** 367–376.
—— 1973. The significance of deformational sequence in discriminating between Precambrian terrains. *Spec. Publ. geol. Soc. S. Afr.* **3,** 45–52.
—— and BOWES, D. R. 1972a. Application of structural sequence to the correlation of Precambrian gneisses, Outer Hebrides, Scotland. *Bull. geol. Soc. Am.* **83,** 107–128.
—— and —— 1972b. Correlation by structural sequence in Precambrian gneisses of Northwestern Britain. *Rep. XXIV Int. geol. Congr.* **1,** 195–200.
HOWIE, R. A. 1964. Some orthopyroxenes from Scottish metamorphic rocks. *Mineral. Mag.* **33,** 903–911.
HULL, E., NOLAN, J., CRUISE, R. J., MCHENRY, A. and HYLAND, J. S. 1890. The Geology of Inishowen. *Mem. geol. Surv. Ireland.*
INGLIS, J. W. 1966. *A study of the granites and associated Lewisian rocks of Loch Laxford, Sutherland.* Univ. Keele Ph.D. thesis (unpubl.).
JEHU, T. J. 1922. The Archaean and Torridonian formations and the later intrusive igneous rocks of Iona. *Trans. R. Soc. Edinb.* **53,** 165–187.
—— and CRAIG, R. H. 1923, 1925, 1926, 1927, 1934. Geology of the Outer Hebrides, Parts I-V. *Trans. R. Soc. Edinb.* **53,** 419–441, 615–641; **54,** 46–89; **55,** 457–488; **57,** 839–874.
JOHNSON, M. R. W. 1960. The structural history of the Moine thrust zone at Loch Carron, Wester Ross. *Trans. R. Soc. Edinb.* **64,** 139–168.
JOHNSTONE, G. S. 1975. The Moine Succession. *In* Harris, A. L., Shackleton, R. M., Watson, J., Downie, C., Harland, W. B. and Moorbath, S. (Editors). *A correlation of Precambrian rocks in the British Isles,* 30–42. *Spec. Rep. geol. Soc. Lond.* No. **6.**
KEPPIE, J. D. 1969. Analysis of a mica subfabric. *Scott. J. Geol.* **5,** 171–186.
KERR, A. 1974. *The mineralogy and petrology of some granulite facies rocks from the Scourie area, Sutherland.* Queens Univ. Belfast Ph.D. thesis (unpubl.).
KHOURY, S. G. 1968a. The structural geometry and geological history of the Lewisian rocks between Kylesku and Geisgeil, Sutherland, Scotland. *Krystalinikum* **6,** 41–78.
—— 1968b. Structural analysis of complex fold belts in the Lewisian north of Kylesku, Sutherland, Scotland. *Scott. J. Geol.* **4,** 109–120.
KING, B. C. 1955. The tectonic pattern of the Lewisian around Clashnessie Bay, near Stoer, Sutherland. *Geol. Mag.* **92,** 69–80.
KROGH, T. E. and DAVIS, G. L. 1971. Zircon U–Pb ages of Archean metavolcanic rocks in the Canadian Shield. *Carnegie Institution Year Book* **70,** 241–242.
KURSTEN, M. 1957. The metamorphic and tectonic history of parts of the Outer Hebrides. *Trans. Edinb. geol. Soc.* **17,** 1–31.
LAMBERT, R. ST J. 1965. Discussion *of* Sabine, P. A. and Watson, J. Isotopic age-determinations of rocks from the British Isles, 1955–64. *Q. J. geol. Soc. Lond.* **121,** 526.
—— 1974. Discussion *of* Evans, C. R. and Lambert, R. St J. The Lewisian of Lochinver, Sutherland; the type area for the Inverian metamorphism. *J. geol. Soc. Lond.* **130,** 149–150.
——, EVANS, C. R. and DEARNLEY, R. 1970. Isotopic ages of dykes and pegmatitic gneiss from the southern islands of the Outer Hebrides. *Scott. J. Geol.* **6,** 214–220.
—— and HOLLAND, J. G. 1972. A geochronological study of the Lewisian from Loch Laxford to Durness, Sutherland, Scotland. *J. geol. Soc. Lond.* **128,** 3–19.
—— and —— 1973. Ca/Y ratios as petrogenetic indicators. *Trans. Am. Geophys. Union* **54,** 1225.
—— and —— 1974. Yttrium geochemistry applied to petrogenesis using calcium—yttrium relationships in rocks and minerals. *Geochim. Cosmochim. Acta* **38,** 1393–1414.
——, MYERS, J. S. and WATSON, J. 1970. An apparent age for a member of the Scourie dyke suite in Lewis, Outer Hebrides. *Scott. J. Geol.* **6,** 221–225.
LISLE, R. J. 1977. The evolution of Laxfordian deformation in the Carloway area, Isle of Lewis, Scotland. *Tectonophysics* **42,** 183–208.
LIVINGSTONE, A. 1976. The paragenesis of spinel- and garnet-amphibole lherzolite in the Rodel area, South Harris. *Scott. J. Geol.* **12,** 293–300.

Lyon, T. D. B. 1973. *A geochronological investigation of the Lewisian rocks of northwestern Scotland.* Univ. Strathclyde Ph.D. thesis (unpubl.).
—— and Bowes, D. R. 1977. Rb–Sr, U–Pb and K–Ar isotopic study of the Lewisian complex between Durness and Loch Laxford, Scotland. *Krystalinikum* **13,** 53–72.
——, Gillen, C. and Bowes, D. R. 1975. Rb–Sr isotopic studies near the major Precambrian junction between Scourie and Loch Laxford, Northwest Scotland. *Scott. J. Geol.* **11,** 333–337.
——, Pidgeon, R. T., Bowes, D. R. and Hopgood, A. M. 1973. Geochronological investigation of the quartzofeldspathic rocks of the Lewisian of Rona, Inner Hebrides. *Q. J. geol. Soc. Lond.* **129,** 389–404.
McCallien, W. J. 1930. The gneiss of Inishtrahull, County Donegal. *Geol. Mag.* **67,** 542–549.
McIntyre, G. A., Brooks, C., Compston, W. and Turek, A. 1966. The statistical assessment of Rb–Sr isochrons. *J. geophys. Res.* **71,** 5459–5468.
Macintyre, R. M., van Breemen, O., Bowes, D. R. and Hopgood, A. M. 1975. Isotopic study of the gneiss complex, Inishtrahull, Co. Donegal. *Sci. Proc. R. Dublin Soc.* A, **5,** 301–309.
McQuillin, R. and Watson, J. 1973. Large-scale basement structures of the Outer Hebrides in the light of geophysical evidence. *Nature Phys. Sci., Lond.* **245,** 1–3.
Max, M. D. 1970. Mainland gneisses southwest of Bangor in Erris, northwest County Mayo, Ireland. *Sci. Proc. R. Dublin Soc.* **A, 3,** 275–291.
Mercy, E. L. P. and O'Hara, M. J. 1965. Websterite from Glenelg, Inverness-shire. *Scott. J. Geol.* **1,** 282–284.
—— 1968. Nepheline normative eclogite from Loch Duich, Ross-shire. *Scott. J. Geol.* **4,** 1–9.
Miller, J. A., Barber, A. J. and Kempton, N. H. 1963. A potassium–argon age determination from a Lewisian inlier. *Nature, Lond.* **197,** 1095–1096.
Miller, J. A. and Brown, P. E. 1965. Potassium–argon age studies in Scotland. *Geol. Mag.* **102,** 106–134.
Moorbath, S. 1969. Evidence for the age of deposition of the Torridonian sediments of north-west Scotland. *Scott. J. Geol.* **5,** 154–170.
—— 1975. Evolution of Precambrian crust from strontium isotopic evidence. *Nature, Lond.* **254,** 395–398.
—— 1976. Age and isotope constraints for the evolution of Archaean crust. *In* Windley, B. F. (Editor). *The Early History of the Earth.* 469–479. Wiley, London.
—— 1977. Ages, isotopes and evolution of Precambrian continental crust. *Chem. Geol.* **20,** 151–187.
—— and Park, R. G. 1972. The Lewisian geochronology of the southern region of the Scottish mainland. *Scott. J. Geol.* **8,** 51–74.
——, Powell, J. L. and Taylor, P. N. 1975. Isotopic evidence for the age and origin of the "grey gneiss" complex of southern Outer Hebrides, Scotland. *J. Geol. Soc. Lond.* **131,** 213–222.
—— and Taylor, P. N. 1974. Lewisian age for the Scardroy mass. *Nature, Lond.* **250,** 41–43.
——, Welke, H. and Gale, N. H. 1969. The significance of lead isotope studies in ancient high-grade metamorphic basement complexes, as exemplified by the Lewisian rocks of northwest Scotland. *Earth Planet. Sci. Lett.* **6,** 245–256.
Moorhouse, S. J. 1976. The geochemistry of the Lewisian and Moinian of the Borgie area, North Sutherland. *Scott. J. Geol.* **12,** 159–165.
Muecke, G. K. 1969. *The petrogenesis of the granulite–facies rocks of the Lewisian of Sutherland, Scotland.* Oxford Univ. D.Phil. thesis (unpubl.).
Mutanen, T. 1976. Komatiites and komatiite provinces in Finland. *Geologi* **28,** 49–56.
Myers, J. S. 1970. Gneiss types and their significance in the repeatedly deformed and metamorphosed Lewisian Complex of Western Harris, Outer Hebrides. *Scott. J. Geol.* **6,** 186–199.
—— 1971a. The late Laxfordian granite-migmatite complex of western Harris, Outer Hebrides. *Scott. J. Geol.* **7,** 254–284.
—— 1971b. Zones of abundant Scourie dyke fragments and their significance in the Lewisian Complex of Western Harris, Outer Hebrides. *Proc. Geol. Assoc.* **82,** 365–378.
Nisbet, H. C. 1961. The geology of North Rona. *Trans. geol. Soc. Glasg.* **24,** 169–189.
O'Hara, M. J. 1961a. Zoned ultrabasic and basic gneiss masses in the Early Lewisian metamorphic complex at Scourie, Sutherland. *J. Petrol.* **2,** 248–276.
—— 1961b. Petrology of the Scourie dyke, Sutherland. *Mineral. Mag.* **32,** 848–865.
—— 1962. Some intrusions in the Lewisian Complex near Badcall, Sutherland. *Trans. Edinb. geol. Soc.* **19,** 201–207.
—— 1965. Origin of ultrabasic and basic gneiss masses in the Lewisian. *Geol. Mag.* **102,** 296–314.
—— 1975. Great thickness and high geothermal gradient of Archaean crust: the Lewisian of Scotland (Extended Abstract). *Int. Conference on Geothermometry and Geobarometry,* Penn. State Univ. Oct. 1975.
—— 1976. High geothermal gradient in Archaean crust. *In* Biggar, G. M. (Editor). *Progress in experimental petrology,* 254. *Nat. Environ. Res. Council, London Publ. Ser. D.* No. 6.
—— 1977. Thermal history of excavation of Archaean gneisses from the base of the continental crust. *J. geol. Soc. Lond.* **134,** 185–200.
Owen, T. R. 1976. *The geological evolution of the British Isles.* Pergamon, Oxford.
Palmer, K. F. 1971. *A comparative study of two Precambrian gneiss areas—the Suportoq region, East Greenland and South Harris, Outer Hebrides—and their bearing on Precambrian crustal evolution.* Univ. Birmingham Ph.D. thesis (unpubl.).

PARK, R. G. 1961. The pseudotachylite of the Gairloch district, Ross-shire. *Am. J. Sci.* **259,** 542–550.
—— 1964. The structural history of the Lewisian rocks of Gairloch, Wester Ross. *Q. J. geol. Soc. Lond.* **120,** 397–425.
—— 1965. Early metamorphic complex of the Lewisian north-east of Gairloch, Ross-shire, Scotland. *Nature, Lond.* **207,** 66–68.
—— 1966. Nature and origin of Lewisian basic rocks of Gairloch, Ross-shire, Wester Ross. *Scott. J. Geol.* **2,** 179–199.
—— 1969a. Structural history of Lewisian rocks of Tollie, Ross-shire. *Scott. J. Geol.* **5,** 293–295.
—— 1969b. Structural correlation in metamorphic belts. *Tectonophysics* **7,** 323–338.
—— 1970a. The structural evolution of the Tollie antiform—a geometrically complex fold in the Lewisian north-east of Gairloch, Ross-shire. *Q. J. geol. Soc. Lond.* **125,** 319–350.
—— 1970b. Observations on Lewisian chronology. *Scott. J. Geol.* **6,** 379–399.
—— 1971. Lewisian chronology. *Scott. J. Geol.* **7,** 184–185.
—— 1973. The Laxfordian belts of the Scottish mainland. *In* Park, R. G. and Tarney, J. (Editors). *The Early Precambrian of Scotland and Related Rocks of Greenland,* 65–76. Univ. Keele.
—— and CRESSWELL, D. 1972. Basic dykes in the early Precambrian (Lewisian) of NW Scotland. *Rep. XXIV Int. geol. Congr.* **1,** 238–245.
—— and —— 1973. The dykes of the Laxfordian belts. *In* Park, R. G. and Tarney, J. (Editors). *The Early Precambrian of Scotland and related rocks of Greenland,* 119–130. Univ. Keele.
—— and —— 1975. Structural evolution of the Laxford front. *Scott. J. Geol.* **11,** 169–170
PEACH, B. N., GUNN, W., CLOUGH, C. T. and GREENLY, E. 1913. The geology of the Fannich Mountains and the country around upper Loch Maree and Strath Broom (Sheet 92). *Mem. geol. Surv. Scotland.*
—— and HORNE, J. 1930. *Chapters on the geology of Scotland.* Oxford Univ. Press, London.
——, ——, GUNN, W., CLOUGH, C. T., HINXMAN, L. W. and TEALL, J. J. H. 1907. The geological structure of the North-West Highlands of Scotland. *Mem. geol. Surv. Scotland.*
——, ——, HINXMAN, L. W., CRAMPTON, C. B., ANDERSON, E. M. and CARRUTHERS, R. G. 1913. The geology of central Ross-shire (Sheet 82). *Mem. geol. Surv. Scotland.*
——, ——, WOODWARD, H. B., CLOUGH, C. T., HARKER, A. and WEDD, C. B. 1910. The geology of Glenelg, Lochalsh and South-east part of Skye. *Mem. geol. Surv. Scotland.*
PHEMISTER, J. 1960. *British Regional Geology—Scotland: The Northern Highlands* (3rd edn.). H.M.S.O. Edinburgh.
PIDGEON, R. T. 1970. Geochronology of the Outer Hebrides. *Scott. J. Geol.* **6,** 412.
—— and AFTALION, M. 1972. The geochronological significance of discordant U–Pb ages of oval-shaped zircons from a Lewisian gneiss from Harris, Outer Hebrides. *Earth Planet. Sci. Lett.* **17,** 269–274.
—— and —— 1978. Cogenetic and inherited zircon U–Pb systems in granites: Palaeozoic granites of Scotland and England. *In* Bowes, D. R. and Leake, B. E. (Editors). *Crustal evolution in northwestern Britain and adjacent regions,* pp 183–220. *Geol. J. Spec. Issue.* No. 10.
—— and BOWES, D. R. 1972. Zircon U–Pb ages of granulites from the Central Region of the Lewisian, northwestern Scotland. *Geol. Mag.* **109,** 247–258.
PIPER, J. D. A. 1976. Palaeomagnetic evidence for a Proterozoic super-continent. *Phil. Trans. R. Soc. Lond.* A, **280,** 469–490.
POWELL, D. W. 1970. Magnetised rocks within the Lewisian of Western Scotland and under the Southern Uplands. *Scott. J. Geol.* **6,** 353–369.
POWER, G. M. and PARK, R. G. 1969. A geochemical study of five amphibolite bodies from the Lewisian of Gairloch, Ross-shire. *Scott. J. Geol.* **5,** 26–41.
PULVERTAFT, T. C. R. 1968. The Precambrian stratigraphy of Western Greenland. *Rep. XXIII Int. geol. Congr.* **4,** 89–107.
RAMSAY, J. G. 1958. Moine–Lewisian relations at Glenelg, Inverness-shire. *Q. J. geol. Soc. Lond.* **113,** 487–520.
—— 1963. Structure and metamorphism of the Moine and Lewisian rocks in the north-west Caledonides. *In* Johnson, M. R. W. and Stewart, F. H. (Editors). *The British Caledonides,* 143–175. Oliver and Boyd, Edinburgh.
—— and GRAHAM, R. H. 1970. Strain variations in shear belts. *Can. J. Earth Sci.* **7,** 786–813.
READ, H. H. 1934. Age problems of the Moine Series of Scotland. *Geol. Mag.* **71,** 302–317.
—— and WATSON, J. 1975. *Introduction to Geology* Vol. 2 *Earth History* Pt. I—*Early Stages of Earth History.* Macmillan, London.
RICHEY, J. E. and THOMAS, H. H. 1930. The geology of Ardnamurchan, North-west Mull and Coll. *Mem. geol. Surv. Scotland.*
ROBERTSON, J. K., RANKIN, J. B., BOWES, D. R. and HOPGOOD, A. M. 1964. The Glasgow University Exploration Society Expedition to Mingulay. *Nature, Lond.* **204,** 25.
RODDICK, J. C., COMPSTON, W. and DURNEY, D. W. 1976. The radiometric age of the Mount Keith Granodiorite: a maximum age estimate for an Archaean greenstone sequence in the Yilgarn block, Western Australia. *Precambrian Res.* **3,** 55–78.
SABINE, P. A. and WATSON, J. V. 1965. Isotopic age-determinations of rocks from the British Isles, 1955–64. *Q. J. geol. Soc. Lond.* **121,** 477–533.
SALOP, L. J. 1977. *The Precambrian of the Northern Hemisphere and general features of early geological evolution.* Elsevier, Amsterdam.

SHERATON, J. W. 1970. The origin of the Lewisian gneisses of Northwest Scotland with particular reference to the Drumbeg area, Sutherland. *Earth Planet. Sci. Lett.* **8,** 301–310.
——, SKINNER, A. C. and TARNEY, J. 1973. The geochemistry of the Scourian gneisses of the Assynt district. *In* Park, R. G. and Tarney, J. (Editors). *The Early Precambrian of Scotland and Related Rocks of Greenland,* 13–30. Univ. Keele.
——, TARNEY, J., WHEATLEY, T. J. and WRIGHT, A. E. 1973. The structural history of the Assynt district. *In* Park, R. G. and Tarney, J. (Editors). *The Early Precambrian of Scotland and Related Rocks of Greenland,* 31–43. Univ. Keele.
SINCLAIR, I. G. L. 1964. *The Lewisian rocks of the island of Tiree, Inner Hebrides.* Univ. St. Andrews Ph.D. thesis (unpubl.).
SMALES, A. A., MAPPER, D., MORGAN, J. W., WEBSTER, R. K. and WOOD, A. J. 1958. Some geochemical determinations using radioactive and stable isotopes. *Proc. Second U.N. Int. Conf. on the Peaceful Uses of Atomic Energy* **2,** 242–248.
SMITH, P. J. and BOTT, M. H. P. 1975. Structure of the crust beneath the Caledonian Foreland and Caledonian Belt of the North Scottish Shelf Region. *Geophys. J. R. astron. Soc.* **40,** 187–205.
SPOONER, C. M. and FAIRBAIRN, H. W. 1970. Strontium 87/Strontium 86 initial ratios in granulite terrains. *J. geophys. Res.* **75,** 6706–6713.
STANDLEY, R. C. 1975. Structural evolution of the Laxford front. *Scott. J. Geol.* **11,** 170–171.
—— 1975b. *The 'Scourie Dyke' suite of the north-west mainland Lewisian of Scotland—with particular reference to the structural geology and geochemistry.* Univ. Keele Ph.D. thesis (unpubl.).
SUN, S. S. and HANSON, G. N. 1975. Evolution of the mantle: geochemical evidence from alkali basalts. *Geology* **3,** 297–302.
SUTTON, J. 1967. The extension of the geological record into the Pre-Cambrian. *Proc. Geol. Assoc.* **78,** 493–534.
—— 1969. Structural history of Lewisian rocks of Tollie, Ross-shire. *Scott. J. Geol.* **5,** 295–296.
—— 1973. The first three-quarters of the geological record in Britain. *In* Park, R. G. and Tarney, J. (Editors). *The Early Precambrian of Scotland and Related Rocks of Greenland,* 9–12. Univ. Keele.
—— 1974. Discussion *of* Evans, C. R. and Lambert, R. St J. The Lewisian of Lochinver, Sutherland; the type area for the Inverian metamorphism. *J. geol. Soc. Lond.* **130,** 149.
—— and DEARNLEY, R. 1964. Discussion *of* Bowes, D. R., Wright, A. E. and Park, R. G. Layered intrusive rocks in the Lewisian of the North-West Highlands of Scotland. *Q. J. geol. Soc. Lond.* **120,** 188–189.
——and WATSON, J. 1951a. The pre-Torridonian metamorphic history of the Loch Torridon and Scourie areas in the north-west Highlands, and its bearing on the chronological classification of the Lewisian. *Q. J. geol. Soc. Lond.* **106,** 241–307.
—— and —— 1951b. Varying trends in the metamorphism of dolerites. *Geol. Mag.* **88,** 25–35.
—— and —— 1959. Structures in the Caledonides, between Loch Duich and Glenelg, north-west Highlands. *Q. J. geol. Soc. Lond.* **114,** 231–257.
—— and —— 1962. Further observations on the margin of the Laxfordian complex of the Lewisian near Loch Laxford, Sutherland. *Trans. R. Soc. Edinb.* **65,** 89–106.
—— and —— 1969. Scourian–Laxfordian relationships in the Lewisian of north-west Scotland. *Geol. Assoc. Can. Spec. Pap.* **5,** 119–128.
TAFT, M. B. 1971. *Structure and tectonic history of the Lewisian Gneiss Complex north-east of Loch Seaforth, Isle of Lewis, Outer Hebrides, Scotland.* Univ. St. Andrews M.Sc. thesis (unpubl.).
—— 1978. Basic minor intrusions in the Lewisian gneisses of southern Lewis, Outer Hebrides. *Scott. J. Geol.* **14** (in press).
TALWANI, M. and ELDHOLM, O. 1972. Continental margin off Norway: a geophysical study. *Bull. geol. Soc. Am.* **83,** 3575–3606.
TARNEY, J. 1963. Assynt dykes and their metamorphism. *Nature, Lond.* **199,** 672–674.
—— 1969. Epitaxic relations between coexisting pyroxenes. *Mineral. Mag.* **37,** 115–122.
—— 1973. The Scourie dyke suite and the nature of the Inverian event in Assynt. *In* Park, R. G. and Tarney, J. (Editors). *The Early Precambrian of Scotland and Related Rocks of Greenland,* 105–118. Univ. Keele.
—— 1976. Geochemistry of Archaean high-grade gneisses, with implications as to the origin and evolution of the Precambrian crust. *In* Windley, B. F. (Editor). *The Early History of the Earth,* 405–417. Wiley, London.
——, DALZIEL, I. W. D. and DE WIT, M. J. 1976. Marginal basin 'Rocas Verdes' complex from S. Chile: a model for greenstone belt formation. *In* Windley, B. F. (Editor). *The Early History of the Earth,* 131–146. Wiley, London.
—— SKINNER, A. C. and SHERATON, J. W. 1972. A geochemical comparison of major Archaean gneiss units from Northwest Scotland and East Greenland. *Rep. XXIV Int. geol. Congr.* **1,** 162–174.
—— and WINDLEY, B. F. 1977. Crustal developments in the Archaean. *IUGS-UNESCO-IGCP Archaean geochemistry. The origin and evolution of Archaean continental crust,* 117–119. Symposium Abstracts, Hyderabad.

TEALL, J. J. H. 1885. The metamorphosis of dolerite into hornblende schist. *Q. J. geol. Soc. Lond.* **41,** 133–144.
TILLEY, C. E. 1936. Eulysites and related rock types from Loch Duich, Ross-shire. *Mineral Mag.* **24,** 331–342.
VAN BREEMEN, O., AFTALION, M. and PIDGEON, R. T. 1971. The age of the granitic injection complex of Harris, Outer Hebrides. *Scott. J. Geol.* **7,** 139–152.
——, BOWES, D. R. and PHILLIPS, W. E. A. 1976. Evidence for basement of late Precambrian age in the Caledonides of western Ireland. *Geology* **4,** 499–501.
——, HALLIDAY, A. N., JOHNSON, M. R. W. and BOWES, D. R. 1978. Crustal additions in late Precambrian times. *In* Bowes, D. R. and Leake, B. E. (Editors). *Crustal evolution in northwestern Britain and adjacent regions*, 81–106. *Geol. J. Spec. Issue* No. 10.
VAN SCHMUS, W. R. 1976. Early and Middle Proterozoic history of the Great Lakes area, North America. *Phil. Trans. R. Soc. Lond.* A **280,** 605–628.
VERNON, R. H. 1970. Comparative grain–boundary studies of some basic and ultrabasic granulites, nodules and cumulates. *Scott. J. Geol.* **7,** 139–152.
VON KNORRING, O. and DEARNLEY, R. 1960. The Lewisian pegmatites of South Harris, Outer Hebrides. *Mineral. Mag.* **32,** 366–378.
WATSON, J. 1965. Lewisian. *In* Craig, G. Y. (Editor). *The Geology of Scotland.* 49–77. Oliver and Boyd, Edinburgh.
—— 1967. Evidence of mobility in reactivated basement complexes. *Proc. Geol. Assoc.* **78,** 211–235.
—— 1968. Post–Scourian metadolerites in relation to Laxfordian deformation in Great Bernera, Outer Hebrides. *Scott. J. Geol.* **4,** 53–67.
—— 1969. The Precambrian gneiss complex of Ness, Lewis, in relation to the effects of Laxfordian regeneration. *Scott. J. Geol.* **5,** 269–285.
—— 1973. Effects of reworking on high–grade gneiss complexes. *Phil. Trans. R. Soc. Lond.* A, **273,** 443–455.
—— 1975. The Lewisian Complex. *In* Harris, A. L., Shackleton, R. M., Watson, J., Downie, C., Harland, W. B. and Moorbath, S. (Editors). *A correlation of Precambrian rocks in the British Isles*, 15–29. *Spec. Rep. geol. Soc. Lond.* No. 6.
—— 1977. The Outer Hebrides: a geological perspective. *Proc. Geol. Assoc.* **88,** 1–14.
—— and LISLE, R. J. 1973. The pre-Laxfordian complex of the Outer Hebrides. *In* Park, R. G. and Tarney, J. (Editors). *The Early Precambrian of Scotland and Related Rocks of Greenland,* 45–50. Univ. Keele.
WESTBROOK, G. K. 1972. Structure and metamorphism of the Lewisian of east Tiree, Inner Hebrides. *Scott. J. Geol.* **8,** 13–30.
WHITTEN, E. H. T. 1951. Discussion *of* Sutton, J. and Watson, J. The pre-Torridonian history of the Loch Torridon and Scourie areas in the North-West Highlands, and its bearing on the chronological classification of the Lewisian. *Q. J. geol. Soc. Lond.* 106, 304–305.
WILLIAMS, D. A. C. and HALLBERG, J. A. 1973. Archaean layered intrusions of the Eastern Goldfields region, Western Australia. *Contrib. Mineral. Petrol.* **38,** 45–70.
WINCHESTER, J. A. 1971. Some geochemical distinctions between Moinian and Lewisian rocks, and their use in establishing the identity of supposed inliers in the Moinian. *Scott. J. Geol.* **7,** 327–344.
—— and LAMBERT, R. ST J. 1970. Geochemical distinctions between the Lewisian of Cassley, Durcha and Loch Shin, Sutherland, and the surrounding Moinian. *Proc. Geol. Assoc.* **81,** 275–301.
WINDLEY, B. F. 1973. Archaean anorthosites; a review with the Fiskenaesset complex, West Greenland as a model for interpretation. *Spec. Publ. geol. Soc. S. Afr.* **3,** 33–46.
—— 1977. *The evolving continents.* Wiley, London.
WOOD, B. J. 1975. Influence of pressure, temperature and bulk composition on the appearance of garnet in orthogneiss—an example from South Harris, Scotland. *Earth Planet. Sci. Lett.* **26,** 299–311.

D. R. Bowes,
Department of Geology,
University of Glasgow,
Glasgow G12 8QQ, Scotland.

# Crustal additions in late Precambrian times

*O. van Breemen, A. N. Halliday, M. R. W. Johnson and D. R. Bowes*

As the subparallel Grenville and Caledonian orogenic belts intersect in northwestern Britain evidence of rock-forming events in late Precambrian times is generally obscured by the products of sedimentation and the effects of tectonism and metamorphic overprinting of the Caledonian orogenic cycle. Seeing back through the Caledonian events can be achieved by (i) Rb-Sr isotopic mineral studies in areas where the overprinting was associated with only small to moderate elevation of temperature, (ii) Rb-Sr isotopic whole-rock studies of large samples and (iii) U-Pb isotopic studies of zircons in igneous rocks that do not have an older radiogenic component. Such studies have been made for rocks of the Annagh Gneiss Complex and Ox Mountains inliers of Co. Mayo, Ireland and of the Morar Division of the Moine assemblage in northwest Scotland.

For the Annagh Gneiss Complex, U-Pb analyses of zircons in two syntectonic granite sheets yield an age of $1070 \pm 30$ m.y. while zircons from a late tectonic granite yield an age of $1000 \pm 30$ m.y. These granites bracket at least two phases of deformation accompanied by amphibolite facies metamorphism. Rb-Sr large sample whole-rock isotopic systems of the host (para) gneisses are not consistent with a history extending back to early-middle Proterozoic times, as is the case for the Lewisian Complex, while U-Pb zircon systems suggest an age of the gneiss-forming event that is not much older than the emplacement of the syntectonic granites. These dated tectonothermal and plutonic events represent part of the Grenville orogeny in the northwestern British Isles.

Rb-Sr large sample whole-rock studies on semipelitic schists from the Ox Mountains and the Morar Division of the Moine assemblage suggest derivation after the Grenville orogeny with $720 \pm 120$ and $700 \pm 100$ m.y. regression ages for the rocks from northwest Scotland. U-Pb zircon analyses of minerals from two pegmatites in northwest Scotland yield slightly older ages. One gives an upper intersection zircon age of $815 \pm 30$ m.y. and a concordant monazite age of $780 \pm 10$ m.y.; the other gives an upper intersection zircon age of $740 \pm 30$ m.y. These ages cannot be interpreted in terms of partial resetting of older ages, and together with geological evidence, confirm the existence of a distinct Morarian tectonothermal orogenic episode which precedes the onset of the main Caledonian orogenic cycle in this part of the crust.

Rb-Sr muscovite and biotite ages in the range of 470–380 m.y. provide urther information concerning the Caledonian thermal regime which has affected all the products of late Precambrian orogenic activity.

## 1. Introduction

The first evidence for the existence of crystalline rocks of late Precambrian age within the Caledonides of northwestern Britain was found in pegmatites associated with psammitic, semipelitic and pelitic schists occurring on the mainland of Scotland, southeast of Skye. These rocks are generally considered to be part of the Moine assemblage which is in tectonic contact, along the Moine thrust zone, with the Precambrian crystalline Lewisian Complex (Figs 1, 6). Rb-Sr ages of up to 770 m.y. for muscovite from the pegmatites provided evidence for the antiquity of the host schists (Giletti *et al.* 1961) but Rb-Sr determinations on coexisting biotites suggested that the isotopic systems of the muscovites had been disturbed by a thermal event at 435±15 m.y. ago, i.e. during the Caledonian orogeny. Pegmatites with similar Rb-Sr muscovite ages were subsequently found by Long and Lambert (1963) to be more extensive than had previously been thought and were interpreted as being "typical secretion products of a regional metamorphism which is inferred to have taken place at the same time" (p. 239), i.e. at 750 m.y. ago. These authors argued for the existence of a NNE–SSW zone containing the products of high-grade Precambrian metamorphism northwest of the Great Glen fault, which died out to the north, in the region of the 550 m.y. old Carn Chuinneag granite (Fig. 1). Subsequently van Breemen *et al.* (1974) confirmed the main conclusions of Long and Lambert (1963) by showing, from Rb-Sr and U-Pb isotopic data, that a lensoid pegmatite concordant with the quartzofeldspathic bands and lenses of migmatised schists had an age of 730 m.y. However they also found evidence for abundant 430±10 m.y. pegmatites emplaced during extensive Caledonian deformation.

Though at this stage there was evidence for an episode distinct from both the Grenville and Caledonian orogenies, Brook *et al.* (1976) reported a 1020±50 m.y. Rb-Sr whole-rock isochron age for the Ardgour granitic gneiss which occurs in the migmatised area of the Moine outcrop (Fig. 1). This age was related to the peak of Grenville metamorphism and migmatisation and, following Stewart (1969), it was argued that all the psammitic, semipelitic and pelitic schists bounded by the Moine thrust to the northwest, and the Great Glen fault to the southeast, had a sedimentary derivation prior to the Grenville orogeny. On this interpretation the Moine metasediments are considered to be older than (i) the uniform psammitic schists (Central Highland Granulites) that occur southeast of the Great Glen fault and which pass apparently conformably into the more pelitic upper Precambrian to Cambrian metasediments of the Dalradian Supergroup and (ii) the undeformed arkosic Torridonian sediments dated by Moorbath (1969) at 970±25 m.y. (Lower Torridonian) and 790±20 m.y. (Upper Torridonian) which lie unconformably on the crystalline Lewisian Complex.

Other recent evidence suggests that crystalline rocks of late Precambrian age are even more widespread. The possibility that at least some of the basement underlying the present outcrop of the Dalradian Supergroup is of late Precambrian age is suggested by a 1310±25 m.y. U-Pb zircon chord upper intersection age from the Ben Vuirig granite in the central Grampian Highlands (Fig. 1; Pankhurst and Pidgeon 1976). In northeast Scotland Sturt *et al.* (1977) report *c.* 700 m.y. Rb-Sr whole-rock isochron ages for gneisses in tectonic juxtaposition with Dalradian metasediments while the 665±17 m.y. Rb-Sr whole-rock isochron age of the associated Portsoy granite indicates that granitic plutonism occurred during the early stages of deposition of the Dalradian Supergroup (Pankhurst 1974). For western Ireland,

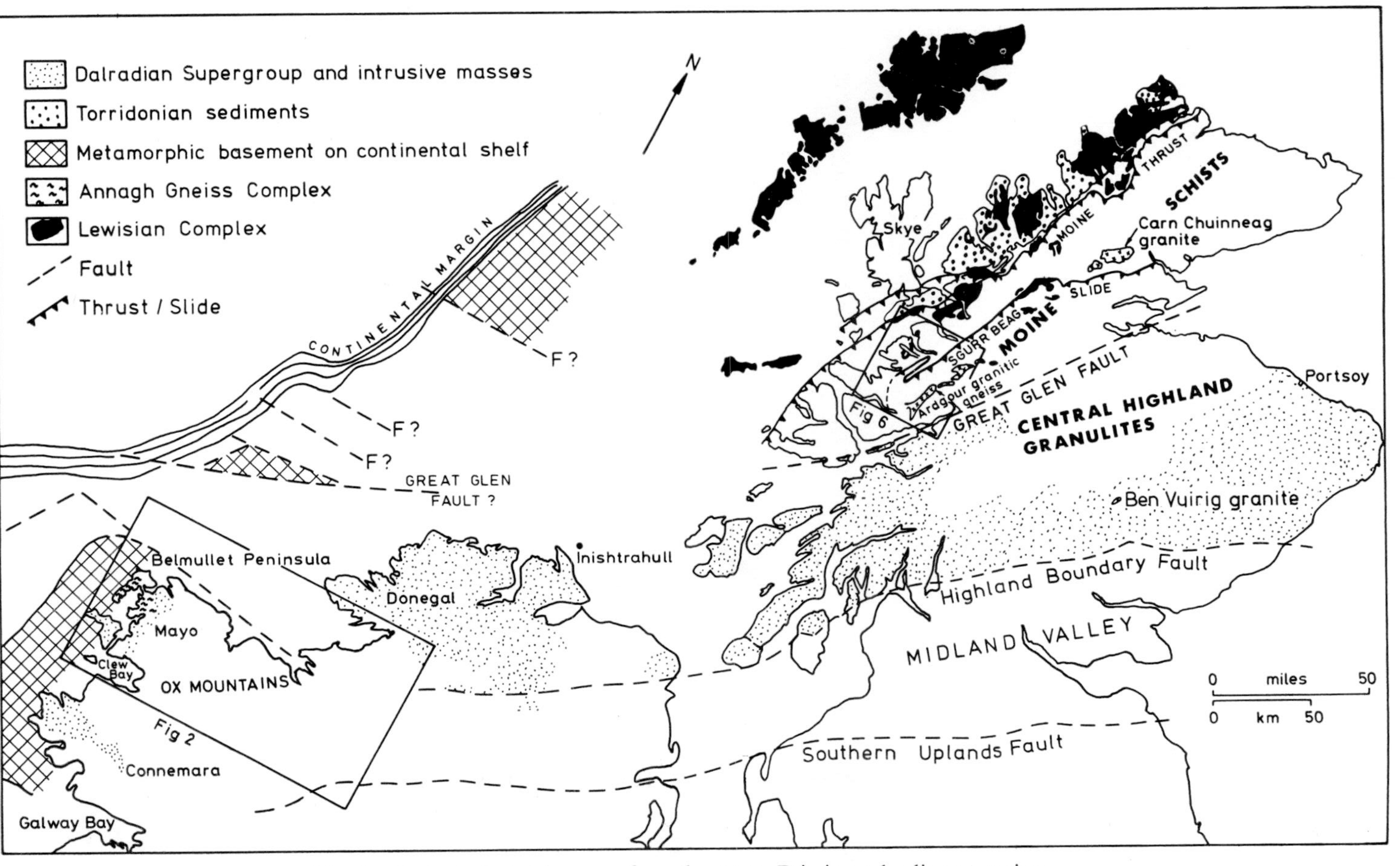


Fig. 1. Geological sketch map of northwestern Britain and adjacent regions.

Phillips *et al.* (1975) present somewhat ambiguous $^{40}Ar/^{39}Ar$ age evidence argued as indicating that granulite facies psammatic rocks in the northeast Ox Mountains are at least 700 m.y. old (Figs 1, 2). Additionally, Sutton and Max (1969) and van Breemen *et al.* (1976) present geological and Rb-Sr whole-rock age evidence which shows that the abundant granite sheets of the Annagh Gneiss Complex on the Belmullet Peninulsa of Co. Mayo are also of late Precambrian age, lying between the limits of *c.* 700 and 1300 m.y. (Figs 1, 2).

In order to evaluate what continental additions have been made in late Precambrian times, it is necessary (i) to separate the products of Caledonian and late Precambrian events, (ii) to decide whether Caledonian radiometric ages are the result of isotopic resetting of late Precambrian rocks, (iii) to assess the extent to which late Precambrian ages may be "mixed ages" due to partial isotopic resetting of older rocks, and (iv) to understand the nature and timing of the late Precambrian events. To this end further Rb-Sr whole-rock and mineral isotopic data, together with U-Pb zircon and monazite isotopic data, are presented here together with a discussion of their significance to the understanding of crustal evolution in the northwest British Isles. The new data are for rocks from both western Ireland (Figs 1, 2) and northwest Scotland (Figs 1, 6), viz. for gneisses and granites of the Annagh Gneiss Complex, semipelitic schists and pegmatites from the Ox Mountains, semipelitic schists from east of Ardnamurchan and pegmatites from Morar. One difficulty of discussing the rocks from northwest Scotland is the use of the term "Moine" for rock assemblages that have been correlated on the basis of lithology and which may not all be of one age. Where possible the structural/stratigraphic units of Johnstone *et al.* (1969, are used (cf. Fig. 6).

Analytical methods follow van Breemen *et al.* (1974) for Rb-Sr studies and Pidgeon and Aftalion (1978) for U-Pb work. All ages have been calculated or recalculated with $\lambda Rb = 1{\cdot}42 \times 10^{-11} yr^{-1}$ and the U decay constants of Jaffey *et al.* (1971). Regression calculations follow York (1969) and all errors are quoted at the $2\delta$ level. Analytical uncertainties for isotope ratios are as follows: $^{87}Rb/^{86}Sr \pm 1{\cdot}4\%$; $^{87}Sr/^{86}Sr \pm 0{\cdot}03\%$; $^{207}Pb/^{235}U \pm 0{\cdot}6\%$ and $^{206}Pb/^{238}U \pm 0{\cdot}4\%$. The analytical data are given in Tables 1 and 2 as are grid references for sample localities. Sample numbers refer to the catalogue of the Isotope Geology Unit, Scottish Universities Research and Reactor Centre, East Kilbride.

## 2. Annagh Gneiss Complex

### 2a. Geological setting

The Annagh Gneiss Complex of western Ireland (Fig. 2; Sutton and Max 1969), which occurs on both the Belmullet Peninsula (Sutton 1972) and on the adjacent mainland (Max 1970), is composed of grey micaceous and dark-coloured hornblendic gneiss of possible sedimentary affinity together with variably foliated granitic gneiss that shows cross-cutting relationships with the grey gneiss. Pink gneiss, resulting from lit-par-lit injection and metasomatism, occurs between the grey and granitic gneisses. This crystalline complex is structurally overlain by as much as 4 km of psammites, which are correlated with the Central Highland Granulites in Scotland (Fig. 1; Max 1970; Crow *et al.* 1971), and an overlying dominantly sedimentary assemblage of the Dalradian Supergroup. On litho-structural grounds, the Annagh Gneiss Complex has been compared with the Lewisian Complex of northwest Scotland (Max 1970) though most of the granite

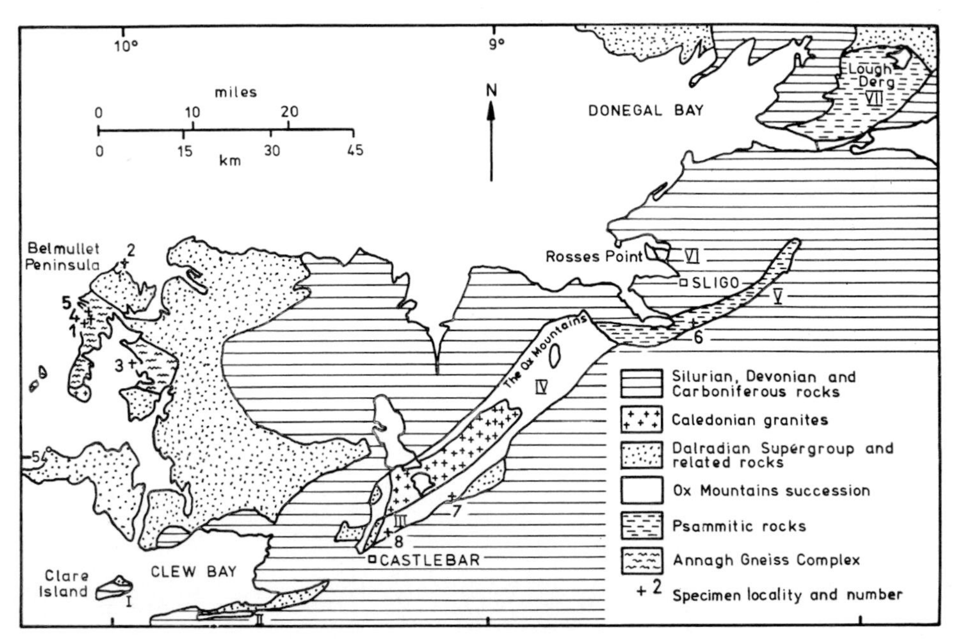


Fig. 2. Geological sketch map of part of western Ireland showing location of dated samples and of inliers of metamorphic rocks: I Clare Island, II Deer Park, III southwest Ox Mountains, IV central Ox Mountains, V northeast Ox Mountains, VI Rosses Point, VII Lough Derg.

sheets have now been shown to be younger than 1300 m.y. (van Breemen *et al.* 1976).

The early tectonothermal history of the grey gneiss includes multiple deformation, amphibolite facies metamorphism and the development of elongate pegmatoid lenses. These quartzofeldspathic masses were boudinaged before the syntectonic emplacement, dominantly as near concordant sheets, of granitic material now seen as foliated gneiss in most parts. Lit-par-lit injection associated with metasomatism resulted in the development of injection complexes, as near Erris Head (cf. van Breemen *et al.* 1976 p. 500). Elsewhere larger and more homogeneous masses occur. Subsequent deformation and metamorphism, up to amphibolite facies, was followed by further late tectonic emplacement of discrete and less extensive granitic to coarse grained quartzofeldspathic pegmatitic masses. Probably synchronous with the intrusion of these small granitic masses was the porphyroblastic growth of plagioclase, microcline, biotite, amphibole, epidote and garnet described by Sutton (1972). There is no published evidence for intense tectonism after the emplacement of these late tectonic granitic masses and before the deposition of the cover rocks and their subsequent deformation and metamorphism during the Caledonian event (Sutton 1972).

## 2b. Zircon morphology

Samples for U-Pb zircon analysis were collected from (i) a 50 cm thick mica-rich layer of the grey gneiss, which shows no obvious metasomatic effects associated with the emplacement of the sheets of foliated syntectonic granite that bound it on either side (Fig. 2, locality 1), (ii) two extensive masses of variably foliated syntectonic

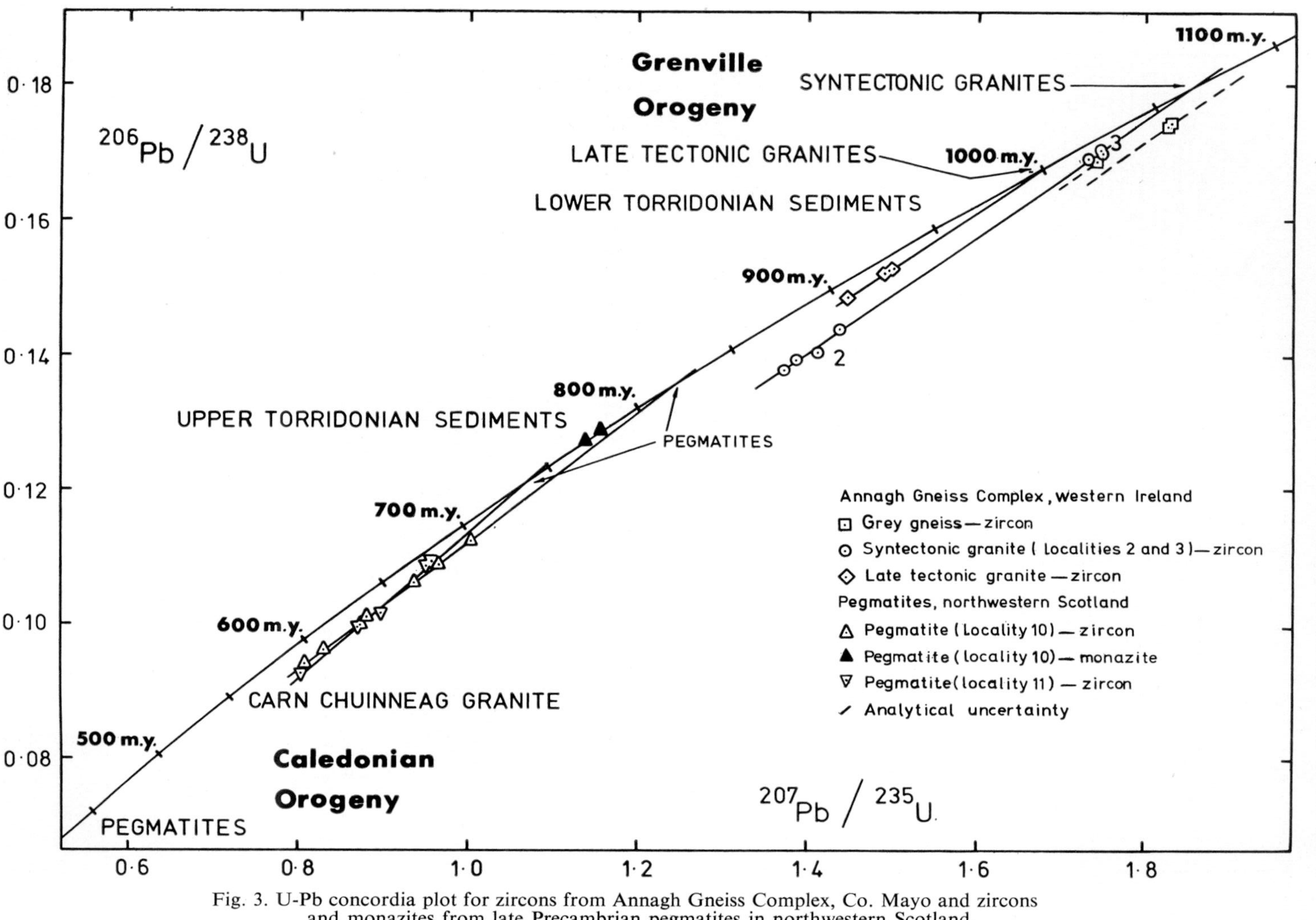


Fig. 3. U-Pb concordia plot for zircons from Annagh Gneiss Complex, Co. Mayo and zircons and monazites from late Precambrian pegmatites in northwestern Scotland.

## Table 1. U–Pb isotopic analyses, western Ireland and northwestern Scotland

| Size (μm) and magnetic fractions | | Pb μg/g | U μg/g | $^{206}Pb/^{204}Pb$ | Atom percent radiogenic lead 206 | 207 | 208 | Atomic ratios $^{207}Pb/^{206}Pb$ | $^{207}Pb/^{235}U$ | $^{206}Pb/^{238}U$ |
|---|---|---|---|---|---|---|---|---|---|---|
| ANNAGH GNEISS COMPLEX, W IRELAND | | | | | | | | | | |
| Grey gneiss (RC 1454): locality 1 (F 642310) | | | | | | | | | | |
| Zircon | | | | | | | | | | |
| - 165 + 142 | NM | 39.6 | 230 | 4721 | 87.04 | 6.632 | 6.328 | 0.07620 | 1.8335 | 0.17451 |
| - 106 + 84 | NM | 38.9 | 226 | 8347 | 87.00 | 6.630 | 6.375 | 0.07621 | 1.8298 | 0.17412 |
| - 45 | NM | 40.3 | 245 | 21201 | 88.44 | 6.624 | 4.937 | 0.07490 | 1.7452 | 0.16898 |
| Syntectonic granite (RC 1319): locality 2 (F 706405) | | | | | | | | | | |
| Zircon | | | | | | | | | | |
| + 142 | NM | 47.3 | 342 | 927 | 89.49 | 6.482 | 4.032 | 0.07244 | 1.4389 | 0.14406 |
| - 142 + 106 | NM | 55.4 | 413 | 1402 | 89.50 | 6.460 | 4.044 | 0.07219 | 1.3894 | 0.13959 |
| - 84 + 61 | NM | 53.2 | 399 | 1583 | 89.00 | 6.433 | 4.564 | 0.07227 | 1.3752 | 0.13799 |
| - 61 | NM | 48.5 | 355 | 1818 | 88.45 | 6.450 | 5.103 | 0.07293 | 1.4138 | 0.14059 |
| Syntectonic granite (RC 1325): locality 3 (F 716244) | | | | | | | | | | |
| Zircon | | | | | | | | | | |
| + 215 | NM | 63.1 | 378 | 3256 | 87.88 | 6.544 | 5.574 | 0.07446 | 1.7518 | 0.17061 |
| - 106 + 84 | NM | 53.8 | 323 | 5372 | 87.81 | 6.555 | 5.635 | 0.07465 | 1.7523 | 0.17022 |
| - 61 + 45 | NM | 56.9 | 343 | 9252 | 87.99 | 6.533 | 5.480 | 0.07425 | 1.7356 | 0.16952 |
| - 45 | NM | 73.0 | 431 | 9895 | 86.83 | 6.528 | 6.643 | 0.07518 | 1.7728 | 0.17101 |
| Late tectonic granite (RC 1320): locality 4 (F 655315) | | | | | | | | | | |
| Zircon | | | | | | | | | | |
| + 215 | NM | 118.6 | 816 | 5378 | 90.53 | 6.429 | 3.039 | 0.07101 | 1.4984 | 0.15302 |
| - 142 + 106 | NM | 115.4 | 793 | 11411 | 90.58 | 6.450 | 2.971 | 0.07121 | 1.5049 | 0.15327 |
| - 61 + 45 | NM | 103.0 | 713 | 10237 | 90.76 | 6.449 | 2.788 | 0.07105 | 1.4940 | 0.15250 |
| - 45 | NM | 92.9 | 659 | 9577 | 90.89 | 6.423 | 2.692 | 0.07067 | 1.4521 | 0.14901 |
| MORAR DIVISION OF MOINE ASSEMBLAGE, NW SCOTLAND | | | | | | | | | | |
| Concordant pegmatite (RC 1399): locality 10 (NM 835925) | | | | | | | | | | |
| Zircon | | | | | | | | | | |
| + 215 | M3° | 441 | 4805 | 3973 | 93.54 | 5.886 | 0.577 | 0.06292 | 0.8666 | 0.09988 |
| - 215 + 165 | M3° | 493 | 5699 | 2816 | 93.51 | 5.838 | 0.650 | 0.06243 | 0.8099 | 0.09409 |
| repeat | | 483 | 5458 | 3448 | 93.56 | 5.859 | 0.584 | 0.06262 | 0.8312 | 0.09626 |
| - 165 + 142 | M3° | 444 | 4775 | 2946 | 93.49 | 5.915 | 0.598 | 0.06327 | 0.8814 | 0.10103 |
| - 142 + 106 | M3° | 402 | 4107 | 3293 | 93.44 | 5.970 | 0.587 | 0.06389 | 0.9375 | 0.10641 |
| - 106 + 61 | M3° | 385 | 3834 | 3477 | 93.46 | 6.000 | 0.543 | 0.06421 | 0.9653 | 0.10903 |
| repeat | | 350 | 3376 | 4814 | 93.50 | 6.030 | 0.468 | 0.06449 | 1.0018 | 0.11266 |
| Monazite | | | | | | | | | | |
| + 165 | | 1552 | 3785 | 5312 | 26.89 | 1.743 | 71.37 | 0.06483 | 1.1386 | 0.12737 |
| - 165 + 61 | | 1704 | 4010 | 3695 | 26.29 | 1.708 | 72.00 | 0.06499 | 1.1570 | 0.12912 |
| Concordant pegmatite (RC 686): locality 11 (NM 803826) | | | | | | | | | | |
| Zircon | | | | | | | | | | |
| + 215 | NM3° | 457 | 5391 | 3246 | 94.01 | 5.922 | 0.0728 | 0.06299 | 0.8050 | 0.09268 |
| + 142* | NM3° | 470 | 5158 | 2890 | 93.97 | 5.962 | 0.0687 | 0.06344 | 0.8714 | 0.0996 |
| - 215 + 165 | NM3° | 502 | 5387 | 2940 | 93.90 | 6.012 | 0.0906 | 0.06403 | 0.8989 | 0.10181 |
| - 106 + 84* | NM3° | 429 | 4300 | 2550 | 93.92 | 5.971 | 0.1124 | 0.06357 | 0.9561 | 0.1091 |
| - 106 + 84 | NM3° | 497 | 4992 | 2101 | 93.93 | 5.954 | 0.1182 | 0.06339 | 0.9503 | 0.10872 |

* From van Breemen *et al.* (1974)

granite (localities 2, 3) and (iii) a late tectonic granite (locality 4). None of the zircons in these rocks has the good euhedral form commonly associated with rocks of magmatic derivation.

In the late tectonic granite (RC 1320) the zircons are generally cloudy with embayments, scallops and pitmarks providing evidence of resorption. The zircons from the syntectonic granite at locality 2 (RC 1319) are of two types, (i) subhedral to anhedral cloudy grains and (ii) clearer, elongate crystals approaching euhedral form, but with some blunting of terminations and numerous inclusions. Those in the syntectonic granite at locality 3 (RC 1325) are also a mixed population. Generally cloudy, irregular or well-rounded grains are associated with fairly well-formed clear

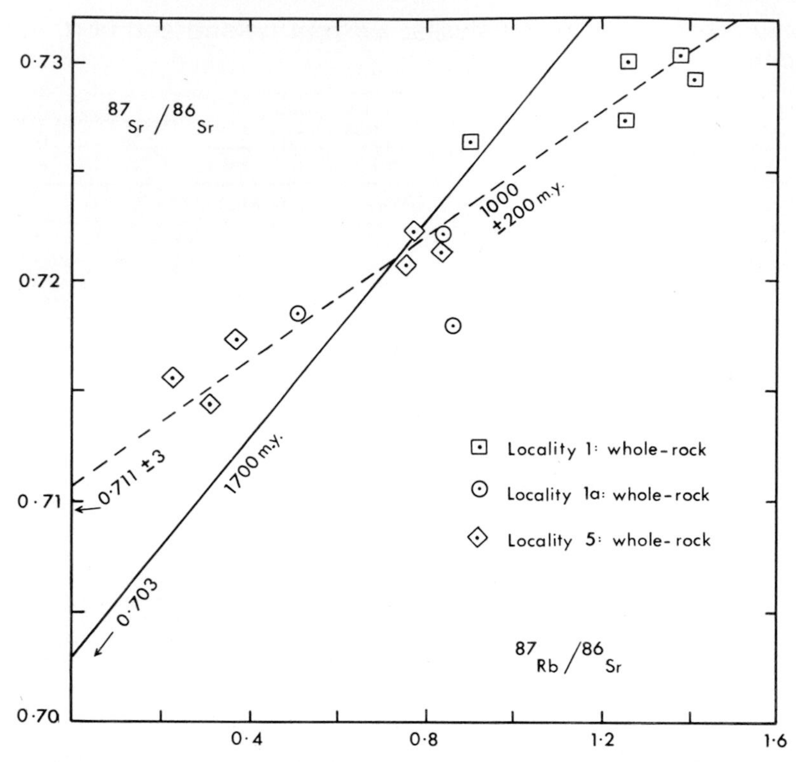


Fig. 4. Rb-Sr whole-rock isochron plot for grey gneiss of the Annagh Gneiss Complex, Co. Mayo; a 1700 m.y. reference line through initial $^{87}Sr/^{86}Sr = 0{\cdot}703$ is shown.

crystals, many of which contain dark rounded cores that have radial cracks extending out into the clear rims. The zircons in the grey gneiss at locality 1 (RC 1454) are similar to those of the granitic sheets, being generally cloudy with the larger grains having irregular forms while the smaller grains are well-rounded.

## 2c. Radiometric data

On a concordia plot the data points for the non-magnetic zircons from both samples of the syntectonic granite are closely grouped and show no direct relationship between size fraction, U content and degree of discordance (Table 1; Fig. 3). A best fit line through the data points for the zircon population from locality 2 passes through the zircon data points for locality 3 and intersects the concordia at $1070 \pm 30$ m.y. Uncertainties have been estimated by drawing lines through the data points for the early granite at locality 2 and the 550 m.y. (early Caledonian) and the 0 m.y. age points along the concordia. A parallel line through the non-magnetic zircon data points from the late granite yields an upper intersection age of $1000 \pm 30$ m.y. where the uncertainties are calculated as above. From the grey gneiss, three non-magnetic zircon fractions have been analysed which plot slightly to the right of the 1070 m.y. chord for the early granite.

The grey gneiss has also been investigated by Rb-Sr whole-rock analysis. A collection of six 4 kg samples from locality 5 (where sheets of syntectonic granite

## Table 2. Rb–Sr isotopic analyses, western Ireland

| Sample | Rb ppm | Sr(total)ppm | $^{87}Rb/^{86}Sr$ | $^{87}Sr/^{86}Sr$ | Mineral — whole-rock age (m.y.) |
|---|---|---|---|---|---|
| | | ANNAGH GNEISS COMPLEX | | | |
| Grey gneiss | | | | | |
| Locality 1 (F 642310) | | | | | |
| RC 1454 | 90.4 | 290 | 0.9041 | 0.72636 | |
| RC 1455 | 111 | 234 | 1.381 | 0.73272 | |
| RC 1456 | 98.0 | 225 | 1.261 | 0.72997 | |
| RC 1459 | 111 | 256 | 1.259 | 0.72729 | |
| RC 1460 | 103 | 211 | 1.412 | 0.72922 | |
| Locality 1a (F 640310) | | | | | |
| RC 1461 | 71.2 | 403 | 0.5116 | 0.71846 | |
| RC 1462 | 121 | 407 | 0.8628 | 0.71795 | |
| RC 1463 | 103 | 355 | 0.8403 | 0.72209 | |
| Locality 5 (F 638338) | | | | | |
| RC 1488 | 135 | 465 | 0.8408 | 0.72136 | |
| RC 1491 | 133 | 506 | 0.7580 | 0.72070 | |
| RC 1492 | 87.0 | 800 | 0.3150 | 0.71434 | |
| RC 1493 | 96.5 | 755 | 0.3703 | 0.71737 | |
| RC 1494 | 134 | 500 | 0.7798 | 0.72226 | |
| RC 1496 | 112 | 1390 | 0.2328 | 0.71557 | |
| | | NE OX MOUNTAINS INLIER | | | |
| Discordant pegmatites | | | | | |
| Locality 6 (G 767300) | | | | | |
| RC 1499 – muscovite | 1110 | 3.37 | 2407 | 16.314 | 455 ± 7* |
| RC 1500 – muscovite | 991 | 4.18 | 1229 | 8.782 | 461 ± 7* |
| | | SW OX MOUNTAINS INLIER | | | |
| Semipelitic schist of Cappagh Formation | | | | | |
| Locality 7 (G 273001) | | | | | |
| RC 1471 | 132 | 132 | 2.904 | 0.74317 | |
| – biotite | 392 | 10.8 | 111.3 | 1.3359 | 384 ± 6 |
| RC 1472 | 117 | 161 | 2.110 | 0.73865 | |
| RC 1473 | 86.8 | 140 | 1.795 | 0.73941 | |
| RC 1474 | 91.1 | 158 | 1.677 | 0.73874 | |
| RC 1475 | 96.3 | 124 | 2.254 | 0.73599 | |
| RC 1476 | 104 | 137 | 2.215 | 0.73726 | |
| RC 1476 | 104 | 137 | 2.206 | 0.73745 | |
| RC 1476 – muscovite | 156 | 135 | 3.359 | 0.74508 | 472 ± 27 |
| repeat | 145 | 122 | 3.455 | 0.74580 | 476 ± 27 |
| RC 1476 – biotite | 357 | 11.5 | 94.12 | 1.2356 | 381 ± 6 |
| Locality 8 (M 166937) | | | | | |
| RC 1465 | 48.7 | 114 | 1.235 | 0.73729 | |
| RC 1466 | 44.5 | 125 | 1.030 | 0.73749 | |
| RC 1467 | 51.4 | 157 | 0.953 | 0.73320 | |
| RC 1468 | 50.1 | 128 | 1.140 | 0.73448 | |
| RC 1469 | 65.0 | 147 | 1.282 | 0.73600 | |
| RC 1470 | 84.4 | 183 | 1.340 | 0.73667 | |

All analyses are of whole-rocks unless otherwise indicated

* Mineral age

are sparse) yield low Rb-Sr ratios and an inconclusive scatter of data points (Table 2; Fig. 4). From the 50 cm thick layer of grey gneiss at locality 1, five 25–60 kg samples were collected. A further three 25 kg samples were collected from another 50 cm thick layer in the grey gneiss *c.* 300 m away (locality 1a). All but two of the data points (for samples from localities 1 and 1a) plot below a 1700 m.y. reference line which has been drawn through a minimum crustal initial $^{87}Sr/^{86}Sr$ ratio of 0·703 (cf. Faure and Powell 1972). A regression line through all 14 data points for the grey gneiss corresponds to an age of 1000±200 m.y.

### 2d. Interpretation

These data indicate that the Annagh Gneiss Complex was involved in deformation, amphibolite facies metamorphism and intrusion of at least two generations of granitic sheets during the Grenville orogeny. Any further conclusions have to be qualified by varying degrees of uncertainty. For instance, as the grey gneiss samples were collected from three separate localities, the *c.* 1000 m.y. Rb-Sr regression line cannot be explained in terms of isotopic homogenisation during metamorphism at that time. If the general alignment of points is not fortuitous, it must reflect a high degree of Sr isotopic homogeneity of the rock assemblage before metamorphism. While this feature could possibly be associated with sedimentary accumulation, the possibility that the grey gneiss was part of the Lewisian Complex (1700 m.y. or older) cannot be dismissed. However this would require Sr isotopic migration on a scale greater than 50 cm, which is unlikely (cf. Hofmann 1975), or a massive introduction of Rb and other alkalies during granite intrusion for which there is no petrographic evidence.

While the upper intersection ages of 1070$\pm$30 m.y. and 1000$\pm$30 m.y., obtained from U-Pb zircon analyses of the syntectonic and late tectonic granitic masses, can be interpreted with confidence as times of emplacement, it is not possible to obtain a precise age of metamorphism of the grey gneiss from U-Pb zircon systems (Fig. 3). The scatter of data points for the gneiss is interpreted in terms of a zircon population, which has an inherited pre-metamorphic lead component, and has suffered lead loss comparable to that of the zircons in the granites. The morphology of the zircons in the granites suggests, however, that they too have had a complex pre-emplacement history.

Despite the qualifications, the radiometric and geological evidence is sufficiently consistent to permit the erection of the following sequence of events for the Annagh Gneiss Complex. The precursors of the grey gneiss (probably sediments) were deposited after the end of the Laxfordian orogenic episode whose products are present in the Lewisian Complex, i.e. after 1700 m.y. ago (cf. Bikerman *et al.* 1975 p. 252). These rocks were metamorphosed under amphibolite facies conditions, intensely deformed and compositionally segregated not long before the pervasive syntectonic emplacement of sheets of granite 1070$\pm$30 m.y. ago. The whole complex then underwent multiple deformation and amphibolite facies metamorphism before the late tectonic emplacement of more granitic bodies 1000$\pm$30 m.y. ago. This igneous activity was accompanied by static growth of minerals in the grey gneiss and marks a late or final phase of the orogenic episode of the Grenville Cycle in the northwestern British Isles.

## 3. Ox Mountains inliers

### 3a. Geological setting

In western Ireland, stretching from Clew Bay to Donegal Bay (Fig. 2) within a region of mainly Palaeozoic sediments, there occur a number of inliers of polymetamorphic rocks. These are the inliers of Clare Island, Deer Park, southwest and central Ox Mountains, northeast Ox Mountains, Rosses Point and Lough Derg. The northeast Ox Mountains and Lough Derg inliers contain thick sequences of psammitic and semipelitic schists which Phillips *et al.* (1975) correlate, on the basis of lithology, with the Moine assemblage of Scotland northwest of the Great Glen fault. However these authors emphasize three notable differences, viz. the presence in the Irish rocks of several lenses of marble, the abundance of garnet-pyroxene pods and the effects of high pressure granulite facies metamorphism. Mineral assemblages there indicate that the metamorphic conditions during the earlier phases of deformation correspond to 800°C and 10 kb. They also suggest a possible Precambrian age for these early episodes of deformation which were followed by further multiple deformation and retrogressive amphibolite facies metamorphism.

A major structural break separates the psammitic schists of the northeast Ox Mountains inlier from the more varied amphibolite facies rocks of the Ox Mountains succession which make up the bulk of the central and southwest Ox Mountains. Long and Max (1977) have proposed a generalised Ox Mountains succession in which the bulk of the lowermost pelites, quartzites and marbles are correlated with part of the Lower Dalradian Group and the overlying quartzites and semipelites are correlated with the Middle Dalradian Group. This sequence has a faulted contact with proximal turbidites of the Upper Dalradian Group. The scheme of Long and Max (1977), which puts all the metasedimentary rocks of the southwest and central Ox Mountains in the Dalradian Supergroup, accords with the views of Currall (1963) and Taylor (1968). It also appears to be consistent with Rb-Sr whole-rock isochron ages of 477±6 m.y. (Pankhurst *et al.* 1976) and 489±19 m.y. (Max *et al.* 1976) for the syn- to late-kinematic Slieve Gamph Complex which intrudes the Ox Mountains succession and is shown on Figure 2 as "Caledonian granite".

A different point of view has been presented by Phillips *et al.* (1975) who point out that the first (Caledonian) schistosity of the greenschist facies Upper Dalradian rocks is superimposed on metamorphic features and deformational fabrics in the Ox Mountain succession which they correlate with the earliest (Precambrian) elements in the psammitic rocks of the northeast Ox Mountains. Long and Max (1977) also record the greater structural and metamorphic complexity of the Ox Mountains succession but question whether the Upper Dalradian Group was originally unconformable on the Ox Mountain succession. In this regard R. M. Shackleton in discussion of similar rocks of the Deer Park Complex (Phillips 1973) suggested that Lower Dalradian rocks might have been affected by metamorphism while the Upper Dalradian rocks were still being deposited.

## 3b. Radiometric data

No attempt was made to date directly the granulite facies psammitic rocks of the northeast Ox Mountains inlier, but two large undeformed pegmatitic bodies 30 m apart (Fig. 2; locality 6) which cut these rocks, yield Rb-Sr ages on muscovite books of 455±7 m.y. (RC 1499—Table 2) and 461±7 m.y. (RC 1500).

The age of the Ox Mountains succession has been investigated using the Rb-Sr whole-rock method. Large samples (25–90 kg) of the uppermost Cappagh Formation (Long and Max 1977) were used in order to detect the existence of any late Precambrian isotopic systems despite the superimposition of a Caledonian thermal event. Six samples were collected from locality 7 along a 25 m road section. At locality 8 three sample pairs (RC 1465–6, 1467–8 and 1469–70) were collected at intervals of 100 m. The spacing within each of these three sample pairs was less than 10 m. The distribution of the Rb-Sr data points on an isochron plot (Table 2; Fig. 5) suggests that even on a scale of metres, complete Sr isotope homogenization has not occurred during any stage of the metamorphism. The data points show a closer overall fit to a 500 m.y. age line than a 1000 m.y. line and one of the data points (RC 1471) plots below the 1000 m.y. reference line drawn through the minimum initial $^{87}Sr/^{86}Sr$ ratio possible at that time (0·703). Rb-Sr muscovite and

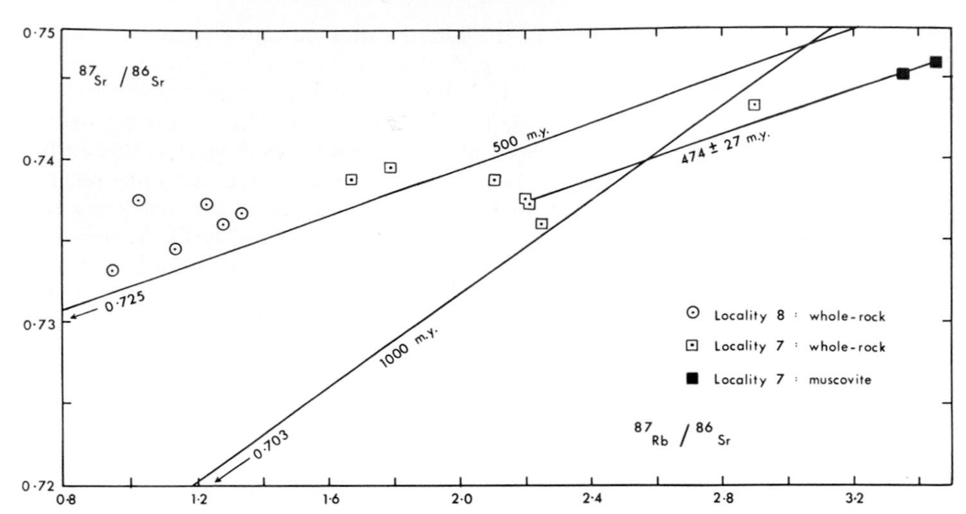


Fig. 5. Rb-Sr isochron plot for the Cappagh Formation of the Ox Mountains succession, western Ireland; 1000 m.y. and 500 m.y. reference lines through initial $^{87}Sr/^{86}Sr = 0{\cdot}703$ and 0·725 respectively are shown.

biotite analyses from sample RC 1476 yield ages of 474±27 m.y. (average of two) and 381±6 m.y., respectively. A biotite analysis from sample RC 1471 yields an age of 384±6 m.y. (Table 2).

## 3c. Interpretation

The pegmatites in the psammitic schists of the northeast Ox Mountains, for which the *c.* 460 m.y. Rb-Sr muscovite ages have been obtained, post-date the deformation associated with granulite facies metamorphism (Phillips *et al.* 1975; A. Molloy personal communication). Hence though the result testifies to a Caledonian event, it does not rule out the existence of a Precambrian event.

In the southwest Ox Mountains, the mean Rb-Sr muscovite and biotite ages of 474±27 m.y. and 383±6 m.y. respectively, are explained as the result of Caledonian heating and subsequent cooling with the 80 m.y. difference in the ages resulting from the difference in blocking temperatures of the two minerals (500±50°C and 300±50°C respectively; Purdy and Jäger 1976). As Rb-Sr muscovite systems are reset under metamorphic conditions in which staurolite grows, and as staurolite is present in these rocks, the muscovite age does does not rule out the existence of a Precambrian event. However, the distribution of the whole-rock data points (Fig. 5) for samples of such large size is not likely to have resulted from thermal overprinting of isotopic systems from a Precambrian event. The crude alignment of points may not have any age significance, but the data tend to negate a Grenville age for the Ox Mountains succession.

# 4. Upper Precambrian rocks in northwestern Scotland

## 4a. Geological setting

The metasedimentary rocks of the Moine assemblage which lie between the

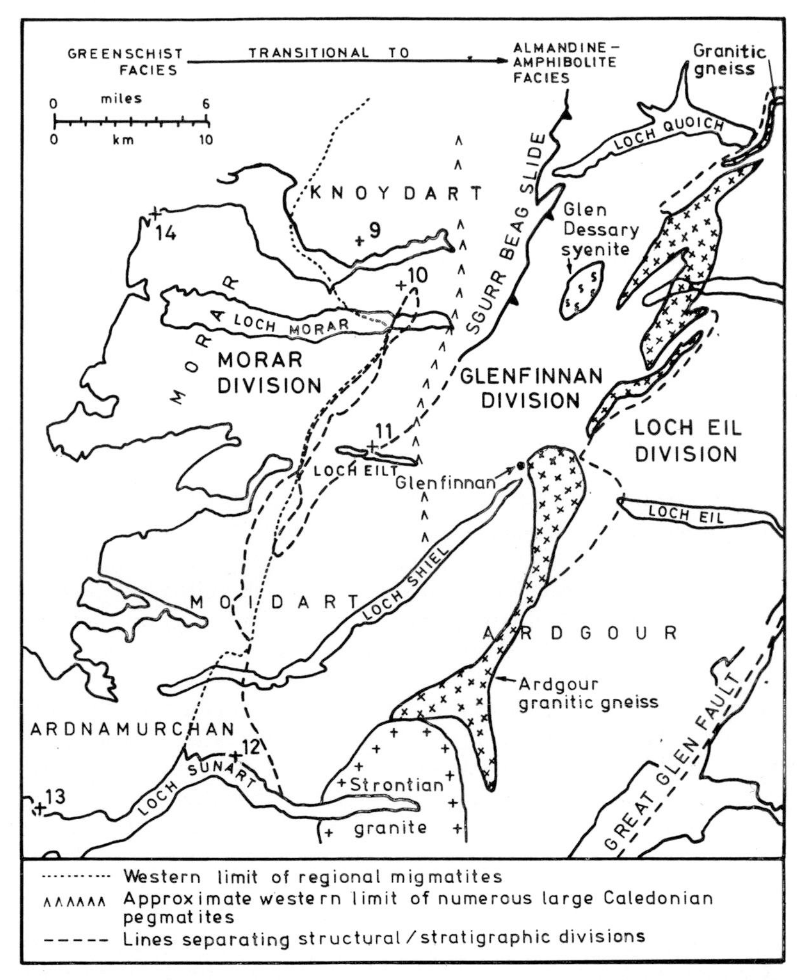


Fig. 6. Geological sketch map of part of northwestern Scotland showing distribution of structural/stratigraphic divisions of the Moine assemblage and location of dated specimens (after Johnstone *et al.* 1969).

Moine thrust and the Great Glen fault (Fig. 1) are generally subdivided into three structural-stratigraphic divisions, the relationships of which are best displayed in the area of Figure 6. From west to east, the Morar Division structurally underlies the Glenfinnan Division which in turn structurally underlies the Loch Eil Division (Johnstone *et al.* 1969).

The stratigraphy of the Morar Division is well defined (Johnstone *et al.* 1969) and ranges from a basal pelite through psammites with semipelitic bands, through striped and pelitic schists to the upper Morar psammites. The occurrence of a basal conglomerate has been taken as evidence that these rocks lie unconformably on Lewisian basement (Peach *et al.* 1910; Ramsay 1958). Over much of the outcrop area of the Morar Division, the metamorphism only attains that of the biotite and almandine metamorphic zones though it generally attains kyanite grade metamorphism in the east (Winchester 1974).

No general stratigraphic sequence has been established for the Glenfinnan Division (Tanner *et al.* 1970). The grade of metamorphism corresponds to the sillimanite zone and evidence of small scale partial melting is ubiquitous (Phemister 1960; Winchester 1974). Rock types consist of pelitic gneisses, semipelites, alternations of pelite, quartzose psammite and quartzite, and layers and lenses of amphibolite, a rock type that is absent from the Morar Division. Contacts with gneisses of the Lewisian inliers (Fig. 1; Moorbath and Taylor 1974) are not clear.

The structurally highest Loch Eil Division consists of uniform psammites with lenses of calc-silicate rocks and in it the foliation is generally flat-lying (Brown *et al.* 1970). A major structural break, the Quoich line, separates the Glenfinnan and Loch Eil divisions (Clifford 1958). According to Winchester (1974) the grade of metamorphism within the Loch Eil Division drops to as low as the almandine zone although neither distinctions between progressive and retrogressive metamorphic effects nor between products of Precambrian or Caledonian events are clearly indicated.

The Ardgour granitic gneiss, for which a 1020±50 m.y. whole-rock Rb-Sr isochron has been reported (Brook *et al.* 1976) extends along the western side of the Quoich line from Ardgour to Loch Quoich. Dalziel (1963) argued that the granitic gneiss was formed by regional migmatisation from the adjacent metasediments partly on the evidence of zircon morphology, which may not be conclusive (Pankhurst and Pidgeon 1976). On the other hand, Gould (1966) showed that the granitic gneiss had a uniform eutectic melt composition and could have been derived from the underlying basement. Unlike the Annagh Gneiss Complex of western Ireland, there are no extensive areas of rocks with compositions intermediate between the granitic gneiss and the pelitic and psammitic host rocks of the Glenfinnan Division. A. L. Harris (in discussion of Winchester 1974) has suggested a tectonic emplacement for this granitic gneiss, in which case it may constitute basement of Grenville age to part or all of the Moine assemblage.

Numerous segregation pegmatites give the granitic gneiss a migmatitic appearance (Harry 1954) and presumably the large samples from which Brook *et al.* (1976) obtained the concordant Rb-Sr isochron included these pegmatites. Hence the large whole-rock systems may have remained isotopically isolated during the segregation of the pegmatites, and the age of the later migmatisation may be considerably younger than the age of granitic gneiss formation. A 555±5 m.y. U-Pb zircon age on one of the segregation pegmatites (unpublished data) would appear to date this event, though as this age is obtained from a lower concordia intercept, it may also relate to a stage of prograde (Caledonian) metamorphism (cf. Gebauer and Grünenfelder 1976).

The possibility that the Ardgour granitic gneiss underwent partial melting after its original formation in Grenville times nonetheless has important implications. At Loch Quoich (Fig. 6), some granitic gneiss appears to cross the Quoich line which has been assumed to be the result of Caledonian tectonism (Johnstone *et al.* 1969). At this locality a narrow sheet of granitic gneiss has gradational contacts with the psammitic schists of the Loch Eil Division, which could be used as evidence to support the view of Brook *et al.* (1976) that these rocks were deposited before the Grenville orogeny. Unpublished isotopic data from this unit is again inconclusive. However the granitic gneiss sheet at Loch Quoich is unmigmatised, while structural evidence suggests that it was formed after the main units of the granitic gneiss in the

Glenfinnan Division (Johnstone *et al.* 1969). Thus the relationships of this apparently complex zone of granitic rocks to the Quoich line and to the juxtaposed Glenfinnan and Loch Eil divisions remains enigmatic and requires further study.

The boundary between the Glenfinnan and Morar divisions is marked by the Caledonian Sgurr Beag slide (Tanner 1971) which is a Caledonian feature and brought up many inliers of Lewisian rocks along its base. This slide generally corresponds to the western margin of sillimanite grade metamorphism (Winchester 1974) that is considered to be of Precambrian age (van Breemen *et al.* 1974). Near the Carn Chuinneag granite (Fig. 1) metamorphic relationships become less clear, partly because of the increasing metamorphic grade of the Caledonian orogeny. However the 550 m.y. old Carn Chuinneag granite was intruded into greenschist facies rocks (Long and Lambert 1963; Shepherd 1973), whilst just south of the Sgurr Beag slide *c.* 750 m.y. old pegmatite appears to have been emplaced into the high grade rocks of the Glenfinnan Division (Long and Lambert 1963). Further south-westwards, pegmatite localities 9, 10 and 11 (Fig. 6), from which *c.* 750 m.y. ages have been reported, are all in rocks of the Morar Division, though van Breemen *et al.* (1974) reported evidence for Precambrian pegmatites in the Glenfinnan Division whose Rb-Sr muscovite systems had been partially disturbed by a Caledonian event. However, the larger pegmatites yielding *c.* 750 m.y. ages may not have been formed during the same event as the smaller pegmatites and migmatitic segregations in the Glenfinnan Division. If the stratigraphic evidence for continuity of sediments in the Morar Division (Powell 1974) is accepted, then it is reasonable to assume that the metamorphism associated with the main fabric of the rocks in the Morar Division (Macqueen and Powell 1977) cannot postdate the intrusion of the late Precambrian pegmatites. It would, for instance, be surprising if the 740 m.y. K-Ar muscovite system (Fitch *et al.* 1969) in the Knoydart pegmatite (Fig. 6, locality 9) had survived the kyanite grade metamorphism there, particularly in view of the recent data of Purdy and Jäger (1976) that the blocking temperature of K-Ar muscovite systems is *c.* 350°C. In addition the pegmatites in the Morar Division appear to have been emplaced late in the local tectonic history (R. St J. Lambert, in Dunning 1972). Thus in order to evaluate whether the metamorphic effects of the Grenville episode extend into rocks of the Morar Division, it is critical to establish whether the pegmatites previously dated at *c.* 750 m.y. are indeed distinct from the Grenville orogeny.

## 4b. U–Pb age data for Precambrian pegmatites

A sample for zircon analysis was collected from a pegmatite at locality 10 (Fig. 6) for which a Precambrian age has been reported by Giletti *et al.* (1961). Seven U-Pb zircon analyses on four magnetic size fractions (RC 1399; Table 1) define a chord (Fig. 3) which corresponds to an upper intersection age of $815 \pm 30$ m.y. Monazite from the same pegmatite yields concordant U-Pb ages of $770 \pm 10$ m.y. and $780 \pm 10$ m.y. ($+165\ \mu$ m and $-165\ \mu$ m size fractions; Table 1; Fig. 3). As the larger size fraction contains inclusions of zircon which have lost lead, the older age of $780 \pm 10$ m.y. is preferred.

A significantly younger U-Pb zircon age of $730 \pm 20$ m.y., in close agreement with Rb-Sr ages on co-existing muscovites, has been reported by van Breemen *et al.* (1974) for a pegmatite from locality 11 (also in the Morar Division). The stubby irregular zircons from this pegmatite appear to be considerably more metamict than the more euhedral elongate crystals from the pegmatite at locality 10. As only two size fractions had previously been analysed, three more analyses are reported in

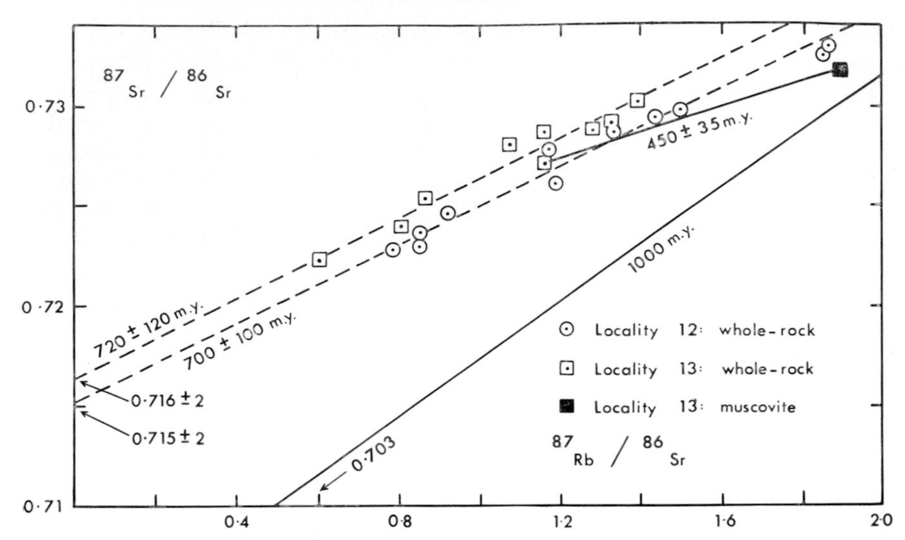


Fig. 7. Rb-Sr isochron plot for semipelites from the Morar Division of the Moine assemblage

Table 1 (RC 686) together with the data from van Breemen *et al.* (1974). A chord through all the data points corresponds to an age of 740±30 m.y. This does not significantly change the previous result.

## 4c. Rb–Sr age data on semipelitic schists of the Morar Division

The critical question of whether the rocks of the Morar Division have been involved in the Grenville orogeny has been approached with Rb-Sr whole-rock and mineral studies. Suites of large (*c.* 25 kg) samples have been collected from localities in the kyanite zone (Fig. 6, locality 12) and in the garnet zone (locality 13). At both localities, generally uniform semipelitic schist extends along 30 m road sections across strike. Two samples of pelitic schist were collected for mineral age determination at locality 14 in the garnet zone.

The Rb-Sr data points for the whole-rock samples from both localities 12 and 13 show considerable scatter (Table 3; Fig. 7). Best fit lines to the data points correspond to ages of 700±100 m.y. (locality 12) and 720±120 m.y. (locality 13). As the scatter of data points is considerably in excess of the analytical uncertainty, the statistical errors quoted here may not give reliable age limits (cf. Brooks *et al.* 1972).

In spite of the fact that Lambert (1969) reported Precambrian biotite K-Ar ages from west Morar, only Caledonian biotite Rb-Sr (430–410 m.y.) and muscovite Rb-Sr ages (*c.* 450 m.y.) were obtained in this study. Uncertainties for the muscovite ages may be slightly larger than the analytical uncertainties quoted, because of the low Rb-Sr ratios and the likelihood of internal sample inhomogeneity. The muscovite Rb-Sr systems are, however, definitely not Precambrian in age.

## 4d. Interpretation

The concordant 780±10 m.y. U-Pb monazite age from the pegmatite at locality 10 cannot be explained by partial Caledonian overprinting of an isotopic system of Grenville age. In the absence of controlled data on the behaviour of U-Pb monazite

Table 3. Rb–Sr isotopic analyses, northwestern Scotland

| Sample | Rb ppm | Sr(total)ppm | $^{87}Rb/^{86}Sr$ | $^{87}Sr/^{86}Sr$ | Mineral — whole-rock age (m.y.) |
|---|---|---|---|---|---|
| MORAR DIVISION OF MOINE ASSEMBLAGE | | | | | |
| Semipelitic schist | | | | | |
| Locality 12 (NM 723639) | | | | | |
| RC 1445 | 111 | 347 | 0.9272 | 0.72456 | |
| RC 1445 - muscovite | 196 | 120 | 4.745 | 0.74837 | 438 ± 10 |
| RC 1445 - biotite | 409 | 7.76 | 167.3 | 1.7050 | 414 ± 6 |
| RC 1446 | 105 | 357 | 0.8560 | 0.72291 | |
| RC 1447 | 141 | 346 | 1.179 | 0.72777 | |
| RC 1447 - muscovite | 202 | 139 | 4.219 | 0.74676 | 439 ± 10 |
| RC 1447 - biotite | 431 | 7.33 | 188.8 | 1.8322 | 413 ± 6 |
| RC 1448 | 110 | 373 | 0.8554 | 0.72358 | |
| RC 1449 | 104 | 381 | 0.7892 | 0.72273 | |
| RC 1450 | 148 | 232 | 1.852 | 0.73254 | |
| repeat | 149 | 230 | 1.869 | 0.73294 | |
| RC 1515 | 122 | 296 | 1.192 | 0.72593 | |
| RC 1516 | 128 | 247 | 1.498 | 0.72965 | |
| RC 1520 | 134 | 269 | 1.440 | 0.72932 | |
| RC 1522 | 127 | 275 | 1.331 | 0.72857 | |
| Locality 13 (NM 596614) | | | | | |
| RC 1432 | 112 | 400 | 0.8094 | 0.72393 | |
| RC 1433 | 113 | 376 | 0.8724 | 0.72537 | |
| RC 1434 | 129 | 319 | 1.167 | 0.72707 | |
| RC 1434 - muscovite | 191 | 292 | 1.897 | 0.73176 | 450 ± 35 |
| RC 1434 - biotite | 419 | 19.2 | 65.75 | 1.1192 | 426 ± 6 |
| RC 1435 | 92.6 | 440 | 0.6100 | 0.72229 | |
| RC 1437 | 120 | 300 | 1.161 | 0.72865 | |
| RC 1437 - biotite | 457 | 18.2 | 75.94 | 1.1768 | 427 ± 6 |
| RC 1438 | 133 | 300 | 1.2859 | 0.72877 | |
| RC 1440 | 119 | 321 | 1.076 | 0.72805 | |
| RC 1441 | 135 | 281 | 1.392 | 0.73016 | |
| RC 1442 | 131 | 284 | 1.333 | 0.72915 | |
| Locality 14 (NM 684977) | | | | | |
| RC 1477 | 72.3 | 439 | 0.4766 | 0.71761 | |
| RC 1477 - muscovite | 194 | 86.2 | 6.550 | 0.75452 | 427 ± 8 |
| repeat | 195 | 86.4 | 6.548 | 0.75474 | 429 ± 8 |
| RC 1477 - biotite | 359 | 21.5 | 49.83 | 1.0082 | 413 ± 6 |
| RC 1480 | 115 | 373 | 0.8905 | 0.72358 | |
| RC 1480 - muscovite | 183 | 122 | 4.340 | 0.74630 | 462 ± 10 |
| RC 1480 - biotite | 419 | 16.3 | 77.67 | 1.1838 | 421 ± 6 |

All analyses are of whole-rocks unless otherwise indicated.

systems during metamorphism, post-Grenville cooling over a period of 200 m.y. to the 500±50°C blocking temperature of Rb-Sr muscovite systems cannot be ruled out. However, this interpretation would appear to be unlikely in an area of variable metamorphic grade where locally the metamorphism has never exceeded the greenschist facies. The 815±30 m.y. (locality 10) and 740±30 m.y. (locality 11) zircon upper intersection ages show a large discrepancy, but they also cannot easily be explained in terms of post-Grenville cooling, particularly as annealing of metamict zircon lattice, which prevents lead loss, can take place at temperatures as low as 350°C (Gebauer and Grünenfelder 1976). Hence the evidence clearly points to the 780±50 m.y. age as being that of pegmatite intrusion, amphibolite facies metamorphism, or more probably both.

The *c*. 700 m.y. Rb-Sr whole-rock regression ages from localities 12 and 13 are not at variance with the above conclusion and it would be very surprising if Sr migration had occurred on a scale greater than the 25 kg samples after the first main metamorphism. This, together with the simple prograde fabric of the rocks at these localities, would tend to suggest a post-Grenville age for the Morar Division. In view of the scatter of data points, this Rb-Sr isotopic evidence alone does not rule out the possibility that the *c.* 700 m.y. regression ages are intermediate between an older age of deposition and a main Caledonian metamorphism. However, this is not consistent with the geological evidence for a dominant Precambrian fabric.

In spite of the fact that the Rb-Sr muscovite and biotite ages for the pelitic schists at localities 12, 13 and 14 are *c.* 450–400 m.y., the stratigraphical continuity in the Morar Division and the textural evidence indicates that the micas grew not later than the formation of the 780±50 m.y. pegmatites. The garnets at locality 13 are generally idioblastic and show the same textural features as garnets that grew during the development of the dominant metamorphic fabric in the schists of Morar and Knoydart (Fig. 6; Macqueen and Powell 1977). The characteristic features of these garnets are (i) threefold zonation, (ii) continuous growth during local $F_2$ as indicated by S-shaped inclusions and (iii) matrix coarsening after garnet growth as indicated by the distribution of inclusions.

The break between Precambrian and Caledonian deformation is put by Powell (1974) between $F_2$ and $F_3$ in the local structural sequence. The later deformation only resulted in crenulation of mica (Macqueen and Powell 1977) something barely evident at locality 13. Thus the fact that the Rb-Sr muscovite ages at localities 12, 13 and 14 are *c.* 450 m.y. must reflect the effects of thermal overprinting during the Caledonian episode. The characteristics of regional, late Caledonian cooling are shown by the time gap of *c.* 25 m.y. between the muscovite and biotite ages at localities 12 and 13 and the occurence of the younger of the biotite ages (414±6 m.y.) at locality (12) which is the further into the Caledonian thermal regime (Dewey and Pankhurst 1970).

## 5. Regional implications

### 5a. Distribution of Grenville rocks

The emerging pattern of radiometric data clearly establishes metamorphic rocks of the Grenville cycle within the Caledonides of the British Isles. The Annagh Gneiss Complex of western Ireland appears to be entirely of Grenville age. The grey (para) gneiss there was probably formed not long before the syntectonic emplacement of abundant granitic sheets, now dominantly granitic gneiss, 1070±30 m.y. ago. This, in turn, was followed by polyphase deformation and metamorphism before the emplacement of discrete discordant masses of granite at 1000±30 m.y., late in the tectonic history of the complex. In northwest Scotland, the Ardgour granitic gneiss dated at 1020±50 m.y. (Brook *et al.* 1976) is likely to fit into a corresponding sequence of events. However, the age of both the host schists of the Glenfinnan Division and the migmatitic activity have yet to be clearly established.

The 1100–1000 m.y. orogenic activity in northwest Britain corresponds closely in time, with that of the Grenville orogeny demonstrated in the inliers of high grade rocks in the Appalachian region of eastern North America (Odom *et al.* 1973), in the Grenville Province of Canada (Krogh and Hurley 1968), in the region of the basal gneisses in the Caledonides of western Norway (Priem *et al.* 1973a), and in the main Grenville/Sveconorwegian belt of southern Scandinavia (Priem *et al.* 1973a, b; Skiold 1976). It is also consistent with the $^{40}Ar/^{39}Ar$ age of 987±5 m.y. from Rockall Bank, reported by Miller *et al.* (1973), that indicates the effects of the Grenville orogeny there. This, together with the recent palaeomagnetic and radiometric evidence that from 980 m.y. to 880 m.y. ago the North American and Baltic

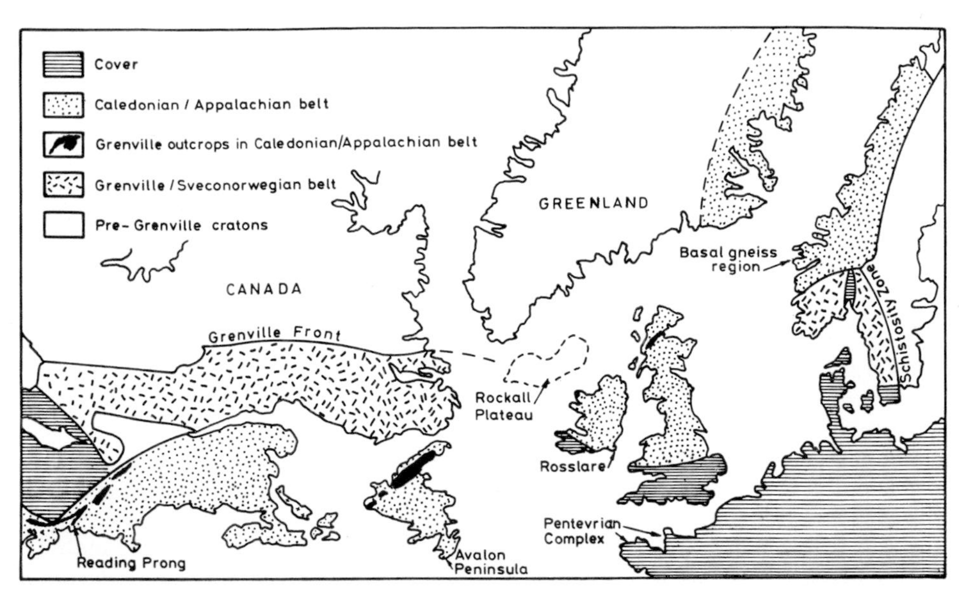


Fig. 8. North Atlantic setting of continental additions in late Precambrian times based on the pre-Mesozoic drift reconstruction of Bullard *et al.* (1965).

Shield regions were juxtaposed (Patchett and Bylund 1977), establishes the continuity of the Grenville belt across the North Atlantic region with considerable certainty (Fig. 8).

The southeastern limit of the Grenville belt in the British Isles is unknown because of the existence of younger overlying rocks. Rb-Sr and K-Ar mineral ages in the range 900–1200 m.y. have been reported from the Pentevrian Complex in Brittany (Leutwein 1968), but both this complex and the Rosslare Complex of southern Ireland appear to be largely of early and middle Precambrian age (Bishop *et al.* 1975; Max 1975). While this may place constraints on the extension of the Grenville belt in a southerly direction, there seem to be no grounds for confining the products of the Grenville orogeny to the Annagh Gneiss Complex of western Ireland and the Ardgour granitic gneiss in northwest Scotland, northwest of the Great Glen fault. The Central Highland Granulites and the metasediments of the Dalradian Supergroup southeast of the fault (Fig. 1), may well be underlain by *c.* 1100–1000 m.y. old basement, particularly as the Annagh Gneiss Complex underlies Dalradian rocks in western Ireland. Such an interpretation is not at variance with the zircon data from the Ben Vuirich granite which intrudes Dalradian rocks. These data can be interpreted as indicating a deep-seated source of *c.* 1300 m.y. old material (Pankhurst and Pidgeon 1976). Alternatively the source region could consist of 1100–1000 m.y. metamorphic rocks containing inherited zircons (from the Lewisian Complex) or may be composite and consist of Lewisian Complex and products of the Grenville orogeny.

Determination of the age of the crystalline basement beneath the Midland Valley of Scotland, which is present as blocks in volcanic vents (Upton *et al.* 1976), and of its equivalent in Ireland (Strogen 1974), will provide valuable information for the reconstruction of continental masses in late Precambrian times. However the available evidence points to the Grenville belt being present under some, and possibly a considerable part, of the British Caledonides and it may project across the Iapetus suture.

## 5b. The post-Grenville–pre-Caledonian period

Elucidation of crustal history in Scotland in the post-Grenville–pre-Caledonian time interval is not a straightforward matter. It could be argued that the Lower Torridonian arkose, for which Moorbath (1969) reports a $970 \pm 25$ m.y. age of diagenesis, represents molasse associated with the epeirogenic uplift stage of the Grenville cycle. However the $790 \pm 20$ m.y. age ascribed to the diagenesis of the Upper Torridonian sediments is not consistent with these rocks also being molasse to the Grenville cycle; neither is their northwesterly and Lewisian provenance (Williams 1969; Moorbath *et al.* 1967) when the evidence for *c.* 800–750 m.y. metamorphic and plutonic events comes from a region to the southeast of the Torridonian outcrop. Stewart (1973 p. 48) pointed out that the Upper Torridonian assemblage "closely resembles some of the thick clastic sequences of the Triassic preserved in fault-bounded basins flanking the North Atlantic" and suggested that the accumulation of these Torridonian rocks was connected with an early opening of Iapetus while Williams (1978) places the beginning of rifting in Newfoundland, with associated igneous activity and accumulation of clastic sequences, at about 800 m.y. ago.

When considering the crustal significance of pegmatite emplacement in northwest Scotland $780 \pm 50$ m.y. ago, it must be pointed out that the plutonic activity is later than the plutonism in southern Scandinavia which continued from before 1000 m.y. until *c.* 850 m.y. (Priem *et al.* 1973a, b; Skiold 1976), i.e. after the tectonothermal activity of the Grenville orogeny. However, the Grenville belt is generally characterized by slow post-orogenic cooling. For example, in the Grenville Province of Canada, Macintyre *et al.* (1967) and Harper (1967) demonstrated that it took until 900 m.y. ago for the blocking temperature of Ar in biotite to be reached (*c.* 300–350°C according to Hanson and Gast 1967). In some of the inliers of Grenville rocks in the Appalachians, blocking temperatures of biotite were not reached until 800–700 m.y. ago while the blocking temperature of hornblende (*c.* 500–550°C according to Hanson and Gast 1967) was not reached until 900 m.y. ago (Dallmeyer 1975a; Dallmeyer *et al.* 1975). These and similar studies in the Appalachians (Dallmeyer 1975b) and southern Scandinavia (Skiold 1976) showing a protracted cooling history argue against a major orogenic episode in the time interval between the Grenville and Caledonian orogenic episodes.

The general agreement of Rb-Sr, U-Pb and K-Ar mineral ages from the late Precambrian pegmatites in northwest Scotland is quite at variance with a cooling hypothesis and points to the existence of a completely separate event at *c.* 800–750 m.y. which the secretion derivation of the pegmatites (Long and Lambert 1963) suggests was tectonothermal. While the Rb-Sr whole-rock data for the Morar Division of the Moine assemblage are not unequivocal, they can also be interpreted as indicative that such an event was tectonothermal. When the structural and fabric evidence are added, an associated set of events indicative of an orogenic episode is suggested which corresponds with what Bowes (1968) referred to as the "Knoydartian orogeny". Subsequently Lambert (1969) used the term "Morarian orogeny" and this term has been used by van Breemen *et al.* (1974) and Johnson (1975).

Different explanations have been given of the nature and significance of this post-Grenville—pre-Caledonian episode. The model of local delayed closure proposed by

Johnson (1975) is not at variance with the hypothesis that the Grenville orogeny resulted from a collision of Himalayan type (Dewey and Burke 1973), but the time span involved (200–250 m.y.) is considerable and substantially longer than for the delayed closure of Iapetus (Dewey and Kidd 1974). The model of episodic activity along a long-standing ocean (Wright 1976) is not consistent with the palaeomagnetic and radiometric evidence of Patchett and Bylund (1977) that the North Atlantic and Baltic Shield regions were juxtaposed 980–880 m.y. ago. An early opening of Iapetus followed by multiphase orogenic activity is consistent with much of the data, with the 780$\pm$50 m.y. pegmatite emplacement being the equivalent of slightly later events in the Avalonian or Celtic episodes recognised elsewhere (Wright 1976). However the apparent geographical localisation of the products of a Morarian episode may better be explained in terms of the opening (and ?closing) of a small basin, or the opening of an ocean to the north. The preferred interpretations place the Morarian episode in an early stage of the overall plate tectonic development of the Caledonian crustal regime, although not necessarily as part of the Caledonian orogenic cycle in the classical sense of this term. Such an interpretation is consistent with the Morar Division of the Moine assemblage showing almost parallel isograds and eastward increasing metamorphic grade resulting from both Caledonian activity and metamorphic activity in late Precambrian times (Fig. 6).

Having established the existence of an orogenic episode between the Grenville and Caledonian orogenic cycles, a major problem awaiting solution is that of the age(s) of the whole Moine assemblage. The data presented here do not support the hypothesis of Brook *et al.* (1976) that all the Moine assemblage between the Great Glen fault and the Moine thrust in Scotland (Fig. 1) is of Grenville age. It could entirely represent a sedimentary accumulation between the Grenville and Morarian orogenic episodes within the same general time as that for the deposition of the Torridonian arkoses (cf. Long and Lambert 1963; Johnson 1965 p. 88). Alternatively parts, such as the migmatitic rocks of the Glenfinnan Division, could represent products of the Grenville Cycle and/or parts could post-date the Morarian episode and show only the effects of the Caledonian orogeny (cf. Soper and Wilkinson 1975), like the Central Highland Granulites southeast of the Great Glen fault. There are similar uncertainties about the age of the various schistose rocks of the Ox Mountains and Loch Derg areas of western Ireland.

## 6. Conclusions

1. An extensive sequence of late Precambrian events between the early stages of the Grenville and Caledonian orogenic cycles is shown in the northwestern British Isles.

2. Dated events in the Annagh Gneiss Complex of western Ireland span a considerable part of the Grenville orogenic episode. Metamorphic activity resulting in the development of (para) gneisses probably only shortly preceded the emplacement of abundant syntectonic granitic sheets 1070$\pm$30 m.y. ago. Subsequent amphibolite facies metamorphism and tectonic activity separated this acidic magmatism from the emplacement of late tectonic granites 1000$\pm$30 m.y. ago. The formation of the Ardgour granitic gneiss in northwest Scotland *c.* 1025 m.y. ago (Brook *et al.* 1976) is interpreted as fitting into this sequence of events.

3. The *c.* 975–850 m.y. time span is unlikely to have included major orogenic activity in the northwestern British Isles, a conclusion supported by both the

palaeomagnetic evidence that the North American and Baltic shields were juxtaposed 980–880 m.y. ago (Patchett and Bylund 1977) and the combined stratigraphic and isotopic evidence of post-orogenic uplift at this time in the major continental masses on either side of the British Caledonides.

4. The deposition *c.* 1000–800 m.y. ago of the Torridonian arkoses in northwest Scotland (Moorbath 1969) may relate to the epeirogenic uplift stage of the Grenville cycle and/or the earliest rifting stage associated with the initiation of a new orogenic cycle.

5. The Morar Division of the Moine assemblage of northwest Scotland is not younger than 815±30—730±20 m.y. old pegmatites but Rb-Sr large sample whole-rock regression ages of 720±120 and 700±100 m.y. on semipelitic rocks are interpreted as indicating that these clastic rocks were not involved in the Grenville orogeny. The regional extent, within the Moine assemblage, of these products of sedimentation, probably deposited during the *c.* 1000–800 m.y. time span, has yet to be demonstrated.

6. A tectonothermal episode—the Morarian episode—which is distinct from both the Grenville and Caledonian orogenic cycles occurred *c.* 800–700 m.y. ago. This episode is related to an early stage of activity in the plate tectonic regime which subsequently produced the British Caledonides and resulted in the intense thermal and tectonic overprinting of the products of late Precambrian times.

*Acknowledgments.* The support and assistance of M. Aftalion, J. Hutchinson, J. Jocelyn, B. E. Leake, W. G. Ross, D. L. Skinner and H. W. Wilson are gratefully acknowledged together with the critical review of the manuscript by P. J. Patchett and the discussions with R. D. Dallmeyer. C. B. Long, M. D. Max and W. E. A. Phillips assisted with the collection of some of the samples from the Belmullet Peninsula. C. B. Long and M. D. Max provided the samples from the southwest Ox Mountains and A. Molloy the samples from the northeast Ox Mountains. The Isotope Geology Unit at the Scottish Universities Research and Reactor Centre is supported by N.E.R.C.

**Note added in proof:**

Brook *et al.* 1977 (*J. geol. Soc. Lond.* **133,** 489–496) give a 1024±96 m.y. Rb-Sr whole-rock regression age for semipelitic schists from the Morar Division of the Moine assemblage from Morar. However, the best fit line joins clusters of two and twelve data points (the latter suggesting a younger age) and is, possibly, less convincing than the *c.* 700 m.y. whole-rock regression ages presented in this paper.

## References


BIKERMAN, M., BOWES, D. R. and VAN BREEMEN, O. 1975. Rb-Sr whole rock isotope studies of Lewisian metasediments and gneisses in the Loch Maree region, Ross-shire. *J. geol. Soc. Lond.* **131,** 237–254.

BISHOP, A. C., ROACH, R. A. and ADAMS, C. J. D. 1975. Precambrian rocks within the Hercynides. *In* Harris, A. L., Shackleton, R. M., Watson, J., Downie, C., Harland, W. B. and Moorbath, S. (Editors). *A Correlation of the Precambrian rocks of the British Isles,* 102–107. *Geol. Soc. Lond. Spec. Rep.* No. 6.

BOWES, D. R. 1968. The absolute time scale and the subdivision of Precambrian rocks in Scotland. *Geol. Fören. Stockholm Förh.* **90,** 175–188.

BROOK, M., BREWER, M. S. and POWELL, D. 1976. Grenville age for rocks in the Moine of north-western Scotland. *Nature, Lond.* **260,** 515–517.

BROOKS, C., HART, S. R. and WENDT, I. 1972. Realistic use of two-error regression treatment as applied to rubidium-strontium data. *Rev. Geophys. Space Phys.* **10,** 551–577.

BROWN, R. L., DALZIEL, I. W. D. and JOHNSON, M. R. W. 1970. A review of the structure and stratigraphy of the Moinian of Ardgour, Moidart and Sunart—Argylle and Inverness-shire. *Scott. J. Geol.* **6,** 309–335.

BULLARD, E., EVERETT, J. E. and SMITH, A. G. 1965. The fit of the continents around the Atlantic. In *A symposium on continental drift* 41–51. *Phil. Trans. R. Soc.* A **258.**

CLIFFORD, T. N. 1958. The stratigraphy and structure of the Kintail district of southern Ross-shire: its relation to the Northern Highlands. *Q. J. geol. Soc. Lond.* **113** (for 1957), 57–92.

CROW, M. J., MAX, M. D. and SUTTON, J. S. 1971. Structure and stratigraphy of the metamorphic rocks in part of Northwest County Mayo, Ireland. *J. geol. Soc. Lond.* **127,** 579–584.

CURRALL, A. E. 1963. The geology of the S.W. end of the Ox Mountains, Co. Mayo. *Proc. R. Ir. Acad.* **63B,** 131–165.

DALLMEYER, R. D. 1975a. $^{40}Ar/^{39}Ar$ age spectra of biotite from Grenville basement gneisses in northwest Georgia. *Geol. Soc. Am. Bull.* **86,** 1740–1744.

—— 1975b. $^{40}Ar/^{39}Ar$ ages of biotite and hornblende from a progressively remetamorphosed basement terrane: their bearing on interpretation of release specta. *Geochim. et Cosmochim. Acta* **39,** 1655–1669.

——, SUTTER, J. F. and BAKER, D. J. 1975. Incremental $^{40}Ar/^{39}Ar$ ages of biotite and hornblende from the northwestern Reading Prong: Their bearing on late Proterozoic thermal and tectonic history. *Geol. Soc. Am. Bull.* **86,** 1435–1443.

DALZIEL, I. W. D. 1963. Zircons from the granitic gneiss of western Ardgour, Argyll: their bearing on its origin. *Trans. Edinb. geol. Soc.* **19,** 349–362.

DEWEY, J. F. and BURKE, K. C. A. 1973. Tibetan, Variscan, and Precambrian basement reactivation: Products of continental collision. *J. Geol.* **81,** 683–692.

—— and KIDD, W. S. F. 1974. Continental Collisions in the Appalachian–Caledonian Orogenic Belt: Variations Related to Complete and Incomplete Suturing. *Geology* **2,** 543–546.

DEWEY, J. F. and PANKHURST, R. J. 1970. The evolution of the Scottish Caledonides in relation to their isotopic age pattern. *Trans. R. Soc. Edinb.* **68,** 361–389.
DUNNING, F. W. 1972. Dating events in the Metamorphic Caledonides: impressions of the symposium held in Edinburgh, September 1971. *Scott. J. Geol.* **8,** 179–192.
FAURE, G. and POWELL, J. L. 1972. *Strontium isotope geology.* Springer-Verlag, Berlin, 188 pp.
FITCH, F. J., MILLER, J. A. and MITCHELL, J. G. 1969. A new approach to radio-isotopic dating in orogenic belts. *In* Kent, P. E., Satterthwaite, G. E. and Spencer, A. M. (Editors). *Time and Place in Orogeny,* 157–195. *Geol. Soc. Lond. Spec. Publ.* No. 3.
GEBAUER, D. and GRÜNENFELDER, M. 1976. U–Pb zircon and Rb-Sr whole-rock dating of low-grade metasediments. Example: Montagne Noire (Southern France). *Contrib. Mineral. Petrol.* **59,** 13–32.
GILETTI, B. J., MOORBATH, S. and LAMBERT, R. St J. 1961. A geochronological study of the metamorphic complexes of the Scottish Highlands. *Q. J. geol. Soc. Lond.* **117,** 233–272.
GOULD, D. 1966. *Geochemical and mineralogical studies of the granitic gneiss and associated rocks of western Ardgour, Argyll.* Univ. of Edinburgh Ph.D. thesis.
HANSON, G. N. and GAST, P. W. 1967. Kinetic studies in contact metamorphic zones. *Geochim. et Cosmochim. Acta* **31,** 1119–1153.
HARPER, C. T. 1967. On the interpretation of potassium-argon ages from Precambrian shields and Phanerozoic orogens. *Earth Planet. Sci. Lett.* **3,** 128–132.
HARRY, W. T. 1954. The composite granitic gneiss of Western Ardgour. *Q. J. geol. Soc. Lond.* **109** (for 1953), 285–309.
HOFMANN, A. W. 1975. Diffusion of Ca and Sr in a basalt melt. *Carnegie Inst. Washington Year Book* **74,** 183–189.
JAFFEY, A. H., FLYNN, K. F., GLENDENIN, L. E., BENTLEY, W. C. and ESSING, A. M. 1971. Precision measurements of half-lives and specific activities of $^{235}$U and$^{238}$ U. *Phys. Rev.* C, **4,** 1889–1906.
JOHNSON, M. R. W. 1965. Torridonian and Moinian. *In* Craig, G. Y. (Editor). *The Geology of Scotland,* 79–113. Oliver & Boyd, Edinburgh.
—— 1975. Morarian Orogeny and Grenville Belt in Britain. *Nature, Lond.* **257,** 301–302.
JOHNSTONE, G. S., SMITH, D. I. and HARRIS, A. L. 1969. Moinian Assemblage of Scotland. *In* Kay, M. (Editor). *North Atlantic—geology and continental drift a symposium,* 159–180. *Mem. Am. Ass. Petrol. Geol.* **12.**
KROGH, T. E. and HURLEY, P.M. 1968. Strontium isotope variation and whole-rock isochron studies, Grenville Province of Ontario. *J. geophys. Res.* **73,** 7107–7125.
LAMBERT, R. St J. 1969. Isotope studies relating to the Pre-Cambrian history of the Moinian of Scotland. *Proc. geol. Soc. Lond.* **1652,** 243–245.
LEUTWEIN, F. 1968. Contribution à la connaissance du précambrian récent en Europe Occidentale et développement géochronologique du Briovérien en Bretagne (France). *Can. J. Earth Sci.* **5,** 673–682.
LONG, C. B. and MAX, M. D. 1977. Metamorphic rocks in the S.W. Ox Mountains Inlier, Ireland; their structural compartmentation and place in the Caledonian orogen. *J. geol. Soc. Lond.* **133,** 413–432.
LONG, L. E. and LAMBERT, R. St J. 1963. Rb-Sr isotopic ages from the Moine Series. *In* Johnson, M. R. W. and Stewart, F. H. (Editors). *The British Caledonides* 217–247. Oliver & Boyd, Edinburgh.
MACINTYRE, R. M., YORK, D. and MOORHOUSE, W. W. 1967. Potassium-argon age determinations in the Madoc–Bancroft area in the Grenville Province of the Canadian Shield. *Can. J. Earth Sci.* **4,** 815–828.
MACQUEEN, J. A. and POWELL, D. 1977. Relationships between deformation and garnet growth in Moine (Precambrian) rocks of western Scotland. *Geol. Soc. Am. Bull.* **88,** 235–240.
MAX, M. D. 1970. Mainland gneisses southwest of Bangor in Erris, northwest County Mayo, Ireland. *Sci. Proc. R. Dubl. Soc.* A, **3,** 275–291.
—— 1975. Precambrian rocks of south-east Ireland. *In* Harris, A. L., Shackleton, R. M., Watson, J., Downie, C., Harland, W. B. and Moorbath, S. (Editors). *A Correlation of the Precambrian rocks of the British Isles,* 97–101. *Geol. Soc. Lond. Spec. Rep.* No. 6.
——, LONG, C. B. and SONET, J. 1976. The geological setting and age of the Ox Mountains Granodiorite. *Bull. geol. Surv. Irel.* **2,** 27–35.
MILLER, J. A., MATTHEWS, D. H. and ROBERTS, D. G. 1973. Rock of Grenville age from Rockall Bank, *Nature Phys. Sci., Lond.* **246,** 61.
MOORBATH, S. 1969. Evidence for the age of deposition of the Torridonian sediments of north-west Scotland. *Scott. J. geol.* **5,** 154–170.
—— and TAYLOR, P. N. 1974. Lewisian age for the Scardroy mass. *Nature, Lond.* **250,** 41–43.
——, STEWART, A. D., LAWSON, D. E. and WILLIAMS, G. E. 1967. Geochronological studies on the Torridonian sediments of north-west Scotland. *Scott. J. Geol.* **3,** 389–412
ODOM, L. A., KISH, S. A. and LEGGO, P. J. 1973. Extension of "Grenville Basement" to the southern extremity of the Appalachians: U-Pb ages of zircons. *Geol. Soc. Am. Abstr. with Programs* **5,** 425.
PANKHURST, R. J. 1974. Rb-Sr whole-rock chronology of Caledonian events in northeast Scotland. *Geol. Soc. Am. Bull.* **85,** 345–350.

Pankhurst, R. J. and Pidgeon, R. T. 1976. Inherited istope systems and the source region pre-history of early Caledonian granites in the Dalradian series of Scotland. *Earth Planet. Sci. Lett.* **31,** 55–68.
——, Andrews, J. R., Phillips, W. E. A., Sanders, I. S. and Taylor, W. E. G. 1976. Age and structural setting of the Slieve Gamph Igneous Complex, Co. Mayo, Eire. *J. geol. Soc. Lond.* **132,** 327–334.
Patchett, P. J. and Bylund, G. 1977. Age of Grenville Belt magnetisation: Rb-Sr and palaeomagnetic evidence from Swedish dolerites. *Earth Planet. Sci. Lett.* **35,** 92–104.
Peach, B. N., Horne, J., Woodward, H. B., Clough, C. T., Harker, A. and Wedd, C. B. 1910. The Geology of Glenelg, Lochalsh and south-east part of Skye. *Mem. geol. Surv. U.K.*
Phemister, J. 1960. The Northern Highlands (3rd ed.) *Brit. reg. Geol., Geol. Surv. and Mus.*, London.
Phillips, W. E. A. 1973. The pre-Silurian rocks of Clare Island, Co. Mayo, Ireland, and the age of the metamorphism of the Dalradian in Ireland. *J. geol. Soc. Lond.* **129,** 585–606.
——, Taylor, W. E. G. and Sanders, I. 1975. An analysis of the geological history of the Ox Mountains inlier. *Sci. Proc. R. Dubl. Soc.* A **5,** 311–329.
Pidgeon, R. T. and Aftalion, M. 1978. Cogenetic and inherited zircon U-Pb systems in granites: Palaeozoic granites of Scotland and England. *In* Bowes, D. R. and Leake, B. E. (Editors). *Crustal evolution in northwestern Britain and adjacent regions,* 183–220. *Geol. J. Spec. Issue* No. 10 (this volume).
Powell, D. 1974. Stratigraphy and structure of the western Moine and the problem of Moine orogenesis. *J. geol. Soc. Lond.* **130,** 575–593.
Priem, H. N. A., Boelrijk, N. A. I. M., Hebeda, E. H., Verdurmen, E. A. Th. and Verschure, R. H. 1973a. Rb-Sr investigations on Precambrian granites, granitic gneisses and acidic volcanics in central Telemark: metamorphic resetting of Rb-Sr whole-rock systems. *Nor. geol. Unders.* **289,** 37–53.
——, ——, ——, —— and —— 1973b. A note on the geochronology of the Hestbrepiggan Granite in West Jotunheimen. *Nor. geol Unders.* **289,** 31–35.
Purdy, J. W. and Jäger, E. 1976. K-Ar ages on rock-forming minerals from the Central Alps. *Mem. Inst. Geol. Mineral. Univ. Padova* **30.**
Ramsay, J. G. 1958. Moine-Lewisian relations at Glenelg, Inverness-shire. *Q. J. geol. Soc. Lond.* **113,** 487–523.
Shepherd, J. 1973. The structure and structural dating of the Carn Chuinneag intrusion. Ross-shire. *Scott. J. Geol.* **9,** 63–88.
Skiold, T. 1976. The interpretation of the Rb-Sr and K-Ar ages of late Precambrian rocks in south-western Sweden. *Geol. Fören. Stockholm Förh.* **98,** 3–29.
Soper, N. J. and Wilkinson, P. 1975. The Moine Thrust and Moine Nappe at Loch Eriboll, Sutherland. *Scott. J. Geol.* **11,** 339–359.
Stewart, A. D. 1969. Torridonian rocks of Scotland reviewed. *In* Kay, M. (Editor). *North Atlantic—geology and continental drift a symposium,* 595–608. *Mem. Am. Ass. Petrol. Geol.* **12.**
—— 1973. 'Torridonian' rocks of western Scotland. *In* Harris, A. L., Shackleton, R. M., Watson, J., Downie, C., Harland, W. B., and Moorbath S. (Editors). *A Correlation of the Precambrian rocks of the British Isles,* 43–51. *Geol. Soc. Lond. Spec. Rep.* No. 6.
Strogen, P. 1974. The sub-Palaeozoic basement in central Ireland. *Nature, Lond.* **250,** 562–563.
Sturt, B. A., Ramsay, D. M., Pringle, I. R. and Teggin, D. E. 1977. Precambrian gneisses in the Dalradian sequence of North-East Scotland. *J. geol. Soc. Lond.* **134,** 41–44.
Sutton, J. S. 1972. The pre-Cambrian rocks of the Mullet Peninsula, County Mayo, Ireland. *Sci. Proc. R. Dubl. Soc.* A **4,** 121–136.
—— and Max, M. D. 1969. Gneisses in the north-western part of County Mayo. Ireland. *Geol. Mag.* **106,** 284–290.
Tanner, P. W. G. 1971. The Sgurr Beag Slide—a major tectonic break within the Moinian of the Western Highlands of Scotland. *Q. J. geol. Soc. Lond.* **126** (for 1970), 435–463.
——, Johnstone, G. S., Smith, D. I. and Harris, A. L. 1970. Moinian stratigraphy and the problem of the Central Ross-shire Inliers. *Geol. Soc. Am. Bull.* **81,** 299–306.
Taylor, W. E. G. 1968. The Dalradian rocks in Slieve Gamph, western Ireland. *Proc. R. Ir. Acad.* B, **67,** 63–82.
Upton, B. G. J., Aspen, P., Graham, A. and Chapman, N. A. 1976. Pre-Palaeozoic basement of the Scottish Midland Valley. *Nature, Lond.* **260,** 517–518.
van Breemen, O., Pidgeon, R. T. and Johnson, M. R. W. 1974. Precambrian and Palaeozoic pegmatites in the Moines of northern Scotland. *J. Geol. Soc. Lond.* **130,** 493–507.
——, Bowes, D. R. and Phillips, W. E. A. 1976. Evidence for basement of late Precambrian age in the Caledonides of western Ireland. *Geology* **4,** 499–501.
Williams, G. E. 1969. Characteristics and origin of a Precambrian pediment. *J. Geol.* **77,** 183–207.
Williams, H. 1978. Geological development of the northern Appalachians: its bearing on the evolution of the British Isles. *In* Bowes, D. R. and Leake, B. E. (Editors). *Crustal evolution in northwestern Britain and adjacent regions*, 1–22. *Geol. J. Spec. Issue* No. 10.

WINCHESTER, J. A. 1974. The zonal pattern of regional metamorphism in the Scottish Caledonides. *J. geol. Soc. Lond.* **130,** 509–524.

WRIGHT, A. E. 1976. Alternating subduction direction and the evolution of the Atlantic Caledonides. *Nature, Lond.* **264,** 156–160.

YORK, D. 1969. Least squares fitting of a straight line with correlated errors. *Earth Planet. Sci. Lett.* **5,** 320–324.

O. van Breemen, A. N. Halliday,
Isotope Geology Unit,
Scottish Universities Research and Reactor Centre,
East Kilbride,
Glasgow G75 0QU, Scotland.

M. R. W. Johnson,
Grant Institute of Geology,
West Mains Road,
Edinburgh EH9 3JW, Scotland.

D. R. Bowes,
Department of Geology,
University of Glasgow,
Glasgow G12 8QQ, Scotland.

# Gravity and magnetic anomalies attributable to basement sources under northern Britain

*D. W. Powell*

Anomalies in both the gravity and magnetic fields on a broad enough scale to have deep basement origins occur within northern Britain. Evidence over Lewisian outcrops is consistent with the generalisation that the intensity of magnetisation in granulites varies inversely with their amphibolitisation. The aeromagnetic anomalies along the L.I.S.P.B. line are used to draw a metamorphic boundary, within the seismic basement layer, that is correlated with granulites. Gravity anomalies to which reference is possible, and from which known non-basement effects have been removed, are restricted to the southwest. The extent to which their interpretation can be made consistent with the velocity structure obtained by L.I.S.P.B. under the Midland Valley and the Southern Uplands is discussed.

## 1. Introduction

Ground magnetic profiles and susceptibility measurements on specimens from some Lewisian outcrops form the basis used in discussing basement magnetisation characteristics. The magnetic anomaly base for the L.I.S.P.B. line interpretation is the smoothed aeromagnetic map with a 13 km wavelength cut-off (Hall and Dagley 1970). Gravity data to which reference is possible are restricted to the west of the region (McLean and Qureshi 1966; Watts 1971; Bott *et al.* 1972; El-Batroukh 1975; Durrance 1976; Hossain 1977).

## 2. Lewisian basement at outcrop

Bott *et al.* (1972) found from gravity profiles across the Ben Stack line in Sutherland, where granite sheets (2·63 to 2·66 gm/cc) separate biotite gneisses (2·69 gm/cc) in the north from granulites (2·78 gm/cc), that the density contrasts must extend

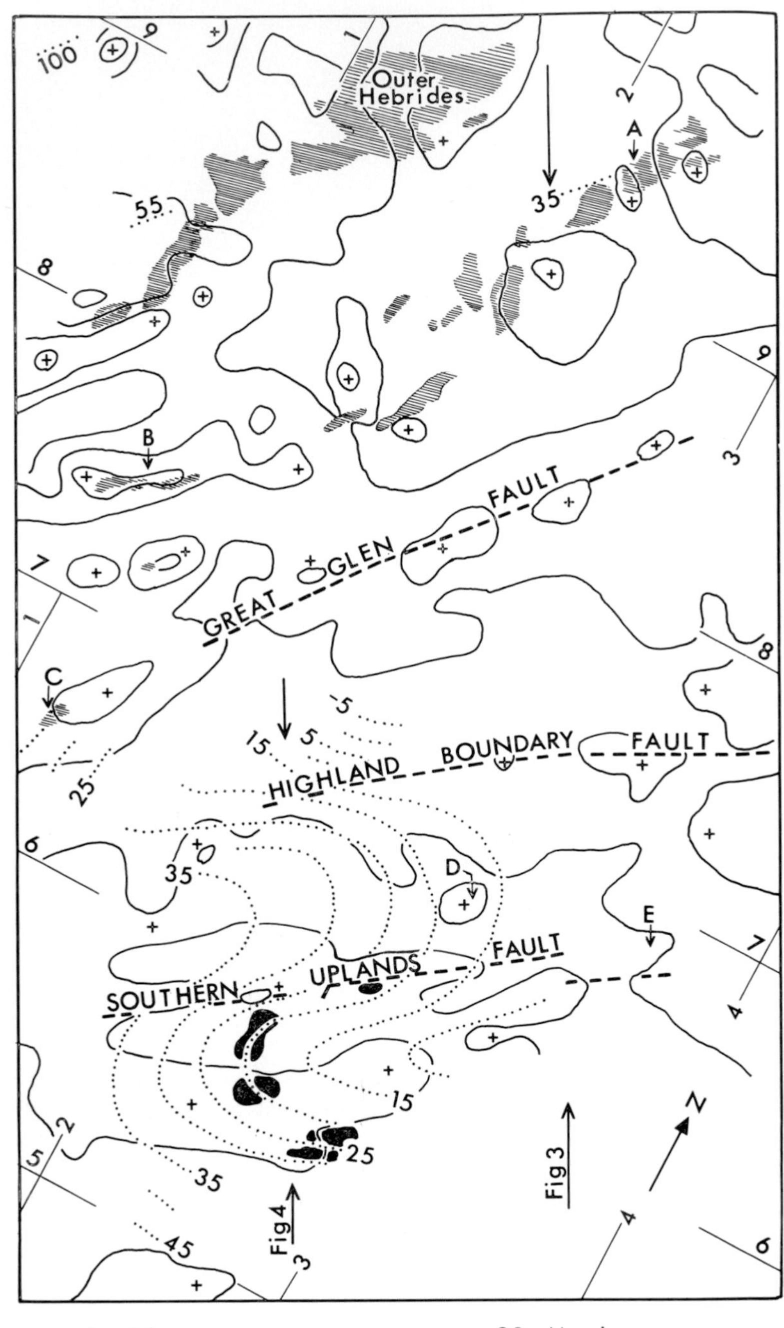

Outer Hebrides
A
B
C
D
E
GREAT GLEN FAULT
HIGHLAND BOUNDARY FAULT
SOUTHERN UPLANDS FAULT
Fig 3
Fig 4
N
100
55
35
25
15
5
-5
45


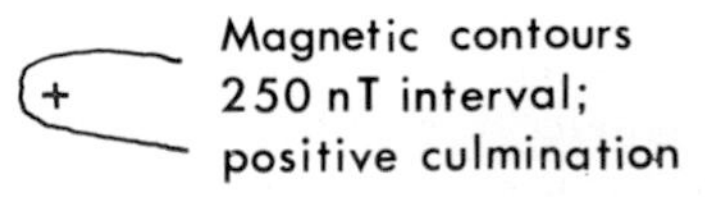

Lewisian outcrop
Granitic plutons (Fig 4)
20 Mgal
Magnetic contours 250 nT interval; positive culmination

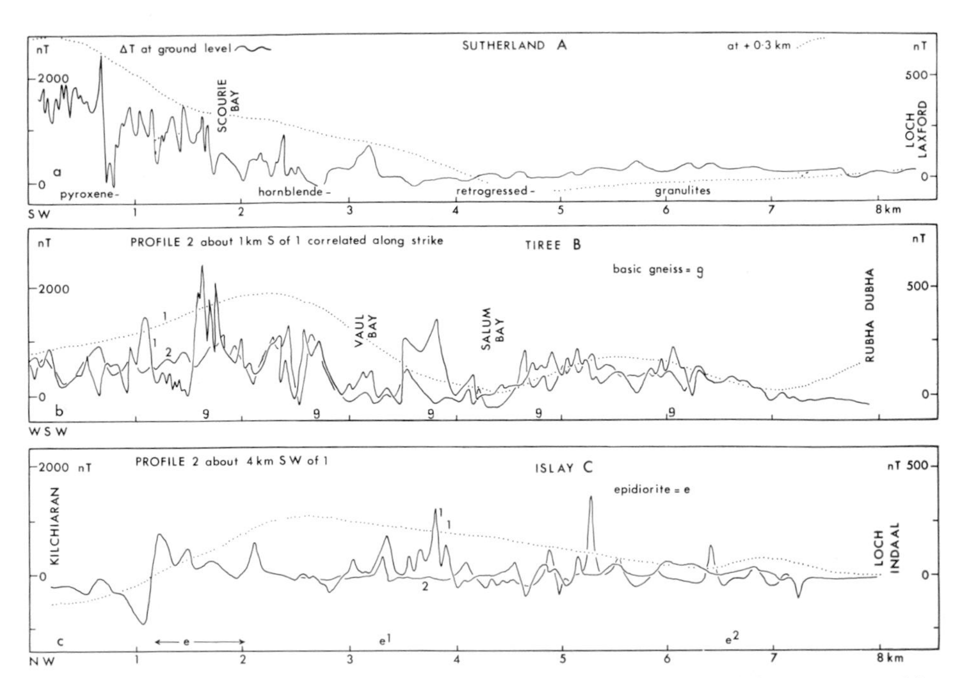


Fig. 2. Magnetometer profiles observed at ground and +0·3 km (dotted lines) over Lewisian outcrops located as A, B and C on Fig. 1. In b and c two parallel ground profiles located near to each other are shown. $e^1$ and $e^2$ are outcrops of epidiorite along profiles 1 and 2.

downwards 3 km and that the denser rocks may approach within a kilometre of the surface under the gneisses. The 500 nT amplitude aeromagnetic anomaly along these profiles is shown to fit a similar model except that the magnetisation contrast, about 0·001 cgs, unlike the density, lies within the granulites. A ground magnetic traverse from Scourie to Laxford (Fig. 2a) confirms this and broadens the basis for the conclusion that magnetite as well as pyroxene is lost when pyroxene granulites are amphibolitised (Powell 1970).

Even stronger aeromagnetic anomalies characterise the Inner and parts of the Outer Hebridean Lewisian. Their occurrence in the southeast part of the Outer Hebridean Lewisian also coincides with pyroxene granulites which are in thrust contact with the western gneisess. Over the Inner Hebrides the most pronounced magnetic field intensity changes occur at probable fault boundaries to the Lewisian outcrops rather than, as in the Outer Hebrides, across boundaries within them. Polarisations around 0·002 cgs are found by fitting step models one or two kilometres high to Inner Hebridean anomalies. Measurements on 95 basic gneiss specimens from eastern Tiree gave a mean intensity of 0·0015 cgs against 0·003 cgs for 167 of other associated rocks, i.e. less than that effective at the Lewisian out-

Fig. 1. Regional magnetic and gravity contours in northern Britain in relation to geological features. Profile locations; A, B and C are magnetic profiles in Fig. 2; D is magnetic anomaly Fig. 4a D; E is vent with granulites (Upton *et al.* 1976). National Grid co-ordinates are marked at edges.

crop boundaries. Most of these rocks show an amphibolite grade of metamorphism superimposed upon a granulite texture and mineral assemblage (Westbrook 1972). The consistent gradational pattern across the ground from strongly to weakly magnetised granulites following their retrogression is not found here (Fig. 2b). Better developed granulite facies rocks do occur in western Tiree (Drury 1972), under the island's biggest amplitude aeromagnetic anomaly, but ground measurements have yet to be made there. This does not, however, subtract from the conclusion that more complex patterns of transition in degree of magnetisation occur here than in Sutherland. Although many possible factors could bring this about, it is postulated for the purpose of the following analysis that the granulite to amphibilite facies control exists, but that it proceeds towards pyroxene in a downward rather than in a horizontal direction.

The rough correlation between aeromagnetic and smoothed ground profiles through an amplitude factor of about four (Figs 2a, 2b) does not obviously apply in Islay (Fig. 2c). Some epidiorites amongst the gneisses at outcrop are strongly magnetised, but must vary as rapidly along strike as they do along the profile or else their effects would be seen more distinctly in the aeromagnetic profiles (+0·3 km). On the other hand, the broad aeromagnetic high is scarcely seen on the ground; it certainly does not increase in amplitude by a factor of four. This suggests that its source is deep compared with the 0·3 km flying height and, as in Tiree, at least admits the possibility of magnetic granulites beneath the surface.

## 3. Concealed granulite basement

The opinion that the 6·4 km/sec. layer under the L.I.S.P.B. profile (Bamford *et al.* 1977) and more generally off northern Scotland (Smith and Bott 1975), correlates with granulites like those in Sutherland, amphibolitised or not, is strengthened by *in situ* velocity measurements on them (Hall and Al-Haddad 1976). Calculation shows that the whole of this layer under the L.I.S.P.B. profile cannot be uniformly magnetised (Fig. 3). The interpretation, therefore, indicates just how much of the layer need be magnetised to match the smoothed aeromagnetic anomalies along the profile. It is implied that the interface, G, within the layer, B–L, separates amphibolitised granulites above from pyroxene granulites below. A source intensity of 0·0025 cgs is necessary for the layer to contain them.

Magnetised granulite is dominant as far south as the basement step at the Great Glen Fault. Apart from the cover thickness, the relationships at the step are magnetically similar to those at the southeastern boundary of the Lewisian outcrop in Islay.

Under the Midland Valley shallow igneous rocks undoubtedly contribute to the filtered magnetic highs and exaggerate the amount of deeper basement sources. The distribution of magnetised basement here is unrelated to the boundary fault structures. Granulite fragments brought to the surface exposure in a vent agglomerate at E (Fig. 1) (Upton *et al.* 1976) over a subdued magnetic high have intensities up to 0·006 cgs. The prominent anomaly at D (Fig. 1 and Fig. 4a) has a pseudogravity effect similar to the Bouguer effect (Fig. 4b) and could represent an inhomogeneity within the basement (+0·45 gm/cc and 0·005 cgs) of, presumably, ultrabasic composition (Fig. 4c). Structural relief on an homogeneous basement (3 km for +0·15 gm/cc and 0·0035 cgs) would be incompatible with the contrasts at the boundary faults (see also next section). A disc shaped basic intrusion (2·75 gm/cc

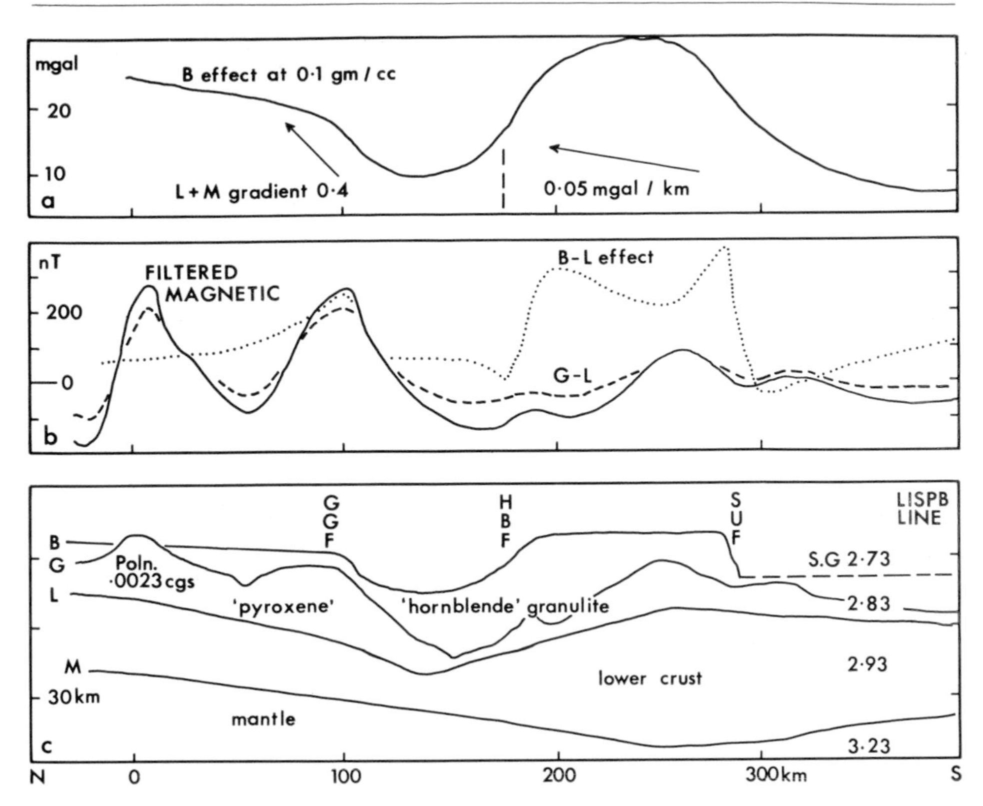


Fig. 3. L.I.S.P.B. profile located on Fig. 1. a. Gravity effect of structure on the top of the "granulite" basement and gravity gradients from the lower crust. b. Magnetic effect of the whole of the "granulite" layer and of a part of it which compares with the filtered magnetic profile. c. Seismic layering (Bamford pers. comm.; Bamford *et al.* 1977) with the physical contrasts applicable to a and b. GGF – Great Glen Fault; HBF – Highland Boundary Fault; SUF – Southern Uplands Fault.

and 0·003 cgs) under physically similar Carboniferous basalts is, however, a possible alternative unrelated to the basement (Fig. 4d).

If a granulite basement exists under the Southern Uplands (Powell 1970), it is more magnetised to the west than under the L.I.S.P.B. line (Fig. 4a and c, anomaly and source A).

## 4. Gravity anomalies and velocity structure

Regional gravity effects from the lower crust appear as the southeast falling gradient over the Hebrides and, probably, the northeast falling gradient throughout the southwest (Fig. 1). The former, inferred from Bouguer culminations (Watts 1971; Bott *et al.* 1972), compares with the 0·4 mgal/km calculated for the lower crustal layers along the L.I.S.P.B. line (Bamford personal communication) (L+M Fig. 3). The latter has no seismic confirmation, being at right angles to the line where the apparent lower crustal effect is reduced to 0·05 mgal/km.

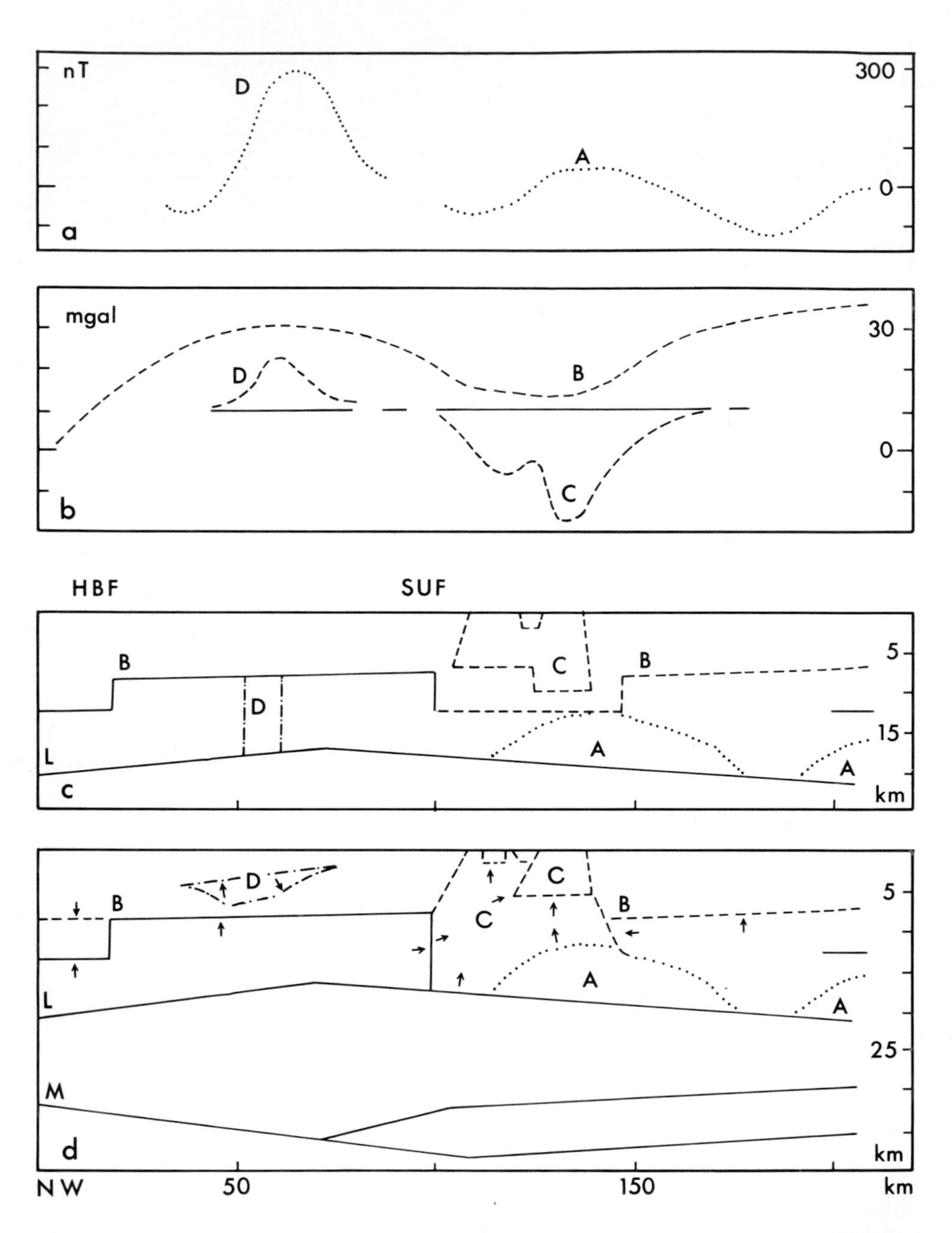


Fig. 4. Western Midland Valley–Southern Uplands profile located on Fig. 1. a. Aeromagnetic anomalies: Southern Uplands (A) and Midland Valley (D of Fig. 1). b. Gravity anomalies: regional (B), granitic plutons (C) and Midland Valley (D). c. and d. Projected L.I.S.P.B. layering (solid lines) with possible alternative potential field sources lettered in correspondence with their respective anomalies. Arrows in d. point in direction of decreasing density. Values in the text. HBF – Highland Boundary Fault; SUF – Southern Uplands Fault.

In Figure 4, showing profiles across the Midland Valley, the calculated effect of the seismic basement uplift (on B; Fig. 3) compares with the inferred gravity regional component from NW to SE (Figs 1 and 4b). If the seismic structure was symmetrical (Fig. 4c), a contrast of 0·1 gm/cc would match the gradients across the Midland Valley boundary faults. The structural amplitude on the basement across the Southern Uplands Fault may, however, be twice that at the Highland Boundary (Fig. 4d) and so require an asymmetrical density distribution. This could be achieved by reducing the contrast at the basement surface (2·78 –2·73 = 0·05 gm/cc) and restoring it across the Highland Boundary Fault by granitic rocks (2·68 gm/cc) against the basement on its northern side (Fig. 4d).

A separation of the gravity effects of the granitic plutons from the regional field has been attempted for the western Southern Uplands (Figs 4b, C, and B respectively) with the intention of minimising the amplitude of the former (El-Batroukh 1975). The maximum depth extent of the granite (7·5 to 10 km at −0·1 gm/cc) is found by neglecting the effect of the generally surrounding tonalite and granodiorite (Fig. 4c, C). Attributing as much of the total mass deficiency as possible to the tonalite (−0·02 to −0·03 gm/cc) at depths less than the seismic boundary L (Fig. 4d, C) gives a granite: tonalite ratio about 1:4. This model is consistent with the possibility that the granulite basement ends at the Southern Uplands Fault and suggests that the magnetic body (Fig. 4d, A) may be part of the batholith. The magnetic source does, however, have a strike extent well beyond the extent of the gravity effect of the body. Gravity calculations based on the seismic layering do not reproduce the southerly rise in the regional gravity field towards the south (Fig. 4b, B). It is supposed in both sections (Figs 4c and d) that the seismic basement under the Pennines is denser than 2·73 gm/cc and extends further north here than on the L.I.S.P.B. line.

## References


BAMFORD, D. NUNN, K., PRODEHL, C. and JACOB, B. 1977. LISPB-III. Upper crustal structure of northern Britain. *J. geol. Soc. Lond.* **133,** 481–488.

BOTT, M. H. P., HOLLAND, J. G., STORRY, P. G. and WATTS, A. B. 1972. Geophysical evidence concerning the structure of the Lewisian of Sutherland, NW Scotland. *J. geol. Soc. Lond.* **128,** 599–612.

DRURY, S. A. 1972. The tectonic evolution of a Lewisian complex on Coll, Inner Hebrides. *Scott. J. Geol.* **8,** 309–334.

DURRANCE, E. M. 1976. A gravity survey of Islay, Scotland. *Geol. Mag.* **113,** 251–261.

EL-BATROUKH, S. I. 1975. *Geophysical investigations on the Loch Doon granite, SW Scotland.* Ph.D. thesis, University of Glasgow.

HALL, D. H. and DAGLEY, P. 1970. Regional magnetic anomalies. An analysis of the smoothed aeromagnetic map of Great Britain and Northern Ireland. *Rep. Inst. geo. Sci.* No. 70/10.

HALL, J. and AL-HADDAD, F. M. 1976. Seismic velocities in the Lewisian metamorphic complex, NW Britain—'in situ' measurements. *Scott. J. Geol.* **12,** 305–314.

HOSSAIN, M. M. A. 1977. *Analysis of the major gravity and magnetic anomalies centred about Bathgate, central Scotland.* M.Sc. thesis, University of Glasgow.

MCLEAN, A. C. and QURESHI, I. R. 1966. Regional gravity anomalies in the western Midland Valley of Scotland. *Trans. R. Soc. Edinb.* **66,** 267–283.

POWELL, D. W. 1970. Magnetised rocks within the Lewisian of western Scotland and under the Southern Uplands. *Scott. J. Geol.* **6,** 353–369.

SMITH, P. J. and BOTT, M. H. P. 1975. Structure of the crust beneath the Caledonian foreland and Caledonian belt of the north Scottish shelf region. *Geophys. J. R. astron. Soc.* **40,** 187–205.

UPTON, B. G. J., ASPEN, P., GRAHAM, A. and CHAPMAN, N. A. 1976. Pre-Palaeozoic basement of the Scottish Midland Valley. *Nature, Lond.* **260,** 517–519.

WATTS, A. B. 1971. Geophysical investigations on the continental shelf and slope north of Scotland. *Scott. J. Geol.* **7,** 189–219.

WESTBROOK, G. K. 1972. Structure and metamorphism of the Lewisian of east Tiree, Inner Hebrides. *Scott. J. Geol.* **8,** 13–30.

D. W. Powell,
Department of Geology,
University of Glasgow,
Glasgow G12 8QQ, Scotland.

# *Part 3*

# Processes of mobile belt evolution

# Ensialic basin sedimentation: the Dalradian Supergroup

*A. L. Harris, C. T. Baldwin, H. J. Bradbury, H. D. Johnson and R. A. Smith*

The Dalradian Supergroup consists of up to 25 km of metasedimentary and volcanic deposits which accumulated between the late Precambrian (Upper Riphean) and Cambro–Ordovician (*c.* 500 m.y.). The depositional history of this succession is analysed in terms of several large-scale, basin-deepening and shallowing sequences (100's m to kilometres thick) which record a progressive upward increase in tectonic instability and contemporaneous basic volcanicity. The dominant source area situated in the northwest supplied mineralogically mature sediment during deposition of the Grampian Group (previously referred to the Moine succession of the Central Highlands), Appin Group and Argyll Group. The emergence of a southern continent is inferred as the most probable source of the mineralogically immature sandstones that are characteristic of the Southern Highland Group. An ensialic basin setting is indicated by the occurrence of high grade metamorphic rocks below the present Dalradian outcrop and below the Midland Valley. The tectonic framework during the evolution of this major, composite basin-deepening sequence was probably dominated by differential fault activity within a marginal basin. A brief comparison is made with a similar major basin-deepening sequence evolving in a marginal, fault-controlled basin, which developed during the Mesozoic and Tertiary history of the North Sea.

## 1. Introduction

In introducing this account of the ensialic setting of the Dalradian Supergroup and of its stratigraphy and sedimentation the various plate tectonic models relevant to Dalradian deposition are examined and the implications of the orogenic events which immediately preceded the accumulation of the Dalradian Supergroup briefly considered.

### 1a. Plate tectonic models

The recent seismic investigation (L.I.S.P.B. experiment) involving a N–S traverse across Britain (Bamford *et al.* 1977) is important in assessing the several plate tectonic models for Grampian–Caledonian evolution that have been proposed in the past ten years. These can be resolved into those in which Dalradian deposition is wholly ensialic and those which require deposition, at least in part, on oceanic crust.

Dewey (1969 fig. 3) showed the Dalradian and Moine rocks as a sedimentary wedge extending from a continental margin, made of Lewisian rocks, on to Iapetus (Proto-Atlantic) oceanic crust. The Moine and Dalradian rocks were regarded as a normal stratigraphical succession of late Precambrian to Cambro–Ordovician age, with the only pre-Caledonian rocks involved in the proposed model of Caledonian orogenesis being inliers of Lewisian gneiss which interleaved with Moine rocks in the North Highlands of Scotland. Lambert and McKerrow (1977) have suggested that the lower parts of the Dalradian Supergroup were deposited on a shelf made of ancient continental rocks including Moine as well as Lewisian assemblages. Their conclusion that the Moine rocks of the North Highlands and the Central Highland Granulites suffered Precambrian orogenesis is at variance with the evidence presented here and their model envisages the possibility that the turbidites of the upper part of the Dalradian Supergroup were deposited on oceanic crust. In correlating the Arenig ending of Dalradian sedimentation and the onset of Barrovian metamorphism in the Grampian orogeny with the subduction of the Iapetus ridge–transform system approximately on the site of the Highland Boundary fault, these authors require lower Ordovician or older rocks below the Midland Valley of Scotland to be on oceanic crust.

Phillips *et al.* (1976) regard the deposition of the Dalradian Supergroup as essentially ensialic, while the work of Graham (1976) has shown that the composition of the basic volcanic sequence of the Tayvallich district in the Southwest Highlands of Scotland is consistent with ensialic volcanism. Phillips *et al.* (1976) suggested that the Ox Mountains in Ireland are the remnant of a continent from which the clasts of high grade metamorphic affinity in Dalradian terrestrial sediments were derived. The view that ancient continental crust extends south of the Highland Border has recently found support in the discovery of fragments of granulite facies material from the Carboniferous Portan Craig vent in Berwickshire (Upton *et al.* 1976).

Wright (1976) envisages deposition of the Dalradian Supergroup on a foundering Morarian island-arc and it may be inferred from his model that the early Great Glen fault formed a tectonically active northwestern margin to the trough which, from late Precambrian until latest Cambrian or early Ordovician times, received Dalradian sediments. The apparent marked contrast in tectonic history of some or all the metamorphic rocks cropping out on either side of the fault lends support to this view. The tholeiitic volcanic rocks of the upper part of the Dalradian sequence are related by Wright (1976) to the opening of a marginal basin, itself related to the presence of an island arc to the southeast. Neither the Morarian nor the Grampian island-arcs proposed are exposed and no contemporary calc-alkaline volcanics are recorded: the presence of these arcs is necessary only to explain 'collision' orogenesis without invoking closure of the whole Proto-Atlantic ocean.

The seismic profiles drawn by Bamford *et al.* (1977) cast considerable doubt on those models that require either deposition of the Dalradian Supergroup on oceanic crust or the presence of oceanic crust below the Midland Valley of Scotland or the Southern Uplands. The inferred presence of high velocity rocks (dense continental crust—?Lewisian complex) at shallow depths below the Midland

Valley is also of great importance in assessing the reality of a continental mass to the south of the Dalradian trough of sedimentation. It is not clear from the profile whether the 'step-up' of the dense material to the south occurs at the present Highland Border zone or whether it lies further to the northwest. If further northwest it may coincide with the Loch Tay fault or significantly, with the northwestern edge of the flat-lying inverted Dalradian rocks which lie in the inverted limb of the Tay Nappe (Shackleton 1958) and the Loch Tay inversion.

## 1b. The southern continent

The existence of rocks having velocities consistent with dense continental crust at shallow depths below the Midland Valley, supports the presence of a continent on the southern side of the Dalradian sedimentary basin. Such an interpretation has hitherto been based on inconclusive evidence like the presence of clasts of metamorphic and granitoid rocks in the younger Dalradian sediments (Phillips *et al.* 1975; Phillips *et al.* 1976 p. 580) that are unlikely to have been derived from the northwest because of the presence there of the extensive Cambro–Ordovician carbonate bank. This somewhat negative evidence, however, is strengthened by the persistent proximal nature of the Southern Highland Group sediments throughout their extensive outcrop and thickness and the survival of abundant fresh, angular clasts of feldspar. These observations make an immediately adjacent source more likely that the distant Lewisian outcrops (cf. Smith 1976).

Nevertheless, the southerly derived currents which deposited the Jura Quartzite (Anderton 1971) are not related to provenance from a southern continent, but to deposition on a tidal shelf connected to an open ocean and adjacent to a landmass to the northwest (Anderton 1976). Much of the evidence from later formations indicates southeasterly accretion of sediments, from a landmass to the northwest, into the oceanic basin envisaged by Anderton (1976) to lie to the southeast. This apparent asymmetrical disposition of the Dalradian stratigraphy and the absence of exposed shallow-water shelf deposits related to the southern continent constitutes evidence opposed to its existence. This asymmetry, however, could be explained by the geometry of the Tay Nappe, itself asymmetrical, which in being transported south may have overridden and concealed part of the continent and all of its shelf deposits, below the flat-lying and inverted rocks of the southern Highlands.

The status of the rocks of the Ox Mountains in Ireland as a fragment of this continent remains an open question, because although Phillips *et al.* (1975) and Pankhurst *et al.* (1976) recognise a pre-Grampian history for most of these rocks, Long and Max (1977) have identified a fairly complete Dalradian succession within the Ox Mountains.

If a continental landmass persisted at the southeastern margin of the trough in which the Dalradian sediments were deposited, vertical movements within the basement must have been an important influence from an early stage. The zone of weakness marking the edge of the continent would be expected to persist through later geological history and possibly this zone is now marked by the Highland Boundary fault. Certainly the present Highland Boundary has been the site of persistent displacements from the lower Palaeozoic to the present day. Several authors (e.g. George 1960) have suggested that the initial throw on the fault was up to the southeast but there is no doubt that that throw was at least partly reversed during the late Silurian or early Devonian. In view of this reversal, the initial upthrow must indeed have been large to explain the persistence of dense continental crust at a high level below the Midland Valley, unless high level dense continental crust had always been there.

Evidence for pre-Devonian control by the basement in the fault zone is clear. The present disposition of the metamorphic rocks shows that the Highland Border zone coincides with low grades of metamorphism, low states of strain, downward-facing structures, steeply inclined rocks and, locally, black shales, cherts and basic volcanic rocks (Highland Border Series). This coincidence cannot be directly or reasonably related to post-Devonian brittle fault movements or to the Devonian movements which, to a great extent, controlled Devonian sediments and perhaps volcanicity (Armstrong and Paterson 1970). To account for the coincidence it is thought that block movements of the basement relatively down to the southeast had already occurred in the late-orogenic phase. These produced the downbending (Highland Border downbend) into the steeply inclined, downward-facing attitude of structures in the Dalradian which were formerly horizontal and facing southeast (Harris *et al.* 1976). Similarly, metamorphic isograds that were formerly flat-lying and marking an upward decrease in metamorphic grade were bent down to become steeply inclined and hence closely spaced in outcrop, causing a rapid decrease in grade towards the Highland Border. The conclusion of Johnson and Harris (1967) that the Highland Border Series has shared the structural history of the adjacent Dalradian rocks and is downward-facing in the axial planes of first structures means that these rocks, too, have been brought down from high levels of the nappe by the down-bending in the Highland Border zone. From their inferred structural position in the nappe they are unlikely to be remnants of oceanic crust and sediments. The pre-Devonian downbend having had this influence on the disposition of the Dalradian rocks, Lower Old Red Sandstone sediments and volcanic rocks were deposited on the eroded downward-facing steeply inclined structures. Movements in the zone that were generally coeval with the Devonian were simply a rejuvenation of the earlier, late orogenic movements and further evidence of the longevity of major zones of weakness in the crust. Similarly, strong control of the structures in the Dalradian rocks by basement blocks is probably an extension of such block movements which controlled certain features of Dalradian sedimentation.

## 2. Stratigraphy

In considering the sedimentary evolution of the Dalradian rocks Harris and Pitcher (1975) envisaged a restricted trough within which Grampian Group ('Moine'—see below) sediments were deposited. The persistence of this trough effectively predetermined the similarly localised outcrop of the Lochaber and Ballachulish subgroups (Fig. 1) but was eventually replaced by a progressively less restricted area in which the overlapping deposition of similar shallow water limestones, carbonaceous shales and orthoquartzites of the Blair Atholl Subgroup took place (Fig. 3). The normally quiescent conditions of deposition of the Appin Group sediments were succeeded by higher energy conditions accompanying the predominantly clastic sedimentation of the Argyll Group, including the late Precambrian glacial episode. It was envisaged that Argyll Group sediments accreted southeastwards, possibly with a migrating axis of maximum sedimentation coinciding with the leading edge of the sedimentary wedge; e.g. proximal turbiditic sandstones of the Crinan Grits in the Islay–Loch Awe region pass

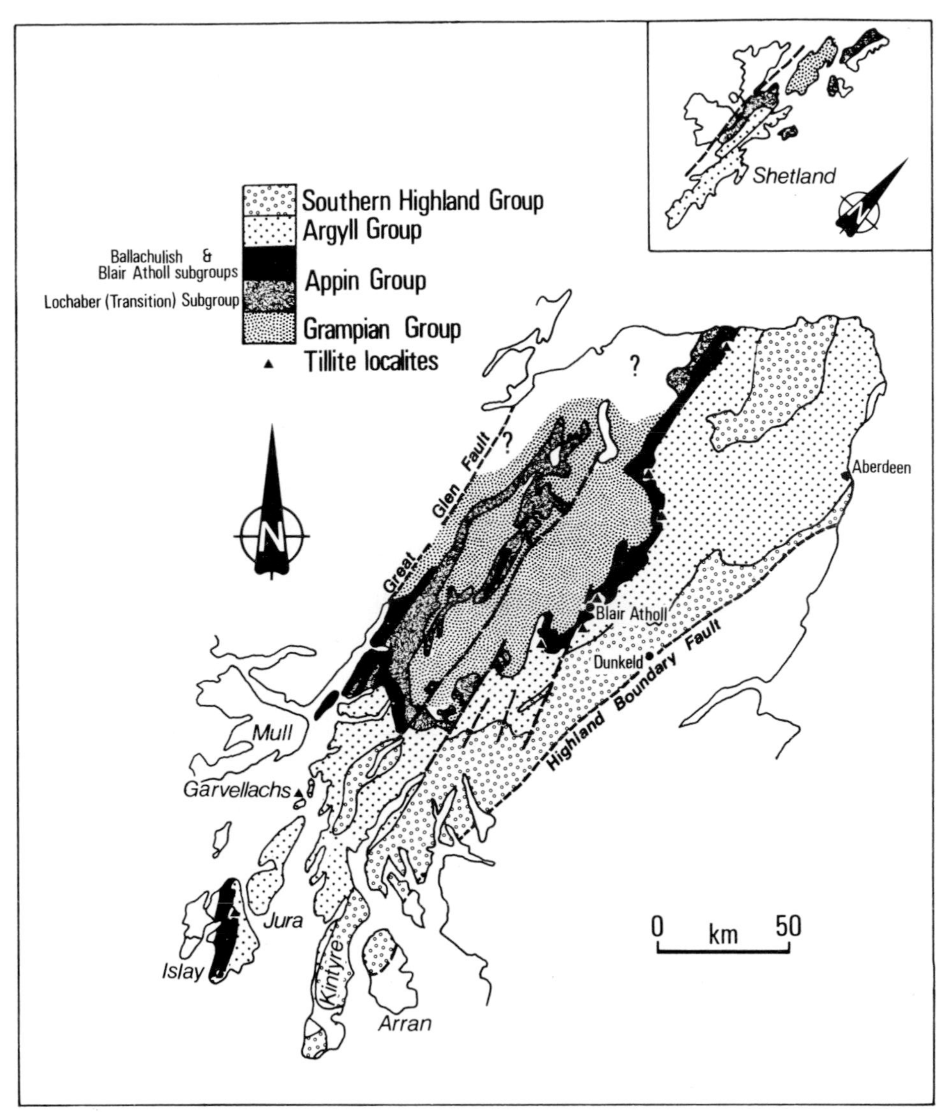


Fig. 1. Distribution of the Grampian, Appin, Argyll and Southern Highland groups in the Dalradian Supergroup of Scotland.

southeastwards into finer grained, more distal deposits (Ben Lui Schist). The provenance of Argyll Group clastic sediments was thought of as being a northwestern continental landmass with northwesterly derived material being subsequently transported axially NE–SW. In a change of provenance probably occurring early in the Cambrian and coinciding with the outbreak of widespread basic volcanicity, a southern continent, of which the Ox Mountains may be a possible remnant, supplied the clasts of high grade metamorphic material that characterize the Southern Highland Group sandstones. This generalised strati-

Table 1. Summary of the lithostratigraphical sub-division of the Dalradian Supergroup

| | Groups | Subgroups | Formations |
|---|---|---|---|
| DALRADIAN SUPERGROUP | SOUTHERN HIGHLAND | (None distinguished) | e.g. Leny Limestone<br>Leny Grit<br>Ben Ledi Grit<br>Aberfoyle Slate |
| | ARGYLL | Tayvallich<br>Crinan<br>Easdale<br>Islay | |
| | APPIN | Blair Atholl<br>Ballachulish<br>Lochaber | |
| | GRAMPIAN | (None distinguished) | e.g. Struan Flags<br>Eilde Flags |
| | | (Base not identified) | |

graphical and sedimentological scheme is now modified below in the light of more recent work with only new stratigraphical information presented in the following parts of this section.

## 2a. "Moine" Central Highland Granulites; the Grampian Group

Sufficient detailed structural and related radiometric work (van Breemen *et al.* 1974; Brook *et al.* 1976, 1977) has been carried out on the Moine assemblage of the North Highlands of Scotland to demonstrate the reality of Precambrian orogenesis—pegmatites yield *c.* 730 m.y. ages (Morarian) and schists and gneisses yield *c.* 1050 m.y. dates (Grenville)—see also van Breemen *et al.* (1978). Profound modification of at least some of the Precambrian rocks during subsequent orogenesis (Tremadoc–Arenig)—the Grampian–Burlingtonian orogeny (Grampian for brevity) of Wright (1976) which affected both the Moine and Dalradian assemblages. Wright restricts the Caledonian orogeny to the end-Silurian and this nomenclature is adopted here.

Demonstration of separate Grenville, Morarian (Lambert 1969—previously named Knoydartian by Bowes 1968) and Grampian events has influenced most recent 'Caledonian' plate tectonic modelling (e.g. Phillips *et al.* 1976; Wright 1976; Lambert and McKerrow 1976). However, this has also produced serious problems of stratigraphical nomenclature which require resolution. The Central Highland Granulites in the Grampian Highlands which Johnstone *et al.* (1969) correlate with Moine rocks in the North Highlands can be shown to pass up by normal sedimentary transition (Bailey 1934; Treagus 1974; Hickman 1975) through the Lochaber (Transition) Subgroup (Harris and Pitcher 1975) into the Ballachulish Subgroup. The existence of a normal sedimentary transition from the Ballachulish into the Blair Atholl Subgroup (Rast and Litherland 1970; Smith and Harris

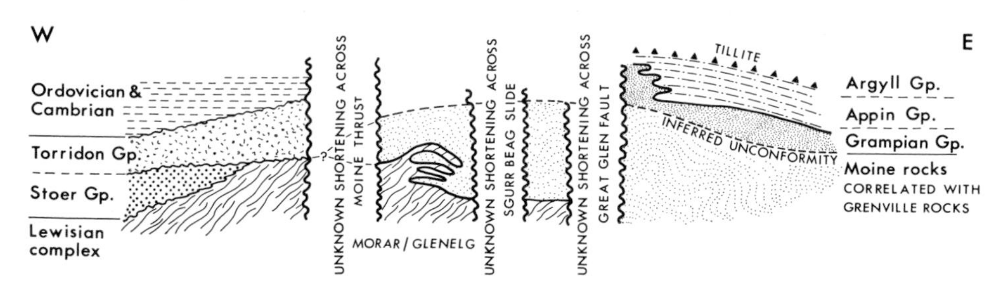


Fig. 2. Relationships between the major Precambrian units across the Scottish NW Highlands.

1976) demonstrates the continuity of the Dalradian succession which can be raced upwards through to the late Precambrian–Cambrian Argyll Group and into the Cambrian (and younger) Southern Highland Group. The evidence that the earliest (Grampian) orogenic events to affect the Dalradian Supergroup were also the first to affect most or all of the rocks of the Grampian Highlands referred to as part of the Moine assemblage throws doubt on the correlation. The Loch Eil (?youngest) division of the Highland Moine rocks lies in structural conformity with the Ardgour granitic gneiss and the extension of that gneiss northwards into Glen Garry, Inverness-shire (see Institute of Geological Sciences 1:50,000 Sheet 62(W) 1976) and the Ardgour granitic gneiss has been dated at 1050 m.y. (Brook *et al.* 1976).

Accordingly the so-called Moine rocks of the Central Highlands are considered to have stronger affinities with the Dalradian assemblage than with the Moine assemblage of north Scotland. Hence it is proposed that those psammitic rocks of the Central Highlands of Scotland which are older than and demonstrated or inferred to be stratigraphically continuous with the Lochaber (Transition) Subgroup be assigned to the Dalradian Supergroup and referred to as the *Grampian Group* (Fig. 1). Such modification to the stratigraphical scheme of Harris and Pitcher (1975) recognises that, like the rest of the Dalradian assemblage, the Grampian Group has no affinity to the Loch Eil or any other division of the Moine assemblage of the North Highlands (Fig. 2). This proposed four-fold subdivision of the Dalradian Supergroup (Table 1) requires the abandonment of the previously used terms Lower, Middle and Upper. All of the rocks formerly assigned to the Central Highland Granulites may not be part of the Grampian Group as autochthonous strata which have suffered Precambrian orogenesis may occur to the southeast of the Great Glen fault and Sturt *et al.* (1977) have produced evidence for allochthonous pre-Grampian gneisses in northeast Scotland. Similarly, Grampian Group rocks which have undergone only Grampian orogenesis may occur to the northwest of the Great Glen (e.g. Soper and Wilkinson 1975).

## 2b. Appin Group

The stratigraphical scheme of Harris and Pitcher (1975 fig. 12) requires some modification. A thick sequence of quartzites, limestones and mica schists in the Ben-y-Ghlo Mountains northeast of Blair Atholl is part of the Ballachulish Subgroup (Smith and Harris 1976). A complete succession without significant structural break can be traced probably from the top of the Ballachulish Limestone to the graphitic schists and limestones of the Blair Atholl Subgroup (Lismore Limestone and Cuil Bay Slate). Mapping (by H.J.B.) to the northwest of Ben Vrackie has shown that, as in Donegal, there are at least two graphitic marble formations in the Blair Atholl Subgroup, and that attempted correlation of these through unexposed terrain would result in spuriously complex outcrop patterns and structural histories. Indeed, work throughout the Blair Atholl district (by H.J.B. and R.A.S.) demonstrates complex lithofacies changes within the Blair Atholl Subgroup with substantial thicknesses of strata showing transition between graphitic schist and marble. It is of interest that the strongly deformed Glen Banvie Series is regarded (by R.A.S.) as either belonging to a stratigraphically low level in the Ballachulish Subgroup (?Leven Schists) (cf. Thomas 1965) or as representing, in severely attenuated form, the whole of the Appin Group over which the Tay Nappe rode.

### 2c. The Argyll and Southern Highland groups

The conventional interpretation of the Perthshire succession of these groups is set out in Table 1 where the lithostratigraphy is represented in what has been regarded as ascending chronological order. The work reported here and represented on the revised Geological Survey Sheet 55E, Pitlochry (in press), shows that each of the lithostratigraphic 'formations', known as Schiehallion Quartzite, Killiecrankie Schist, Carn Mairg Quartzite and Ben Eagach Schists, should be interpreted as a sedimentary facies and that they are at least partly the lateral equivalent of one another. Although believing that this is the case in part of Perthshire and realising that our conclusions imply the lateral equivalence of not only formations but subgroups, we are reluctant at this stage to propose any formal changes to the existing scheme until the application of our model has been tested elsewhere in the central Highlands. The sedimentary model that we are proposing derives largely from recent detailed mapping (by H.J.B.) and is incorporated in part of section 3.

## 3. Sedimentation

Attention is focused on the distribution of lithofacies in time and space and the interpretation of lithofacies sequences. Constraints, however, are placed on the sophistication of sedimentary environment interpretation by the degree of tectonic and metamorphic obliteration of primary structures, the lack of chronostratigraphical control and the emphasis inevitably placed on the least deformed area. Although the gross lithostratigraphy of the Scottish and Irish Dalradian rocks shows remarkable lateral persistence of lithofacies (Harris and Pitcher 1975) it should be emphasized that these are not necessarily time equivalents and as such large-scale lateral facies variations should always be interpreted with extreme care.

The main aim of this section is to impose some sedimentological interpretation on to the existing subgroup lithostratigraphy, attempting to establish and correlate major changes in basin development. This has been achieved by recognizing large-scale lithofacies sequences apparently recording gradational changes in environmental conditions. Two types of sequence are identified: transgressive and regressive sequences refer to relatively small scale events, typically within and between individual formations (several 10's to 100's m thick), while basin-deepening and basin-shallowing sequences are of a larger scale and typically involve changes on the scale of one or more subgroups (several 100's m to kilometres thick). Basin-deepening and shallowing constitute parts of multiple basin-fill sequences and their identification permits the development of the Dalradian basin to be analysed on various scales at differing levels of detail (Fig. 3).

## 3a. Grampian Group

Little sedimentological detail is available and any facies interpretations must remain at best highly speculative. Hickman (1975) suggests that the Eilde Flags of the Lismore–Glen Roy area comprise a generally coarsening-upwards sequence in excess of 1 km thick. The sediments were originally silts and muds interbedded with poorly sorted, immature, locally pebbly, quartz and feldspathic sands, with cross stratification, ripple marks, grading and erosional features indicating deposition in shallow water, probably marine, environments. Mud cracks and rain pits suggest the presence locally of either shoreline or terrestrial environments interdigitating with marginal marine facies. While a spatial facies change is indicated by the relative concentration of quartzite and schistose lithologies southwards and westwards towards Glen Coe and Onich, the environmental implications cannot be specified.

## 3b. Appin Group

In the Appin area the boundary between the Grampian and Appin groups is shown to be significantly diachronous (Hickman 1975 fig. 2). The Grampian Group Eilde Flags, as a near shore or paralic(?) facies, is shown by Hickman (*op. cit.*) to migrate progressively northwestwards with sub-littoral mud facies following sympathetically. The three cross bedded, feldspathic quartzites (Eilde, Binnein and Glen Coe quartzites) thin northeastwards and pass laterally into schists. Northward directed palaeocurrents derived from cross bedding are inferred to represent a major longshore component of sediment transport analogous to that of the younger Jura Quartzite (Anderton 1976). The southwestward increase in thickness of the quartzites is accompanied by increasing feldspar content and Hickman (*op. cit.*) suggests that these marine quartzites may pass southwards into deltaic facies.

The gradual replacement of Eilde Flags marginal facies by sub-littoral muds (now schists) is more or less complete by the base of the Glen Coe Quartzite and the first stage of what amounts to a phase of basin deepening is completed by the kilometre of fining-upwards dark psammitic schists which comprise the lower part of the Leven Schist Formation. The psammitic schists themselves grade upwards into grey-green phyllites and mica schist with thin intercalations of cream limestones. The blanket geometry of this facies and the pale colouration of the limestones, which indicate deposition in freely circulating water, suggests an open marine setting for this part of the Leven Schist Formation but one which is differentiated from its lower counterpart by the progressive diminution of terrigenoclastic sediments. This upper part of the Leven Schist therefore heralds a major phase of carbonate accumulation recorded by the Ballachulish Limestone which itself appears to represent a carbonate build up only slightly 'diluted' by terrigenoclastic sediments. The latter eventually terminates and the carbonate facies gives way by transition into the pyritiferous black slate facies of the Ballachulish Slates which probably accumulated in stagnant basinal environments.

Terrigenous inputs gradually return in the form of thin locally graded quartzites

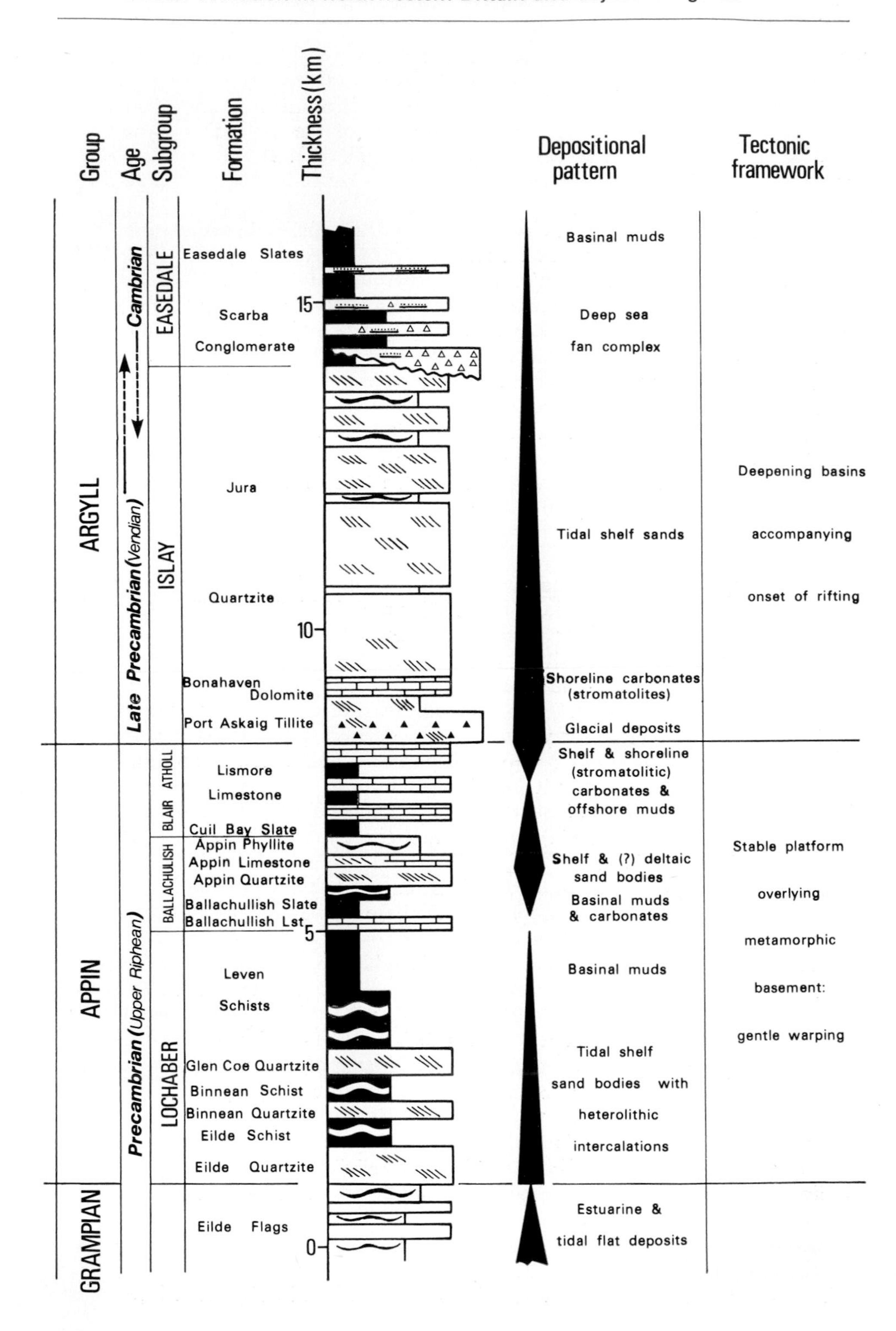

Group
Age
Subgroup
Formation
Thickness (km)
Depositional pattern
Tectonic framework
ARGYLL
APPIN
GRAMPIAN
Late Precambrian (Vendian)
Cambrian
Precambrian (Upper Riphean)
EASEDALE
ISLAY
BLAIR ATHOLL
BALLACHULISH
LOCHABER
Easedale Slates
Scarba
Conglomerate
Jura
Quartzite
Bonahaven Dolomite
Port Askaig Tillite
Lismore Limestone
Cuil Bay Slate
Appin Phyllite
Appin Limestone
Appin Quartzite
Ballachullish Slate
Ballachullish Lst
Leven Schists
Glen Coe Quartzite
Binnean Schist
Binnean Quartzite
Eilde Schist
Eilde Quartzite
Eilde Flags
15
10
5
0
Basinal muds
Deep sea fan complex
Tidal shelf sands
Shoreline carbonates (stromatolites)
Glacial deposits
Shelf & shoreline (stromatolitic) carbonates & offshore muds
Shelf & (?) deltaic sand bodies
Basinal muds & carbonates
Basinal muds
Tidal shelf sand bodies with heterolithic intercalations
Estuarine & tidal flat deposits
Deepening basins accompanying onset of rifting
Stable platform overlying metamorphic basement: gentle warping

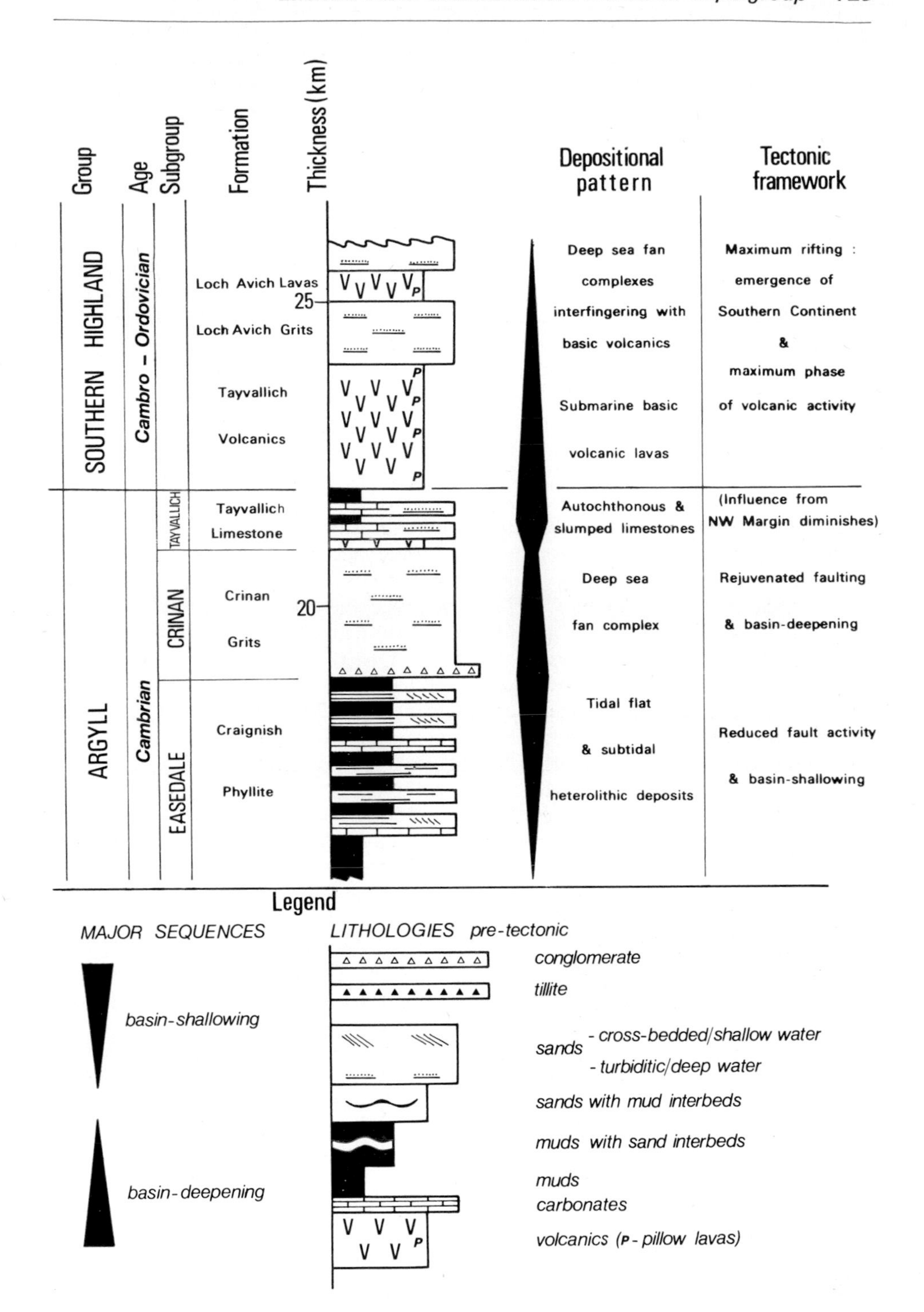


Fig. 3. Lithostratigraphical section through the Dalradian Supergroup (drawn to scale) emphasising the major depositional features and their inferred tectonic framework.

which pass into the cross bedded, rippled and graded pebbly quartzites of the Appin Quartzite Formation. Facies changes within the formation follow the pattern previously developed in the Lochaber Subgroup quartzites, thinning towards the northeast while maturity decreases and grain size increases towards the southwest. Again deltaic facies are suggested to develop in the southwest (Hickman 1975). A similar increase in sand content is recorded in the rippled Appin Phyllites which interdigitate with the open circulation, shallow marine carbonates of the Appin Limestone Formation.

Additional lateral facies changes within the Ballachulish Subgroup are recognized from other areas. Litherland (1970) recognized facies changes, non-sequences and, eastwards from the Appin district, a significant thinning of the Ballachulish Subgroup. However, stratigraphical work in Perthshire (Smith and Harris 1976) demonstrates the lateral persistence of Ballachulish Subgroup rocks and their northeastwards continuation through Blair Atholl towards Braemar. Thus, the restriction of Ballachulish Subgroup rocks to a sub-basin which was 'inherited' from the time of deposition of the Lochaber Subgroup (Harris and Pitcher 1975) is no longer necessary. Palaeogeographically therefore, the Dalradian trough need not have been significantly subdivided or constricted at this time.

The open marine oxidised sediments of the Appin Phyllite and Limestone formations contrast with the overlying black slate and dark limestones which form the Cuil Bay Slate and Lismore Limestone formations of the Appin region. On geochemical grounds Hickman (1975) considered these to be shallow marine or lagoonal with the slate facies possibly constituting pro-delta deposits marginal to the (?)deltaic quartzite facies of the Appin Quartzite and Phyllite formations.

Thus a basin-deepening to basin-shallowing couplet is recognized in the Appin region with the deepening phase commencing with the Grampian Group and terminating within the Ballachulish Slate Formation. Hence the upwards transition from open basinal facies through deltaic(?) to lagoonal facies records a significant basin shallowing phase which culminated in the deposition of the widespread and possibly locally emergent Vendian Tillites, the lower boundary of which may well constitute a sedimentological break as recognized in other North Atlantic regions (e.g. Føyn 1937; Reading and Walker 1966).

## 3c. Argyll Group

Sedimentation in the Argyll Group can be conveniently summarized in several basin-deepening and basin-shallowing sequences (Fig. 3). These large scale sequences were, however, punctuated by numerous smaller scale fluctuations in sea level comprising transgressions and regressions.

Deposition of the Port Askaig tillites (*c.* 750 m thick, Spencer 1971) includes several complex facies deposited in various glacial sub-environments (massive tillites, reworked marine tillites, outwash sands etc.) but essentially deposition occurred in continental and, or, marginal marine environments. The upward transition through Member 5 of the tillite succession (formerly the Lower Fine Grained Quartzite) and into the succeeding Bonahaven Dolomites represents continued shallow water sedimentation within tidally influenced shoreline environments (Klein 1970; Spencer and Spencer 1972) that were accumulating clastic and mixed clastic–carbonate sediments respectively. This type of transition is widely recorded in other post-Vendian tillite sequences (e.g. Reading and Walker 1966; Edwards 1972). In this instance inferred tidal flat-dominated sedimentation continued uninterrupted with the return of dolomites presumably recording a change mainly in climatic and, or, chemical conditions. The numerous small scale fining and coarsening upward sequences within these deposits record minor changes in shoreline configuration which, based on comparison with modern and other similar ancient environments, probably reflect relatively small scale progradational periods separated by minor transgressions.

One of the more fundamental changes in the Argyll Group sedimentation accompanied the influx and rapid accumulation of cross bedded sands and pebbly sands of the Jura Quartzite (>5 km thick). Anderton (1976) argues that these shallow water clastic sediments in the Islay–Jura region were mainly deposited in a tide-dominated subtidal shelf environment. The intercalation of beach-like sediments, however, implies periodic episodes of shoreline progradation followed by repeated, small scale transgressions. The sharp junction with the Bonahaven Dolomites reflects a major transgression when carbonate–clastic shore line sedimentation was replaced by more offshore, open marine conditions accompanied by the transport of generally coarser grained sediment. In the Islay–Jura–Scarba region there appears to have been a well defined sediment transport path which persisted throughout the time of deposition of the Jura Quartzite. The dominant palaeocurrent modes are directed towards the northeast, a trend which is paralleled by a decrease in mean grain size, increased proportions of fine grained (silt and mud) sediment and decrease in thickness (>5 km on Jura to 0·50 km at Ardmucknish Bay). In the context of a tidal shelf environment Anderton (1976) suggested that these dominant northeastward directed currents were flowing parallel to a shoreline, possibly within a partially enclosed, elongate sea. The major source area comprised a landmass situated in the northwest and possibly west while partial basin constriction, inferred as a requirement for generating the necessary powerful tidal currents, may have been provided by offshore ridges or swells (submarine?) to the southeast. Thus, the latter positive features are suggested as having separated the Jura Quartzite shelf sea from a deep basin to the southeast (Anderton 1976). The southwest to northwest facies changes need not necessarily represent an increase in water depth but could simply represent a decrease in competence of the tidal currents. However, the persistence of this trend, throughout deposition of the Jura Quartzite, and the repeated facies changes in overlying deposits, when compared with modern tidal seas (e.g. Stride 1963; Belderson and Stride 1966; Kenyon and Stride 1970) suggest that deeper water conditions probably did prevail in the northeast.

A major break in depositional conditions occurred at the top of the Jura Quartzite and the base of the Easdale Subgroup; a boundary that has been tentatively correlated with the base of the Cambrian (Harris and Pitcher 1975). This initially comprised a relatively widespread return to fine grained sedimentation with the deposition of the Jura Slates, particularly on Islay and south Jura. This was rapidly succeeded by the Scarba Conglomerate, mainly comprising mass flow and turbidite deposits which fine and interfinger northwards, through the Islay–Jura archipelago towards Kerrera (Baldwin and Johnson 1977), with the Easdale Slates. The cessation of tidal currents and the predominance of density-driven currents implies a sudden deepening and increase in sea bed gradient. Concomitant uplift is suggested by the relatively large (2–5 cm diameter) moderate to well rounded quartz and quartzite pebbles which characterize the relatively thin Scarba Conglomerate deposits on southwest Islay. The sands become finer grained through Jura and the northern islands where they are mainly coarse and very coarse grained sandstones. A marked change in facies coincides with a sedimentary rotational slump sheet in North Jura (Anderton 1977). Extending northwards from this inferred slump is a series of debris flows comprising slumped blocks of lithified, fine grained sediments (siltstones, mudstones and limestones) and sandstones with similar intraformational clasts and numerous structures widely recognized as indicative of turbidity currents (graded beds, convolute lamination, partial Bouma sequences, amalgamations and channelling). The best development of debris flows within this facies occur on Scarba and South Lunga. When traced northwards the size and frequency of intraformational clasts decreases, individual sandstone units are thinner and become more sharply defined against thicker intercalations of dark siltstones and mudstones. It is possible that the distal turbidite-like facies in the Easdale Slates on Kerrera are the basinward lateral equivalents of the Scarba Conglomerate (Baldwin and Johnson 1977).

Basinal sedimentation characterized the laterally equivalent and partially younger Easdale Slates; these relatively quiescent conditions being only rarely interrupted by thin, sporadic sand incursions from turbidity currents. These conditions were most widely developed in the Luing–Seil area and to the north, while shallower and more active conditions prevailed to the south in the partially laterally equivalent Port Ellen Phyllites on Islay (Anderton 1975). This period appears to coincide with the maximum development of a large scale basin deepening sequence which began at the base of the Argyll Group; a sequence which comprises some 6·5 km of sediments (Fig. 3).

The remainder of the Easdale Subgroup is replaced by a widespread regressive sequence. Basinal muds are replaced upwards by sandier heterolithic deposits in the Port Ellen Phyllites and the localized development of limestones and associated current reworked sandstones e.g. Degnish Limestones on Shuna and Degnish Point (Baldwin and Johnson 1977). This shallowing continues in the widespread Craignish Phyllites in which gypsum pseudomorphs indicate periodic emergence (Anderton 1975). These extremely variable deposits comprise alternations of variously bedded and laminated siltstones, sandstones and limestones, some of which were deposited in tidal flats and subtidal environments (Anderton 1975). Large scale cross bedding in the Shira limestone of the Loch Awe district also supports these conclusions (Borradaile 1973). Thus, the top of the Easdale Subgroup in the Islay–Jura to Loch Awe region is everywhere represented by very shallow water, probably marine-dominated conditions, which marks the top of the first major basin-shallowing sequence in the Argyll Group (Fig. 3).

It has been mentioned in section 2 that the outcrop pattern in Perthshire requires at least partial lateral equivalence of formations and subgroups. The clean, white, well sorted Schiehallion Quartzite that was probably deposited in shallow water, gives way upwards and laterally to the sands, silts and muds of the poorly sorted Killiecrankie Schist. Either or both formations may have fed fluxoturbidite deposits (Carn Mairg Quartzite) which periodically invaded euxinic areas in which carbonaceous mud (Ben Eagach Schist) was slowly accumulating and which were otherwise largely starved of coarse clastic sediments, a situation closely comparable to that of equivalent deposits in the Islay–Jura region. Locally, the rocks otherwise characterised by the mixed lithologies of the Killiecrankie Schist contain mappable bands of dark graphitic mica schist or pebbly graded quartzites. On the western flanks of Ben Vrackie the Ben Eagach Schist facies is locally absent and the succession between the Killiecrankie and Ben Lawers Schists is occupied by lenticular outcrops of pebbly quartzite interpreted as individual channel deposits now viewed in cross section. The local absence of the Carn Mairg

Quartzite facies on the northern slopes of Sron Mhor was interpreted by Sturt (1961) as a tectonic slide, but the low state of strain of these rocks suggests that the Carn Mairg Quartzite was locally not deposited. Even in Glen Lyon where the Geological Survey (1905) mapped thick Carn Mairg Quartzite some of the outcrops marked merely represent a few pebbly quartzite beds within a succession that is essentially of black graphitic mica schist. Returning to the Islay–Tayvallich–Loch Awe region, the Easdale Subgroup is sharply succeeded by the Crinan Subgroup the base of which coincides with a deepening and concomitant deposition of thick bedded, proximal turbidites of the Crinan Grits. Laterally, into the Highland Border region, this facies is replaced by thinner bedded, finer grained and considerably more strongly deformed deposits comprising the Ben Lui Schists and while transport directions are not available from these deposits, this facies change is interpreted as a basinward transition from relatively proximal submarine fan-type deposits in the west (Crinan Grits) to more distal deposits in the east (Ben Lui Schists). However, investigations on the role of pressure solution during the tectonic deformation of coarse clastic sediments (e.g. Harris *et al.* 1976) suggests that large proportions (20% or more) by volume are removed from such rocks during the early stages of orogenesis; micas and feldspars preferentially survive as insoluble residue. Thus, although *prima facie* spatially related increasing mica in schists of clastic origin may indicate increasing distality in the original sediments (e.g. Crinan Grits →Ben Lui Schist), this increase may be the result of increasing deformation-induced pressure solution. It is likely that some of the Ben Lui and Pitlochry Schists of the Pitlochry district have been passively enriched in mica in this way, and were originally coarser clastic rocks.

The base of the Tayvallich Subgroup represents a sudden change from deep sea clastic sediments to limestone sedimentation and contemporaneous basic volcanicity (Gower 1973). Autochthonous limestones (e.g. Inishowen) accumulated in relatively shallow shelf environments. However, in many Scottish localities the limestones were redeposited from turbidity currents and occasionally alternated with periods of volcanic extrusion. There is a complex interfingering of these facies with volcanic deposits (Gower 1977) comprising pillow lavas, hyaloclastites and airfall tuffs. Although this interval is of variable thickness (up to *c.* 1·2 km) the amount of limestone is relatively constant at around 100 m thick. Thus, the main variation in thickness is determined by the proportion of black mudstone and volcanic material. A shallower environment than the underlying deposits is envisaged by virtue of the autochthonous limestones. Since they are widespread it may be concluded that there was shallowing at the base of the Tayvallich Subgroup. However, turbidite sedimentation continued to be an important mechanism of redepositing shelf (?)limestones in deeper water.

In summary, the Argyll Group in Scotland broadly comprises four repeated, large scale basin deepening and shallowing sequences (Fig. 3) during the deposition of which there was a widespread change from high energy, shallow water conditions at the base to deeper water, tectonically less stable conditions towards the top. Part of this group is probably synchronous with Lower Cambrian shelf deposits on the northwest foreland (Harris and Pitcher 1975). In terms of facies the Lower Cambrian Quartzites were deposited in a similar tide-dominated, shelf environment to that of the Jura Quartzite, but shelf conditions prevailed throughout Cambrian times in northwest Scotland. The major basal Cambrian transgression resulting in shelf sedimentation on the foreland may correlate with one of the two major upward deepening sequences (discussed later).

Palaeogeographic reconstructions must remain tentative without sophisticated time control. However, several repeated facies patterns in the Argyll Group are consistent with a major landmass to the northwest and flanked to the southeast by a broad, shallow shelf during the time of deposition of the Islay Subgroup. A deeper basin lay still further to the southeast.

Continued basin deepening through the remainder of the Argyll Group and into the Southern Highland Group saw the replacement of relatively stable shelf environments by more unstable deeper water environments.

## 3d. Southern Highland Group

This group remains undivided into subgroups and in terms of thickness and sedimentological significance is certainly comparable with an individual subgroup. As yet detailed sedimentological and lithofacies interpretations are unavailable but are in progress (C.T.B. and A.L.H).

Detailed analysis of the sandstone–slate alternations in the Dunkeld area does perhaps provide the basic outlines of a facies model. Both the Birnam and Dunkeld Grits are comprised of thick sequences (>2000 m) of coarse turbidites (cf. Harris 1972). Within these members both lenticular and tabular fining up-thinning up units reaching to 15 m thick are recognized. The 'tabular' sequences commence with 4–5 m thick pockets of amalgamated, pebbly, immature sandstones. Individual beds have subdued lenticular geometries due to the presence of small scale scours, loading structures and distinct channel forms only a few metres wide. These units grade up into rippled, thinly bedded clean sandstone and mudstone alternations that have a high degree of lateral continuity. The lenticular units are composed of amalgamated coarse and subordinate pebbly amalgamated graded sandstones which may either pass laterally into flat, parallel bedded sandstones, or pinch out to be replaced by thin bedded and slate facies. These lenticular units vary in width from 10 m to over 30 m and in thickness from about 3 m to 10 m. In numerous cases the erosional contacts of these coarse grained lenticular sands with the adjacent fine grained facies exhibit a spectrum of soft sediment deformation structures ranging from gently undulating through loaded to characteristically slumped relationships.

The fine grained facies (slates) vary from pure mudstones to laterally continuous two or three metres thick units of interbedded mudstones and ripple cross and parallel laminated coarse siltstones.

The structures, small scale sequences and particularly local sandstone geometries are comparable with lower slope and inner zones of modern deep water submarine fans (cf. Walker and Mutti 1973; Kruit *et al.* 1975). Lithological variations on both small scales (cm to few m) and medium scale (up to 15–20 m thick) may relate to individual channel filling, overtopping and lateral migration phases of sedimentation and largely represent spatial changes in sub-facies type (cf. Mutti 1977). At a still larger scale concentrations of numerous fining-up units form the composite grit members and their alternation with sandstone deficient slates possibly represents phases of large and small scale fan progradation interspersed with phases of fan inactivity or abandonment. Thus, fan sandstones interdigitate with either overbank and, or, outer fan facies of the same fan or from adjacent fans. Such a facies interpretation supports the contention of Harris (*op. cit.*) that the slate–grit contacts may well be diachronous.

At a more general level the basin fill sequences of the Southern Highland Group suggest a striking persistence of facies at any one place. Superficially 'distal' slate facies increase in relative importance towards the southwest but the thick slate sequences of Banffshire which are correlated with the sandstone–slate development of Perthshire and Argyll (Harris and Pitcher 1975) may be allochthonous (Sturt *et al.* 1977). However, a pattern of increased variability and perhaps bathymetry may be implied from both the northeastward decrease in carbonate intercalations and the relative increase in slates in the same direction.

Superimposed on this deep sea fan-type sedimentation pattern are contemporaneous basic volcanic extrusions, several of which display pillow lavas (e.g. Borradaile 1973). This appears to represent a progressive increase in volcanic activity, developed initially in the upper Argyll Group, which was concomitant with an increase in the tectonic instability of the basin. Perhaps the most striking manifestation of this increasing tectonic instability is the change in sandstone mineralogy; similar turbidites in the Crinan Grits are relatively quartz rich (>80%) with rather low feldspar proportions (Borradaile 1973) whereas Southern Highland Group sandstones in Scotland and their lateral equivalents in Ireland contain abundant fresh, angular fragments of potassium feldspar and metamorphic and granitic rock fragments (e.g. Borradaile 1973; Harris 1972). Indeed in the Inishowen region of Ireland, the Fahan Grits include arkoses with more than 25% of large, fresh feldspar clasts (Roberts 1973). Such a dramatic change in mineralogy suggests derivation from a different source area.

## 4. Discussion and summary of basin development

Dalradian sedimentation spanned a period of approximately 300 m.y. extending from the late Precambrian or Upper Riphean (Grampian and Appin Groups), through Vendian tillites (Islay Subgroup) and Lower Cambrian deposits (Tayvallich Subgroup; Downie *et al.* 1971; Downie 1975), and terminated in late Cambrian and, or, early Ordovician times (*c.* 500 m.y.). During this period the basin was filled by a sequence dominated by clastic sediments with subordinate carbonate and basic volcanic intercalations, which after deformation comprises a stratigraphical succession with a maximum thickness of about 25 km. This sequence accumulated at a rate of approximately 80 m/million years, a rate which undoubtedly necessitated large and rapid vertical movements. Since the Dalradian Supergroup is presumed to rest on continental crust comprising an initially relatively stable metamorphic basement, probably consisting of Grenville and Lewisian rocks, it seems likely that subsidence was controlled by basement faults. Although an immense stratigraphical thickness is involved, it should be stressed that this varies considerably from place to place (Pitcher and Shackleton 1966; Cambray 1969; Spencer 1971; Borradaile 1973; Gower 1973; Hickman 1975; Harris and Pitcher 1975, Anderton 1975, 1976) thereby supporting the likelihood of separate fault-bounded troughs and highs within the Dalradian basin.

Although the succession spans the late Precambrian–Cambrian boundary, the base of the Cambrian is presently only imprecisely defined, occurring between the Port Askaig Tillite and the lower Cambrian Tayvallich Limestone (Downie *et al.* 1971; Harris and Pitcher 1975). Since many North Atlantic regions, including northwest Scotland, record a major transgression at the base of the Cambrian (cf. Swett and Smit 1972) it seems likely that one of the two basin-deepening sequences in the Argyll Group may be the product of the same event: (1) Port Askaig Tillite→Jura Quartzite→Scarba Conglomerate–Easdale Slate, or (2)

Craignish Phyllites→Crinan Grits. The first sequence could record the diachronous migration of tidal shelf sand facies from the Dalradian trough (Jura Quartzite, Anderton 1976) onto the northwest Foreland shelf in the Lower Cambrian (Eriboll Quartzite, Swett *et al.* 1971), followed by continued deepening in the Dalradian trough. The second sequence, however, marks a more major facies change with the final disappearance of widespread shallow marine conditions which were replaced by deep water clastic and contemporaneous basic volcanic deposition. At present there is insufficient evidence to permit a positive preference for either of these two possibilities.

Throughout the time of deposition of the Dalradian the basin was flanked to the northwest by a source area supplying relatively mature sediment. This source was most active during deposition of the Grampian, Appin and lower Argyll groups (Hickman 1975; Chenevix-Trench 1975; Anderton 1976) and remained the dominant source area for Lower Cambrian quartzites (Swett and Smitt 1972). A local southerly source area has been invoked by Anderton (1976) during Jura Quartzite times but the emergence of a possible southern continental area is not apparent until during deposition of the Southern Highland Group. This is suggested by the influx of abundant, fresh metamorphic, granitic and large potassium feldspar fragments, within proximal-like turbidite beds of inferred deep sea fan origin which persist along the whole length of the Dalradian outcrop (cf. Roberts 1973; Borradaile 1973; Harris 1972). In view of the extensive shelf area accumulating mature Cambro–Ordovician sediments it seems unlikely that this relatively immature sediment could have been supplied from the northwestern source area. Further support for a southern continent is provided from deep seismic investigation (Bamford *et al.* 1977), and is consistent with the plate tectonic models of both Phillips (1976) and Wright (1976). The persistence of old continental crust at a high level to the south of the present Dalradian outcrop in Scotland is more consistent with the proposals of Phillips *et al* (1976). Foundering of blocks within a marginal basin may have produced the sedimentation patterns in the Dalradian Supergroup, but if the marginal basin was associated with a southern island arc and northerly subduction of oceanic crust then it is difficult to account for the absence of calc-alkaline volcanic rocks within the Dalradian Supergroup. It is more in keeping with the available evidence to suggest that the Dalradian deposition was taking place in a subsiding ensialic basin separated from the major southern ocean, if such existed at this time, by an area of continental crust which periodically at least served as a source of metamorphic and granitoid debris. If a major ocean to the south of the Dalradian basin were still expanding in the late Precambrian–Cambrian, northerly directed subduction of oceanic crust need not have been initiated during Dalradian deposition. The foundering of ensialic blocks and sporadic contemporaneous basic volcanism in the Dalradian basin may have been the product of local instabilities within the continental crust marginal to an expanding ocean, a situation perhaps comparable to the Mesozoic and Tertiary history of the North Sea during the opening of the North Atlantic Ocean.

Because of the quiescent conditions necessary for the accumulation of the Cambro-Ordovician carbonate succession, Grampian underthrusting from the northwest (cf. Phillips *et al.* 1976) is unlikely to have contributed to the Upper Dalradian syndepositional instability alluded to above. Nevertheless, Grampian crustal shortening (and hence thickening) possibly by underthrusting, as envisaged by Phillips *et al.* (1976), may have produced buoyancy to the northwest, and hence generated a gradient for the transport of the Tay (gravity) Nappe. 'Collision' in

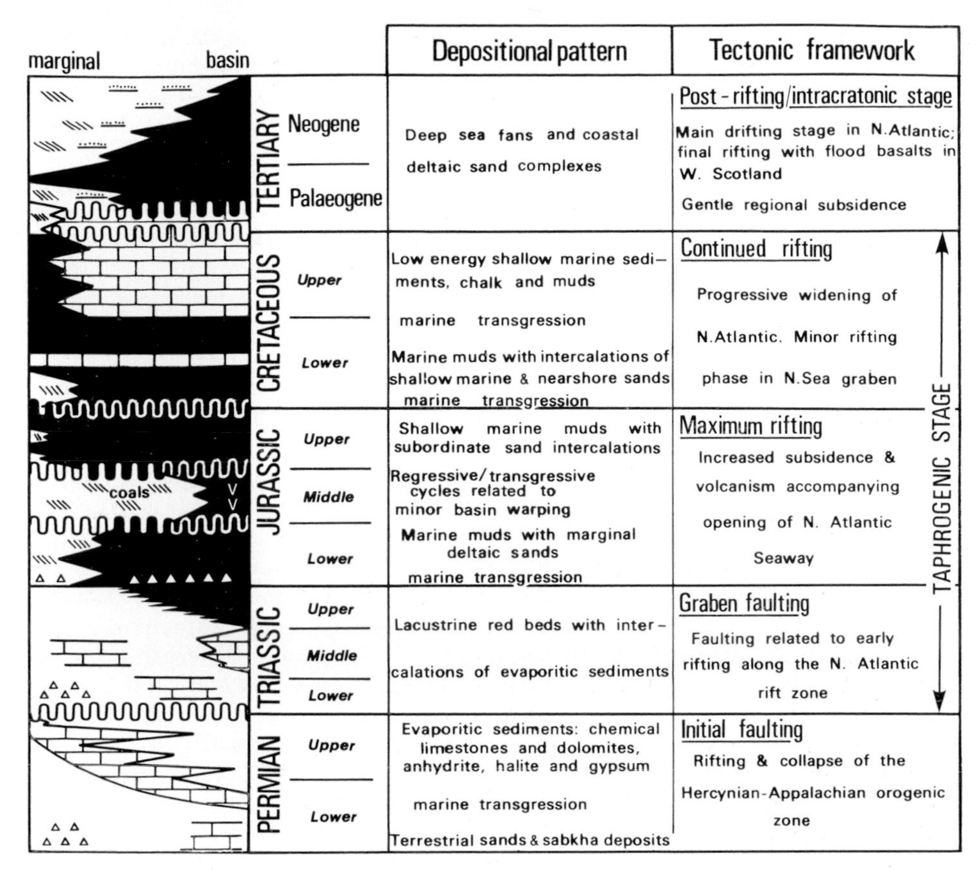


Fig. 4. Depositional pattern and tectonic framework of the Mesozoic and Tertiary deposits of the North Sea (section redrawn after Ronnevik *et al.* 1975). Depositional pattern based on Pegrum *et al.* 1975; tectonic framework based on Ziegler, P. A. 1975 and Ziegler, W. H. 1975. Legend as Fig. 3; crenulated line refers to major tectonic breaks.

the sense envisaged by Wright (1976) is probably unnecessary to explain Grampian orogenesis of the Dalradian rocks; regional crustal shortening of Grampian age may be accomplished along zones of high strain such as those cropping out in the North Highlands, viz. the Hebridean thrust, the Moine thrust, the Sgurr Beag slide (Tanner *et al.* 1970) and analogous structures, and possibly the Great Glen fault.

If a southern ocean was expanding contemporaneously with Dalradian deposition, its tectonic development should be recorded in stratigraphical sequences from any adjacent marginal basins. Thus the various sedimentary sequences documented earlier may reflect not only the changing tectonic framework of the Dalradian basin but also that of an adjacent expanding ocean. The initial phases of basin development comprised superimposed large scale basin-shallowing and basin-deeping sequences (Fig. 3), several hundred metres thick, which during Grampian, Appin and lower Argyll Group times consisted mainly of shallow marine clastic sediments (well sorted sands and carbonaceous muds) with subordinate carbonate deposits. These sequences were superimposed on smaller scale transgressive and regressive periods, and are interpreted as reflecting oscillations in the relative rates of sediment accumulation and subsidence governed by tectonic processes. The basin-deepening transition from the Easdale to Crinan subgroups documents the beginning of more tectonically active conditions and the initiation of widespread deep sea fan-like turbidite deposition (Crinan Grits). Increased subsidence was presumably linked with extensional stresses (?rifting) accompanied by contemporaneous basic volcanism represented as extrusive lavas and high level intrusions intercalated with other proximal turbidite sands (Southern Highland Group) and autochthonous and resedimented carbonates (Tayvallich Subgroup). Following deposition of the Tayvallich Subgroup the increase in volcanic activity was accompanied by a major change in sandstone mineralogy and textural maturity, related to the emergence of a southern source area. On the basis of all this evidence, the development of the Dalradian basin is interpreted to have occurred during a progressive increase in tectonic activity and the rock assemblage to represent a major basin-deepening sequence (Fig. 3).

Comparison can be made with a similar basin-deepening sequence which encompasses the Mesozoic and Tertiary deposits of the North Sea (Fig. 4). This sequence records a complex basin-fill which occurred during a period of increasing tectonic activity in response to the opening of the North Atlantic Ocean and the Alpine orogeny (Ziegler, P. A. 1975; Ziegler, W. H. 1975). Although the subsequent thermal and tectonic activity did not reach the intensity which characterized and is recorded in the development of the Dalradian Supergroup, the basin-fill from continental deposits (Permian and Triassic) through shallow water clastic rocks and carbonates with contemporaneous basic volcanic rocks (Jurassic and Cretaceous) into deeper water clastic sediments (Tertiary) is superficially similar. The time span of the North Sea basin of *c.* 270 m.y. is also comparable to that of the Dalradian basin (*c.* 300 m.y.) but in the former case an upper stratigraphical thickness of approximately 10 km gives a much lower sedimentation rate of 27 (cf. 80) m/million years. This is more comparable to sedimentation rates experienced in more stable marginal basins during late Precambrian to Cambro-Ordovician times: in North Norway, for example, a continuous stratigraphical sequence extends from the Upper Riphean (base *c.* 810 m.y.) to Tremadocian (*c.* 500 m.y.) accumulated at a rate of 17 m/million years (Johnson *et al.* 1978). However, the broad comparison of the basin–fill sequences in the Dalradian Supergroup and the Mesozoic and Tertiary assemblages of the North Sea supports the suggestion that the former probably accumulated in a continental marginal basin, possibly during extensional tectonics.

It is concluded that there is no balance of evidence which unequivocally points to any one of the many plate tectonic models proposed or to the particular disposition of late Precambrian–Cambrian continental and oceanic crust in the British region. Indeed, it has recently been pointed out (Kent 1977) that there was dramatic fault control, presumably through the differential movement of basement segments, of Mesozoic sedimentation, on a world wide basis. This largely took place *before* the inception of the major modern oceans. Thus, if the analogy of the depositional environment and history of the Mesozoic of the North Sea to that of the Dalradian Supergroup is rigorously applied, one may be forced to conclude that the Iapetus Ocean did not come into existence until fault-controlled deposition of the Dalradian was already complete. The similarity of the sedimentation pattern in the Dalradian Supergroup with late Precambrian–Cambrian sedimentation patterns elsewhere in the North Atlantic region (Banks *et al.* 1971; Caby 1972; Harland

and Gayer 1972; Swett and Smit 1972; Siedlecka 1975; Bjørlykke *et al.* 1976; Nystuen 1976; Johnson *et al.* 1978) may bear out the analogy with the Mesozoic, and lead to the conclusion that sedimentation of this age largely preceded the development of the lower Palaeozoic oceans.

*Acknowledgments.* H.D.J. is grateful for a post-doctoral research fellowship in the School of Environmental Sciences at the University of East Anglia. One part of this contribution (H.J.B. and R.A.S.) derives from a contract between the Department of Geology, Liverpool University and I.G.S. for mapping Sheet 55E (Pitlochry). We thank Ruth Johnson for kindly typing the manuscript.

## References

ANDERTON, R. 1971. Dalradian palaeocurrents from the Jura Quartzite. *Scott. J. Geol.* **7,** 175–178.

—— 1974. *Middle Dalradian Sedimentation in Argyll with particular reference to the Jura Quartzite, Scarba Conglomerate and Craignish Phyllites.* Univ. of Reading Ph.D. thesis.

—— 1975. Tidal flat and shallow marine sediments from the Craignish Phyllites, Middle Dalradian, Argyll, Scotland. *Geol. Mag.* **112,** 337–340.

—— 1976. Tidal-shelf sedimentation: an example from the Scottish Dalradian. *Sedimentology* **23,** 429–458.

—— 1977. A field guide to the Dalradian rocks of Jura, Argyll. *Scott. J. Geol.* **13,** 135–142.

ARMSTRONG, M. and PATERSON, I. B. 1970. The Lower Old Red Sandstone of the Strathmore region. *Rep. Inst. geol. Sci.* No. 70/12, 23 pp.

BAILEY, E. B. 1934. West Highland Tectonics: Loch Leven to Glen Roy. *Q. J. geol. Soc. Lond.* **90,** 462–525.

BALDWIN, C. T. and JOHNSON, H. D. 1977. The Dalradian rocks of Lunga, Luing and Shuna. *Scott. J. Geol.* **13,** 143–154.

BAMFORD, D., NUNN, K., PRODEHL, C. and JACOB, B. 1977. LISPB—III. Upper crustal structure of northern Britain. *J. geol. Soc. Lond.* **133,** 481–488.

BANKS, N. L., EDWARDS, M. B., GEDDES, W. P., HOBDAY, D. K. and READING, H. G. 1971. Late Precambrian and Cambro–Ordovician Sedimentation in East Finnmark. *Norg. geol. Unders.* No. 269, 197–236.

BELDERSON, R. H. and STRIDE, A. H. 1966. Tidal current fashioning of a basal bed. *Mar. Geol.* **4,** 237–257.

BJØRLYKKE, K., ELVSBORG, A. and HØY, T. 1976. Late Precambrian sedimentation in the central sparagmite basin of South Norway. *Norsk. geol. Tidsskr.* **56,** 233–290.

BORRADAILE, G. J. 1973. Dalradian structure and stratigraphy of the northern Loch Awe district, Argyllshire. *Trans. R. Soc. Edinb.* **69,** 1–21.

BOWES, D. R. 1968. The absolute time scale and the subdivision of Precambrian rocks in Scotland. *Geol. Fören. Stockholm Förh.* **90,** 175–188.
BROOK, M., BREWER, M. S. and POWELL, D. 1976. Grenville age for rocks in the Moine of north-western Scotland. *Nature, Lond.* **260,** 515–517.
——, POWELL, D. and BREWER, M. S. 1977. Grenville events in rocks of the Northern Highlands, Scotland. *J. geol. Soc. Lond.* **133,** 489–496.
CABY, R. 1972. Preliminary results of mapping in the Caledonian rocks of Canning Land and Wegener Halvø, East Greenland. *Grønlands geol. Unders.* **48,** 21–38.
CAMBRAY, F. W. 1969. The Kilmacrenan succession east of Glenties, Co. Donegal. *Proc. R. Ir. Acad.* **47B,** 291–302.
CHENEVIX-TRENCH, J. R. 1975. *The Lower Dalradian rocks of the Fort William and Creeslough areas.* Univ. of Newcastle-upon-Tyne Ph.D. thesis.
DEWEY, J. F. 1969. Evolution of the Appalachian–Caledonian orogen. *Nature, Lond.* **222,** 124–129.
DOWNIE, C. 1975. Precambrian of the British Isles: Palaeontology. *In* Harris, A. L. *et al.* (Editors). *A correlation of Precambrian rocks in the British Isles,* 113–115. *Spec. Rep. geol. Soc. Lond.* No. 6.
——, LISTER, T. R., HARRIS, A. L. and FETTES, D. J. 1971. A palynological investigation of the Dalradian rocks of Scotland. *Rep. Inst. geol. Sci.* No. 71/9, 30 pp.
EDWARDS, M. B. 1972. *Glacial, interglacial and postglacial sedimentation in a Late Precambrian shelf environment, Finnmark, North Norway.* Univ. of Oxford D.Phil. thesis.
FØYN, S. 1937. The Eocambrian Series of the Tana district, Northern Norway. *Norsk. geol. Tidsskr.* **2,** 65–164.
GEORGE, T. N. 1960. The stratigraphical evolution of the Midland Valley of Scotland. *Trans. geol. Soc. Glasg.* **26,** 32–107.
GOWER, P. J. 1973. *The middle-upper Dalradian boundary with special reference to the Loch Tay Limestone.* Univ. of Liverpool Ph.D. thesis.
—— 1977. Dalradian rocks of the west coast of the Tayvallich Peninsula. *Scott. J. Geol.* **13,** 125–134.
GRAHAM, C. M. 1976. Petrochemistry and tectonic significance of Dalradian metabasaltic rocks of the SW Scottish Highlands. *J. geol. Soc. Lond.* **132,** 61–84.
HARLAND, W. B. and GAYER, R. A. 1972. The Arctic Caledonides and earlier Oceans. *Geol. Mag.* **109,** 289–314.
HARRIS, A. L. 1972. The Dalradian rocks at Dunkeld, Perthshire. *Bull. geol. Surv. Gt Br.* **38,** 1–10.
——, BRADBURY, H. J. and MCGONIGAL, M. H. 1976. The evolution and transport of the Tay Nappe. *Scott. J. Geol.* **12,** 103–113.
—— and PITCHER, W. S. 1975. The Dalradian Supergroup. *In* Harris, A. L. *et al.* (Editors). *A correlation of Precambrian rocks in the British Isles,* 52–75, *Spec. Rep. geol. Soc. Lond.* No. 6.
HICKMAN, A. H. 1975. The stratigraphy of late Precambrian metasediments between Glen Roy and Lismore. *Scott. J. Geol.* **11,** 117–142.
JOHNSON, H. D., LEVELL, B. K. and SIEDLECKI, S. 1978. Lithostratigraphy and structure of some late Precambrian sedimentary rocks in East Finnmark, N. Norway and their relation to the Trollfjord-Komagelv fault on Varanger Peninsula. *J. geol. Soc. Lond.*
JOHNSON, M. R. W. and HARRIS, A. L. 1967. Dalradian–?Arenig relations in parts of the Highland Border, Scotland and their significance in the chronology of the Caledonian Orogeny. *Scott. J. Geol.* **3,** 1–16.
JOHNSTONE, G. S., SMITH, D. I. and HARRIS, A. L. 1969. Moinian Assemblage of Scotland. *In* Kaye, M. (Editor). *North Atlantic—geology and continental drift,* 159–180. *Mem. Am. Ass. Petrol. Geol.* **12.**
KENT, P. E. 1977. The Mesozoic development of aseismic continental margins. *J. geol. Soc. Lond.* **134,** 1–18.
KENYON, N. H. and STRIDE, A. H. 1970. The tide-swept continental shelf sediments between the Shetland Isles and France. *Sedimentology* **14,** 159–173.
KLEIN, G. de V. 1970. Tidal origin of a Precambrian quartzite—the Lower Fine-Grained Quartzite (Middle Dalradian) of Islay, Scotland. *J. sed. Petrol.* **40,** 973–985.
KRUIT, C., BROUWER, J., KNOX, G., SCHOLLNBERGER, W. and VLIET, A. 1975. Une excursion aux cones d'allurions en eau profonde d'age Tertiaire près de San Sebastian (Province de Guipuzcoa, Espagne). *Guide pour l'excursion A–23, IX^e Congrès International de Sedimentologie, Nice (France)* 77 pp.
LAMBERT, R. ST J. 1969. Isotope studies relating to the Pre-Cambrian history of the Moinian of Scotland. *Proc. geol. Soc. Lond.* **1652,** 243–245.
—— and MCKERROW, W. S. 1977. The Grampian Orogeny. *Scott. J. Geol.* **12,** 271–292.
LITHERLAND, M. 1970. *The stratigraphy and structure of the Dalradian rocks around Loch Creran, Argyll.* Univ. of Liverpool Ph. D. Thesis.
LONG, C. B. and MAX, M. D. 1977. Metamorphic rocks in the SW Ox Mountains Inlier, Ireland; their structural compartmentation and place in the Caledonian orogen. *J. geol. Soc. Lond.* **133,** 413–432.

MUTTI, E. 1977. Distinctive thin-bedded turbidite facies and related depositional environments in the Eocene Hecho Group (south-central Pyrenees, Spain). *Sedimentology* **24,** 107–131.

NYSTUEN, J. P. 1976. Facies and sedimentation of the late Precambrian Moelv Tillite in the eastern part of the Sparagmite Region, southern Norway. *Norg. geol. Unders.* No. 329, 1–70.

PANKHURST, R. J., ANDREWS, J. R., PHILLIPS, W. E. A., SANDERS, I. S. and TAYLOR, W. E. G. 1976. The age and structural setting of the Slieve Gamph Igneous Complex, Co. Mayo, Eire. *J. geol. Soc. Lond.* **132,** 327–334.

PEGRUM, R. M., REES, G. and NAYLOR, D. 1975. *Geology of the North-West European Continental Shelf.* Graham Trotman Dudley Ltd., London. 225 pp.

PHILLIPS, W. E. A., TAYLOR, W. E. G. and SANDERS, I. S. 1975. An analysis of the geological history of the Ox Mountains inlier. *Sci. Proc. R. Dubl. Soc.* **5A,** 311–329.

——, STILLMAN, C. J. and MURPHY, T. 1976. A Caledonian plate tectonic model. *J. geol. Soc. Lond.* **132,** 579–609.

PITCHER, W. S. and SHACKLETON, R. M. 1966. On the correlation of certain Lower Dalradian successions in north-west Donegal. *Geol. J.* **5,** 149–156.

RAST, N. and LITHERLAND, M. 1970. The correlation of the Ballachulish and Perthshire (Iltay) Dalradian successions. *Geol. Mag.* **107,** 259–272.

READING, H. G. and WALKER, R. G. 1966. Sedimentation of Eocambrian tillites and associated sediments in Finnmark, northern Norway. *Palaeogeogr. Palaeoclimatol. Palaeoecol.* **2,** 177–212.

ROBERTS, J. C. 1973. The stratigraphy of the Middle and Upper Dalradian succession between Glengad Head and Greencastle, Inishowen, Co. Donegal. *Proc. R. Ir. Acad.* **73B,** 151–164.

RONNEVIK, H., BERGSAGER, E. I., MOE, A., ØVREBØ, O., NAVRESTAD, T. and STANGENES, J. 1975. The Geology of the Norwegian continental shelf. *In* Woodland A. W. (Editor). *Petroleum and the continental shelf of north-western Europe,* **1** *Geology*, 117–128, Applied Science Publishers, Barking, Essex, England.

SHACKLETON, R. M. 1958. Downward-facing structures of the Highland Border. *Q. J. geol Soc. Lond.* **113,** 361–392.

SIEDLECKA, A. 1975. Late Precambrian stratigraphy and structure of the north-eastern margin of the Fennoscandian Shield (East Finnmark–Timan Region). *Norg. geol. Unders.* No. 316, 313–348.

SMITH, A. G. 1976. Plate tectonics and orogeny: a review. *In* Briden, J. C. (Editor). Ancient Plate Margins. *Tectonophysics* **33,** 215–285.

SMITH, R. A. and HARRIS, A. L. 1976. The Ballachulish rocks of the Blair Atholl district. *Scott. J. Geol.* **12,** 153–157.

SOPER, N. J. and WILKINSON, P. 1975. The Moine Thrust and Moine Nappe at Loch Eriboll, Sutherland. *Scott. J. Geol.* **11,** 339–359.

SPENCER, A. M. 1971. Late Precambrian glaciation in Scotland. *Mem. geol. Soc. Lond.* **6,** 100 pp.

—— and SPENCER, M. O. 1972. The Late Precambrian–Lower Cambrian Bonahaven Dolomite of Islay and its stromatolites. *Scott. J. Geol.* **8,** 269–282.

STRIDE, A. H. 1963. Current swept floors of the southern half of Great Britain. *Q. J. Geol. Soc. Lond.* **119,** 175–199.

STURT, B. A., 1961. The geological structure of the area south of Loch Tummel. *Q. J. geol Soc. Lond.* **117,** 131–156.

——, RAMSAY, D. M., PRINGLE, I. R. and TEGGIN, D. E. 1977. Precambrian gneisses in the Dalradian sequence of NE Scotland. *J. geol. Soc. Lond.* **134,** 41–44.

SWETT, K., KLEIN, G. de V. and SMIT, D. E. 1971. A Cambrian tidal sand body—the Eriboll Sandstone of northwest Scotland: an Ancient–recent analogue. *J. Geol.* **79,** 400–415.

—— and SMITT, D. E. 1972. Palaeogeography and Depositional Environments of the Cambro–Ordovician shallow-marine Facies of the North Atlantic. *Geol. Soc. Am. Bull.* **83,** 3223–3248.

TANNER, P. W. G., JOHNSTONE, G. S., SMITH, D. I. and HARRIS, A. L. 1970. Moinian stratigraphy and the problem of the central Ross-shire inliers. *Geol. Soc. Am. Bull.* **81,** 299–306.

THOMAS, P. R. 1965. *The structure and metamorphism of the Moinian rocks in the Glen Garry, Glen Tilt and adjacent districts of Scotland.* Univ. of Liverpool Ph.D. thesis.

TREAGUS, J. E. 1974. A structural cross-section of the Moine and Dalradian rocks of the Kinlochleven area, Scotland. *J. geol. Soc. Lond.* **130,** 525–544.

UPTON, B. J. G., ASPEN, P., GRAHAM, A. and CHAPMAN, N. A. 1976. Pre-Palaeozoic basement of the Scottish Midland Valley. *Nature, Lond.* **260,** 517–518.

VAN BREEMEN, O., PIDGEON, R. T. and JOHNSON, M. R. W. 1974. Precambrian and Palaeozoic pegmatites in the Moines of northern Scotland. *J. geol. Soc. Lond.* **130,** 493–507.

——, HALLIDAY, A. N., JOHNSON, M. R. W. and BOWES, D. R. 1978. Crustal additions in late Precambrian times. *In* Bowes, D. R. and Leake, B. E. (Editors). *Crustal evolution in northwestern Britain and adjacent areas,* 81–106. *Geol. J. Spec. Issue* No. 10.

WALKER, R. G. and MUTTI, E. 1973. Turbidite facies and facies associations. *Turbidites and deep-water sedimentation. Soc. Econ. Paleont. Miner.*, Short Course Notes, 119–157.

WRIGHT, A. E. 1976. Alternating subduction direction and the evolution of the Atlantic Caledonides. *Nature, Lond.* **264,** 156–160.

ZIEGLER, P. A. 1975. North Sea basin history in the tectonic framework of North Western Europe. *In* Woodland, A. W. (Editor). *Petroleum and the continental shelf of north-western Europe* **1**, *Geology*, 131–148. Applied Science Publishers, Barking Essex, England.

ZIEGLER, W. H. 1975. Outline of the geological history of the North Sea. *In* Woodland, A. W. (Editor). *Petroleum and the continental shelf of north-western Europe*. **1**, *Geology*, 165–187. Applied Science Publishers, Barking, Essex, England.

Anthony L. Harris,
Jane Herdman Laboratories of Geology,
University of Liverpool,
Liverpool L69 3BX, England.

Christopher T. Baldwin and Howard D. Johnson,
School of Environmental Sciences,
University of East Anglia,
Norwich NR4 7TJ, England.

Harry J. Bradbury,
Department of Geology,
Cambridgeshire College of Arts and Technology,
Collier Road,
Cambridge CB1 2AJ, England.

Richard A. Smith,
Geological Survey of N. Ireland,
20 College Gardens,
Belfast BT9 6BS, Northern Ireland.

# Faunal provinces and the Proto-Atlantic

*N. Spjeldnaes*

Our knowledge of the distribution of early Palaeozoic faunas in the Atlantic area is rapidly increasing. The faunal provinces and their changing distribution become progressively clearer, and are of special value for interpreting the plate movements in the Atlantic. The later (Jurassic) opening of the Atlantic did not follow the original Proto-Atlantic (Iapetus) seam, but left fragments on the opposite plates. By nature of the material, the biogeographical data are not as precise as others, but they put constraints on models based on geophysical and structural evidence, and are important for dating the sequence of surface events. An active feedback between the biological and physico-chemical studies is needed in order to get a realistic model for the closing of the Proto-Atlantic (Iapetus) Ocean. The biogeographical data indicate that: (1) the closing of the Proto-Atlantic was completed at the end of the Ordovician although the plates may not have been in actual contact, but the intervening ocean had ceased to act as a barrier to the dispersal of marine benthonic faunas; (2) the European plate moved from higher to lower latitudes towards the comparatively stable North American plate; (3) the South American plate moved from lower to higher latitudes, towards Africa, closing the southern Proto-Atlantic by the end of the Ordovician; (4) there is no strong evidence for a major split between north and south Europe, or between Europe and Africa during these plate movements. The pattern of faunal provinces in this region may be better explained by climatic zoning than by large scale plate movements.

## 1. Introduction

The distribution of fossil faunas and floras has been used extensively in the discussion of former crustal movements. The main arguments for Wegener's original continental drift hypothesis were to a great extent based on the distribution of fossils. So was the competing hypothesis of land-bridges, and this illustrated both the importance and weakness of the biological material in deciding on mechanisms in crustal movements.

At present the emphasis is more on geophysical methods, which are certainly highly important and indispensable, especially for revealing the mechanisms of crustal movements. The intention of this paper is to stress the value of the combination of geophysical and biological methods, as they complement and supplement one another. Although the biological methods are numerically less precise than the geophysical ones, they are important as an independent check on the physical processes postulated, and may give a good time sequence of the surface expressions of the crustal movements. The biological material could be used either as a simple dating device or as a fossil guide to the crustal movements. They can also be used in a more sophisticated way to interpret movements and changes in geography more directly involving methods from ecology and biogeography. The concept which has been most effective is that of faunal provinces.

## 2. Faunal provinces

In the Lower Palaeozoic, the faunal provinces which are useful for the discussion of crustal movements are those of the shallow marine shelf seas. Land and freshwater life were not established, and the deep water faunas are too few and difficult to evaluate. It is therefore important in the following discussion to bear in mind that the provinces are more or less wide strips of shallow water around the continents, supplemented with the fauna of outlying volcanic islands.

The faunal provinces (in the sense of Ekman 1953) are areas characterized by a typical fauna, which is discriminated from that of the neighbouring provinces by composition, percentage of endemic forms and by its position in the climatic zones. The areas are typically of continent or sub-continent size. Large continents, which range over more than one climatic zone, will normally have discrete faunal provinces for each zone. Very often continents have different provinces on different sides. The distinctiveness of marine faunal provinces depends on their geographical location and history. The recent west Pacific tropical province, which is found on the west coast of North and South America, is distinguished from the west Atlantic tropical province (mainly in the Caribbean) by a high degree of endemism at the species level, but a low endemism on the generic level. This is easily explained by the fairly recent (a few million years) separation of these two provinces by the Isthmus of Panama.

The east and west Atlantic tropical provinces on the other hand show a high endemism not only on the species level, but also on the generic and family level, reflecting the longer time since the opening of the Atlantic in the Cretaceous. The east Atlantic province shows relations to the Indo-Pacific one, since the communication through the Tethys was open longer.

In the North Atlantic there is a good separation between the tropical and boreal provinces, but the Arctic one is uniform on both sides, since the almost continuous shelves around the Arctic Ocean permit a continuous exchange of faunas.

By using recent analogues like the examples demonstrated here, it is possible to identify the climatic zone of a given area at a particular time in the geological record. Discussion of the criteria for defining the climatic zone in which the faunal province belonged is given by Spjeldnaes (1961). It is also possible to define roughly how long two provinces in the same climatic zone have been separated. It is evident from the examples given and from the imprecision of the fossil data that such datings will not be accurate, and they should preferably be expressed as "short", "longer", and "very long" time rather than in millions of years.

One of the complexities of the study of faunal provinces is that the difference between two long separated provinces in the same climatic zone is as strong in terms of endemics as two adjoining provinces from different climatic zones along the same continent. By using ecological methods it is often possible to identify the two as either belonging to the same, or to two different, climatic zones. This is done by a study of the composition of the faunas and the enclosing sediments, or by a study of the boundary between the two provinces. Transitional faunas and interfingering due to small scale fluctuations in climate are common if the two provinces are on the same continent, but will normally not be found, or be of a special, easily identifiable nature if they were separated by an ocean.

## 3. Faunal provinces and crustal movements

Modern reconstructions of the plate movements in Europe and Africa (e.g. Whittington and Hughes 1972) suppose a wide Tethys ocean between northern Europe, and southern Europe and Africa. These models are based on palaeomagnetic results, indicating an early Palaeozoic pole in Northwest Africa, and on the striking difference in faunas between North and South Europe. A closer examination reveals that the palaeomagnetic evidence is not entirely reliable. There are considerable uncertainties both as to methodology, sample quality, and sample dating, which reduces the value of the pole determinations. The number of samples is also surprisingly small for such a far-reaching conclusion. The faunal provinces are well defined, but an ecological analysis (Spjeldnaes 1961; Dean 1966) indicates that the South European–African province was a polar one, whereas the North European was a boreal to temperate one. A study of the boundary region (Spjeldnaes 1978) in Wales, Belgium, and the Carnic Alps shows interfingerings and faunal exchanges, which indicate that the two provinces adjoined the same continent, and that the differences between them were due to differences in climate and not to a separating ocean.

The tectonic processes in the Mediterranean may not be of the same type as the large scale plate movements known from the Atlantic, but can rather be illustrated in terms of rotating and translative movements of small plates, such as indicated by Illies (1975). These small plates are individually beyond the resolution power of the rather crude biogeographical methods and have to be identified by geophysical and structural geological methods. Moreover, the sequence of faunas is often rather complicated, and it is in many cases difficult to see from which plate individual genera originated.

It has been suggested by Neuman (1972) that some of the faunas in this region come from isolated volcanic islands which acted as centres of development and dispersal. By later plate movements these islands then became part of the consolidated continent. Even if the rapid speciation on isolated volcanic islands is best documented from the terrestrial fauna, such as from the Galapagos Islands, it is possible that they could also have acted as centres of rapid development for marine benthos. The mechanism for this is, however, somewhat more complicated than for the finches of the Galapagos Islands, and requires greater separating distances and more time.

The porambonitids occur in Europe from the middle part of the Lower Ordovician, and develop a number of endemic species and subgenera throughout the Ordovician. In the late Ordovician they spread to the south European province, probably indicating a warming of the climate. It is not known if the sparse incursions of the family into the American continent proper are due to exchange with the Baltic region in Europe, or whether they came from the hypothetical spreading centres within the Proto-Atlantic ocean, or from somewhere else.

The family Christianidae (with the monotypic genus *Christiania*) seems to have originated in the Baltic–Scandinavian region at about the Arenig–Llandvirn boundary. The family was endemic to this region until the early Caradoc, when it suddenly spread to eastern North America (including Scotland). It is found in Kazakhstan and Alaska from beds of similar or somewhat younger age. It is noticeable that it did not spread to England or other parts of the European plate at this time. Later, in the Ashgill the family spread to England and Belgium. An occurrence in the Carnic Alps is somewhat doubtful, as the species resembles the old Caradoc or even Scandinavian Llandeilo species, but the age of the beds is given as Caradoc–Ashgill. The family became extinct at the end of the Ordovician.

The suborder Clitambonitidina was, with the exception of the family Polytechiidae, basically a Baltic group with numerous endemic genera in the Lower and Middle Ordovician. Some of the genera, like *Kullervo, Vellamo*, and perhaps *Atelelasma* migrated into North America during the Caradoc, and other genera are found sporadically in other parts of the European plate.

By using this type of data it is possible to build up an admittedly crude picture of the faunal distributions, and the directions of migrations. There are grave uncertainties in these reconstructions because it is not always easy to identify the exact time and place of the origin of a group. This may be due to taxonomic problems, but more often to the incompleteness of the record. Very often new discoveries, especially from the border zone between the two plates, give supplementary data which supports Neuman's (1972) idea that many of the genera involved originated on volcanic islands between the continental masses, or in areas which are now lost by subduction.

In spite of this, there is sufficient material from well studied animal groups to give a model which is internally consistent, and which can be used to check the geophysical data. The distribution of the faunal provinces is based on studies of the pelagic forms, mainly graptolites, cephalopods, and conodonts, and among the benthic forms the brachiopods and trilobites are the most important ones. Gastropods and ostracods also give important information, and the bryozoans are potentially of great value, because of their great sensitivity to ecological changes. Unfortunately, the bryozoan faunas in the Atlantic regions have not yet been adequately described, and the few remarks on this group made below are based on the author's unpublished studies of European Ordovician bryozoan faunas.

The fact that few of the "European" immigrants to North America got established permanently indicates that the waves of immigration in this direction were induced by climatic changes, and that most of the new forms perished when the climate changed back to normal. Only a few species managed to adapt to the new conditions, and survived. The "American" immigrants into Europe had a much higher rate of survival, and this rate increases with time throughout the Ordovician. This is taken to indicate that Europe moved towards lower latitudes, closer to America, and that the conditions on the European plate became progressively less hostile to the American warm water forms.

When comparing the biogeographical and geophysical evidence it must be remembered that the distance recorded by the faunal provinces is that of effective difference between the faunas. This difference is—to begin with—not the difference between either the coastlines or the continental masses, but between the shelves. The 200 m depth contour will be a reasonable approximation. What is measured is not actual distance in kilometres along a great circle, but the barriers to the dispersal of the faunas. This will depend on the free larval lifetime of the species, which was certainly different among the Ordovician species and genera, as it is in the marine benthonic fauna today. Since most of the benthonic animals in the Ordovician had larvae which were either passively drifting or at least not very effective swimmers, the dispersal would depend heavily on the oceanic current systems. This may have led to a situation where fairly short distances represented unsurmountable barriers to migrations, and also the opposite situation where favourable currents could carry benthonic larvae, and also more or less passive pelagic ones (such as the graptolites and nautiloids) over considerable oceanic distances. Some of the relevant palaeo-oceanographic models have been discussed by Ross (1975, 1976), Williams (1976) and Spjeldnaes (1978).

The details of the current pattern will depend not only on the latitudinal position of the continents, but also on details in their topographic configuration. It will be especially important if continents are present at the Equator, blocking the equatorial warm water currents and deflecting them towards higher latitudes. Movements of continental parts of plates obliquely to the oceanic circulation pattern (which is mostly latitudinal), and especially movements into and out of the equatorial zone will cause considerable changes in the current patterns and, accordingly, in the dispersal potential of the benthonic faunas.

In the early Palaeozoic, the Equator ran through the North American continent, in what is today a N–S direction. This meant that North America was placed squarely at the Equator and therefore deflected the warm equatorial currents towards higher latitudes, much like the present Gulf current. This may have resulted in cold counter currents, like the present Greenland current. Because of our lack of detailed knowledge of the coastal topography of the former North American continent, it is impossible to reconstruct the current pattern in detail, but it may be concluded that it was of a type which would respond rapidly to minor differences in distribution of the water masses or topography, giving great changes in climate.

The sensitivity to changes in oceanic and atmospheric circulation was probably not equal during the whole of the early Palaeozoic. The absence of polar ice caps during the Cambrian and Silurian (Spjeldnaes 1961, 1978) probably resulted in a much reduced intensity of oceanic and atmospheric circulation. Large bottom areas became stagnant, explaining the widespread distribution of black shales in the Cambrian and Silurian, and this would have influenced the mobility of the faunas. The reduced circulation would, by reducing the current intensities, automatically reduce the effective distance of migration. On the other hand, the broader and more diffuse climatic zones would make migration easier, and the distribution areas of faunal provinces larger.

This is seen in the present North Atlantic fauna, where the Arctic faunas on both sides communicate easily, whereas the temperate boreal ones are effectively separated by the wide ocean. If the climatic zoning was less sharp, the boreal-temperate faunas may also have been able to use the northern migration routes.

In the Ordovician, the presence of polar ice caps resulted in a circulation regime similar to the present ones, and this created some extra barriers for migration by increasing the definition of the climatic zones, but at the same time the increased intensity of circulation gave increased opportunities for migration with oceanic currents.

## 3a. Chronology of crustal movements from faunal evidence

There is no faunal evidence for the opening of the Proto-Atlantic. The first Cambrian faunas in the region are well separated into faunal provinces of a type that suggests a definite oceanic barrier between them. The hypothesis that the two plates were joined before the Cambrian rests therefore solely on geophysical and structural evidence.

In the Cambrian, the faunal provinces remained separate with remarkably little faunal exchange (Cowie 1971). Only a few special groups, such as the agnostid trilobites were cosmopolitan and common to both provinces. There are striking internal differences, especially in the North American Cambrian, with a warm-looking fauna in the cratonic deposits in the continental interior, and more temperate-looking ones in the thicker sedimentary sequences along the continental margins. It has been much debated whether these differences are due only to bottom conditions, or if they are climatically conditioned—or a combination of both (Palmer 1972, 1973; Lochman-Balk 1971).

The movements of South America are considerably less well known (Spjeldnaes 1978), but the available evidence indicates that a south Proto-Atlantic existed in Cambrian and early Ordovician time, and was closed during late Ordovician to Silurian times. The palaeomagnetic data are few, but do not contradict this model.

This is based on the observations that the Lower Ordovician of South America is similar to that at intermediate to low latitudes (Serpagli 1973), and that traces of a glaciation are found in the uppermost Ordovician to basal Silurian, resembling the conditions in Africa. This may be interpreted as a movement of South America from intermediate to high latitudes, thereby closing the south Proto-Atlantic.

## 3b. Crustal movements in the North Atlantic region

The gross outline of the crustal movements in the North Atlantic region is well known. The Proto-Atlantic (Iapetus), which may have opened in the late Precambrian, closed during the late Ordovician to Silurian and the present Atlantic opened from the Jurassic onwards along a suture which did not follow the old one. North of a line from Newfoundland to West Ireland there are remnants of the old "America" on the European side—in Ireland, Scotland, West and North Norway, and Spitsbergen. South of Newfoundland there are remnants of "Europe" along the east coast of North America, and Florida belonged originally in southern Europe or Africa.

Part of the evidence for this model comes from studies of the faunal provinces—the "American" remnants in Europe and vice versa were identified because of the characteristic faunas, and this is not only the basis of the present plate tectonic interpretations of the genesis of the Atlantic, but was used extensively by Wegener before the concept of plate tectonics had developed.

In the present situation, the faunal provinces therefore support the geophysical picture, and the further discussion of the faunal provinces will be concentrated on the faunal evidence for the timing of the crustal movements, and what can be deduced from the latitudinal positions of the individual plates at various times.

Another aspect of the palaeoclimatic pattern, which may be of help in interpreting the crustal movements, is the difference in contrast along the Proto-Atlantic suture. The difference in climate, supposedly corresponding to a difference in latitude, increases towards the present South, from northern Scandinavia and Greenland to south Appalachia and south Europe–Africa. The fossil evidence from the metamorphic region of the southern Appalachians is not very good, but the Ordovician in subsurface Florida is definitely of the type found from Brittany to Morocco. The Ordovician of the Appalachian belt appears to have been parallel to the Equator and does not show a corresponding change in climate. Also, there are no South European (polar) immigrants in the faunas in the southern part of the Appalachian belt.

This situation indicates that either the closing of the Proto-Atlantic has a rotational element to it, so that the original distance between the plates was larger in the south than in the north, or that a greater amount (about 500 km wide) of plate was lost due to subduction in the south, or to a combination of these two processes.

The first wave of supposed European forms in North America consisted mainly of *Porambonites* and *Ingria*, which are found in west North America, in Nevada, and California (Cooper 1956). This migration may not have been directly linked to the closing of the Proto-Atlantic. European-looking faunas from younger parts of the Ordovician are found both in California (Rohr *et al.* 1974) and Alaska (Ross and Dutro 1966; Patton and Dutro 1969), and as long as the palaeobiogeographical problems surrounding these faunas are unsolved, it may be safest not to draw any conclusions from this wave of Whiterock age. It should also be noted that contemporaneous faunas in east North America and northwest Europe are strictly endemic and show very few signs of faunal exchange.

The first undoubted wave of faunal exchange took place in the early Middle Ordovician, in a time interval corresponding to the transition between the zone of *Nemagraptus gracilis* and that of *D. peltifer*. This corresponds to the low Porterfield (Arline, Little Oak, and Boutetort Formation) in the southern Appalachians (Cooper 1956), and to the Derfel Limestone in North Wales (Whittington and Williams 1955), the Lower Chasmops Shale (4b$\alpha$) in the Oslo Region, Norway, and the Kukruse stage and its equivalents in the Baltic area. In Europe this has been named the Kukruse wave (Spjeldnaes 1978).

In terms of brachiopods, which is the best studied group, the European forms found in North America are typified by *Christiania* and *Kullervo*, and typical American forms found in Europe are *Ptychoglyptus* and *Anisopleurella*. A number of other genera are common, but their history is not quite clear, and it is difficult to tell in which province they originated. This group comprises genera such as *Nicolella, Skenidioides, Reuschella, Sowerbyella, Palaeostrophomena, Oepikina* and *Leptaena*. Some of these genera are not only common to the two provinces, but they are also represented by very similar, if not identical, species. Most of these common forms fade away rather rapidly, and even the immediately succeeding beds have considerably higher endemism in both provinces.

The next wave is found in the upper Caradoc, corresponding to the Trenton (Oranda Formation) in North America, and is especially manifest in the Vasalemma fauna in the Baltic, with its equivalents. This wave—the Vasalemma wave, is also accompanied by a distinct warming of the climate in northwest Europe, as indicated by the presence of bioherms and warm water carbonates. Also after this wave, the diversity diminishes, and the endemism increases, but a considerable number of the American forms "catch on" and give rise to new endemics.

At the very end of the Ordovician, the *Hirnantia* fauna is becoming almost cosmopolitan (Cocks and Price 1975), but, except for the Quebec Province (political, not faunal), it does not seem to appear in North America (Lesperance 1974). In the rest of the brachiopod faunas there is an increased exchange, and in the Lower Silurian the endemism is low, and there are not only common genera, but even common species. This appears to such an extent that it may be concluded that at

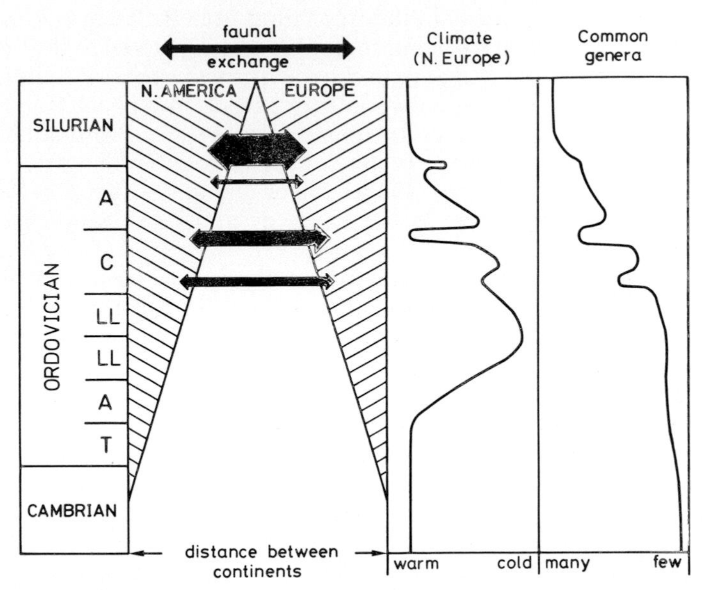


Fig. 1. Diagram illustrating the author's model of the faunal development in the north Proto-Atlantic during the Ordovician. The model is based mainly on the distribution of brachiopods, trilobites, bryozoans, nautilids and graptolites, supplemented by sedimentological observations.

least from the Middle Llandovery, the Proto-Atlantic had ceased to be a barrier to the distribution of the benthonic faunas.

The trilobites, which are also well known and which have been discussed by Whittington (1973), Whittington and Hughes (1972, 1973) and Fortey (1975), show a similar picture with a strong endemism in the Lower Ordovician, decreasing rather rapidly in the early Middle Ordovician, after which there is no appreciable provinciality. There is a clear dominance of American forms migrating to Europe over European forms migrating the opposite way.

The bryozoans have not yet been adequately described, neither from eastern North America nor from northwest Europe. From the author's unpublished studies it appears that the oldest Ordovician fauna (which is somewhat older in Europe being Arenig than in America—Whiterock) is strictly endemic. The European faunas from Llanvirn, Llandeilo, and lower Caradoc appear to be highly endemic and of low diversity, but a single find from South Wales (Spjeldnaes 1963) may indicate that the diversity was locally fairly high, and that there were some genera in common with the American faunas. The Kukruse wave resulted not only in a strong influx of American forms in northwest Europe, but also in a high diversity of endemic forms.

The diversity, which may have been connected to special ecological features, declined rapidly after the Kukruse wave, and the Vasalemma wave brought a new influx of American forms. At this time there were fewer endemic genera in Europe, and even a few species common between the two provinces.

The Upper Ordovician bryozoans in Europe are not too well known, and the faunas differ considerable from the well known Maysville and Richmond faunas in

North America. In spite of this, there is an increasing number of American forms found, not only in the already rather "American" Porkuni fauna in Esthonia, and of the Boda Limestone in Sweden, but even in the previously polar province in southern Europe and North Africa.

In the Silurian, our rather limited data indicate that there was very little endemism on the generic level between the two provinces.

The graptolites, representing a pelagic group, show high endemism in the Lower Ordovician, and partly also in the lower part of the Middle Ordovician, but already the *Nemagraptus gracilis* fauna is largely cosmopolitan, and the later graptolite faunas in the Ordovician and Silurian in the two areas do only show trivial differences which can hardly be described as provincial.

The cephalopods show also a considerable degree of endemism in the Lower Ordovician, but from the Llandeilo on there was an influx of American forms (Sweet 1958), and during all the remainder of the Ordovician, the European fauna was dominated by American forms and had rather few endemic native forms.

All this adds to a consistent picture of the European plate moving from higher latitudes towards the comparatively stable American plate, which was at the Equator. As could be expected, the first animal group to "bridge the gap" was the mobile, pelagic cephalopds which, in addition to active swimming, had the possibility of passive spread by empty shells with ocean currents like the recent *Nautilus*. The planktonic graptolites were next, followed by the trilobites, representing the vagrant benthos (including also a number of actively swimming forms).

The sessile benthos lagged behind, as would be expected, and the provinciality was not completely extinguished before the end of the Ordovician.

Also the climatic development based both on the faunas and the sediments supports this model. The shelf carbonates of the Baltic Lower Ordovician may at first glance be taken to indicate warm water, but as shown by Jaanusson (1972) they were indicative of an intermediate and not a warm climate.

As mentioned above, the Silurian does not show any appreciable provinciality, and there are even some common species, especially among the brachiopods and corals, and a number of pairs of species which are so close that it may be suspected that the main difference is that the material was described by different scientists and from different continents.

## 3c. The present boundaries between the provinces

There are two types of difficulties that may confuse the boundaries between the provinces, as exposed now.

It is necessary to be very careful, especially when gathering data from the older literature, in order to be certain that the material in question comes from the right province. This is especially so when discussing the areas on the "wrong" present continent. Sometimes it becomes apparent that genera, which were first described in Europe, really originated in North America, and vice versa. Such problems of a taxonomic and nomenclatorial nature are further complicated by the incomplete and often inhomogeneous nature of the description of the faunas. This is particularly so with the bryozoans, but also other important groups have not been subject to modern comprehensive treatment. This is the major obstacle for a detailed and precise use of the biogeographical methods on the crustal movements in the North Atlantic region.

The other type of difficulty stems from the actual geological conditions. Much of the critical evidence has been lost due to subduction, particularly of course the very shelves where the faunas lived. Much of the terrain has been metamorphosed almost beyond recognition, and tectonic mixing has taken place, which may confuse originally clear-cut boundaries.

Sometimes the superstructure of the plates, including fossiliferous sediments, was detached from the basement by decollement tectonics and compressed by folding or thrusting. This is common in the Scandinavian Caledonides, where the precise boundary between the superstructure material from the two plates may be difficult to distinguish because it has been telescoped together by the superficial tectonics.

This may lead to doubts as to which plate the fossils belong. Hughes *et al.* (1975) have reported the trinucleid trilobite *Stapeleyella* from the American plate, based on the occurrence of *S.*(?) *forosi* from the Trondheim region in Norway. Most of this region did certainly belong to the American plate, but there are indications that the southeast parts of it contain beds indigenous to the European plate. The genus *Stapeleyella* is almost exclusively restricted to this plate, and its occurrence could instead be used to indicate that the beds in which it was found really belonged to the European plate.

The factual evidence is tenuous indeed in this particular instance. Only one specimen has ever been found of *S.*(?) *forosi*, and that was not collected *in situ*, but from the roofing slate of an old church. The quarry from which the slate was supposedly taken has been identified (Størmer, in Kiaer 1932), and demonstrates the ambiguity of some of the biogeographical evidence, after it has been exposed to geological influences for 450 million years.

## 4. Conclusions

The study of faunal provinces may contribute to the understanding of the crustal movements in the North Atlantic area during the early Palaeozoic. The evidence bears on the effective distance between the continents, and the latitudinal positions of the plates. The effective distance is not the direct geographical distance, but represents degrees in the barrier effect of the intervening ocean. It can be either a complete barrier, an unstable filter, or no barrier. With the last, this does not mean that there was no salt water between the continental plates, but merely that the remaining water bodies did not represent a barrier to migration of the benthic marine fauna.

In considering the latitudinal position of the continental parts of the plates, it should also be considered that the precision of the biological methods is about $\pm 10°$, and that longitude can only be estimated through indirect methods with considerably less accuracy.

The results relevant to the closing of the Proto-Atlantic from the study of the faunal provinces seem to be mainly:

(i) Throughout the Cambrian there was an effective oceanic barrier, comparable to that between the boreal-temperate provinces in the Atlantic today. Due to the supposed lower intensity of circulation, the real distance between the continents may have been smaller.

(ii) During the Ordovician the distance between the two continental masses decreased from full oceanic distance in the early Ordovician to no effective distance in the late Ashgill.

(iii) Since the Equator remained almost stationary over the North American continent, and the European one moved from higher to lower latitudes during the

Ordovician, it appears that probably the European–African plate was the active one and the American one was the passive.

The main interest in the use of faunal provinces for the study of crustal movements, particularly in the North Atlantic, is that it provides an independent method by which the mainly geophysical structural evidence can be tested. Even if modern physical methods are highly sophisticated and have considerable numerical precision, there are comparatively few direct observations on the actual processes going on in the deeper part of the crust and in the mantle. In order to test the validity of the models, it is therefore useful to have a set of independent methods by which they can be checked. The theory of plate tectonics has also been fruitful in the development of biogeography, particularly regarding the development of the faunal provinces, and a feedback system of mutual benefit should be further developed. Because of the different approaches, both of method, and of data obtained, the two approaches are also to a large extent complementary, and may profitably be used together in the study of crustal movements.

## References


COCKS, L. R. M. and PRICE, D. 1975. The biostratigraphy of the upper Ordovician and lower Silurian of south-west Dyfed, with comments on the *Hirnantia* fauna. *Palaeont.* **18,** 703–724.

COOPER, G. A. 1956. Chazyan and related Brachiopods. *Smithsonian Misc. Coll.* **127.**

COWIE, J. W. 1971. The Cambrian of the North American Arctic Regions. *In* Holland, C. H. (Editor). *Lower Palaeozoic Rocks of the World* **1,** 325–383. Wiley, London.

DEAN, W. T. 1966. The Lower Ordovician Stratigraphy and Trilobites of the Landeyran Valley and the neighbouring district of the Montaine Noire, South-Western France. *Bull. Brit. Mus. (Nat. Hist.) Geology.* **12,** 247–353.

EKMAN, S. 1953. *Zoogeography of the Sea.* Sidgwick & Jackson. London.

FORTEY, R. A. 1975. Early Ordovician trilobite communities. *Fossils and Strata* **4,** 331–352.

HUGHES, C. P., INGHAM, J. K. and ADDISON, R. 1975. The morphology, classification and evolution of the Trinucleidae (Trilobita). *Phil. Trans. R. Soc.* B. **272,** 537–607.

ILLIES, H. 1975. Intraplate tectonics in stable Europe as related to plate tectonics in the Alpine system. *Geol. Rundschau* **64,** 677–699.

JAANUSSON, V. 1972. Aspects of carbonate sedimentation in the Ordovician of Baltoscandia. *Lethaia* **6,** 11–34.

LESPERANCE, P. J. 1974. The Hirnantian fauna of the Percé area (Quebec) and the Ordovician–Silurian boundary. *Am. J. Sci.* **274,** 10–30.

LOCHMAN-BALK, C. 1971. The Cambrian of the Craton of the United States. *In* Holland, C. H. (Editor). *Lower Palaeozoic of the World.* **1,** 79–167. Wiley, London.

NEUMAN, R. B. 1972. Brachiopods of Early Ordovician Volcanic Islands. *24th Intern. Geol. Congr.* **7,** 297–301.

PALMER, A. R. 1972. Problems of Cambrian Biogeography. *24th Intern. Geol. Congr.* **7,** 310–315.

—— 1973. Cambrian Trilobites. *In* Hallam. A. (Editor). *Atlas of Palaeobiogeography,* 3-11. Elsevier, Amsterdam.

PATTON, W. W. Jr. and DUTRO, J. T. Jr. 1969. Preliminary report on the Paleozoic and Mesozoic sedimentary sequence on St. Lawrence Island, Alaska. *U.S. Geol. Survey, Prof. Paper* 650–D, D138–D143.

ROHR, D. M., BOUCOT, A. J. and POTTER, A. W. 1974. Age corrections for some Silurian localities in northern California. *J. Paleont.* **48,** 413–414.

ROSS, R. J. Jr. 1975. Early Paleozoic trilobites, sedimentary facies, lithosphaeric plates and ocean currents. *Fossils and Strata* **4,** 307–329.

—— 1976. Ordovician Sedimentation in the Western United States. *In* Bassett, M. G. (Editor). *The Ordovician System.* Univ. of Wales Press, and National Museum of Wales, 73–105.

—— and DUTRO, J. T. Jr. 1966. Silicified Ordovician Brachiopods from East-Central Alaska. *Smithsonian Misc. Coll.* 149.

SERPAGLI, E. 1973. Carbonati de tipo bahamitico nell ordoviciano inferiore delle precordillera argentina e relative osservationa paleoclimato logiche. *Atti Naz. Lincei, Rendic. Sci. Fis. Mat. Nat. ser. VIII,* 55, no. 5.

SPJELDNAES, N. 1961. Ordovician Climatic Zones. *Norsk Geol. Tidsskr.* **41,** 45–77.

—— 1963. Some silicified Ordovician fossils from South Wales. *Palaeontology* **6,** 254–263.

—— 1978. Lower Palaeozoic Palaeoclimatology. *In* Holland C. H. (Editor). *Lower Palaeozoic rocks of the World* **3,** Wiley, London (in press).

STØRMER, L. 1932. Trinucleidae from the Trondheim area, with contributions by Th. Vogt. *In* Kiaer, J. The Hovin Group in the Trondheim Area. *Skr. Vidensk. Akad. Mat.-Naturvid. Kl. 1932*, no. **4,** 169–175.
SWEET, W. C. 1958. The Middle Ordovician of the Oslo Region, Norway 10. Nautiloid Cephalopods. *Norsk Geol. Tidsskr.* **38,** 1–176.
WHITTINGTON, H. B. 1973. Ordovician Trilobites. *In* Hallam A. (Editor). *Atlas of Palaeobiogeography*, 13–18. Elsevier, Amsterdam.
—— and HUGHES, C. P. 1972. Ordovician geography and faunal provinces deduced from trilobite distribution. *Phil. Trans. R. Soc.* B. **263,** 235–278.
—— and —— 1973. Ordovician trilobite distribution and geography. *Spec. Paper Palaeontology* **12,** 235–240.
—— and WILLIAMS, A. 1955. The Fauna of the Derfel Limestone of the Arenig District, North Wales. *Phil. Trans. R. Soc.* B. **238,** 397–429.
WILLIAMS, A. 1973. Distribution of Brachiopod Assemblages in relation to Ordovician Palaeogeography. *Spec. Paper Palaeontology* **12,** 241–269.
—— 1976. Plate Tectonics and Biofacies evolution as factors in Ordovician Correlation. *In* Bassett, M. G. (Editor). *The Ordovician System.* University of Wales Press and National Museum of Wales, 29–66.

Nils Spjeldnaes,
Department of Palaeoecology,
The University,
Aarhus, Denmark.

# Geology of a continental margin 1: the Ballantrae Complex

*B. J. Bluck*

---

The Ballantrae Complex, although containing most of the components of an ophiolite, has an unproven sequence. There is evidence of continual overlap between igneous and metamorphic events. The structural complexity of the ophiolite may be related either to a series of obductions, or the incorporation, during a single obduction, of major rock units from a range of geographical locations. The ophiolite was obducted over a marginal basin of Middle Arenig age, and the sediments of this basin record the Middle Arenig part of the obduction history.

## 1. Introduction

The history of research into the Ballantrae Volcanic Complex has been given by Bailey and McCallien (1952). An early interpretation by Geikie (1866), that the whole Ballantrae sequence was metamorphic in origin was corrected by an enlightened Bonney (1878) who demonstrated it to be mainly igneous. There followed the detailed descriptions of Peach and Horne (1899), with comments by Teall, and the meticulous petrographic work of Balsillie (1932, 1937). Bailey and McCallien (1957) attempted to see a stratigraphy in the complex and Bloxam and Allen (1960) initiated more critical appraisal of aspects of the metamorphism. The next major change in thinking came with the recognition that ophiolite was possibly a part of the oceanic crust, and the discovery of Ordovician obducted crust in Newfoundland —the strike continuation of the Caledonides. This prompted a series of papers (e.g. Church and Gayer 1973; Dewey 1974), which drew analogies between the Ballantrae ophiolite and other well exposed examples, and also placed the Ballantrae ophiolite in a major earth structural context.

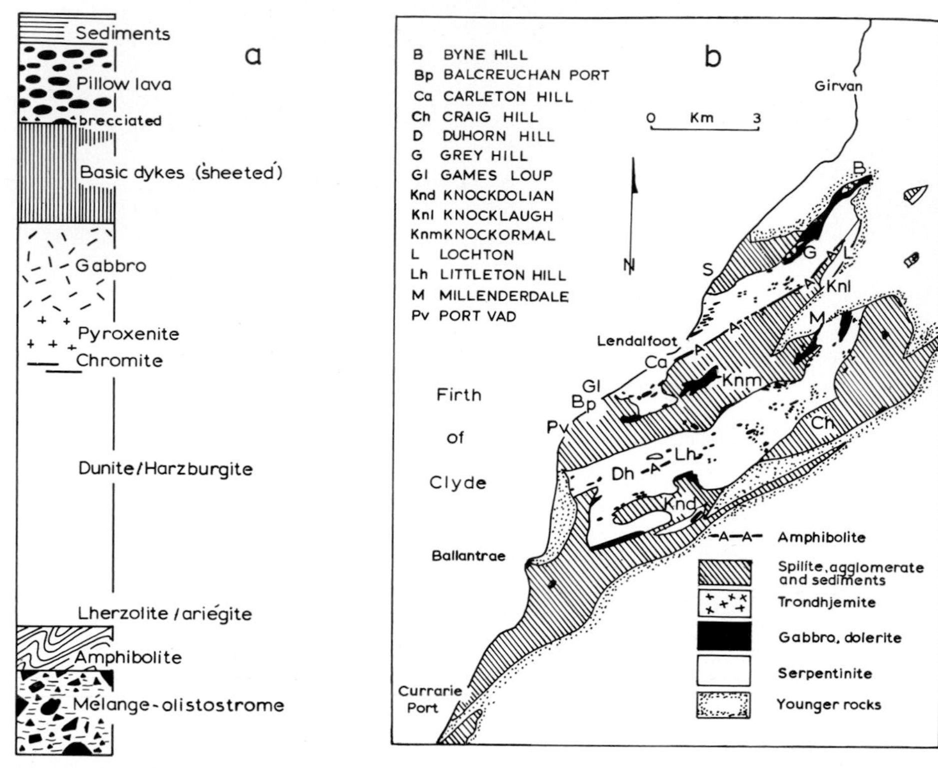


Fig. 1. (a) section of an ideal ophiolite; (b) simplified map of the Ballantrae Complex.

## 2. Description of rocks

### 2a. Spilitic lavas

Spilitic lavas, most of which are pillowed, crop out abundantly in the Ballantrae Complex (Fig. 1). They range in texture from the porphyritic type of Stockinray, Pinbain Hill, Port Vad and Knockdolian through "normal" fine grained varieties to the sub-variolitic tachylyte commonly found in conglomerates and agglomerates. Compositionally they have representatives of quartz tholeiites, tholeiites and basic andesites. The chemistry of the lavas suggested to Bloxam and Lewis (1972 p. 136) a possible island-arc origin, whilst Wilkinson and Cann (1974) distinguished hot-spot basalts (Bennane Head–Balcreuchan Port), island-arc tholeiites (Balcreuchan Port–Games Loup) and representatives of ocean-floor basalts in the metamorphic rocks of Knockormal and Knocklaugh.

The spilites interstratify and, for example on the shore north of Pinbain, intertongue with conglomerate and possible agglomerate. Rocks, thought to be agglomerate by Peach and Horne (1899 p. 449) and Bailey and McCallien (1957 p. 38) interstratify with lava on Knockdolian, where fragments in the agglomerate are sometimes >2 m in diameter. Brecciation of the spilite occurs at other localities, and a fine example is seen in the ash sequence on the northern side of Pinbain Bay.

Bailey and McCallien (1957 p. 37) recognised two spilitic assemblages, the

Knockdolian Spilitic Group and the Downan Point Spilitic Group, separated by the serpentinite. Mendum (1968 p. 261) divided the sequence into three units, the lowest one of which was Arenig in age and separated from the following Phenocrystic unit by an unconformity. The Phenocrystic unit is in turn succeeded unconformably by the Newer Spilitic unit, both these units being of uncertain age. However Walton (1961 pp. 68–71) demonstrated that some of the spilites are interstratified with sediments younger than Arenig at Currarie Point (within the Downan Point spilite outcrop), and Kelling (1961 p. 43) found volcanic rocks interbedded with sediments of Llandeilo age in the Rhinns of Galloway. Whittington (*in* Williams *et al.* 1972 p. 51) pointed out that at Craighead cherts, interstratified with pillow lavas, have yielded Llandeilo conodonts to Lamont and Lindstrom (1957). On the other hand, Peach and Horne (1899 p. 439) record Middle Arenig red shales interbedded with pillow lava at Balcreuchan Port, and Middle Arenig blue shales in the agglomerates which are interstratified with the Pinbain Hill spilite assemblage.

There is evidently some range in the ages of the spilites, and need for a more thorough investigation into their age and stratigraphy in relation to differences in their chemistry (i.e. are basalts ascribed to a "hot-spot" origin of the same general age as those regarded as of island-arc type?).

## 2b. "Sheeted dyke complex"

Dykes intruded into foliated gabbros are well exposed in a series of knolls 500 m east of Millenderdale Farm. The gabbros have a steeply dipping foliation which, although variably oriented, has a general NE–SW trend. It comprises hornblende augen enclosed by saussurite (albite, analcite, prehnite and chlorite) or pyroxene rimmed with hornblende associated with andesine and labradorite (Balsillie 1932 p. 118). The foliated gabbros are cut by dykes up to 3 m wide and which occupy about 25% of the exposures (cf. Dewey 1974 p. 943). The dykes have a variable orientation but with a fairly strong mode in a NNE–SSW direction, and are often faulted or otherwise have a tortuous course through the gabbro. Although many dykes have no chilled margins (Balsillie 1932 p. 118), some do. They enclose masses of foliated gabbro and although the foliation in the main gabbro mass is cut by the dykes, examples of foliation bending to conform to the orientation of the dyke wall are quite common. The dykes are mainly beerbachites (Bloxam 1955 p. 334), some hornblendic, others containing enstatite or hypersthene.

Evidently the dykes post-date the foliation, but the absence of chilled margins in some dykes, and the bending of foliation at the dyke margins suggests the gabbro to have been plastic at the time of dyke intrusion (Balsillie 1932 p. 119; see also Bonatti *et al.* 1975 who consider this as one way of explaining foliated gabbros of the Mid-Atlantic ridge). The origin of the gabbro, its foliation, the dyke intrusion, and their granulitization may therefore have belonged to the same igneous and deformational episode.

These outcrops have been interpreted as part of a sheeted dyke complex (Church and Gayer 1973 p. 501; Dewey 1974 p. 943), but the proportion of dykes to country rocks is considerably lower than is found in other ophiolites (e.g. 90–100% of the sheeted dyke complex is dyke rock at Troodos—Moores and Vine 1971 p. 450). Dyke intrusion in a tectonically active regime where gabbroic rocks were forming and cooling might well fit the model of a spreading ridge, but on the exposed evidence the outcrop is merely a gabbro cut by a few dykes. Dykes are also uncommon in the spilite sequence. Hence a sheeted dyke complex is either poorly represented or not present at all in the Ballantrae ophiolite.

## 2c. Unfoliated gabbros and associated rocks

Unfoliated gabbros are well exposed in a series of hills running southwestwards from Byne Hill to Grey Hill. These gabbros, chilled against the enclosing serpentinite, comprise two main types (Bloxam 1968 p. 108), (i) an ophitic olivine gabbro of plagioclase feldspar, clinopyroxene (partly replaced by brown hornblende), olivine and bronzite, and (ii) dioritic hornblende gabbro, transitional between gabbro and diorite. The feldspars are in stages of albitization, and the hornblende, replacing pyroxene, is distinctive in being mainly green (Balsillie 1932 p. 116; Bloxam 1968 p. 108). These gabbroic intrusions are sometimes leucocratic and grade into irregular masses of anorthosite, which are now mostly albitized.

Within the serpentinite north of Lendalfoot, a gabbro-pegmatite sheet (Bonney's Dyke) is well exposed on the shore. This rock, fractured or flasered at its margin, carries inclusions of the local bastite serpentinite and diallage pegmatite. As with the previously described gabbros, this one has uralitized pyroxenes, and the feldspars are replaced by prehnite and pectolite.

On Byne Hill and Grey Hill trondhjemites occur with the gabbros, often with a dioritic zone of variable thickness between. Bloxam (1968 pp. 110–111) has divided them into hornblende- and biotite-bearing varieties. They contain secondary quartz and albite.

These rocks resemble in composition, texture and alteration those rocks found below the sheeted dyke complexes of other ophiolites (e.g. Reinhardt 1969 p. 6 records gabbros with brown hornblende from Oman and albitization of more calcic feldspar is recorded from Troodos gabbros by Moores and Vine 1971 p. 454). Myashiro *et al.* (1970) record gabbros with brown hornblende from the Mid-Atlantic ridge so the gabbros at least have a composition and alteration similar to those found on ocean-ridges.

## 2d. Serpentinites and associated ultrabasic rocks

The serpentinite lies in two major outcrops, interpreted by Bailey and McCallien (1957 fig. 2) to represent a single outcrop disposed about a plunging fold. Lherzolite-serpentinite is the most abundant rock type, but harzburgite and dunite are also found (Balsillie 1932 pp. 112–115). Church and Gayer (1973 p. 500), however, suggest that much of the serpentinite was originally harzburgite-dunite. Bloxam and Allen (1960) record ceylonite-bearing wehrlite and ariégites from Knockormal and Church and Gayer (1973 p. 500), an amphibolitic ariégite from Knocklaugh. Fresh pyroxenites occur near Knocklaugh (Balsillie 1932 p. 114) and Knockormal (Bloxam and Allen 1960 p. 14). Picrite occurs in some fairly extensive exposures in the southern serpentine belt (Balsillie 1932 p. 112).

Church and Gayer (1973 p. 500) and Dewey (1974 p. 945) have drawn a comparison between these rocks and those occurring in other better exposed ophiolites. They regard the lherzolite-ariégite as upper mantle in origin, and the wehrlites, clinopyroxenites and picrites as cumulates forming at the top of the upper mantle, beneath the gabbros (Fig. 1).

## 2e. Amphibolite and related rocks

Amphibolites are recognized in two general localities. The first is within the southern serpentinite in a discontinuous series of outcrops near Littleton Hill (Balsillie 1932 p. 123). Here the amphibolites are scapolite-bearing and form part of a group of sheared granulitic pods extending from Duhorn Hill to Craig Hill. The origin of these rocks is uncertain, but may be linked with the development of the beerbachites (p. 153). The second is along the northwesterly-dipping contact between serpentinite and sediments or volcanic rocks between Carleton Hill and Loch Lochton (Anderson 1936 fig. 2) where a structural thickness of 30–50 m represents the metamorphic aureole of the northern serpentinite. This is thinner than the aureole beneath the White Hills and Bay of Islands sheets of Newfoundland (Williams and Smyth 1973 p. 600), the reduced thickness possibly being due to structural thinning in what is clearly a zone of considerable dislocation.

In this contact zone there is a general gradation from sheared volcanic and sedimentary rocks, to epidote-chlorite foliated rocks, to pyroxene and garnet-bearing amphibolite which, in some cases, alternates with the epidote-chlorite schists. The ultrabasic rocks at their contact with the amphibolite have a mylonitic texture and Balsille (1932 p. 114) records a tremolite schist from this junction near Knocklaugh. In the greenschist rocks the banding, defined by the growth of chlorite and epidote, is roughly parallel to the outcrop of the amphibolite. The foliation has been deformed by at least two fold phases. Accordingly, as with the Newfoundland examples of Williams and Smyth (1973), the main phase of mineral growth preceded the folding. The shear belt is cut by undeformed dykes presumed to belong to the Ballantrae igneous event (Anderson 1936 p. 541).

Amphibolites occur at the bases of many ophiolite sequences, e.g. the Bay of Islands Complex of Newfoundland, the Semail Complex of Oman and the Vourinos Complex of Greece (see Dewey and Bird 1971 p. 3183; Williams and Smyth 1973 p. 616). In Ballantrae, as in Newfoundland there is a decreasing metamorphic grade away from the serpentinite. Church and Gayer (1973 p. 500) have suggested that, as with the Newfoundland example, these amphibolites were produced during the obduction of hot mantle onto sedimentary and volcanic rocks. However, it is possible that the metamorphism was induced by the frictional heat of overthrusting (see Graham and England 1976; Woodcock and Robertson 1977).

## 2f. Other metamorphic rocks

The southern serpentinite belt contains a number of masses of beerbachite (pyroxene, hornblende-bearing metadolerites, Balsillie 1932, 1937; Bloxam 1955). They occur in a belt extending from Littleton Hill to Millenderdale (Bloxam 1955 fig. 1), often show a schistosity consistent within an outcrop, but are not uniform from one outcrop to the next (Balsillie 1932 p. 122). They occasionally show chilled margins when traced to the enclosing serpentinite, and this serpentinite/beerbachite contact is often a zone of shearing. Early workers (Teall *in* Peach and Horne 1899; Tyrrell 1909) considered them to be basic rocks thermally metamorphosed by the serpentinite, whilst Balsillie (1932) and Bailey and McCallien (1952), amongst others, considered them to be intrusions which had suffered some metamorphism whilst still hot. Bearing in mind the observed shearing of the serpentinite at the margins of the intrusions, the absence of metamorphism in the surrounding serpentinite (Balsillie 1932 p. 122) and the diversity in the orientation of the schistosity, they could well be tectonic inclusions in the serpentinite, having been formerly intrusions into the original peridotite.

Rodingite dykes in serpentinite and along the contact between serpentinite and gabbro have been described by Bloxam (1954, 1968 pp. 116–119). Glaucophane schist found at Knockormal (Balsillie 1937) shows a transition to spilite in greenschist facies (Bloxam and Allen 1960 p. 6). The presence of glaucophane schist, requiring a moderate temperature and (when accompanied by lawsonite, for example) a high pressure regime in which to form, has presented a problem in the interpretation of

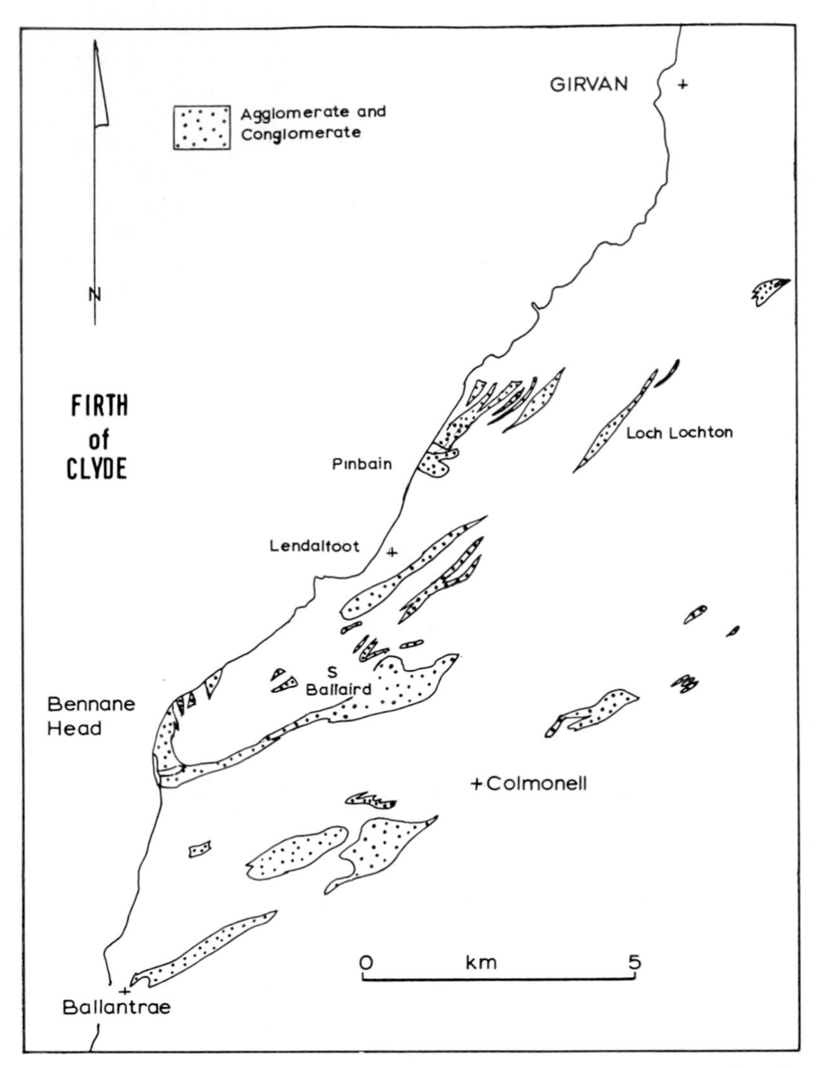


Fig. 2. Distribution of rudaceous sediments associated with the Ballantrae Complex (conglomerates and agglomerates together with breccias of uncertain origin).

some ophiolites. According to some workers, blueschists imply a previous history of subduction, so that there is a structural problem in getting the glacuophane schist into the obducted pile (Coleman 1972 p. 23). In the Knockormal instance Lambert and McKerrow (1976 p. 277) see no need for "substantial subduction" (the meaning of this is uncertain) and suggest metamorphism in the range 3–5·5 kb. Coleman (1971, 1972) suggests that tectonic overpressures produced during obduction may also be responsible for bluecschist formation, although this might not amount to more than 1 kb (Brace *et al.* 1970). In the Ballantrae ophiolite it has already been suggested that the obduction was accompanied by the development of garnet amphibolites (produced at fairly high temperatures). Either the Ballantrae ophiolite

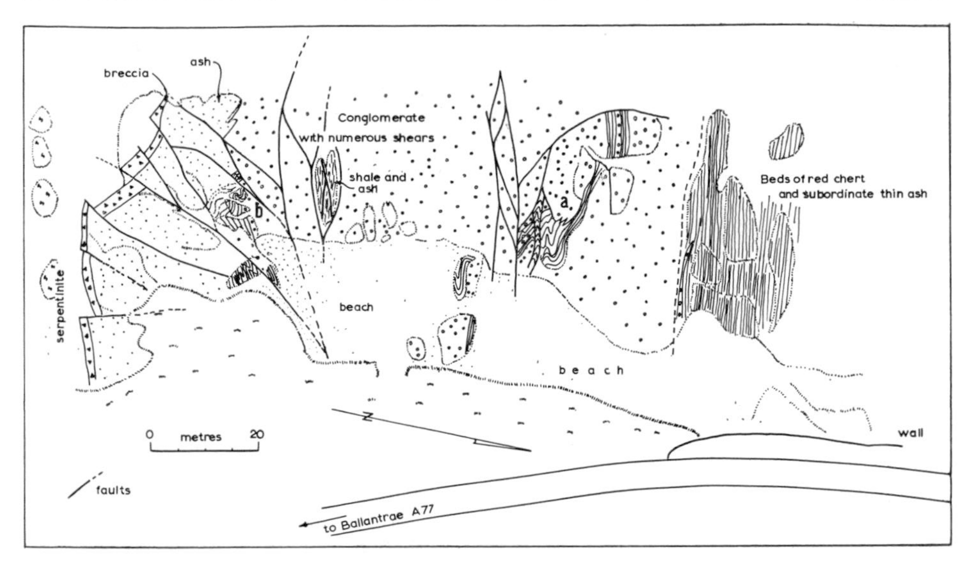


Fig. 3. Plane-table map of the foreshore, south of Bennane Head [NX 091861]: a = interstratification of chert and conglomerate; b = locality of Fig. 4.

has had a more complex history than previously thought, with possibly more than one phase of obduction, or the already obducted ophiolite has been affected by a later period of thrusting during which time the blueschist developed (see Church and Gayer 1973 p. 507). Whatever may be the case, if the glaucophane schist pebbles found in the Middle Arenig conglomerates at Pinbain (p. 158) are derived from the Knockormal and related occurrences, then all the possible subduction, obduction and subsequent thrusting will have been completed by Middle Arenig time.

## 2g. Ophiolitic olistostromes and associated sediments

Rocks, previously considered to be agglomerates at Bennane Head, Pinbain and S. Ballaird (Fig. 2) have been reinterpreted as olistostromes and mélanges by Church and Gayer (1973 p. 503) and Dewey (1974 p. 945). These olistostromes, made up of angular and sub-rounded clasts up to 7 m in diameter, have a matrix of rock fragments and minerals as in normal poorly sorted conglomerates of this kind. They have little internal structure, but sometimes contain streaked out and deformed black shales (e.g. at Pinbain) suggesting that soft sediment flowage and folding took place during the emplacement of the conglomerate. At Bennane Head the conglomerates interstratify and intertongue with cherts which are themselves often slumped (Figs 3, 4, 5). Where the cherts are interbedded with sandstones soft sediment boudins are well developed. Planes of discontinuity which cut out parts of the chert sequence are probably slide planes developed in the unconsolidated chert sediments. At Pinbain, large clasts are totally enclosed in a shale matrix (Fig. 6), and to the north of Pinbain Burn, conglomerates intertongue with brecciated and deformed black shale.

The structure and texture of the conglomerates suggest a mass-flow origin. Their continual interbedding with cherts and black shales suggest the movement of coarse sediments into environments quiet enough to allow these fine grained sediments to accumulate. The conglomerates, sometimes containing very large clasts, probably

Fig. 4. Slumping in cherts at Bennane Head (cf. Fig. 3b).

needed quite steep slopes down which to flow and are thought to have been formed when major sediment masses were displaced into deeper water.

The varied composition of these coarse sediments was noted by Balsillie (1937 p. 32) at S. Ballaird where the occurrence of clasts of amphibolite, gabbro, granulites, serpentinite and spilite was recorded. Bailey and McCallien (1957 p. 47) record an assemblage of glaucophane and epidote schists, diallage schists, gabbro, black shale and spilite as clasts at Pinbain. In addition, Church and Gayer (1973 p. 503) found limestones, chert and pyroxenite at this exposure. To this list may be added greywacke, amphibolite, granulite, diorite, and various dolerites. The conglomerates at Bennane Head have yielded a more restricted range of clast types: spilites, cherts, greywacke and a range of carbonate rocks some of which have undergone severe deformation. Bailey and McCallien (1957 p. 41) recognised picotite in some of these carbonate clasts. Other olistostromes, previously mapped in as agglomerate (Geol. Surv. Scotland 1 inch Sheet 7) have, in addition to the clasts already mentioned, yielded albitized andesite. Almost every rock type found in the presently exposed ophiolite can be identified as a clast in these conglomerates. Evidently at the time of conglomerate deposition the whole ophiolite was undergoing erosion, even to the amphibolite and glaucophane schist.

Peach and Horne (1889) found graptolites of Middle Arenig age in most of the shales which interstratify or lie adjacent to the conglomerates. From this information it is clear that the ophiolite was being eroded in Middle Arenig times, and that far travelled oceanic sediments, with a wide range of ages, are not present.

The shales and cherts are therefore considered to have accumulated in a marginal sea, the Ballantrae part of which was fairly short lived. During Middle Arenig times it was overridden by the rising ophiolite. The sea-floor in front of this rising mass

Fig. 5. Interstratification of radiolarian chert (a) and conglomerate (b); Bennane Head.

was uplifted, and newly laid sediments slid down the tilting surface, only to be again deformed by the advancing front. As the nappe rose, so major mass flow deposits moved off into the basin.

With the possible exception of the Bennane Head sequence (but see below), all the olistostromes have an exceptionally wide range of clast types, coming from practically every part of the ophiolite. There are a number of explanations for this:

(i) The ophiolite sequence was fully exposed by the time it had reached Ballantrae, i.e. given no structural complication in its upper surface, the sedimentary record of a rising and advancing nappe of ophiolite would have its topmost unit (sediments and pillow lavas) contribute clasts to the first-formed conglomerates. Progressive additions of clasts from deeper in the ophiolite pile would occur in the conglomerates deposited in front of the advancing nappe and, in an undeformed sequence, a horizontal ordering of clast type would occur in the conglomerates from few clast types (mainly from the top of the ophiolite), to many clast types (including ones from the base).

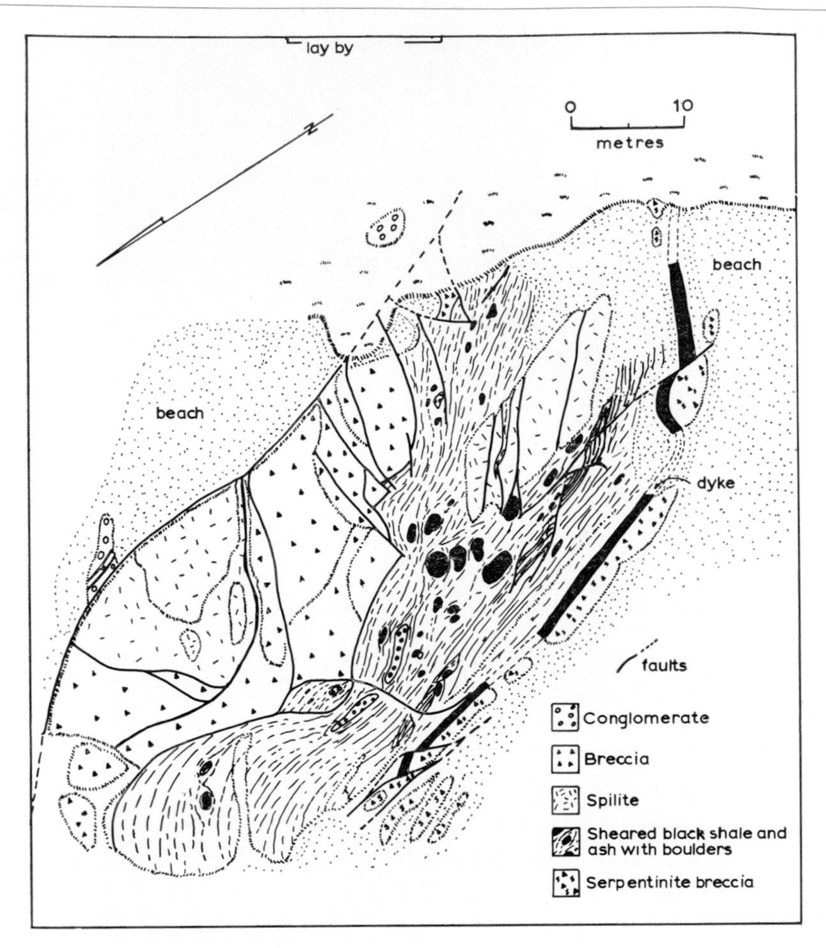


Fig. 6. Part of plane-table map of Pinbain, south of Pinbain Burn [NX 137915], illustrating the occurrences of some of the types of conglomerates.

(ii) The ophiolite was compound, being composed of slices of more than one ophiolitic complex.

(iii) The ophiolite was structurally complex and a considerable re-arrangement of its units had taken place by the time it reached Ballantrae. Along with (ii) this would present a varied outcrop on the face of the ophiolite from which to derive a polymictic conglomerate.

(iv) At the time of obduction the ophiolite might have had a variable topography, perhaps due to faulting. Talus (see Heezen and Rawson 1977) accumulated in front of these faults, would then be mixed when erosion of the obducted plate commenced.

The discovery of conglomerate fragments within the olistostromes, together with pebbles of greywacke favour explanations (ii), (iii) and (iv). The present day structural complexity of the Ballantrae ophiolite, with its glaucophane schist, favours (ii) and (iii).

Other mapped "agglomerates" (Fig. 2) are massive breccias, and investigations are now underway to determine if they accumulated at the foot of fault scarps ((iv) of above).

## 3. Conclusions

Along with many other ophiolites (Dewey 1976) the one at Ballantrae appears to have formed in, or adjacent to, a marginal basin (cf. Church and Gayer 1973), over which it was obducted. From a consideration of the structure of the ophiolite and the type of debris in the conglomerates, it is probable that the ophiolite was compound or at least structurally complex at the time of its obduction.

## References


ANDERSON, J. G. C. 1936. Age of the Girvan–Ballantrae serpentine. *Geol. Mag.* **73,** 535–545.

BAILEY, E. B. and McCALLIEN, W. J. 1952. Ballantrae problems: historical review. *Trans. Edinb. geol. Soc.* **15,** 14–38.

—— and —— 1957. The Ballantrae serpentine, Ayrshire. *Trans. Edinb. geol. Soc.* **17,** 33–53.

BALSILLIE, D. 1932. The Ballantrae Igneous Complex, South Ayrshire. *Geol. Mag.* **69,** 107–131.

—— 1937. Further observations on the Ballantrae Igneous Complex, South Ayrshire. *Geol. Mag.* **74**, 20–33.

BONATTI, E., HONNOREZ, J., KIRST, P. and RADICATI, F. 1975. Meta-gabbros from the Mid-Atlantic Ridge at 06°N: contact-hydrothermal-dynamic metamorphism beneath the axial valley. *J. Geol.* **83,** 61–78.

BLOXAM, T. W. 1954. Rodingite from the Girvan–Ballantrae Complex, Ayrshire. *Mineral. Mag.* **30,** 525–528.

—— 1955. The origin of the Girvan–Ballantrae beerbachites. *Geol. Mag.* **92,** 329–336.

—— 1968. The petrology of Byne Hill, Ayrshire. *Trans. R. Soc. Edinb.* **68,** 16–19.

—— and ALLEN, J. B. 1960. Glaucophane-schist, eclogite, and associated rocks from Knockormal in the Girvan–Ballantrae Complex, South Aryshire. *Trans. R. Soc. Edinb.* **64,** 1–29.

—— and LEWIS, A. D. 1972. Ti, Zr and Cr in some British pillow lavas and their petrogenetic affinities. *Nature, Phys. Sci., Lond.* **237,** 134–136.

BONNEY, T. G. 1878. On the serpentines and associated igneous rocks of the Ayrshire coast. *Q. J. geol. Soc. Lond.* **34,** 769–785.

BRACE, W. F., ERNST, W. G. and KALLBERG, R. W. 1970. An experimental study of tectonic overpressure in Franciscan rocks. *Geol. Soc. Am. Bull.* **81,** 1325–1338.

CHURCH, W. R. and GAYER, R. A. 1973. The Ballantrae ophiolite. *Geol. Mag.* **110,** 497–510.

COLEMAN, R. G. 1971. Plate tectonic emplacement of upper mantle peridotites along continental edges. *J. geophys. Res.* **76,** 1212–1222.

—— 1972. Blueschist metamorphism and plate tectonics. *Proc. Int. geol. Congr.* **24** (2), 19–26.

DEWEY, J. F. 1974. Continental margins and ophiolite obduction: Appalachian Caledonian System. *In* Burk, C. A. and Drake, C. L. (Editors). *Geology of Continental margins,* 933–950. Springer-Verlag, New York.

—— 1976. Ophiolite obduction. *Tectonophysics* **31,** 93–120.

—— and BIRD, J. M. 1971. Origin and emplacement of the ophiolite suite: Appalachian ophiolites in Newfoundland. *J. geophys. Res.* **76,** 3179–3206.

GEIKIE, J. 1866. On the metamorphic Lower Silurian rocks of Carrick, Ayrshire. *Q. J. geol. Soc. Lond.* **22,** 513–534.

GRAHAM, C. M. and ENGLAND, P. C. 1976. Thermal regimes and regional metamorphism in the vicinity of overthrust faults: an example of shear heating and inverted metamorphic zonation from Southern California. *Earth Planet. Sci. Lett.* **31,** 142–152.

HEEZEN, B. and RAWSON, M. 1977. Visual observations of contemporary erosion and tectonic deformation on the Cocos Ridge crest. *Mar. Geol.* **23,** 173–196.

KELLING, G. 1961. The stratigraphy and structure of the Ordovician rocks of the Rhinns of Galloway. *Q. J. geol. Soc. Lond.* **117,** 37–75.

LAMBERT, R. ST J. and MCKERROW, W. S. 1976. The Grampian Orogeny. *Scott. J. Geol.* **12,** 271–292.

LAMONT, A. and LINDSTROM, M. 1957. Arenigian and Llandeilian cherts identified in the Southern Uplands of Scotland by means of conodonts, etc. *Trans. Edinb. geol. Soc.* **17,** 60–70.
MENDUM, J. R. 1968. Unconformities in the Ballantrae volcanic sequence. *Trans. Leeds geol. Soc.* **7,** 261–264.
MIYASHIRO, A., SHIDO, F. and EWING, M. 1970. Crystallisation and differentiation in the abyssal tholeiites and gabbros from mid-ocean ridges. *Earth Planet. Sci. Lett.* **7,** 361–365.
MOORES, E. M. and VINE, F. J. 1971. The Troodos Massif, Cyprus and other ophiolites as oceanic crust: evaluation and implications. *Phil. Trans. R. Soc.* A **268,** 433–466.
PEACH, B. N. and HORNE, J. 1899. The Silurian rocks of Great Britain 1, Scotland. *Mem. Geol. Surv. U.K.*
REINHARDT, B. M. 1969. On the genesis and emplacement of ophiolites in the Oman Mountains geosyncline. *Schweiz. mineral. petrogr. Mitt.* **49,** 1–30.
TYRRELL, G. W. 1909. A new occurrence of picrite in the Ballantrae District and its associated rocks. *Trans. geol. Soc. Glasg.* **13,** 283–290.
WALTON, E. K. 1961. Some aspects of the succession and structure in the Lower Palaeozoic rocks of the Southern Uplands of Scotland. *Geol. Rundsch* **50,** 63–77.
WILKINSON, J. M. and CANN, J. R. 1974. Trace elements and tectonic relationships of basaltic rocks in the Ballantrae igneous complex, Ayrshire. *Geol. Mag.* **111,** 35–41.
WILLIAMS, A., STRACHAN, I., BASSETT, D. A., DEAN, W. T., INGHAM, J. K., WRIGHT, A. D. and WHITTINGTON, H. B. 1972. A correlation of Ordovician rocks in the British Isles. *Geol. Soc. Lond. Spec. Rep.* No. 3.
WILLIAMS, H. and SMYTH, W. R. 1973. Metamorphic aureoles beneath ophiolite suites and Alpine peridotites: tectonic implications with West Newfoundland examples. *Am. J. Sci.* **273,** 594–621.
WOODCOCK, N. H. and ROBERTSON, A. H. F. 1977. Origin of some ophiolite related metamorphic rocks of the "Tethyan" belt. *Geology.* **5,** 373–376.

B. J. Bluck,
Department of Geology,
University of Glasgow,
Glasgow G12 8QQ, Scotland.

# Geology of a continental margin 2: middle and late Ordovician transgression, Girvan

*J. K. Ingham*

The Ballantrae Volcanic Group was deformed and uplifted, perhaps as obducted sheets, by mid-Ordovician times for conglomerates, limestones and greywackes of ?Upper Llanvirn–mid-Caradoc age are markedly transgressive northwestwards across a Ballantrae Volcanic Group "basement". The middle Ordovician to lower Silurian cover was deposited on an unstable slope or slopes probably paralleling an active subduction trench to the southeast. There is constant intertongueing of conglomerate wedges with thick proximal and distal turbidite sequences, together with some limestones, siltstones, mudstones and black shales. Slumping and other evidence of instability is widespread. Faunas are mainly graptolitic but there are numerous shelly horizons, much of the material washed in or slump-derived and there are some deep water indigenous faunas of a highly specialised nature.

## 1. Introduction

The essentially ophiolitic Ballantrae rocks were already severely deformed and had accreted or were in process of accreting to the northwestern plate margin by middle Ordovician times for the middle to upper Ordovician sedimentary cover in the Girvan area is markedly transgressive northwards or northwestwards across the Ballantrae "basement". An active subduction trench became established to the south of the Girvan district, probably in the vicinity of the line of the Southern Uplands fault (McKerrow *et al.* 1977) and there a thick succession, mainly of greywacke and shale, was built up during middle and late Ordovician times which sequentially imbricated and became progressively accreted to the new plate margin.

## 2. The Stinchar Valley succession

At Girvan the oldest, post-Ballantrae sediments are found in the southern part of

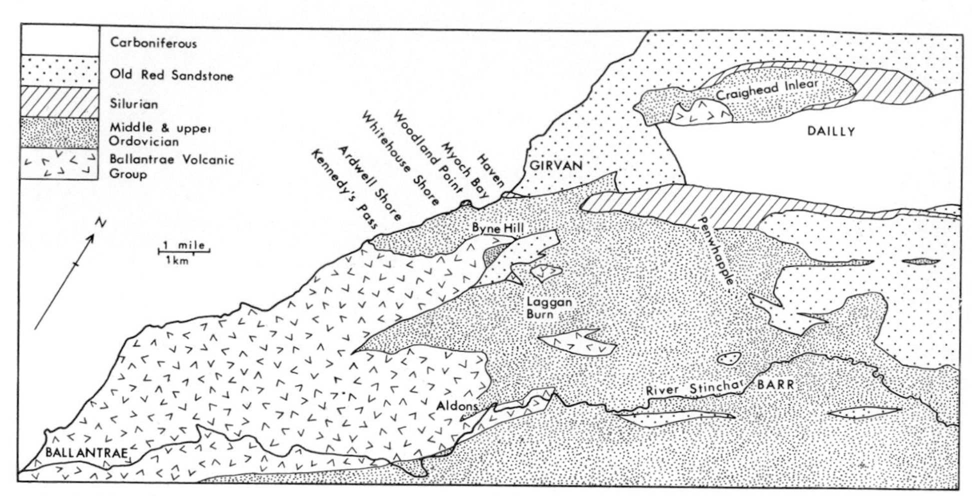


Fig. 1. Simplified geological map of the Girvan district.

the area, in the vicinity of the Stinchar Valley (Figs 1, 2). The succession there begins with the Kirkland Conglomerate, at least 250 m in thickness which, together with other Ordovician conglomerates in the area, was interpreted by Williams (1962) as a slide conglomerate, probably deposited against a contemporaneous normal fault or faults throwing down to the southeast. At its summit occurs a locally developed impure limestone, the Auchensoul Limestone, whose sparse shelly fauna has largely yet to be described. On the northern side of the Stinchar Valley the Kirkland Conglomerate passes upwards into the calcareous, sometimes pebbly sandstone of the Confinis Flags, some 45 m thick, which contain a rich shelly fauna (Williams 1962; Tripp 1962) and these in turn are succeeded by approximately 60 m of Stinchar Limestone, also richly fossiliferous locally (Williams 1962; Tripp 1967). The overlying Didymograptus superstes Mudstone reaches a maximum of 40 m in thickness and locally has an abundant shelly fauna at its base which is calcareous (Williams 1962; Tripp 1976). These beds are succeeded by the thickest known development of the Benan Conglomerate (up to 640 m), another suspected slide conglomerate which has, in places, scoured deeply into the D. superstes Mudstone and locally rests on Stinchar Limestone. In the vicinity of Aldons Quarry, the Kirkland Conglomerate and the Confinis Flags are not developed and here it is the pebbly lower part of the Stinchar Limestone which rests directly on Ballantrae rocks.

Along the southern side of the Stinchar Valley a different development is seen consisting for the most part of greywackes, conglomerates and mudstones (the Traboyack and Albany divisions). Locally, the Albany Mudstone Formation is calcareous and has yielded the most southerly shelly faunas in the Girvan area (Williams 1962; Tripp 1965; Ross and Ingham 1970) which indicate a correlation with the Stinchar Limestone to the north.

As Williams (1962 pp. 58–62) pointed out, the shelly faunas of the Stinchar Valley area in particular bear little resemblance to approximately contemporaneous faunas in the Anglo-Welsh region but show very great similarities to middle Ordovician shelly faunas from the Appalachian belt of eastern North America. It

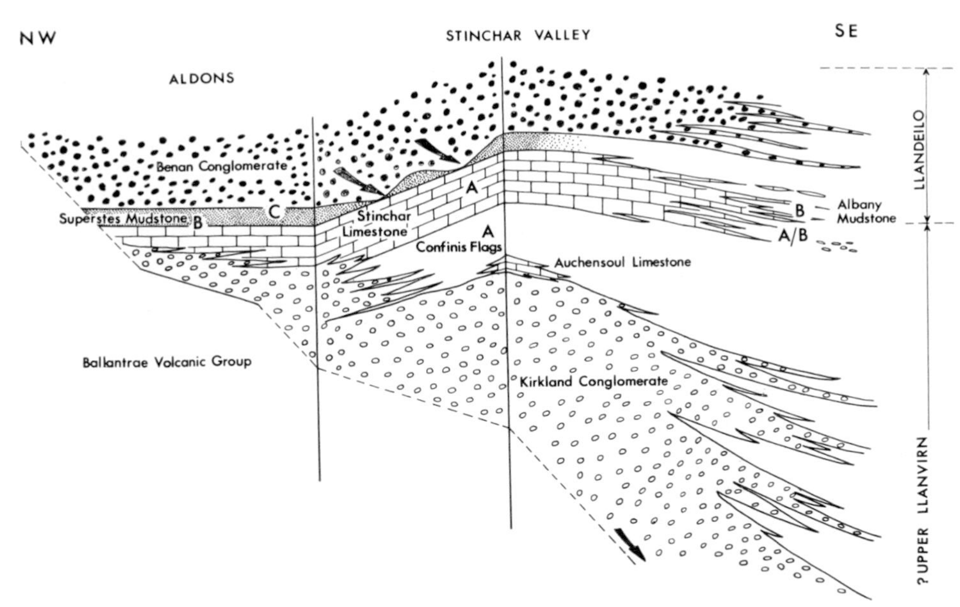


Fig. 2. Schematic fence diagram of the Stinchar Valley middle Ordovician succession (not to scale); capital letters explained in the text.

is not surprising therefore that shelly communities recognised albeit in the lower Ordovician of Spitzbergen (Fortey 1975a, 1975b), which was part of the American Province, should find some general application in the Girvan area. Fortey recognised three benthic trilobite associations characterising increasingly deep environments together with an association of pelagic forms whose widespread distribution was not controlled by water depth. Decreasing endemicity with increasing depth is a particularly important observation Fortey drew from his data. His shallowest benthic community, dominated by cheirurid and illaenid trilobites, has its closest Girvan parallel in the faunas of the Confinis Flags and the Stinchar Limestone (A in Fig. 2) whereas a rather deeper water benthos was dominated by nileid trilobites and such faunas are particularly characteristic of the upper part of the Albany Mudstone to the south of the River Stinchar and the basal Superstes Mudstone further northwest (B in Fig. 2). His deepest, olenid community is not represented in the Stinchar Valley: its place is perhaps taken by an impoverished fauna of inarticulate brachiopods and graptolites in the main part of the Superstes Mudstone(C in Fig. 2). Thus, not only is the transgressive nature of the Stinchar Valley succession reflected in local stratigraphical detail, but the faunas show evidence of progressive deepening, with diachronous effect, from southeast to northwest. The culmination of this local transgressive phase is the very thick Benan Conglomerate wedge which, as already noted, has, in places, channelled through the Superstes Mudstone.

## 3. The age of the Stinchar Valley succession

In spite of the great faunal dissimilarity between the middle Ordovician at Girvan and the Anglo-Welsh area, the Stinchar Valley sequence has been regarded as Caradoc in age (Williams 1962). This was primarily because the *Nemagraptus gracilis* Zone, which has supposed correlatives in the Superstes Mudstone, was for a number of years interpreted as the lowest graptolite zone represented in the Caradoc Series. Addison (*in* Williams *et al.* 1972 pp. 35–36) proved this not to be the case in the Llandeilo and adjacent areas in South Wales and showed that whereas only a small portion of the *N. gracilis* Zone is represented in the lowest part of the type Caradoc Series in Salop, the bulk of the zone is of Llandeilo age and its base probably lies within the Lower Llandeilo of the type area. The *Glyptograptus teretiusculus* Zone, previously equated with the whole of the Llandeilo Series, has thus been relegated to its lowest part. This situation was anticipated by Bergström (1971 p. 110) who, on the basis of his researches on conodonts in Europe and North America, would go further. A conodont zonal scheme, initially established in Sweden and linked with varying degrees of precision to the Ordovician graptolite zones, has been widely recognised in the Atlantic region and its application, particularly in parts of the Appalachians and also in Britain, has provided correlations which are, in terms of traditional assessments, controversial.

Bergström recorded a fauna characteristic of the *Eoplacognathus lindstroemi* Subzone of the *Pygodus serrus* Zone from the highest pre-type Llandeilo Ffairfach Group. In Sweden, this correlates with a lower part of the *Glyptograptus teretiusculus* graptolite zone, the implication being that the Upper Llanvirn Series includes not only the *Didymograptus murchisoni* Zone but a substantial part of the *G. teretiusculus* Zone. Macrofaunal evidence appears to be somewhat contradictory (Bassett, *in* Williams *et al.* 1972 p. 33) and the position is not yet settled but Bergström considers that the important *Pygodus serrus*/*P. anserinus* zonal boundary approximates closely to the Llanvirn/Llandeilo boundary. This correlation affects the interpretation of the age of the Girvan succession for the conodont zonal boundary just noted falls within the Stinchar Limestone. Thus, on Bergström's assessment, the lower part of the Stinchar Limestone, the Confinis Flags and the Kirkland Conglomerate are likely to be late Llanvirn in age (Table 1).

Williams (1962 p. 58) draws particular attention to the close brachiopod faunal similarity, even at specific level, with the Pratt Ferry and Little Oak formations of Alabama and it is perhaps significant that at Pratt Ferry itself, Bergström (1971 p. 117) has indicated that the *P. serrus*/*P. anserinus* zonal boundary lies within the Pratt Ferry Formation, although this is not necessarily the case throughout Alabama as both formations seem to be diachronous. Both the Pratt Ferry and Little Oak formations, together with the Arline Formation in eastern Tennessee, were correlated by Cooper (1956) with his Porterfield Stage, whose type area is in southern Virginia, on the basis of their very similar brachiopod assemblages and this correlation was followed by Williams. Bergström's interpretation, on the basis of conodont zonation, is very different. He considers that many of the widespread middle Ordovician limestone formations of the southern Appalachians are, in general, diachronous upwards towards the northwest and that the type Porterfield Stage, of approximately middle Llandeilo to early Caradoc age is younger than the previously equated horizons of Alabama and Tennessee. This controversy is not yet resolved.

A late Llanvirn age for the lower part of the Stinchar Valley succession appears to be in conflict with the youngest age recorded for the Ballantrae Volcanic Group in the Craighead Inlier at Girvan. Lamont and Lindström (1957 p. 65) noted a conodont fauna from Craighead Hill in cherts said to be infolded with pillow lavas which

Table 1. A correlation of the type series with graptolite and conodont zonations in the middle part of the Ordovician System

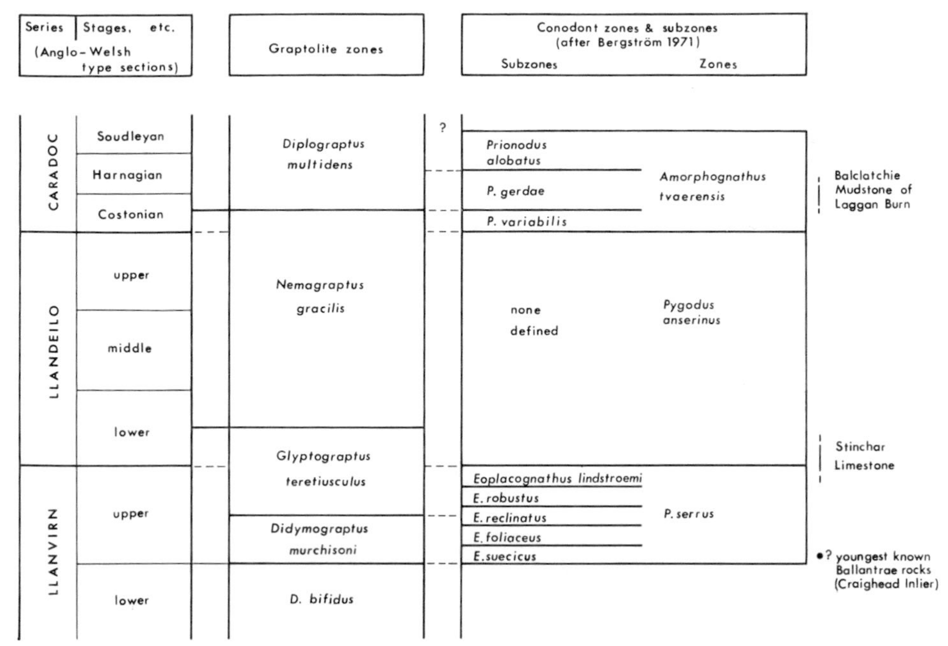

| Series | Stages, etc. (Anglo-Welsh type sections) | Graptolite zones | Conodont subzones (after Bergström 1971) | Conodont zones (after Bergström 1971) | |
|---|---|---|---|---|---|
| CARADOC | Soudleyan | *Diplograptus multidens* | ? *Prionodus alobatus* | *Amorphognathus tvaerensis* | |
| | Harnagian | | *P. gerdae* | | Balclatchie Mudstone of Laggan Burn |
| | Costonian | | *P. variabilis* | | |
| LLANDEILO | upper | *Nemagraptus gracilis* | none defined | *Pygodus anserinus* | |
| | middle | | | | |
| | lower | *Glyptograptus teretiusculus* | | | Stinchar Limestone |
| LLANVIRN | upper | | *Eoplacognathus lindstroemi* | *P. serrus* | |
| | | | *E. robustus* | | |
| | | *Didymograptus murchisoni* | *E. reclinatus* | | |
| | | | *E. foliaceus* | | |
| | | | *E. suecicus* | | • ? youngest known Ballantrae rocks (Craighead Inlier) |
| | lower | *D. bifidus* | | | |

they dated as "Llandeilian". The fauna is dominated by *Periodon aculeatus*, an important species in the *Pygodus serrus* and *P. anserinus* zones. Its common occurrence at Craighead, without species of *Pygodus*, could indicate a basal *P. serrus* zonal age, i.e. basal Upper Llanvirn in terms of Bergström's conodont zonal correlation, but the evidence is not conclusive. Nevertheless, there may be little, if any, time gap between the youngest Ballantrae rocks of Craighead and the lowest part of the Stinchar Valley succession.

## 4. The coastal and inland sequences north of the Tormitchell thrust

The Ordovician rocks of the Girvan district have been foreshortened by asymmetrical open folding of late or post-Silurian age but more importantly by large thrusts of the same age whose slightly undulating planes dip to the southeast. To the north of the Tormitchell thrust (Williams 1959, 1962) the Stinchar Valley succession below the Benan Conglomerate is not represented. Moreover, the great thickness of the Benan Conglomerate seen to the north of the Stinchar Valley and in the vicinity of Aldons Farm is already reduced to about 60 m over the crest of the Aldons anticline. In the coastal area, where the Benan Conglomerate is the basal formation of the middle and upper Ordovician sedimentary sequence it reaches about 180 m in thickness. Here it is overlain by approximately 90 m of fossiliferous greywackes—

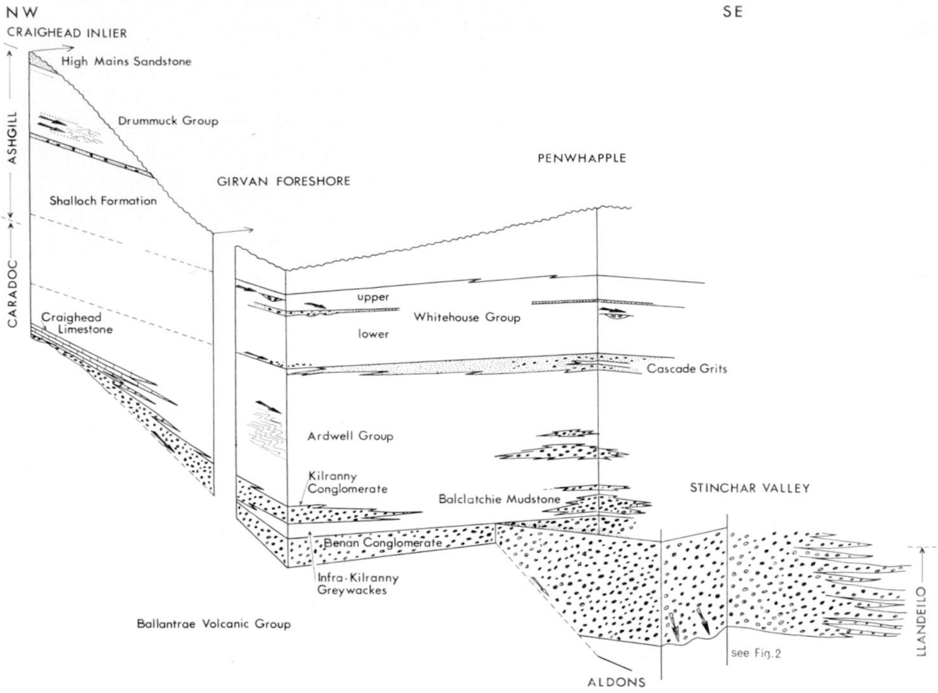

Fig. 3. Schematic fence diagram of the middle and upper Ordovician sections in the Girvan coastal and inland areas, together with that in the Craighead Inlier.

the Infra-Kilranny Greywackes of Williams (1962)—and these in turn are overlain by another conglomerate wedge, the Kilranny Conglomerate, up to 150 m thick, which is of the same general type as the Benan Conglomerate. These are unsorted conglomerates for the most part, the predominantly igneous cobbles usually not being in contact but supported by chloritic sand and silt. There is imbrication evident in many layers (Hubert 1966 p. 678) suggesting deposition, at least of the Kilranny Conglomerate, by currents which flowed to the southeast. The Ardwell Flags which follow have a thickness in excess of 1200 m along the coast and they rest with apparent discordance on the Kilranny Conglomerate—Henderson's unconformity (Henderson 1935). This supposed unconformity may not be real for the basal Ardwell Flags are seen to rest with seeming overlap in a channel at the summit of the conglomerate. Moreover, there is evidence of movement, not necessarily tectonic, both at the junction and within the basal Ardwell Flags. The Ardwell Flags therefore may have moved to some extent over the Kilranny Conglomerate, perhaps penecontemporaneously. The Ardwell Flags along the foreshore consist for the most part of fine greywackes with laminated muds and silts. Some horizons show much penecontemporaneous brecciation and intraformational folding (Henderson 1935; Hubert 1965) with *in situ* stromatolitic algal mats at one horizon (Bluck, personal communication). Whether the depositional environment for the Ardwell Flags was wholly neritic (Hubert 1966) or at least partly a downslope turbidite regime (Kuenen 1953; Williams 1962) has still to be resolved.

Along the southwestern and central sections of the Ardwell foreshore, the Ardwell Flags are thrown into spectacular cascading folds which Williams (1959 pp. 631–634, 663–664) interpreted as representing a local tectonic phase within the general Caledonide deformation. New information (Bluck, personal communication) reveals that, whereas some features, such as conjugate and reverse faulting related to the disposition of the fold axes, suggest tectonism, there are a number of fold-related features which are not of tectonic origin. These include sand flow at fold closures and, in at least one instance, the emplacement of an axial planar sandstone dyke. Added to this the fact that the fold belt is stratigraphically confined and can be traced some distance inland is suggestive of a major slump sheet, perhaps of partly consolidated sediment. If this is the case, and when the general steep northwesterly dip in the coastal area is taken into account, the direction of down-slope slumping would have been from the northwest.

Inland, the Ardwell Flags reach an estimated 1400 m in thickness and contain a number of local conglomerate wedges, some thick. In Penwhapple Burn, where the Ardwell Flags succession is much foreshortened by thrusting, the beds are underlain by the dark Balclatchie Mudstone and conglomerates. These beds have yielded a trilobite fauna dominated by raphiophorids and remopleuridids together with a brachiopod fauna including many inarticulates. A possible slope environment is suggested. At the Laggan Burn locality, calcareous horizons in a Balclatchie succession have yielded a rich and well-preserved graptolite fauna of early Caradoc age which was monographed by Bulman (1945–1947). Many of the graptolites show current orientation. Bergström (1971 p. 114) noted a conodont fauna from this locality, also suggesting an early Caradoc age and indicative of the *Amorphognathus tvaerensis* Zone.

Williams (1962 p. 32) showed that the Balclatchie sequence of the inland sections is equivalent to the Infra-Kilranny Greywackes, the Kilranny Conglomerate and at least 45 m of the basal Ardwell Flags in the foreshore area. The fauna of the Ardwell Flags is predominantly graptolitic and orthocone fragments are relatively common. One well known "upper Balclatchie", shelly horizon low in the unit at Ardmillan Braes, 1 km east of Ardwell Farm has yielded a rich brachipod fauna and many well preserved, usually complete trilobites representing, however, few genera. These fossiliferous beds are of limited stratigraphical distribution and the preponderance of complete trilobite specimens is suggestive of rapid accumulation, perhaps under the influence of strong bottom currents transporting mud and silt, together with the associated benthos, some distance from its original site. Similar mass transport is known to have occurred at various levels higher in the Ordovician at Girvan (see below). The Ardwell graptolite faunas indicate a late middle or late Caradoc age for the top of the unit. In Penwhapple Burn and the surrounding area, the summit of the Ardwell Flags is marked by the Cascade Grits and Conglomerates which have interbedded graptolitic mudstones. On the foreshore, the same graptolite fauna is known from the Whitehouse Shore, adjacent to Ardwell Bay. Here, striped mudstones are interbedded with sandstone tongues and these probably represent the Cascade horizon inland. Lapworth (1882 pp. 597, 606) mistakenly included these beds as the basal division of the Whitehouse Group: their recognition as Ardwell Flags negates Williams (1962 p. 41) supposition that some 300 m of Ardwell Flags are absent in the vicinity of Ardwell Bay owing to major faulting.

On the Whitehouse Shore the Ardwell Flags are followed conformably by the lithologically diverse Whitehouse Group. One of the most recently published stratigraphical and sedimentological summaries of this division was given by Hubert (1966 p. 683) but in this work no account was taken of important strike and reverse faults affecting the foreshore sections. Consequently, there are correlation inconsistencies and thickness inaccuracies in his assessment.

The Lower Whitehouse Group has three, as yet un-named formations. The lowest division, with an estimated thickness of 60 m or more, is a "limestone flysch" succession (Dzulynski 1964 p. 249) as noted by Hubert (1966) in which, particularly in the lower part, graded sandy limestones alternate with unfossiliferous green shales. The coarser portion of each graded unit is commonly cross- or dune-bedded (Hubert 1966 pp. 684–685) and consists of terrigenous debris of volcanic and igneous rocks, presumably mainly derived from a Ballantrae "basement", together with some metamorphic material and shell debris. The basal bed, resting on the Ardwell Flags, is particularly coarse and probably best described as a breccia. Many of the angular fragments are of locally derived Ardwell Flags and were probably broken up by strong bottom currents. Higher in each graded unit, current ripples are common. Some graded units are multiple. Cross-bed and flute orientation data collected by Hubert indicate deposition by currents flowing from the west and northwest. Hubert was of the opinion that much of the Girvan succession was deposited in shallow water and he preferred not to use the term "turbidite". However, it is not known whether the environment was deep water or otherwise although the palaeoslope was presumably inclined to the northwest. Indentifiable shelly material in the graded limestones is rare but the trilobite *Tretaspis*, of the *ceriodes* species group, has been found. This group is characteristic of the late Caradoc Upper Chasmops Limestone in the Oslo region of Norway, similar horizons in Sweden and also from high Actonian and Onnian (latest Caradoc) strata in the Welsh Borderlands and northern England.

The limestone flysch unit is followed on the foreshore by a sandstone flysch division although the junction between the two is somewhat gradational. Much of this division, which as yet has yielded no fossil material, consists of thin, but very persistent grey graded sandstone beds interleaved with green micaceous shales. Flutes indicate a current derivation from the northeast and the succession may represent a distal turbidite sequence deposited parallel to the northwestern margin of the sedimentary basin. This division has a maximum known thickness of 45 m. Its summit is nowhere seen because of major reverse faulting which brings it into juxtaposition with Upper Whitehouse beds.

The third and uppermost division of the Lower Whitehouse Group was not known to Hubert. It is absent from the Whitehouse Shore for reasons given above although parts of it are exposed in the lower reaches of Myoch Burn, adjacent to Myoch Bay. The best section is that seen in Penwhapple Burn where about 90 m of striped grey mudstones and siltstones with some sandstone beds lie below the Upper Whitehouse Group. The lithology is similar to that of much of the Ardwell Flags and is locally rich in graptolites indicating the *Pleurograptus linearis* Zone. In Penwhapple Burn there occurs one thick, slumped mudstone conglomerate containing both graptolites and shelly fossil debris, presumably derived from up-slope locations.

The Upper Whitehouse Group contains several well-defined, albeit diachronous divisions and, for the most part, consists of green and reddish silty mudstones. At or near the base of the division in the foreshore area is a variably thick pebbly sandstone and mudstone conglomerate unit. A considerable degree of channelling occurs locally at the base of some of the sub-units of this bed and the sandstones commonly contain flat pebbles of locally derived green shale. Cross-bed data from the sandstones (Hubert 1966 pp. 688, 690) indicate deposition by strong, scouring bottom currents from the northwest. The overlying reddish and green silty mudstones, together with the sandstone and conglomerate, total approximately 30 m in thickness. An upward sequence in the mudstones typically consists of green silty mudstones with impersistent thin sandstones in the lower part, alternating green and reddish silty mudstones followed by similar beds which are predominantly reddish in colour. These colour divisions are markedly diachronous both within and at the top of the unit and the boundaries show an apparent northeasterly general descent. Sandstone dykes occur in places. Ripple and cross-lamination in the thin sandstones indicates current directions from the northwest.

The reddish beds, particularly near the top of the unit contain the remains of a large indigenous fauna which was unknown to Lapworth (1882) and which will be described elsewhere, together with a complete stratigraphical revision. The fauna consists mostly of bizarre trilobites of which both blind and large-eyed forms are the dominant species. (For a list of genera, see Ingham and Williams *in* Bassett *et al.* 1974 p. 58.) Important among the large-eyed forms are the probably pelagic cyclopygids, telephinid and remopleuridid. Some of the blind forms, such as the raphiophorids and agnostids may also have been pelagic, but others, like the dionidids and some specialised trinucleids, were benthic animals of restricted environmental distribution. Many of the forms occur together in down-slope deposits in the late Ordovician of the mid-Wales basin. The restricted brachiopod fauna from the reddish mudstones belongs to the *Foliomena* community (D. A. T. Harper, personal communication) which was regarded by Sheehan (1973 p. 60), following Berry (1972), as belonging to a relatively deep benthic marine life zone. This community, dominated by trilobites, has been regarded in the Silurian of Britain as intermediate between the *Clorinda* community and the graptolite facies (Ziegler *et al.* 1968 p. 23). Hubert (1966 pp. 690–691) preferred to regard these Upper Whitehouse beds as being of shallow water prodeltaic origin. In the writer's opinion this thesis is untenable and the facies conforms to that recognised in red-bed sequences by Ziegler and McKerrow (1975) in the Silurian of the British Isles, Norway and eastern North America where it was identified in deep shelf, basin and ocean-floor environments. The existence of red-bed sequences other than in non-marine situations was attributed to derivation from oxidised source material and rapid burial in quiet marine environments which precluded reduction. The writer regards the depositional environment of the bulk of the Upper Whitehouse sediments as having been the subsiding unstable slopes of a submarine fan.

Locally, there are green sandstone and siltstone wedges intertongueing with the upper part of the reddish mudstones both in the foreshore area and at an equivalent horizon in Penwhapple Burn. In the latter section, the occurrence of a *Tretaspis* of the *ceriodes* species group again indicates a late Caradoc age for this part of the succession.

In the Myoch Bay section, the upper part of the reddish mudstones pass laterally into a thin, mainly dark grey graptolitic shale which has yielded a fauna containing *Dicellograptus complanatus* although the bulk of the fauna is more typical of the *Pleurograptus linearis* Zone. Above this, a relatively thin succession of calcareous shales and siltstones has yielded typical *D. complanatus* zonal forms. Within this succession at Myoch Bay is a local, calcareous mudstone conglomerate lens, 0 to 5 m in thickness, whose matrix contains an up-slope fauna, dateable in Anglo-Welsh terms as being of Pusgillian age and thus belonging to the lower part of the Ashgill

Series. The conglomerate contains many local mudstone clasts, some of them contorted and was deposited rapidly in a local scour channel by means of sediment movement from an up-slope location. On the Whitehouse Shore, only the uppermost (3 m to more than 15 m) of these high Upper Whitehouse beds rest on the reddish mudstones and there has been local intermixing of mudstones at the junction, again indicating penecontemporaneous mass movement of sediment. In one section on the Whitehouse Shore, it is estimated that up to 12 m of strata, represented elsewhere, are missing at this junction. The following Shalloch Formation is lithologically very persistent in detail over wide areas and clearly a degree of topographic stability had been achieved at this level. Topographic restoration for the late Upper Whitehouse sediments, prior to the large scale movement of material, indicates an apparent palaeoslope to the northeast, i.e. parallel to the present outcrop pattern. This compares with the apparent direction of downwards diachronism noted above. The actual palaeoslope direction was probably to the east or east-southeast.

The highest Upper Whitehouse beds resting on the reddish mudstones in the Whitehouse Shore outcrop consist of grey and green striped mudstone and shale, containing small flakes of similar rocks at some levels, followed by shales with lensing cross-bedded limestones in shallow scour channels. The limestones consist for the most part of comminuted shelly debris and the cross-beds indicate a current derivation from the southeast. The shales contain *D. complanatus* Zone graptolites and a trilobite and brachipod fauna in thin seams similar to that in the underlying reddish mudstones, although additional genera are present. It is from this horizon that the bulk of the Gray Collection (British Museum, Natural History), used in early monographs, was derived.

The Shalloch or Barren Flags Formation begins with a thin, grey mudstone unit which is persistent over a wide area. Above this, some 330 m of flysch-like sandstones with greenish and grey shales extends to the unconformable base of the Silurian in the Myoch Bay/Haven section, although further southwest half of this thickness has been removed below the Silurian overstep. Within the width of Woodland Point, over 12 m of beds are cut out at this junction. The sandstones of the Shalloch Formation are particularly persistent and individual beds can be correlated between the foreshore and Penwhapple Burn. Grading is rare, although horizontal lamination and current-ripple sequences are common. Convolute bedding also occurs and there are numerous sandstone dykes in the lower part of the sequence. At several horizons, thin graded limestones also occur with, in one case, recognisable low Ashgill shelly debris. Graptolites are known from two levels. *D. complanatus* has been found at 9 m from the base and *D. anceps* at about 180 m. According to Hubert (1966 pp. 691–694) the current direction changes within the Shalloch Formation from a southwesterly to a predominantly northeasterly derivation and this correlates with a change from greenish to predominantly grey shales, reflecting the difference in derivation of material.

## 5. The Craighead Inlier

The lowest post-Ballantrae Ordovician sediments in the Craighead Inlier are the Craighead Limestones with their associated basal conglomerates. Williams (1962) described in detail these predominantly shallow water rocks, which were deposited over a very irregular volcanic basement, some within wave base. A variety of limestone lithologies is present, ranging from brecciated algal limestones to sandy and rubbly pelmatozoan limestones and calareous shales, all of them interdigitating in a complex manner. The Kiln Mudstone tongue, near the top of the limestone sequence, has yielded rich faunas which were monographed by Tripp (1954) and Williams (1962). Williams correlated the Craighead Limestones with the upper part of the Ardwell Flags to the south and suggested that they equated with a horizon perhaps as high as the Cascade Grits. However, the fact that the limestones are overlain by at least 300 m of siltstones and shales containing an Ardwell graptolite fauna (the Plantinhead Flags) suggests that a rather lower Ardwell horizon is represented.

There is no outcrop of any part of the Whitehouse Group or the lower part of the Shalloch Formation in the Craighead Inlier owing to the disposition of a large east–west fault extending from the northern side of Craighead Hill. The upper part of the Shalloch Formation is known, however, and probably includes beds younger than those known from either the foreshore area or from Penwhapple Burn. It is overlain by the Drummuck Group, some 400 m thick, which has a thin basal sandstone and conglomerate, the Auldthorns Conglomerate of Lamont (1935) and this is followed by the Quarrel Hill Mudstones (lower Drummuck Group). Trilobite, brachiopod and sponge faunas were described by Lamont (1935) primarily from small exposures on Quarrel Hill but recently, Forestry Commission excavations at Glenmard Quarry have produced extensive exposures. These show convincingly that the Quarrel Hill Mudstones mainly comprise a series of major mud flow sheets with exquisitely preserved shelly debris and more complete material, gathered together in pockets. Neither the sediment nor the fauna is indigenous to the locality and is presumably derived from an unknown upslope location. The fauna is dated as low middle Ashgill (mid Cautleyan) age.

The mudstones and siltstones of the middle part of the Drummuck Group are poorly exposed and have yielded few fossils but the upper part, comprising the Ladyburn Mudstones and Starfish Beds, have produced extensive faunas of trilobites, brachiopods, molluscs, echinoderms and some current oriented graptolites. Most specimens, whether from the sandy Starfish Beds or from the mudstones, are complete and Goldring and Stephenson (1972), in a study of selected starfish beds, considered that the fauna was entombed by turbulence which stirred up sediment and animals and redeposited them with some concentration of the fauna. They ascribed the disturbance to a shallow, turbulent marine environment (pp. 615–616) but judging from evidence of slope instablity in other parts of the Ordovician succession at Girvan, it is conceivable that down-slope currents and turbulence may be involved here also. The Ladyburn Mudstones are poorly bedded and thus resemble much of the Quarrel Hill Mudstone in Glenmard Quarry but direct evidence of sediment flow is lacking. The age of these high Drummuck beds is late mid Ashgill (high Rawtheyan Stage). By Ashgill times, marine faunas generally are less provincial than in middle Ordovician times and although this also applies to the Girvan succession, the faunas still contain elements of American aspect which are not found in contemporaneous Anglo-Welsh faunas. Among the trilobites, *Cryptolithus* species and *Flexicalymene* of the *meeki* group, are notable examples.

The Drummuck Group is followed conformably by the High Mains Sandstone of latest Ordovician (Hirnantian) age. Outcrops are limited by the basal Silurian unconformity but a rich *Hirnantia* brachiopod fauna has been collected and described (Lamont 1935) and this is probably of shallow water origin.

## 6. The basal Silurian unconformity

Reference has already been made to the pronounced southerly overstep of the basal Silurian, of low Llandovery age, in the foreshore area. Here, neither highest Shalloch Formation, Drummuck Group nor High Mains Sandstone have representatives preserved. The degree of overstep is reflected in the Craighead Inlier to the northeast where the stratigraphical break is slight. The basal Silurian unit in the Craighead Inlier is the Mulloch Hill Conglomerate (Ladyburn Conglomerate of Cocks and Toghill 1973), whereas the local, thick scour channel Craigskelly Conglomerate, of approximately the same age, is the basal bed at the Haven. There is no basal conglomerate either in Penwhapple Burn or on much of Woodland Point. It may be that the unconformity is not a subaerial one but reflects a shallowing and subsequent down-slope channelling in earliest Silurian times.

Slope instability was still an important factor in the Silurian regime at Girvan. The Quartz Conglomerate tongue at the base of the Scart (= Saugh Hill) Grits at the Haven, for example, has visibly scoured into the underlying low Llandovery Shales of the Woodland Formation, with much contemporaneous deformation of the unconsolidated strata and intrusion of shale laminae by the conglomerate (see Kuenen 1953 pp. 41–42).

## 7. Conclusions

The overall picture of post-Ballantrae Lower Palaeozoic stratigraphy and sedimentation is one of progressive northwesterly transgression over a subsiding, structurally complex volcanic terrain. There is much evidence of local rapid subsidence linked with slope instability and the general regime deduced by Williams (1962) for the middle Ordovician is probably a true reflection of the broad environmental pattern.

Active, low-angle normal faulting in the basement, paralleling the palaeogeographical grain and throwing down towards the southeast seems to have been the controlling factor in the formation of slope basins and the accumulation of downslope and laterally transported debris. This regime, which persisted from middle Ordovician through early Silurian times in the Girvan area is a reflection of the existence of an ancient continental margin in southern Scotland.

## References

BASSETT, D. A., INGHAM, J. K. and WRIGHT, A. D. 1974. *Field excursion guide to type and classical sections in Britain.* The Palaeontological Association Ordovician System Symposium. Birmingham 1–66.

BERGSTRÖM, S. M. 1971. Conodont biostratigraphy of the Middle and Upper Ordovician of Europe and eastern North America. *Geol. Soc. Am. Mem.* **127,** 83–157.

BERRY, W. B. N. 1972. Early Ordovician bathyurid province lithofacies, biofacies and correlations—their relationship to a proto-Atlantic Ocean. *Lethaia* **5,** 69–83.

BULMAN, O. M. B. 1945–47. A monograph of the Caradoc (Balclatchie) graptolites from limestones in Laggan Burn, Ayrshire. *Palaeont. Soc. Monogr.,* Parts **1-3**, 1–42 (1945); 43–58 (1946); i-ix, 59–78 (1947).

COCKS, L. R. M. and TOGHILL, P. 1973. The biostratigraphy of the Silurian rocks of the Girvan district, Scotland. *J. geol. Soc. Lond.* **129,** 209–243.

COOPER, G. A. 1956. Chazyan and related brachiopods. *Smiths. Misc. Coll.* **127,** Part 1 (text), 1–1024; Part 2 (plates), 1025–1245.

DZULYNSKI, S. 1964. Flysch facies. *Ann. Soc. Geol. Pol.* **34,** 245–266.

FORTEY, R. A. 1975a. Ordovician trilobite communities. *Fossils and Strata* **4,** 339–360.

—— 1975b. The Ordovician trilobites of Spitzbergen II. Asaphidae, Nileidae, Raphiophoridae and Telephinidae of the Valhallfonna Formation. *Nor. Polarinst. Skr.* No. **162,** 1–207.

GOLDRING, R. and STEPHENSON, D. G. 1972. The depositional environment of three starfish beds. *Neues Jahrb. Geol. Paläont. Mh.* 611–624.

HENDERSON, S. M. K. 1935. Ordovician submarine disturbances in the Girvan district. *Trans. R. Soc. Edinb.* **63,** 487–509.

HUBERT, J. F. 1966. Sedimentary history of upper Ordovician geosynclinal rocks, Girvan, Scotland. *J. Sediment. Petrol.* **36,** 677–699.

KUENEN, P. H. 1953. Graded bedding, with observations on Lower Palaeozoic rocks of Britain. *K. Ned. Akad. Wet. Afd. Nat. Verh.* **20,** 1–47.

LAMONT, A. 1935. The Drummuck Group, Girvan; a stratigraphical revision with descriptions of new fossils from the lower part of the group. *Trans. geol. Soc. Glasg.* **19,** 288–334.

—— and LINDSTRÖM, M. 1957. Arenigian and Llandeilian cherts identified in the Southern Uplands of Scotland by means of conodonts, etc. *Trans. geol. Soc. Edinb.* **17,** 60–70.

LAPWORTH, C. 1882. The Girvan succession. *Q. J. geol. Soc. Lond.* **38,** 537–666.

MCKERROW, W. S., LEGGETT, J. K. and EALES, M. H. 1977. Imbricate thrust model of the Southern Uplands of Scotland. *Nature* **267,** 237–239.

ROSS, R. J. and INGHAM, J. K. 1970. Distribution of the Toquima–Table Head (Middle Ordovician Whiterock) faunal realm in the northern hemisphere. *Bull. geol. Soc. Am.* **81,** 393–408.

SHEEHAN, P. M. 1973. Brachiopods from the Jerrestad Mudstone (Early Ashgillian, Ordovician) from a boring in southern Sweden. *Geol. et Palaeont.* **7,** 59–76.

TRIPP, R. P. 1954. Caradocian trilobites from mudstones at Craighead Quarry, near Girvan, Ayrshire. *Trans. R. Soc. Edinb.* **62,** 655–693.

—— 1962. Trilobites from the "Confinis" Flags (Ordovician) of the Girvan District, Ayrshire. *Trans. R. Soc. Edinb.* **65,** 1–40.

—— 1965. Trilobites from the Albany Division (Ordovician) of the Girvan District, Ayrshire. *Palaeontol.* **8,** 577–603.

—— 1967. Trilobites from the upper Stinchar Limestone (Ordovician) of the Girvan district, Ayrshire. *Trans. R. Soc. Edinb.* **67,** 43–93.

—— 1976. Trilobites from the basal *superstes* Mudstones (Ordovician) at Aldons Quarry, near Girvan, Ayrshire. *Trans. R. Soc. Edinb.* **69,** 369–423.

WILLIAMS, A. 1959. A structural history of the Girvan district, S.W. Ayrshire. *Trans. R. Soc. Edinb.* **63,** 629–667.

—— 1962. The Barr and Lower Ardmillan Series (Caradoc) of the Girvan district, south-west Ayrshire, with descriptions of the brachiopoda. *Mem. geol. Soc. Lond.* No. **3,** 1–267.

——, STRACHAN, I., BASSETT, D. A., DEAN, W. T., INGHAM, J. K., WRIGHT, A. D. and WHITTINGTON, H. B. 1972. A correlation of Ordovician rocks in the British Isles. *Geol. Soc. Lond. Spec. Rep.* No. **3,** 1–74.

ZIEGLER, A. M., COCKS, L. R. M. and BAMBACH, R. K. 1968. The composition and structure of Lower Silurian marine communities. *Lethaia* **1,** 1–27.

—— and MCKERROW, W. S. 1975. Silurian marine red beds. *Am. J. Sci.* **275,** 31–56.

J. K. Ingham,
Hunterian Museum,
University of Glasgow,
Glasgow G12 8QQ, Scotland.

# Geology of a continental margin 3: gravity and magnetic anomaly interpretation of the Girvan–Ballantrae district

*D. W. Powell*

A map of the residual Bouguer gravity and vertical magnetic component anomalies of the Girvan–Ballantrae district is presented. The low density and high susceptibility of the serpentinite are reflected in the anomalies over its northern outcrop. The cross-section of the serpentinite body is calculated from its magnetic effect. An observed positive gravity anomaly, modified by correction for the pseudo-gravity effect of the serpentinite, is interpreted as an extensive body of older basic rocks shallowly underlying the southern serpentinite outcrop.

## 1. Introduction

The magnetic field contours of Figure 1 are based on about 2000 vertical component readings with supplementary control on the lateral extent of some anomalies from total field readings. The station spacing is variable; across anomalies it is normally at 50 m intervals along lines less than 1 km apart. The 250 nT contour interval, from an arbitrary datum of 1000 nT, is dictated by the scale of the map and the detail required to outline weaker anomalies. Interpretation is mainly concerned with the distribution of serpentinite.

The gravity field contours are based on about 500 readings reduced to sea level, against IGF (1930), as terrain corrected Bouguer anomalies with a 2·72 gm/cc density. A linear regional, shown on the map, has been subtracted. Stations lie at 100 to 500 m intervals along roads and tracks, more sparsely elsewhere. The source of the central high [Nat. Grid 2195.5895] is the chief problem requiring explanation. The physical properties of the rocks are indicated in Table 1. A plot of the magnetic susceptibility aganst density (Fig. 2) for rocks of the area shows a roughly inverse relationship which goes some way towards accounting for the absence of magnetic anomalies over the main gravity high and their concentration over the gravity low northwest of it.

## 2. Discussions of anomalies

The features of the residual gravity map in relation to geological boundaries are:

1. 'Lows' over serpentinite (northern belt) and near Nat. Grid position [217.587] due to a −0·2 gm/cc contrast with the volcanic rocks.

2. Gradients on the flanks of the main 'high'. (a) The gravity gradient from [217.591] curving to [2140.5885] is associated with the fault shown between the serpentinite belts (Fig. 1). Probably the outcrop of older basic rock which straddles this gradient at the northern locality is related to a concealed dense body causing the gravity effect. If so, the fundamental structure does not directly join this outcrop to that of similar rocks near Lendalfoot: rather these outcrops are related to en échelon faults. (b) Approximately along the east and southeast margins of the 2 km wide volcanic block, over which the 'high' culminates, the volcanic block is in part faulted against sediments with a −0·06 gm/cc contrast. Alone, this requires a lava thickness of 7 km to produce the observed amplitude but the computed gradients are then too small. To match the gradients a contrast $\geqslant$ 0·1 gm/cc and a thickness $\geqslant$ 2·5 km seem necessary. Towards the northeast corner of the 'high' [220.592] there seems no correlation with the observed geology.

It is concluded that, for the present purpose, the volcanics and sediments can be grouped together, i.e., positive contrasts imply older basic rocks, negative ones serpentinite.

The principal features on the magnetic map relatable to geological units are:

1. The line of anomalies along its south and southeast limit which are caused by the greywacke containing serpentinous detritus, principally from [217.582] to [225.591], and lavas, principally from [209.582] to [215.585]. These are both members of a younger series than the ophiolite and faulted against it.

2. Relatively minor anomalies due to igneous pebble conglomerates north of the Upper Stinchar fault zone from [220.589] to beyond the map at [225.592].

3. The general lack of anomalies over the volcanic rocks indicating that low susceptibility volcanics predominate within the ophiolite complex. Exceptions are near [215.595] where some non-porphyritic spilites are found with high susceptibilities and near [2095.5865] where agglomerates may be the source of the magnetic variation. Elsewhere between the serpentinite belts, anomalies are more likely to be due to serpentinite in fault zones than to volcanic rocks.

4. Anomalies over the serpentinites with amplitudes of 2000 nT confirming that, under their northern outcrop particularly, the susceptibility is frequently over 0·006 cgs. Lower, less variable field values near [217.593] are consistent with a specimen from there with a susceptibility of less than 0·005 cgs and a density of 2·83 gm/cc (note the 'saddle' between the gravity 'lows' on Fig. 1).

To the northeast beyond its main outcrop, serpentinite (susceptibility over 0·006 cgs) forms lenticular bodies rising into cover rocks where traversed by fault zones. Some intensity variations within the main mass are also across shear zones, but their complex pattern is generally contoured arbitrarily. Best defined are those in

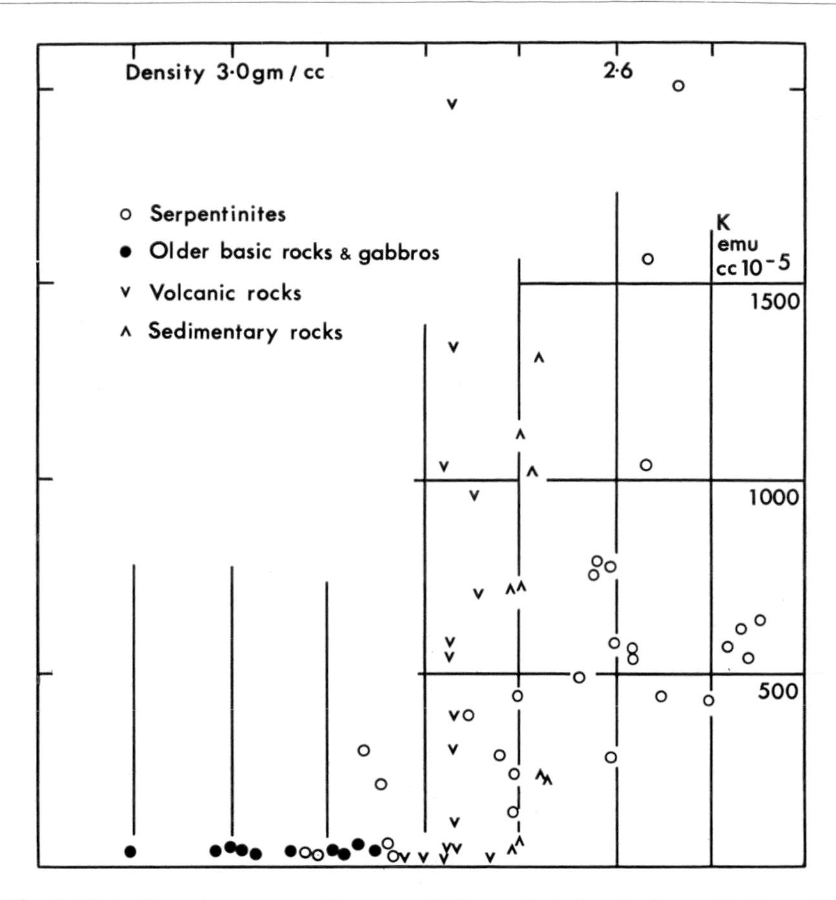


Fig. 2. Plot of specimen magnetic susceptibility against density for rocks from the Girvan–Ballantrae district.

Table 1

| *Rock Type* | *Density gm/cc* | | *Susceptibility emu/cc* | |
|---|---|---|---|---|
| | *Mean* | *Range* | *Mean* | *Range* |
| Serpentinites | 2·56 | 2·43 to 2·92 | 0·006 | 0·0001 to 0·03 |
| Volcanic rocks | 2·77 | 2·74 to 2·85 | 0·0025 | 0·0001 to 0·03 (sub-area dependent) |
| Older basic rocks & gabbros | 2·95 | 2·80 to 3·10 | 0·00025 | 0·0001 to 0·003 |
| Sedimentary rocks | 2·71 | 2·66 to 2·73 | 0·002 | 0·0001 to 0·013 (sub-area dependent) |

the eastern half of the southern serpentinite. A minor gravity 'low' [216.587] coincides with part of one. Apart from these, neither strong magnetic anomalies nor gravity 'lows' occur over this belt. Its northern margin is marked by a weak but persistent magnetic 'low'. The possibilities are either that this serpentinite exhibits weak contrasts towards its surroundings (Fig. 2), or it is relatively thin. Although the mineralogical possibility of the first correlation has not been specifically studied, the choice of interpretation requires some discussion of it.

Derivation from a harzburgite parent (essentially olivine with orthopyroxene) with a density of 3·2 gm/cc indicates that under serpentinisation when the density has become, say, 2·73 gm/cc:

$$2{\cdot}73 = (\text{vol}\,\%\,(\text{OL}+\text{OPX})\times 3{\cdot}2) + (\text{vol}\,\%\,\text{serpentine} \times 2{\cdot}45) + (\text{say}\ 3\,\%\ \text{Fe-oxides} \times 5{\cdot}0).$$

This suggest that two thirds of the volume is serpentine. Since the ratio OL/OPX, estimated from the 45–46% $SiO_2$ in the serpentinite analyses (Balsillie 1932), adjusted as water-free, is around 2 to 1 it is concluded that this density (2·73 gm/cc) represents a point where serpentine has replaced olivine but not pyroxene. The relative resistance of the latter is seen in many specimens. The possibility of serpentinite showing little magnetisation contrast with volcanic rocks in this part of the area requires it, like most of the lavas between the two belts, to be weakly magnetised (say less than 0·001 cgs). This seems hardly possible if all the olivine has been altered since it probably carried half or more of the total FeO in the parent rock. Petrographic evidence shows micron size iron oxide granules, mainly magnetite from susceptibility, outlining former olivine grain boundaries and much finer dust in the serpentine of their cores. $En_{75}$ with $Fo_{95}$, dividing FeO fairly equally between the two for a 1:2 ratio, is about the Mg:Fe partitioning limit with a co-existing pair (Ramberg and De Vore 1951). Also, there is no evidence that any of the main mass of harzburgite has survived in a less serpentinised state.

It is concluded that the southern serpentinite is more likely to be thin than less than 66% serpentinised. Weakly magnetised rocks underlie it. The gravity high suggests that these are dense and stretch beyond the serpentinite outcrop under the adjacent volcanic rocks. The older basic rocks have the required properties and their sporadic distribution (Fig. 1) is consistent with this possibility.

## 3. Quantitative models

Two models are considered. Both commenced with fitting a serpentinite body to a magnetic anomaly. Its gravity effect was calculated and added to the Bouguer residual field. A dense body representing older basic rocks was adjusted to match this.

The first magnetic model (0·005 cgs) needed to be 3 to 4 km thick and, fitted to the simpler aeromagnetic profile, was found to be too laterally extensive to fit the ground magnetic anomaly. Its gravity effect was 12 mgal and it transformed the observed gravity high so that its crest was shifted to mid-way between the serpentinite outcrops. The second magnetic model (Fig. 3) (0·008 cgs) is a reasonable fit to the anomalies at both levels. Its gravity effect barely removes the low over the northern serpentinite. Whilst open to some improvement, no fundamental change is to be expected unless the basic assumptions are altered.

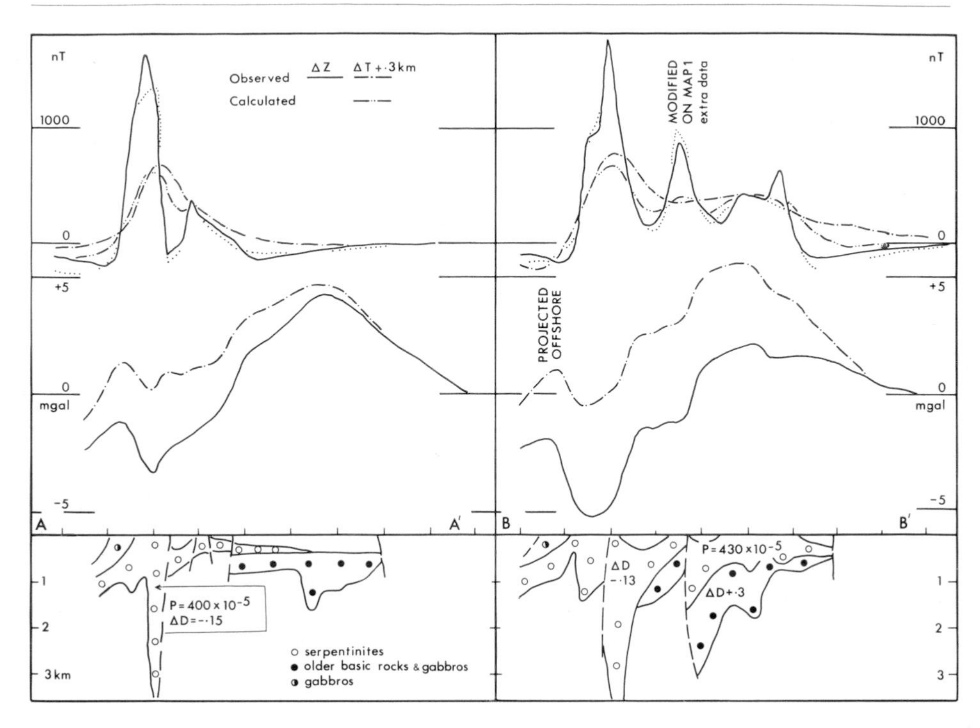


Fig. 3. Observed and calculated magnetic and gravity anomalies along profiles A and B shown in Fig. 1. The pseudo-gravity effect of the serpentinite has been added to the residual gravity profile (solid line) to produce the profile on which the dense body is modelled. P = polarisation contrast in cgs; △D = density contrast in gm/cc.

Since laboratory measurements indicated that the older basic rocks have velocities over 6·0 km/sec. compared with 5·0 km/sec. for volcanic rocks and even less for serpentinite, field refraction measurements were made in an attempt to find high velocity material beneath the northern end of the southern serpentinite. The results were inconclusive because not only was there no evidence of a concealed high velocity layer within a kilometre of the surface but there was also no evidence that the older basic rocks at outcrop had distinctive velocities.

*Acknowledgment.* I thank Dr. J. Hall for making the refraction measurements.

## References

BALSILLIE, D. 1932. The Ballantrae igneous complex, south Ayrshire. *Geol. Mag.* **69**, 107–131.
RAMBERG, H. and DE. VORE, G. 1951. The distribution of $Fe^{++}$ and $Mg^{++}$ in coexisting olivines and pyroxenes. *J. Geol.* **59**, 193–220.

D. W. Powell,
Department of Geology,
University of Glasgow,
Glasgow, G12 8QQ, Scotland.

# Cogenetic and inherited zircon U–Pb systems in granites: Palaeozoic granites of Scotland and England

*R. T. Pidgeon and M. Aftalion*

Zircon U–Pb isotopic systems (84 new analyses) are described for 24 Caledonian granites from the Northern and Grampian Highlands, the Midland Valley, the Southern Uplands, the Lake District and Leicestershire. Zircons from pre-tectonic granites from the Grampian and Northern Highlands, including the Carn Chuinneag and Ben Vuirich granites, the Ardgour granite gneiss and the granite at Vagastie Bridge, have complex U–Pb systems which indicate the presence of a significant component of *c*. 1600 m.y. old zircon. Zircon U–Pb systems from most of the Newer granites (440$\pm$10 m.y.) from the Grampian and Northern Highlands also indicate the presence of inherited 'old' zircon. None of the granites investigated from the Midland Valley (Distinkhorn), the Southern Uplands or the Lake District (except Eskdale) and Leicestershire show isotopic evidence of inherited zircons. This suggests that either these granites formed under different conditions from those to the north, and all old zircon xenocrysts were dissolved, or that older zircon xenocrysts were not present in these granites, which were presumably derived from new Palaeozoic crust. We favour the second explanation and tentatively mark the Highland Boundary fault as the fundamental structure separating a dominantly 'Proterozoic' basement to the north from a mainly middle Palaeozoic basement to the south. Age determinations based on the present U–Pb results indicate a common age of *c*. 560 m.y. for the Ardgour granite gneiss and the Carn Chuinneag granite. They also support previous views that the main Grampian metamorphism in the Highlands was completed 475 m.y. ago and that the major influx of Newer granites occurred 400$\pm$10 m.y. ago. The *c*. 405 m.y. age for the foliated granite at Vagastie Bridge presents a problem as this is in apparent conflict with previous conclusions on the age of the regional $D_2$ deformation in this part of the Moine assemblage.

## 1. Introduction

Most of the granites of the British Caledonides were emplaced in late Silurian to early Devonian time although granite episodes are known to span much of the history of the Caledonian orogeny. Despite differentiation and contamination, the granites retain a chemical memory of their source and isotopic studies of elements

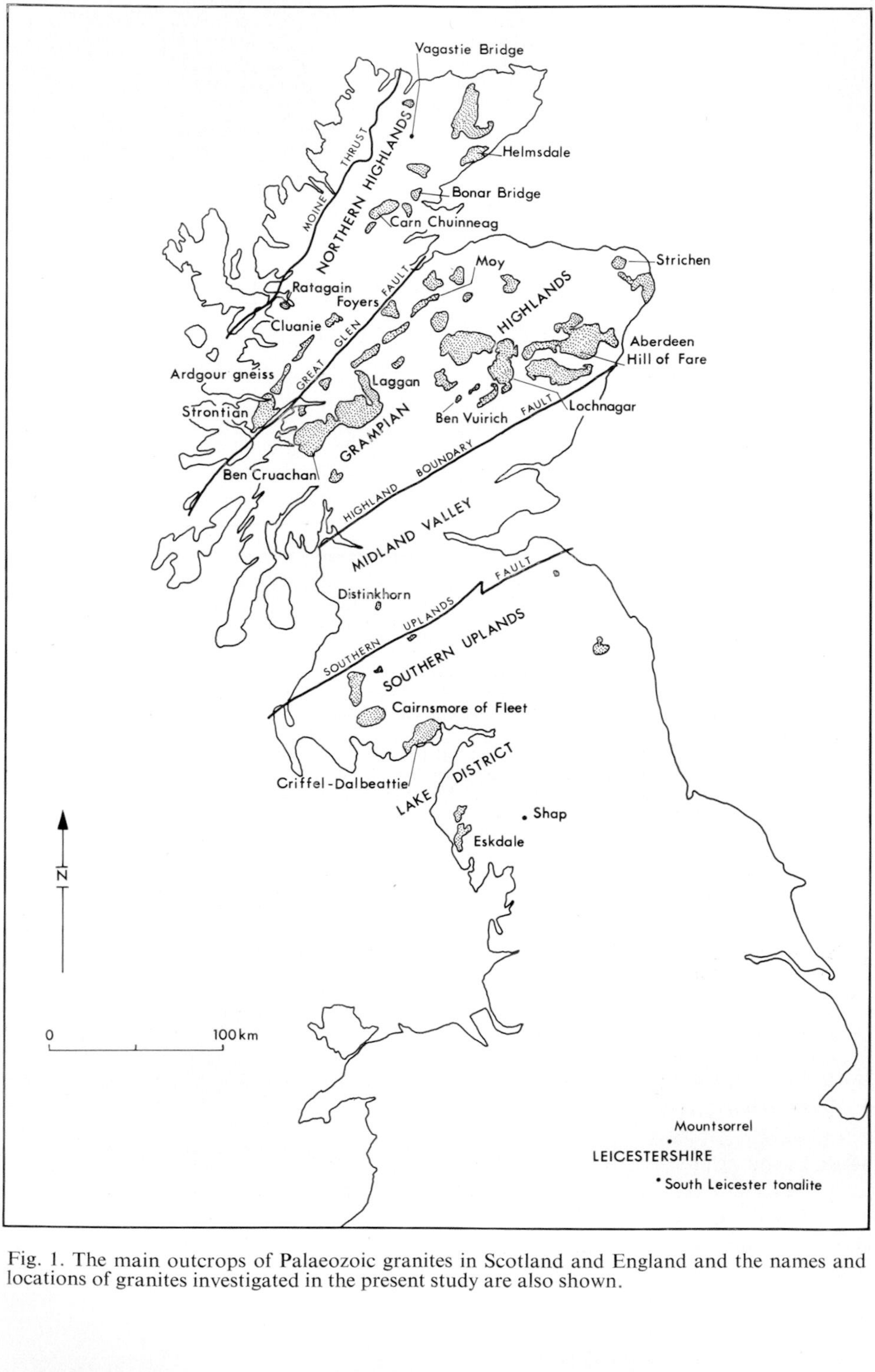

Fig. 1. The main outcrops of Palaeozoic granites in Scotland and England and the names and locations of granites investigated in the present study are also shown.

such as Sr, Pb, O and H yield information on the processes of granite formation and the nature of the source rocks. Zircon is important for under favourable circumstances it may, at least partially, survive anatexis and granite emplacement. Zircon xenocrysts in intrusive granites can be detected through their U–Pb isotopic systems, and in this paper we describe the distribution of inherited and cogenetic zircon in north British granites and examine the implications of the results on the age and origin of the granites and the nature of the granite source rocks.

The term granite is used in a general sense to cover granitoid rocks such as granite, granodiorite, tonalite etc. We report 84 new zircon U–Pb analyses, and analyses of a monazite and a sphene, from a total of 22 Caledonian granites from the Northern Highlands to central England (Fig. 1). Many important granites have not been investigated and generally we have investigated only one sample from each complex. Consequently we have only limited information on the comparative behaviour of zircon U–Pb systems between the different rock types comprising a single complex and on U–Pb systems of zircons from granite xenoliths and no information at present on the zircons from the Moinian and Southern Uplands sediments. Nevertheless, this study points to the significance of zircon U–Pb systems as valuable isotopic indicators of processes of granite formation and of the nature of granite source rocks.

## 1a. Methods

Laboratory work was performed at the Scottish Universities Research and Reactor Centre using the chemical techniques described by Krogh (1973). Isotope ratios were measured on a modified AEI–GEC MS–12 mass spectrometer with digitised data output and on-line processing using a Data General Nova computer. Lead was analyzed using the silica gel technique. Uranium was loaded on $Ta_2O_5$ with $H_3PO_4$ on a single rhenium filament and analysed as $UO^+$. Results for the NBS radiogenic lead standard (SRM 983) over the latter period of this work were:— $^{207}Pb/^{206}Pb = 0{\cdot}07122 \pm 0{\cdot}00010$, $^{208}Pb/^{206}Pb = 0{\cdot}01373 \pm 0{\cdot}00010$ and $^{204}Pb/^{206}Pb = 0{\cdot}00039 \pm 0{\cdot}00004$. Repeat analyses are reported for seven zircon samples and the monazite. The zircon repeats cannot be considered true duplicates as they represent small aliquots from zircon fractions known to be inhomogenous. This is shown in Figure 3 where a large component of the variation in the repeat analyses (samples 14, 15, 16) can be attributed to sample inhomogeneity. Over the six year period of the study analytical precision steadily improved and earlier measurements are not, as a rule, as precise as later ones. This has caused difficulties in assigning errors as it has not been possible to reassess every measurement in the course of writing this paper. So to construct error boxes and calculate regressions we have used an overall uncertainty of $\pm 0{\cdot}3\%$ in $^{207}Pb/^{206}Pb$ and $\pm 0{\cdot}6\%$ in Pb/U. Where possible the position of chords has been calculated by the method of least squares, using a correlation coefficient of 0·7. The intersections of the chord with concordia were then determined by allowing the slope of the chord to vary about its centroid by plus and minus its two sigma limits.

The constants used in this paper are as follows: $\lambda^{235}U = 0{\cdot}98485 \times 10^{-9} y^{-1}$; $\lambda^{238} = 0{\cdot}155125 \times 10^{-9} y^{-1}$; $^{238}U/^{235}U$ was taken as 137·88. Correction for the common lead in the zircon analyses was made using the assumed composition of $^{206}Pb/^{204}Pb = 18{\cdot}1$, $^{207}Pb/^{204}Pb = 15{\cdot}5$ and $^{208}Pb/^{204}Pb = 36{\cdot}8$. For Rb–Sr and K–Ar ages we have used $\lambda_\beta = 1{\cdot}39 \times 10^{-11} y^{-1}$ for $^{87}Rb$ and $\lambda_e = 0{\cdot}584 \times 10^{-10} y^{-1}$, and $\lambda_\beta = 4{\cdot}72 \times 10^{-10} y^{-1}$ for $^{40}K$.

## 2. Previous observations on zircon U-Pb systems in granites

Ideally, the U–Pb ages of cogenetic zircons from granitic rocks should be concordant i.e. the $^{206}Pb–^{238}U$ ages should be the same (within the experimental uncertainty) as the $^{207}Pb–^{235}U$ ages. In general, this is rarely so for the U–Pb systems are usually disturbed or 'discordant'. In discordant systems the $^{206}Pb–^{238}U$ and $^{207}Pb–^{235}U$ ages are invariably too low, indicating that the zircons have leaked radiogenic lead or accumulated uranium at some time since initial crystallization.

Silver (1963) showed that cogenetic zircon suites from individual samples of granite had variable uranium contents and that the relative degree of isotopic discordance was a function of the uranium (+Th) concentrations. Armed with this knowledge Silver (1963) and Silver and Deutsch (1963) were able to divide their cogenetic zircon suites into a number of size and magnetic fractions with different uranium contents and different degrees of isotopic discordance. On a concordia plot the plotted Pb/U ratios of these zircon fractions were seen to define a line (discordia). The upper intersection of the discordia with concordia, at the point of zero discordance, gives the age of the zircon (Wetherill 1956). The lower intersection with concordia indicates the age of the event responsible for disturbing the zircon U–Pb isotopic system. This applies for a single episodic disturbance. Where the zircons have undergone a number of isotopic disturbances or have lost lead by slow continuous diffusion (Tilton 1960; Wasserburg 1963) the interpretation is more complex. An important feature of this behaviour is that the $^{207}Pb/^{206}Pb$ apparent ages are never older than the true age of the granite.

In strong contrast to this are discordant U–Pb patterns of zircons from a number of granites which point to the influence of *older inherited zircon*. In these systems the $^{207}Pb/^{206}Pb$ ages and the Pb/U ages are generally 'older' than the *known* age of granite emplacement. Some examples of intrusive granites with clearly demonstrated inherited zircon components are the Vire–Carolles Granite from Normandy (Pasteels 1970), the Bergell Granite from the Southern Alps (Gulson and Krogh 1973), the Idaho Batholith (Grauert and Hofmann 1973) the Gunpowder Granite from Maryland (Grauert 1973) ,and the Carn Chuinneag granite (Pidgeon and Johnson 1974) and the Ben Vuirich granite (Pankhurst and Pidgeon 1976) from the Scottish Highlands. The points of non-magnetic zircon size fractions from these granites frequently show a generally linear correlation on a concordia plot (e.g. Grauert 1973 and Pankhurst and Pidgeon 1976 and Figs 2–5 in this paper). In these systems the lower intersections with concordia indicate (approximately) the age of granite emplacement and the discordia lines trend backward in time. To distinguish this behaviour from the previously described 'normal' zircon discordance we have adopted the terminology of Gulson and Rutishauser (1976) and refer to such linear discordia patterns as 'reverse discordia'.

The lower intersection of a reverse discordia will closely approximate the age of the granite if the zircons have remained closed system to U–Pb migration since emplacement. However we have already noted that cogenetic zircons are rarely concordant. Consequently we would expect that zircon points defining a reverse discordia would, in general, have also experienced some isotopic disturbance subsequent to granite emplacement. It has been observed that later disturbances of such zircon U–Pb systems tend to rotate the discordia or distort it into a curve (Gulson and Rutishauser 1976) or to disturb the linearity of U–Pb points, especially in the case of relatively high uranium 'magnetic' zircon (Krogh and Davis 1972; Pidgeon and Hopgood 1976).

An excellent example of the complexities of behaviour of zircon U–Pb systems in different rock types from a single granite complex is described in a study of the Carn Chuinneag granite (Pidgeon and Johnson 1974) mentioned below.

## 3. Zircon samples and their isotopic systems

In this section we outline the granite samples, the zircon morphology and the analytical results on zircon size and magnetic fractions. The pre-tectonic granites show deformational fabrics indicating that they were emplaced before the major Caledonian orogeny whose peak is considered by Dewey and Pankhurst (1970) to have been a short Arenig episode. These granites, such as the Carn Chuinneag and Ben Vuirich granites, are included in Read's (1961) early granites. The post-tectonic granites show little or no evidence of tectonism and were emplaced after the major Caledonian orogeny being the Newer and Late Granites of Read (1961).

### 3a. The pre-tectonic granites

(*i*) *The Carn Chuinneag granite–Inchbae granite* is the largest of the pre-kinematic granites. It is a complex body and has been described by Harker (1962), Johnson and Shepherd (1970) and Shepherd (1973). Johnson and Shepherd (1973) inferred that the main regional metamorphism and the $D_2$–$D_4$ structures in Moinian of northern Ross-shire and Sutherland post-date the intrusion of the Carn Chuinneag granite. The behaviour of zircon U–Pb systems from different rock types of the granite has been described by Pidgeon and Johnson (1974) who reported zircon U–Pb analyses on three separate samples of the main granite type (the Inchbae rock of Harker 1962), one sample of the minor more alkaline Lochan a' Chairn rock and two samples of the restricted riebeckite granite gneiss.

Zircons from two biotite granites, the Inchbae and Lochan a' Chairn rocks, contain excess radiogenic lead but none has been found in a third granite, the riebeckite gneiss. The excess radiogenic lead is attributed to zircon xenocrysts which have partially survived the formation of the granite magma and been incorporated in newly crystallizing zircon. The Inchbae results suggest $1500 \pm 200$ m.y. as the age of the source rocks (Fig. 2). The absence of excess radiogenic lead in the riebeckite gneiss zircons is attributed to either the progressive exclusion of zircon xenocrysts during the formation of the Lochan a' Chairn rock and riebeckite gneiss magmas or to a decrease in the stability of the zircon with increasing alkalinity of the magma. Zircons from the Inchbae rock have not been isotopically disturbed since granite emplacement. On the other hand zircons from the finer-grained Lochan a' Chairn rock have been slightly disturbed and the riebeckite gneiss zircons have experienced a strong recent isotopic disturbance. This suggests a correlation between granite type and the stability of the zircon U–Pb isotopic systems, though the mechanism is not understood. The stability of the Rb–Sr whole-rock systems can also be correlated with rock type. Whereas the Inchbae and Lochan a' Chairn rocks have not been isotopically disturbed since granite emplacement the Rb–Sr whole-rock systems of the riebeckite gneiss were strongly disturbed *c.* 425 m.y. ago.

(*ii*) *The Ardgour granitic gneiss* has a transitional contact with the surrounding high-grade migmatic Moine metasediments and most workers consider this to be a metasomatic derivative of the Moine metasediments (Harry 1953). Dalziel and Johnson (1963) showed that the gneiss formed after the regional $F_1$ folds but before

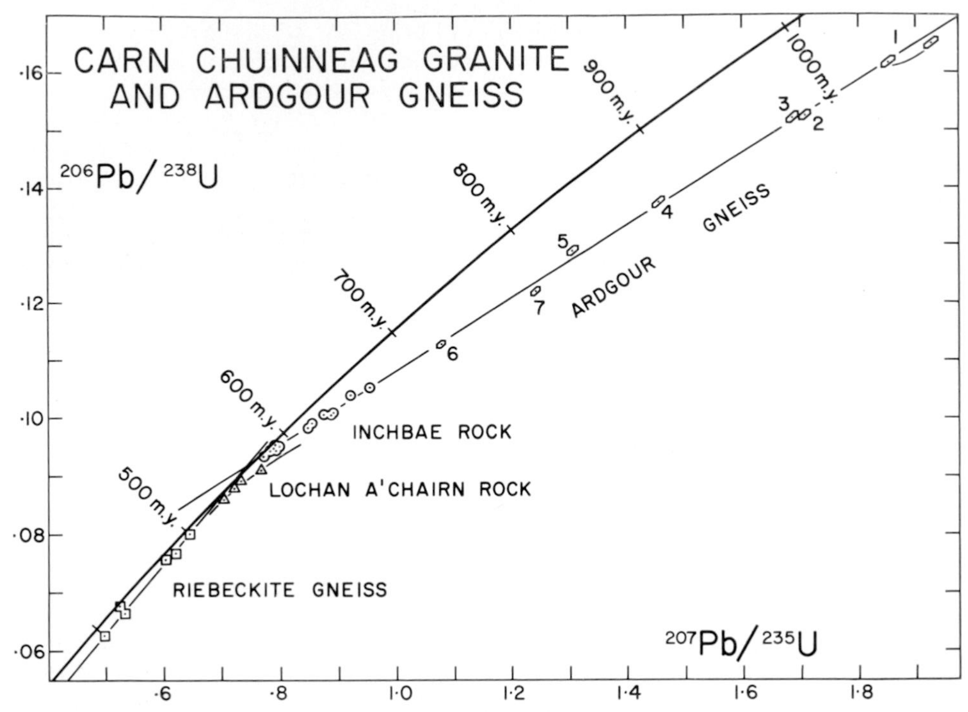


Fig. 2. Concordia plot showing the correlation of zircon points from the Ardgour gneiss (RC 705, number 1–7) and the Inchbae rock (RC137, 138 and 317) of the Carn Chuinneag granite on a single reverse discordia with intersections with concordia of 568 ± 3 m.y. and 1544 ± 26m.y. The lower intersection of this chord corresponds with the upper intersection of a 'normal discordia' through the riebeckite gneiss zircon points.

the major $F_2$ folds. Mercy (1963) however was struck by the marked magmatic features of the gneiss chemistry and speculated that this body could be a 'distinct kind of magma emplaced among high-grade metamorphic rocks'.

We report U–Pb analyses on zircons from a single sample of gneiss (RC705) from location 6 of van Breemen *et al.* (1974) [National Grid NH919800] just east of Glenfinnan. The rock is a biotite (with minor muscovite) bearing gneiss. Oriented biotites are red-brown with strong pleochroic halos around large zircons. Quartz makes 20% of the rock and plagioclase is the dominant feldspar. Garnet and subhedral apatite are accessory.

Zircons are invariably rounded and vary from needle-like (length/breadth ratios up to 6) to squat almost equidimensional oval and irregular shaped grains. A number of crystals are completely free of inclusions and many others contain rare, small, colourless, irregular shaped and needle-like bodies. A number of squat crystals are clouded with alteration. A few have elongate central cavities and rare grains have irregular zircon cores. No internal zoning was observed. Dalziel (1963) interpreted the rounded zircons as evidence that the gneiss formed from Moine metasediments without passing through a truly magmatic stage, In out view these morphological features are not sufficiently diagnostic to preclude a magmatic stage. For instance these zircons are remarkably similar to zircons from the intrusive Ben Vuirich granite (Pankhurst and Pidgeon 1976).

## Table 1. The pre-tectonic granites

| Zircon Fraction (size in microns) | Pb µg/g | U µg/g | $^{206}Pb/^{204}Pb$ | Atom percent radiogenic lead 206 | 207 | 208 | Atomic Ratios $^{207}Pb/^{206}Pb$ | $^{207}Pb/^{235}U$ | $^{206}Pb/^{238}U$ |
|---|---|---|---|---|---|---|---|---|---|
| **Ardgour gneiss (RC705)** | | | | | | | | | |
| 1. + 116µ NM1° | 50.1 | 304.3 | 3585 | 84.42 | 7.017 | 8.564 | .08312 | 1.852 | .1616 |
| repeat | 53.5 | 318.1 | 6545 | 84.54 | 7.157 | 8.305 | .08466 | 1.928 | .1651 |
| 2. −116 + 106µ NM1° | 47.9 | 309.7 | 5885 | 84.85 | 6.893 | 8.257 | .08123 | 1.708 | .1525 |
| 3. −106 + 84µ NM1° | 55.2 | 360.1 | 4130 | 85.31 | 6.861 | 7.833 | .08043 | 1.685 | .1520 |
| 4. −84 + 61µ NM1° | 50.3 | 365.0 | 11,000 | 85.90 | 6.596 | 7.502 | .07679 | 1.456 | .1375 |
| 5. −61 + 45µ NM1° | 51.1 | 401.0 | 2214 | 86.92 | 6.393 | 6.685 | .07355 | 1.307 | .1288 |
| 6. −45µ NM1° | 46.3 | 421.2 | 8300 | 88.23 | 6.102 | 5.663 | .06916 | 1.075 | .1128 |
| 7. −84 + 61µ M8° | 52.1 | 426.7 | 2082 | 85.77 | 6.349 | 7.884 | .07403 | 1.243 | .1218 |
| **Vagastie Bridge granite (RC662)** | | | | | | | | | |
| 8. + 142µ NM1° | 44.7 | 548.5 | 3325 | 86.06 | 5.684 | 8.252 | .06603 | .7420 | .08148 |
| 9. −106 + 84µ NM1° | 50.9 | 602.0 | 5000 | 85.49 | 5.693 | 8.812 | .06660 | .7719 | .08406 |
| 10. −61 + 45µ NM1° | 57.9 | 764.4 | 11190 | 84.61 | 5.222 | 10.16 | .06172 | .6336 | .07445 |
| 11. −45µ NM1° | 62.2 | 867.9 | 6935 | 83.34 | 4.849 | 11.81 | .05819 | .5565 | .06937 |
| 12. sphene | 12.8 | 193.9 | 158.1 | 88.06 | 4.972 | 6.972 | .05635 | .5256 | .06766 |
| **Vagastie Bridge granite (RC663)** | | | | | | | | | |
| 13. + 116µ NM1° | 65.0 | 728.2 | 5620 | 86.19 | 5.986 | 7.825 | .06945 | .8562 | .08941 |
| 14. −106 + 84µ NM1° | 79.5 | 830.6 | 4927 | 86.11 | 6.264 | 7.628 | .07152 | .9610 | .09581 |
| repeat | 76.1 | 807.9 | 8960 | 86.16 | 6.292 | 7.542 | .07303 | .9505 | .09440 |
| 15. −84 + 61µ NM1° | 81.7 | 829.5 | 8787 | 86.00 | 6.371 | 7.631 | .07409 | 1.006 | .09845 |
| repeat | 78.2 | 776.5 | 6902 | 85.74 | 6.417 | 7.841 | .07485 | 1.036 | .10036 |
| 16. −61 + 45µ NM1° | 78.7 | 908.4 | 7160 | 85.71 | 5.867 | 8.420 | .06844 | .8148 | .08634 |
| repeat | 78.5 | 895.0 | 11,940 | 85.67 | 5.958 | 8.367 | .06954 | .8379 | .08738 |

U–Pb systems of the seven analysed size and magnetic fractions (Table 1) show a regular increase in isotopic discordance with increasing uranium concentration and decreasing size—in accordance with observations on cogenetic zircon suites (e.g. Silver and Deutsch 1963). On a concordia plot (Fig. 2) the Pb/U points fall on a single chord with intersections with concordia at $574 \pm 30$ m.y. and $1556 \pm^{73}_{66}$ m.y. The degrees of discordance are extreme as all points fall in the lower half of this chord.

This chord is coincident, in slope and lower intersection, with the reverse discordia through zircon points from the Inchbae rock of the Carn Chuinneag granite (Fig. 2) We interpret this as strong evidence of a relationship between these rocks and discuss this later.

(*iii*) *The Ben Vuirich granite* is one of a number of small stocks which outcrop in the Dalradian in the neighbourhood of Pitlochry, Perthshire. It has a well developed foliation and was intruded before the peak of metamorphism ($M_2$ of Rast 1963; $M_3$ of Johnson 1962) which is locally of kyanite grade. Bradbury *et al.* (1976) regarded the granite as having been emplaced in the Dalradian between the second and third deformation episodes of a Caledonian $D_1$–$D_4$ sequence.

The sample is a uniform K–feldspar, oligoclase, biotite gneiss with accessory sphene and garnet. Zircons are light yellow to brown and generally rounded. Pankhurst and Pidgeon (1976) refer them to Dalziel's (1963) groups B and C, referring respectively to euhedral grains whose crystal edges and corners are slightly rounded and to well rounded grains. Pankhurst and Pidgeon (1976) reported U–Pb analyses on six size and magnetic fractions from this rock. Pb/U points are strongly discordant and define a single chord with intercepts with concordia of $514^{+6}_{-7}$ m.y. and $1316^{+26}_{-25}$ m.y. Geological constraints suggest that the lower intersection records the post $F_2$, pre-$M_3$ emplacement age. The upper intersection reflects the presence of old zircon xenocrysts which have survived in the granite magma without complete isotopic resetting. The ultimate source of these xenocrysts could be a metamorphic basement complex which formed about 1320 m.y. ago.

(*iv*) *Vagastie Bridge granite*. Two pieces (RC662 and RC663) were taken from a lens of foliated augen granite in Strath Vagastie [National Grid NC534272] in Sutherland (Soper 1971; Soper and Brown 1971). They were hornblende bearing gneisses with dominant plagioclase and microcline phenocrysts, minor chloritized biotite and accessory sphene and subhedral apatite. The presence of euhedral hornblende and sphene suggests these phases crystallized from the granite magma; as well as this we can find no compelling petrographic evidence for correlating the granite with hybrid phases of the Carn Chuinneag granite as suggested by Soper (1971). The Vagastie Bridge granite possesses the regional $D_2$ B–tectonite fabric and, like the Carn Chuinneag granite, has a pre-$D_2$ age (Soper 1971; cf. Fergusson 1978).

The zircon populations are composite and show a variation in morphology from euhedral prismatic forms with length/breadth ratios up to 6, to rounded, indented and generally irregular shaped squat grains with cracked and cloudy interiors. A number of euhedral grains have light brown irregularly shaped central zircon cores surrounded by colourless euhedrally zoned zircon rims (Fig. 8).

U–Pb analyses on four zircon size fractions and one sphene from RC662 and four size fractions (three repeated) from RC663 are given in Table 1. On a concordia plot (Fig. 3) the Pb/U points are strongly discordant but fall on a single reverse discordia which intersects concordia at 405±11 m.y. and 1576±100 m.y. The single sphene point is essentially concordant and lies precisely on the point of lower intersection of the zircon discordia and concordia. This demonstrates that the zircons have not undergone a significant 'recent' isotopic disturbance and we interpret the lower intersection as the age of granite formation and emplacement at which time euhedral sphene crystallised and 1600 m.y. old zircon xenocrysts were encased in rims of new zircon.

## 3b. The post tectonic granites: Grampians and the Northern Highlands

(*i*) *The Helmsdale granite* sample (RC972) [National Grid ND028158] is a medium grained leucogranite composed mainly of tabular altered plagioclase, anhedral poorly twinned K–feldspar and quartz with accessory chloritised biotite, iron ore and a trace of muscovite. The zircon population is complex. Many grains are irregular with rounded corners and terminations. A few grains are euhedral. A number of sub-hedral grains have central zircon cores with rims of colourless zoned zircon. Inclusions and central holes were observed in a minority of grains.

Three zircon size fractions were analysed (Table 2). The uranium concentrations increase and the Pb/U ratios decrease as the size fractions become finer. On a

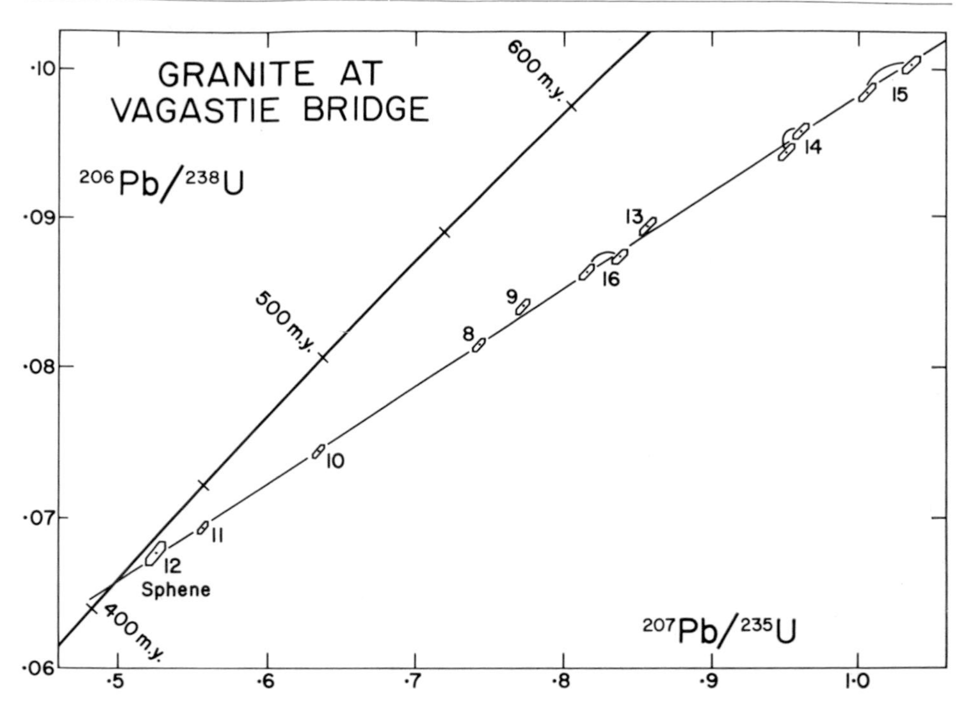


Fig. 3. Concordia plot showing zircon points from granite samples RC662 (points 8–12 from Table 1) and RC663 (points 13–16). Error boxes are drawn around individual points. A chord through these points, including the sphene point (12), intersects concordia at 405 ± 11 m.y. and 1576 ± 100 m.y.

concordia plot the three Pb/U points are strongly discordant but well correlated and lie on a single reverse discordia with a poorly defined lower intersection with concordia at *c.* 420 m.y. and a very uncertain upper intersection somewhere in the vicinity of 2050 m.y. (Fig. 10). The biotite K–Ar age of 400 ± 15 m.y. (401 ± 14 m.y. and 397 ± 14 m.y., Miller and Brown 1965) is in agreement with the zircon lower intersection age and indicates that the zircon U–Pb systems remained essentially closed since granite formation *c.* 410 m.y. ago. The reverse trend of the zircon discordia indicates the presence of zircon xenocrysts with an average age of *c.* 2050 m.y. We relate these to the oval and irregular shaped grains and cores observed in the zircon population.

(*ii*) *The Bonar Bridge granite* sample (RC976) [National Grid NH663913] is a medium grained leucogranite composed of poikilitic K–feldspar, plagioclase laths, minor chloritized biotite and subhedral apatite. The zircon population is complex consisting generally of rounded to anhedral squat grains with a minority of sharply euhedral forms. Many grains appear to consist of rounded and irregular shaped, cloudy, central zircon cores encased in clear, zoned zircon rims sometimes with an overall euhedral shape. Inclusions are not common. Like the Helmsdale granite the yield of zircons was small and only 120 mg was recovered from a 26 kg sample.

The three analysed size fractions (Table 2) were relatively high in uranium but only a limited dispersion of uranium concentration and discordance was achieved (Fig. 4). Nevertheless the spread is sufficient to define a reverse discordia with

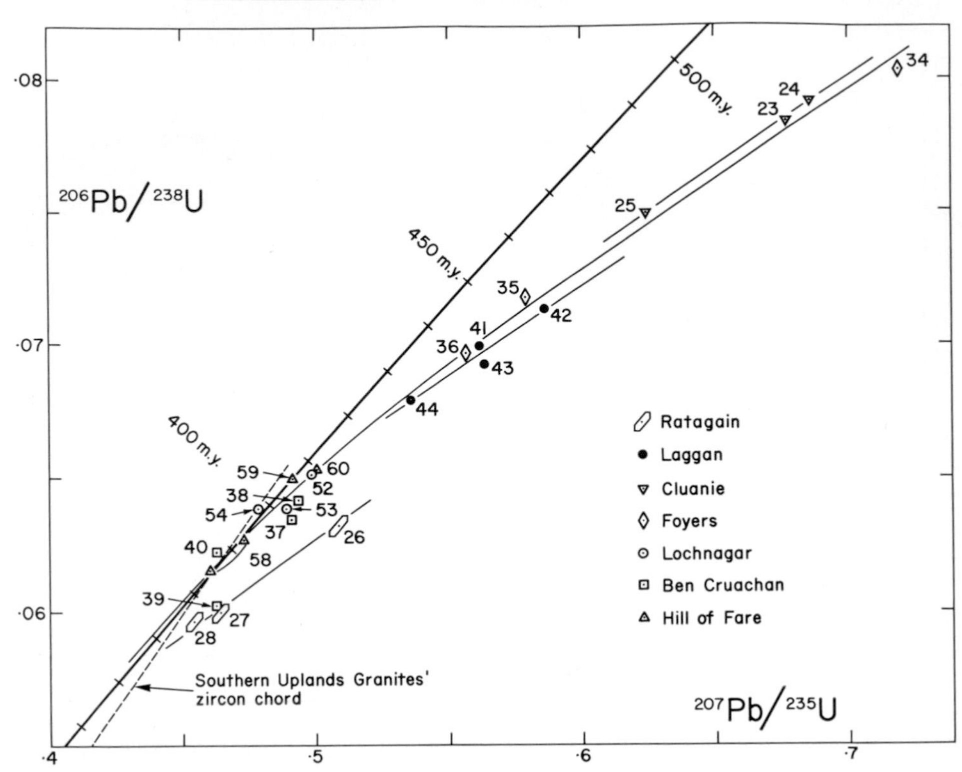


Fig. 4. Concordia plot showing zircon points from some of the Newer granites from the Grampian and Northern Highlands. All granites have some evidence of an inherited 'old' zircon component in their zircon U–Pb systems.

uncertain intersections with concordia at *c.* 400 m.y. and approximately 1400–1800 m.y. There seems little doubt that these old zircon xenocrysts are the irregular and rounded grains observed in the zircon population.

(*iii*) *The Cluanie granite* sample (RC721) [National Grid NH175104] is a granodiorite. The main constituents are plagioclase, with sharply twinned laths and 'blurry' twinned irregular crystals with complex oscillatory zoning, quartz, K–feldspar, minor chloritised hornblende and accessory subhedral apatite and sphene. The zircon population is complex and consists of a few euhedral crystals with sharp terminations, a number of blunted forms with rounded interfacial angles, and abundant rounded, irregular shaped, cloudy grains. A number of grains have rounded zircon cores surrounded by zoned rims of 'new' zircon. Inclusions are rare.

Three size fractions were analysed (Table 2). These showed only a small spread in uranium concentrations and degree of isotopic discordance but sufficient to show that the Pb/U points fall on a single discordia (Fig. 10) with a lower intersection with concordia at *c.* 417 m.y. and a very uncertain upper intersection at *c.* 1000 m.y. The lower intersection is interpreted as indicating the approximate age of granite emplacement. The reverse discordia demonstrates the presence of a significant component of old zircon xenocrysts in the Cluanie granite.

# Table 2. Post-tectonic granite—Grampian and Northern Highlands

| Zircon Fraction (size in microns) | Pb μg/g | U μg/g | $^{206}Pb/^{204}Pb$ | Atom percent radiogenic lead | | | Atomic Ratios | | |
|---|---|---|---|---|---|---|---|---|---|
| | | | | 206 | 207 | 208 | $^{207}Pb/^{206}Pb$ | $^{207}Pb/^{235}U$ | $^{206}Pb/^{238}U$ |
| **Helmsdale Granite (RC972)** | | | | | | | | | |
| 17. + 84μ NM | 71.3 | 590 | 756 | 83.53 | 7.776 | 8.691 | .09309 | 1.508 | .1175 |
| 18. -84 + 61μ NM | 77.6 | 719 | 1183 | 84.26 | 7.376 | 8.396 | .08720 | 1.269 | .1056 |
| 19. -61μ NM | 82.1 | 817 | 8665 | 85.12 | 7.107 | 7.767 | .08349 | 1.144 | .0994 |
| **Bonar Bridge Granite (RC976)** | | | | | | | | | |
| 20. + 84μ NM | 112.1 | 1436 | 510 | 87.38 | 5.851 | 6.766 | .06696 | .7324 | .07932 |
| 21. -84 + 61μ NM | 134.4 | 1695 | 1839 | 87.02 | 5.838 | 7.144 | .06709 | .7423 | .08024 |
| 22. -61μ NM | 117.8 | 1472 | 1327 | 86.06 | 5.862 | 8.073 | .06811 | .7518 | .08006 |
| **Cluanie Granite (RC721)** | | | | | | | | | |
| 23. -106 + 84μ NM | 69.4 | 923 | 3787 | 89.66 | 5.621 | 4.717 | .06270 | .6777 | .07839 |
| 24. -84 + 61μ NM | 71.6 | 943 | 3480 | 89.52 | 5.629 | 4.848 | .06288 | .6859 | .07911 |
| 25. -45μ NM | 67.9 | 946 | 2734 | 89.86 | 5.430 | 4.705 | .06042 | .6243 | .07493 |
| **Ratagan Granite (RC720)** | | | | | | | | | |
| 26. + 84μ NM | 28.3 | 379 | 4230 | 72.80 | 4.248 | 22.95 | .05836 | .5084 | .06319 |
| 27. -61 + 45μ NM | 32.2 | 460 | 3250 | 73.79 | 4.142 | 22.06 | .05613 | .4641 | .05998 |
| 28. -45μ NM | 34.7 | 497 | 1516 | 73.62 | 4.064 | 22.31 | .05520 | .4542 | .05967 |
| **Strontian Granite (RC1258)** | | | | | | | | | |
| 29. + 106μ NM | 32.9 | 450 | 4855 | 79.56 | 4.436 | 16.00 | .05576 | .5200 | .06763 |
| 30 -45μ NM | 42.1 | 577 | 6675 | 77.59 | 4.315 | 18.09 | .05561 | .5042 | .06576 |
| **Xenolith in Strontian Granite (RC1259)** | | | | | | | | | |
| 31. + 106μ NM | 37.9 | 501 | 5143 | 77.36 | 4.298 | 18.34 | .05556 | .5206 | .06794 |
| 32. -84 + 61μ NM | 54.8 | 668 | 6294 | 72.09 | 3.999 | 23.91 | .05547 | .5099 | .06660 |
| 33. -45 NM | 54.2 | 684 | 4066 | 72.54 | 4.031 | 23.42 | .05557 | .5115 | .06675 |
| **Foyers Granite (RC981)** | | | | | | | | | |
| 34. + 142μ NM | 17.2 | 195 | 2611 | 78.45 | 5.096 | 16.45 | .06496 | .7197 | .08035 |
| 35. -106 + 84μ NM | 17.4 | 214 | 1565 | 76.14 | 4.458 | 19.40 | .05855 | .5791 | .07174 |
| 36. -45μ NM | 21.7 | 267 | 2192 | 73.82 | 4.277 | 21.89 | .05793 | .5564 | .06965 |
| **Ben Cruachan Granite (RC724)** | | | | | | | | | |
| 37. + 106μ NM | 23.1 | 343 | 1495 | 80.94 | 4.547 | 14.51 | .05617 | .4913 | .06343 |
| 38. -106 + 84μ NM | 26.0 | 385 | 526 | 79.73 | 4.578 | 15.69 | .05740 | .4959 | .06265 |
| repeat | 26.3 | 382 | 1071 | 80.41 | 4.490 | 15.10 | .05584 | .4938 | .06413 |
| 39. -84 + 61μ NM | 34.3 | 524 | 1510 | 79.19 | 4.411 | 16.40 | .05570 | .4630 | .06029 |
| 40. -45μ NM | 49.8 | 741 | 591 | 79.90 | 4.310 | 15.79 | .05395 | .4633 | .06229 |
| **Laggan Granite (RC723)** | | | | | | | | | |
| 41. + 142μ NM | 29.2 | 399 | 1948 | 82.15 | 4.786 | 13.06 | .05825 | .5615 | .06991 |
| 42. -106 + 84μ NM | 35.5 | 477 | 1345 | 82.25 | 4.903 | 12.84 | .05960 | .5860 | .07130 |
| 43. -61 + 45μ NM | 38.7 | 531 | 1075 | 81.67 | 4.821 | 13.51 | .05903 | .5631 | .06918 |
| 44. -45μ NM | 47.3 | 660 | 2794 | 81.49 | 4.660 | 13.85 | .05719 | .5356 | .06792 |
| **Moy Granite (RC722)** | | | | | | | | | |
| 45. -142 + 106μ NM | 105.0 | 775 | 1066 | 86.88 | 7.223 | 5.891 | .08313 | 1.569 | .1368 |
| 46. -106 + 84μ NM | 106.9 | 795 | 1075 | 86.98 | 7.276 | 5.740 | .08366 | 1.569 | .1361 |
| 47. -84 + 61μ NM | 105.8 | 844 | 1086 | 87.06 | 7.241 | 5.695 | .08317 | 1.454 | .1268 |
| 48. -61 + 45μ NM | 110.5 | 914 | 1024 | 87.07 | 7.203 | 5.726 | .08273 | 1.396 | .1224 |
| 49. -45μ NM | 60.5 | 525 | 1578 | 88.05 | 6.947 | 4.997 | .07889 | 1.283 | .1179 |
| 50. -106 + 84μ M1° | 140.7 | 1426 | 353 | 88.05 | 6.785 | 5.167 | .07706 | 1.073 | .1010 |
| repeat | 137.6 | 1346 | 420 | 87.91 | 6.969 | 5.116 | .07927 | 1.142 | .1045 |
| 51. -106 + 84μ M5° | 125.0 | 1516 | 227 | 88.33 | 6.719 | 4.946 | .07667 | .888 | .0847 |
| **Lochnagar Granite (RC982)** | | | | | | | | | |
| 52. + 142μ NM | 32.7 | 467 | 4732 | 80.16 | 4.453 | 15.38 | .05554 | .4989 | .06513 |
| 53. -106μ NM | 45.6 | 648 | 3240 | 78.15 | 4.343 | 17.51 | .05557 | .4893 | .06386 |
| 54. -45μ NM | 64.5 | 931 | 8120 | 78.08 | 4.313 | 17.61 | .05524 | .4788 | .06387 |
| **Aberdeen Granite (RC871)** | | | | | | | | | |
| 55. + 142μ NM | 55.6 | 595 | 3748 | 88.69 | 6.293 | 5.020 | .07096 | .9422 | .09630 |
| 56. -106 + 84μ NM | 59.0 | 640 | 2513 | 88.93 | 6.275 | 4.791 | .07056 | .9271 | .09529 |
| 57. -61μ NM | 57.4 | 632 | 11,830 | 88.68 | 6.260 | 5.060 | .07059 | .9105 | .09354 |
| **Hill of Fare Granite (RC872)** | | | | | | | | | |
| 58. + 142μ NM | 101.6 | 1606 | 6040 | 85.23 | 4.665 | 10.11 | .05474 | .4731 | .06268 |
| repeat | 100.5 | 1622 | 5211 | 85.47 | 4.639 | 9.90 | .05427 | .4606 | .06155 |
| 59. -106 + 84μ NM | 97.3 | 1484 | 9000 | 85.29 | 4.680 | 10.03 | .05487 | .4915 | .06496 |
| 60. -61 + 45μ NM | 79.8 | 1201 | 2957 | 84.50 | 4.699 | 10.80 | .05561 | .5005 | .06527 |
| **Strichen Granite (RC1283)** | | | | | | | | | |
| 61. + 106μ NM | 67.5 | 889 | 8500 | 89.97 | 5.717 | 4.310 | .06350 | .6967 | .07951 |
| 62. -84 + 61μ NM | 62.5 | 870 | 7870 | 90.22 | 5.323 | 4.457 | .05900 | .6135 | .07542 |
| 63. -45μ NM | 55.1 | 771 | 7910 | 89.26 | 5.189 | 5.552 | .05814 | .5948 | .07419 |
| 64. monazite | 917 | 1193 | 3113 | 8.539 | .4861 | 90.97 | .05693 | .5943 | .07571 |
| repeat | 954 | 1221 | 44700 | 8.500 | .4800 | 91.02 | .05648 | .5968 | .07663 |

(*iv*) *The Ratagain granite* sample (RC720) [National Grid NG901198] is a quartz monzonite with zoned plagioclase laths, poikilitic, anhedral, untwinned K–feldspar, minor green-brown biotite, euhedral hornblende and sphene and accessory iron ore and apatite.

The zircon population is inhomogeneous. In general crystals are stubby with a few inclusions and vary from an abundance of sharply euhedral crystals, showing high order pyramids and fine, multiple, euhedral zoning, to irregular shaped, cloudy grains. A few euhedral grains appear to consist of a euhedral outer rim surrounding a rounded zircon core.

Analyses of three size fractions (Table 2) show that the zircons are relatively high in thorium, with $^{208}$Pb accounting for 22% of the radiogenic lead. On a concordia plot (Fig. 4) the three Pb/U points are almost concordant but form a short reverse discordia intersecting the concordia at *c.* 365 m.y. and at a very uncertain point in excess of 1000 m.y. The reverse trend indicates a trace of old inherited radiogenic lead in the zircon U–Pb system and we attribute this to the observed rounded and irregular zircon grains and cores. In view of other age results on the Newer granites we consider that the lower intersection age is too low to be accepted as the age of the granite without independent confirmation. We interpret this as having been lowered from the true age by a minor, relatively recent loss of radiogenic lead.

(*v*) *The Strontian granite* sample (RC1258) [National Grid NM850610] is a medium grained tonalite consisting of plagioclase, frequently with complex oscillatory zoning, quartz, euhedral hornblende, with inclusions of small apatite rods, brown biotite and accessory euhedral sphene and apatite.

A few zircons are sharply euhedral though in general terminations are blunted and most grains have irregular and rounded forms. Zircons commonly contain a few small colourless rod-shaped inclusions and many grains are cracked and cloudy. Faint euhedral zoning was seen in about half the crystals.

Analyses of the coarsest and finest size fractions are recorded in Table 2. The zircons are nearly concordant. The $^{207}$Pb/$^{206}$Pb apparent ages of the two zircons fractions are 443 and 437 m.y. respectively and on a concordia plot (Fig. 5) the two points fit a normal discordia which intersects concordia at 425–445 m.y. and close to zero. There is no indication of an inherited component of older zircon. Miller and Brown (1965) reported three isotope dilution K–Ar biotite ages of 421 $\pm$ 19 m.y., 407 $\pm$ 18 m.y. and 381 $\pm$ 17 m.y. for the Strontian granite. We tend to disregard the 381 m.y. result as the $K_2O$ content was rather low at 4·74%. The two older ages are in reasonable agreement with the zircon U–Pb age.

(*vi*) *Xenolith in the Strontian granite* is a typical rounded amphibolite xenolith (RC1259) taken near the location of the granite samples. It is finer grained than the granite and consists of clumps of granoblastic amphibolite dispersed in a matrix of plagioclase, hornblende and biotite with a few phenocrysts of plagioclase with complex zoning. A few euhedral hornblendes and sphenes were also present in the matrix.

The zircons are rounded and corroded either squat grains or elongate irregular-shaped rods. Very few crystals with sharp terminations were observed. Most crystals contain a few variable inclusions including rounded red-stained bodies. Faint euhedral zoning was observed in a number of grains.

The three analysed size fractions (Table 2) were similar in uranium content to

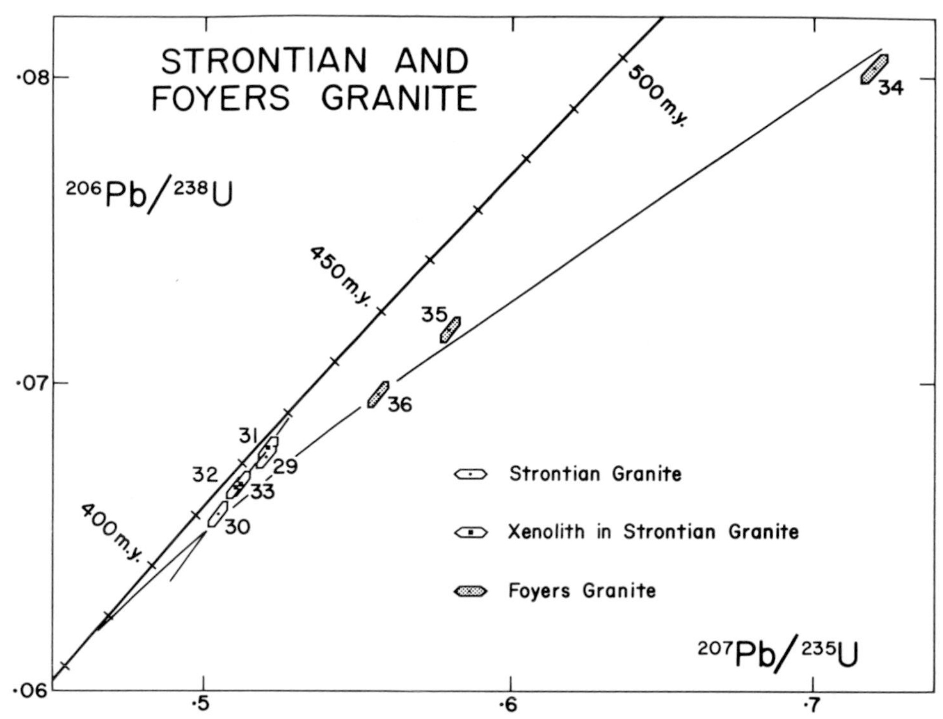


Fig. 5. Concordia plot showing zircon points from the Strontian (RC1258) and Foyers (RC981) granites and a xenolith (RC1259) from the Strontian granite. Whereas the Foyers granite zircons (34–36) have a significant component of 'old' inherited zircon there is no indication of an old inherited component in zircons from the Strontian granite (29–30) or its xenolith (31–33). These zircons behave as a slightly discordant cogenetic suite and define a chord with an upper intersection with concordia of approximately 430 m.y.

the granite zircons and on a concordia plot (Fig. 5) the Pb/U points fall precisely on the short chord through the granite zircons intersecting the concordia at $435 \pm 10$ m.y. and essentially zero million years. This shows that the xenolith has no significant component of inherited 'older' zircon and that the xenolith zircons crystallised during crystallization of the granite magma.

(*vii*) *The Foyers granite* sample (RC981) [National Grid NH548204] is a medium grained granodiorite consisting of finely twinned plagioclase, with complex oscillatory zoning, quartz, untwinned K–feldspar, hornblende, rarely subhedral and also as granoblastic aggregates, biotite, with inclusions of apatite and euhedral sphene.

Zircons are generally stubby. A number have sharp, often high order bipyramids though the majority have rounded and irregular shapes. Small transparent rod-shaped inclusions or bubbles and irregular shaped bodies are fairly common inclusions and a few elongate crystals have central cavities. A few crystals show euhedral zoning.

The uranium contents of the three analysed size fractions (Table 2) are low (195–267 ppm) relative to other zircons from Newer granites. The Pb/U points on a concordia plot (Fig. 5) have a useful dispersion and define a single reverse discordia with a lower intersection with concordia of approximately 385 m.y. and an upper intercept in excess of 1000m.y. Brown *et al.* (1968) reported biotite K–Ar ages of 392±18 m.y. and 409±18 m.y. for the Foyers granite. This suggests that the zircons have undergone a minor recent isotopic disturbance which has slightly rotated the discordia and depressed the lower intersection with concordia. Nevertheless the trend of the discordia clearly demonstrates the presence of a component of old inherited zircon.

(*viii*) *The Ben Cruachan granite* sample (RC724) [National Grid NM039333] is a medium grained granodiorite composed of plagioclase, including larger crystals with complex oscillatory zoning, quartz, minor microcline, hornblende and biotite with associated sphene and iron ore and accessory apatite.

The zircons are uniformly squat grains. A number are sharply euhedral but the majority have blunted terminations and rounded facial angles. Crystals generally have one or two rod-like or irregular shaped inclusions though a number of grains are completely clear. Euhedral internal zoning was observed in a few grains.

Although the uranium concentrations of the four analysed zircon size fractions varied from 343 to 741 ppm (Table 2) they showed very little spread in isotopic dicordance. U/Pb points from the two coarsest fractions fall close to concordia at *c.* 405 m.y. The two finer fractions fall slightly below this suggesting that these points have experienced a relatively recent, minor, isotopic disturbance. This age of *c.* 405 m.y. is in accord with biotite K–Ar ages of 387±6 m.y. and 399±8 m.y. (Brown *et al.* 1968) for the Cruachan complex and fits the Lower Old Red Sandstone age indicated on geological grounds (Roberts 1966). A trace component of inherited lead in the zircon population of the present sample of Ben Cruachan granite is indicated by the suggestion of a 'reverse' trend of a discordia through the zircon points (Fig. 4) so the U–Pb age is interpreted as a maximum.

(*ix*) *The Laggan granite* sample (RC723) [National Grid NM371809] is a medium grained granodiorite consisting of plagioclase, with complex oscillatory zoning, poorly twinned K–feldspar, quartz, biotite, hornblende in discrete euhedral grains and smaller subhedral grains in finer grained amphibolitic clumps. Euhedral sphene and iron ore are accessory.

The zircons are squat prismatic crystals. A majority are sharply euhedral. The rest are irregularly blunted and rounded and sometimes terminated by high order bipyramids. The majority of grains contain a few transparent rod-shaped inclusions and lesser amounts of other unidentified inclusions occur. A few grains are clouded with alteration and many crystals show euhedral zoning.

Analyses of four zircon size fraction (Table 2) show a spread in uranium content from 339 ppm in the coarsest fraction to 660 ppm in the finest. On a concordia plot the Pb/U points fall almost at the bottom of a reverse discordia which intersects the concordia at 400±10 m.y. and in excess of 1000 m.y. (Fig. 4). The lower intersection age of 400±10 m.y. agrees with ages reported for other Newer granites (see Brown *et al.* 1968). Consequently we do not believe the zircons have been significantly isotopically disturbed since granite emplacement. The distribution of points along a reverse discordia is clear evidence of a component of old (>1000 m.y.) inherited zircon in the Laggan granite.

(*x*) *The Moy granite* sample (RC722) [National Grid NH726260] is a two mica granite composed of plagioclase, which in some crystals shows complex oscillatory zoning, microcline, quartz, muscovite and minor biotite and garnet.

Very few zircons show sharply euhedral forms. In general the population is composed of rounded and irregular shaped grains with few inclusions but with internal staining. Some grains have light yellow-brown oval shaped centres surrounded by rims of thin colourless zircon which thicken, and show fine euhedral zones, at the bipyramidal terminations.

Analyses of five non-magnetic and two magnetic zircon fractions (Table 2) show a spread in uranium concentration from 775 ppm in the coarsest non-magnetic fraction to 1516 ppm in the most magnetic fraction. These also show a considerable spread in isotopic discordance and, on a concordia plot (Fig. 10), the distribution of Pb/U points reveals an exceptional influence of old inherited radiogenic lead. The data points scatter about a 'reference' reverse discordia which intersects the concordia at *c.* 300–350 m.y. and 15–1600 m.y. The points extend up to 25% along the length of the discordia indicating that a considerable portion of the zircon is inherited zircon xenocrysts. This may be related to the round and irregular zircon grains previously described. This being the case it follows that very little new zircon grew during consolidation of the granite magma. The scatter of zircon points about the reference discordia suggests possible inhomogeneities in the source zircons. We discuss this later in the paper. Independent age measurements on this granite are not available. However from other age results on Newer granites we conclude that the lower intersection age of 300–350 m.y. is too low and that the zircon U–Pb systems have experienced a relatively recent, isotopic disturbance.

(*xi*) *The Lochnagar granite* sample (RC982) [National Grid 340900] is a coarse grained granite composed of untwinned K–feldspar, quartz, plagioclase, generally with complex oscillatory zoning, biotite and minor hornblende, associated euhedral sphene and accessory iron ore and apatite.

The zircons vary from rare elongate forms, with length/breadth ratios of up to 7, to generally stubby to almost round grains. Many crystals are sharply euhedral though others are rounded and a number have scalloped forms and irregular shapes. Many crystals have internal staining. Numerous small colourless rod-shaped, probably apatite, inclusions are common and a number of crystals have sinuous cavities.

Uranium concentrations in the three analysed size fractions vary from 467 ppm in the coarsest fraction to 931 ppm in the finest (Table 2). However the Pb/U points are almost concordant at 400–415 m.y. (Fig. 4) and in general agreement with the age of many Newer granites (Brown *et al.* 1968) and we conclude that this age is close to the true age of granite emplacement. Thus the zircons have remained closed U–Pb systems since granite emplacement. The distribution of zircon U–Pb points, although almost concordant, suggests a small, hardly significant component of old radiogenic lead which may result in a slight overestimate of the age of emplacement.

(*xii*) *The Aberdeen granite* sample (RC871) was taken from the Dyce quarry [National Grid NJ886135]. It is a two mica granite consisting mainly of microcline, quartz, plagioclase with faint complex zoning, red-brown biotite with distinctive pleochroic halos, muscovite and accessory apatite.

Zircons are generally stubby though a few elongate grains have length/breadth ratios up to 7. Forms are generally blunted and rounded and vary from completely irregular shapes to well developed pointed bipyramids. Grains are frequently

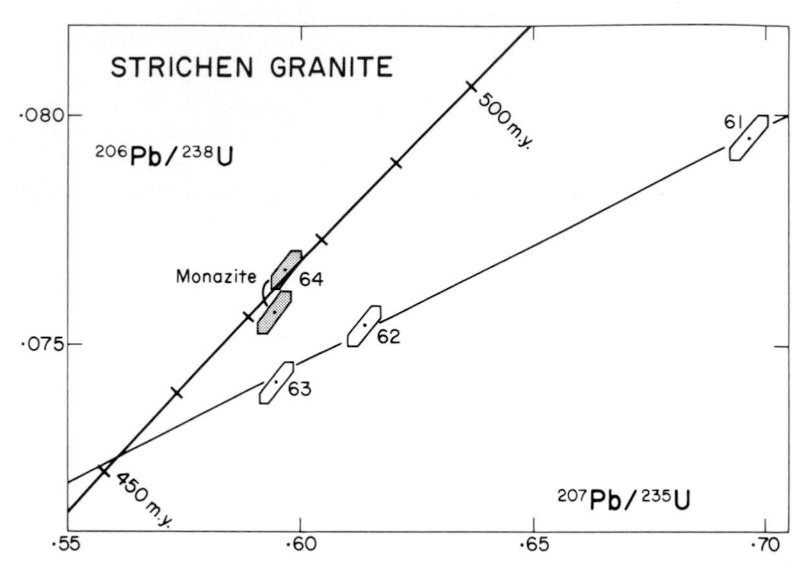


Fig. 6. Concordia plot showing zircon (61–63) and monazite (64) points from granite sample RC1283. The concordant monazite points indicate a granite age of 575 ± 5 m.y. The intersection of the zircon chord at *c*. 450 m.y. is attributed to a relatively recent disturbance of the zircon U–Pb systems.

inclusion-free though a number contain a sprinkling of small colourless rods and many have a core of yellowish rounded zircon surrounded by a rim of clear zoned zircon which thickens at the terminations producing pointed bipyramids.

Analyses of their size fractions (Table 2) produced only a limited spread in uranium (595–640 ppm) and a corresponding restricted range of discordance. Nevertheless, on a concordia plot (Fig. 10) the points give clear evidence of a component of old inherited zircon. The three points define a short segment of a reverse discordia which on extrapolation intersects the concordia in the vicinity of 3–400 m.y. and in excess of 1000 m.y. Unfortunately the zircon results give no direct age information.

(*xiii*) *The Hill of Fare granite* sample (RC872) [National Grid NJ711006] is a leucogranite composed of microcline, quartz, finely twinned plagioclase frequently showing complex oscillatory zoning, minor brown biotite and accessory iron ore.

Most of the zircons have rounded and irregular shapes though a number of euhedral forms were present. Almost all euhedral grains consist of a core of cloudy, brownish, rounded zircon encased in a rim of colourless euhedrally zoned zircon. Crystals have a sprinkling of rod-shaped colourless inclusions.

The uranium concentrations of the three analysed size fractions (Table 2) varied from 1606 ppm in the coarsest to 1201 ppm in the finest fraction which is the reverse of the previously observed distribution of uranium with zircon size. On a concordia plot (Fig. 4) the Pb/U points are nearly concordant at *c*. 400 m.y. However there is an indication that points 59 and 60 contain a minor component of inherited zircon (Fig. 4) so the determined age should be a maximum. Point 60 in particular falls on the extension of several reverse discordia through inherited zircon systems from other Newer granites. The presence of inherited zircon is also indicated by the rounded zircon cores.

(*xiv*) *Strichen granite* sample (RC1283) [National Grid NJ911605] is a medium grained biotite granite composed mainly of microcline, plagioclase with normal and occasionally complex zoning, red-brown biotite with prominent pleochroic halos, rare, possibly secondary muscovite and accessory subhedral apatite.

The zircon population contains a number of elongate euhedral crystals (length/breadth ratios of 5–10) which contain a sprinkling of minute, colourless, rod-shaped and bubble-like bulbous inclusions. The majority of grains however are euhedral, rounded and frequently irregular squat-shaped grains. A number of the more euhedral zircons contain rounded zircon cores.

The uranium content of the three analysed zircon size fractions (Table 2) decreases with decreasing zircon size. On a concordia plot (Fig. 6) the three Pb/U points are strongly discordant but are aligned on a chord with a lower intersection with concordia at *c*. 450 m.y. and an upper intersection well in excess of 1000 m.y. This clearly demonstrates the presence of a significant component of old inherited zircon in the Strichen granite. The lower intersection age could be interpreted as the age of granite emplacement. However the monazite Pb/U points (Table 2), also shown on Figure 6, are essentially concordant at $475 \pm 5$ m.y. The uncertainty in the Rb–Sr isochron age of the Strichen granite of $455 \pm 22$ m.y. (Pankhurst 1974) is too great for the result to be used to indicate whether the zircon lower intersection age or the concordant monazite age is more likely to be correct. Late isotopic disturbance of zircon in granites is common and the reverse discordia could have been slightly rotated by the effects of such an event and intersect the concordia at apparent ages which are too low. On the other hand it is very unlikely that the monazite has an old inherited component. This is also evident from the concordancy of the monazite point, which confirms that monazite has remained a closed U–Pb system since emplacement. Monazite has proved to be a reliable indicator of age (e.g. Köppel and Grünenfelder 1975) and we interpret the concordant monazite age as the age of emplacement of the Strichen granite.

## 3c. The post tectonic granites: the Midland Valley and Southern Uplands

(*i*) *The Distinkhorn granite* sample (RC965) [National Grid NS599350] is a medium grained tonalite. In general plagioclase is sharply twinned with little or no zoning though a minority of crystals have blurred twins and complex oscillatory zones. The brown biotite and hornblende (sometimes biotite rimmed) contain inclusions of euhedral apatite.

The zircons are broken crystal fragments. In the coarser grain sizes many of the fragments can be seen to be pieces of what were large sharply euhedral zircon crystals terminated by high order bipyramids. Inclusions are rare but a minority of crystals have central cavities.

Three zircon size fractions were analysed (Table 3). On a concordia plot the three Pb/U points define a normal discordia which intersects the concordia at $390 \pm 6$ m.y. and close to zero million years (Fig. 7). The $-142+106\mu$ fraction is nearly concordant at 394 m.y. and the $^{207}Pb/^{206}Pb$ apparent ages for the three fractions vary from 383 to 398 m.y. There is no suggestion of excess radiogenic lead

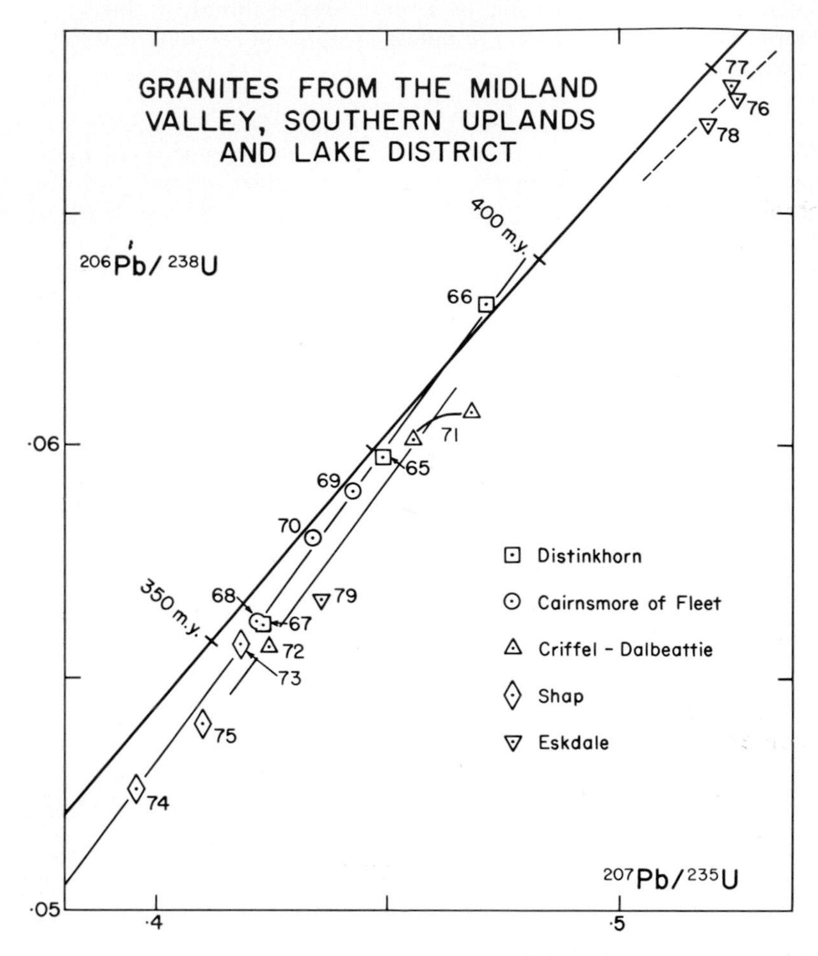


Fig. 7. Plot showing distribution of zircon U–Pb points from granite samples 65–79 (Table 3). A chord through zircon points from the Distinkhorn, Cairnsmore of Fleet and Shap granites intersects concordia at 390 ± 6 m.y. The Criffel–Dalbeattie granite appears to be slightly older. Zircon points from the Eskdale granite indicate a minor component of inherited older zircon.

and we interpret the upper intersection age of 390 ± 6 m.y. as the age of granite emplacement.

(*ii*) *The Cairnsmore of Fleet granite* sample (RC848) [National Grid NX546753] is a two mica granite with large poikilitic microcline, sharply twinned, normally zoned, plagioclase, red-brown biotite, minor muscovite and accessory apatite and epidote.

The zircon population is inhomogeneous and zircons range in size from squat grains to elongate crystals with length/breadth ratios as great as six. Most zircons are euhedral though a number of blunted and irregular shaped grains occur. A number of euhedral zircons contain numerous fine needle and rod-shaped inclusions and have multiple euhedral zones. Many other euhedral zircons consist of cores of

## Table 3. Post-tectonic granite—Midland Valley, Southern Uplands, Lake District and Leicestershire

| Zircon Fraction (size in microns) | Pb µg/g | U µg/g | $^{206}Pb/^{204}Pb$ | Atom percent radiogenic lead | | | Atomic Ratios | | |
|---|---|---|---|---|---|---|---|---|---|
| | | | | 206 | 207 | 208 | $^{207}Pb/^{206}Pb$ | $^{207}Pb/^{235}U$ | $^{206}Pb/^{238}U$ |
| **Midland Valley and Southern Uplands** | | | | | | | | | |
| **Distinkhorn Granite (RC965)** | | | | | | | | | |
| 65. + 165µ NM1° | 35.9 | 562 | 2220 | 80.40 | 4.384 | 15.22 | .05453 | .4493 | .05976 |
| 66. −142 + 106µ NM1° | 37.2 | 556 | 1940 | 81.03 | 4.398 | 14.57 | .05427 | .4717 | .06303 |
| 67. −61 + 45µ NM1° | 30.0 | 507 | 2850 | 81.66 | 4.463 | 13.87 | .05465 | .4233 | .05617 |
| **Cairnsmore of Fleet Granite (RC848)** | | | | | | | | | |
| 68. + 106µ NM1° | 97.0 | 1750 | 2230 | 87.19 | 4.750 | 8.060 | .05448 | .4221 | .05620 |
| 69. −106 + 84 µ NM1° | 89.0 | 1534 | 1465 | 87.48 | 4.758 | 7.765 | .05440 | .4428 | .05903 |
| 70. −61µ NM | 101.2 | 1753 | 1842 | 86.49 | 4.693 | 8.820 | .05426 | .4340 | .05801 |
| **Criffel - Dalbeattie Granite (RC846)** | | | | | | | | | |
| 71. −84 + 61µ NM1° | 80.1 | 1293 | 962 | 83.52 | 4.582 | 11.90 | .05486 | .4552 | .06017 |
| repeat | 80.9 | 1290 | 902 | 83.29 | 4.660 | 12.05 | .05596 | .4687 | .06075 |
| repeat | 79.3 | 1291 | 673 | 83.78 | 4.456 | 11.76 | .05318 | .4382 | .05976 |
| 72. −45µ NM1° | 102.0 | 1771 | 646 | 83.22 | 4.598 | 12.18 | .05525 | .4242 | .05568 |
| **Lake District** | | | | | | | | | |
| **Shap Granite (RC967)** | | | | | | | | | |
| 73. + 142µ NM1° | 63.5 | 1072 | 3406 | 81.02 | 4.400 | 14.58 | .05428 | .4174 | .05577 |
| 74. −106 + 84µ NM1° | 77.8 | 1397 | 2930 | 81.33 | 4.439 | 14.23 | .05458 | .3957 | .05253 |
| 75. −61µ NM1° | 83.0 | 1442 | 2433 | 80.74 | 4.447 | 14.81 | .05508 | .4102 | .05400 |
| **Eskdale Granite (RC968)** | | | | | | | | | |
| 76. −84 + 61µ NM1° | 33.5 | 492 | 1500 | 85.26 | 4.813 | 9.926 | .05645 | .5253 | .06749 |
| 77. −61 + 45µ NM1° | 38.2 | 566 | 2385 | 86.31 | 4.840 | 8.848 | .05607 | .5240 | .06777 |
| 78. −45µ NM1° | 39.9 | 603 | 2485 | 86.90 | 4.889 | 8.207 | .05626 | .5192 | .06692 |
| 79. −61 + 45µ M1° | 80.0 | 1466 | 605 | 89.32 | 4.975 | 5.703 | .05570 | .4355 | .05670 |
| **Leicestershire** | | | | | | | | | |
| **Mountsorrel Granodiorite (RC995)** | | | | | | | | | |
| 80. + 106µ NM1° | 29.0 | 379 | 471 | 73.54 | 4.134 | 22.32 | .05621 | .5062 | .06531 |
| 81. −84 + 61µ NM1° | 37.7 | 482 | 528 | 71.58 | 4.012 | 24.41 | .05605 | .5026 | .06502 |
| 82. −61 + 45µ NM1° | 39.7 | 497 | 824 | 70.66 | 3.953 | 25.59 | .05595 | .5065 | .06565 |
| 83. −45µ NM1° | 51.2 | 694 | 482 | 70.88 | 3.967 | 25.15 | .05596 | .4690 | .06078 |
| 84. −61 + 45µ M1° | 57.2 | 782 | 429 | 71.21 | 3.983 | 24.81 | .05594 | .4660 | .06041 |
| **South Leicester Tonalite (RC997)** | | | | | | | | | |
| 85. −84 + 61µ NM1° | 55.6 | 434 | 3280 | 47.08 | 2.631 | 50.29 | .05588 | .5381 | .06983 |
| 86. −61 + 45µ NM1° | 57.4 | 442 | 3308 | 46.33 | 2.592 | 51.08 | .05596 | .5369 | .06959 |

unzoned, cloudy, rounded zircon with inclusions, surrounded by an inclusion free, rim of zoned zircon.

The uranium contents of the three analysed size fractions (Table 3) are the highest encountered in the present study—slightly greater than that found in zircons from the Hill of Fare granite. On a concordia plot (Fig. 7) the three Pb/U points are seen to fall precisely on the Distinkhorn granite zircon discordia with an upper intersection with concordia at 390±6 m.y. There is no indication of inherited zircon and the age agrees with the K–Ar biotite age of 390±6 m.y. reported by Brown *et al.* (1968). We conclude that the zircon upper intersection age of 390±6 m.y. is the age of granite emplacement.

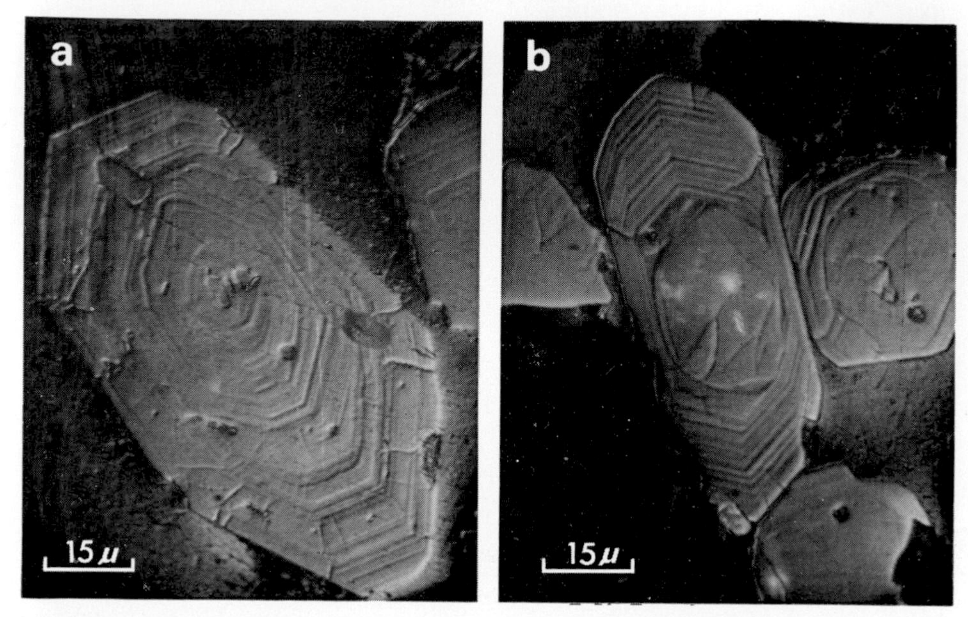


Fig. 8. Polished grain mount of zircons (−85+60 micron size fractions) from (a) The Criffel–Dalbeattie granite (RC846) showing fine, multiple, euhedral, concentric zones which are interpreted as providing a record of an essentially continuous 'single stage' history of zircon growth in the granite melt. (b) The Vagastie Bridge granite (RC662); showing two grains with anhedral, rounded central zircon cores, without zones, encased in rims of zircon showing fine, multiple, euhedral growth zones. Note the zonal thickening which indicates enhanced longitudinal over lateral zircon growth. The central cores are interpreted as corroded 'old' zircon xenocrysts which have survived anatexis and formed nuclei for new zircon (with characteristic euhedral zones) growth during crystallization of the granite melt.

(*iii*) *The Criffel–Dalbeattie granite* sample (RC846) [National Grid NX838612] is a medium to coarse grained leucogranite consisting of orthoclase phenocrysts, quartz, sharply twinned plagioclase and chloritized biotite.

Most zircons are sharply euhedral with multiple euhedral zones thickening at the terminations which are generally high order bipyramids (Fig. 8). A few grains consist of a core of unzoned, rounded, stained and fractured zircon encased in a colourless, euhedrally-zoned zircon rim. Generally grains contain a sprinkling of small transparent rod-shaped inclusions and other irregular shaped bodies.

Two zircon size fractions were analysed (Table 3). The large variation in the repeats of the $-84+61\mu$ fraction is attributed to the relatively high common lead contents of these fractions. On a concordia plot (Fig. 7) the Pb/U points are slightly discordant but fall essentially on a 0–410 m.y. discordia. The isotopic systems show no evidence of excess radiogenic lead. The upper intersection age of $406\pm15$ m.y. (error visual estimate only) agrees with K–Ar biotite ages of $397\pm8$ m.y. and $391\pm8$ m.y. (Brown *et al.* 1968) and we interpret this as the age of granite emplacement. We relate the slight discordance of the two fractions to a recent minor isotopic disturbance.

### 3d. Post-tectonic granites: the Lake District and Leicestershire

(*i*) *The Shap granite* sample (RC967) [National Grid NY555085] is coarse grained with phenocrysts of orthoclase, quartz, altered normally zoned plagioclase, chloritized brown biotite and accessory iron ore, sphene and apatite. Biotite, feldspar, sphene and apatite are generally euhedral.

The zircons are mainly euhedral, with sharp euhedral, multiple, growth zones. A minority of rounded grains are also present. A few euhedral grains appear to be composed of rounded unzoned zircon cores surrounded by clear, euhedrally zoned zircon rims. The zircons contain a few small rod-shaped (apatite) and other inclusions.

The Pb/U ratios of the three analysed fractions are slightly discordant (Table 3). Nevertheless on a concordia plot the points fall on the same discordia, with intersections with the concordia at $390 \pm 6$ and 0–50 m.y., like the Distinkhorn and Cairnsmore of Fleet granites (Fig. 7). The Shap granite has been the subject of numerous age determinations and Brown *et al.* (1968) reported a mean of 21 Rb–Sr and K–Ar determinations on biotite of $393 \pm 11$ (standard deviation) m.y. The zircon upper intersection age of $390 \pm 6$ m.y. is in good agreement with this average value though a direct comparison is difficult to make because of decay constant uncertainties. However on this evidence we conclude that the Shap granite zircons have no significant component of excess radiogenic lead and that the upper intersection age is the age of granite emplacement.

(*ii*) *The Eskdale granite* sample (RC968) [National Grid SD113944] is a red-brown biotite bearing granite with heavily altered plagioclase. K–feldspar shows graphic intergrowth with quartz. Biotite is largely recrystallized to smaller flakes and garnet and apatite are accessory.

A number of zircons are euhedral. Others have rounded and irregular shapes. Grains are generally stubby and are unzoned or very faintly zoned. They are frequently free of inclusions, though small round and rod-shaped bodies occur in a few grains. No zircon cores were observed.

Three non-magnetic size fractions were analysed (Table 3). On a concordia plot (Fig. 7) the three non-magnetic Pb/U points are slightly discordant but closely grouped near *c.* 425 m.y. A line joining this group of points with the more discordant magnetic zircon point intersects the concordia at $460 \pm 10$ m.y. and close to zero million years. Unfortunately the discordance spread of the non-magnetic points is insufficient to independently define the chord and the trend of the line is completely dependent on the magnetic point (Table 3, No. 79). The chord indicates that this granite was significantly older than the Shap and the Southern Uplands granites. However disregarding the magnetic zircon point, the remaining zircon points could also be interpreted as indicating a minor component of older inherited zircon in the Eskdale granite zircon population.

A dashed line on Figure 7 shows a hypothetical reverse discordia through these points to indicate the possible effects of an inherited component. Morphological examination of the zircon shows no evidence of older zircon. However, Brown *et al.* (1964) reported a 'mean' total volume, biotite, K–Ar age of 383 m.y. for the Eskdale granite. Whereas this result cannot be accepted as the definitive age determination of this granite it does indicate that the Eskdale granite was probably emplaced at about the same time as the Shap and Southern Uplands granites. On this evidence it is suspected that the zircons from the Eskdale granite have a minor component of inherited zircon though further measurements will have to be made to resolve this question.

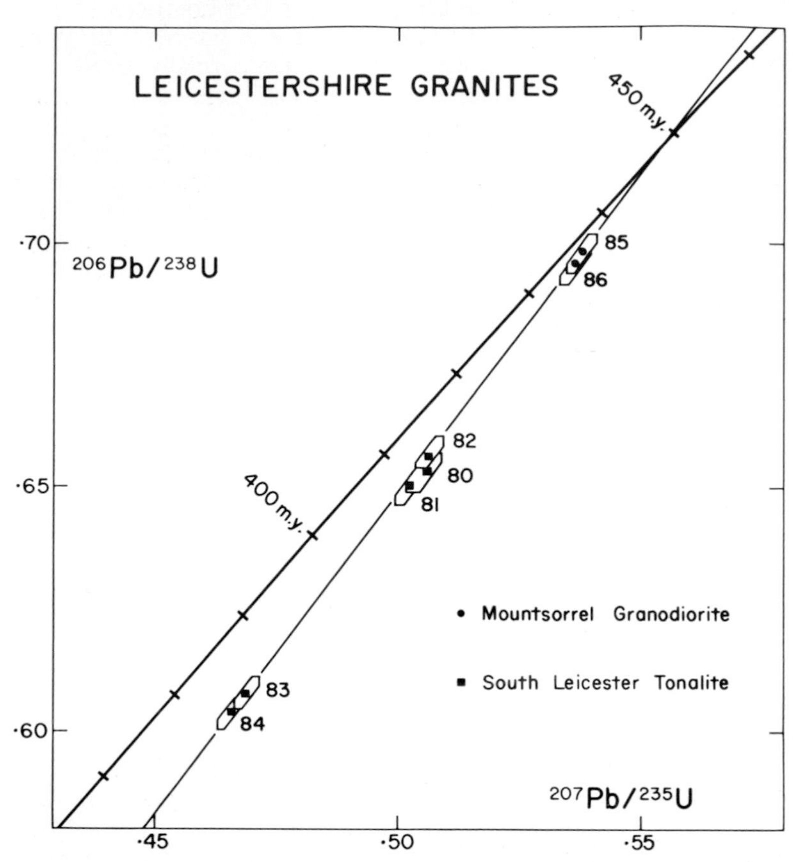


Fig. 9. Concordia plot showing zircon points from granite samples RC995 (85 and 86) and RC997 (81–84). A chord throught the points intersects concordia at $452 \pm ^{8}_{6}$ m.y. (errors are 2 $\sigma$) and at approximately zero million years.

(*iii*) *The Mountsorrel granodiorite* sample (RC995) [National Grid SK570152] is from the Cocklaw Quarry in a boss *c*. 3 km in diameter near Mountsorrel and about 10 km from Leicester. It consists of plagioclase with strong oscillatory zoning, quartz, orthoclase, chloritized biotite and hornblende, and euhedral sphene. Finer grained aggregates of highly altered euhedral hornblende and plagioclase also occur.

The zircons are stubby, prismatic, frequently sharply euhedral crystals, though a number of rounded, corroded and irregular shaped grains occur. The zircons are variable in their inclusion contents. Some are free of inclusions but generally grains have a sprinkling of fine rod-shaped bodies. Zoning is rare and no cores of potentially older zircon were observed.

The five analysed non-magnetic and magnetic size fractions have higher than usual common lead (Table 3). Nevertheless on a concordia plot Pb/U points are only slightly discordant and fall a single discordia chord with intersections with the concordia at 452±8 m.y. and 0–50 m.y. (Fig. 9). There is no indication in the U–Pb

muscovite pegmatites) (van Breemen *et al.* 1974). To the southeast in the Grampian Highlands the second regional deformation is considered to have occurred in Arenigian times (Dewey and Pankhurst 1970). There is no independent evidence to indicate a post 405±11 m.y. $D_2$ deformation in the Caledonides.

The possibility that the zircon reverse discordia and the sphene have been affected by some form of metamorphic 'resetting' is very unlikely in view of our experience with zircon U–Pb systems from the Carn Chuinneag granite (Pidgeon and Johnson 1974) which have retained the granite emplacement age, in agreement with Rb–Sr measurements (Long 1964), despite being overprinted by what appears to be the same $D_2$ deformation as found at Vagastie Bridge. Also the possibility of explaining the zircon lower intersection age in terms of a significant 'recent' lead loss, similar to that observed in a number of Caledonian granites such as Cairnsmore of Fleet and Criffel–Dalbeattie, has already been discounted on the evidence of the coincidence of the concordant sphene point with the point of lower intersection of the zircon reverse discordia and concordia (Fig. 3). Consequently, we see no reason to doubt the zircon U–Pb age of 405±11 m.y. as the true age of the Vagastie Bridge granite.

## 5. Episodes of granite emplacement: the post-tectonic granites

### 5a. Previous geochronological results

Most of the Caledonian granites were emplaced after the main metamorphism and tectonism during late Silurian to early Devonian time.

Although Read (1961) classified these according to their modes of emplacement with the last or permitted granites emplaced as ring complexes by cauldron subsidence, only rarely can field relationships be used to indicate the ages of the granites. Thus Brown *et al.* (1968) found that it was not possible, on the basis of isotopic ages, to distinguish forceful from permitted granites and they did not consider this classification to be entirely satisfactory in discussing the variety of intrusive phenomena which the granites display. Brown *et al.* (1964), Miller and Brown (1965) and Brown *et al.* (1968) reported K–Ar ages, mainly on biotites, for a number of Caledonian granites from the Lake District to the Northern Highlands. They concluded that the majority of the Newer Granites were emplaced 400±10 m.y. ago. Strictly, these ages are 'cooling ages' dating the closure of biotite to argon loss as the granite cools. Dewey and Pankhurst (1970) point out that in the Central Highlands, where the regional blocking temperature for argon retention in biotite is not reached until after granite emplacement, the biotite K–Ar ages of granites reflect the regional cooling and could be interpreted as minimum ages. Such biotite K–Ar ages for the Foyers, Strontian and Cluanie granites could be minimum ages. Hornblende K–Ar ages on a few granites such as Rogart, Ballachulish, Ben Cruachan, Garabal Hill and Criffel–Dalbeattie are in general agreement with the biotite K–Ar ages. Rb–Sr results reported for a few Newer granites, such as Garabal Hill (Summerhayes 1966) and Shap (Kulp *et al.* 1960), are also in broad agreement with the K–Ar ages. However, Pankhurst (1974) has reported Rb–Sr isochron ages of *c.* 460 m.y. for a number of small granites, the Strichen, Kennethmont, Aberchirder, and possibly Longmanhill, in Aberdeenshire. Pankhurst proposes that these granites were emplaced during a late Ordovician magmatic episode following $F_3$ folding and that the Kennethmont, Strichen and Aberchirder granites may actually be part of a large batholithic mass underlying the Boyndie syncline. Pankhurst (1974) interpreted the high initial $^{87}Sr/^{86}Sr$ values (0·714 to 0·717) as indicating that the granites were derived from a deep, underlying crustal layer, possibly Lewisian gneiss and not from the mantle or subducted oceanic lithosphere. Bell (1968) reported biotite whole rock Rb–Sr ages of $385 \pm 8$ m.y. (initial $^{87}Sr/^{86}Sr = 0{\cdot}718 \pm 0{\cdot}004$) for the Bennachie, Peterhead and Lochnagar granites.

## 5b. Complexities in zircon U–Pb systems

The use of zircon U–Pb systems to determine the age of granite emplacement is seriously affected in a number of granites by the presence of inherited zircon. For instance, only a very approximate estimate of the age of emplacement can be obtained from the small dispersion of zircon points situated well up the reverse discordia for the Aberdeen and Bonar Bridge granites (Fig. 10). In addition, there is the possibility that the zircon U–Pb systems have been open to minor lead loss since emplacement. This is clearly the case for the Ratagain granite and is suspected for the Strichen, Foyers and Moy granites (Fig. 10). Other granites with linear (reverse) discordias such as Helmsdale, Cluanie and Laggan (Fig. 10) give lower intersections with concordia between 400 to 420 m.y. and are thought to have experienced little or no post emplacement isotopic disturbance.

## 5c. Correlation of the Strontian and Foyers granites

The zircon U–Pb systems of the Strontian granite and its xenolith show no evidence of old inherited zircon but behave as a slightly discordant cogenetic zircon suite with an upper intersection of a short discordia with concordia of $435 \pm 10$ m.y. (Fig. 5). We interpret this as the age of emplacement of the granite.

In contrast to this the zircon U–Pb systems of the Foyers granite show clear evidence of an inherited component of old zircon (Fig. 5). We have previously concluded that the zircon U–Pb system has been slightly disturbed since granite emplacement and that the reverse discordia does not indicate the age of granite emplacement. K–Ar ages of $392 \pm 18$ m.y. and $409 \pm 18$ m.y. suggest that the Foyers granite was emplaced significantly later than the Strontian granite though there is the possibility that these are minimum ages related to the regional cooling (Dewey and Pankhurst 1970). Clearly, a K–Ar age measurement of a hornblende from the Foyers granite is needed to enable a meaningful comparison of the ages of these two granites.

However, we believe that the presence of old inherited zircon in the Foyers granite, compared to its absence in the Strontian granite and its xenolith, indicates fundamental differences in the source rocks of the two granites and is strong evidence that the Foyers and Strontian granites were not originally the same body (cf. Kennedy 1946).

## 5d. Granites of the Midland Valley, Southern Uplands and Lake District

Zircon U–Pb systems from the Distinkhorn, Cairnsmore of Fleet and Shap granites behave as slightly discordant cogenetic suites with U–Pb points defining a single discordia with an upper intersection with concordia of $390 \pm 6$ m.y. and a lower intersection of 0–50 m.y. (Fig. 7). The two Criffel–Dalbeattie zircon points appear to indicate a slightly older age of $406 \pm 15$ m.y. (as only two points were measured the error estimate is subjective and relatively large). At present we are not prepared to use the zircon data points from the Eskdale granite as indicators of the age of this body. The points fall well above the chord through the other Southern Uplands granites and it is suspected that this granite has a component of old inherited zircon.

Lambert and Mills (1960) consider the Shap granite was intruded in the early part of the interval upper Silurian to upper Devonian. The Southern Uplands granites were emplaced in folded Ordovician and Silurian sediments and Blyth (1955) suggests their age is lower Devonian. The Distinkhorn granite intrudes sandstone immediately below fossiliferous lower Devonian (Mercy 1960). On this evidence the zircon U–Pb age of $390 \pm 6$ m.y. for the Distinkhorn, Criffel–Dalbeattie and Shap granites can be related to the lower to middle Devonian.

### 5e. Post-tectonic granite activity

The zircon U–Pb age of the Strichen granite ($475 \pm 5$ m.y.) supports the conclusion of Pankhurst (1974) that a major igneous event took place in the northwest Highlands *c.* 470 m.y. ago, probably in association with the emplacement of the younger gabbros during the Arenig Grampian orogeny. This supports the view of Dewey and Pankhurst (1970) that all deformation and metamorphism in this region was completed within a relatively short climactic episode between $501 \pm 17$ m.y. and $475 \pm 5$ m.y.

The zircon results also support the conclusions of Brown *et al.* (1968) that the majority of Newer granites were emplaced $400 \pm 10$ m.y. ago. There is an indication from the present work that the majority of Midland Valley, Southern Uplands and Lake District granites ($390 \pm 6$ m.y.) are slightly but significantly younger than most of the Grampian granites (400 – 415 m.y.) though we are not entirely confident that the latter have not been influenced by a trace component of old inherited zircon.

## 6. The distribution and significance of old inherited zircon in Caledonian granites

Granites with significant components of old inherited zircons include the Ardgour granite gneiss, Carn Chuinneag, Ben Vuirich, Vagastie Bridge, Ratagain, Helmsdale, Bonar Bridge, Cluanie, Laggan, Moy, Aberdeen, Strichen and Foyers granites. Granites which contain a trace of inherited zircon include Ben Cruachan, Lochnagar, Hill of Fare and Eskdale. These granites, except for Eskdale, are restricted to the Grampian and Northern Highlands. Granites with no isotopic evidence of an inherited component, which include Strontian, Cairnsmore of Fleet, Distinkhorn, Criffel–Dalbeattie, Shap and Mountsorrel, and the South Leicester tonalite, occur mainly in the Midland Valley, the Southern Uplands and south into England. The boundary separating granites with inherited zircon to the north from granites without inherited zircon to the south appears to be at or a little north of the Highland Boundary fault (Figs 1, 11).

Possible explanations for this distribution are:

1. Zircon xenocrysts, although originally present, were unstable or excluded under the conditions of formation and emplacement of granites in the southern 'non orogenic' Caledonides.
2. The old inherited zircons derive from country rock xenoliths incorporated into the rising granite magmas. 'Old' zircon is present in Moine and Dalradian sediments in the north but absent in sediments in the south.

3. The granite source rocks south of the Highland Boundary fault are fundamentally different from those to the north. In the south the granite source rocks were dominantly new 'Silurian–Ordovician' crust whereas granite source rocks in the Grampian and Northern Highlands were dominantly Precambrian basement.

The reasons for the survival of zircon in some granite melts and not others is not understood. Work on the Carn Chuinneag granite (Pidgeon and Johnson 1974) suggests that magma composition is an important factor. Inherited zircon is present in the calcic Inchbae granite but is lower in amount or absent in the associated alkaline granites. The broad similarity of many of the present granite samples, from north and south of the Highland Boundary fault, to the Inchbae granite suggests that in general the Palaeozoic granite magmas are favourable to zircon survival—provided that magma temperatures (and $P_{H_2O}$) are similar to those of the Inchbae granite magma. Higher temperatures of magma formation might cause complete zircon dissolution. However to explain the distribution of inherited zircon in these terms would require consistently higher magma temperatures for granites south of the Highland Boundary fault compared to temperatures in granite magmas north of the fault. We are not aware of any evidence to support this and consider this unlikely.

Another possible explanation for the absence of old inherited zircon in the granites south of the Highland Boundary fault is that these zircons have been removed during emplacement.

It is generally agreed that at the source the normal product of anatexis is a mush of crystals plus granitic melt (Winkler 1967; Wyllie 1971). The proportion of melt to residual crystals (including zircons) depends mainly on the source rock composition, the temperature, and the water content. White and Chappell (1977) consider that as the magma moves upwards 'the melt may progressively separate or free itself from its residue' (p. 12), the final granite being a record of the extent of this process. The detailed chemical results necessary to establish such 'unmixing' trends are not available for the Caledonian granites. However, the brief descriptions of granite samples and their zircons provides qualitative information on the presence of residual source material including zircon xenocrysts.

Plagioclase with complex oscillatory zoning, a definite indicator of restite (White and Chappell 1977) is present in the Cluanie, Laggan, Moy, Stricken, Ben Cruachan, Strontian, Lochnagar, Distinkhorn and Mountsorrel granites. Of these, the first five contain old inherited zircon U–Pb systems whereas the remainder have no significant isotopic evidence of inherited zircon. Plagioclase in the Ratagain, Helmsdale, Foyers, Bonar Bridge and Vagastie Bridge granites is sharply twinned and normally zoned suggesting little if any restite and yet all these granites have isotopic evidence of inherited zircon. The Helmsdale, Bonar Bridge and Hill of Fare granite samples are leucogranites and might prove to be essentially 'restite free' minimum melts. Nevertheless, Bonar Bridge and Helmsdale granites have zircon U–Pb systems which indicate a significant component of old inherited zircon.

Fig. 10. Composite concordia plot showing: A. The relationship of the zircon reverse discordia of the pre-tectonic and Newer granites with discordant zircon U–Pb systems from the Lewisian of the foreland. B. Zircon U–Pb systems of Dalradian metasediments and the Galway granite and the Thorr adamellite of the Donegal granite from Ireland. C. Zircon U–Pb systems of Caledonian granites in Scotland north of the Highland Boundary fault. D. Zircon U–Pb systems of zircons from Caledonian granites in Scotland and England south of the Highland Boundary fault.

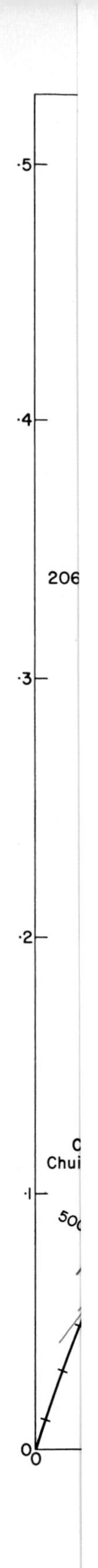
·5
·4
206
·3
·2
Chui
·1
0
0

Therefore, at this stage of the study we have not been able to discover a relationship between the occurrence of old inherited zircon and the presence or absence of identifiable restite material. Certainly, there is no evidence that melt–residua separation processes have controlled the distribution of old zircon xenocrysts in the Caledonian granites.

Only in one case, the Strontian granite, have we made a study of zircon U–Pb systems from a sample of granite and one of its xenoliths. This granite intrudes Moine schists and gneisses and, at the sample location in Glen Tarbet, contains numerous basic xenoliths. The studied xenolith was an amphibolite, which suggests this was primary residua of the granite melt rather than country rock. This view is further supported by the zircon U–Pb results which show no isotopic evidence of inherited zircon in the tonalite or the xenolith. So, although intruded into gneiss known to be at least 740 m.y. old (van Breemen *et al.* 1974) the tonalite has not assimilated significant amounts of country rock, and consequently old zircon. On the other hand the Cluanie hornblende granodiorite, a similar rock, also emplaced in Moine gneisses has definite isotopic evidence of old inherited zircon. We have no reason to suspect that the nature and general contents of country rock xenoliths (Moine and Dalradian) in these granites are significantly different and we conclude that most xenoliths are probably of deep seated origin, as suggested by White and Chappell (1977), and that assimilation of xenoliths of country rock has not played a major role in governing the presence or absence of old inherited zircon in the Caledonian granites.

Finally as we have argued that the restricted occurrence of old zircon xenocrysts in granites from the Grampian and Northern Highlands cannot be explained in terms of granite forming processes, separation of zircon xenocrysts during emplacement or contamination with country rock, we interpret the regional pattern as the result of fundamental differences in the granite source rocks.

## 7. The nature of the granite source rocks

### 7a. The 15–1600 m.y. 'upper intersection age' of the reverse discordias

The upper intersection with the concordia of the reverse discordia is in excess of 1000 m.y. for the Carn Chuinneag and Ardgour gneiss (1550 m.y.), the Ben Vuirich granite (1316$\pm$26 m.y.), the Vagastie Bridge granite (1576$\pm$100 m.y.), the Moy granite (15–1600 m.y.), the Helmsdale granite (*c.* 2050 m.y.), the Strichen, Ratagain, Foyers, Laggan, Aberdeen and Cluanie granites. These reverse discordia are shown on Figure 10 together with discordia trends of zircon suites from Lewisian gneisses and granites (from van Breemen *et al.* 1971; Pidgeon and Aftalion 1972; Pidgeon and Bowes 1972; and Lyon *et al.* 1973). Figure 10 shows that the general trend of the granite zircon reverse discordias is towards the 15–1600 m.y. interval of the concordia curve. Notable exceptions to this are the Ben Vuirich granite, with an upper intersection age of 1316$\pm$26 m.y. and the Helmsdale granite, with an upper intersection age of *c.* 2050 m.y.

This trend of discordia towards the 15–1600 m.y. interval is present in the pre-tectonic (560 m.y. old) and post tectonic granites and appears to be independent of granite age. Also, with the above exceptions, this general trend of discordia is consistent for granites throughout the region and appears to be independent of granite type.

These relationships provide important information on the nature of the granite source rocks. For instance, it is obvious from a comparison of the general trend of the granite reverse discordia and the zircon U–Pb systems of Lewisian rocks shown on Figure 10 that the granite zircons cannot have been derived directly from *c*. 2700 m.y. old 'Scourian' (or Badcallian) gneisses and granulites which occur immediately west of the Caledonide belt and are infolded with Moine metasediments. A probable exception to this is the Helmsdale granite where the 2050 m.y. value can be explained in terms of a complex source region which contains a significant proportion of zircons of Scourian age intermixed with younger, possibly 15–1600 m.y. old zircons. Possible source rocks which could produce the inherited zircon U–Pb discordance patterns of the remaining bodies mentioned are Moine and, or, Dalradian metasediments or Proterozoic crust formed essentially 1600 m.y. ago—and possibly related to the Laxfordian igneous and metamorphic episodes of the Lewisian.

## 7b. Moine and Dalradian metasediments as the granite source rocks

At present there are no zircon U–Pb results on Moine metasediments and the only zircon U–Pb study of Dalradian metasediments has been made on the Connemara Schists and Bens Quartzite from Connemara, Western Ireland (Pidgeon 1969) where the U–Pb systems of zircon size fractions from these two metasediments were strongly discordant. However on a concordia plot the zircon points were well correlated and intersected the concordia at 400–500 m.y. and 1475–1675 m.y. (Bens Quartzite) and *c*. 400 m.y. and *c*. 1290 m.y. (Connemara Schist) (Fig. 10). Pidgeon (1969) commented that 'Without additional zircon sample measurements . . . it is not possible to distinguish between the following explanations of the results:
1. The linear trends or 'chords' through the data-points are due to a strong isotopic disturbance(s) during the Dalradian metamorphism of simple original zircon populations. This requires that the schist and quartzite zircons were from separate source rocks 1290 m.y. and 1475–1675 m.y. old respectively.
2. The linear trends are the results of artificially averaging by sizing a zircon population of unknown original complexity which was isotopically disturbed during the Dalradian $M_1$ metamorphism, and possibly by other preceding events. On a concordia plot linear trends from size splits of individual samples would have intersections with the concordia with no unique geological significance'.

The results of this one study raise the possibility that the melting of such sedimentary material to form granite magma could result in a pattern of inherited zircon hardly distinguishable from that presently observed. However, until the zircon U–Pb systems of Moine and Dalradian metasediments in Scotland are investigated it is not possible to confirm or deny the potential of these rocks as uniform sources of 14–1600 m.y. old zircons. Nevertheless, we would argue that these rocks are not a likely source of the Caledonian granites on the basis of the distribution of rock types and the absence of any identified correlation of inherited zircon with granite type.

Chappell and White (1974) recognised two distinct granite types in southeast Australia which they called I (igneous) and S (sedimentary) types. S types are those granitoids derived from the ultrametamorphism of rocks that have been through weathering processes (sedimentation) on the earth's surface. I types are those generated from source rocks derived directly from pre-existing rocks of igneous origin (White and Chappell 1977). The extensive chemical data required to distinguish these granite types is not available for the Caledonian granites. However, the occurrence of key minerals such as hornblende, indicating I types and two micas including a red-brown biotite, indicating S types, may be of value in recognising I and S granite types. Probable S types such as Strichen, Aberdeen and the Ardgour granite gneiss occur in the Grampians and Northern Highlands and have older inherited zircons and might well be derived from Moine and Dalradian sediments. However, hornblende-bearing probable I types granites such as Cluanie and Foyers also occur in the northern region and these also contain inherited old zircon.

Possible I type granites such as Shap and Distinkhorn are also found in the southern region as are possible S types. The Eskdale granite, which has a probable small component of inherited zircon, has the appearance of an S type and possibly other muscovite granites (e.g. Cairnsmore of Fleet) in the Southern Uplands without inherited zircon may turn out to belong to this category.

In conclusion therefore, the boundary separating granites with old zircons to the north from those without to the south is not an I–S line in the sense of White *et al.* (1976) as I granite types occur north and south of the Highland Boundary fault and both I and S type granites in the Grampian and Northern Highlands contain old inherited zircon xenocrysts. This argues against Dalradian and Moine sediments as source rocks for the magmas of these granites as anatexis of sediments would be expected to lead a dominance of hornblende-free S type granitoids (White and Chappell 1977) in the Grampians and Northern Highlands.

## 7c. Proterozoic crust as source rocks for the Caledonian granites in the Grampians and Northern Highlands

The general consistency of the upper intersection ages of reverse discordia of 15–1600 m.y., the lack of a relationship between this presence of inherited zircon and granite type, and the similarity of granite types in the Grampians and Northern Highlands with others elsewhere in the Caledonides, leads us to propose that a major component of the source rocks of those Caledonian granites with old inherited zircon are crustal rocks formed *c.* 1600 m.y. ago. The possibility that these source rocks are 'Scourian' crust which has been completely reconstituted during Laxfordian times cannot be entirely ruled out. Certainly the zircon U–Pb systems from the Helmsdale and Ratagain granites retain isotopic evidence of $>2000$ m.y. old zircon, thereby indicating the presence of a possible Scourian component.

The 1640–1900 m.y. old South Harris injection complex (Myers 1971) and igneous complex (Jehu and Craig 1927; Dearnley 1963) may be an expression of the *c.* 1600 m.y. old crust forming event as, although formed in a region dominated by gneiss with 2700 m.y. old Scourian ages (Pidgeon and Aftalion 1972) the zircon U–Pb systems of the Harris granites and the tonalite (unpublished results) have no isotopic evidence of a component of Scourian zircon (van Breemen *et al.* 1971).

Although the presence of inherited zircon indicates Proterozoic source rocks it does not preclude the possibility that this material has been mixed with younger (Middle Palaeozoic in this case) rocks during formation of the Newer granites.

The position of the zircon U–Pb points on the reverse discordia cannot be used to provide quantitative estimates of the proportion of old-inherited to newly-formed zircon without taking into account:

1. The complexity of the inherited zircon population.
2. The uranium concentration of the inherited zircon relative to that of newly-formed zircon.

3. The nature and degree of isotopic discordance sustained by the inherited zircons before and during crystallization of the granite magma.
4. The susceptibility of inherited zircon relative to newly-formed zircon to post granite isotopic disturbance.

Morphological examination, electron probe and fission track analysis can assist in resolving the relative amounts of old and new zircon, and in measuring the uranium concentrations, but we know no way of adequately assessing the complexity of the original source zircons or of estimating the discordance of inherited zircon at the time of granite emplacement.

## 7d. The granite source rocks in the Midland Valley, the Southern Uplands and England

On the basis of conclusions reached in the preceding sections, we interpret the absence of old zircon xenocrysts from the granites from the Midland Valley and south into England as evidence that these granites were derived from new middle Palaeozoic crust.

This conclusion is supported by the results of a preliminary investigation of common lead in Caledonian granites from the Highlands and Southern Uplands (Blaxland *et al.* in prep.). These authors found that granites from the Southern Uplands contained lead compatible with a Palaeozoic crust whereas granites in the Highlands had common lead indicative of significantly older source rocks. An equivalent study of strontium isotopic ratios in Caledonian granites is not available at present.

The Eskdale granite from the Lake District is an exception as this has an apparent component of older zircon. The source rocks of this granite maybe different from those of the other Southern Uplands granites. This granite has affinities with White and Chappell's (1977) S (sedimentary) granite types and may represent a mixture of deep crustal source material with overlying sediment containing older detrital zircon.

## 7e. The boundary between Proterozoic and Palaeozoic granite source rocks

The patterns of inherited zircon in the pre- and post tectonic granites suggest a boundary between Palaeozoic and largely Proterozoic granite source rocks beneath the region immediately north of the Highland Boundary fault. They also suggest a possible overlap of the dominantly Palaeozoic granite source rocks within the time intervals between emplacement of the Ben Vuirich granite and the invasion of the Newer granites.

The presence of Proterozoic granite source rocks beneath the southwest Moines *c.* 560 m.y. ago is indicated by the zircon U–Pb systems of the Ardgour gneiss and Carn Chuinneag granite and the Ben Vuirich granite results suggest that these source rocks extended essentially as far south as the Highland Boundary fault *c.* 510 m.y. ago. However, the zircon U–Pb systems from some of the larger Newer granite masses emplaced just north of the Highland Boundary fault *c.* 400 m.y. ago, such as Ben Cruachan, Lochnager and Hill of Fare, contain only minor components of inherited Proterozoic zircon. It is possible that much of the older zircon has been removed by one or other of the processes already discussed. Alternatively, these granites could have been derived from dominantly Palaeozoic source rocks, mixed with minor amounts of Proterozoic material. In view of the significant component of Proterozoic zircon in Newer granites further to the north, such as

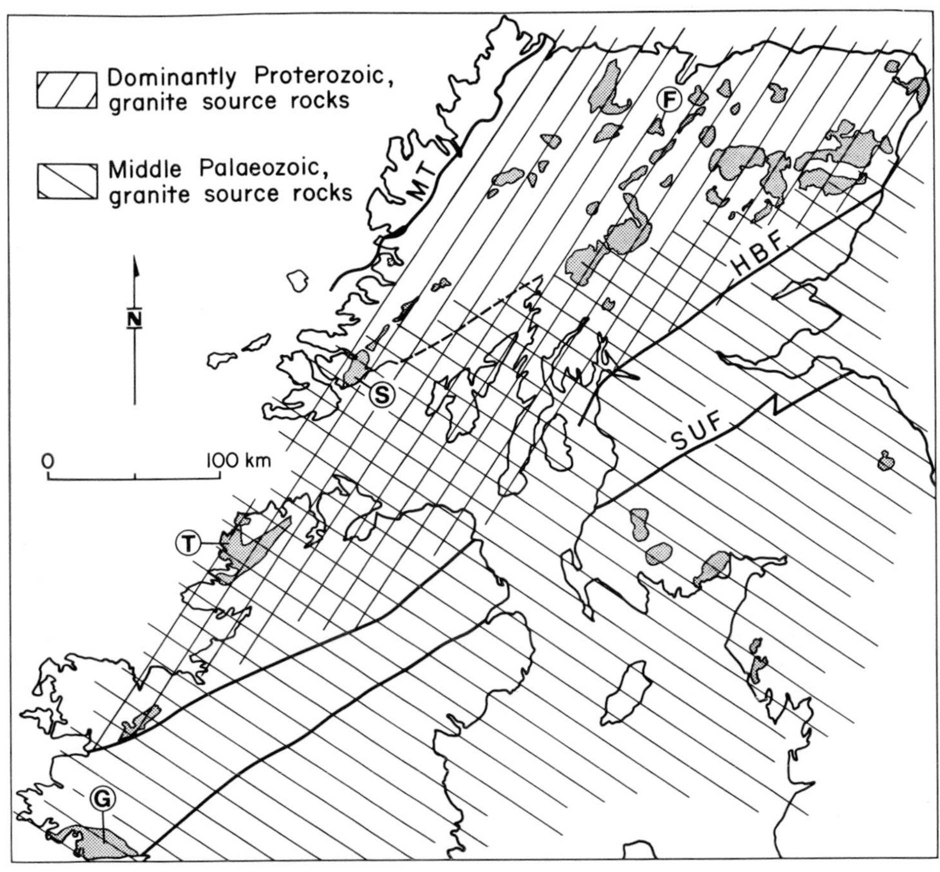


Fig. 11. Reconstruction of the northern British Isles (after Brown and Hughes 1973) showing the proposed distribution of granite source rocks in the early Devonian. The locations of the Highland Boundary fault (HBF) and Southern Uplands fault (SUF) in Ireland are taken from Leake (1963). S—Strontian granite. F—Foyers granite. T—Thorr adamellite. G—Galway granite.

Cluanie, Laggan, Foyers and Aberdeen, we favour this interpretation which further suggests that the proportion of Palaeozoic granite source rocks decreases northward.

The Strontian granite is important in this regard as it has no isotopic evidence of Proterozoic zircons in either the granite sample or the enclosed xenolith. Clearly, this needs further testing by analyzing zircons from additional samples of other phases of the granite but tentatively we interpret this as the earliest of the granites derived from dominantly Palaeozoic source rocks. The geographical location of this granite introduces a difficulty however, as a significant sinistral movement of the Great Glen fault (Kennedy 1946; Winchester 1973) would require that the Strontian granite was generated in a region which we propose consisted of dominantly Proterozoic granite source rocks. However, the zircon U–Pb isotopic systems in the granite could be explained by postulating a dextral movement of the fault (Holgate 1969; Brown and Hughes 1973). A dextral movement such as that proposed by

Brown and Hughes (1973) has been used in the reconstruction on Figure 11. Under this model the Strontian granite is moved south to a position approximately in line with the other Newer granite masses with little isotopic evidence of inherited Proterozoic zircon. The estimated area of mixed granite source rocks is shown on Figure 11.

A possible extension of these source region limits into Ireland (Fig. 11 ) is suggested by the presence of a minor component of inherited zircon in the Donegal granite (unpublished results) and the absence of inherited zircon in the Galway granite (Pidgeon 1969).

## 7f. The origin of the middle Palaeozoic granites

Recent plate tectonic models have linked granite formation with northward dipping (e.g. Dewey 1971), southward dipping (e.g. Jeans 1973) and northward and southward dipping (e.g. Mitchell and McKerrow 1976) subduction zones located in the vicinity of the Midland Valley. Conclusions from the present work have an important bearing on a number of these models. For instance, we find no evidence to suggest continuing granite activity from the Arenig metamorphism to the Devonian as suggested by Dewey and Pankhurst (1970) and Dewey (1971) or for a linear trend of granite ages, as suggested by Phillips *et al.* (1976). Also, we find no evidence of Precambrian granite source rocks south of the Highland Boundary fault as might be expected from models involving the collision of two Precambrian continental masses (e.g. Dewey 1969 fig. 2). On the other hand, we do find evidence of a fundamental division in the granite source rocks between dominantly *c.* 1600 m.y. material north of the Highland Boundary fault and middle Palaeozoic source rocks to the south. The zone of mixing of these granite source rocks could be interpreted as a suture and the possible location of an extinct northward dipping subduction zone; however, such a model cannot explain the southern continuation of Silurian–early Devonian granites. Granites of this age have been found beneath the Lower Carboniferous cover in Yorkshire (Dunham 1974) and may extend further south. The Leicestershire granites could represent an earlier phase. In any case it appears that a widespread invasion of middle Palaeozoic granites intruded Moine and Dalradian sediments in the Highlands and Lower Palaeozoic rocks south of the Highlands, and that south of the Highlands the granites show no indication of having been derived from older source rocks.

At present we see no direct link between this widespread granite activity and subduction during the Arenig or Grampian orogeny which terminated 50 m.y. earlier (Pankhurst 1974; Lambert and McKerrow 1976). Rather, we are led to consider the possibility that the middle Palaeozoic granites are the surface expression of an extensive crustal thickening by an unknown 'sub-crustal accretion' process. The addition of 'new' mantle-derived granitic material to the crust has been widely advocated (e.g. Hurley *et al.* 1962; and review by Boettcher 1973). This process could account for the normal thickness of continental crust underlying the British Isles (Powell 1971; Bamford and Prodehl 1977).

*Acknowledgments.* The results were gathered in the course of a six year research programme on the evolution of the Scottish Caledonides. During this time we have benefitted from many valuable field trips and stimulating discussions with scientists from the Scottish Universities, the Geological Survey and elsewhere. In particular, we acknowledge our indebtedness to M. R. W. Johnson, J. Shepherd and D. Wilson of Edinburgh, M. Munro and P. E. Brown of Aberdeen, E. Stevens of St. Andrews,

B. E. Leake of Glasgow, A. L. Harris and H. J. Bradbury of Liverpool, R. J. Pankhurst of the Antarctic Geological Survey, H. W. Wilson, O. van Breemen and J. Jocelyn of the Scottish Universities Research and Reactor Centre and recent discussions with B. W. Chappell and A. J. R. White from the Australian National and La Trobe Universities. P. E. Brown and A. J. R. White kindly criticised the manuscript. We also gratefully acknowledge the Natural Environment Research Council for support over the period of this work.

## References

BAMFORD, D. and PRODEHL, C. 1977. Explosion seismology and the continental crust-mantle boundary. *J. geol. Soc. Lond.* **134,** 139–851.

BELL, K. 1968. Age relations and provenance of the Dalradian Series of Scotland. *Geol. Soc. Am. Bull.* **79,** 1167–1194.

BLAXLAND, H. B., AFTALION, M. and VAN BREEMEN, O. Preliminary Pb isotopic measurements from feldspars of late Caledonian granites: Evidence for the crust underlying Scotland (in preparation).

BOETTCHER, A. L. 1973. Volcanism and orogenic belts—the origin of andesites. *Tectonophysics* **17,** 223–240.

BRADBURY, H. J., SMITH, R. A. and HARRIS, A. L. 1976. 'Older' granites as time-markers in Dalradian evolution. *J. geol. Soc. Lond.* **132,** 677–684.

BROOK, M., BREWER, M. S. and POWELL, D. 1976. Grenville age for rocks in the Moine of north-western Scotland. *Nature, Lond.* **260,** 515–517.

BROWN, G. C. and HUGHES, D. J. 1973. Great Glen Fault and timing of granite intrusion on the Proto-Atlantic continental margins. *Nature Phys. Sci., Lond.* **244,** 129–132.

BROWN, P. E., MILLER, J. A. and GRASTY, K. L. 1968. Isotopic ages of Late Caledonian granitic intrusions in the British Isles. *Proc. Yorks. geol. Soc.* **36,** 251–276.

——, —— and SOPER, N. J. 1964. Age of the principal intrusions of the Lake District. *Proc. Yorks. geol. Soc.* **34,** 331–342.

CHAPPELL, B. W. and WHITE, A. J. R. 1974. Two contrasting granite types. *Pacific Geol.* **8,** 173–174.

DALZIEL, I. W. D. 1963. Zircons from the granitic gneiss of Western Ardgour, Argyll; their bearing on its origin. *Trans. Edinb. geol. Soc.* **19,** 349–362.

—— and M. R. W. JOHNSON 1963. Evidence for geological dating of the granitic gneiss of Western Ardgour. *Geol. Mag.* **100,** 244–254.

DEARNLEY, R. 1963. The Lewisian Complex of South Harris. *Q. J. geol. Soc. Lond.* **119,** 243–307.
DEWEY, J. F. 1969. Evolution of the Appalachian/Caledonian orogen. *Nature, Lond.* **222,** 124–129.
—— 1971. A model for the Lower Palaeozoic evolution of the southern margin of the early Caledonides of Scotland and Ireland. *Scott. J. Geol.* **7,** 219–240.
—— and PANKHURST, R. J. 1970. The evolution of the Scottish Caledonides in relation to their isotopic age pattern. *Trans. R. Soc. Edinb.* **68,** 361–389.
DUNHAM, K. C. 1974. Granite beneath the Pennines in North Yorkshire. *Proc. Yorks. geol. Soc.* **40,** 191–194.
FERGUSSON I. W. 1978. Structural age of the Vagastie Bridge Granite. *Scott. J. Geol.* **14,** 89–92.
GRAUERT, B. 1973. U–Pb isotopic studies of zircons from the Gunpowder Granite, Baltimore County, Maryland. *Ann. Rep. Dept Terr. Magn. Year Book* **72,** 288–290.
—— and HOFMANN, A. 1973. Old radiogenic lead components in zircons from the Idaho Batholith and its metasedimentary aureole. *Ann. Rep. Dept Terr. Magn. Year Book* **72,** 297–299.
——, HÅNNY, R. and SOPTRAJANOVA, G. 1974. Geochronology of a polymetamorphic and anatectic gneiss region: The Moldanubicum of the area Lam–Deggendorf, Eastern Bavaria, Germany. *Contr. Mineral. Petrol.* **45,** 37–63.
GULSON, B. L. and KROGH, T. E. 1973. Old lead components in the young Bergell Massif, south-east Swiss Alps. *Contr. Mineral. Petrol.* **40,** 239–252.
—— and RUTISHAUSER, H. 1976. Granitisation and U–Pb studies of zircons in the Lauterbrunnen Crystalline Complex. *Geochem. J.* **10,** 13–23.
HARKER, R. I. 1962. The older ortho-gneisses of Carn Chuinneag and Inchbae. *J. Petrol.* **3,** 215–237.
HARRY, W. T. 1953. The composite granitic gneiss of Western Ardgour, Argyll. *Q. J. geol. Soc. Lond.* **109,** 285–308.
HOLGATE, N. 1969. Palaeozoic and Tertiary transcurrent movements on the Great Glen Fault. *Scott. J. Geol.* **5,** 97–139.
HURLEY, P. M., HUGHES, M., FAURE, G., FAIRBAIRN, H. W. and PINSON, W. H. 1962. Radiogenic strontium–87 model of continent formation. *J. Geophys. Res.* **67,** 5315–5334.
JEANS P. J. F. 1973. Plate tectonic reconstruction of the south Caledonides of Great Britain. *Nature, Phys. Sci., Lond.* **245,** 120–122.
JEHU, T. J. and CRAIG, R. M. 1927. Geology of the Outer Hebrides, Part IV. South Harris. *Trans. R. Soc. Edinb.* **55,** 457–488.
JOHNSON, M. R. W. 1962. Some time relations of movement and metamorphism in the Scottish Highlands. *Geol. Mijnb.* **42,** 121–142.
—— and SHEPHERD, J. 1970. Notes on the age of metamorphism of the Moinian. *Scott. J. Geol.* **6,** 228–230.
KENNEDY, W. Q. 1946. The Great Glen Fault. *Q. J. geol. Soc. Lond.* **102,** 41–72.
KÖPPEL, V. and GRÜNENFELDER, M. 1971. A study of inherited and newly formed zircons from paragneisses and granitised sediment of the Strona-Ceneri-zone (Southern Alps). *Schweiz. Mineral. Petr. Mitt.* **51,** 385–409.
—— and —— 1975. Concordant U–Pb ages of monazite and xenotime from the Central Alps and the timing of the high temperature Alpine metamorphism. *Schweiz. Mineral. Petr. Mitt.* **55,** 129–132.
KROGH, T. E. 1973. A low-contamination method for hydrothermal decomposition of zircon and extraction of U and Pb for isotopic age determinations. *Geochim. Cosmochim. Acta* **37,** 485–494.
—— and DAVIS, G. L. 1972. The effect of regional metamorphism on U–Pb systems in zircon and a comparison with Rb–Sr systems in the same whole rock and its constituent minerals. *Carnegie Inst. Year Book* 72, 601–610.
KULP, J. L., LONG, L. E., GRIFFIN, A. A., MILLS, R., LAMBERT, R. ST J., GILETTI, B. J. and WEBSTER, R. K. 1960. Potassium–argon and rubidium–strontium ages of some granites from Britain and Eire. *Nature, Lond.* **185,** 495–497.
LAMBERT, R. St J. and MILLS, A. A. 1960. Some Critical points for the Palaeozoic Time scale from the British Isles. *Annals of the N.Y. Acad. Sci.* **91,** 378–389.
—— and MCKERROW, W. S. 1976. The Grampian Orogeny. *Scott. J. Geol.* **12,** 271–292.
LEAKE, B. E. 1963. Location of the Southern Uplands Fault in Central Ireland. *Geol. Mag.* **100,** 420–423.
LONG, L. E. 1964. Rb–Sr chronology of the Carn Chuinneag intrusion, Ross-shire, Scotland. *J. Geophys. Res.* **69,** 1589–1597.
—— and LAMBERT, R. St J. 1963. Rb–Sr isotope ages from the Moine series. *In:* Johnson, M. R. W. and Stewart, F. H. (Editors). *The British Caledonides.* 217–239. Oliver and Boyd, Edinburgh.
LYON, T. D. B., PIDGEON, R. T., BOWES, D. R. and HOPGOOD, A. M. 1973. Geochronological investigation of the quartzofeldspathic rocks of the Lewisian of Rona, Inner Hebrides. *J. geol. Soc. Lond.* **128,** 389–404.
MACGREGOR, M. 1937. The western part of the Criffel–Dalbeattie igneous complex. *Q. J. geol. Soc. Lond.* **93,** 457–486.

MERCY, E. L. P. 1960. Caledonian igneous activity. *In:* Craig, G. Y. (Editor). *The Geology of Scotland,* 229–267. Oliver and Boyd, Edinburgh.
—— 1963. The geochemistry of some Caledonian Granitic and metasedimentary rocks. *In:* Johnson, M. R. W. and Stewart, F. H. (Editors). *The British Caledonides,* 189–215. Oliver and Boyd, Edinburgh.
MENEISY, M. Y. and MILLER, J. A. 1963. A geochronological study of the crystalline rocks of Charnwood Forest, England. *Geol. Mag.* **100,** 507–523.
MILLER, J. A. and BROWN, P. E. 1965. Potassium argon age studies in Scotland. *Geol. Mag.* **102,** 106–134.
MITCHELL, A. H. G. and MCKERROW, W. S. 1975. Analogous evolution of the Burma Orogen and the Scottish Caledonides. *Geol. Soc. Am. Bull.* **86,** 305–315.
MYERS, J. S. 1971. The Late Laxfordian granite–migmatite complex of Western Harris, Outer Hebrides. *Scott. J. Geol.* **7,** 254–284.
PANKHURST, R. J. 1974. Rb–Sr whole-rock chronology of Caledonian events in Northeast Scotland. *Geol. Soc. Am. Bull.* **85,** 345–350.
—— and PIDGEON, R. T. 1976. Inherited isotope systems and the source region pre-history of early Caledonian Granites in the Dalradian Series of Scotland. *Earth Planet. Sci. Lett.* **31,** 55–68.
PASTEELS, P. 1970. Uranium–lead radioactive ages of monazite and zircon from the Vire–Carolles Granite (Normandy). A case study of zircon–monazite discrepancy. *Eclogae geol. Helv.* **63,** 231–237.
PHILLIPS, W. E. A., STILLMAN, C. J. and MURPHY, T. 1976. A Caledonian plate tectonic model. *J. geol. Soc. Lond.* **132,** 579–609.
PIDGEON, R. T. 1969. Zircon U–Pb ages from the Galway granite and the Dalradian of Connemara, Ireland. *Scott. J. Geol.* **5,** 375–392.
—— and AFTALION, M. 1972. The geochronological significance of discordant U–Pb ages of oval-shaped zircons from a Lewisian gneiss from Harris, Outer Hebrides *Earth Planet. Sci Lett.* **17,** 269–274.
—— and BOWES, D. R. 1972. Zircon U–Pb ages of granulites from the Central Region of the Lewisian, northwestern Scotland. *Geol. Mag.* **109,** 247–258.
—— and HOPGOOD, A. M. 1975. Geochronology of Archaean gneisses and tonalites from north of the Frederikshåbs isblink, S. W. Greenland. *Geochim. Cosmochim. Acta.* **39,** 1333–1346.
—— and JOHNSON, M. R. W. 1974. A comparison of zircon U–Pb and whole-rock Rb–Sr systems in three phases of the Carn Chuinneag granite, northern Scotland. *Earth Planet. Sci. Lett.* **24,** 105–112.
POWELL, D. W. 1971. A model for the Lower Palaeozoic evolution of the southern margin of the Early Caledonides of Scotland and Ireland. *Scott. J. Geol.* **7,** 369–372.
RAST, N. 1963. Structure and metamorphism of the Dalradian rocks of Scotland. *In:* Johnson, M. R. W. and Stewart, F. H. *The British Caledonides,* 123–142. Oliver and Boyd, Edinburgh.
READ, H. H. 1961. Aspects of Caledonian magmatism in Britain. *Lpool and Manchr geol. J.* **2,** 653–683.
ROBERTS, J. L. 1966. Ignimbrite eruptions in the volcanic history of the Glencoe Cauldron subsidence. *Lpool and Manchr geol. J.* **5,** 173–184.
SHEPHERD, J. 1973. The structure and structural dating of the Carn Chuinneag intrusion, Ross-shire. *Scott. J. Geol.* **9,** 63–88.
SILVER, L. T. 1963. The relation between radioactivity and discordance in zircons. *Nucl. Geophys.* **1075,** 34–39.
—— and DEUTSCH, S. 1963. Uranium–lead isotopic variations in zircons: A case study. *J. Geol.* **71,** 721–758.
SOPER, N. J. 1971. The earliest Caledonian structures in Moine thrust belt. *Scott. J. Geol.* **7,** 241–253.
—— and BROWN, P. E. 1971. Relationship between metamorphism and migmatization in the northern part of the Moine Nappe. *Scott. J. Geol.* **7,** 305–325.
SUMMERHAYES, C. P. 1966. A geochronological and strontium isotope study on the Garbal Hill-Glen Fyne igneous complex, Scotland. *Geol. Mag.* **103,** 153–165.
TILTON, G. R. 1960. Volume diffusion as a mechanism for discordant lead ages. *J. Geophys. Res.* **65,** 2933–2945.
VAN BREEMEN, O., AFTALION, M. and PIDGEON, R. T. 1971. The age of the granitic injection complex of Harris, Outer Hebrides. *Scott, J. Geol.* **7,** 139–152.
——, PIDGEON, R. T. and JOHNSON, M. R. W. 1974. Precambrian and Palaeozoic pegmatites in the Moines of northern Scotland. *J. geol. Soc. Lond.* **130,** 493–507.
WASSERBURG G. J. 1963. Diffusion processes in lead–uranium systems. *J. Geophys. Res.* **68,** 4823–4846.
WETHERILL, G. W. 1956. Discordant uranium–lead ages. *Trans. Am. Geophys. Union* **37,** 302–326.
WHITE, A. J. R. and CHAPPELL, B. W. 1977. Ultrametamorphism and granitoid genesis. *Tectonophysics* **43,** 7–22.
——, WILLIAMS, I. S. and CHAPPELL, B. W. 1976. The Jindabyne thrust and its tectonic, physiographic and petrogenetic significance. *J. Geol. Soc. Australia* **23,** 105–112.

WINCHESTER, J. A. 1973. Pattern of regional metamorphism suggests a sinistral displacement of 160 km along the Great Glen Fault. *Nature, Phys. Sci., Lond.* **241,** 81–84.
WINKLER, H. G. F. 1967. *Petrogenesis of metamorphic rocks.* Springer Verlag, Berlin, 237 pp.
WYLLIE, P. J. 1971. *The Dynamic Earth.* John Wiley, New York. 416 pp.

R. T. Pidgeon,
Research School of Earth Sciences,
Australian National University,
Canberra, Australia
(formerly at Scottish Universities Research and Reactor Centre).

M. Aftalion,
Isotope Geology Unit,
Scottish Universities Research and Reactor Centre,
East Kilbride,
Glasgow G75 OQU, Scotland.

# Granite emplacement: the granites of Ireland and their origin

*B. E. Leake*

The Irish Caledonian granites fall into three age groups: about 500 m.y. (e.g. the Ox Mountains and the Connemara ortho-granites), 460 m.y. (Oughterard Granite) and 415 m.y. (Galway, Leinster, Newry, etc.) and are spatially associated with known major NE–SW lineaments, both slides and faults. These ages closely agree with the times of major uplift in the Scottish and Irish Dalradian and Moines as recorded by K-Ar ages. It is postulated that movement, especially uplift, on these major lineaments that extended right through the crust to the mantle, brought hot mantle material against the lower crust and generated and concentrated granitic magma from partial melting of the lower crust. The Sr isotope data are compatible with derivation from a previously highly metamorphosed granulite facies crust, presumably therefore of Grenville or Lewisian age. The late (415 m.y.) phase is probably connected with major sinistral transcurrent faulting in the British Isles. The granites partly control the present elevation of the Irish crust whereas sedimentary basins adjoining the plutons may be partly formed by subsidence due to lateral migration of granite. The overall result of granite emplacement is an increased density stratification of the crust whereby lighter material is moved to higher levels leaving a more basic and denser residue at the base of the crust and thus assisting stabilisation and cratonic development.

## 1. Introduction

A constantly repeated feature of continental crustal evolution is the reoccurrence of granitic intrusions, both early, syn and most characteristically, late or long after orogenic movements. There is almost no substantial part of the continental crust that is totally devoid of granite intrusions but it is quite remarkable that they are totally absent from the oceanic crust. This single fact strongly suggests that granites are derived from continental crust by melting and not from the mantle, whose composition is generally accepted to be ultrabasic and thus unsuitable to provide large volumes of acid magma.

This account outlines the major features of the Irish Caledonian granites and suggests a mechanism to explain the otherwise puzzling occurrence of granites up to 100 m.y. after the metamorphic and structural climax of the Caledonian orogeny. The occurrence of the granites is spatially and genetically linked to major crustal lineaments, either slides or deep faults, that sliced through the crust to the mantle and caused uplift and brought hot mantle material against lower crustal matter that melted to give granite magma. The episodic nature of these movements is shown by age data on substantial segments of the crust in northwest Britain and the ages obtained coincide with that of major granite emplacement epochs. Even the present elevation of the crust seems to have been influenced by Caledonian granites and there seems to be no substantial bodies of buried granite in Ireland or off the coast of Ireland.

The granites of Ireland occur peripherally around the edge of the island with only minor occurrences in the centre of Ireland (e.g. the contaminated granodiorite to quartz dioritic Crossdoney Granite, near Cavan; Skiba 1952) and these minor intrusions are not considered. It is suggested that the sedimentary basins of central

Table 1. Ages and Initial $^{87}Sr/^{86}Sr$ ratios of Irish granites

| | K-Ar | U-Pb | Rb-Sr | Rb-Sr minus K-Ar | $(^{87}Sr/^{86}Sr)_i$ |
|---|---|---|---|---|---|
| Main Donegal | 378 ± 7 (1) | | 498 ± 5 (2) | 120 | 0·708 (2) |
| Carnsore | 350 ± 10 (3) | | 555 ± 6 (4) | 205 | 0·709 (4) |
| Ox Mtns | | | 500 ± 18 (5) | | 0·707 (5) |
| Ox Mtns | | | 487 ± 6 (6) | | 0·706 (6) |
| Oughterard | | | 469 ± 7 (7) | | 0·701 ?(7) |
| Rosses | 384 ± 8 (1) | | 430? (2) | 46 | |
| Leinster | 392 ± 8 (1) | | 428 (8) | 36 | 0·704–0·706 (7) |
| Newry | | | 399 ± 21 (10) | | 0·706 (10) |
| Corvock | 376 ± 14 (11) | | 388 ± 19 (11) | | |
| Galway | 278 ± 10 (3) | 415 ± 15 (12) | 407 ± 12 (7) | 129 | 0·704 (7) |
| Omey | | | 410 ± 17 (7) | | 0·706 (7) |
| Roundstone | | | 418 ± 80 (7) | | |
| Inish | | | 427 ± 8 (7) | | 0·704 (7) |
| Mourne Mtns | 58 ± 4 (13) | | | | |

$\lambda = 1{\cdot}39 \times 10^{-11}\ yr^{-1}$ for $Rb^{87}$

References given by numbers in parentheses thus: (1) Brown *et al.* 1968; (2) Leggo *et al.* 1969; (3) Leutwein, F. 1970; (4) Leutwein *et al.* 1972; (5) Max *et al.* 1976; (6) Pankhurst *et al.* 1976; (7) Leggo *et al.* 1968; (8) Lambert and Mills 1961; (9) O'Connor and Brück 1976; (10) O'Connor 1976; (11) Moorbath *et al.* 1968; (12) Pidgeon 1969; (13) Evans *et al.* 1973.

Ireland and off-shore marginal to the granites are genetically linked with nearby granite emplacement and peripheral subsidence near to the intrusions.

Throughout the account all Rb-Sr ages are based on the decay constant for $^{87}$Rb of $1{\cdot}39 \times 10^{-11}$ yr$^{-1}$.

## 2. Episodic ages of the Irish granites

The Irish granites are principally of Caledonian age and fall into two or more probably three discrete age ranges, 500±10 m.y.; *c.* 460 m.y. and 415±15 m.y. (Table 1). The older ages come from the Donegal, the Ox Mountains and the Connemara migmatites, while the younger (415 m.y.) main post-orogenic phase involved the Leinster batholith, the Galway batholith and its satellitic granites of Roundstone, Inish and Omey; the Newry, Corvock and possibly some of the Donegal granites. The Oughterard Granite in Connemara may have been emplaced shortly after 500 m.y. or most likely is about 469±7 m.y. old, falling into an intermediate group.

In Scotland these three granite phases are well known e.g. Ben Vuroch and Dunfallendy Hill granites, Rb-Sr age 491–514 m.y. (Pankhurst and Pidgeon 1976) and the Aberdeenshire younger granites 412±7 m.y. (Bell 1968) with a major phase at 460 m.y. in Aberdeenshire (Pankhurst 1974) and possibly northern England in the Threlkeld microgranite (Rb-Sr age, 471±16 m.y.; Wadge *et al.* 1974) that is not generally recognised in Ireland except for perhaps the Oughterard Granite. The final major phase of Caledonian granite emplacement at 415±15 m.y. was widespread throughout the British Isles from Mount Sorrel, Leicester (433±17 m.y. Cribb 1975) and Shap (420±11 m.y. Lambert and Mills 1961) to many bodies such as the Cheviot Hills Granite whose geological relations and rather younger K-Ar ages suggest that they were probably intruded at the same period. Likewise there was near synchronous intrusion in the Appalachians, e.g. Charlotte Belt, southeast of the Inner Piedmont Belt with a NE–SW line of plutons about 414–385 m.y. (Fullagar 1971), in southwest New Brunswick and southeast Nova Scotia, e.g. 417±38 m.y. (Cormier and Smith 1973). Accordingly, the Irish Caledonian granites were the result of processes that acted synchronously over a wide area of continental crust, i.e. these processes were not local but controlled by major widespread crustal events.

The Carnsore Granodiorite appears to be older (a poor Rb-Sr isochron suggests *c.* 550 m.y.; Luetwein *et al.* 1972) and was emplaced into the Rosslare complex in southeast Ireland.

The only substantial non-Caledonian Irish granite is the Mourne Mountains body that is about 58 m.y. old (Evans *et al.* 1973) and is clearly related to the Tertiary episode of igneous activity connected with the breakup of the crust in the northwest Atlantic which provoked major outpouring of basaltic magma.

The above ages are based on the information available in early 1977 but it must be appreciated that the ages of the Leinster and Donegal granites are subject to substantial uncertainty and the latter may belong to the 415 m.y. phase.

## 3. The granites of Donegal

The granites of Donegal and their aureoles have been described in detail and at a

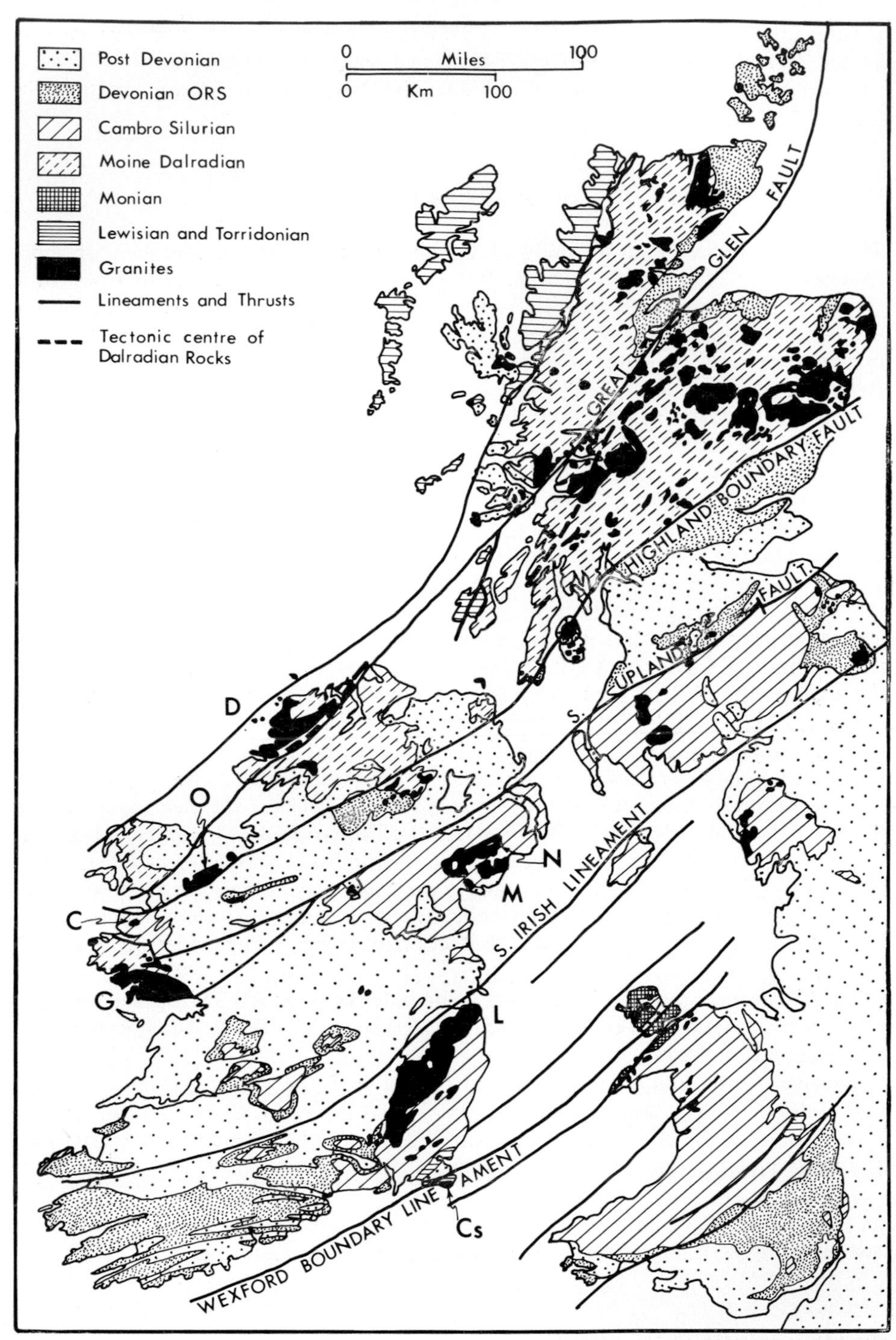


Fig. 1. Distribution of major lineaments and granitic bodies in the British Isles. D = Donegal Granites; O = Ox Mountain Granodiorites; C = Corvock Granodiorites; G = Galway Granites; Cs = Carnsore Granite; M = Mourne Mountains Granite; N = Newry Granodiorite.

length in numerous publications listed in the definitive account of Pitcher and Berger (1972). Repetition or extended summary of this work would merely duplicate the available literature and yet no account of the Irish granites should fail to draw on the mass of extremely high quality work carried out and still underway on the Donegal granites.

Donegal contains eight granites distinguished for their variety of final emplacement modes. The Thorr and Fanad granite were emplaced by reactive stoping, the Ardara and Toories pluton by diapiric action, the Rosses and Barnesmore granite by cauldron subsidence, the Main Donegal Granite by magmatic wedging complicated by intense deformation. The Trawenagh Bay Granite, is rather an enigma, possibly being emplaced by piecemeal stoping but the whole petrogenesis of this granite is clouded by the apparently incompatible time-relationships between it and the Main Donegal Granite which it adjoins. This might be resolved if the Trawenagh Bay Granite were emplaced as a late phase of the Donegal Granite, after some but not all of the foliation and fabric in the Donegal Granite was imposed.

The Main Donegal Granite is remarkably close to and parallel with the major Knockateen slide for most of its outcrop with the northeast termination of the granite almost running into this slide. The Knockateen slide separates the northwest facing Creeslough succession of northwest Donegal from the southeast facing Kilmacrenan succession with the Lough Foyle synform, the lateral equivalent in Ireland of the Iltay boundary slide and the Loch Awe structure of the southwest Highlands of Scotland. It is therefore an early fundamental major tectonic break that in much later (post-granite emplacement, post-Silurian) times was nearly coincident with the major transcurrent faulting that is exemplified by the Leannan Fault system with its 40 km of lateral displacement and substantial (several km), but unknown, vertical movement. Accordingly it is the writer's view that the slab of crust along which the Main Donegal Granite was intruded was a major zone of crustal weakness along one or more fundamental lineaments that extended through most or all of the crust. Renewed or continuing movements in this zone were responsible for imposing the fabric of the Main Donegal Granite on that body. The fact, emphasised by Pitcher and Berger (1972 p. 359) that there is little evidence for fault movements pre-dating the granite emplacement is not surprising because nearly all such faults would later be re-used by continuing movements after the emplacement of the granites and their early history would be usually undecipherable in the field. Nevertheless, Pitcher and Berger (1972 p. 73) postulated "considerable thrusting . . . at a comparatively late stage in the structural history, though clearly pre-dating the emplacement of the granites". When this is considered along with the difficulties of identifying the Knockateen slide as an *exposed* tectonic break, the quality of the exposure in much of the critical ground outside the granites, the form of the Main Donegal Granite and the surprisingly unfaulted nature of the granite boundaries (except the Barnesmore pluton), it is clear that emplacement guided by a major tectonic lineament is a viable hypothesis. Work on the Main Donegal aureole suggests that the aureole to the northwest of the granite was significantly cooler than that to the southeast but there are differing views as to whether the hotter situation in the southeast predated the granite emplacement. The possibility of different tectonic levels being involved on each side of the granite is feasible and although difficult to confirm might have arisen from later faults moving both laterally and vertically.

The Rb-Sr isochron age of the Main Donegal Granite at 498$\pm$5 (Leggo *et al.* 1969) is no less than 100 m.y. older than the K-Ar ages of hornblendes in the Thorr and Ardara granites (392±8 m.y.) or micas in the Main Donegal Granite (357–382 m.y.) (Lambert 1966; Brown *et al.* 1968). This is far too long for a simple cooling pattern and suggests that either the Rb-Sr age is too old or that the Main Donegal Granite and probably the remaining plutons, were crystallised and partly emplaced about 500 m.y. ago and cooled only slowly at depths which the metamorphic aureoles suggest were about 8–9 km. These depths would therefore, with a geothermal gradient of 20–25° C/km, be above the threshold of about 200°C below which argon diffusion ceases. About 400 m.y. ago major faulting could have uplifted and rapidly cooled the rocks to low (<200°C) temperatures resulting in freezing of the K-Ar ages. The widespread occurrence of K-Ar ages of 400±25 m.y. in the Caledonian granite suite of Britain is believed to be therefore only partly the result of crystallisation and intrusion at that time being also the consequence of major fault-induced cooling by uplift of plutons already emplaced into the upper crust.

It is interesting to notice that the main concentration of late Caledonian granites in Scotland (Fig. 1) is curiously near to the extension of the Ardrishaig anticline, the root of the Tay Nappe (Roberts 1974), almost along strike from the Donegal Granite. Whether this early structure was re-activated later and was partly responsible for the siting of the major granite belt of the Grampians is uncertain until more Rb-Sr isochrons are available to compare with the K-Ar ages. In many ways, the form of the Donegal granites is more consistent with emplacement in the late 415 m.y. phase which is about the age of the main Grampian granites. This would link it more positively with the Leannan Fault.

## 4. The granites of Connemara

### 4a. Granite gneisses

The oldest Connemara Granite is the variably foliated K feldspar gneiss belonging to the Migmatitic Quartz Diorite Gneiss of the Connemara Migmatites. These granite ortho-gneisses (Fig. 2) contain quartz, andesine, K feldspar, chloritised biotite and hornblende. They crystallised very late in the Migmatitic Quartz Diorite Gneiss sequence being late differentiates of the magmatic series (Leake 1970). Because of the migmatitic nature of the gneiss complex, which is intimately involved with metasediment in some areas, and possibly also because of low temperature retrogression, several attempts to obtain reliable isochrons have proved unsuccessful. Hornblendes from the granite gneisses have yielded 435±17 m.y. by K-Ar analyses (Leggo *et al.* 1966) but this must be a late cooling age as the gneisses pre-date the Oughterard Granite and are thought to follow shortly after the Connemara basic and ultrabasic rocks (which they inject) and must therefore be younger than 510±10 m.y. based on U-Pb in zircons in the basic rocks (Pidgeon 1969).

The Connemara Migmatites represent a major incursion of water-rich basic and ultrabasic magma into the Dalradian rocks along a major E–W movement zone by which a strong foliation was imposed on the gneisses as they crystallised and the early peridotite and hornblende-rich gabbros were tectonically torn apart in the more trondjemitic Quartz Diorite Gneiss. The volume of granite gneiss is minor compared with these other rock types—perhaps 10–15% and crystallisation of the granite gneiss seems to have outlasted the penetrative movements (ascribed to $F_3$) so that some of the granite gneisses are hardly foliated. The main distribution of these late granite gneisses in the migmatite complex (Fig. 2) is concentrated in a zone running east from Cashel for several kilometres which suggests that the last magma fractions of the magmatic episode were selectively concentrated in certain critical zones by tectonic activity (Leake 1970) and probably also migrated upwards to now eroded levels. The tectonic reconstructions are suggestive of a "root-zone" through which magma moved upwards in the crust and water was concentrated being latterly dispersed into the adjoining metasediments to the north and carrying with it it elements enriched in the late stage magma fractions. These elements were fixed in the peripheral metasediments which have suffered major metasomatism for up to 5 km from the zone of migmatites (Senior and Leake 1978).

## 4b. The Oughterard Granite

This granite is quite unlike the remaining Caledonian granites of Ireland as the detailed account of Bradshaw *et al.* (1969) makes evident. The granite occurs in two main vaguely circular bodies, in the west, the Tullaghboy mass which is faulted apart, and in the east the Oughterard intrusion. The two bodies are joined by a most irregular dyke (Fig. 2). The Tullaghboy mass, which is highly irregular and often xenolithic with country rock fragments, passes into a swarm of small granite pods, veins, sills, and dykes that extend for several kilometres to the west in the Connemara Schists and also in the Migmatitic Quartz Diorite Gneiss. It is this scattered pod-like occurrence and the lack of a single simple intrusion that marks the granite off from most other plutons. The intrusions largely lie astride the $F_4$ Connemara antiform being clearly later than both the $F_3$ and $F_4$ folds which they truncate. There are no pushing or distension structures and the granite has been emplaced partly by dissolving its way into the country rock and partly by dilation. There is no detectable contact aureole and it seems as if the magma may have been emplaced before the regional metamorphism had completely declined.

Leggo *et al.* (1966) showed that the minerals in the granite closed to Rb-Sr diffusion $469 \pm 7$ m.y. and that the youngest likely whole rock isochron was $539 \pm 37$ m.y. but this age was based on 9 of the 12 samples analysed, there being no single isochron. Subsequent geochronological work (Pidgeon 1969) made this age most unlikely because the Oughterard Granite is clearly much younger than the basic and ultrabasic rocks dated at $510 \pm 10$ m.y. by U-Pb measurements on zircons (Pidgeon 1969). In places the granite follows fault lines that cross the foliation of the schists and gneisses thus emphasising its rather late age. That the granite is severely chloritised, sericitised and in places sheared, suggests that the reason the samples fail to plot on a single isochron could be due to loss of $^{87}Sr$ from chloritised biotite which is ubiquitous. Accordingly it would now seem more appropriate to exclude the three samples that have the lowest $^{87}Sr/^{86}Sr$ and on this basis an isochron using the remaining 9 samples gives an age close to the 469 m.y. whole rock mineral isochron. This is therefore now considered to be the most likely age of the Oughterard Granite. The initial ratio of $^{87}Sr/^{86}Sr$ was 0·710. This granite is therefore not only unusual in its occurrence but also in its age, no other Irish granites being known of this type and age.

The mineralogy is rather monotonous with quartz, oligoclase, microcline microperthite, chlorite after biotite, and a little muscovite. Hornblende is restricted to a small intrusion while garnet is rare. Chemically it has high $SiO_2$ (74%), $Na_2O$ slightly in excess of $K_2O$, CaO about 1% and high Ba.

## 4c. The Galway batholith

Modern detailed mapping on this body began in 1958 when Wright (1964) commenced his study of the Carna district and was subsequently followed by systematic

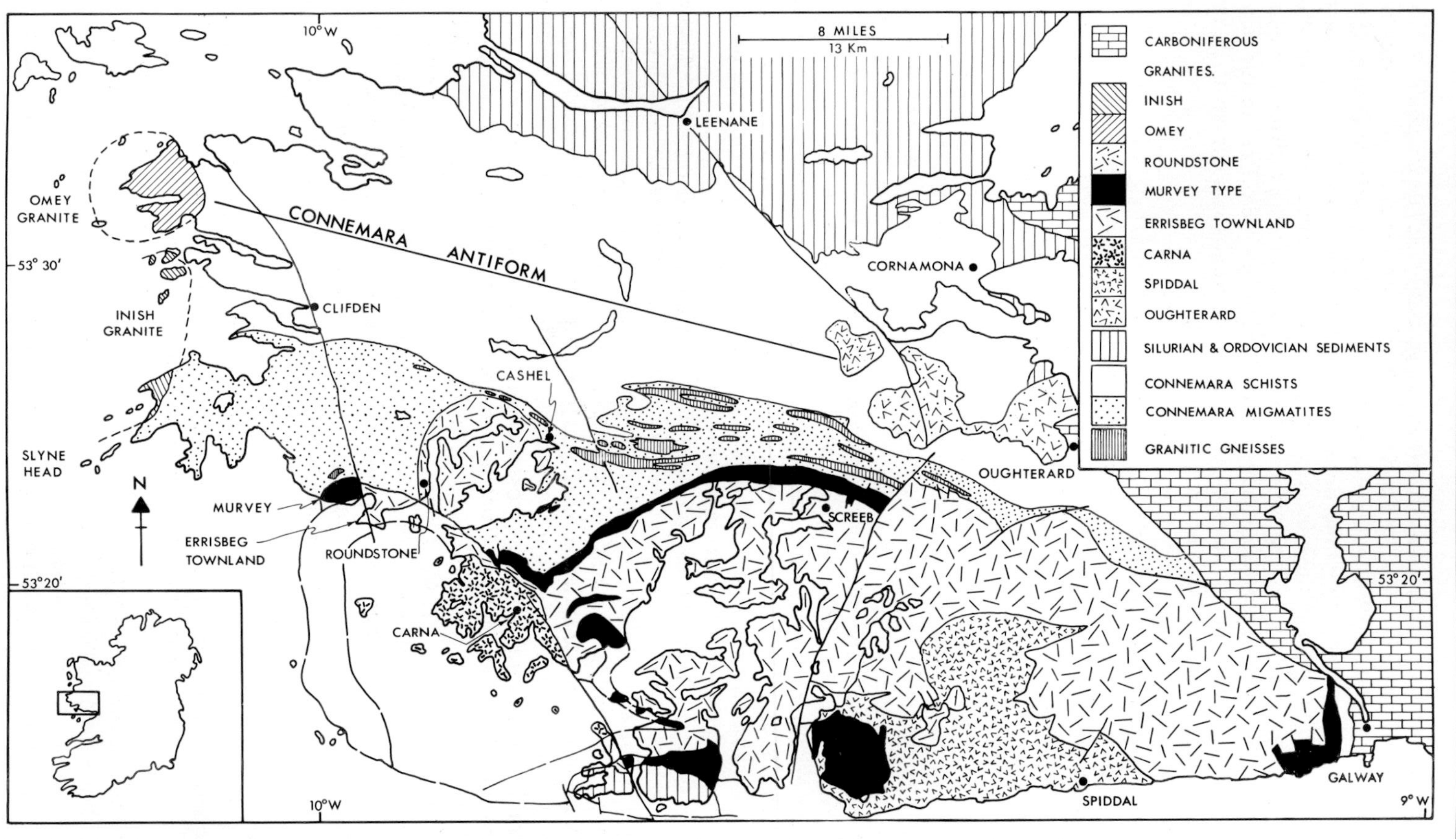


Fig. 2. The Connemara Granites with the central part of the Galway Granite adapted from Max *et al.* (1978).

coverage of parts of the batholith to the east of Carna by Aucott (1965, 1968, 1970), Claxton (1968, 1969, 1970, 1971), Lawrence (1968), Coats and Wilson (1970) and Plant (1968) while Leake (1968) worked west of Carna and in 1974 described the crystallisation history of the western part of the Galway Granite, the whole batholith having been dated at 407±12 m.y. by a Rb-Sr isochron (Leggo *et al.* 1966) and 415±15 m.y. by U-Pb measurements on zircons (Pidgeon 1969). Recently Max *et al.* (1977, 1978) have completed a rapid survey of the unmapped parts of the batholith, including some under-sea exposures on the western seaboard, and have synthesised the available information to give a valuable preliminary overall picture which, with the detailed mapping referred to, forms the basis of Figure 2.

The batholith lies on the assumed direct extension of the Southern Uplands Fault across Ireland, a major splay of which probably passed through the ground now occupied by the batholith to emerge in the Skerd Rocks, southwest of the granite where a major pre-granite fault brings the South Connemara Series, of supposed Ordovician age, against the Connemara basic and ultrabasic rocks (Leake 1963; Max and Ryan 1975).

The batholith is generally well-exposed, particularly in the west and a good deal of fine detail has been mapped. There are two, or possibly, three intrusion centres (Max *et al.* 1978) but a porphyritic orthoclase-microcline adamellite, the Errisbeg Townland Granite, is common to both centres which are those of Carna and Galway–Kilkieran, there being insufficient detailed mapping to determine whether the latter centre is really composed of two partly coalesced domes. Contiguous to the Errisbeg Townland Granite (ETG) near Roundstone is a major satellitic plug that forms the adamellitic Roundstone Granite, the nature of the contact between the ETG and the Roundstone Granite being obscured by Bertraboy Bay.

(i) *The Carna dome.* The western part of the batholith, the Carna dome, has a ring structure with a central granodiorite (Carna Granite) containing irregular vertical bands of K feldspar-richer material that forms the Cuilleen variety. In places the Carna Granite passes over a 300 m transition zone with gradually increasing numbers of K feldspar phenocrysts into the ETG which has only a little hornblende compared with the Carna variety. The ETG possesses excellent primary magmatic layers, marked by variations in the proportions of biotite with hornblende against feldspar with quartz, by current bedding and by graded bedding with biotite-rich bottoms and quartz-K feldspar rich tops. The layering almost always dips outwards from the Carna centre, dipping northwards in the north at Errisbeg Townland, westwards in the west on Croaghnakeela Island and southwards near Lettermullan in the south. It is quite clear that this layering, which now dips commonly between 20° and 45° but reaches up to 75°, must have formed in a near flat position such as described from the superbly exposed granites and syenites of south Greenland by Emeleus (1963) and Ferguson and Pulvertaft (1963). Consequently the layering must have been tipped up by the upward rise of the centre of the Carna dome, a circumstance that suggests that the emplacement of the Carna Granite into its present position was later than the crystallisation of the ETG despite the more potassic and higher silica contents of the ETG. The transition zone is capable of differing interpretations but based on traverses across part of the Carna dome with exhaustive chemical analyses of rocks and microprobe analyses of biotite, Leake (1974) showed that the Carna granite had crystallised from the outside inwards and it would seem that it largely crystallised and was then emplaced into its present position by nearly vertical concentric upward movement. This is deduced from vertical igneous layers produced first by flow-sorting and then later by local remobilisation of the almost crystallised granite, the remobilisation layers being clearly distinguishable from the settling layers in the ETG because they lack grading, current bedding and occasionally cross earlier xenoliths. The outer edge of the Carna Granite is then considered to have been partly K feldspathised by the largely crystallised ETG which differentiated outwards, upwards and away from the Carna Granite. The alternative of a cauldron sinking Carna Granite floundering into ETG is possible geochemically but fails to explain the tipped up layering in the ETG and is therefore rejected.

There is much detailed evidence recording progressive movements during and after the crystallisation of the granite in the Carna dome and throughout the batholith. Although country rock xenoliths are almost totally absent throughout the batholith, small (2 cm to 2 m, occasionally larger) rounded microdiorite xenoliths are common and these must be of early igneous crystallisation and have been broken, dispersed and carried up by the granite. They cannot be country rock material because they are non-metamorphic and do not match any of the country rocks which are mostly high grade metamorphic rocks. Their wide distribution in the granites makes it difficult to envisage an earlier major diorite intrusion being involved as it would need to underlie almost the whole batholith, moreover the fine grain size is only matched in the crystallisation of dykes; none of the granites has such a fine grain size. On Croaghnakeela Island, there is a swarm of E–W trending dykes up to 10 m wide with fine grained chilled margins and coarsely crystallised centres, so coarse that they resemble granite and some of these dykes can be walked along into broken up and disoriented xenoliths; clearly disturbed by later movement of the granite that earlier had fractured, allowed the dykes to intrude and then chilled the dyke magma. Such a remobilisation process carried out at depth might explain the microdiorite xenoliths and would certainly imply a great deal of upward transport and irregular flow.

The Galway batholith is significantly different from that of the Leinster pluton in that the latter, like most intrusions, has the oldest and most basic rocks at the periphery whereas the Galway batholith has the most acid and latest members at the margin. The roof and margins of the batholith often have a fine grained, aphyric, acidic (73–76% $SiO_2$), biotite-poor albite granite, the Murvey Granite which at Murvey has a final phase that contains garnets of spessartine-almandine composition. This is a situation that puzzled the author for many years as an igneous body can only lose heat from the outside and crystallisation from the edge would be expected with a chilled margin. How could this mineralogically and chemically late differentiate crystallise at the edge?

A theory to explain the observations was developed in Leake (1974). The Murvey Granite occurs due to a combination of accumulated late magmatic differentiate on top of the ETG and infilling of space left by stoped blocks. The geochemistry shows that the ETG differentiated towards the Murvey Granite composition both in rock and mineral compositions so that at Murvey, where the roof dips northwards at 50°–60°, and north of Carna where the roof dips at values down to 20°, the Murvey Granite lies on top of layered ETG which differentiated upwards to give the Murvey Granite through a prominent zone of layers. The lack of chilled margins could be explained by stoping away of the early chilled granite on sunk blocks of country rock. Layered accumulation is not alone an adequate explanation however as the texture of the Murvey Granite becomes increasingly aplitic indicating faster crystallisation the nearer the country rock and the further from the ETG.

The additional evidence was provided by a strip of the margin at Errisbeg Townland where Murvey Granite was absent and ETG came directly against the previously (500 m.y.) metamorphosed basic and ultrabasic rocks that form most of the northern country rock. Large (900 x 100 m) slabs of ultrabasic rock have been arrested in the process of being prised off into the granite and would have dropped into the magma and sunk without reaction if the process had continued (Leake 1974). The ETG veins that were wedging off the rock change along their length into Murvey Granite with local garnets (described in Leake 1968), topaz and tourmaline. Clearly the Murvey Granite formed by "draining" or segregation out of the residual ETG magma into the low pressure space ("vacuum") left behind by the prised off blocks. Early stoping would simply involve mass movement of the ETG into the vacated space but increasingly as the ETG solidified it would be able only to supply residual low temperature uncrystallised magma—the Murvey Granite. Thus a process very similar to that which produces aplites would have operated. Aplites are normally produced by dilation during the last stages of consolidation of a granite. Cracking results in the residual acid magma draining into the crack where it rapidly crystallises giving the characteristic aplitic texture. In the Galway Granite, such aplites are common and in the ETG they have a very similar composition to the Murvey Granite (whereas those in the Carna Granite are less differentiated). The aplitic texture of the Murvey Granite would thus be explained and also the absence of aplite in the garnetiferous marginal Murvey Granite—no further residual magma could be extracted.

Thus the Murvey Granite is a later differentiate of the ETG magma, with which granite it is always a contact, and crystallised in the space left by late stoping of the margin. Where Garnetiferous Murvey Granite occurs (e.g. at Murvey) this is due to the terminal stoping that drew the ultimate magma out of the normal Murvey Granite. This is rather rare but is matched by the albite aplites in the normal Murvey Granite which often contain garnets whereas those in the remaining granites of the batholith are garnet-free. A fuller detailing of the evidence for the above view is given in Leake (1974).

It then follows that the Murvey Granite outcrops east and southeast of Carna must be near to the roof, a view supported by Max *et al.* (1978) who suggested that this zone be called the Kilkieran septum. This defines the boundary between the Carna and Galway–Kilkieran domes.

(ii) *The Galway–Kilkieran dome*. This dome has non-aplitic garnet-free Murvey Granite along the northern boundary only as far as the Shannawona fault (Claxton 1971; Leake and Leggo 1963; Plant 1968) which sinistrally moves the margin of the granite over 2 km with unknown but substantial vertical displacement—probably several kilometres. This fault is believed to lift up on the east side exposing a much deeper level of the batholith that has the 1 km wide Shannawona Granite at the edge with the Shannaweelaun type inside, both of which are much more basic than the Murvey and ETG types (Plant 1968) and have a distinct foliation reminiscent of that produced by crystallisation during near vertical movement which continued in places after most of the granite had crystallised. The Shannaweelaun Granite averages only 62–63 % $SiO_2$ being lower than the Carna type (68 %) which is otherwise close to the composition of the Shannaweelaun variety. As yet the distributions of the Shannaweelaun and Shannawona types, are largely unmapped. There is a pocket of largely non-aplitic Murvey granite and containing only garnets in rare aplites, in the corner of the batholith at Galway, but otherwise the main occurrences of the Murvey-type granite are in the south.

According to Max *et al.* (1978) a large part of the Galway–Kilkieran dome in the south is made of the Spiddal Granite which is similar to the Carna Granite but has sharp, probably intrusive and certainly highly irregular contacts with the ETG. This irregular magmatic contact contrasts with the ETG–Carna Granite relations which suggest cooler, more consolidated emplacement of the Carna Granite as a more nearly crystallised body than the Spiddal Granite. The evidence then points to the Galway–Kilkieran dome being either earlier than the Carna dome or presenting a deeper cross-section or both. In both domes the central granodiorite, the Carna or Spiddal Granite, was intruded or emplaced into the ETG and there is marginal Murvey-type Granite though the northeastern portion of the batholith is clearly different and probably has very early granite crystallised during substantial upward movement.

The supposedly disrupted dykes, the central emplacement of Carna and Spiddal granites, the outward dipping of current and graded bedding, the vertical movement zones identified in the vertical foliation of the northeastern portion of the batholith and the vertical layers within the batholith, all provide a picture of progressive upward movement of material in the centres of the domes that constitute the batholith.

## 4d. Satellitic plutons of the Galway batholith

The satellitic bodies include the Roundstone, Omey and Inish granites (Table 1 and Fig. 2), all dated about 415 m.y. with minor occurrences of three granites (probably one body at depth) near Letterfrack in northern Connemara (Townend 1966). The Roundstone Granite has not been studied in detail but is a circular adamellitic plug with a diameter of 7 km and frequently has a faulted contact. Natural contacts show no sign of chilling. A marginal quartz-rich facies that is foliated parallel to the contact is sometimes present and this presumably formed by upward movement of the nearby solid magma plug. Hornfelsing has been detected within 50 m of the contact. Further details are given by Leake (1970) and Evans and Leake (1970).

The circular Omey pluton (Fig. 2) in northwest Connemara has been described in detail by Townend (1966). It is a vertical plug of porphyritic adamellite, which carries, particularly in the southeast, a little hornblende. It was emplaced near the axial plane of the Connemara antiform. A number of intrusive phases culminated with a central intrusion and in the west with a marginal band of fine adamellite that is the most acidic and alkali of the granite phases. This is similar to the Murvey Granite in the Galway batholith. Radial aplites emphasise the symmetrical nature of the final stresses acting on this body. There is a complex aureole that Ferguson and Harvey (1978) have investigated and a gas-breccia plug on the southeast side of the intrusion that the same authors are examining.

The Inish Granite occurs on the western seaboard of Connemara (Fig. 2) and the available exposures have been mapped by Townend (1966) and Leake (unpublished). This granite has an aplitic and microcline microperthitic pegmatitic marginal zone with graphic granite. The zone is about 300 m wide and is choked with disoriented country rock xenoliths and passes outwards into a complex vein network that injects the country rock and then into a fringe zone of concordant mixed pegmatite and aplite veins that gradually become less common over a distance of up to 1 km. The xenolith-choked aplopegmatitic facies is truncated inwards by a uniform xenolith-free adamellite that on Inishturk and Turbot Island, at the north end of

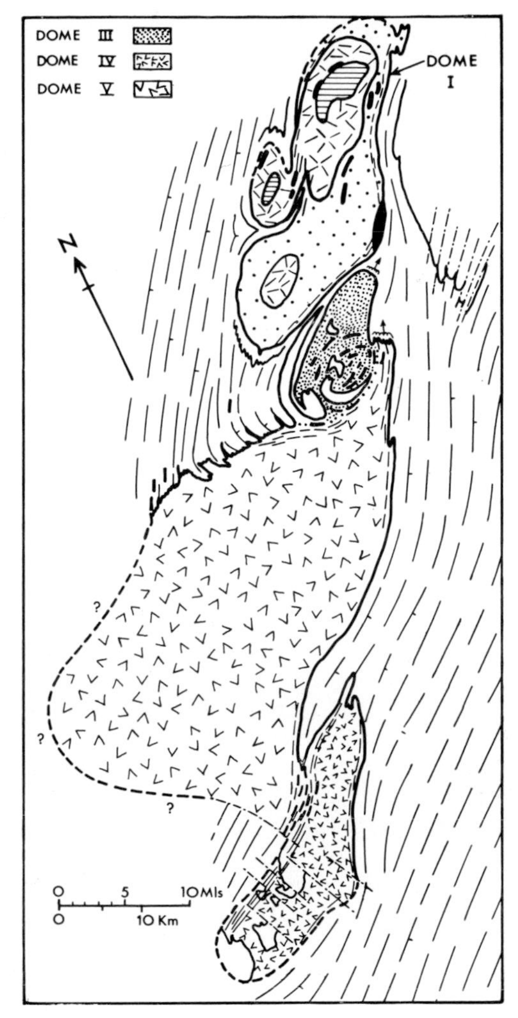


Fig. 3. The domal arrangements of the Leinster batholith after various publications by Brindley detailed in text.

the intrusion, contains a little hornblende. Radiating into the country rock on the SE side with SE strike is a swarm of granite and plagioclase porphyry dykes that cross the concordant granite veins but do not cross the uniform adamellite, stopping short at the adamellite contact. Spectacular hornfelsing is observed at Knock with andalusite replacing regional garnet.

## 5. The Leinster Granite

The Leinster Granite, the largest batholith in the British Isles, is a compound body with five dome-like units (Fig. 3) probably emplaced by distension and stoping with lateral and upward heaving into a Lower Palaeozoic succession. There are comprehensive summaries of the geology available (Brindley 1969, 1973; Brindley and Gupta 1974; Brück 1974; Brück and O'Connor 1977) including the associated appinitic rocks (Brindley 1970) which pass outwards into a lamprophyre swarm (McArdle 1974). Zones of granitic sheets flank the batholith in the country rocks to the southeast and Brindley (1973) regards the Rockabill Granite, lying 6 km off the north Dublin coast (Brindley and Kennan 1972) as belonging to a zone of minor granitic intrusions lying northwest of the main batholith. Radiometric ages of $386 \pm 6$ and $392 \pm 9$ m.y. based on K-Ar mineral analyses and a single Rb-Sr biotite age of $430 \pm 12$ m.y. have been obtained by Kulp *et al.* (1960), Lambert and Mills (1961) and Brown *et al.* (1968).

The northern part of the batholith contains Domes I and II and is, with Dome III, the best exposed and has been studied in the most detail. There are four principal concentrically-arranged granite types in Dome I and II ranging from early quartz diorite or granodiorite through porphyritic microcline and muscovite-bearing adamellites to equigranular adamellite. According to Brück and O'Connor (1977), the oldest and most basic types occur peripherally or as disc-shaped inclusions up to 2 km long in the porphyritic adamellite which itself is centrally intruded by the equigranular adamellite while subsequent deuteric alteration has changed the centres of Domes I and II. The granite varieties run concordant with the eastern sides of Domes I, II, and III, which dip moderately eastwards (35°–68°), but the western sides are steep or overturned with inward dips. The Domes are partly separated by septa of country rock, some of which pass into xenoliths in the granite. The aureole is usually narrow, less than 3 km, and contains andalusite, staurolite, almandine and biotite as scattered porphyroblasts with local chloritoid and sillimanite, indicating modest pressures during metamorphism—perhaps 3–4 kb or a depth of about 10 km. The picture is one of successive pulses of pushing magma each intruded into the centre of the preceding magma in agreement with the concentric disposition and the rather sharp contacts between the different granite types.

Dome III has the most complex internal structure with many circularly-disposed septa of country rock which are believed (Brindley 1973) to be concordont prongs of country rock pendents that follow a circular flow-form and descend from the roof that is still partly preserved on the mountain summits. A mechanism to produce a horizontal circular flow of granite magma is difficult to conceive, the more especially as the roof pendents lie parallel to the layering and not at the high angle that would be necessary if these had acted in a paddlewheel fashion with a vertical axis of rotation. Again these observations suggest vertical upward movement of magma with presumably some upward pushing of the roof, some lateral distension, some stoping and some lateral wedging of pieces of country rock as shown by tonguing of the magma into the country rock. The presence of a cataclastic foliation, especially near the margins, shows that pushing continued when the early peripheral rocks were completely crystallised.

The present outcrop is near to the roof which except for the northern end generally dips southwards at 15°–25°, being at about 750 m elevation in the centre of the exposed batholith.

Between Domes IV and V is a fascinating 6 km wide zone of foliated fine biotite diorites and adamellites to coarse quartz diorites with a plethora of country rock septa, aplites showing variable degrees of deformation, granite gneisses, sillimanite-bearing schists and pegmatites with quartz, muscovite and microcline and sometimes spodumene, beryl, tourmaline and garnet (Brindley 1974).

O'Connor and Brück (1976) give results that suggest that the earliest phase, the grandiorite, probably had an initial ratio of $^{87}Sr/^{86}Sr$ of 0·704 whereas the late Porphyritic Microcline type had 0·707 so that significant differences occur but the explanations require more precise data and more indication of whether there exists a gradation from one ratio to another or not.

## 6. Other granites

The Carnsore Granodiorite occupies a small area in southeast Ireland around Carnsore Point and has a marginal biotite-rich facies. The age of *c.* 550 m.y. (Leutwein *et al.* 1972) indicates an early Caledonian phase of granite activity that is not dealt with here.

The Ox Mountains granodiorite contains evidence of its relatively early age in that is has been penetratively deformed and the fabrics can, according to Max *et al.* (1976) be traced into the fabric of the envelope metasediments, the intrusion being syn to late kinematic with the second phase of deformation. There is a general tendency to give more acidic types to the southeast.

The Newry Granodiorite differs from the remaining Caledonian granites of Ireland in having a pronounced negative gravity anomaly, in being genetically associated with ultramafic, diorite and monzonitic rocks and in not being linked with any known lineament. According to Dr Meighan, his work with Dr Neeson shows that the ultramafic to granodiorite series is a differentiation series with possible minor contamination by crustal assimilation.

The Corvock Granite (Fig. 1) lies south of Clew Bay and about 3 km south of the serpentinite zone of the supposed extension of the Highland Boundary Fault and has recently been restudied by Dr T. Kelly who has kindly allowed me to include some of his results. The intrusion is a thin sheet roughly 25 km² in area but less than 1 km thick as determined from gravity studies. The body is a vertically differentiated biotite granodiorite that passes upwards into a more acid, alkali-rich granite and was probably intruded upwards from the south and spread out in a thin sheet northwards. Contamination by country rock is negligible but there is a distinct thermal aureole. The emplacement of the granite was preceded by two sets of path-locating cracks which formed pre-granite feldspar porphyry dykes geochemically related to the granite. The first formed parallel to the granite margin that was to be and the second followed radial cracks, though both sets cut the main slatey cleavage in the Ordovician and Silurian country rocks and were themselves cut by the granite which was emplaced along fractures partly determined by the earlier dykes.

## 7. Tectonism and granite emplacement

The major Irish granites are spatially associated with various NE–NW lineaments (Fig. 1); the Donegal Granite with the Knockateen slide and subsequently the Leannan Fault system; the Leinster Granite with the south Irish lineament (Rast 1969 p. 207; Gardiner 1975); the Ox Mountains Granodiorite and possibly the Corvock Granite with the Highland Boundary Fault extensions in Ireland; the Connemara orthogneisses with a strong E–W movement zone; the Galway Granite with the extension of the Southern Uplands Fault in Ireland (Leake 1963; Max and Ryan 1975) and possibly the Carnsmore Granodiorite with an early expression of the Wexford Boundary lineament (Gardiner 1975), leaving only the Newry Granite not associated with some previously postulated lineament. This is not to imply that each granite is precisely and exactly controlled in its position of final consolidation (i.e. where it is now) by the surface expression of these lineaments but that generation and emplacement in the crust were genetically related to these major deep faults and possibly even the alignment of the major granite bodies was influenced by these deep fractures.

The hypothesis is strongly supported by the remarkable agreement of the timing of the granite magmatism and that of the freezing of argon diffusion, as recorded by K-Ar ages, in the metamorphic rocks. These K-Ar ages are primarily a measure of marked uplift and cooling by movement on major deep faults.

Thus the oldest K-Ar ages in the Dalradian rocks are parallel and close to the Highland Boundary Fault (Harper 1967) being 503$\pm$6 m.y. or coincident with the 500 m.y. granite episode. The Highland Boundary Fault is itself also remarkably parallel to the fringe of the downward facing Aberfoyle anticline which suggests that this fault has syntectonic Caledonide affinities and must involve substantial vertical movement (Leake 1977; Harris *et al.* 1978). This is consistent with the evidence from the Ox Mountains of the syn-D2 emplacement of the narrow elongated Ox Mountain Granodiorite at 500$\pm$18 m.y. (Max *et al.* 1976) although of course subsequent continuing movements on these fractures have faulted the granite itself. It is therefore perfectly consistent that the Main Donegal Granite with a Rb-Sr age of 498$\pm$5 m.y. should lie almost along the syntectonic Knockateen slide. This 500 m.y. uplift episode is also recorded in the Ingletonian of northern England (Rb-Sr ages, 505$\pm$7 m.y., O'Nions *et al.* 1973).

The 500 m.y. granite phase was accompanied by the emplacement of major basic and ultrabasic intrusions in Connemara (U-Pb age, 510$\pm$10 m.y., Pidgeon 1969) and Aberdeenshire (Rb-Sr age, 501$\pm$17 m.y. on the basic rocks with 500$\pm$24 m.y. for their aureoles; Pankhurst 1970) and possibly in Tyrone. Clearly this episode involved substantial tapping of the mantle and gave in Connemara a wide range of peridotites, hornblendic gabbros and quartz diorite gneisses culminating in potash feldspar ortho-gneisses whose ages must be slightly younger than the early basic phase dated at 510$\pm$10 m.y. by U-Pb in zircons (Pidgeon 1969).

In the Dalradian of both Connemara and Scotland the main concentration of K-Ar ages is around 450$\pm$15 m.y. (Moorbath *et al.* 1968; Dewey and Pankhurst 1970) overlapping the 460$\pm$10 m.y. Rb-Sr age granites of northeast Scotland and possibly the Oughterard Granite of Connemara.

The adoption of a slightly larger value for $^{87}$Rb decay would make this agreement even closer and the different isotopic systems would be expected to close such that the K-Ar age is slightly younger than the Rb-Sr age.

In north Connemara major faulting about this time let down the Connemara Schists to the north by at least 5 km in Ordovician times alone (Dewey *et al.* 1970) and by this means the thick Ordovician and Silurian sedimentation of the Murrisk trough was possible, the Murrisk trough lying between the Highland Boundary Fault extension along the south side of Clew Bay and the Doon Rock and Derry Faults on the north edge of Connemara. Perhaps contemporaneously the Oughterard Granite was emplaced and may indeed have contributed to the relative uplifting of the Connemara massif. The downwarping and faulting of the Murrisk trough continued into Silurian times and in many ways this trough has similarities in the Midland Valley of Scotland except the latter is dominated at the present surface by Upper Palaeozoic rocks. The stratigraphical continuity of the Dalradian succession from Donegal to Connemara seems to rule out any possibility of oceanic crust intervening in the ground between these two areas which must be underlain by Dalradian rocks.

Accordingly, a pattern emerges of granite emplacement and synchronous uplift of segments of the orogen, uplift certainly in places known to be connected with major faulting and this picture is confirmed by the timing of the latest Caledonian granite phase at 415±15 m.y. which corresponds with major uplift of the Moines in which K-Ar ages peak at 410–430 m.y. (Dewey and Pankhurst 1970) and probably with the age of the Moine Thrust (van Breemen, personal communication). Possibly these movements were the early precursors of the massive transcurrent movements that fractured and sinistrally displaced the Scottish Highlands, and include the Great Glen and Leannan fault systems. Recently Morris (1976a) has suggested from palaeomagnetic measurements that during Devonian times enormous sinistral tear faulting parallel to the Caledonide–north Appalachian orogenic belt moved Britain 1800±800 km and that this was linked with a major 30° clockwise rotation of the whole of the Connemara–Murrisk region (Morris 1976b) thus explaining the marked swing in Caledonian strike from its normal NE trend to E–W in the west of Ireland. These movements probably spread over a considerable period (>25 m.y.) and there must have been some vertical movement though it should be noted that if the British Isles has been moved substantially to the northeast then the Grenville belt in North America would not then project across to Britain.

Associated in time with the late granite emplacement was the outpouring in Scotland and northern England of lower Devonian volcanic material which includes olivine basalts, andesites, dacites, and rhyolites (e.g. Gandy 1975; Groome and Hall 1974) though it is clear that many, if not all, of the volcanic rocks were derived from quite separate magmas from those giving the nearby plutonic intrusions of approximately the same age (Groome and Hall 1974). The location of lower Devonian volcanics in the Midland Valley of Scotland seems to show a spatial association with the Highland Boundary and Southern Uplands Faults but it is not clear to what extent this is an accidental rather than a genetic connection.

Accordingly, the siting and timing of granite emplacement in Ireland is postulated to be largely controlled by major deep faulting which is responsible for triggering the granite episodes so that although the lower crust would reach partial melting temperatures during the metamorphic climax, granite emplacement is not restricted to the metamorphic climax nor is emplacement a continuous process from 500 to 400 m.y. ago. Major deep faulting would serve to promote partial melting of the lower crust by (i) raising segments which might begin to melt isothermally due to pressure reduction; (ii) initiating and funnelling basic magma intrusion into the crust; (iii) lifting hotter upper mantle or bottom crustal material against slightly higher and colder lower crust and thus continuously heating and then melting the latter; (iv) providing zones of active movement that would facilitate aggregation of the partial melt into distinct gobules of magma that (v) would move upwards as plutons in the weakened zones of the crust (Fig. 4). The crustal segments that the British Isles is split into (see map, Dunning and Max in Harris *et al.* 1975), mainly by major lineaments, contain many sedimentary basins between the granite segments presumably due to relative sinking (lateral migration of magma to zones of intrusion and extrusion?) or less active rising than the lower density granite-containing segments

The vertical displacement of the Moho and the movement of mantle material against crustal matter has been described by Krestnikov and Nersesov (1964) and

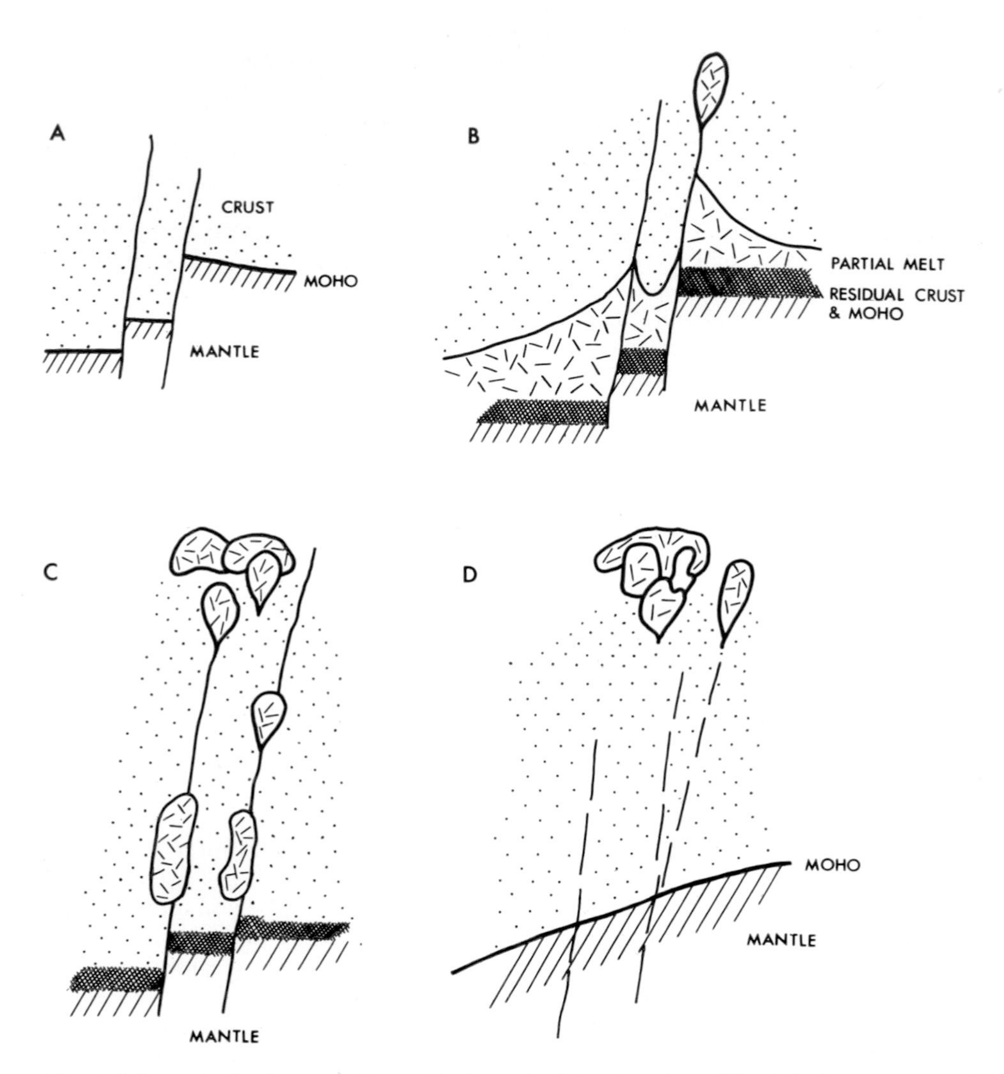


Fig. 4. Schematic development (A to D) of granite intrusions by uplift of the lower crust and mantle causing partial melting of the rather dry lower crust and upward concentration of magma along or near to major lineaments.

the massive transcurrent movements described above and the vertical movements in north Connemara must have involved mantle movement. Deep lineaments provide an explanation for the otherwise puzzling occurrence of major granite emplacement up to 100 m.y. after the metamorphic climax, a circumstance found in a great many orogenic belts. They also explain why some granites have K-Ar ages up to and over 100 m.y. younger than the Rb-Sr or U-Pb age and they provide a link between the magmatic episode and the orogenic and post-orogenic tectonic adjustments of the country rock.

## 8. Form and thickness of the Irish granites

The general form of the Irish granites is ellipsoidal with variation from a sub-spherical shape such as the Ardara pluton through variously elongated ellipsoids such as the Leinster, Galway and Main Donegal granites possess to sheet-like forms such as the Barnesmore and Corvock granites seem to show. The tendency to form dome-like masses in now known to be much more pronounced than previously thought because it is now clear that the Galway batholith has a dual dome structure, the Leinster Granite is a series of separate domes and the seaward limits of the Inish, Fanad, and Toories granites indicate domal structures that probably have several variants through to a plug-like shape which is the most likely form of the Rosses, Omey and Roundstone granites, all of which are probably late in the 415 m.y. granite episode.

The gravity evidence of Bott and Smithson (1967), Bott (1974) and Young (1974) shows that most, perhaps even all, the Irish granites, extend to little more than 10 km in depth and many are much thinner. Thus Young (1974) has shown that the Ardara diapir, which is 8 km in diameter at the surface must narrow downwards and only extends about 5 km in depth having a drop-like shape. The Rosses Granite has a similar thickness while the Main Donegal Granite, which extends for 40 km from NE to SW and is commonly 5 km wide, fits the gravity evidence best with a thickness of only 4 km and is in places even thinner than this (Young 1974). Likewise the Fanad pluton has a maximum thickness of less than 4 km while unpublished gravity measurements on the Corvock Granite (kindly communicated by Dr T. Kelly) suggest that the granite is probably not more than 1 km thick. As yet there are no published thicknesses for the Leinster and Galway batholiths but the gravity anomaly map of Ireland, published by the Dublin Institute for Advanced Studies as Geophysical Bulletin No. 32 (1974), suggests that the Galway batholith is between 4 and 8 km thick but the Leinster mass is probably thicker, perhaps the thickest of all the Irish granites and may reach 10 km.

The tendency of the gravity interpretations to favour rather thin granites compared with their horizontal extents suggests that the shapes preserved are not the forms that would have carried the intrusions upwards through the crust. They have consolidation shapes rather than a transport geometry. Ramberg's (1970) experiments model the situation exactly with blebs and pipes of magma rising vertically to spread out laterally as the upward motion is arrested, presumably by increasing solidification of the magma and increasing viscosity of the country rock so that it is distorted to allow the passage of the magma with increasing difficulty so that stoping becomes the more effective means of inching the granite upwards, but even this removes heat from the magma as the cold blocks slide through to the floor of the chamber. It is important to appreciate the difference between a transportation geometry and a consolidation geometry because otherwise the Grout (1945)–Fyfe (1970) model of bubble movement of granite magma through the crust must be rejected because nearly all granites have steep outward sloping contacts whereas the bubble model requires that inward slopes should be much more commoner than they appear to be.

## 9. Petrogenesis

It is generally now appreciated that in order to preserve a granite magma from crystallising and to enable it to be emplaced in the upper crust, the conditions of melting must be almost anhydrous with water being obtained almost entirely by the melting of mica or perhaps hornblende. Any water-rich melting merely results in a magma that rapidly crystallises on upward movement, whereas a dry magma can, as it rises, continue to melt partially dissolved material.

The single fact that basalts are practically anhydrous although water should be enriched in any liquid phase obtained by partially melting the mantle, demonstrates, according to Bottinga and Allègre (1976), that the Upper Mantle has a maximum of 0·1% water. This rules out deviation of magnesian andesite (tonalite) magma by partial melting of the mantle because such tonalite magmas can only be obtained under nearly water saturated conditions (Nicholls 1974). In view of the anhydrous nature of all probable lower crust or mantle parents, water saturated magma production is unrealistic anyway.

In agreement with this, Maaløe and Wyllie (1975) have shown that the sequence of crystallisation observed in a typical biotite granite can only be obtained with less than 1·2% $H_2O$ in the magma and that crystallisation is probably between 1100°C and 700°C. Richardson and Powell (1976) have shown that temperatures of 1100°C were probably reached at the base of a 50 km thick crust during the time of Dalradian metamorphism in Scotland even if heat flow from the mantle was normal, i.e. about 0·8 HFU. The production of granite magmas by partially melting the lower crust about 500 m.y. is therefore feasible and if higher heat flows from the mantle are assumed then heating would be more rapid and temperatures higher for a thinner crust.

Recent progress in experimental petrology has tended to confirm that granite magmas can only be obtained in quantity from the crust because partial fusion of a rock must give a magma that is in equilibrium with the residual minerals left in the parent source rock at the time of melting and for a wide range of granitic to granodioritic compositions, over a wide range of temperature, pressure and $pH_2O$, quartz and plagioclase are on the liquidus—minerals not found as residual minerals in partially melted peridotite. Thus Wyllie *et al.* (1976) have argued that the residual minerals of primary magmas derived from a peridotitic mantle must include some of the following minerals, olivine, orthopyroxene, clinopyroxene, garnet, spinel or calcic plagioclase. Olivine is totally absent in magmas containing up to 5% $H_2O$ and derived from a wide range of granitic to granodioritic composition. Thus on this evidence a mantle origin is unlikely.

Brown (1977) has however recently argued against a crustal origin for the Cordillerian granites of the Andes, partly on the grounds that the thickness of the crust in the Central Andes, which reaches a maximum of 70 km (James 1971) could only be obtained by underplating from mantle-derived material as "evidence for crustal shortening in this region is slight". However, the nature and age of much of this thick volume of crustal rock is quite uncertain, except, as James (1971) argues, it is unlikely to be basic material. Whether it is young or old in age is quite uncertain and it has not yet been shown that the lower crust in this region differs significantly from the usual lower crust which shows, based on seismic reflection results, a layered structure indicative of metamorphic rocks rather than plutonic rocks (Smith and Brown 1977). Crustal thickening even if due to granite magma, might have been caused by subducted continental matter belonging to a now-destroyed continent (Pitcher personal communication).

The restrictions of granites to areas of continental crust and their total absence from the oceanic floor strongly suggests that the origin of granitic magma lies in the crust even recognising that it is now well established (e.g. Anderson 1967; Sipkin and Jordan 1976) that the mantle under the continents has seismic features that distinguish it from the mantle under the oceans. These seismic differences are not such as to indicate extremely high water or salic contents in the sub-continental mantle. According to Taylor (1977), the fact that granitic plutons possess $\delta^{18}O$ about $+10$ to 13, with D/H values of the biotite and hornblende between, indicates that the magma has been derived by melting of high $\delta^{18}O$ metasediments or altered volcanic rocks and that the primary ("magmatic") water in granite batholiths, as distinct from the secondary water involved in meteoric-hydrothermal convection systems, is ultimately derived by dehydration, and, or, partial melting of the lower crust or subducted lithosphere. Such conclusions are in agreement with the discovery of Precambrian zircons in some Caledonian granites as reported in this volume by Pidgeon and Aftalion (1978), though in fairness it must be added that we do not really know whether there are zircons in the mantle and if there are what age they would record. It is quite feasible however, that if zircons do exist in that mantle they would have a similar age (1600 m.y.) to that obtained by Pidgeon and Aftalion because a good deal of evidence, based on Rb-Sr "error-chrons" (Brooks *et al.* 1976a, 1976b) and Pb-Pb work (Sun and Hanson 1975) suggests that the mantle has been closed to major chemical redistribution for the last 1500 m.y. Nevertheless the most likely source of the zircons would appear to be crustal material as ultrabasic rocks are normally very poor in zircons and zirconium.

It has been common to cite the rather low $^{87}Sr/^{86}Sr$ ratios of many granites (e.g. Galway Granite, 0·7044, Leggo *et al.* 1966; Leinster Granite, 0·704–0·706; O'Connor and Brück 1976; Newry Granodiorite, 0·706, O'Connor 1976) as evidence of a mantle derivation on the basis that higher values would be expected for crustally derived material. The opinion that crustally derived magma should have high initial $^{87}Sr/^{86}Sr$ is largely based on the assumption that the rock being melted is average crust with Rb/Sr of 0·10–0·20 whereas it seems much more likely that most granites are derived from the lower continental crust which numerous studies (e.g. Lambert and Heier 1968; Sheraton *et al.* 1973; Field and Clough 1976) indicate must be severely depleted in Rb compared with the upper crust. Thus Hurley (1968) considered the lower continental crust to have an *average* $^{87}Sr/^{86}Sr$ of 0·704 with 15 ppm Rb and 500 ppm Sr, giving a Rb/Sr ratio of only 0·03, closely comparable with typical supposed mantle values. Quite clearly such a composition, 400–500 m.y. ago, could have yielded a $^{87}Sr/^{86}Sr$ at least as low as that observed for the Galway Granite remembering that an *average* of 0·704 implies some values below this. Agreed that our knowledge of this lower crust is scanty because it is doubtful whether such material that immediately overlies the Moho is anywhere exposed, but nevertheless pyroxene granulites are the nearest known approach to such bottom crustal rocks and they are severely depleted in Rb and are similar to the lower crustal composition deduced by Hurley (1968).

If a pyroxene granulite type basement is considered the most probable parental material for the Irish granites, then clearly the most likely *known* source would be Lewisian of Scourian (2800 m.y.) type but we do not know what was the nature of the deep-seated metamorphosed Grenville basement (van Breeman *et al.* 1976) or indeed the *lower* Laxfordian basement (granulites beneath amphibolites?) and these could also be parental sources. It is relevant that the Grenville metamorphism in North America is characteristically granulite facies. While it is not certain that many pyroxene granulites have obtained their Rb, K, U, Th depleted compositions as result of metamorphism rather than being an original feature, nevertheless the favoured view is that metasomatic loss by incipient partial melting or movement of material at high temperature has taken place. This implies that in order to obtain low $^{87}Sr/^{86}Sr$ in Caledonian granites a *previous* episode of partial melting or metasomatism was necessary in order to deplete the lower crust of Rb and this would need to have taken place sometime well back into the Precambrian depending upon the model adopted for the partial melting e.g. whether isotopic homogenisation between the different minerals in the parent rock took place upon partial melting or whether biotite was preferentially retained or melted without complete isotopic homogenisation. Biotite is particularly critical because Rb preferentially enters biotite and thus the $^{87}Sr/^{86}Sr$ of biotite increases rapidly compared with that of other minerals but since Sr is reluctant to enter mica lattices, high temperature metamorphism will encourage $^{87}Sr$ to migrate and this might leave the rock preferentially to $^{86}Sr$, which is more safely ensconced in feldspar or hornblende. Thus preferential loss of $^{87}Sr$ with lowering of the $^{87}Sr/^{86}Sr$ ratio is conceivable as a result of very high grade metamorphism. Metasomatic enrichment of $^{87}Sr/^{86}Sr$ by solutions leaching $^{87}Sr$ out of positions in crystal structures where $^{87}Sr$ is unstable without melting temperatures being reached have been described by Blaxland *et al.* (1976) and van Breemen *et al.* (1975) so that $^{87}Sr$ depletion is not a novel concept. A previous episode of partial melting in pre-Caledonian times is necessary otherwise a Rb-rich and high $^{87}Sr/^{86}Sr$ granite would be produced by a single episode of partial fusion.

Although exposed pyroxene granulites are similar to the predicted lowermost continental crust it seems likely that many typical granulites contain more quartzo-feldspathic material than is present in material just above the Moho, a view certainly in accordance with the seismic properties of the deepest crust. Lambert and Heier's (1968) work strongly suggests that depletion in Rb in granulite facies rocks is greatest in high pressure granulites. Such rocks are poor in biotite while any K feldspar is impoverished in Rb but may contain substantial amounts of Sr. The depletion of Rb in K feldspar is presumably linked to the large ionic if Rb (1·47Å) compared with K (1·33Å) from which it follows that under high pressure Rb is discriminated against relative to K even more than at lower pressure. This is part of the process of adjusting element proportions to those most stable in the mantle, where K/Rb is very high. In addition, the principal Sr-bearing phases, plagioclase, which will only accept minor Rb (commonly 5–20 ppm Rb compared with several hundred ppm Sr), and hornblende which will accept rather more (*c.* 50 ppm) are commonly accompanied by pyroxene and garnet which are almost Rb-free. More over, experimental evidence (e.g. Drake and Weill 1975) shows that Sr favours plagioclase over the liquid phase at all geologically relevant temperatures whereas Rb favours the liquid phase so that partial melting would indeed accentuate any metasomatic Rb loss. Consequently the formation of pyroxene granulite 1000–3000 m.y. ago would involve major Rb loss but Sr retention, as documented by Sheraton *et al.* (1973) with resulting low Rb/Sr and slow increase of $^{87}Sr/^{86}Sr$ and would provide a suitable parental material to provide subsequent granite magma with low $^{87}Sr/^{86}Sr$ provided further substantial melting was provoked.

Because the formation of such a pyroxene granulite will normally involve partial or even nearly complete dehydration which will at least partly be a further means of transporting the removed elements, the rock will not subsequently melt easily to yield a second magma and a substantial increase in temperature will be required. If Oxburgh's (1969) postulate that there is a continuous slow outward diffusion of water from the mantle independent of magmatic activity is accepted, then partial re-hydration of granulite might occur and such an effusion could also carry mantle-derived K, Rb and Sr into the crust metasomatically (e.g. pyroxene → hornblende → biotite as seen in many granulites). This would be part of the general process of progressive loss of K and Rb from the mantle to the crust, as postulated by many workers. Although the D/H evidence puts restrictions on the likely proportions of primary juvenile waters derived from the mantle that occurs in granites when they crystallise in the upper crust, it would be unlikely that at the temperature and pressure found at the base of the crust, the Moho could be a barrier preventing metasomatic transfer of material and this is a further factor that could cause modification of the $^{87}Sr/^{86}Sr$ ratio. Such partial re-hydration, though only trivial, would assist by markedly lowering partial melt temperatures.

Some means of episodically melting granite out of the granulite basement is needed and this mechanism must allow for a concentration of granite magma by lateral migration because the quantity of granite that can be derived from a granulite basement in any one place is severely limited and would hardly provide plutons 2–10 km thick. Ramberg (1970) has shown experimentally that lateral migration of quite thin layers of low density material into even *healed* fracture zones is adequate to concentrate low density material and form a rising pluton and thus the petrogenesis is linked to major deep fractures as already outlined. Perhaps the association of sinking sedimentary basins peripheral to many major granitic intrusions is a reflection of crustal adjustment following lateral movement of granite magma out of the deep basement.

Finally the question might be asked as to whether average basic and acid pyroxene gneisses contain enough K etc. to provide granite magma by partial melting. If a spherical pluton 5 km in diameter of average granitic composition is obtained by partial melting of a cylinder of lower crust parallel to the lineament and with radius $2\frac{1}{2}$ times 5 (i.e. 12·5 km) and half-length $3\frac{1}{2}$ times 5 (i.e. 17·5 km or 35 km in full-length) then only 0·35% of the volume needs to be abstracted to give the granite bubble providing lateral migration of the partial melt is possible up to distances along the length of $3\frac{1}{2}$ times the 5 km bubble diameter—values that seem compatible with Ramberg's (1970) experiments. Such a trivial abstraction is easily made even from basic gneisses of the type described by Sheraton *et al.* (1973) with under 1% $K_2O$ etc. with only quite minor changes in composition so that the remnant material has no unusual composition. Quantitatively, there is no problem therefore in obtaining the needed granitic constituents out of the lower crust even after it has suffered previous granulite facies metasomatism, though of course such a small volume loss is inadequate to explain basinal subsidence alone.

## 10. Granites and crustal evolution

Although it is now well recognised that many of the major batholithic complexes of tonalite to granite are connected with destructive continental margins such as the circum-Pacific belt it is not yet clear exactly how the Caledonian granites of Britain fit the now numerous subduction models proposed for the British Isles e.g. Dewey and Pankhurst (1970); Lambert and McKerrow (1976); Phillips *et al.* (1976). Hall (1971, 1972) has attempted to find systematic regional variation in the geochemistry of the Caledonian granites but unlike the western U.S.A., no clear pattern of progressive change of K across the orogenic belt has yet emerged. Preliminary work, because much more data is needed (Hall 1971) does suggest small overall differences between Caledonian granites and those of the Variscan and Alpine orogenic belts that are ascribed by Hall (1971) to different geothermal gradients in each belt and perhaps levels of uplift and erosion of the crust. As yet, the granite picture does not enable discrimination between the different proposed subduction models but the implications of Hall's deductions are that the origin of the granites concerned lies in the crust and depends on its metamorphic style.

Granite episodes are a means by which the crust is regenerated and stabilised. Material is melted or purged out of the lower crust and rises into the upper crust under the influence of its low density and mobility and the upper crust is uplifted and partly eroded. The depleted lower crust, or restite, increased in density and refractory elements, sinks down into a dense residue tending to chemical and physical equilibrium with the upper mantle. Thus density stratification and consequent stabilisation and eventually cratonisation are promoted; a view confirmed by the seismic properties of the crust. After one, two or three periods of partial melting, the lower crust becomes much less able to provide any new granite magma and any further granites will only arise from emplacement of hot basic magma into the middle or upper crust or differentiation from the basic magma and such is the picture of the Tertiary granites, represented in Ireland by the Mourne Granite with its high initial $^{87}Sr/^{86}Sr$ ($>0{\cdot}710$) ratio and its temporal association with the basaltic outpourings about 60 m.y. ago. Yet again this episode is related to deep crustal fracturing connected with the pulling apart of northwest Britain and Greenland but probably with a dominance of dilatation not present in the Caledonian espisodes.

It is remarkable that except for a possible buried extension of the Leinster Granite to the southwest and a possible shallow buried granite south of Castleisland in southwest Ireland, the gravity anomaly map of Ireland (Dublin Inst. Advanced Studies, 1974) indicates that there are no substantial masses of buried granite in Ireland. This means that all the large Irish granites have consolidated close to the present exposed land surface. Even more remarkable is the fact that several of the granites have their roofs within 2 km of the present land surface. Thus the Leinster Granite has a roof at about 500 m sloping downwards to the southwest while the roof of the Galway granites, as deduced from the moderate outward dips of the margins and the existence of the Kilkieran septum, is within 2 km of the present surface. According to Pitcher and Berger (1972) the Thorr Granite and the Main Donegal Granite have their roofs within 2 or at the most 3 km of the present surface. Similar relationships are observed elsewhere e.g. the Lake District batholith (Bott 1974) and the Cornish batholith. Also relevant is the apparent scarcity of substantial bodies of granite offshore on the continental shelves and in the North Sea basin and several Irish granites that occur along the coast e.g. Fanad, Toories, Inish and Galway granites (Evans and Whittington 1976; Bremner and Leake 1977), all extend only a short distance away from the shore.

As the Irish granites consolidated at depths of between 3 and 10 km in the crust, as deduced from their metamorphic aureoles, while the crust is at least 30 km thick and may have been as much as 50 km thick in Caledonian times, it follows that granite magma rises through most of the crust and consolidates in quite a narrow band near the surface, perhaps ultimately being arrested by cooling outwardly and inwardly (by the rain of cold stoped xenoliths), aided finally by the influence of groundwater. But if this narrow zone in the crust is subsequently to be more or less coincident with the *future* land surface it suggests that this elevation is partly controlled over major segments of the crust by the granites themselves. Possibly the adjoining sedimentary basins result from less positive uplift giving relative

LEAKE, B. E. and LEGGO, P. J. 1963. On the age relations of the Connemara Migmatites and the Galway Granite, west of Ireland. *Geol. Mag.* **100,** 193–204.
LEGGO, P. J., TANNER, P. W. G. and LEAKE, B. E. 1969. Isochron study of Donegal Granite and certain Dalradian rocks of Britain. *In* Kay, M. (Editor). *North Atlantic—Geology and Continental drift. Am. Assoc. Petrol. Geol. Memoir* **12,** 354–362.
LEUTWEIN, F., SONET, J. and MAX, M. D. 1972. The age of the Carnsore Granodiorite. *Geol. Surv. Ireland Bull.* **1,** 303–309.
MAALØE, S. and WYLLIE, P. J. 1975. Water content of a granite magma deduced from the sequence of crystallisation determined experimentally with water saturated conditions. *Contrib. Mineral. Petrol.* **52,** 175–191.
MAX, M. D. and RYAN, P. 1975. The Southern Uplands Fault and its relation to the metamorphic rocks of Connemara. *Geol. Mag.* **112,** 610–612.
——. LONG, C. B. and SONET, J. 1976. The geological setting and age of the Ox Mountains Granodiorite. *Geol. Surv. Ireland Bull.* **2,** 27–35.
——, LONG, C. B., KEARY, R., RYAN, P. D., GEOGHEGAN, M., O'GRADY, M., INAMDAR, D. D. and MCINTYRE, T. 1977. Preliminary report on the geology of the NW approaches to Galway Bay and part of its landward area. *Geol. Surv. Ireland. Rep. Series* RS 75/3 44 pp.
—— LONG, C. B. and GEOGHEGAN, M. 1978. The Galway Granite and its setting. *Geol. Surv. Ireland Bull.* (in press).
MOORBATH, S., BELL, K., LEAKE, B. E. and MCKERROW, W. S. 1968. Geochronological studies in Connemara and Murrisk, Western Ireland. *In* Hamilton, E. I. and Farquhar, R. M. (Editors). Radiometric dating for geologists. Interscience, London, 259–298.
MORRIS, W. A. 1976a. Transcurrent motion determined palaeomegnetically in the northern Appalachians and Caledonides and the Acadian Orogeny. *Canad. J. Earth Sci.* **13,** 1236–1243.
—— 1976b. Paleomagnetic results from the Lower Palaeozoic of Ireland. *Canad. J. Earth Sci.* **13,** 294–304.
MCARDLE, P. 1974. A Caledonian lamprophyre swarm in south-east Ireland. *Sci. Proc. R. Dublin Soc.* **A, 5,** 117–122.
NICHOLLS, I. A. 1974. Liquids in equilibrium with peridotitic mineral assemblages at high water pressures. *Contrib. Mineral. Petrol.* **45,** 289–316.
O'CONNOR, P. J. 1976. Rb-Sr whole-rock isochron from the Newry Granodiorite, NE Ireland. *Sci. Proc. R. Dublin Soc.* **A, 5,** 407–413.
—— and BRÜCK, P. M. 1976. Strontium isotope ratios for some Caledonian Igneous rocks from Central Leinster, Ireland. *Geol. Surv. Ireland, Bull.* **2,** 69–77.
O'NIONS, R. K., OXBURGH, E. R., HAWKESWORTH, C. J. and MACINTYRE, R. M. 1973. New isotopic and stratigraphical evidence on the age of the Ingletonian: probable Cambrian of northern England. *J. geol. Soc. Lond.* **129,** 445–452.
OXBURGH, E. R. 1969. The deep structure of orogenic belts—the root problem. *In* Kent, P. E. *et al.* (Editors). *Time and Place in Orogeny*, 251–273. *Geol. Soc. Lond. Spec. Publ.* No. 3.
PANKHURST, R. J. 1970. The geochemistry of the basic igneous complexes. *Scott. J. Geol.* **6,** 83–107.
—— 1974. Rb-Sr whole-rock chronology of Caledonian events in NE Scotland. *Geol. Soc. Am. Bull.* **85,** 345–350.
PHILLIPS, W. E. A., STILLMAN, C. J. and MURPHY, T. 1976. A Caledonian plate tectonic model. *J. geol. Soc. Lond.* **132,** 579–609.
PIDGEON, R. T. 1969. Zircon U-Pb ages from the Galway Granite and the Dalradian, Connemara, Ireland. *Scott. J. Geol.* **5,** 375–392.
—— and AFTALION, M. 1978. Cogenetic and inherited zircon U-Pb systems in granites—Palaeozoic granites of Scotland and England. *In* Bowes, D. R. and Leake, B. E. (Editors). *Crustal evolution in northwestern Britain and adjacent regions*, 183–220. *Geol. J. Spec. Issue,* No. 10.
PITCHER, W. S. and BERGER, A. R. 1972. *The geology of Donegal.* Wiley, London.
PLANT, A. G. 1968. *The geology of the Leam–Shannawona district, Connemara, Ireland.* Ph.D. thesis, University of Bristol.
RAMBERG, H. 1970. Model studies in relation to intrusion of plutonic bodies. *In* Newall, G. and Rast, N. (Editors). *Mechanism of Igneous Intrusion. Geological J. Special Issue* No. 2, 261–286.
RAST, N. 1969. Orogenic belts and their parts. *In* Kent, P. E. *et al.* (Editors). *Time and Place* in *Orogeny,* 197–213. *Geol. Soc. Lond. Spec. Publ.* No. 3.
RICHARDSON, S. W. and POWELL, R. 1976. Thermal causes of the Dalradian metamorphism in the central Highlands of Scotland. *Scott. J. Geol.* **12,** 237–268.
ROBERTS, J. L. 1974. The structure of the Dalradian rocks in the SW Highlands of Scotland. *J. geol. Soc. Lond.* **130,** 93–125.
SENIOR, A. and LEAKE, B. E. 1978. Regional metasomatism and the geochemistry of the Dalradian metasediments of Connemara, Western Ireland. *J. Petrol.* **19.**
SHERATON, J. W., SKINNER, A. C. and TARNEY, J. 1973. The geochemistry of the Scourian gneisses of the Assynt district. *In* Park, R. G. and Tasney, J. (Editors). *The early Precambrian of Scotland and related rocks of Greenland*, 13–30. Univ. of Keel.

SIPKIN, S. S. and JORDON, T. H. 1976. Lateral heterogeneity of the Upper Mantle determined from the travel times of Multiple ScS. *J. geophysical Res.* **81,** 6307–6320.
SKIBA, W. 1952. The contact phenomena on the NW of the Crosscloney complex, Co. Cavan. *Trans. Edinb. geol. Soc.* **15,** 322–345.
SMITHSON, S. B. and BROWN, S. K. 1977. A model for the lower continental crust. *Earth Planet. Sci. Lett.* **35,** 134–144.
SUN, S. and HANSON, G. N. 1975. Evolution of the mantle: Geochemical evidence from alkali basalt. *Geology* **3,** 297–302.
TAYLOR, H. P. JR. 1977. Water-rock interactions and the origin of water in granitic batholiths. *J. geol. Soc. Lond.* **133,** 509–558.
TOWNEND, R. 1966. The geology of some granite plutons from western Connemara, Co. Galway. *Proc. R. Ir. Acad.* **65B,** 157–202.
VAN BREEMEN, O., HUTCHINSON, J. and BOWDEN, P. 1975. Age and origin of the Nigerian Mesozoic Granites: A Rb-Sr isotopic study. *Contrib. Mineral. Petrol.* **50,** 157–172.
—— and BOWES, D. R. and PHILLIPS, W. E. A. 1976. Evidence for basement of late Precambrian age in the Caledonides of western Ireland. *Geology* **4,** 499–501.
WADGE, A. J., HARDING, R. R. and DERBYSHIRE, D. P. F. 1974. The Rb-Sr age and field relations of the Thelkeld microgranite. *Proc. Yorks. geol. Soc.* **40,** 211–222.
WRIGHT, P. C. 1964. The petrology, chemistry and structure of the Galway Granite of the Carna area, County Galway. *Proc. R. Ir. Acad.* **63B,** 239–264.
YOUNG, D. G. G. 1974. The Donegal Granite—a gravity analysis. *Proc. R. Irish Acad.* **74B,** 63–73.

B. E. Leake,
Department of Geology,
University of Glasgow,
Glasgow G12 8QQ, Scotland.

# Sedimentation in a late orogenic basin: the Old Red Sandstone of the Midland Valley of Scotland

*B. J. Bluck*

The Old Red Sandstone rocks of the Midland Valley of Scotland comprise a thick sequence of 'red beds' deposited in the final stage of the Caledonian orogenic cycle. The Lower Old Red Sandstone is a conglomerate–lithic arenite assemblage, partly interstratified with acid to basic volcanic rocks. It was deposited in at least two basins: the *Strathmore basin* in the northwest, and the *Lanark basin* in the southwest. Both basins have axes parallel with the Highland Boundary and Southern Uplands faults, which respectively bound one of their margins. Both basins are longitudinally filled from the northeast. The Strathmore Basin began in the northeast and extended to the southwest so that it has a fill which overlaps and thins in a southwesterly direction.

The Upper Old Red Sandstone rocks are partly derived from the Lower over which they unconformably lie. The basin, with a roughly E–W axial orientation, is dominated by an easterly palaeocurrent flow. It began in the Clyde region and was initially extended by successive normal faults, which were the response to a sinistral shift on the Highland Boundary fault. The Upper Old Red Sandstone in being thinner, finer grained, devoid of volcanic activity, and in having quartz-arenites and numerous caliche beds, is thought to have been laid down in comparative tectonic stability.

## 1. Introduction

Rocks of Old Red Sandstone age occur widely in Scotland. Although outliers of Old Red Sandstone sedimentary rocks lie scattered over the Highlands, major outcrops bound their eastern margin from Caithness to Aberdeenshire (Fig. 1). These outcrops form the western margin of a major basin, the Orcadian basin, which extends to the east to include Orkney. To the west of the Highlands, small outcrops occur near Oban, but to the south extensive outcrops are found in the Midland Valley, which are regarded, together with those of the Southern Uplands, as part of the Caledonian basin.

The Old Red Sandstone, a continental facies of the Devonian, has three divisions:

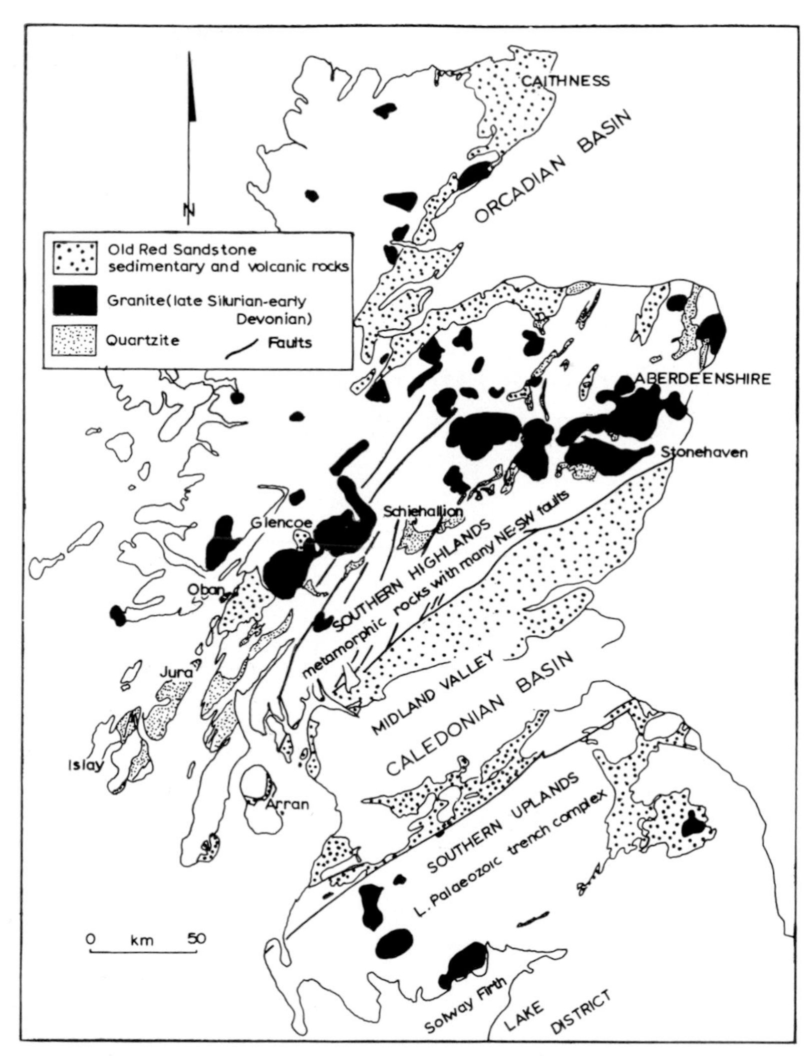


Fig. 1. Location of Scottish Old Red Sandstone basins, Newer Granites and some of the NE–SW 'Tay set' of faults.

Lower, Middle and Upper (Murchison 1859). With the discovery of Lower Old Red Sandstone spores in the Orcadian basin (Collins and Donovan 1977), the suspicions of Westoll (1951) have been confirmed, and all three divisions are to be found there. However, in the Caledonian basin the Lower Old Red Sandstone is unconformably overlain by the Upper. With such sparse organic remains with which to date the beds, it has been thought possible that the Middle Old Red Sandstone could be represented in one or both of these divisions. However, Richardson (1967) considered the spores from the Strathmore Group (at the top of the Lower Old Red Sandstone) to be early Devonian in age. The Upper Old Red Sandstone in the Caledonian basin is relatively thin, and passes upward conformably into the

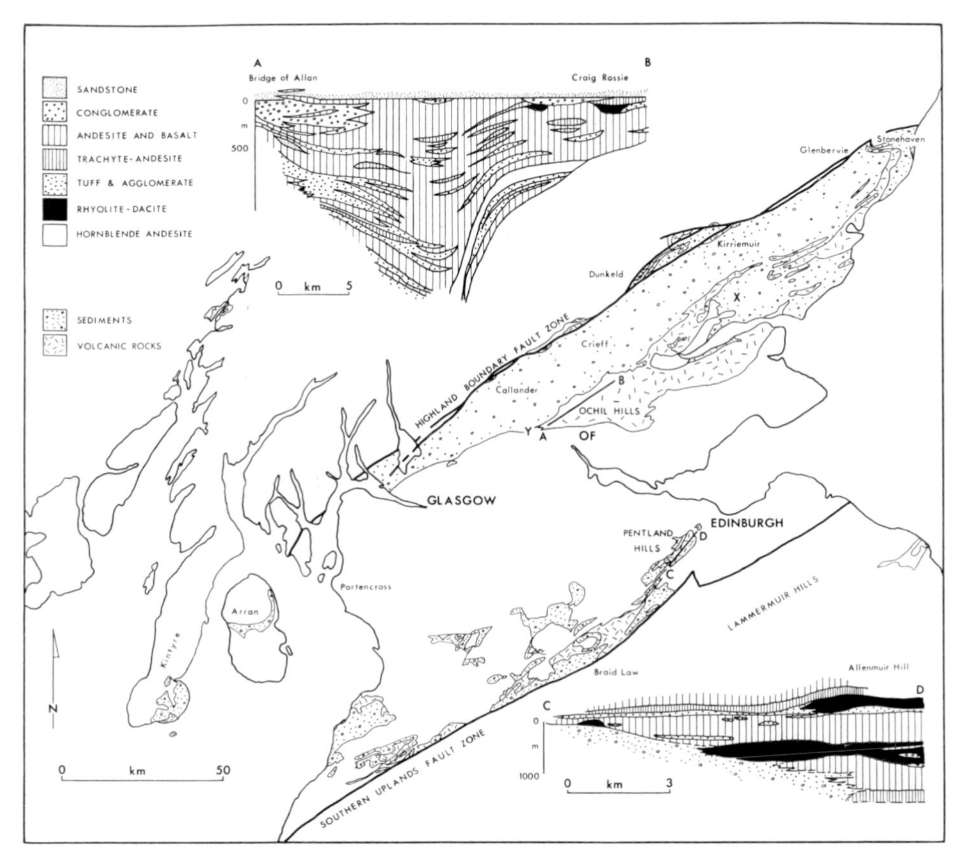


Fig. 2. Distribution of Lower Old Red Sandstone outcrops in the Midland Valley of Scotland, together with a general locality map for places referred to in the relevant parts of the text. Section A–B is of the Ochil lava pile (Francis *in* Francis *et al.* 1970 fig. 4); section C–D for the Pentland Hills (Mykura 1960 fig. 2); X, Y locate areas where the lavas of the Ochil–Sidlaw sequence inter-tongue with sediments; OF=Ochil fault.

Carboniferous: in places the basal part of the sequence could be Middle Old Red Sandstone in age.

There are abundant Late Silurian–Early Devonian plutonic rocks in Scotland (Brown *et al.* 1968). Some are linked with acidic to basic volcanic activity (Bailey 1960; Roberts 1974) and in the region around Oban there was a major development of Lower Old Red Sandstone andesites—the Lorne Plateau Lavas, which rest on rocks of Downtonian age (Waterston 1965 p. 282).

The Old Red Sandstone igneous activity has been related to a northwest subduction in the region south of the Southern Uplands (Dewey 1969, 1971). However, as Dewey (1969 p. 125) points out, there is a substantial gap between the end of subduction and the beginning of Old Red Sandstone extrusive and intrusive igneous activity. Phillips *et al.* (1976) explain this by suggesting that a triple junction moved southwestwards along the subduction zone to the south (part of which is thought to have existed in the Solway region), terminating volcanic activity on the northern border as it migrated. By early Devonian times this migration had travelled 980 km to the southwest. Those rocks to the north of the subduction zone, with Devonian

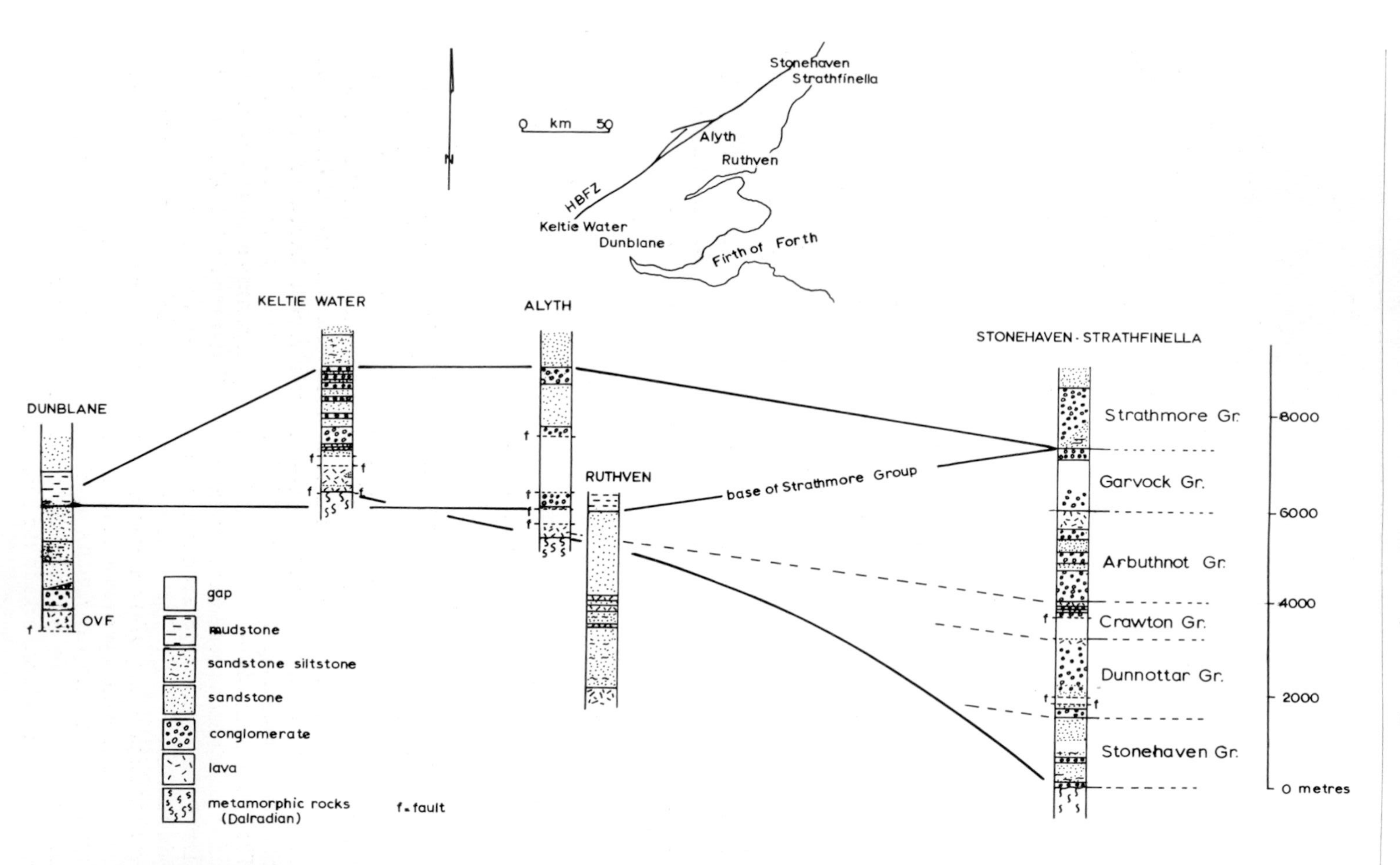


Fig. 3. Comparative sections through the Lower Old Red Sandstone along the northwestern and southeastern limbs of the Strathmore syncline; OVF = Ochil Volcanic Formation (based on Armstrong and Paterson 1970).

igneous activity were then displaced dextrally by some 980 km along the Solway line to their present position opposite the Lake District. This wrench movement is thought to have taken place in Devonian times (Phillips *et al.* 1976 p. 595).

The Scottish Old Red Sandstone was deposited after major orogenic events had taken place in the Southern Uplands, and it overlaps in time with the intrusion of many granite batholiths. The Midland Valley occupied a position between the earlier Grampian orogen to the north (Lambert and McKerrow 1976) and the Caledonian orogen to the south. It is bounded by laterally continuous faults one of which, the Highland Boundary fault, posesses a serpentinite screen. Whatever might have been its early history, the major recognisable event in the Midland Valley was the deposition of the Old Red Sandstone and to a very large extent, the later structural history of the Midland Valley and the surrounding uplifting blocks is written in the Old Red Sandstone rocks. These rocks thus provide a case history of sedimentation in a late orogenic basin.

## 2. The Lower Old Red Sandstone

Rocks belonging to the Lower Old Red Sandstone almost totally occur in two strips of outcrop lying adjacent to the Highland Boundary and Southern Uplands faults (Fig. 2—both faults are really fault zones, particularly the Highland Boundary fault). The northern outcrops are fairly continuous in the main part of the Midland Valley, formed by the elongate Strathmore syncline, and the less continuous Sidlaw anticline (Figs 2, 5), both of which lie parallel with the Highland Boundary fault. The southerly exposures are sporadic, and show evidence of greater tectonic activity being sharply folded and often having fractures running roughly parallel with the Southern Uplands fault.

### 2a. Northern exposures

The sequence along the northern margin of the Midland Valley comprises red and drab coarse conglomerates, lithic arenites and subordinate mudstones and shales together with interstratified and partly intertonguing acidic to basic volcanic rocks.

(i) *Stratigraphy.* The full, known sequence of Lower Old Red Sandstone rocks is best exposed in the northeastern part of the Midland Valley where it is estimated to be 9 km thick (Armstrong and Paterson 1970 p. 20). This sequence of mainly conglomerate, sandstone and interstratified lava (Fig. 3), thins towards the southwest (Fig. 4), although in the southwestern end of the Sidlaw anticline the base of the partly coeval lava pile is not seen (Fig. 2). The thinning of the sequence along the northwestern limb of the Strathmore syncline is thought to have been achieved by overstep in a southwesterly direction. A previously described dacitic lava (Campbell 1913 p. 943; Allan 1928 pp. 66–71), occurring between Dunkeld and Glenbervie (Fig. 2) has been reinterpreted as an ignimbrite (Paterson and Harris 1969). The ignimbrite forms the local base of the Old Red Sandstone at Dunkeld where it rests on Dalradian rocks. Further to the northeast, at Glenbervie, the ignimbrite is thought to lie at the top of the Crawton Group, and is considered to be at roughly the same horizon as the Crawton Lavas on the coast (Fig. 3; Armstrong and Paterson 1970 p. 10). Along this northern edge of the Midland Valley the Stonehaven, Dunnottar and Crawton groups are therefore overstepped in the distance between Stonehaven and Dunkeld (Fig. 3).

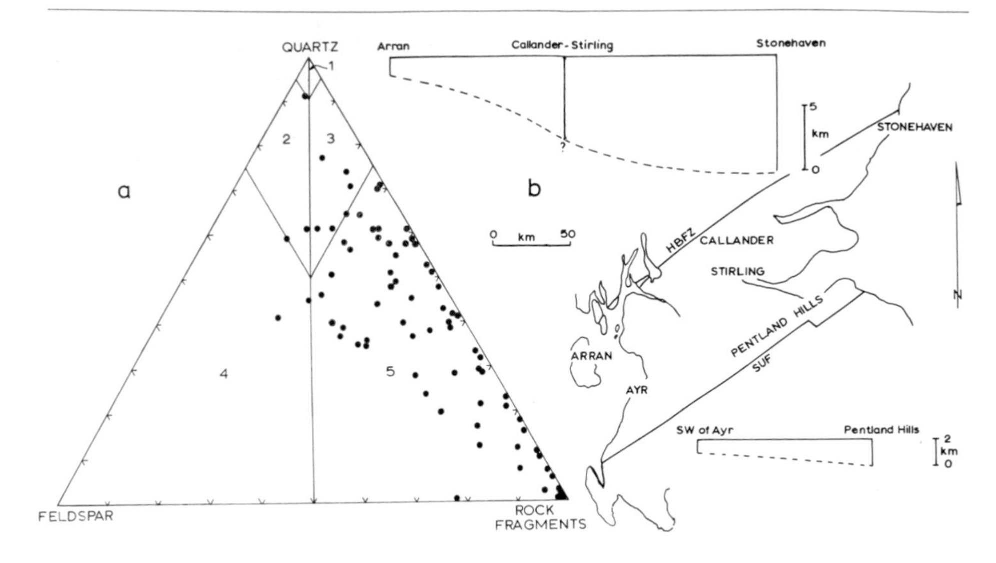


Fig. 4 (a). Modal composition of sandstones from the Midland Valley west of a line from Callander to the Pentland Hills; 1 quartz arenite, 2 subarkose, 3 sublithic arenite, 4 arkosic arenite, 5 lithic arenite; 76 samples, 400 counts per thin section, by point counting.
(b) Areal variations in the thickness of Lower Old Red Sandstone sequences.

Rocks of the Arbuthnot Group rest on Dalradian rocks north of Crieff (Armstrong and Paterson 1970 fig. 3) and rocks assignable to this group are also thought to occur as far southwest as Kintyre (Morton 1976 fig. 2.9). If this latter lithostratigraphic correlation is correct, and given that Strathmore Group rocks occur in this southwestern part (as indicated by Armstrong and Paterson 1970 fig. 1), then the continued thinning of strata to the southwest (Fig. 4b) is not due to overstep.

Overstep of this kind cannot be demonstrated on the southeast limb of the Strathmore syncline as the base of the lavas in the Ochils is not seen. However, the oldest dated beds, at some distance from the present Highlands in the southwest, are the Sandy's Creek Beds, exposed at Portencross. These beds are Breconian or Dittonian in age (Downie and Lister 1969 p. 199). Although the base of the sequence is not exposed here, lithologically similar beds are found low in a sequence which rests on Dalradian rocks in Arran (the second division of the Basal Quartz Sandstones of Friend *et al.* 1963 p. 380). If the Arran beds are to be correlated with the Sandy's Creek Beds as Downie and Lister (1969) suggest, then the lower part of the Stonehaven succession is missing. This is in keeping with concept of a southwesterly overstep. Unfortunately the biostratigraphy is not sufficiently precise to suggest the number of groups in the Lower Old Red Sandstone which might be missing here.

Despite the southwesterly thinning of the sequence the lithology shows little variation. Along the Highland border, thick conglomerates are interstratified with sandstones, with many of the conglomerate units comprising single cycles of coarse clasts at the base grading upwards into fine clasts, and eventually into the overlying sandstones. In the axial region of the Strathmore syncline, siltstones, mudstones and shales are very well developed in the Strathmore Group and in the Garvock Group of the Dunblane region. Some of these fine grained units exceed 1000 m in thickness.

On the southeastern limb of the Strathmore syncline the sediments are generally finer grained than those to the northwest, and are partly interstratified with a major lava pile which forms the Ochils. Although along the Highland border lava flows occur intermittently from Stonehaven (Figs 2, 3) to Arran (Gunn 1903) and tuffs with evidence of vents occur in Kintyre (Friend and MacDonald 1968), there is no exposed lava sequence as thick as that of the Ochils. Here at least 3000 m of variable lavas poured out during Lower Old Red Sandstone times. In residual outcrop the lavas are thickest in the region of Tillicoultry (Francis *in* Francis *et al.* 1970 p. 24) and although they are terminated to the south by the Ochil fault, the andesitic lavas which occur north of the Highland Boundary fault at Crieff and Callander have been interpreted by Allan (1940 p. 183) and George (1974 p. 14) as a northerly extension of the Ochil-Sidlaw lavas. In that event they would probably underlie the Strathmore syncline. Allan (1940 p. 183), in the area north of the Highland Boundary fault at Crieff, found xenoliths of Dalradian-type in the lava which may well have come from local pipes cutting the Dalradian rocks north of the Highland Boundary fault, but beneath the present Devonian outcrop. The alternatives are that either the lavas came from sources in the southern Highlands (Campbell 1913 p. 950) or that the floor of the Midland Valley in the vicinity of Crieff is made of Dalradian-type rocks. There appears to be little evidence for volcanic pipes or fissures of this type and age in the southern Highlands, although there are some granite plutons to the north and northwest which at higher levels may have been accompanied by lavas, as at Glencoe. Northerly derived conglomerates contain considerable amounts of quite variable volcanic detritus (p. 257).

The age of the Ochil–Sidlaw lavas is difficult to precisely determine. To the northeast they intertongue with sandstones and shales of Arbuthnot Group (Fig. 2–X); to the southwest and at the top they intertongue with sandstones belonging to the Garvock Group (Fig. 2–Y).

(ii) *Sediment dispersal.* Most early workers recognised the Highlands as being the source of sediment to the Old Red Sandstone of the adjacent Strathmore syncline. Comparative sections taken from the northwestern and southeastern limbs of this structure clearly demonstrate a fining of sediments towards the south (Fig. 3). Similarly, conglomerates high in the Strathmore Group, northeast of Callander, thin in a southeasterly direction (Fig. 5a—section A–B). Here also a thick sandstone and siltstone succession on the southwestern limb of the syncline is replaced partly by conglomerates to the northwest. Directional structures mapped in these Highland border conglomerates by Wilson (1971) and Morton (1976) demonstrate that, for the region between Crieff and Kintyre, they dispersed towards the southeast, in a direction roughly perpendicular to the trend of the Highland Boundary fault.

The sandstones with which the conglomerates interstratify and intertongue have quite variable cross bed dip directions, but in the region between Crieff and Kintyre there is a general cross stratal dip towards the southwest (Wilson 1971 figs 125, 130; Morton 1976 fig. 7.4; personal observations). Conglomerates associated with the Ochils are dispersed north and northwestwards from them and sometimes appear to occupy valleys cut into the underlying lavas (Forsyth *in* Francis *et al.* 1970 p. 71; Wilson 1971).

In the region southwest of Crieff, there is evidence of a trough which was bounded by the Highlands to the northwest and at least partly by the Ochils to the southeast. The sediment dispersed from these lateral sources as well as along the trough axis. Preliminary calculations suggest that as much as 70% of the sediment was axial.

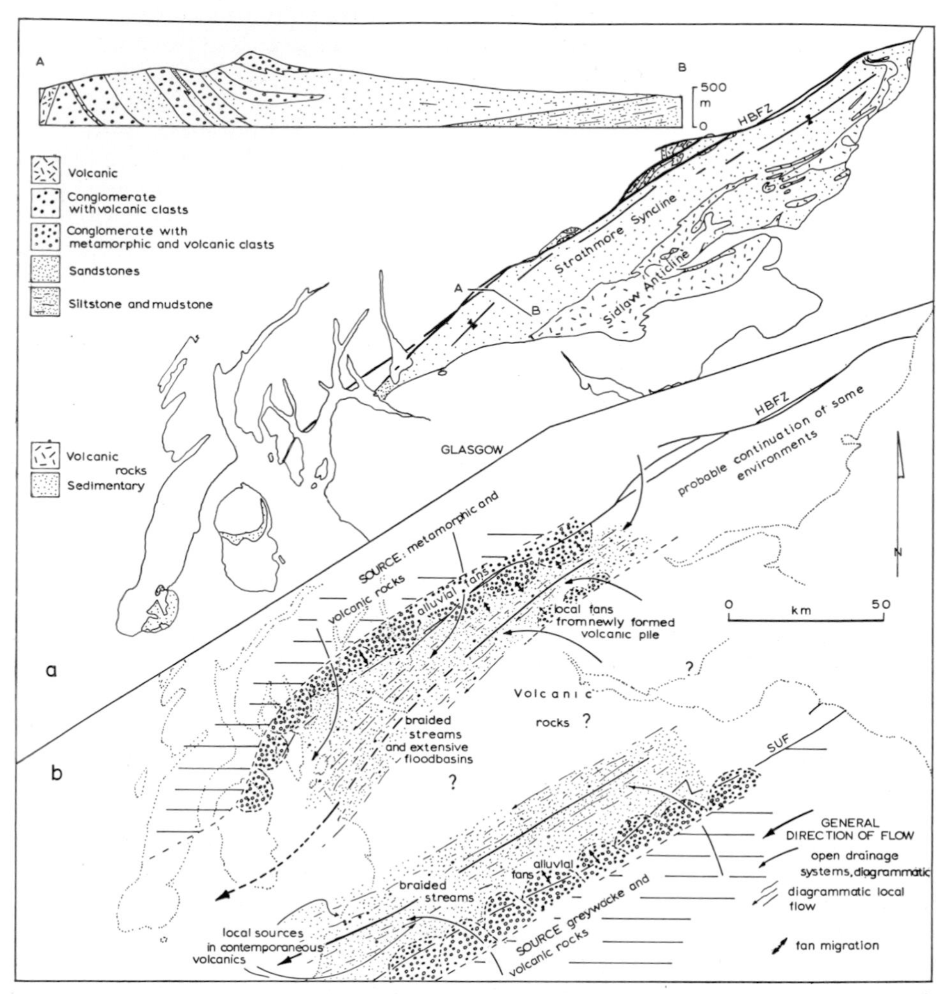


Fig. 5 (a). Distribution of folds in the northern Lower Old Red Sandstone outcrops, and a section A–B taken from the Geol. Surv. 1:50,000 sheet 39 W (Stirling) 1974.
(b) Palaeogeographical map of Lower Old Red Sandstone times, based partly on the work of McGiven (1967); Wilson (1971) and Morton (1976). The distribution of volcanic centres is omitted, as they possibly continue under the Strathmore syncline and in areas to the south. However, with the exception of the area from Loch Lomond to the Clyde, volcanic eruptions occurred at sometime almost everywhere in the Lower Old Red Sandstone; the two main known centres are the Ochil and Pentland hills. HBFZ = Highland Boundary fault zone; SUF = Southern Uplands fault zone.

However, experience in Recent fault bounded basins, analogous to these Old Red Sandstone basins, suggest the possibility that the ultimate sediment sources of this southwesterly-dispersed, axial sediment are the trough margins (see Figs 16, 17). There may have been times when the comparatively restricted fan drainage basins of the southern Highlands captured parts of the large drainage basins which would have existed in the Highlands to the north and so the lateral fans might have become major "open" fluvial systems draining into and through the axial trough

(Fig. 5b). The alternative is to derive the axial sediment from a region of uplift along strike to the northeast in the North Sea, but within the area of the projected Midland Valley. As illustrated in Figure 18, an uplift in this general area is probable.

(iii) *Composition*

CONGLOMERATES. The conglomerates contain clasts mainly of volcanic, hypabyssal, and metamorphic provenance. In general there is a sequence of clast types within the conglomerates: the lower beds, where they rest on Dalradian rocks, are breccias with Dalradian clasts (Allan 1940 p. 175; Friend *et al.* 1963 p. 393; Friend and MacDonald 1968 p. 268). At Stonehaven, the lowest beds (Stonehaven Group) contain clasts dominated by Highland Border Series rocks at the base, and acid and basic igneous rocks further up the sequence (Campbell 1913 p. 931). Above this, conglomerates with abundant quartzite clasts are common in the Stonehaven area (Dunnottar Group—Campbell 1913 p. 937), Arran (Friend *et al.* 1963 fig. 7) and Kintyre (Friend and MacDonald 1968 p. 268), but volcanic clasts occur in this position in the central areas, between Crieff and Kirriemuir (Allan 1928, 1940). The conglomerates above this are rich in volcanic clasts, but towards the top of the sequence rocks of local Dalradian type appear. In the northeast, Campbell (1929 p. 555) records sillimanite and staurolite "gneiss" whilst further to the southwest vein quartz, schist and schistose grit are found (Allan 1929 p. 555; Friend *et al.* 1963 fig. 7.).

This sequence of clast types in the conglomerates has led Campbell (1913 p. 953) and Allan (1929 p. 555) to invoke the following sequence of events: (1) erosion of Dalradian rocks, particularly in the west, and erosion of an extensive cover of Highland Border rocks in the northeast, (2) volcanic activity covering the source area with lavas, which were then eroded to supply debris to the Midland Valley, and (3) cessation of volcanic activity and re-exposure of the Dalradian rocks beneath.

This general sequence, useful though it is, ignores some problems in provenance. The quartzite clasts, some of boulder size, are ubiquitous in the Lower Old Red Sandstone and older rocks exposed elsewhere in the Midland Valley. They are very well rounded, often better rounded than the associated clasts of schistose grits and volcanic rocks. They petrographically resemble the middle Dalradian quartzites of Islay, Jura (according to Friend *et al.* 1963) and Perthshire, although where rocks become very quartz-rich it is difficult to find distinctive characteristics in them. However, it is improbable that boulders of quartzite travelled the 50 km from Jura to Arran, or even the 30 km from Schiehallion to Crieff, and as they are middle Dalradian rocks they would not normally be expected to have originally overlain the upper Dalradian sequence of the southern Highlands.

From the extreme rounding of these quartzite clasts it is clear that they have to be at least second cycle, as their coarseness precludes the possibility of them having acquired high rounding by travelling a great distance from their source in a single fluvial cycle. This must therefore mean that the southern Highlands were partly covered with conglomerates containing quartzite clasts during this time (Bluck 1969 p. 714). These clasts may ultimately have come from further north (Jura–Schiehallion), having accumulated in conglomerates which formed either slightly earlier than, or during Lower Old Red Sandstone times, i.e. there may have been basins of Lower Old Red Sandstone deposition in the southern Highlands which were eventually reworked into the Midland Valley sequence.

SANDSTONES. Most sandstones are lithic arenites (Fig. 4a). The lithic clasts, not

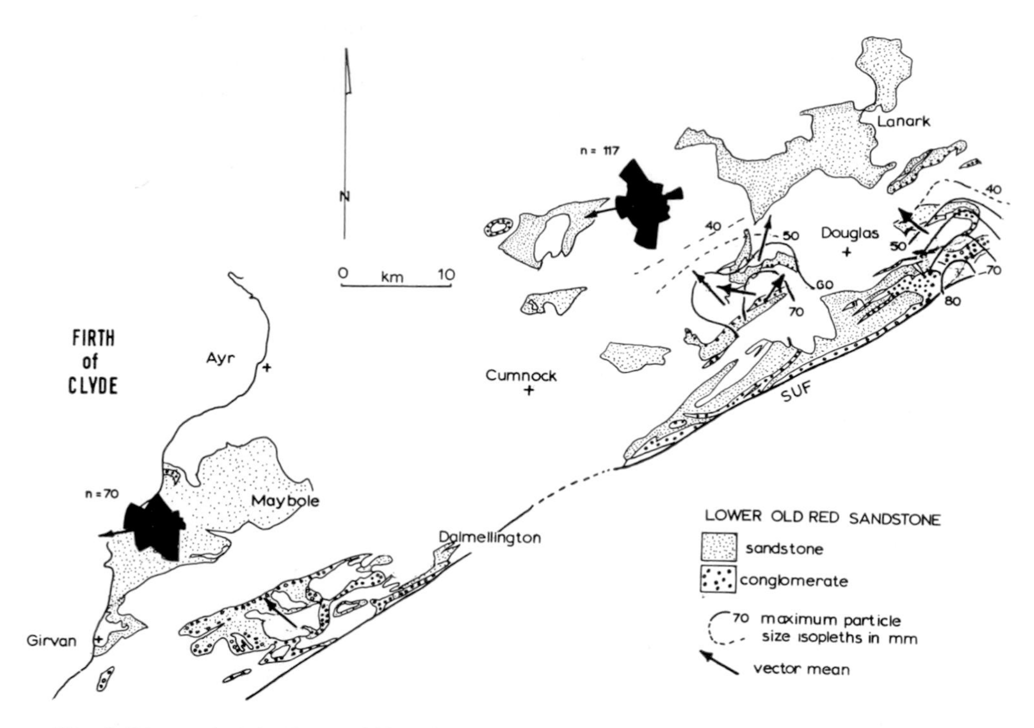


Fig. 6. Dispersal of the Lower Old Red Sandstone sediments along part of the southern margin of the Midland Valley (the Lanark basin). Isopleths and palaeocurrent data in the northeast after McGiven (1967 fig. 41); palaeocurrents elsewhere along the Southern Uplands fault (SUF) are from the lava conglomerate; data in rose diagrams from the sandstone in all the outliers.

always easily identified because of post-depositional decay, are dominantly volcanic, even in the sandstone matrix to conglomerates with metamorphic clasts. Metamorphic rock fragments do occur in some sandstones.

## 2b. Southern exposures

The Lower Old Red Sandstone rocks of the southern margin of the Midland Valley partly crop out between the Southern Uplands fault and a number of parallel faults to the north of it, and partly as inliers, the two largest of which are centred on Lanark and Maybole (Figs 2, 6). Once again the sequence comprises red and drab conglomerates and lithic arenites which partly interfinger and interstratify with lavas of basalt, andesite, dacite and rhyolite.

(i) *Stratigraphy*. The sequence is best known and probably thickest in the Pentland Hills, where Mykura (1960 fig. 2) has demonstrated that some 2000 m of volcanic rocks pass southwestwards into a thinner sequence of conglomerates and sandstones (Fig. 1, C–D). Further to the southwest in the area just southwest of Cumnock, a basal "Greywacke Conglomerate" and associated sandstones are overlain by lavas, sandstones and then by conglomerates with lava clasts in a succession which may be as much as 1700 m thick. Further to the southwest, in the area around Dalmellington, sedimentary rocks of conglomerate and sandstone are overlain by lavas to reach a total thickness of 1200 m (Fig. 3), while in the Maybole inlier, the sequence is about 1500 m thick.

The lithostratigraphy of the kind established for the northern margin of the Midland Valley is not found here, discontinuity of outcrop being a hindering factor.

(ii) *Dispersal.* Units of conglomerate thin towards the northwest, and in the same direction there is a decrease in particle size (McGiven 1967 fig. 41 demonstrated this for the "Greywacke Conglomerate"; see also Fig. 6). The "Lava Conglomerate" also shows a northwesterly thinning, and directional structures within the conglomerates, or thin sandstone lenses interstratified with them indicate a dispersal in the same northwesterly direction (Fig. 6).

The sandstones which interstratify with the distal ends of the conglomerate, and which occur in the Lanark and Maybole inliers, have cross strata which dip to the southwest, once again showing the strong axial flow parallel to the Southern Uplands fault.

(iii) *Composition*

CONGLOMERATES. The conglomerates contain clasts of greywacke, lava, chert, quartzite, and many hypabyssal rock types. Although there is considerable variability, the sequence of clast types run generally from greywacke at the base to lava at the top of the succession. The greywacke clasts resemble more the Silurian than the Ordovician rocks of the Southern Uplands (MacGregor and MacGregor 1948 p. 20; McGiven 1967 p. 155), but chert and pillow lava may well be from the Ordovician assemblage.

SANDSTONES. Most sandstones are lithic-arenites with an abundance of volcanic clasts. Even the sandstones of the Lanark and Maybole inliers have an abundance of these grains, which must either come via open drainage systems from the volcanic areas to the south, or from the volcanic fields within the basin area and possibly to the northeast.

## 2c. Palaeogeography and basin evolution

The essential palaeogeography (Fig. 5) is one where the Highlands and Southern Uplands shed debris which laterally fills two elongate basins both of which have a longitudinal flow to the southwest. Whilst most workers agree that the Devonian Midland Valley was flanked by older rocks which provided a source for the sediments, there are conflicting views about the nature and position of the basin source boundaries. Kennedy (1958 p. 112) considered that the Highland Boundary and Southern Uplands faults substantially defined the basin margin, whilst George (1960 pp. 48, 65–66) gave many examples where the Old Red Sandstone oversteps both faults and suggested a basin of Old Red Sandstone accumulation extending into the Highlands to the north and the Southern Uplands in the south.

Accepting that the Old Red Sandstone basin of sedimentation was not limited by these two major faults, the question left to resolve is whether it was a major downwarp centred somewhere in the Midland Valley (George 1960 fig. 12), or parts of it (e.g. Allan 1940 p. 186 considered the Highland Boundary fault to be the centre of downwarp for the area around Crieff), or whether there was a fault control on sedimentation by faults other than the Highland Boundary and Southern Uplands faults.

(i) *Nature and age of Highland Boundary and Southern Uplands fault movement.* No substantial evidence has been offered which would demonstrate lateral movement on these faults (cf. Anderson 1947 p. 511), but there is good evidence for vertical displacements of thousands of metres, as Old Red Sandstone rocks lie faulted against Dalradian rocks. On Lomondside, Upper Old Red Sandstone is downfaulted to the north by the Highland Boundary fault, so that the main southerly downthrow affecting the Lower Old Red Sandstone was pre-Upper Old Red Sandstone in age (Anderson 1947 p. 509). No evidence for the age of movement has been observed to the northeast along this fault, but as it is not likely to be of Upper Old Red Sandstone age, a pre-Upper Old Red Sandstone movement is probable. Similar kinds of evidence exist for the Southern Uplands fault, where in the Lammermuir Hills, south of the Pentland Hills, Upper Old Red Sandstone oversteps Lower Old Red Sandstone to lie on lower Palaeozoic rocks to the south of the Southern Uplands fault, or its equivalent (George 1960 pp. 60–63). On this evidence, and the presence of an unconformity between Lower and Upper Old Red Sandstone in the Midland Valley, a Middle Old Red Sandstone age for the faulting, folding and rift-like structure has been proposed (e.g. George 1960 p. 66).

A Middle Old Red Sandstone age for the major fault movement presents some problems. If the structure of the Lower Old Red Sandstone along the Highland Boundary fault is regarded as being dominantly monoclinal (however, see Allan 1928 fig. 3; 1940 figs 1, 4) then the pre-Upper Old Red Sandstone throw must have been considerable, up to 9 km. If this were to have taken place in Middle Old Red Sandstone times, then the yield of sediment to the (relatively) subsiding Midland Valley would have been substantial and a thick sequence of Middle Old Red Sandstone sediment would have accumulated there. Instead, the Lower Old Red Sandstone rocks within the Midland Valley were themselves undergoing erosion at this time. Disposing of the sediment eroded off the Highlands before the laying down of the Upper Old Red Sandstone is therefore something of a problem if the faulting is to be of this Middle Old Red Sandstone age.

On the other hand, in order to accumulate up to 9 km of Lower Old Red Sandstone rocks, a considerable vertical displacement is required. Within this sequence there are coarse, alluvial fan-type conglomerates which not only require persistent subsidence but also continual steep slopes in the source areas to ensure a copious supply of coarse debris out into the basin. Moreover, the alternation of conglomerate beds with sandstones (Fig. 3) suggests repeated rejuvenation of the uplift. Fanglomerates of this coarseness and structure, when in thicknesses of this order, require sharply defined areas of uplift and downwarp which are most easily obtained by faulting. Such structures would need to lie parallel with the depositional strike of the conglomerates, and would very likely parallel the longitudinal basin axes both of which, in this Lower Old Red Sandstone instance, are parallel to the Highland Boundary and Southern Uplands faults (Fig. 5b).

Conglomerate beds thicken, become more abundant and often contain frameworks of boulders where the succession is traced towards the Highland Boundary and Southern Uplands faults (Figs 3, 5, 6), so that the basin edge cannot have been far from the present fault positions.

Faults of the type which may bound basins of this kind are illustrated in Figure 7. According to Kennedy (1958 p. 117), the Highland Boundary fault was a major topographic feature, and Figure 7a represents his interpretation where the fault is marginal to the source, the basin and the outcrop. The interpretation of George (1960) is represented by Figure 7b, where the main fault defines the outcrop margin for most of its length, but there are preserved transgressions over it (i.e. it does not

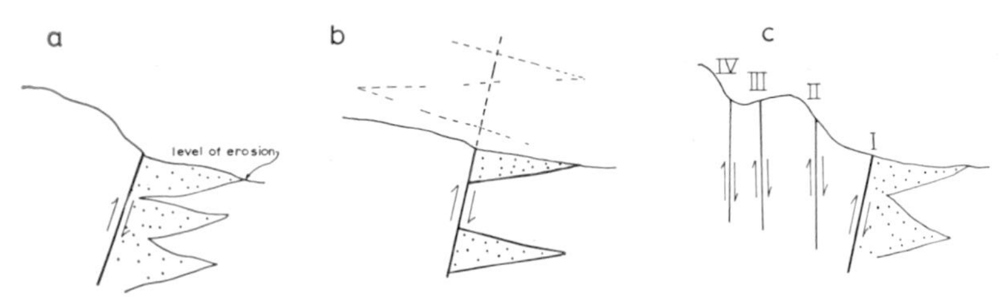


Fig. 7. Illustrating some of the faults which may bound sedimentary successions of the kind discussed here. Faults are regarded in three ways: outcrop bounding, where they are responsible for preserving the sedimentary basin (b); source bounding, where they control the uplift of the source blocks (c, II to IV); and basin margin, where they determine the margin of the basin (ignoring occasional overlaps into the source area (c, I) ). Where all the fault types—basin, source and outcrop—are embodied in one fracture is shown in (a). The bounding faults to the Midland Valley may have, over their length or over periods of time, behaved like any one of these faults illustrated.

define the basin margin). In the interpretation here adopted the fault may have behaved differently depending upon position along the fracture, but is most likely to be represented by Figure 7c where sourcewards of the main fracture, there were relatively minor related faults that were responsible for the source uplift, and even basins of deposition within the Highlands (e.g. between III and IV of Fig. 7c). These basins may have been subsequently tapped of their sediments which were then brought over the main preserving fracture. Location of the secondary fractures in older rocks bordering the Southern Uplands and Highland Boundary faults is both difficult and speculative. Firstly they may be in the strike direction of the existing Dalradian and Southern Uplands lower Palaeozoic rocks and secondly, if they dip into the main fault (Fig. 7c) then most would disappear as uplift proceeds and those remaining would have a fraction of the throw of the parent fault (I of Figure 7c). It is possible that, for the Highland area, fractures of or related to the Tay set (Fig. 1) were the secondary fractures that controlled uplift. Some of these occur quite close to the Highland Boundary fault. In the Southern Uplands there are many NE–SW fractures which may have been responsible for the uplifts required there.

The broad lithostratigraphy of the Old Red Sandstone of the Strathmore Basin permits further interpretation of the basin history. If the stratigraphic relations north of the Highland Boundary fault can be extended to the area immediately to the south, then the succession in the Strathmore syncline would have been deposited in a basin which, through time, gradually extended in a southwesterly direction (Fig. 18), being filled by sediments transported in a similar direction (cf. Bryhni 1964). In such a case the throw of the fault(s) would decrease towards the southwest, where the succession is thinner (Fig. 4b), finer grained and in an extensive outcrop is seen to rest unconformably on Dalradian rocks in Kintyre. In the area where the Highland Boundary fault splits, and its course is difficult to detect, the elongate fault bounded basin appears to have opened out. This area southwest of Loch Lomond may have been a hinge where, probably due to a considerable splaying out of the Highland Boundary fault, the deeply subsiding, elongate Strathmore basin stopped its southwesterly migration (see Figs 4b, 18).

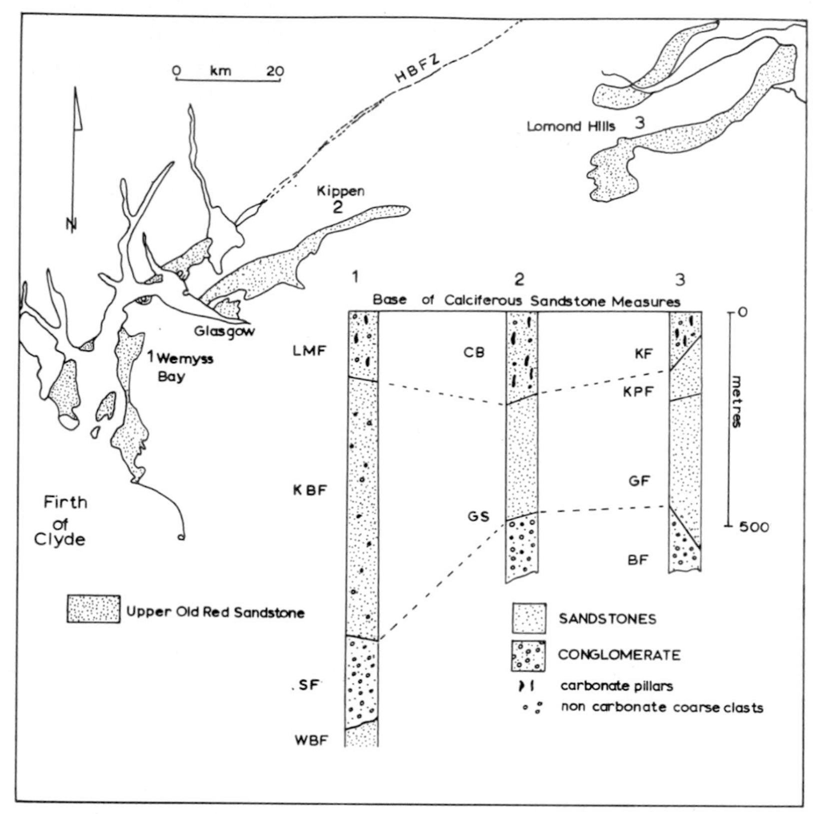


Fig. 8. Comparative sections through the Upper Old Red Sandstone. Kippen section, west of Stirling (after Francis, *in* Francis *et al.* 1970 pp. 97–114), Lomond Hills (after Chisholm and Dean 1974 fig. 9). WBF = Wemyss Bay Formation; SF = Skelmorlie Formation; KBF = Kelly Burn Formation; LMF = Leap Moor Formation; GS = Gargunnock Sandstone; CB = Cornstone Beds; BF = Burnside Formation; GF = Glenvale Formation; KPF = Knox Pulpit Formation; KF = Kinnesswood Formation. HBFZ = Highland Boundary fault zone

## 3. The Upper Old Red Sandstone

The Upper Old Red Sandstone contrasts with the Lower in its general finer grain size (fewer, and finer conglomerates), reduced thickness, change from lithic arenites to sublithic and quartz arenites, presence of many cornstones and absence of volcanic rocks.

### 3a. Stratigraphy

In contrast to the Lower, the succession is most complete and thickest in the Firth of Clyde–Glasgow region where it may exceed 1000 m. Here the sequence has been divided into four units (Fig. 8), the upper three of which are traceable northeastwards as far as the Lomond Hills (Chisholm and Dean 1974 fig. 9). This simplified four fold division is not traceable southeastwards (Fig. 9). In that direction first the Wemyss Bay Formation, then the major conglomerate division is rapidly lost just south of Skelmorlie, where it is thought to have disappeared against a contemporaneous fault. Such a disappearance is also envisaged for the basal

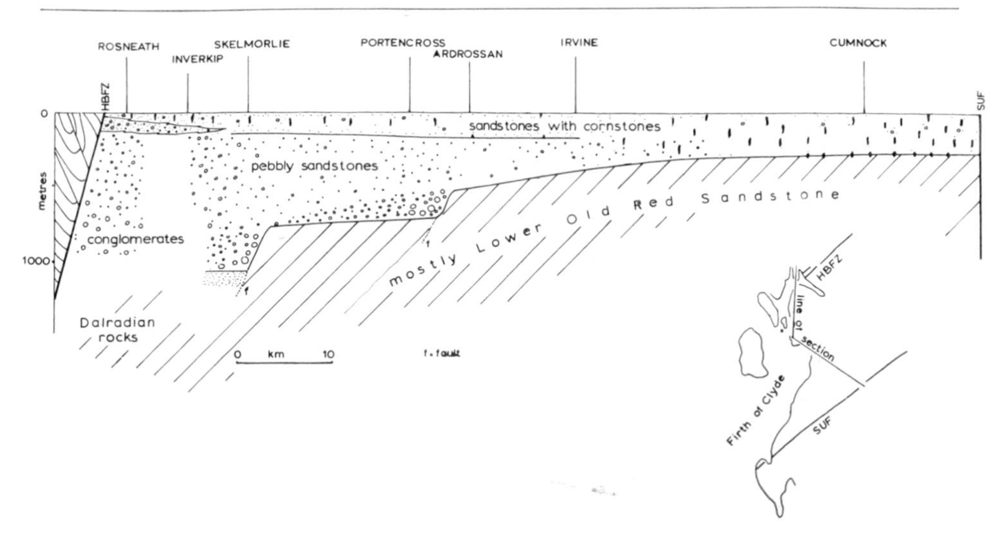


Fig. 9. Provisional section through Upper Old Red Sandstone in the western end of the Midland Valley. HBFZ = Highland Boundary fault zone; SUF = Southern Uplands fault.

conglomerate at Portencross (Bluck 1967), which here probably rests on Lower Old Red Sandstone. Further to the southeast (Fig. 9), the Upper Old Red Sandstone does not have conglomerates with the coarse grain size and distinctive structure of those at Skelmorlie and Portencross. In this southern area it often has a cornstone (fossil caliche) at its base (Fig. 10; cf. Burgess 1960, fig. 4).

Compared with the succession in the Wemyss Bay district, the lower sandstone (the Wemyss Bay Formation) dies out both to the northeast and southeast and the thick wedges of conglomerate and overlying sandstones are replaced to the southeast, where, either by overlap or facies change or both, cornstone beds dominate the succession (Fig. 9).

## 3b. Sediment dispersal

The coarsest sediments in the Upper Old Red Sandstones are found in the northwestern part of the Midland Valley where, with respect to the other cross stratal and grain size data, they appear to occupy the upstream end of roughly E–W trough (Fig. 11). Outside the Midland Valley, in the eastern Southern Uplands, a roughly southeasterly flow is evident. Palaeocurrent measurements have been taken from the conglomerate, pebbly sandstone and sandstone with cornstone beds which correspond to the lower, middle and upper divisions (Fig. 12; cf. Fig. 9). In the absence of a good chronostratigraphic framework, the intervals chosen are wide so as to minimise any possibility of complications due to diachronism.

The cross bed dip direction for the very lowest beds, the Wemyss Bay Formation, are not recorded in Figure 12 but, on the basis of 32 readings, they have a bipolar distribution, with the larger mode dipping towards the NNW. The lower division (Fig. 12) has cross stratal dips which show some divergence but on the whole indicate a northeasterly flow direction (with the exception of the data around Loch Lomond, which shows a southwesterly direction of transport). The data from around Kippen are taken from Read and Johnson (1967 figs 4, 6) who do not divide readings from the conglomerates and overlying sandstones.

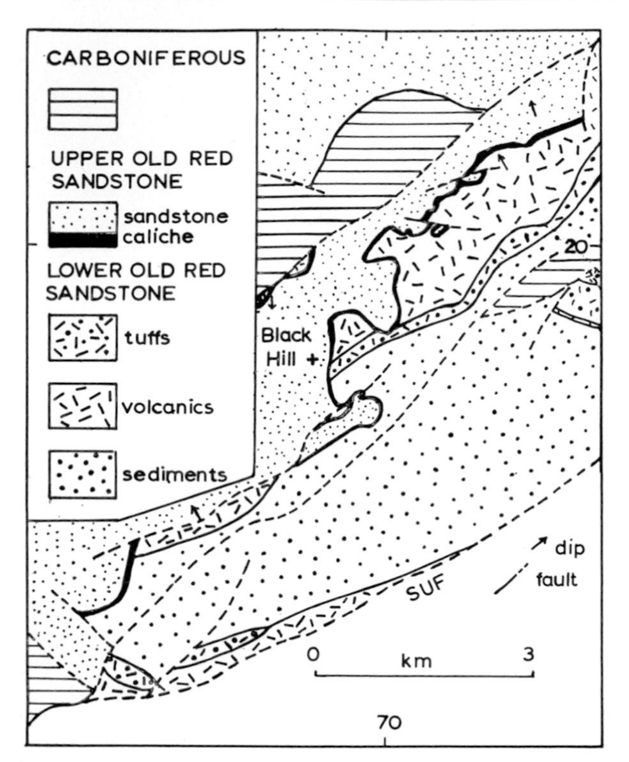


Fig. 10. The basal unconformity of the Upper Old Red Sandstone just north of Sanquhar, where a caliche soil is developed on the exposed Lower Old Red Sandstone surface. Based on Geol. Surv. 1″ Sheet 15 (Sanquhar). SUF = Southern Uplands fault.

The middle part of the sequence shows a regional change to a easterly and northeasterly dispersal again with exception of the data from Loch Lomond to Kippen. The upper part of the sequence continues this trend with a general southeastly dispersal, but with some exceptions.

## 3c. Composition

(i) *Conglomerates.* The conglomerates have a strong areal variation in composition. In the southern exposures clasts of vein quartz, quartzite, andesite-basalt and greywacke are common indicating the derivation from the Southern Uplands and the Lower Old Red Sandstone of that general area. On the Clyde coast, coarse conglomerates at Portencross (Fig. 9) contain clasts of vein quartz, quartzite and sandstones (Bluck 1969 figs 7, 9), partly derived from the underlying Lower Old Red Sandstone. A little further north, the Skelmorlie conglomerates contain clasts of vein quartz, quartzite, red sandstone, volcanic rocks and rocks of Dalradian-type (Bluck 1969, figs 8, 9). These two compositionally discrete conglomerates belong to different dispersal shadows; dispersal shadows of less discrete composition also occur in the conglomerates which lie along the Highland Boundary fault from South Bute to Loch Lomond (Fig. 12). Here the dominant clast type is of a low grade metamorphic kind, matching some of the Dalradian rocks cropping out to the north. In the areas around Loch Lomond, schistose grit and greywacke are common in Dalradian outcrop and as clasts in the conglomerate, but at Rosneath

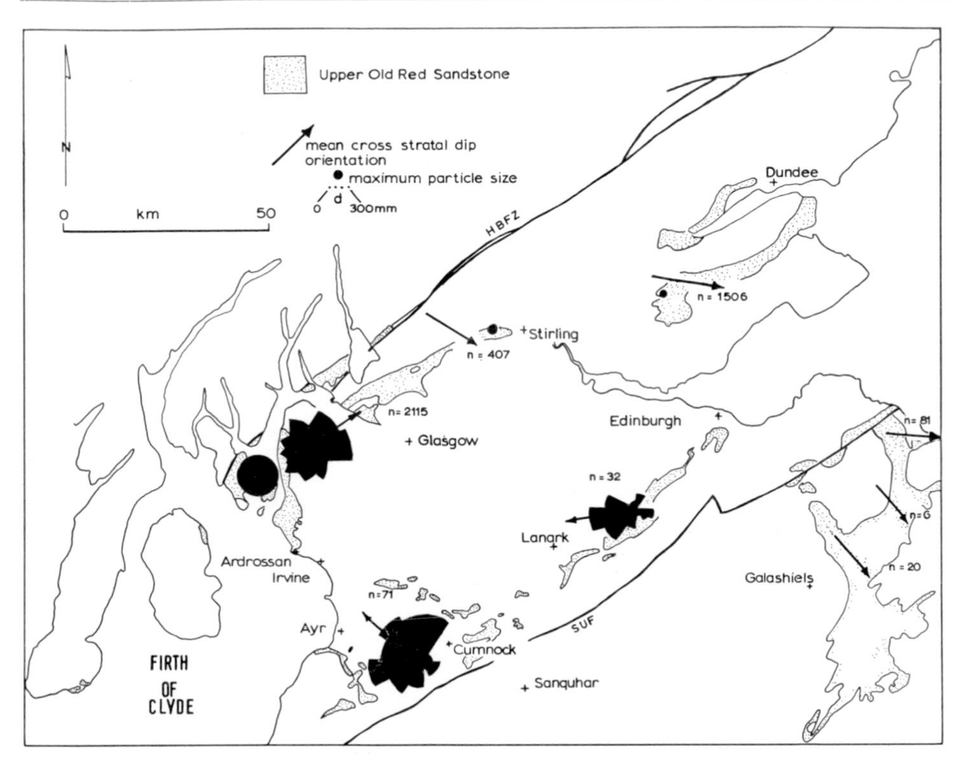


Fig. 11. Directional structures and an average maximum clast size for the basal part of the Upper Old Red Sandstone. Isolate arrows show observations of other workers: data from Stirling (Kippen) area come from Read and Johnson 1967 figs 4, 6; Lomond Hills (Chisholm and Dean 1974 table 2; Southern Uplands, Paterson *et al.* 1976 table 1). n = number of observations, d = diameter of clast in millimetres. HBFZ, SUF as in other figures.

the conglomerates contain clasts of schists and igneous rocks. In the areas of Toward Point and South Bute, clasts of schistose material are accompanied by slate, phyllite and epidiorite, again reflecting changes in the local Dalradian rocks to the north.

The conglomerates of the Kippen and Lomond Hills areas contain vein quartz, quartzite, and low grade metamorphic rocks, together with lava derived from the Lower Old Red Sandstone (Francis *in* Francis *et al.* 1970 p. 99); Chisholm and Dean 1974 p. 11).

Sections through the conglomerate show an upward change in composition with vein quartz and quartzite increasing at the expense of less stable clasts. Although this change may be attributed to the inherently small size of vein quartz, which in a sequence getting finer would therefore increase, a similar (albeit out of phase) change in sandstone (the data of Figure 13) argues for other causes (p. 270).

(ii) *Sandstones.* The sandstone composition of the Upper Old Red differs appreciably from that of the Lower. The conglomerates of the lower part of the Upper Old Red Sandstone contain interstratified sandstones which are lithic arenites or sublithic arenites (Figs 13, 14a) and contain many more metamorphic rock fragments than the Lower Old Red Sandstones. There is an upward change to sublithic

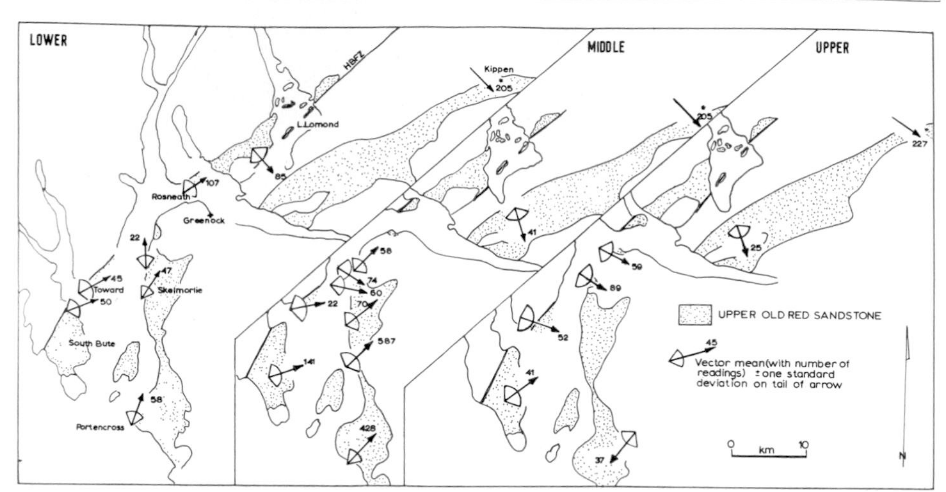


Fig. 12. Cross stratal dip orientation for the Upper Old Red Sandstone of the western Midland Valley. The lower, middle and upper divisions correspond respectively to the conglomerate, pebbly sandstone and sandstone subdivisions of Figure 9. HBFZ as in other figures.

arenites (pebbly sandstones of the middle part of the sequence) and sublithic arenites, subarkoses and quartz arenites in the sandstones with cornstones of the upper (Fig. 14b, c). A corresponding increase in feldspar content (at the expense of rock fragments) has also been recorded from the Lomond Hill sequence by Chisholm and Dean (1974 fig. 5).

## 3d. Palaeogeography and basin evolution

The Upper Old Red Sandstone accumulated in a large terrestrial basin under the dominant influence of alluvial regimes. The palaeogeography is illustrated in Figure 15 using the same three intervals as in Figure 12, to allow for the probability of considerable diachronism.

It is considered that the earliest sediments accumulated in the area around Wemyss Bay where the sandstones of the Wemyss Bay Formation crop out extensively. The base of this formation is not seen, and in the normal cycle of Upper Old Red Sandstone kind, beds of this type would occur at the top of a sequence beginning with conglomerates. If this were to be the case then a considerable thickness of sediments may underlie this general area, something that the geophysical results of Qureshi (1970 pp. 481, 502) also suggest is likely. The beds contain large-scale cross strata and it is possible that they may be partly aeolian. From the Wemyss bay area, a new basin of sedimentation began to form and gradually extend, probably by progressive marginal downfaulting (Fig. 15). This postulate is based on the occurrence of the two conglomerate wedges of Skelmorlie and Portencross (Fig. 9) both of which have evidence of alluvial fan deposition. Sediments containing clasts of Portencross-type overlie beds with clasts of the Skelmorlie-type so providing evidence of the later age of the Portencross dispersal shadow and the faulting which initiated and controlled it.

A fairly coarse conglomerate, probably of basal kind, runs from Lomondside to South Bute. At Rosneath it is coarse with structures of proximal alluvial fan type, and has a distinctive clast composition. This conglomerate is overlain by a finer one

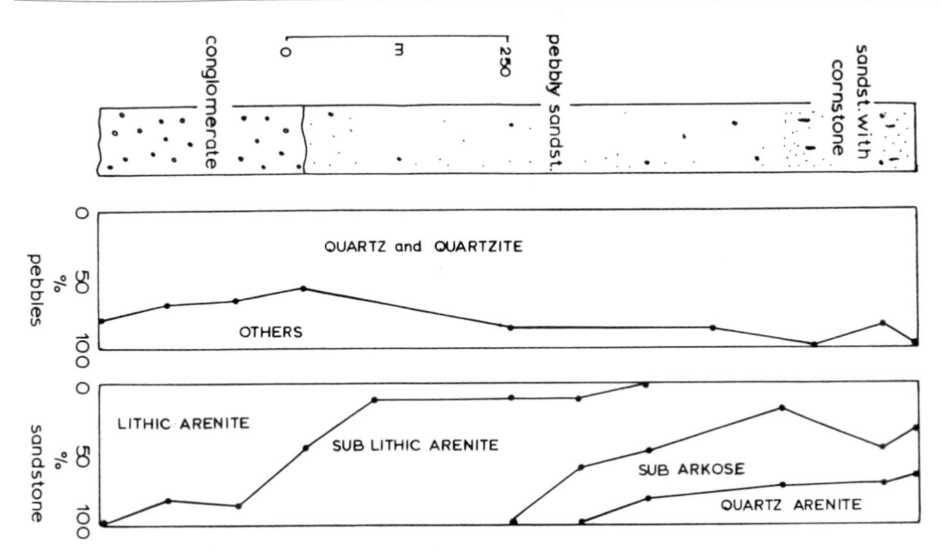


Fig. 13. Vertical variation in composition of conglomerates and sandstones for a section in Skelmorlie–Wemyss Bay area; at least 100 pebble counts per locality in conglomerate. With the exception of the basal beds, thin section data for sandstones are taken from nested samples (4–8) within ±5m stratigraphically of point shown on diagram.

containing assemblages of clasts similar to those found at Toward and South Bute. The South Bute succession begins with a coarse conglomerate of proximal alluvial fan structure; the one beginning the succession on Toward is largely finer and of more distal type. The Rosneath, Toward and South Bute coarse basal conglomerates all share roughly the same palaeocurrent direction, but on the evidence given could not have formed at the same time: the conglomerates at Rosneath are proximal and differ in composition. As clasts of South Bute–Toward type overlie the Rosneath proximal conglomerates, the Rosneath sequence is thought to be the earlier. The South Bute–Toward coarse conglomerates are thought to belong to the same dispersal shadow with the downstream Toward conglomerates being the finer. By analogy with the east Clyde coast, the Rosneath dispersal shadow is thought to have been divided from the South Bute–Toward one by a fault which was active during the sedimentation of the Rosneath succession with a later fault probably providing the relief for the South Bute sequence (Fig. 15). The postulated faults are thought to trend perpendicular to the local palaeocurrent direction and this, in addition to consideration of location and age, suggests the form of basin migration as outlined in Figure 15 (lower).

The controlling factors on the Lomondside sequences have not yet been examined in detail, but coarse, proximal type conglomerates occur in this sequence, and a southerly downthrowing fault, contemporaneous with sedimentation, located just north of the outcrop, seems likely.

During the middle part of the Upper Old Red Sandstone, the sediments in the Clyde area are much finer, and do not show the same changes in thickness which characterizes some of the underlying conglomerates (Fig. 9). Contemporaneous faulting is not thought to have been active in the area during this time. Braided stream deposits, some of which were major sediment cones, dominate the record, and this sediment overlapped the basal conglomerates as the basin of accumulation

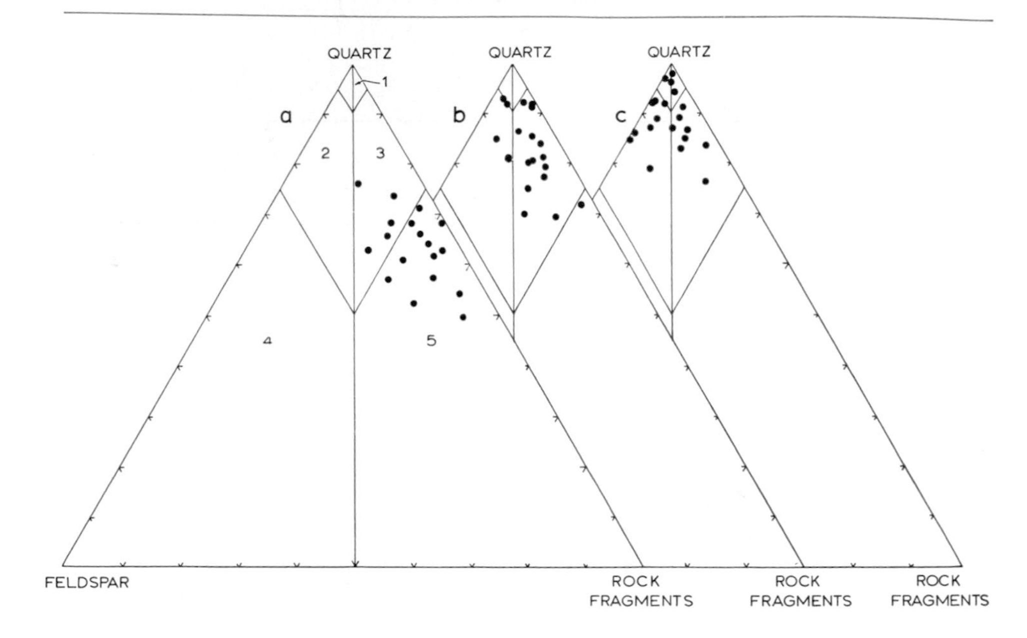


Fig. 14. Modal compositions for the Upper Old Red Sandstone. a, sandstones in the lower (conglomerate) beds; b, sandstones in the pebbly sandstone beds; c, sandstones in the sandstones with cornstone (upper). 1, 2 etc. as for Figure 4. Total 57 samples, 400 counts per thin section, by point counting.

failed to subside as fast as the rate of sediment supply. This extension of the basin of accumulation implies that uplift of the source, outside the present area of study, was then on a scale larger than it had previously been, and preliminary investigations into the structure of the sediment deposited during this time suggest an increase in the depth of the streams which laid down the sediment.

The composition of the sediments of this middle part of the sequence is more uniform than previously. There are many rounded clasts, but also, in some localities, fairly angular Dalradian ones which are probable first cycle. Southwest of the present area of study, and between the generally accepted line of the Highland Boundary and Southern Uplands faults, Dalradian-type rocks were probably exposed.

Sedimentation is thought to have begun in the southern part of the Midland Valley at this time, probably by ephemeral streams draining the Southern Uplands (Fig. 15, middle). Here, mature cornstone profiles at least 4 m thick (Eyles 1949 p. 48; Burgess 1960) are common through the sequence, suggesting a low rate of subsidence in an area receiving little sediment. The Southern Uplands are therefore considered to be mature uplands at this time, with a comparatively condensed sequence forming in front of them. There would appear to be little activity on the Southern Uplands fault during this period.

Important environmental changes took place in the upper part of the Upper Old Red Sandstone. The basin probably further expanded and fining upward cycles of the type laid down by meandering streams partly replace the braided stream deposits below. These cycles, often less than 3 m thick and having quite thin cross strata, are probably the product of quite small streams. The abundant development

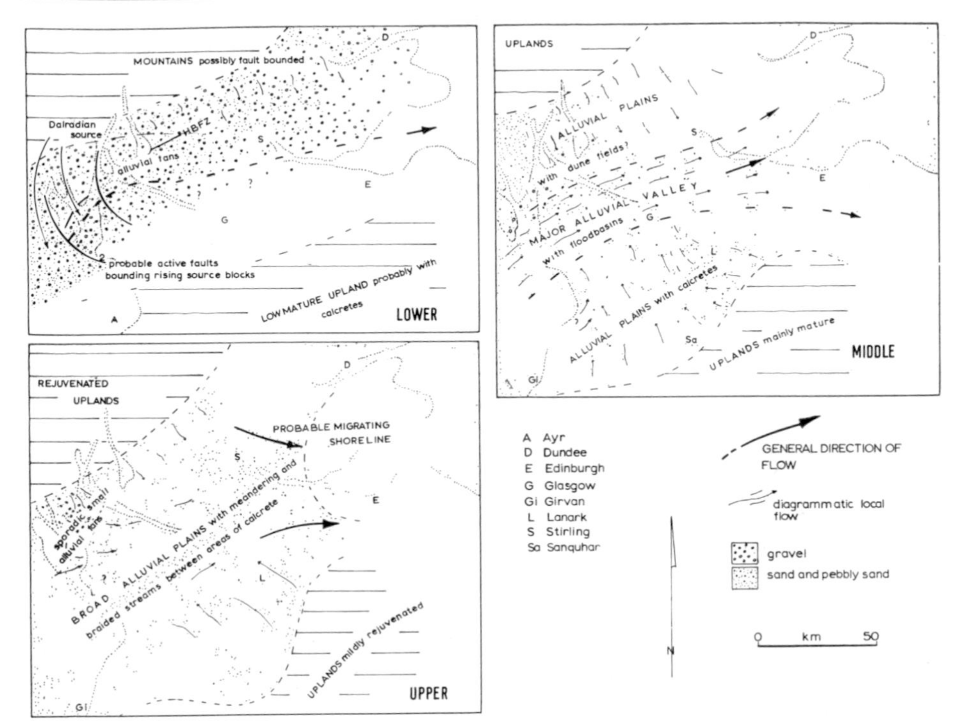


Fig. 15. Palaeogeographical reconstruction of Upper Old Red Sandstone times. Lower = conglomerate beds; middle = pebbly sandstones; upper = sandstones with cornstones (see Fig. 9). 1, 2 are faults which initially controlled much of the sedimentation and these are thought to have migrated southwards.

of caliche horizons probably indicates a reduction in the rate of sedimentation which, together with evidence for smaller streams suggests a reduction in the area and/or activity of the source. There is some evidence in the top of the sequence of intertidal sedimentation. Here there are sheets of flat or low-angled beds, some with bioturbation (*Skolithos*) and conglomerates with highly rounded clasts (cf. Chisholm and Dean 1974 p. 15). Large-scale cross strata, and fairly structureless sandstones with well rounded grains may indicate aeolian conditions in parts of the sequence.

Towards the end of Old Red Sandstone times conglomerates with angular Dalradian clasts, and a proximal-type structure, appear in the regions near the Highland Boundary fault. These may reflect a short lived rejuvenation of the Highlands to the north (Fig. 15, upper).

There is an upward change in sandstone composition throughout the Old Red Sandstone sequence which may reflect a fundamental change in tectonism. The Lower Old Red Sandstone is immature, being mainly a lithic arenite with up to 100% rock fragments and feldspar (Fig. 5) mostly derived from igneous sources, but even within this sequence, at the western end of the Midland Valley, there is an upward change to a more quartz-rich lithic arenite.

The Upper Old Red Sandstone records a fairly sudden change to sandstones richer in quartz and mostly of a low grade metamorphic provenance. Although a change in source may have brought about this increase in quartz, the sediments continue to increase in maturity upward in the sequence despite a constant meta-

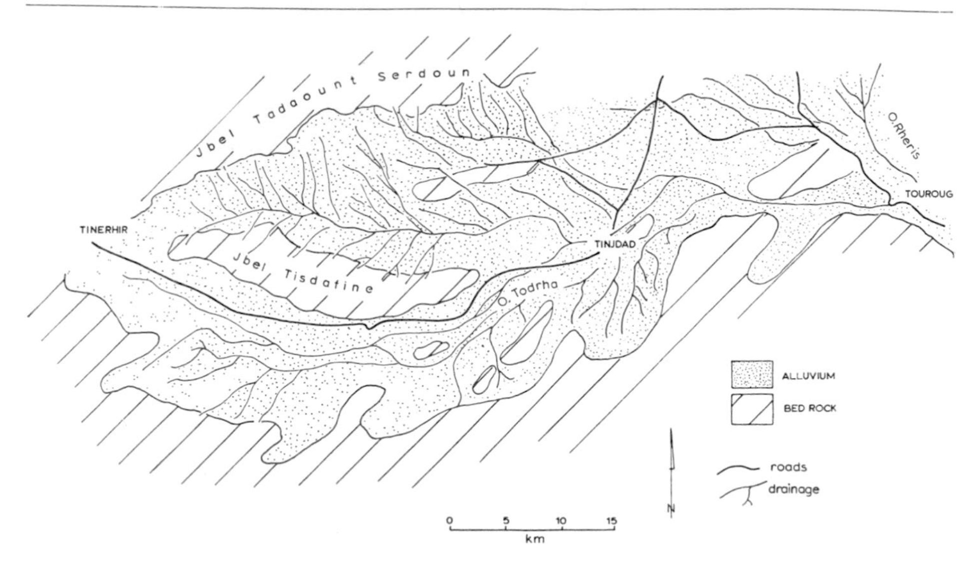


Fig. 16. Morphology of a small embayment or depression along the southern margin of the Atlas in southern Morocco. The basin is situated in the major fault zone running from Agidir in Morocco to Gabes in Tunisia. The Oued Todrha longitudinally fills the embayment with sediment which, like the lateral fill to the north, come off the Atlas. The Todrha emerges from its extensive Atlas drainage basin at Tinerhir.

morphic provenance (Figs 13, 14). This compositional maturing is accompanied by a general decrease in grain size of the sedimentary sequence, a shift in environment to meandering rather than braided streams, and an increase in the abundance of caliche. Whilst the decrease in grain size might indicate a more distal sediment, the latter two, together, suggest a decrease in sediment supply and general conditions of stability. Such caliche soil profiles have been shown to develop in areas where sediment supply is low compared with rate of soil growth, and mature caliche profiles are found in areas with a low rate of sedimentation (Steel 1974).

The Upper Old Red Sandstone sequence is a major upward fining, upward maturing sequence, which records the initiation and completion of a small, late orogenic-type basin. It began as a basin where there was small scale local sediment accumulations and was followed by a period where a major source flooded sediment into the basin. It ended as a stable basin with a subdued source.

## 4. Tectonic setting of the Old Red Sandstone

Kennedy (1958 p. 116) regarded the Old Red Sandstone as the molasse of the Caledonian orogeny, comparing it with the Tertiary Alpine molasse of the Swiss Plain and the Siwaliks of the Himalyan front (cf. Anderson 1947 p. 506). He envisaged it as having accumulated in a rift valley bounded by reverse faults (Kennedy 1958 fig. 2, p. 118).

Mitchell and McKerrow (1975) interpreted the Old Red Sandstone of the Midland Valley as the upper (fluvial) part of a major sedimentary basin the lower part of which was marine. The filling is thought to have begun from the northeast where, on the southern margin of the Midland Valley, the transition from marine to

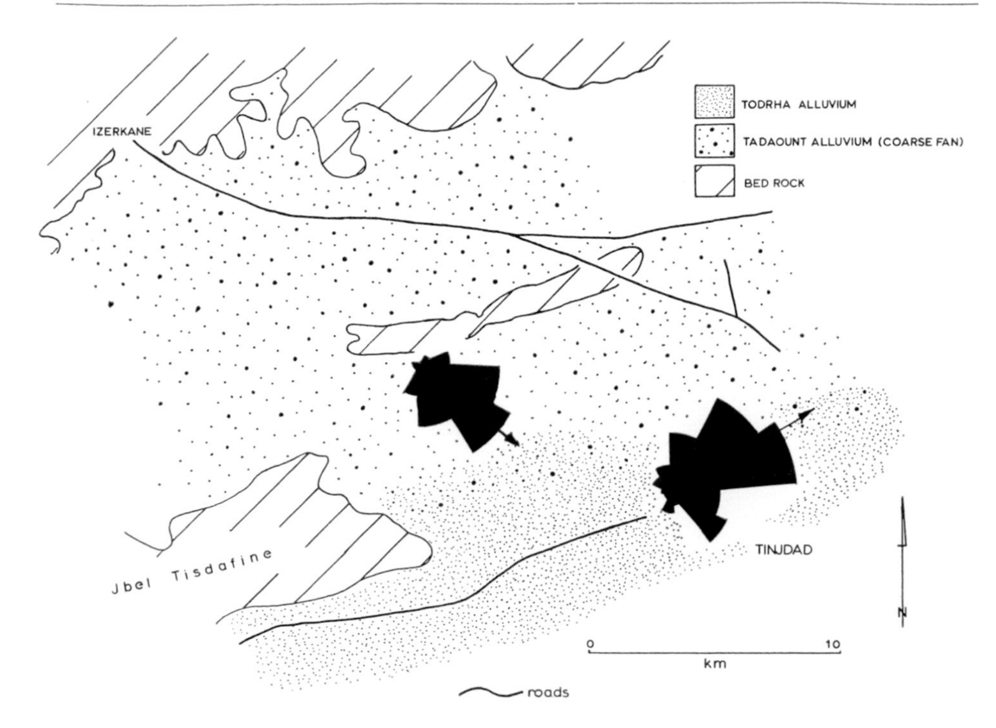


Fig. 17. Cross stratal dips for the various sediments in the Tinjdad embayment. The coarse lateral fill, and the finer longitudinal fill has its analogy in the Lower Old Red Sandstone of the Midland Valley.

non-marine is earlier than it is to the southwest (*op. cit.* p. 308). They drew analogy between the longitudinal fill of the Midland Valley and the filling of the Central Burma Lowlands by the Irrawaddy–Chindwin river systems.

The Old Red Sandstone cannot be entirely regarded as the late stage deposits of the orgenies which formed the mountains to the north and south of the Midland Valley. In this sense it is not the molasse as conceived by Kennedy (1958), since the uplift of the Highlands to the north began 505±10 m.y. ago (Dewey and Pankhurst 1970) some 100 m.y. before the beginning of deposition of the Old Red Sandstone with the main uplift of the Dalradian rocks being 470–430 m.y. ago. The evidence on uplift age is equivocal for the Southern Uplands; there are indications of uplift as early as late Llandovery times in the Pentland Hills but the northerly dispersal of conglomerate did not take place until Wenlock times (McGiven 1967 p. 99). The Old Red Sandstone of the southern margin of the Midland Valley may therefore represent the late stage, molasse-type sediments of the Southern Uplands orogenic phase. It does, however, share with the Old Red Sandstone of the Strathmore basin, a basin orientation, a palaeogeographical, and to some extent a petrographical evolution. These facts suggest that the structural events which caused the development of the northern belts of the Midland Valley Old Red Sandstone may also have controlled the southern.

Palaeogeographical and palaeocurrent data suggest the Old Red Sandstone to have been deposited in a series of basins which resemble the embayments found along the major, South Atlas fault zone running from Agidir almost to Gabes in

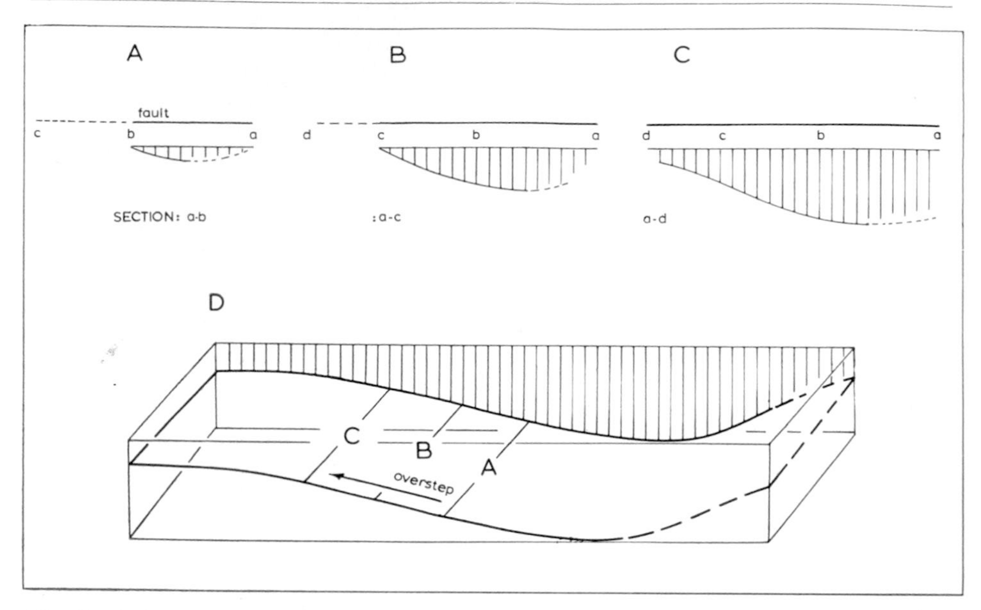


Fig. 18. A, B, C shows the extension of a fault bounded basin by the lateral migration of a fault a b c d. The original basin has become the locus of thick sediment accumulation and in the direction of fault migration the sediment pile shows overstep onto the basin floor (D). This illustrates the deposition of the Lower Old Red Sandstone in the Strathmore basin, assuming the overstep suggested north of the Highland Boundary fault, applies to the ground to the south.

Tunisia. There, as instanced near Tinjdad (Fig. 16) most of the sediment fill to a basin, whether it is axial or lateral, comes from the Atlas Range to the north (Fig. 17). Although the palaeogeography and palaeocurrents may be controlled by the topography of the basin, its geometry, stratigraphy and evolution is controlled by tectonics. In these latter respects the Old Red Sandstone basins resemble those found in major fault zones such as the Dead Sea (Quennell 1958), Hawke Bay, New Zealand (Kingma 1958) and the San Andreas (Crowell 1974 p. 229). Here the uplift and erosion of the source rocks may not belong to the same orogenic cycle which originally created and folded them.

The geometry and stratigraphy of the Lower Old Red Sandstone in the Strathmore syncline is outlined in Figure 18, where it can be seen that deposition along a southwestwards migrating fault is postulated. At present it is not possible to be more specific about the structural regime in which this basin formed, but it is highly likely to involve strike-slip movement on the Highland Boundary fault.

The Upper Old Red Sandstone basin began forming where the Lower Old Red Sandstone migrating hinge zone stopped. Here a series of normal faults developed, which are thought to have migrated southwestwards in response to a southwesterly-growing fracture (Fig. 19A, B). The geometry and stratigraphy outlined in Figure 19D (cf. Fig. 9) again resembles basins described by others as occurring in major fault zones (see Kingma 1958 fig. 2). In this instance a pull apart along the Highland Boundary fault seems likely. At position b on Figure 19A there would be sediment dispersal from the wall of fault 1 and the Highland Boundary fault wall (a-b), so that the dip direction of the cross strata should be an average of these effects and have a high standard deviation. Although based on only few observations, the cross

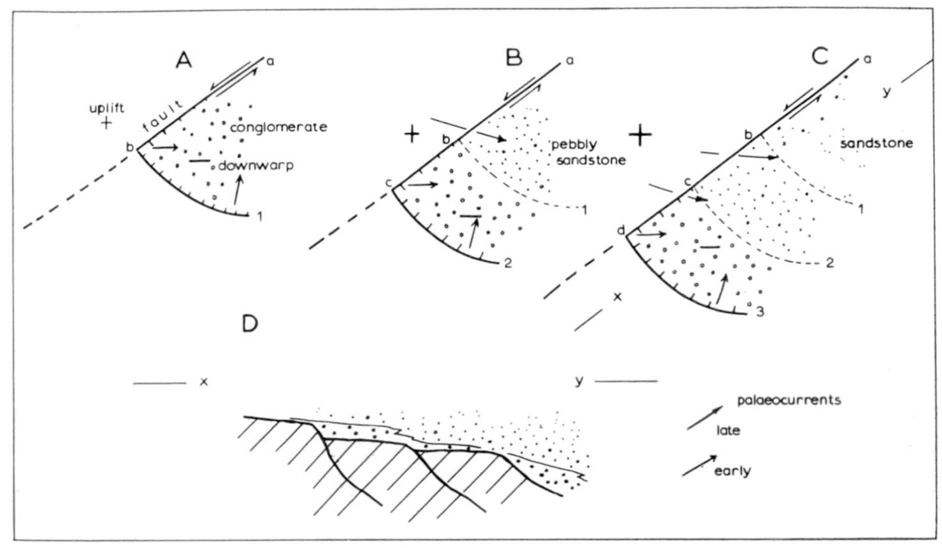


Fig. 19. A, a sinistral fault, a–b, extends the basement on its right moving side and initiates normal faulting (1). Further growth of the fault takes place to c as the basin continues to develop in the region of 1. Another fracture is formed as a result of extension, and the basin further subsides (B, C). D = section X–Y (cf. Fig. 9).

strata taken from the proximal facies along the Highland Boundary fault do have a higher standard deviation than those taken from the Portencross and Skelmorlie areas. The change in the dispersal of sediment upward in the sequence (Fig. 12 middle and upper), requires uplift to occur on the northwestern side of the fault and intensify to the southwest. With this being a sinistral fault, there should be a complimentary uplift on the northwestern side which should grow in the ground a little southwestwards from the present basin position. Such an uplift would be needed to produce the strong E to NE dispersed sediment yield required in the middle of the Upper Old Red Sandstone (see Fig. 12).

## 5. Conclusions

The deposition of the Old Red Sandstone was a late orogenic event the timing of which largely followed, and partly overlapped intrusive and extrusive igneous activity of the type normally associated with a descending plate. Phillips *et al.* (1976) have provided a possible solution to the problem of a pre-Old Red Sandstone Iapetus closure (and consequent cessation of subduction) and the Old Red Sandstone igneous activity by dextrally shifting Scotland from the southwest where, in the terms of their "oblique subduction" model, subduction during Old Red Sandstone times had taken place.

The Old Red Sandstone sedimentary basins have a sedimentary strike and elongation parallel to the structural grain of the adjacent rocks. In the Strathmore basin there is a time gap between the orogeny in the adjacent source rocks and the inception of the basin, so that a similarity in the orientation of the growing basin and the structural orientation of the source gives a spurious impression of a unity in the development of them both. This similarity in orientation is brought about by the reactivation of an older fracture, the Highland Boundary fault. The pre-Old Red Sandstone age for this fracture is suggested by the discovery of serpentinite fragments in nearby Lower Old Red Sandstone conglomerates (Allan 1928 p. 74) and a serpentinite conglomerate associated with Highland Border rocks at Balmaha (cf. Anderson 1947 p. 496). The serpentinite in these instances is thought to have been derived from the existing serpentinite screen in the Highland Boundary fault.

The Old Red Sandstone comprises a major cycle beginning with the development of coarse conglomerates and volcanic rocks, and ending with the widespread development of sandstones and caliche-soils. This upward change is accompanied by increasing quartz clasts and grains in both conglomerates and sandstones, which in the sandstones produces a change from lithic arenites at the base to quartz arenites at the top. Although this change could be a function of a switch in provenance from volcanic terrane (Lower) to metamorphic terrane (Upper Old Red Sandstone), within the latter, the maturing of the sediments continues under the same provenance. This suggests sediment maturing as a response to the establishment of a more subdued tectonic regime through Old Red Sandstone times: the upward increase in caliche, the decrease in grain size of the rocks and the absence of volcanic activity all indicate this upward change to stability.

Within the Lower Old Red Sandstone, the upper beds (the Strathmore Group of the Strathmore basin) have no lava flows and are commonly quite fine grained, indicating a more subdued tectonic regime. But the folding during Middle Old Red Sandstone times rejuvenated the system, and the ensuing Upper Old Red Sandstone is a major fining upward unit in which conglomerates with lithic-arenites are replaced by quartz arenites at the top. This upward change in the complete Upper Old Red Sandstone sequence is interpreted as the result of the paring down of the rejuvenated source.

Although faulting had a dominant influence on sedimentation in both Lower and Upper Old Red Sandstone, the type and degree of fault control was different in each case. The nature of the fault control is not well known for the Lower Old Red Sandstone of the southern (Lanark) basin. The northern (Strathmore) basin is filled with a southwesterly overlapping and thinning succession; the axial palaeocurrents also flow in this same direction. The basin began in the northeast and gradually extended southwestwards; it was bordered on the northwestern side by the Highland Boundary fault, to the northwest of which smaller faults are thought to have been actively uplifting the source. Subsidence in the basin lessened towards the southwest, where in the Clyde region the Highland Boundary fault splits, and (in Kintyre) there is a major northerly overstep. This broadly defined area is then regarded as a hinge zone.

Sedimentation of the Upper Old Red Sandstone began in the Clyde region, mainly as a result of a sinistral movement on the Highland Boundary fault. Rocks on the southeastern margin of the fault responded by normal faulting. By this means successive faults, and the basin they created, migrated southwestwards along the line of the Highland Boundary fault. The oldest sediment, preserved where the faulting began, is overstepped away from this area as the basin gradually extended outwards. At this time the rocks, folded during Middle Old Red Sandstone times, were undergoing erosion: there are clasts in the Upper Old Red Sandstone which have been derived from the Lower.

Whilst the tectonic evolution of the basin controlled its geometry and stratigraphy,

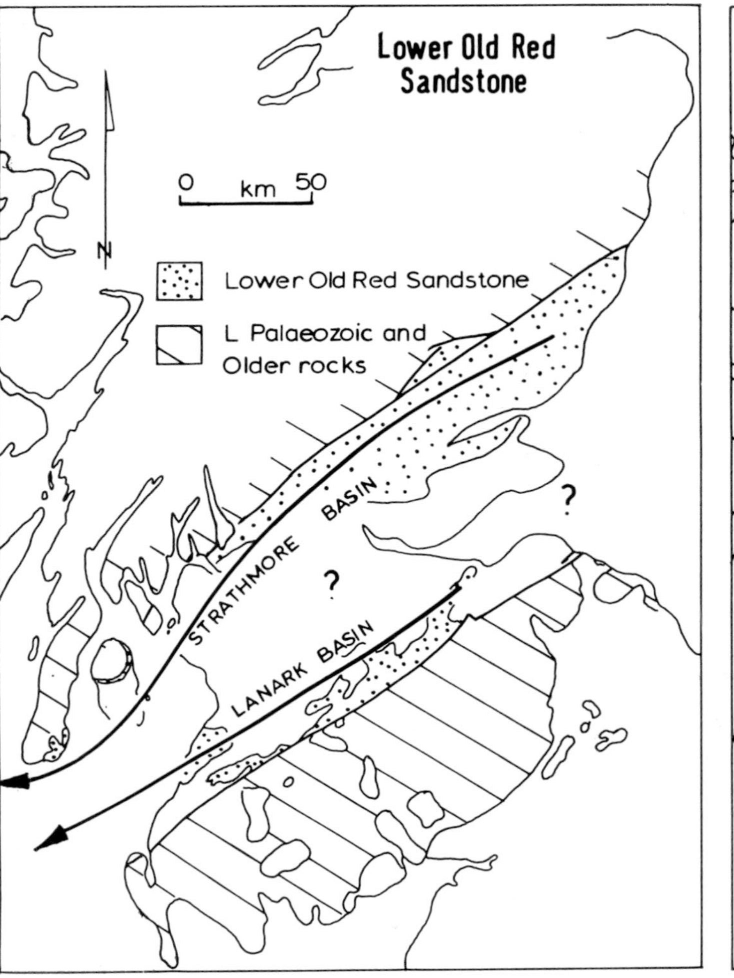


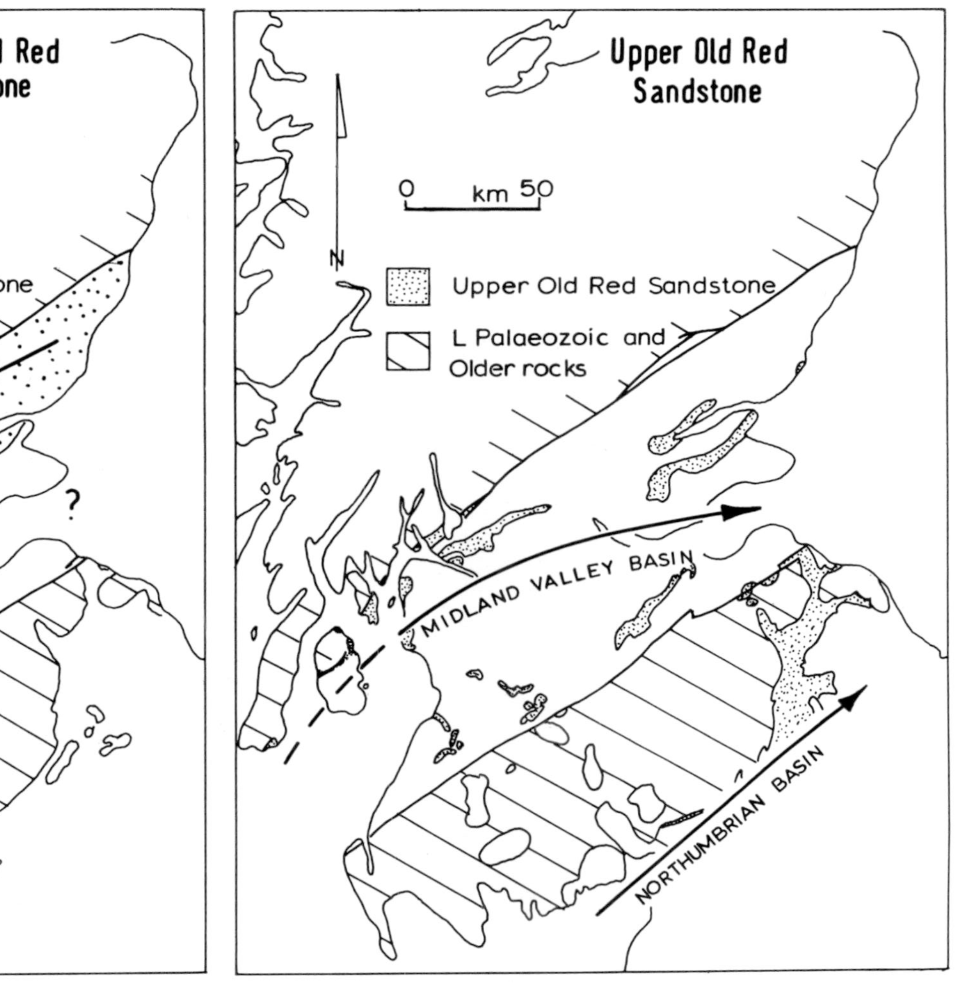


Fig. 20. A summary diagram of the dip of major longitudinal basins in Lower and Upper Old Red Sandstone times.

the palaeocurrents were controlled by the morphology of the surface of accumulation. In this respect the basins appear to resemble the embayments which are strung out along the South Atlas fault. In the region Midland Valley, the embayments faced towards the southwest, suggesting a positive area in the region northeast of the present Midland Valley (Fig. 20). In Upper Old Red Sandstone times the general palaeoslope had almost reversed, and dipped to the east. A corresponding easterly flow for the generally parallel Northumbrian basin has been demonstrated by Leeder (1973 fig. 10).

Faults which may bound basins have been divided into three kinds: outcrop bounding, where the fault is merely responsible for the termination and preservation of a sedimentary basin; basin bounding, where the edge of the basin of sedimentation is defined by the fracture; and source bounding, where the uplift is controlled by the fault. During the deposition of the Old Red Sandstone the marginal faults to the Midland Valley behaved as all three at various times and in some instances the faults performed all three functions at once.

*Acknowledgments.* I thank Drs McGiven, Morton and Wilson for permission to use some of their unpublished information.

## References

ALLAN, D. A. 1928. The geology of the Highland Border from Tayside to Noranside. *Trans. R. Soc. Edinb.* **56,** 57–88.

—— 1929. A preliminary account of certain Lower Old Red Sandstone Conglomerates in Perthshire and Forfar. *Rep. Br. Ass. Adv. Sci.* **96,** 555.

—— 1940. The geology of the Highland Border Glen Almond to Glen Artney. *Trans. R. Soc. Edinb.* **60,** 171–193.

ANDERSON, J. G. C. 1947. The geology of the Highland Border: Stonehaven to Arran. *Trans. R. Soc. Edinb.* **61,** 479–515.

ARMSTRONG, M. and PATERSON, I. B. 1970. The Lower Old Red Sandstone of the Strathmore Region. *Rep. Inst. geol. Sci.* No. 70/12, 24 pp.

BAILEY, E. B. 1960. The geology of Ben Nevis and Glen Coe (Sheet 53). 2nd Ed. *Mem. geol. Surv. U.K.*

BLUCK, B. J. 1967. Deposition of some Upper Old Red Sandstone conglomerates in the Clyde area: A study in the significance of bedding. *Scott. J. Geol.* **3,** 139–167.

—— 1969. Old Red Sandstone and other Palaeozoic conglomerates of Scotland. *In* Kay, M. (Editor). *North Atlantic—geology and continental drift: a symposium,* 711–723. *Mem. Am. Assoc. Petrol. Geol.* **12.**

BROWN, P. E., MILLER, J. A. and GRASTY, R. L. 1968. Isotopic ages of the Caledonian granitic intrusions in the British Isles. *Proc. Yorks geol. Soc.* **36,** 251–276.

BRYHNI, I. 1964. Migrating basins on the Old Red continent. *Nature, Lond.* **202,** 384–385.

BURGESS, I. C. 1960. Fossil soils in the Upper Old Red Sandstone of South Ayrshire. *Trans. geol. Soc. Glasg.* **24,** 138–153.

CAMPBELL, R. 1913. The geology of south-eastern Kincardineshire. *Trans. R. Soc. Edinb.* **48,** 923–960.

—— 1929. The composition of the conglomerates of the Downtonian and Lower Old Red Sandstone of the Stonehaven district. *Rep. Br. Ass. Adv. Sci.* **96,** 554–555.

CHISHOLM, J. I and DEAN, J. M. 1974. The Upper Old Red Sandstone of Fife and Kinross: a fluviatile sequence with evidence of marine incursions. *Scott. J. Geol.* **10,** 1–30.
COLLINS, A. G. and DONOVAN, N. 1977. The age of two Old Red Sandstone sequences in Southern Caithness. *Scott. J. Geol.* **13,** 53–57.
CROWELL, J. C. 1974. Sedimentation along the San Andreas fault, California. *In* Dott, R. H. and Shaver, R. H. (Editors). *Modern and Ancient Geosynclinal sedimentation,* 292–303. *Spec. Publs Soc. econ. Paleontol. Mineral.* **19.**
DEWEY, J. F. 1969. Evolution of the Appalachian/Caledonian orogen. *Nature, Lond.* **222,** 124–129.
—— 1971. A model for the Lower Palaeozoic evolution of the southern margin of the early Caledonides of Scotland and Ireland. *Scott. J. Geol.* **7,** 219–240.
—— and PANKHURST, R. J. 1970. The evolution of the Scottish Caledonides in relation to their isotopic age pattern. *Trans. R. Soc. Edinb.* **68,** 361–389.
DOWNIE, C. and LISTER, T. R. 1969. The Sandy's Creek beds (Devonian) of Farland Head, Ayrshire. *Scott. J. Geol.* **5,** 193–206.
EYLES, V. A., SIMPSON, J. B. and MACGREGOR, A. G. 1949. Geology of Central Ayrshire. *Mem. geol. Surv. U.K.*
FRANCIS, E. H., FORSYTH, I. H., READ, W. A. and ARMSTRONG, M. 1970. The geology of the Stirling District. *Mem. geol. Surv. U.K.*
FRIEND, P. F., HARLAND, W. B. and HUDSON, J. D. 1963. The Old Red Sandstone and the Highlands boundary in Arran, Scotland. *Trans. Edinb. geol. Soc.* **19,** 363–425.
—— and MACDONALD, R. 1968. Volcanic sediments, stratigraphy and tectonic background of the Old Red Sandstone of Kintyre, W. Scotland. *Scott. J. Geol.* **4,** 265–282.
GEORGE, T. N. 1960. The stratigraphical evolution of the Midland Valley. *Trans. geol. Soc. Glasg.* **24,** 32–107.
—— 1974. The geology of the Stirling region. *In* Timms, D. (Editor). *The Stirling region,* 5–29. *Br. Ass. Adv. Sci., Stirling.*
GUNN, W. 1903. The geology of North Arran, South Bute, and the Cumbraes, with parts of Ayrshire and Kintyre. *Mem. geol. Surv. U.K.*
KENNEDY, W. Q. 1958. The tectonic evolution of the Midland Valley of Scotland. *Trans. geol. Soc. Glasg.* **23,** 107–133.
KINGMA, J. T. 1958. Possible origin of piercement structures, local unconformities, and secondary basins in the Eastern Geosyncline, New Zealand. *N.Z. J. Geol. Geophys.* **1,** 269–274.
LAMBERT, R. ST J. and MCKERROW, W. S. 1976. The Grampian Orogeny. *Scott. J. Geol.* **12,** 271–292.
LEEDER, M. 1973. Sedimentology and palaeogeography of the Upper Old Red Sandstone in the Scottish Border Basin. *Scott. J. Geol.* **9,** 117–144.
MCGIVEN, A. 1967. Sedimentation and provenance of post-Valentian conglomerates up to and including the basal conglomerate of the Lower Old Red Sandstone in the southern part of the Midland Valley of Scotland. *Univ. Glasg. Ph.D. thesis.*
MACGREGOR, M. and MACGREGOR, A. G. 1948. The Midland Valley of Scotland. *Br. reg. Geol. Geol. Surv. and Mus.,* Lond.
MITCHELL, A. H. G. and MCKERROW, W. S. 1975. Analogous Evolution of the Burma Orogen and the Scottish Caledonides. *Bull. geol. Soc. Am.* **86,** 305–315.
MORTON, D. J. 1976. Lower Old Red Sandstone sedimentation in the North–West Midland Valley and North Argyll areas of Scotland. *Univ. Glasg. Ph.D. thesis.*
MURCHISON, R. I. 1859. On the succession of the older rocks in the northernmost counties of Scotland. *Q. J. geol. Soc. Lond.* **15,** 145–151.
MYKURA, W. 1960. The Lower Old Red Sandstone rocks of the Pentland Hills. *Bull. geol. Surv. G.B.* **16,** 131–155.
PATERSON, I. B. and HARRIS, A. L. 1969. Lower Old Red Sandstone ignimbrites from Dunkeld, Perthshire. *Rep. Inst. geol. Sci.* No. 69/7, 7 pp.
——, BROWNE, M. A. E. and ARMSTRONG, M. 1976. Upper Old Red Sandstone palaeogeography. *Scott. J. Geol.* **12,** 89–91.
PHILLIPS, W. E. A., STILLMAN, C. J. and MURPHY, T. 1976. A Caledonian plate tectonic model. *J. geol. Soc. Lond.* **132,** 579–609.
QUENNELL, A. M. 1958. The structural and geomorphic evolution of the Dead Sea Rift. *Q. J. geol. Soc. Lond.* **114,** 1–24.
QURESHI, I. R. 1970. A gravity survey of a region of the Highlands Boundary Fault in Scotland. *Q. J. geol. Soc. Lond.* **125,** 481–502.
READ, W. A. and JOHNSON, S. R. H. 1967. The sedimentology of sandstone formations within the Upper Old Red Sandstone and lowest Calciferous Sandstones Measures west of Stirling, Scotland. *Scott. J. Geol.* **3,** 242–269.
RICHARDSON, J. B. 1967. Some British Lower Devonian spore assemblages and their stratigraphic significance. *Rev. Palaeobot. Palynol.* **1,** 111–129.
ROBERTS, J. L. 1974. The evolution of the Glencoe cauldron. *Scott. J. Geol.* **10,** 269–282.

STEEL, R. J. 1974. Cornstone (fossil caliche)—its origin, stratigraphic, and sedimentological importance in the New Red Sandstone, western Scotland. *J. Geol.* **82,** 351–369.
WATERSTON, C. D. 1965. Old Red Sandstone. *In* G. Y. Craig (Editor). *Geology of Scotland,* 269–308. Oliver and Boyd, Edinburgh.
WESTOLL, T. S. 1951. The vertebrate-bearing strata of Scotland. *Rep. Int. geol. Congr.* **24** (11), 5–21.
WILSON, A. C. 1971. Lower Devonian sedimentation in the north–west Midland Valley of Scotland. *Univ. Glasg. Ph.D. thesis.*

B. J. Bluck,
Department of Geology,
University of Glasgow,
Glasgow G12 8QQ, Scotland.

# *Part 4*

# Cratonic igneous activity, mineralization, tectonism and sedimentation

# Igneous activity in a fractured craton: Carboniferous volcanism in northern Britain

*E. H. Francis*

Lasting instability in the Carboniferous North America–North Europe craton is shown by almost continuous, though sporadic, volcanism related to localised hinge-line tectonism expressed by variations in sediment thickness. Similarities shown by this relationship throughout northern Britain are more significant than local differences. Thus, though there are contrasts in the volume of eruptive rocks and in structural alignments on either side of the Southern Uplands massif, all areas share a common pattern of volcanic location, mechanism and magmatic type. Activity is invariably concentrated along folds, hinges or lines of renewed faulting, being excluded from thick basinal sequences of sediments; in each area it consists dominantly of early lavas followed by later pyroclastics; and magma composition is virtually restricted to inter-related alkaline basalts with minor local acid differentiates. The pattern was interrupted in the late Westphalian–early Stephanian, when a single episode of tholeiitic intrusion was associated with a system of E–W fractures which cut discordantly across the earlier structures. Alkaline volcanism then became re-established, but in association with new hinge-line controls which mainly followed northwesterly trends. The ages and distribution of the volcanic rocks are difficult to reconcile with the concept of mid-plate hot-spot partial melting of the Upper Mantle. Nor are they obviously related to subduction during the closure of the mid-European ocean and Tethys to the south or the Uralides suture to the east. These flanking orogenies nevertheless may be correlated tentatively with stress patterns set up within the craton, though up to late Westphalian times at least they were no more influential than eustatic controls in repeatedly reactivating early structural lineaments. The later Stephanian orogeny is represented in part by E–W fractures along which tholeiites were emplaced and as this suite of intrusions can be traced as far east as south Sweden a more direct correlation between orogeny and plate collision can be inferred.

## 1. Introduction

It is convenient for the purpose of this paper to define northern Britain as lying between the Highland Boundary fault and the Wales–Brabant ridge, for there is a unity in the pattern of Carboniferous volcanism throughout that region. The Highlands to the north include few sediments or igneous rocks of proven Carboniferous age, while southwest England is excluded because the composition and associations of the volcanic rocks there indicate emplacement in a separate environment. The volcanic rocks also form a unity corresponding to the now unfashionable concept of a petrographical province. For that reason the youngest rocks are included in this account, though they may be earliest Permian in age according to the current, still not wholly satisfactory, stratigraphical and radiometric age dating.

The pattern of the volcanism is relatively simple. It consists almost entirely of alkaline basalts; and though felsic differentiates were locally produced in Scotland during early phases of maximum volumetric effusion, basaltic varieties of alkaline magma remained available for eruption at one or more scattered centres throughout the period. Central complexes are unknown, but in addition to plugs related to necks, minor intrusions are common being of two kinds. One, closely linked both geographically and chemically to the extrusive rocks, consists almost exclusively of sills or sill-complexes up to 120 m thick. The other is represented by intrusions of quartz dolerite and tholeiite emplaced virtually by a single magmatic pulse which interrupted—but did not end—the long-lasting alkaline sequence.

The distribution of volcanic rocks, summarized in Figures 1 and 2, is derived from a literature too extensive to list here. Most Scottish occurrences, however, are listed in reviews by Francis (1965, 1967) which have since been augmented by Upton (1969), Read and Elliott (*in* Francis *et al.* 1970), Davies (1974), McAdam (1974), Livingstone and McKissock (1974), Whyte and MacDonald (1974) and Craig and Hall (1975), while Macdonald (1975) and Macdonald *et al.* (1977) have provided modern reviews of their petrochemistry. For England the bibliography listed by Francis (1970) has since been supplemented by Leeder (1974, 1976) and Poole (1977), while Irish occurrences are covered by Ashby (1939), Schultz and Sevastopulo (1965) and Strogen (1973a, b). In the sections which follow an attempt is made to reappraise some of this data in the light of recent and various hypotheses of Carboniferous global tectonics.

## 2. Volcanic stratigraphy

In his review of Scottish Carboniferous–Permian volcanicity, MacGregor (1948) outlined the following eruptive sequence: (i) effusion of alkaline basalts with local felsic differentiates during the Dinantian and Namurian, (ii) quiescence during the Westphalian and Stephanian, (iii) tholeiitic intrusion in the late Carboniferous or early Permian, (iv) renewed effusion of alkaline basalt in Permian times. He recognised that because the two magma types were derived by partial melting at different levels within the upper mantle, their alternation demanded a complex tectono-volcanic explanation.

An attempt by Francis (1965) to simplify this sequence by assuming all the alkaline volcanism to be pre-tholeiitic and pre-Stephanian has been invalidated; plants interbedded with the Mauchline Volcanic Group of Ayrshire, once thought to be Westphalian D in age (Mykura 1965) are now reassigned to earliest Permian (Wagner, *in* Smith *et al.* 1974 p. 23). The evidence now available suggests a volcanic history which was more, rather than less, complicated than that postulated by MacGregor. New sections, mainly from boreholes, have substantially extended the stratigraphical range of activity as expressed by lavas and tuffs interbedded with sediments, while radiometric (mainly whole rock K-Ar) dating has provided a better framework for assessing the age of the intrusions than had previously been

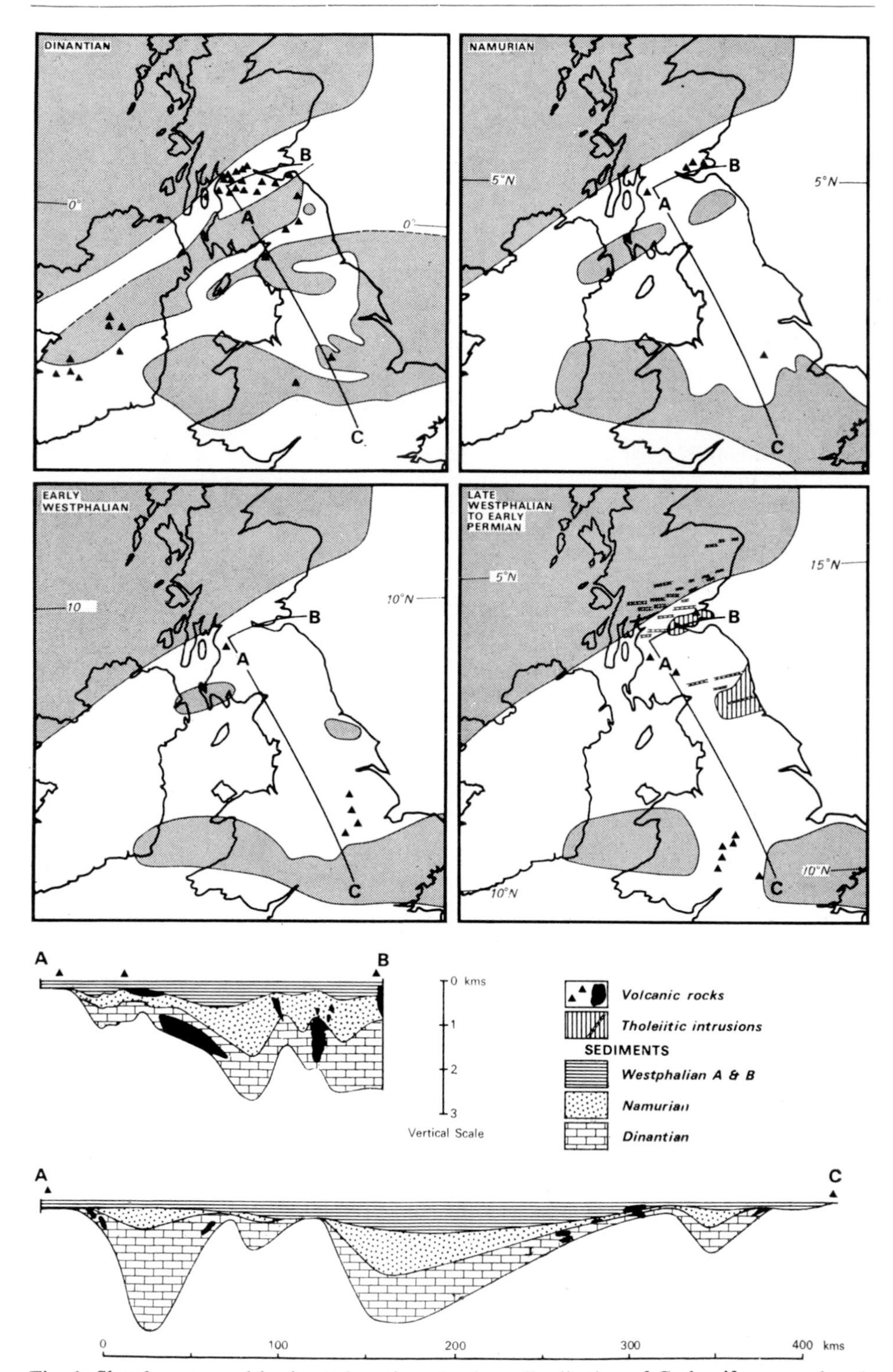


Fig. 1. Sketch maps and horizontal sections to show distribution of Carboniferous and early Permian volcanic rocks and their relationship to the main structural controls.

offered by cross-cutting relationships. The time scale of Francis and Woodland (1964) modified by Fitch *et al.* (1970) is followed below, assuming the constants obtaining then and at the time of writing.

## 2a. Extrusive rocks

Lavas and tuffs within known sedimentary successions are the most unequivocal indices relating activity to time and this evidence forms the basis of Figure 2. This shows volcanism to have been most widespread and voluminous in the Dinantian and that the bulk of the extrusive rocks of that age were lavas. In most areas activity ceased before the end of Dinantian times and was never renewed. In Ayrshire, Fife and the English Midlands, however, interbedded tuffs testify to continuing explosive activity from groups of small short-lived centres. Collectively they span most of Namurian times. Continuity was subsequently maintained in Ayrshire in the form of the Passage Group lavas of late Namurian to early Westphalian age. Simultaneously, and continuing to the end of Westphalian A times there was activity in the east Midlands, where the volcanic rocks first noted in the oilfields are being established in current coalfield exploration boreholes to consist of basaltic lavas and tuffs forming the most extensive Carboniferous volcanic field south of the Scottish Midland Valley. Also simultaneously, tuffs with rare lavas were accumulating off the coast of east Fife (and almost certainly onshore at Largo Law too), but there the activity continued almost to the end of Westphalian B times. Apart from the presence of shards in the Etruria Marls of the west Midlands, indicating slightly later (Westphalian B/C boundary) activity, the stratigraphic column provides no evidence of subsequent eruption until the Mauchline Volcanic Group was formed in Ayrshire and neighbouring counties in early Permian times. There are no recognisable extrusive rocks of Westphalian C, D and Stephanian ages, but activity nevertheless continued throughout the period, judging by radiometric dates obtained from alkaline intrusions.

## 2b. Intrusions

Apart from plugs and other minor intrusions sited at eruptive centres, the alkaline intrusions are represented almost entirely by dolerite sills or sill-complexes. They are rarely seen to be fed by, or otherwise related to dykes and they are closely associated geographically with the necks and extrusive rocks, being restricted in their distribution to the areas of Carboniferous sedimentation. The association is one of the lines of evidence which has led Francis (1968) to suppose that some of the ephemeral Silesian ash-cones were probably subaqueous, having been initiated by phreatic activity from the roofs of sills intruded into thick local accumulations of wet low density Carboniferous sediments.

MacGregor (1948) used cross-cutting evidence to establish that there were at least two ages of alkaline dolerite sill emplacement in Ayrshire and that the necks which fed the Permian Volcanic Group were younger than both. Radiometric analyses of such sills and related neck intrusions throughout the craton, however, give a range of ages which is almost as wide as that of the consanguineous extrusive rocks. They assume particular importance where they offer the only evidence of volcanism at times or in places where the stratigraphical column is inadequate.

Examples by De Souza (1974) include late Namurian, Westphalian and Stephanian ages for some of the sills and plugs cutting early Dinantian strata in East Lothian, and Namurian–Westphalian and Permian ages for plugs cutting early Dinantian rocks along the northern margin of the Northumberland trough. In east

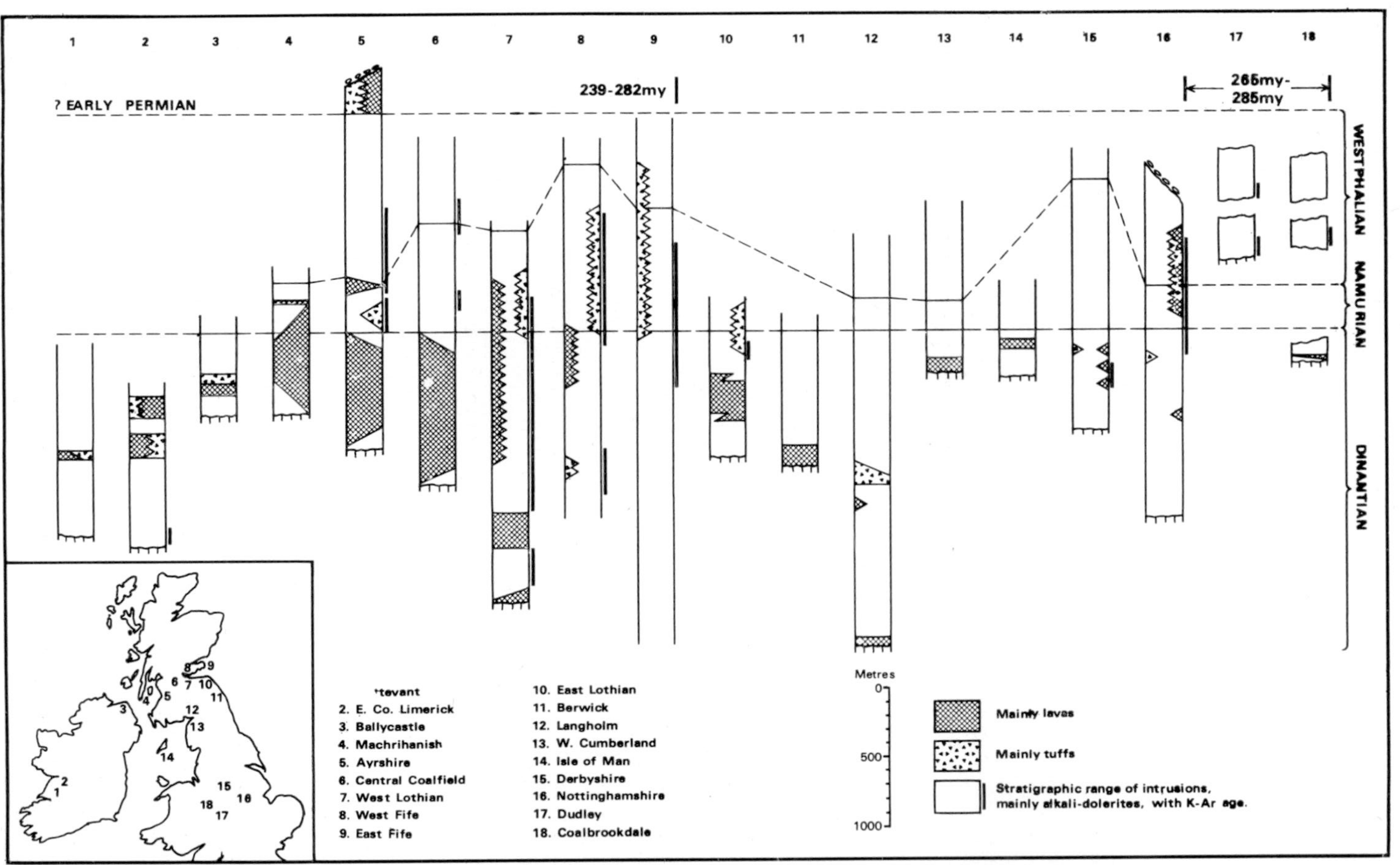


Fig. 2. Generalized vertical sections to show the stratigraphical distribution of extrusive rocks and the range of some of the principal alkaline intrusions.

Fig. 3. Sketch map showing quartz dolerite and tholeiite sills and dykes in northern Britain and southern Sweden.

Fife, Fitch *et al.* (1970) obtained Permian ages of 274±5 m.y. for plugs, and comparable ages have now been obtained by Forsyth and Rundle (1978) from plugs and sills in the same area, together with Stephanian and even late Westphalian ages. It is interesting to note that while on the one hand no extrusive equivalents have been preserved, on the other hand neither are there any Namurian radiometric dates from necks or sills, though tuffs are known to be intercalated in Namurian sequences in neighbouring outcrops and boreholes (Forsyth and Chisholm 1977). It seems likely that the originating Namurian necks are not exposed, or else they contain no crystalline rocks suitable for radiometric analysis.

In the west Midlands of England, Fitch *et al.* (1970) obtained Permian ages of 265±5 m.y. from one group of alkaline basaltic sills cutting Westphalian strata and Stephanian ages of 295 m.y. for another. The last agrees well with the age of 298±6 m.y., also believed to be Stephanian, obtained by Rundle (*in* Poole 1977) on the Wales–Brabant massif farther south. In the east Midlands, the 302 m.y. age given by the sill cutting Westphalian A sediments indicates a Westphalian A age corresponding to the adjacent extrusive rocks. There seems, however, to have been a slightly earlier (308±10 m.y.) Westphalian episode in the west Midlands both on the evidence of the age of the Barrow Hill Basalt (Fitch *et al.* 1970) and interbedded shards nearby (see above).

In summary, therefore, the alkaline intrusions appear to indicate that Westphalian volcanism was more widespread than has been established at outcrop or in drilling and that, moreover, activity continued both into late Westphalian and Stephanian though no extrusive equivalents have been preserved. Taking this in conjunction with the stratigraphy established by the known lavas and tuffs, therefore, there seem to have been few occasions during the Carboniferous or early Permian when there was not activity at one locality or other within the north British part of the craton. Moreover, the ages of about 295 m.y. indicate that there was no significant hiatus in the availability of alkaline basalt, though that time corresponds with the single episode of tholeiite intrusions.

The quartz-dolerite and tholeiite intrusions differ from the alkaline rocks as much in their dynamic setting as in their petrochemistry. Though the two main sill complexes, namely the Midland Valley Sill of southeast Scotland and the Whin Sill of northeast England (Figs 1 and 3) are comparable to the alkaline sills in thickness (up to 120 m) they have no extrusive equivalents. By way of further contrast, they are linked with, and fed by, dykes which are emplaced along E–W fractures extending beyond the areas of Carboniferous sedimentation. One of the problems posed by these rocks is that they are restricted in occurrence to an E–W belt about 200 km wide covering southern Scotland and northern England. It is here suggested that this belt continues eastward for at least 1000 km. Extrapolating beyond Denmark, where it is assumed to be concealed by the overlying cover of younger deposits, it can next be identified in southern Sweden where it forms a swarm of dolerite dykes trending WNW or NW (Fig. 3). The dykes, which are associated with sills, were only tentatively described as quartz dolerites by Hjelmqvist (1939) and assumed by him to be related to the British rocks, though it is clear from their petrographic description that they include such diagnostic features as micrographic intergrowth of quartz and feldspar. More conclusive, however, is that they are now determined as having an average radiometric age of 294±4 m.y. (Klingspor 1976).

## 3. Structural framework

The Carboniferous craton can be regarded as having taken form some time after the final closure of the Proto-Atlantic—an event marked, according to McKerrow and Ziegler (1972) by the middle Devonian Acadian orogeny which can be equated, in turn, with the middle ORS deformation in the Midland Valley (Dewey and Kidd 1974). This new global tectonic view accords with the demonstrations by Kennedy (1958) and George (1960) that the Midland Valley first became recognisable as a graben structure in mid-Devonian times. By the start of the Carboniferous, therefore, the region had become part of the southern marginal shelf of the North America–North Europe craton, flanked to the north by a landmass which included the Highlands. This landmass continued to supply sediments to the variably subsiding shallow seas and deltas which covered the shelf throughout the period. Because Figures 1 and 2 are designed to show where volcanic activity was located, they should not be allowed to obscure the fact that most parts of the craton contain thick complete sequences of Carboniferous sediments wholly uninterrupted by lava or tuff.

The line of the Proto-Atlantic suture has been variously equated with the Highland Boundary Fault, the Southern Uplands Fault and the northern edge of the Northumberland trough. In terms of Carboniferous events, however, the exact position is perhaps less important than the recognition that the suturing initiated a series of NE fractures including the Midland Valley itself, as well as rifts and strike-slips within it, and, farther south, the Northumberland trough—another graben according to Leeder (1974). In this northern part of the region the fractures seem likely to have begun to operate as syndepositional controls as early as lower ORS times as shown by such structures as the Strathmore basin (Armstrong and Paterson 1970). Their continuing effects on Carboniferous sedimentation is demonstrated by rapid thickness variations over such fractures on the Kerse Loch, Inchgotrick and Dusk Water faults in the western part of the Midland Valley (McLean 1966; Mykura 1967) and the syndepositional troughs and axes of uplift farther east (Fig. 1).

In the southern part of the region, too, sedimentation patterns indicate the influence of hinge-lines, though the northeasterly Caledonoid trends are replaced by an E–W Armorican grain (George 1969). It can now be argued that this change may reflect progressively increasing distance from the Proto-Atlantic suture, though an additional factor, as noted by Bott (1967) and Leeder (1976), is that these deep-seated fault controls tend to be determined by the presence of early Devonian calc-alkaline plutons that were themselves emplaced as legacies of the late Caledonian orogeny. They underlie parts of the Southern Uplands, Cumbrian, Alston, Askrigg and Wales–Brabant massifs (Fig. 5). By way of further, though minor contrast with the northern NE-trending structures, the fractures in the south tend to be fewer and mainly normal in character, while the syndepositional troughs tend to be fewer, more extensive and lower in amplitude. It is notable that the southernmost basins, the Widmerpool, Gainsborough and Eden "gulfs" of Kent (1966), have WNW alignments parallel to the Hercynian Front. Both these and the structures farther north waned in influence as the period progressed, however, and there is little evidence of their continued activity by late Westphalian times.

Although Simpson (1962) rejected the concept of Variscan orogenic phases, either or both of Stille's early Namurian (Sudetic) and mid to late Westphalian or Stephanian (Asturic) phases have been invoked in more recent Carboniferous plate tectonic models (Burrett 1972; McKerrow and Ziegler 1972; Laurent 1972; Johnson 1973; Dewey and Burke 1973). The existence of the earlier of these two phases is not supported by the depositional evidence. Differential movements along the hinge-lines did give rise to unconformities and discordances, but they are local. Away from the hinge-lines into the centres of the syndepositional troughs and basins, the cyclic sequences form a sedimentation pattern which can be entirely explained in terms of eustatic transgression and regression throughout Dinantian, Namurian and early to mid-Westphalian times (Calver 1971; Ramsbottom 1973).

The case for one or more later (?Asturic) phases is better founded. In the Mauchline basin the succession, though apparently concordant, lacks the whole of the Stephanian (Smith *et al.* 1974). There is no Stephanian elsewhere in Scotland and even if Poole's (1975) recognition of the stage in the English Midlands is accepted, it is only in the Warwickshire Coalfield that there is a seemingly uninterrupted sequence up into the Permian. Elsewhere there is angular unconformity between Westphalian and Permian, indicating orogeny, uplift and erosion.

The E–W fractures associated with tholeiite emplacement clearly represent one tectonic phase at about 295 m.y. It was followed by another phase represented by the NW–SE system of fractures which Mykura (1967) claims to have exerted control over Permian accumulations in the Mauchline basin. Hall (1974) extrapolates the system southeastwards through Sanquhar and Thornhill into the Vale of Eden and suggests that it began to replace the NE structures as a synsedimentary control during the later part of Carboniferous times. McLean (1978) includes the system in a "Clyde Belt", another component of which is the Stranraer–Liverpool Bay–Cheshire basin linear system. The NW systems and the E–W tholeiites and fractures seem to be mutually exclusive in the areas which they traverse.

## 4. Relation of structure to volcanism

A close relationship between the contemporaneously active northeasterly structures and the alkaline volcanism is exemplified by the lines of necks along the Campsie fault (Craig and Hall 1975), along the northern hinge-line of the Northumberland trough (Leeder 1976) and by the complex history of both volcanism and movement along the Ardross fault (Francis and Hopgood 1970). It is apparent, too, from the thickness variations in the Clyde Plateau Lavas (Johnstone 1965; Hall 1974) and from the fact that throughout the craton volcanic activity was located in areas shown by sedimentation patterns to have been on crests or immediate flanks (hinge-lines) of rigid blocks or axes of relative uplift (Fig. 1). In Ireland, too, volcanic centres are concentrated at, and aligned along the margin of the syn-depositional Shannon trough (Strogen, personal communication).

The volcano-tectonic relationship changed during the single phase of tholeiitic intrusion in the late Carboniferous. As noted above, the Midland Valley and Whin sills were themselves restricted to the Carboniferous sedimentary basins, but were fed by E–W dykes emplaced along normal faults which cut all the earlier NE structures and also extend beyond the basins. More significantly, this phase ended the long history of northeasterly controls. Although alkaline volcanism became re-established again related to syndepositional structures—the new northwesterly alignments indicate an entirely new stress system in latest Carboniferous and early Permian times. It is interesting to note that if McLean's (1978) Stranraer–Cheshire basin line is extrapolated southeastward it explains the later alkaline volcanism of the west Midlands as well as the penecontemporaneous breaching of the Wales–Brabant massif.

Anderson (1951) construed the E–W phase in terms of a brief period of tension following a long earlier period of compression and stress. This explanation accords well with some of the purely structural evidence. The earlier and continuing NE fractures, for instance, do indeed provide evidence of compressional stresses in having transcurrent components which contrast with the normal and presumably extensional faulting of the late Carboniferous tholeiitic phase. The stress sequence appears to fit less well, however, with the syndepositional troughs and basins, and these suggest to Leeder (1972) an originating tensional stress field oriented approximately NW–SE relative to present poles. It seems similarly difficult to reconcile maximum compression with maximum eruption of alkaline lavas. Macdonald *et al.* (1977) therefore propose that the production of the large volumes of Dinantian lavas during a first thermal cycle was facilitated by regional tensional stress and normal faulting. The decreased availability of alkaline magmas during a second (Namurian to Permian) thermal cycle is then ascribed to dominantly compressional stresses relieved by intermittent tension with continued normal faulting. These authors recognise, however, that this sequence does not explain the tholeiitic suite of intrusions. One difficulty is that on their reading, the E–W faulting expressed only one of a succession of phases of extensional stress, yet the fractures traverse pre-existing structures and also tapped magma sources lying at higher levels in the mantle than any of those which provided alkaline magma before or afterwards. A further problem is that the second thermal cycle must, on the evidence reviewed here, embrace not only the regime of NE structural controls preceding the tholeiites, but also the NW controls which followed them.

Until there is more positive evidence bearing on this problem it is here suggested that Anderson's interpretation of early compressional stresses might after all be reconciled with maximum alkaline lava effusion if those stresses were both intermittent and tangential to the hinge-controls—a concept which would be compatible with some of the plate tectonic models discussed on p. 290.

## 5. Petrochemistry

After the reviews of Tomkeieff (1937) and MacGregor (1937, 1948) little was published on Scottish Carboniferous petrochemistry until Macdonald's (1975) modernisation of the rock classification and nomenclature and his recognition that the Scottish Dinantian suite consists of mildly undersaturated alkaline basalts. These he assigns to several geographically distinct lineages and suggests that chemical variation within them is related to crustal fractionation within high-level magma chambers. Derivation is by melting of mantle peridotite at depths ranging from 50 to 100 km according to the degree of silica undersaturation, according to Macdonald *et al.* (1977) who follow Macdonald (1975) in inferring an increase in undersaturation with time. The few available analyses of Permian rocks certainly supports the view that they, at least, are more undersaturated, but that the change is progressive throughout Namurian and Westphalian cannot be said to have been conclusively established because there are not enough analysable specimens in the dominantly pyroclastic accumulations of these ages (the similar shortage of Namurian and Westphalian rocks suitable for radiometric age-dating is discussed on p. 284). It is unfortunate, too, that there is so little comparable detailed petrochemistry allied to age dating for the English rocks that it is not yet possible to construct an integrated petrogenetic model for the region as a whole.

Trace element data, though limited, accords well with the geodynamic evidence establishing the magmatism as being typically mid-plate and continental. Though Higazy (1952) is concerned with a single alkaline dolerite sill and Livingstone and McKissock (1974) with a single borehole lava sequence, their results agree with average values for continental alkaline dolerites given by Bell and Doyle (1971). Similarly, the K-Rb ratios together with the Sr and Zr values given by Dunham and Kaye (1965) for the Little Whin Sill, agree with continental tholeiite ratios quoted by Condie *et al.* (1970); they can be more confidently extrapolated to the other Scottish and English tholeiites in view of their probably consanguineous, if not actually comagmatic, relationship.

The demonstration by Dunham and Kaye (*op. cit.*) that the Little Whin Sill lies mainly between the fields of alkaline and tholeiitic basalts on alkali-silica plots leads them to claim that the sill represents the parental magma of the Great Whin Sill. This is supported by the work of Harrison (1968) who finds that the latter itself falls partly within Kuno's (1960) field of high alumina basalt and suggests that the parent magma was potentially olivine-bearing, becoming oversaturated and differentiated. Transitional plots are also indicated by a late Namurian or early Westphalian sill in Derbyshire (Harrison, *In* Stevenson and Gaunt 1971) and the late Westphalian or Stephanian Barrow Hill basalt in the west Midlands (Francis 1970). The question raised by such transitional types is whether there may be a generic link between the alkaline and tholeiitic rocks of the region. For the bulk of the tholeiites, i.e. those having a uniform 295 m.y. age, the balance of the evidence is to the contrary. Their tectonic setting and the lack of extrusive equivalents (or any other markedly differentiated variants) suggest derivation by partial melting at a higher level (perhaps 15 to 35 km depth according to Green *et al.* 1967) and under a different stress regime than the alkaline rocks.

## 6. Plate tectonic models

Up to eight lithospheric plates, four continental and four oceanic, are identified in the literature dealing with Carboniferous plate tectonics over the past five years.

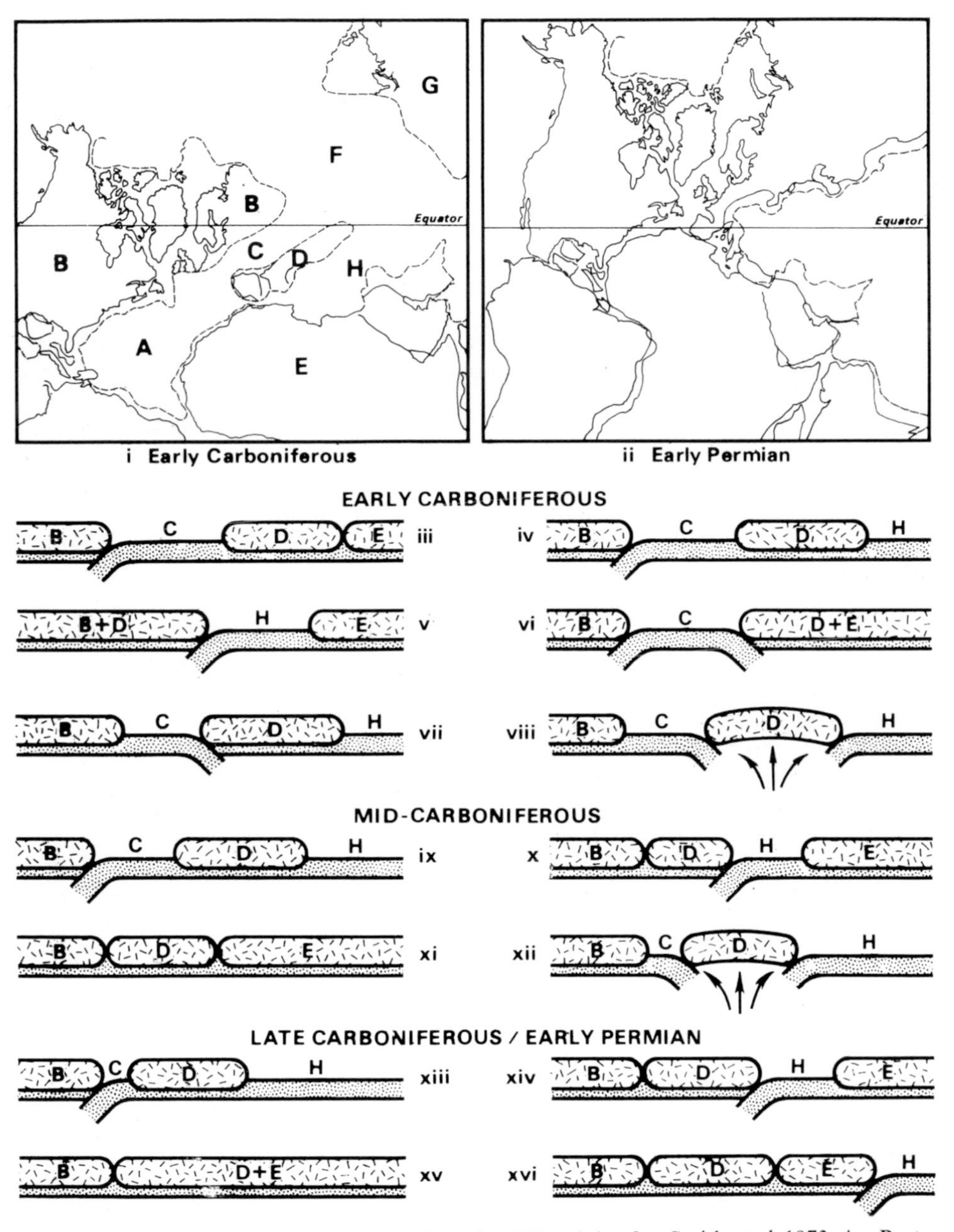


Fig. 4. Palaeogeographic plate reconstructions: i and ii mainly after Smith *et al.* 1973. A = Proto South Atlantic or Phoibic Ocean: B = North America–North Europe continent: C = Mid-European or Rheic Ocean; D = South Europe Continent or microplate assemblage: E = African continent: F = Pleionic Ocean: G = Siberian Platform. The plate tectonic models shown in iii–xvi are included in various combinations by Burrett 1972; Cosgrove 1972; Floyd 1972; Nicolas 1972; Johnson 1973; Riding 1974; Dewey and Burke 1974; Badham and Halls 1975; Leeder 1976; Lorenz 1976; Lorenz and Nicholls 1976.

They are shown in Figure 4 together with some of the principal permutations of uncoupling, subduction and suturing which have been postulated.

Evidence has already been advanced on p. 286 to indicate that plate-collision, if any, was not reflected in the Dinantian, Namurian or early to mid-Westphalian sequences of the main sedimentary basins. It cannot be assumed with certainty that the localised discordances and unconformities at the sedimentation hinge-lines were equally unrelated to plate movement. It is impossible, however, to make specific correlation between any phase of collision on the one hand with local structure or local volcanic activity on the other. If there was any other influence exerted by collision or suturing (e.g. of the Rheic Ocean) much before Stephanian times it can only have been in the periodic exertion of N–S compression. The alignment of the southernmost troughs of the craton tends to support this view. So does the general northward drift implied by the geomagnetic palaeogeography and the note by Lorenz (1976) that the forces driving the south Europe microplates against adjacent plates was not very great. Lorenz' (*op. cit.*) evidence that the south European microplates were rotated during this period may further explain an alternation of extensional stress with compression.

The late Westphalian–Stephanian orogeny is more obviously related to suturing regardless of whether or not it marked the final closure of the Rheic Ocean. Although these forces must have been essentially compressional there is need to accommodate the initial E–W tensional faulting associated with the tholeiitic intrusion. As the belt in southern Sweden swings toward the southeast, parallel to the Tornquist Line, two explanations for temporary crustal extension offer themselves. One is implied by the rotation of the Lorenz model mentioned above, especially as the Tornquist Line is an integral part of that model. The other is the possible effect of tangential partial collision involved in the Uralides suturing farther to the east, a closure already in operation though not completed until Permian or Triassic times according to Hamilton (1970). Neither Rheic nor Uralides suturing provide obvious explanations for the northwesterly structures which were initiated, presumably by extension, in the late Carboniferous and continued into the Permian. It may be more feasible to correlate these with the formation of the pre-Triassic Rockall trough (McLean 1978).

Another critical question raised by the plate tectonic models and shown in Figure 4 is whether there was a subduction zone extending northwestwards beneath North Britain and, if so, how might it have been related to the volcanism. Models which do postulate northwesterly subduction fall within three groups, all correlating it to some degree with the late Devonian and early Carboniferous greenstones of southwest England. Only one of these (Johnson 1973) extrapolates the subduction farther north to include the volcanism discussed here. Johnson postulates northwesterly consumption of a mid-European ocean (the Rheic Ocean of McKerrow and Ziegler 1972) which, by inference, continued throughout the Carboniferous into early Permian times. The argument is not in itself invalidated by his further assumption—subsequently contradicted by Floyd (1972) on geochemical grounds—that the southwest England greenstones were part of an oceanic ophiolite suite.

A second group, including Burrett (1972), Riding (1974) and Badham and Halls (1975) also assume a northwesterly dipping subduction, but believe it to relate only to the southwestern greenstones so that the process ended early in Carboniferous times. Both Riding and Badham and Halls agree that there was later Carboniferous subduction of a Tethyan plate (Fig. 4) and this they have in common with the third group represented by Floyd (1972), Cosgrove (1972) and Nicolas (1972) who do not recognise a Rheic Ocean at any time during the Carboniferous, but envisage instead a northwestward subduction of Tethys throughout the period.

What militates most against Johnson's correlation between north British volcanism and Rheic subduction is the nature of the volcanic rocks and their distribution both in time and space. Moreover, data on their geochemistry, admittedly sparse, gives no indication of the sort of linear variation which might be expected if the rocks were generated at various levels of a NW-dipping Benioff Zone. Instead, for instance, the Permian basalts of southwest England are richer in potash than their Scottish counterparts, rather than poorer.

The models here preferred, therefore, are those of Laurent (1972), Lorenz and Nicholls (1976) and Lorenz (1976), together with those elements from the proposals by Badham and Halls (1975) and Burrett (1972) which postulate southeasterly subduction of a Rheic Ocean beneath a southern European continental plate or aggregate of microplates. This preference is strengthened by the apparently close relationship between subduction and the south European intrusive domes, acid to intermediate volcanism (including extensive ignimbrite flows) and intensive hydrothermal alteration, all of which, as emphasised by Lorenz and Nicholls, contrast starkly with the more alkaline and titanium-rich rocks of North Britain.

Much of the current literature on mid-plate volcanic activity ascribes magmatism to mantle plumes. These, however, are as difficult to reconcile with the evidence presented here as is subduction. As noted by Bailey (1977) plumes have fixed locations in the mantle. Activity generated by them should therefore change location with time to match the movement of the overriding continental plates. It can be seen from Figure 4, however, that the distribution of Carboniferous volcanism does not fit with such an interpretation. Even if the large number of Dinantian centres could be reconciled with plumes there is no movement of volcanic focus with time. On the contrary, activity repeatedly renewed throughout the period at certain fracture-controlled centres even though the palaeogeographic reconstruction of Smith *et al.* (1973) shows the region to have drifted through some 15° of latitude across the equator between early Carboniferous and early Permian time.

An alternative to plumes which seems to fit better with the evidence of drift is the membrane tectonics theory advanced by Turcotte and Oxburgh (1973). Conformity of the plate to slight changes in the earth's curvature is assumed thereby to engender either tensional or compressional stresses in various parts of the plate, as well as corresponding thermal stresses. The analogy cannot be carried too far because the inhomogeneous composition of the Carboniferous craton and its irregular shape do not match the necessarily idealized model advanced by Turcotte and Oxburgh (1973). In principle, however, the known amount of latitudinal drift could adequately account for the complex history of tension, compression and magma propagation indicated by the rocks.

The suggestion by Russell (1971) that the Atlantic may have begun to open as early as in Carboniferous times has little support from the supposed N–S fracturing quoted and even less from critical discussion (Leeder 1972). The evidence of NW structural controls during the late Carboniferous and early Permian as quoted by Hall (1974) is less equivocal, though his belief that they too are related to the initiation of Atlantic rifting requires further substantiation (see also McLean 1978). Even if they are so related, thereby explaining the late alkaline volcanism along the Stranraer–Cheshire–West Midlands line and the Mauchline–Sanquhar–Thornhill line, it is not clear what such fracture have to do with penecontemporaneous volcanism in Fife.

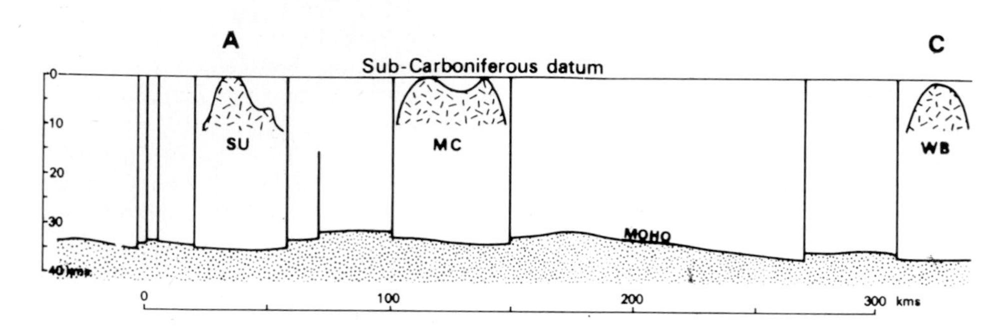


Fig. 5. Diagrammatic section approximately along the line A–C of Fig. 1 with the base of the Carboniferous drawn to horizontal datum and the Moho trace derived by corresponding adjustment to the section given by Bamford *et al.* 1976.

## 7. Discussions and conclusions

During the Carboniferous the region formed part of a continental plate—a craton of a kind, though not perhaps as stable as the strictest definition of that term might demand. The instability was closely related to, if not caused by, an inhomogeneity imparted during the Devonian plate-suturing when the deep-seated NE fracturing in the north was accompanied by the emplacement of granitic plutons throughout the region (Fig. 5). Although the hinge-lines correlated with these earlier structures gave rise to local unconformities and discordances, the regional pattern of sedimentation up to late Westphalian or Stephanian times at least is explained in terms of eustatic transgression and regression rather than by externally imposed orogeny.

This is opposed to the proposition that either the fracturing or the related volcanism were initiated or reactivated by subduction or by plate collisions identifiable as orogenic phases. Indeed, even if it were possible to recognise the so-called Sudetan (pre-Namurian) or Erzgebirgian (pre-Westphalian) orogenies in the sedimentary sequences there do not seem to have been penecontemporaneous volcanic outpourings to match them. Having rejected plumes, subduction and collision as mechanical agents, misgivings are also here expressed as to Bott's (1971, 1976) theory of hot creep in lower continental crustal levels, for though it explains extensional fracturing in the brittle upper levels it seems specifically to rule out a continuity of fracturing at depth and to deny access to magma rising from the mantle. The part played by the fractures in such ascent through continental cratons is well established by modern analogy, though there is little sign of arching or updoming (Baker and Wohlenberg 1971; Bailey 1974) in the British Carboniferous sequences.

A single alternative model would be to assume a combination of regional epierogenic uplift with local isostatic uplift engendered (Bott 1967) by the granitic plutons. This, added to the northward drift of the craton as indicated by palaeomagnetism may have been sufficient to have caused rifting, lithospheric thinning and increased thermal gradients in the hinge zones to give repeated interaction between fracture and magmatism. The application of the membrane tectonic principle (Turcotte and Oxburgh 1973) fits well as an additional rather than as a necessary factor.

If plate collisions surrounding the craton had any influence on the internal volcanism and faulting only two lines of evidence seem to be relevant. One is the WNW alignment of the southernmost deposition and troughs of the region which may reflect the gradual closure of a Rheic Ocean to the south. The other is the tholeiitic intrusion episode at 295 m.y., approximating perhaps to the start of an Asturian orogeny. The E–W normal fractures which presumably tapped magma at relatively high levels in the mantle, may perhaps have been formed in response to oblique collision. If so it is not clear whether it was with one of the rotated south European microplates or was even related to closure of the more distant Uralides suture to the east. Oblique collision may also explain the alternation of tensional and compressional stresses indicated by volcanic and tectonic relationships throughout the period.

*Acknowledgments*. This work was completed while the author was on the staff of I.G.S. and permission to publish has been given by the Director.

## References

ANDERSON, E. M. 1951. *The dynamics of faulting* (2nd Edit.). Oliver and Boyd, Edinburgh.
ARMSTRONG, M. and PATERSON, I. B. 1970. The Lower Old Red Sandstone of the Strathmore region. *Inst. Geol. Sci. Rep.* **70/12,** 23 pp.
ASHBY, D. F. 1939. The geological succession and petrology of the Lower Carboniferous volcanic area of Co. Limerick. *Proc. Geol. Assoc.* **50,** 324–330.
BADHAM, J. P. N. and HALLS, C. 1975. Microplate tectonics, oblique collisions and the evolution of the Hercynian orogenic systems. *Geology* **3,** 373–376.
BAILEY, D. K. 1974. Continental rifting and alkaline magmatism, *In* Sørenson, H. (Editor). *The Alkaline Rocks*, 148–159. Wiley, London.
—— 1977. Lithosphere control of continental rift magmatism. *J. geol. Soc. Lond.* **133,** 103–106.
BAKER, B. H. and WOHLENBERG, J. 1971. Structure and evolution of the Kenya Rift Valley. *Nature, Lond.* **229,** 538–542.
BAMFORD, D., FABER, S., JACOB, B., KAMINSKI, W., NUNN, K. PRODEHL, C., FUCHS, K., KING, R. and WILLMORE, P. 1976. A lithospheric seismic profile in Britain—I. Preliminary results. *Geophys. J. R. astron. Soc.* **44,** 145–160.
BELL, K. and DOYLE, R. J. 1971. K-Rb relationships in some continental alkaline rocks associated with the East African Rift System. *Geochem. Cosmochim. Acta.* **35,** 903–916.
BOTT, M. H. P. 1967. Geophysical investigations of the northern Pennine basement rocks. *Proc. Yorks. geol. Soc.* **36,** 139–168.
—— 1971. Evolution of young continental margins and formation of shelf basins. *Tectonophysics* **11,** 319–327.
—— 1976. Formation of sedimentary basins of graben type by extension of continental crust. *Tectonophysics* **36,** 77–86.

BURRETT, C. F. 1972. Plate tectonics and the Hercynian orogeny. *Nature, Lond.* **239,** 155–156.
CALVER, M. A. 1971. *Westphalian marine faunas in northern England and adjoining areas.* Ph.D. thesis, Univ. of Reading.
CONDIE, J. C., MACKE, J. E. and REIMER, T. O. 1970. Petrology and geochemistry of early Precambrian greywackes from the King Tree Group, South Africa. *Bull. geol. Soc. Am.* **81,** 2758–2776.
COSGROVE, M. 1972. Geochemistry of the potassium-rich Permian volcanic rocks of Devonshire, England. *Contr. Mineral. Petrol.* **36,** 155–170.
CRAIG, P. M. and HALL, I. H. S. 1975. The Lower Carboniferous rocks of the Campsie-Kilpatrick area. *Scott. J. Geol.* **11,** 171–174.
DAVIES, A. 1974. The Lower Carboniferous (Dinantian) Sequence at Spilmersford, East Lothian, Scotland. *Bull. geol. Surv. Gt Brit.* **45,** 1–38.
DE SOUZA, H. A. F. 1974. *Potassium-argon ages of Carboniferous igneous rocks from East Lothian and the south of Scotland.* M.Sc. thesis, Univ. of Leeds.
DEWEY, J. F. and BURKE, K. C. 1973. Tibetan, Variscan and PreCambrian basement reactivation: products of continental collision. *J. Geol.* **81,** 683–692.
—— and KIDD, W. S. F. 1974. Continental collisions in the Appalachian-Caledonian orogenic belt: variations related to complete and incomplete suturing. *Geology* **2,** 543–546.
DUNHAM, A. C. and KAYE, M. J. 1965. The petrology of the Little Whin Sill, County Durham. *Proc. Yorks. geol. Soc.* **35,** 229–276.
FITCH, F. J., MILLER, J. A. and WILLIAMS, S. C. 1970. Isotopic ages of British Carboniferous rocks. *C. R. 6me Congr. Int. Stratigr. géol. Carbonif.,* Sheffield 1967 **2,** 771–789.
FLOYD, P. A. 1972. Geochemistry, origin and tectonic environment of the basic and acid rocks of Cornubia, England. *Proc. geol. Assoc.* **83,** 385–404.
FORSYTH, I. H. and CHISHOLM, J. I. 1977. The geology of East Fife. *Mem. geol. Surv. U.K.*
—— and RUNDLE, C. C. 1978. The age of the volcanic and hypabyssal rocks of east Fife. *Bull. geol. Surv. Gt Brit.* **60.**
FRANCIS, E. H. 1965. Carboniferous and Carboniferous-Permian Igneous Rocks, *In* Craig, G. Y. (Editor). *The Geology of Scotland.* Oliver and Boyd, Edinburgh and London, 309–357.
—— 1967. Review of Carboniferous-Permian volcanicity in Scotland. *Geol. Runds* **57,** 219–246.
—— 1968. Effect of sedimentation on volcanic processes, including neck-sill relationships, in the British Carboniferous. *Proc. 23rd Int. geol. Congr. Prague,* 1968 **2,** 163–174.
—— 1970. Review of Carboniferous volcanism in England and Wales. *J. Earth Sci.* (Leeds.) **8,** 41–56.
——, FORSYTH, I. H., READ, W. A. and ARMSTRONG, M. 1970. The geology of the Stirling District. *Mem. geol. Surv. U.K.*
—— and HOPGOOD, A. M. 1970. Volcanism and the Ardross Fault, Fife, Scotland. *Scott. J. Geol.* **6,** 162–185.
—— and WOODLAND, A. W. 1964. The Carboniferous Period. *In* Harland, W. B. (Editor). *The Phanerozoic Time-Scale:* a Symposium. *Q. J. geol. Soc. Lond.* **120S,** 221–232.
GEORGE, T. N. 1960. The stratigraphical evolution of the Midland Valley. *Trans. geol. Soc. Glasg.* **24,** 32–107.
—— 1969. British Dinantian Stratigraphy. *C. R. 6me Congr. Int. Stratigr. géol. Carbonif.,* Sheffield 1967 **1,** 193–218.
GREEN, T. H. GREEN, D. H. and RINGWOOD, A. E. 1967. Origin of high-alumina basalts and their relationship to quartz tholeiites and alkali basalts. *Earth Planet. Sci. Lett.* **2,** 41–51.
HALL, J. 1974. A seismic reflection survey of the Clyde Plateau Lavas in north Ayrshire and Renfrewshire. *Scott. J. Geol.* **9,** 255–279.
HAMILTON, W. 1970. The Uralides and the motion of the Russian and Siberian platforms. *Geol. Soc. Amer. Bull.* **81,** 2553–2576.
HARRISON, R. K. 1968, Petrology of the Little and Great Whin Sills in the Woodland Borehole, Co. Durham. *Bull. geol. Surv. Gt Brit.* **28,** 38–54.
HIGAZY, R. A. 1952. The distribution and significance of the trace elements in the Braefoot Outer Sill, Fife. *Trans. Edinb. geol. Soc.* **15,** 150–186.
HJELMQVIST, S. 1939. Some post-Silurian dykes in Scania and problems suggested by them. *Sver. Geol. Unders.* C. 430.
JOHNSON, G. A. L. 1973. Closing of the Carboniferous sea in western Europe. *In* Tarling, D. H. and Runcorn, S. K. (Editors). *Implications of Continental Drift to the Earth Sciences.* Academic Press, London **2,** 843–850.
JOHNSTONE, G. S. 1965. The volcanic rocks of the Misty Law-Knockside Hills district, Renfrewshire. *Bull. geol. Surv. Gt Brit.* 53–64.
KENNEDY, W. Q. 1958. The tectonic evolution of the Midland Valley of Scotland. *Trans. geol. Soc. Glasg.* **23,** 106–133.
KENT, P. E. 1966. The structure of the concealed Carboniferous rocks of north-eastern England. *Proc. Yorks. geol. Soc.* **35,** 323–352.

KLINGSPOR, I. 1976. Radiometric age-determination of basalts, dolerites and related syenite in Skåne, southern Sweden. *Geol. Fören. Stockh. Fordh.* **98,** 195–215.

KUNO, H. 1960. High-alumina basalt. *J. Petrol.* **1,** 121–145.

LAURENT, R. 1972. The Hercynides of south Europe, a Model. *24th Int. geol. Congr.* **3,** 363–370.

LEEDER, M. R. 1972. North-south geofractures in Scotland and Ireland. *Scott. J. Geol.* **8,** 283–291.

—— 1974. The origin of the Northumberland Basin. *Scott. J. Geol.* **10,** 283–296.

—— 1976. Sedimentary facies and the origins of basin subsidence along the northern margin of the supposed Hercynian Ocean. *Tectonophysics* **36,** 167–179.

LIVINGSTONE, A. and MCKISSOCK, G. M. 1974. The geochemistry of the igneous rocks in the Lower Carboniferous (Dinantian) sequence at Spilmersford, East Lothian, Scotland. *Bull. geol. Surv. Gt Brit.* **45,** 47–61.

LORENZ, V. 1976. Formation of Hercynian subplates, possible causes and consequences. *Nature, Lond.* **262,** 374–377.

—— and NICHOLLS, I. A. 1976. The Permo–Carboniferous basin and range province of Europe. An application of plate tectonics. *In* Falke, H. (Editor). *Continental Permian in Central, West and South Europe.* NATO ASI Series, C22, 313–342. Reidel, Dordrecht, Holland.

MCADAM, A. D. 1974. The petrography of the igneous rocks in the Lower Carboniferous (Dinantian) at Spilmersford, East Lothian, Scotland. *Bull. geol. Surv. Gt Brit.* **45,** 39–46.

MACDONALD, R. 1975. Petrochemistry of the early Carboniferous (Dinantian) lavas of Scotland. *Scott. J. Geol.* **11,** 269–314.

——, THOMAS, J. E. and RIZZELLO, S. A. 1977. Variations in basalt chemistry with time in the Midland Valley province during the Carboniferous and Permian. *Scott. J. Geol.* **13.**

MACGREGOR, A. G. 1937. The Carboniferous and Permian volcanoes of Scotland. *Bull. volcan.* **27,** 1, 41–58.

—— 1948. Problems of Carboniferous–Permian volcanicity in Scotland. *Q. J. geol. Soc. Lond.* **104,** 133–153.

MCKERROW, W. S. and ZIEGLER, A. M. 1972. Palaeozoic oceans. *Nature Phys. Sci. Lond.* **240,** 92–94.

MCLEAN, A. C. 1966. A gravity survey in Ayrshire and its geological interpretation. *Trans. R. Soc. Edinb.* **66,** 239–265.

—— 1978. Evolution of fault–controlled ensialic basins in northwestern Britain. *In* Bowes, D. R. and Leake, B. E. (Editors). *Crustal evolution in northwestern Britain and adjacent regions,* 325-346 *Geol. J. Spec. Issue* No. 10.

MYKURA, W. 1965. The age of the lower part of the New Red Sandstone of south-west Scotland. *Scott. J. Geol.* **1,** 9–18.

—— 1967. The Upper Carboniferous rocks of south-west Ayrshire. *Bull. geol. Surv. Gt Brit.* **26,** 23–98.

NICOLAS, A. 1972. Was the Hercynian orogenic belt of Europe the Andean type? *Nature, Lond.* **236,** 221–223.

POOLE, E. G. 1975. Correlation of the Upper Coal Measures of Central England and adjoining areas and their relationship to the Stephanian of the Continent. *Bull. Soc. belge Géol.* **84,** 57–66.

—— 1977. Stratigraphy of the Steeple Aston Borehole, Oxfordshire. *Bull. geol. Surv. Gt Brit.* **57,** 1–85.

RAMSBOTTOM, W. H. C. 1973. Transgressions and regressions in the Dinantian: a new synthesis of British Dinantian stratigraphy. *Proc. Yorks. geol. Soc.* **39,** 567–608.

RIDING, R. 1974. Model of the Hercynian fold belt. *Earth Planet Sci. Lett.* **24,** 125–135.

RUSSELL, M. J. 1971. North–south geofractures in Scotland and Ireland. *Scott. J. Geol.* **8,** 75–84

SCHULTZ, R. W. and SEVASTOPULO, G. D. 1965. Lower Carboniferous rocks near Tulla, Co. Clare. *Scient. Proc. R. Dubl. Soc.* **2,** 153–162.

SIMPSON, S. 1962. Variscan orogenic phases, *In* Coe, K. (Editor). *Some aspects of Variscan fold belts.* Manchester University Press, 65–73.

SMITH, A. G., BRIDEN, J. C. and DREWRY, G. E. 1973. Phanerozoic world maps. *In* Hughes, N. P. (Editor). *Organisms and Continents through time. Palaeontol. Spec. Pap.* **12,** 1–42.

SMITH, D. B., BRUNSTROM, P. G. W., MANNING, R. I., SIMPSON, S. and SHOTTON, F. W. 1974. *A correlation of Permian rocks in the British Isles.* Spec. Rep. *Geol. Soc. Lond.* **5,** 1–45.

STEVENSON, I. P. and GAUNT, G. D. 1971. Geology of the country around Chapel en le Frith *Mem. geol. Surv. U.K.*

STROGEN, P. 1973a. Brecciated lavas from County Limerick and their significance. *Geol. Mag* **110,** 351–364.

—— 1973b. The volcanic rocks of Carrigogunnel area, Co. Limerick. *Scient. Proc. R. Dubl. Soc.* **5,** 1–26.

TOMKEIEFF, S. I. 1937. Petrochemistry of Scottish Carboniferous-Permian igneous rocks. *Bull. volcan.* **2,** 59–87.

TURCOTTE, D. L. and OXBURGH, E. R. 1973. Mid-plate tectonics. *Nature, Lond.* **244,** 337–339.

UPTON, B. G. J. 1969. *Field excursion guide to the Carboniferous volcanic rocks of the Midland Valley of Scotland.* Geological Society of Edinburgh.

WHYTE, F. and MACDONALD, J. G. 1974. Lower Carboniferous vulcanicity in the northern part of the Clyde Plateau. *Scott. J. Geol.* **10,** 187–198.

E. Howel Francis,
Department of Earth Sciences,
University of Leeds,
Leeds LS2 9JT England.

# Mineralization in a fractured craton

*M. J. Russell*

The largest base-metal deposits in Ireland were formed early in early Carboniferous times (Courceyan). They are considered to be by-products of convective systems that involved Carboniferous seawater, then occupying pore spaces in the upper crust, and driven by heat left over from the Caledonian orogeny with the metals leached mainly from Caledonian geosynclinal rocks. These hydrothermal systems may have been engendered as permeability increased in response to east–west relative tension. The conduits were probably sited at fracture intersections, in some cases where north–south geofractures were crossed by easterly- or northeasterly-trending faults. Towards the close of the Carboniferous, tension finally caused the lithosphere to fail mainly along the Caledonian grain. Hot mantle diapirs rose into 'breaks' in the lithosphere and extensive partial melting produced basaltic magma, often tholeiitic, that intruded into the crust in Britain, the North Sea, Scandinavia, and Newfoundland. Phlogopite broke down under these conditions and fluorine was released into silicate melts and concentrated in residual magmas. Fluorite deposits were formed in late crystallizing intrusions or in related vein deposits in Norway, Sweden and England. At the same time, in the early Permian, the lithosphere was probably separating by ocean-floor spreading between Greenland and northwest Europe.

## 1. Introduction

The study of ore deposits of hydrothermal origin can provide information about the thermal and tectonic state of the crust and/or upper mantle at the time of ore formation. Earlier papers in this volume show how the crust developed on the North Atlantic margins up to the formation of the Caledonian late orogenic basins. Here it is shown that the geothermal gradient remained high in the Carboniferous and the lithosphere was therefore relatively thin. This was taken advantage of by tensional stresses that probably caused Rockall trough, the Faeroe–Shetland channel and the east Norwegian Sea (the Proto-North Atlantic) to open by ocean-floor spreading in early Permian times.

## 2. Early Carboniferous mineralization

### 2a. Irish ore deposits

The post-Caledonian ore deposits in Ireland are composed of ±lead±zinc± copper±barite and occupy Carboniferous limestones of Courceyan (Tournaisian) age. The largest deposits in central Ireland are clearly pre-Arundian (pre-Viséan) and in some cases demonstrably sedimentary or syn-diagenetic (Russell 1974, 1975; Anon. 1975; Coomer and Robinson 1976). The metals and some of the sulphur were probably derived from underlying lower Palaeozoic geosynclinal sediments (Russell 1968). The thermal energy driving the hydrothermal systems (presumably involving Carboniferous sea water) may have been heat left over from the Caledonian orogeny. Leake (1978) has pointed out that the late Caledonian granites, which are of the order of 415 m.y. old, occur around the margins of Ireland. They would have effectively removed heat from the lower crust whereas the apparent lack of granite in most of the central Irish plain implies that heat was conducted upwards at a relatively slow rate to be used to drive the Courceyan hydrothermal systems. The fact that the ore deposits are approximately the same age (*c.* 355 m.y.) suggests a tectonic triggering to the systems. A possibility explored previously (Russell 1968, 1973) is that the underlying controls to the mineralization are N–S geofractures formed as the first response to the tension that led to the eventual separation of the plates in this region (Fig. 1). The largest deposit, Navan, was discovered in 1970 on the southerly extrapolation of the trace of the N–S trending Kingscourt basin (see Russell 1968, 1972). The local structural controls are reactivated E–W to northeasterly-trending faults. Movement on these faults was concomitant with mineralization (Russell 1975) and may have been due to adjustments to the north of the southeasterly-subducting lithosphere (Russell 1976).

### 2b. Irish-type ore deposits in Scotland

Lead-zinc deposits in Scotland occur in veins in Caledonian rocks and consequently their age is uncertain. However, the two largest, Leadhills and Tyndrum, may have been generated in the same way and at the same time as the Irish ores. K-Ar dating of wall rock clays (Ineson and Mitchell 1974) gives a Carboniferous age for both deposits although these could be minimum ages (Mitchell and Krouse 1971).

### 2c. Irish-type ore deposits in Nova Scotia

Sulphide deposits in Lower Carboniferous rocks in Nova Scotia are very similar to the Irish ores (Russell 1968). Here again the underlying rocks are geosynclinal and are of early Palaeozoic age. Although the two largest lead-zinc deposits, Walton and Gays River, have been described as epigenetic (MacEachern and Hannon 1974; Boyle *et al.* 1976) their fine grained and intimately mixed mineralogies are strongly suggestive of syndiagenetic deposition.

## 3. Further evidence for tension and a high geothermal gradient in Carboniferous times

Intrusion and extrusion of alkali basalt was a feature of the Carboniferous period in Scotland (MacDonald 1975; Francis 1978), England, Ireland (Strogen 1977), Nova Scotia and New Brunswick (Rose *et al.* 1970). MacDonald (1975) has shown that

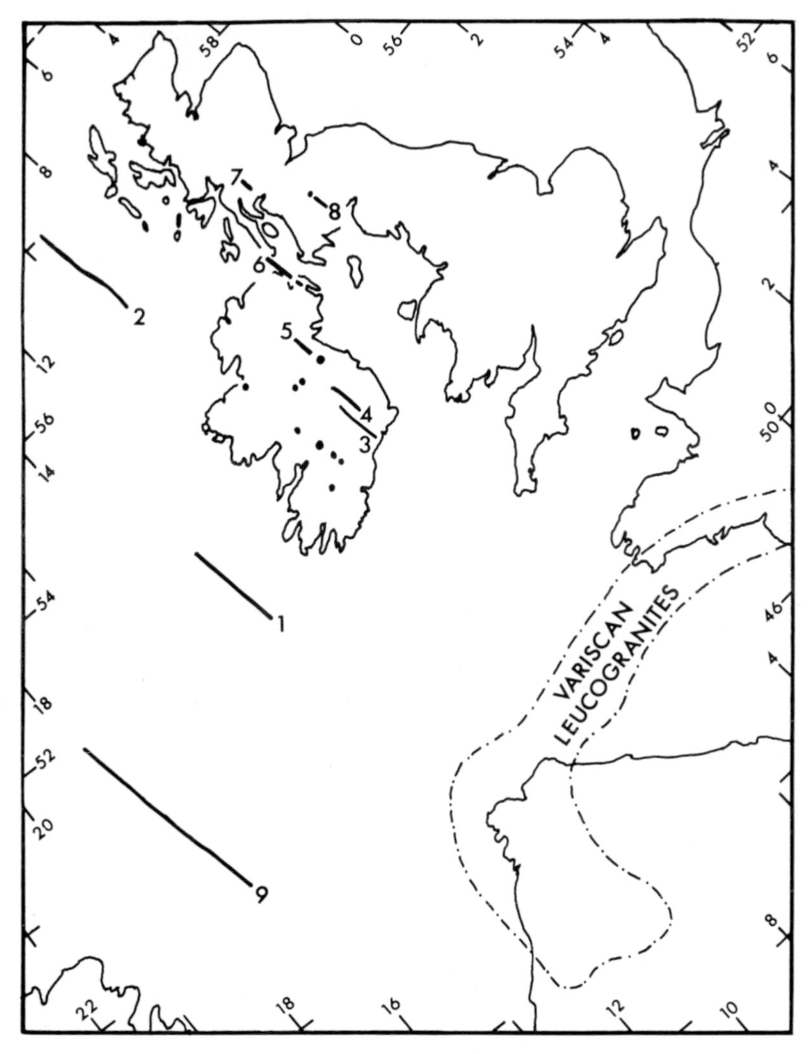


Fig. 1. Oblique Mercator projection of a reconstruction of part of the North Atlantic region for the Carboniferous period (Smythe *et al.* 1978). Ore deposits of Carboniferous age with $>10^4$ tonnes of metal shown as black dots. N–S structures: (1) trace of linear Free-Air gravity low (Gray and Stacey 1970), (2) continental margin bathymetry trace (five of the ore deposits in Lower Carboniferous Limestone occur on the line of its southerly extrapolation—Russell 1972), (3) fault and Namurian basin (Geological Survey map of Ireland, 1,750,000 3rd edn 1962), (4) high permeability zone (Wright 1976), (5) Kingscourt basin and Navan orebody, both of early Carboniferous age (Jackson 1956, Russell 1972), (6) unusual Tertiary dyke trend and N–S trending vein of lead–zinc at Newtownards, Northern Ireland (Caston 1975), (7) north Loch Lomond and minor lead–zinc deposit at Tyndrum (Russell 1972), (8) Thornhill basin and Leadhills deposit (MacGregor 1944), (9) bathymetry trace of the northwest Newfoundland continental margin; Variscan leucogranites after Arthaud and Matte (1975).

the volcanism in Scotland became progressively more undersaturated and occurred on a much reduced scale as time progressed. This was presumably due to cooling and magma being derived from partial melts at greater depths. This post-orogenic activity shows that there was still a relatively high geothermal gradient in

Carboniferous times on the site of the Caledonian–Appalachian orogen. Relative tension is also to be presumed from basin subsidence at the same time (Leeder 1974; Russell 1976).

## 4. Late Carboniferous—early Permian mineralization

Although there are a variety of ore types of this age, fluorite is the common denominator. This is in marked contrast to the Lower Carboniferous ores which are largely devoid of the mineral. The fluorite occurs intimately associated with granites in Newfoundland and Norway and as veins in Norway, Sweden and northern England (Fig. 2).

### 4a. Newfoundland

Large concentrations of vein fluorite in the St Lawrence granite represent the earliest mineralization of this type in the margins of the North Atlantic. The granite is dated at 326±10 m.y. (Bell and Blenkinsop 1975; all Rb-Sr ages recalculated for $\lambda\ ^{87}Rb = 1{\cdot}42 \times 10^{-11}y^{-1}$). Teng and Strong (1976) have shown that the St Lawrence peralkaline granite is the product of extreme differentiation. It contains 20–40% quartz and 30–60% orthoclase, some plagioclase and minor amounts of fluorite, riebeckite, aegerine, biotite, magnetite and haematite. The hypersolvus granite was probably produced by limited partial melting of anhydrous continental crust and involved an earlier granite. A crustal influence is supported by the high initial $^{87}Sr/^{86}Sr$ ratio of 0·722±0·005 (Bell and Blenkinsop 1975). Teng and Strong (1976) point out that peralkaline granites are typical of rifting environments and suggest that limited fractional melting at the base of the continental crust due to an increase in thermal gradient would be capable of producing peralkaline or alkaline granite magma (Luth *et al.* 1964; Bailey and Schairer 1966). The St Lawrence granite is one of several of similar age known in Newfoundland (Bell and Blenkinsop 1977). Significant in this context is the Newport granite in East Newfoundland which crosscuts unfoliated dolerite dykes trending between N and NNE. Jayasinghe and Berger (1976) suggest that these dykes relate to an early rifting episode comparable to East Greenland in early Tertiary times. The relationship between the dolerites and the Newport granite lends credence to the views of Teng and Strong (1976).

### 4b. Norway

Hysingjord (1971, reported in Sørensen 1975) has found the border zones of the alkali granite in the Ramnes cauldron in the Oslo graben to contain up to 20% fluorite and the plutonism has been dated at 270±2 m.y. (recalculated from Heier and Compston 1969). The earliest igneous activity in the Oslo region was extrusion of alkali basalt, one flow of tholeiite and flows of rhomb-porphyry. Sundvoll (1976) has dated this activity at about 293±10 m.y. Heier and Compston (1969) have shown that, in contrast to the St Lawrence granite in Newfoundland, the Ramnes monzonites and granite along with igneous intrusions from other cauldrons, could not have formed by crustal anatexis (initial $^{87}Sr/^{86}Sr = 0{\cdot}7041 \pm 0{\cdot}0002$). A mantle source for the Oslo felsic plutons now finds support in the gravity work of Ramberg (1976) who has shown that a mass of dense rock underlies the Oslo graben. This mass is interpreted to be the gabbro from which the Oslo igneous series was produced by crystal fractionation.

Fluorite also occurs associated with intrusive rocks in the Baerum and Nittedal

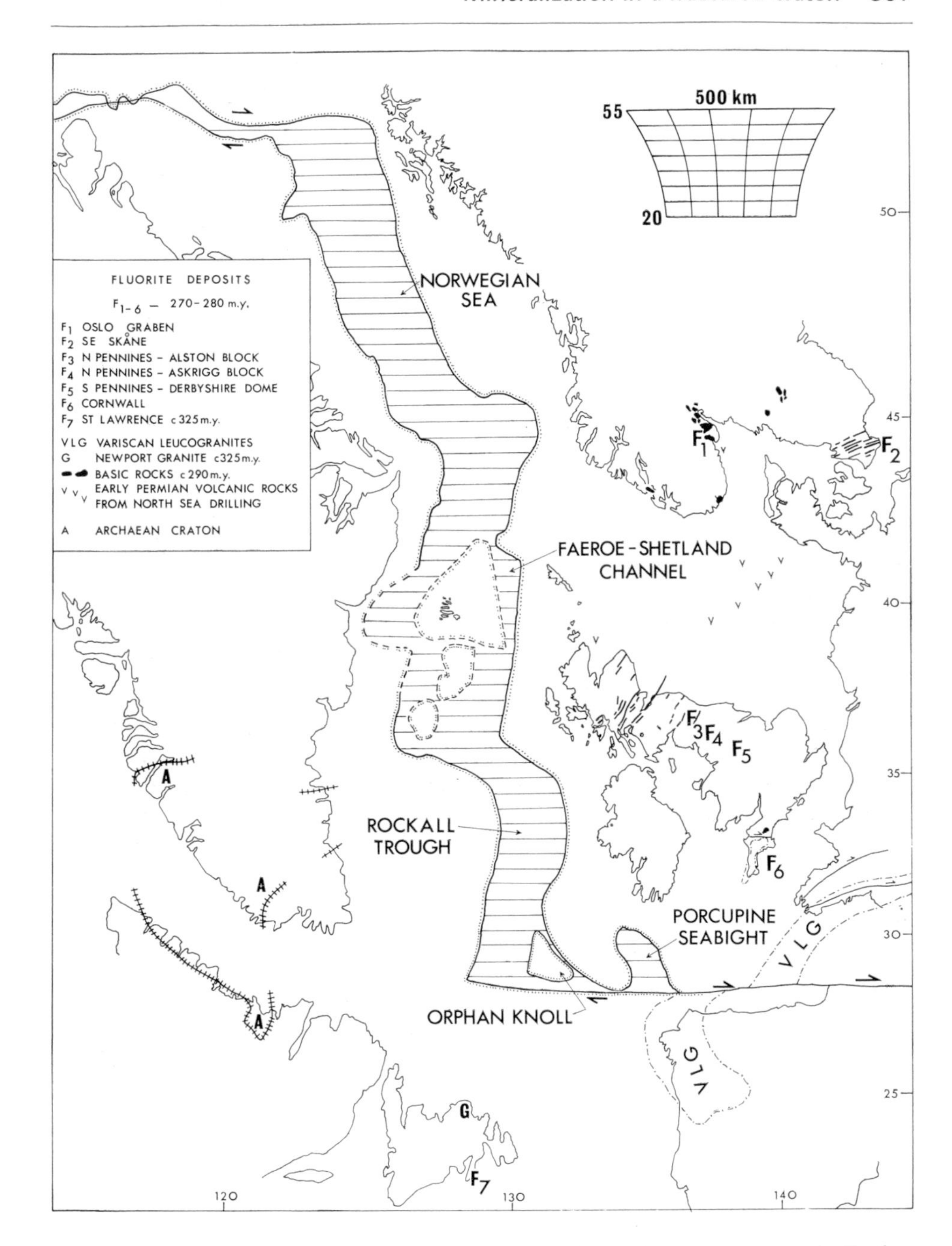


Fig. 2. Oblique Mercator projection of the northern North Atlantic region prior to the Tertiary opening and the formation of the Bay of Biscay. The position of Iberia relative to Europe is similar to that favoured by Le Pichon *et al.* (1971). Movement along the proto-Bay of Biscay transform fault postdated the Variscan leucogranties and predated the Triassic basins in France. The pre-Tertiary (pre-anomaly 24) presumed ocean (comprising Rockall trough, Faeroe–Shetland channel, and the Norwegian Sea) is opened by this movement. Basic rocks after Priem *et al.* 1968; Miller *et al.* 1970; Halvorsen 1970; Faerseth *et al.* 1976; Klingspor 1976 and Sundvoll 1976; North Sea volcanic rocks after Kent 1975 and Ziegler 1976; Archaean craton of Greenland after Bridgewater *et al.* 1973; reconstruction after Smythe *et al.* 1978.

cauldrons (Oftedahl 1960; B. Larsen personal communication). It is also present in veins at Lassedalen to the west of the Oslo graben (Willms 1975) and at Gjerpenfeltet (Kaspersen 1976) to the south of the graben.

### 4c. Southern Sweden

Late Carboniferous–early Permian movements near the northwestern end of the Tornquist line has permitted the intrusion of both dolerites dated at 294±4 m.y. (Klingspor 1976—K-Ar) and brecciated fluorite-quartz veins (Wickmann *et al.* 1963; Magnusson 1973; Vokes and Gale 1976).

### 4d. Northern England

Sawkins (1966) estimated there to have been 20 million tonnes of fluorite deposited in veins in Carboniferous limestones and sandstones in the northern Pennines. Dunham (1959) calculated the total galena content to have been six million tonnes and the sphalerite tonnage a little lower. The veins are post-quartz dolerite intrusion (*c.* 295 m.y.—K-Ar; Fitch and Miller 1967) but probably predate the Yellow Sands of early Permian age in which Versey (1925) records detrital grains of fluorite and barite. Although there is an obvious spatial association with the tholeiitic intrusions, Dunham (1949) pointed out that fluorite was not one of the minerals developed in the well defined hydrothermal stage of the tholeiitic magmatism. Bott and Masson-Smith (1957) favoured a granitic source for the mineralization and although a granite of the appropriate age was not encountered in subsequent drilling (Dunham *et. al.* 1965), work by Smith (1974) supports an igneous, probably granitic, derivation as the fluorite is enriched in Y, Ce and La and there are anomalous concentrations of Zr in wallrock alteration zones (Ineson 1969). Moreover Smith and Phillips (1975) estimate that the veins were filled within 1000 years, which implies an expulsion of fluids from crystallizing magma rather than a long lived goethermal system. The veining took advantage of tensional structures formed in response to the doming in the northern Pennines which took place shortly after the intrusion of the quartz dolerite sills and dykes (Dunham 1949).

Significant fluorite deposits also occur in the Askrigg block of the north Pennines and the Derbyshire dome in the southern Pennines. Although a relationship to igneous activity is tenuous, Mitchell and Krouse (1971) have argued for an origin of fluorite and lead and zinc in the Askrigg block related to anatexis of the lower crust in the Permo-Carboniferous interval. In Derbyshire the first and probably the most significant mineralizing episode has been dated at 270 m.y. by Ineson and Mitchell (1973). It therefore presumably has a similar origin to that of the northern Pennines mineralization, although the source is more distant judging from the lower temperatures determined from fluid inclusion studies and the Y proportions in the fluorite (Smith and Hirst 1974).

Fluorite is one component of the tin-copper-lead-zinc mineralization in Cornwall of approximately the same age. This mineralization is generally considered to relate to the Armorican orogeny.

## 5. Models for fluorite mineralization

Vokes (1973) suggested that the mineralization in the Oslo region may have been related to early rifting in the northern North Atlantic and Sawkins (1976), while concurring, added the north Pennine deposits to this period of tectonic activity.

There is a growing body of evidence favouring rifting towards the end of the Carboniferous and in the early Permian times involving ocean-floor formation between Greenland and northwestern Europe (Russell 1976; McLean 1978; Smythe *et al.* 1978). A single model for the generation of fluorite is not possible. Accordingly models for the two districts in which there is the most information are constructed and applied to the other districts.

### 5a. St Lawrence, Newfoundland

In late Carboniferous times tension in the lithosphere was focused along the Caledonian–Appalachian mountain belt. Minor lithosphere separation under Newfoundland allowed "asthenospheric" mantle to rise and partial melting to take place. Some basic magma was intruded (Jayasinghe and Berger 1976) and possibly extruded on to the continental crust (Teng and Strong 1976). More extensive intrusion into lower crustal levels enabled heat transfer and caused some anhydrous partial melting so producing hypersolvus peralkaline magmas (see Bailey and Schairer 1966). Biotite in lower crustal gneisses was rendered unstable due to very low water pressures and $H_2O$ and HF released from hydroxyl sites entered the melt (Gavrilin *et al.* 1972; Bailey, J. C. 1977). Peralkaline magma rose due to buoyancy and stoping and increased its water and fluorine content. The HF:$H_2O$ ratio would have been higher in a peralkaline melt than in those of orogenic calc-alkaline type which result from fusion of relatively wet geosynclinal sediments. As HF is more soluble than water in silicate melts and lowers the melting temperature more, the fluorine became concentrated in the residual melt (Wyllie and Tuttle 1961). On crystallization of this residual melt, calcium fluoride was precipitated as a late stage accessory and the remainder entered the vapour phase and streamed upwards precipitating along a cooling gradient in fissures produced by contraction of the cooling granite (Teng and Strong 1976).

### 5b. Oslo graben, Norway

Towards the end of Carboniferous times, mantle diapirs rose into the lithosphere as a response to relative tension (see Bailey, D. K. 1977). This would have caused general heating in the uppermost mantle and extensive partial melting of mantle material produced basic magma that was "ponded" below the crust in the Oslo district. This basic mass gave rise to updoming in the region (Ramberg 1976). Some of the magma broke through to the surface to give flows of basalt and rhomb-porphyry. Under the influence of relative tension the dome finally collapsed and the Oslo graben formed. A high degree of partial melting of a phlogopite-lherzolite upper mantle (cf. Flower 1969) would have released fluorine from the (OH, F) site in phlogopite. Differentiation by crystal fractionation to monzonites, syenites and eventually peralkaline granites would have had the effect of concentrating fluorine in the residual melts. Fluorite either precipitated as a primary mineral or escaped into the vapour phase to crystallize in veins some distance from the magmatic source.

### 5c. Northern Pennines, England

It is envisaged that the generation of fluorite deposits in the northern Pennines was similar to that of the Oslo district. At about the same time as the intrusion and extrusion of basic rocks around Oslo, quartz dolerites were intruded in the northern Pennine region (295 m.y. ago). Here too there was doming but in this case local rifting did not follow. Again alkali granites with high fluorine contents were probably produced either by differentiation or crustal anatexis or both. These granites could have intruded the older granites in the northern Pennines so explaining the steep geothermal gradient at the time. Fluorite escaped in the vapour phase and was precipitated in open veins along a cooling gradient (Smith and Phillips 1975). The lead, zinc and barite may have been leached out of relatively deep levels of the Caledonian granite (Pb average 37 ppm—Holland 1967) by hydrothermal waters convectively driven by the cooling early Permian granite. The fluorite deposits to the south probably originated at the same time and in a similar way but the magmatic source was more distant (p. 302). Nevertheless the fluorite provides an indication of the geographical extent of partial melting in the upper mantle around the Permo-Carboniferous interval.

### 5d. Skåne, South Sweden

As in the northern Pennines there is a close relationship in time and space between dolerite dykes and fluorite-lead-zinc deposits and presumably the deposits were produced in a similar way.

### 5e. Cornwall, England

The fluorite deposits in Cornwall have generally been related to the Armorican granites. Just what the effect or influence of relative tension was in the area is difficult to decipher.

## 6. Tectonic environment

As the Greenland–Svalbard and proto-Bay of Biscay faults are small circles to a common pole it is considered that they accommodated transform motion at the same time (Smythe *et al.* 1978). According to a map produced by Arthaud and Matte (1975, fig. 8), dextral movement of about 150 km took place between the intrusion of Variscan leucogranites in Iberia and France and the formation of basins in early Triassic times. One way of explaining this movement is as a response to the opening of Rockall trough and its topographical and structural neighbours, the Faeroe-Shetland channel and the east Norwegian sea (the pre-anomaly 24 oceanic crust). The position of Spain relative to France favoured by Le Pichon *et al.* (1971) implies that the Rockall trough was fully open prior to the formation of the Bay of Biscay (Fig. 2). Russell (1972, 1976) has previously suggested that the opening of the pre-Tertiary Atlantic began at the same time as the intrusion and extrusion of basaltic magma in late Carboniferous times. The opening probably took place in early Permian times so allowing the Boreal sea to flood parts of northwestern Europe at the beginning of late Permian times. McLean (1978) also shows that Rockall trough was probably in existence before the development of the Permo-Trias marginal troughs on the continental margin of the British Isles.

## 7. Synthesis

The Caledonian–Appalachian orogen had stabilized by about 400 m.y. ago but it remained a zone of potential lithospheric weakness owing to the orogenic grain and high heat content. Post-orogenic basaltic magmatism derived from limited partial melting of the upper mantle was a feature of the early Carboniferous, especially in Scotland, but continued intermittently in the Late Carboniferous in Britain, Ireland

and the Maritimes of Canada. Mineralizing systems driven by heat left over from the Caledonian–Appalachian orogeny were generated in Carboniferous sea water that percolated down into the orogen leaching metals in its path. The uprising hydrothermal fluids took advantage of E–W and northeasterly-trending faults, and there is some evidence in favour of N–S geofractures, produced by E–W tension, as the underlying control. Tension continued throughout Carboniferous times causing basin formation and eventually the lithosphere began to fail. Mantle diapirs rose into the breaks and partial melting produced basic magma first in Newfoundland, then suddenly and coevally on a large scale in Britain, the North Sea and Scandinavia (*c.* 295 m.y. ago), implying that the strength of the lithosphere in tension was being generally exceeded. The main zone of failure was between Greenland and northwest Europe. Elsewhere some doming took place over isolated mantle diapirs (e.g. northern Pennines and Oslo graben), and a high degree of partial melting of phlogopite-lherzolite mantle produced fluorine-rich magmas. In the case of the Oslo region, the dome eventually collapsed under the influence of tension and the graben was formed. Even where the magma failed to reach the surface, the extent of the partial melting can be surmised from the occurrence of magmatic hydrothermal fluorite deposits (Fig. 2). Large scale convective systems in pore waters were engendered by some of the Permian felsic intrusions and important lead, zinc and barite deposits were generated and are found associated with some of the fluorite deposits.

The opening of the proto-North Atlantic was probably completed by the end of the early Permian times and the Zechstein Sea flooded part of north central Europe at the beginning of the late Permian. Some of the heat introduced into the lower crust in the early Permian remained. It eventually drove hydrothermal systems, involving late Permian sea-water, that were small in comparison to those of the north Pennine-type of the early Permian (Hirst and Smith 1974), but large enough to introduce metals into the Zechstein Sea through hot springs and deposit lead and zinc in the Marl slate.

This phase of ocean-floor spreading was short lived and it was to be another 200 m.y. before the northern North Atlantic was to open in earnest. Again basic dykes and peralkaline intrusions were a prominent feature of the splitting, and fluorite and base metal mineralization are represented in the Mourne Mountains and Skye granites.

*Acknowledgments.* I thank A. Hall, B. Larsen, M. R. Leeder, R. MacDonald, R. M. Macintyre, C. J. Morrissey, E.-R. Neumann, F. W. Smith, D. K. Smythe and B. Sundvoll for discussions and help.

## References

ANON. 1975. Tara—Ireland's major lead/zinc prospect still involved in lease talks. *Min. Mag.* July 1975, 14–19.

ARTHAUD, F. and MATTE, P. 1975. Les decrochements tardi–Hercyniens du sud–ouest de l'Europe. Geometrie et essai de reconstitution des conditions de la deformation. *Tectonophysics* **25**, 139–171.

BAILEY, D. K 1977. Lithosphere control of continental rift magmatism. *J. geol. Soc. Lond.* **133,** 103–106.
—— and SCHAIRER, J. F. 1966. The system $Na_2O$–$Al_2O_3$–$Fe_2O_3$–$SiO_2$ at 1 atmosphere, and the petrogenesis of alkaline rocks. *J. Petrol.* **7,** 114–170.
BAILEY, J. C. 1977. Fluorine in granitic rocks and melts: a review. *Chem. Geol.* **19,** 1–42.
BELL, K. and BLENKINSOP, J. 1975. The geochronology of eastern Newfoundland. *Nature, Lond.* **254,** 410–411.
—— and —— 1977. Geochronological evidence of Hercynian activity in Newfoundland. *Nature, Lond.* **265,** 616–618.
BOTT, M. H. and MASSON-SMITH, D. 1957. The geological interpretation of a gravity survey of the Alston Block and the Durham Coalfield. *Q. J. geol. Soc. Lond.* **113,** 93–117.
BOYLE, R. W., WANLESS, R. K. and STEVENS, R. D. 1976. Sulphur isotope investigation of the barite, manganese and lead-zing-copper-silver deposits of the Walton–Cheverie area, Nova Scotia. Canada. *Econ. Geol.* **71,** 749–762.
BRIDGEWATER, D., WATSON, J and WINDLEY, B. F. 1973. The Archaean craton of the North Atlantic region. *Philos. Trans. R. Soc. Lond.* A **273,** 493–512.
CASTON, G. F. 1975. Igneous dykes and associated scour hollows of the North Channel, Irish Sea. *Mar. Geol.* **18,** 77–85.
COOMER, P. G. and ROBINSON, B. W. 1976. Sulphur and sulphate–oxygen isotopes and the origin of the Silvermines deposits, Ireland. *Mineral. Deposita* **11,** 155–169.
DUNHAM, K. C. 1949. Geology of the Northern Pennine orefield 1. *Mem. geol. Surv.* U.K.
—— 1959. Non-ferrous mining potentialities of the northern Pennines. In *Future of non-ferrous mining in Great Britain and Ireland,* 204–232. *Inst. Min. Met.*, London.
——, DUNHAM, A. C., HODGE, B. C. and JOHNSON, G. A. 1965. Granite beneath Viséan sediments with mineralization at Rookhope, northern Pennines. *Q. J. geol. Soc. Lond.* **121,** 383–414.
FAERSETH, R. B., MACINTYRE, R. M. and NATERSTAD, J. 1976. Mesozoic alkaline dykes in the Sunnhordland region, western Norway: ages, geochemistry and regional significance. *Lithos* **9,** 331–345.
FITCH, F. J. and MILLER, J. A. 1967. The age of the Whin Sill. *Geol. J.* **5,** 233–250.
——,—— and WILLIAMS, S. C. 1970. Isotopic ages of British Carboniferous rocks. *C. R. 6e Congr. Strat. Geol. Carbonif.* **2,** 771–789.
FLOWER, M. F. J. 1969. Phlogopite from Jan Mayen Island (North Atlantic). *Earth Planet. Sci. Lett.* **6,** 461–466.
FRANCIS, E. H. 1978. Igneous activity in a fractured craton: Carboniferous volcanism in northern Britain. *In* Bowes, D. R. and Leake, B. E. (Editors). *Crustal evolution in northwestern Britain and adjacent regions*, 279–296. *Geol. J. Spec. Issue* No. 10.
GAVRILIN, R. D., AGAFONNIKOVA, L. S. and SAVINOVA, E. N. 1972. Behaviour of fluorine in the initial stages of granitization. *Geochem. Int.* **9,** 180–185.
GRAY, F. and STACEY, A. P. 1970. Gravity and magnetic interpretation of Porcupine Bank and Porcupine Bight. *Deep-Sea Res.* **17,** 467–475.
HALVORSEN, E. 1970. Palaeomagnetism and the age of the younger diabases in the Ny-Hellesund area, S. Norway. *Nor. Geol. Tidsskr.* **50,** 157–166.
HEIER, K. S. and COMPSTON, W. 1969 Rb-Sr isotopic studies of the plutonic rocks of the Oslo region. *Lithos* **2,** 133–145.
HIRST, D. M. and SMITH, F. W. 1974. Controls of barite mineralization in the Lower Magnesian Limestone of the Ferryhill area, County Durham. *Trans. Inst. Min. Metall.* **83B,** 49–55.
HOLLAND, J. G. 1967. Rapid analysis of the Weardale granite. *Proc. Yorks. geol. Soc.* **36,** 91–113.
INESON, P. R. 1969. Trace-element aureoles in limestone wallrocks adjacent to lead–zinc–barite–fluorite mineralization in the northern Pennines and Derbyshire ore fields. *Trans. Inst. Min. Metall.* **78B,** 29–40.
—— and MITCHELL, J. G. 1973. Isotopic age determination on clay minerals from lavas and tuffs of the Derbyshire ore field. *Geol. Mag.* **6,** 501–512.
—— and —— 1974. K-Ar isotopic age determinations from some Scottish mineral localities. *Trans. Inst. Min. Metall.* **83B,** 13–18.
JACKSON, J. S. 1956. The Carboniferous succession of the Kingscourt outlier with notes on the Permo–Trias. *Univ. Coll. Dublin Ph.D. thesis.*
JAYASINGHE, N. R. and BERGER, A. R. 1976. On the plutonic evolution of the Wesleyville area, Bonavista Bay, Newfoundland. *Can. J. Earth Sci.* **13,** 1560–1570.
KASPERSEN, P. O. 1976. En malmgeologisk undersøkelse av flusspatmineraliseringen i Gjerpenfeltet nord for Skien. *Univ. Trondheim* thesis.
KENT, P. E. 1975. Review of North Sea basin development. *J. geol. Soc. Lond.* **131,** 435–468.
KLINGSPOR, I. 1976. Radiometric age-determination of basalts, dolerites and related syenite in Skåne, southern Sweden. *Geol. Fören. Stockh. Förh.* **98,** 195–215.
LEAKE, B. E. 1978. Granite emplacement: the granites of Ireland and their origin. *In* Bowes, D. R. and Leake, B. E. (Editors). *Crustal evolution in northwestern Britain and adjacent regions,* 221–248. *Geological J. Special Issue* No. 10.
LEEDER, M. R. 1974. The origin of the Northumberland basin. *Scott. J. Geol.* **10,** 283–296.

LE PICHON, X., BONNIN, J., FRANCHETEAU, J. and SIBUET, J. -C. 1971. Une hypothèse d'évolution tectonique du golfe de Gascogne, In *Histoire structurale du golfe de Gascogne, tome* **2**, Publications de l'institut Français du Pétrole, Paris, VI, 11–1.

LUTH, W. C., JAHNS, R. H. and TUTTLE, O. F. 1964. The granite system at pressures of 4 to 10 kilobars. *J. geophys. Res.* **69,** 759–773.

MACDONALD, R. 1975. Petrochemistry of the Early Carboniferous (Dinantian) lavas of Scotland. *Scott. J. Geol.* **11,** 269–314.

MACEACHERN, S. B. and HANNON, P. 1974. The Gays River discovery. *Can. Min. J.* **95** (4), 77–79.

MACGREGOR, A. G., 1944. Barytes in central Scotland. *Wartime pamph. geol. Surv. U.K.*

MCLEAN, A. C. 1978. Evolution of fault-controlled ensialic basins in northwestern Britain. *In* Bowes, D. R. and Leake, B. E. (Editors). *Crustal evolution in northwestern Britain and adjacent regions*, 325–346. *Geol. J. Spec. Issue* No. 10.

MAGNUSSON, N. H. 1973. *Malm i Sverige* **1**, *Mellersta och södra Sverige*. Almqvist and Wiksell, Stockholm.

MITCHELL, R. H. and KROUSE, H. R. 1971. Isotopic composition of sulfur and lead in galena from the Greenhow–Skyreholme area, Yorkshire, England. *Econ. Geol.* **66,** 243–251.

OFTEDAHL, C. 1960. Permian rocks and structures of the Oslo Region. *In* Holtedhal, O. (Editor). *Geology of Norway,* 298–343. *Nor. geol. Unders.* **208.**

PRIEM, H. N., MULDER, F. G., BOELRIJK, N. A., HEBEDA, E. H., VERSCHURE, R. H. and VERDURMEN, E. A. 1968. Geochronological and palaeomagnetic reconnaissance survey in parts of central and southern Sweden. *Phys. Earth Planet. Inter.* **1,** 373–380.

RAMBERG, I. B. 1976. Gravity interpretation of the Oslo Graben and associated igneous rocks. *Nor. geol. Unders.* **325,** 1–194.

ROSE, E. R., SANFORD, B. V. and HACQUEBARD, P. A. 1970. Economic minerals of southeastern Canada. *In* Douglas, R. J. W. (Editor). *Geology and Economic Minerals of Canada,* 305–364. *Geol. Surv. Can., Econ. Geol. Rep.* No. 1.

RUSSELL, M. J. 1968. Structural controls of base metal mineralization in Ireland in relation to continental drift. *Trans. Inst. Min. Metall.* **77B,** 117–128.

—— 1972. North–south geofractures in Scotland and Ireland. *Scott. J. Geol.* **8,** 75–84.

—— 1973. Base-metal mineralization in Ireland and Scotland and the formation of Rockall Trough. *In* Tarling, D. H. and Runcorn, S. K. (Editors). *Implications of continental drift to the earth sciences* **1,** 581–597. Academic Press, London.

—— 1974. Manganese halo surrounding the Tynagh ore deposit, Ireland: a preliminary note. *Trans. Inst. Min. Metall.* **83B,** 65–66.

—— 1975. Lithogeochemical environment of the Tynagh base-metal deposit, Ireland, and its bearing on ore deposition. *Trans. Inst. Min. Metall.* **84B,** 128–133.

—— 1976. A possible Lower Permian age for the onset of ocean floor spreading in the northern North Atlantic. *Scott. J. Geol.* **12,** 315–323.

SAWKINS, F. J. 1966. Ore genesis in the North Pennine orefield, in the light of fluid inclusion studies. *Econ. Geol.* **61,** 385–401.

—— 1976. Metal deposits related to intracontinental hot spot and rifting environments. *J. Geol.* **84,** 653–671.

SMITH, F. W. 1974. Factors governing the development of fluospar orebodies in the North Pennine orefield. *Univ. Durham Ph.D. thesis.*

—— and HIRST, D. M. 1974. Analysis of trace elements and fluid inclusions in fluorite from the Ardennes Massif. *An. Soc. geol. Belg.* **97,** 281–285.

—— and PHILLIPS, R. 1975. Temperature gradients and ore deposition in the North Pennine orefield. *Fortschr. Minerals.* **52,** 491–494.

SMYTHE, D. K., KENOLTY, N., RUSSELL, M. J. and SCRUTTON, R. A. 1978. Re-interpretation of the early history of the Rockall Trough and the northern North Atlantic, and its economic implications. *(In press).*

SØRENSEN, R. 1975. The Ramnes cauldron in the Permian of the Oslo region, southern Norway. *Nor. geol. Unders.* **321,** 67–86.

STROGEN, P. 1977. The evolution of the Carboniferous volcanic complex of southeast Limerick, Ireland. *J. geol. Soc. Lond.* **133,** 409–410.

SUNDVOLL, B. 1976. Rb/Sr relationship in the igneous rock of the Oslo palaeorift. *4th Eur. Colloq. on Geochron. Abstr.* Amsterdam, 89.

TENG. H. C. and STRONG, D. F. 1976. Geology and geochemistry of the St. Lawrence peralkaline granite and associated fluorite deposits, southeast Newfoundland. *Can. J. Earth Sci.* **13,** 1374–1385.

VERSEY, H. C. 1925. The beds underlying the Magnesian Limestone in Yorkshire. *Proc. Yorks. geol. Soc.* **20,** 200–214.

VOKES, F. M. 1973. Metallogeny possibly related to continental break-up in Southwest Scandinavia. *In* Tarling, D. H. and Runcorn, S. K. (Editors). *Implications of continental drift to the earth sciences,* **1,** 573–579. Academic Press, London.

—— and GALE, G. H. 1976. Metallogeny relatable to global tectonics in southern Scandinavia. *Geol. Ass. Can. Spec. Pap.* **14,** 413–441.

WICKMANN, F. E., BLOMQVIST, N. G., GEIJER, P., PARWEL, A., UBISCH, H. V. and WELIN, E. 1963. Istotopic constitution of ore lead in Sweden. *Ark. Mineral. Geol.* **11,** 193–257.

WILLMS, J. 1975. *Fluorit-Vorkommen in Telemark, Südnorwegen.* Diplomarbeit, Hamberg 1975.
WRIGHT, G. R. 1976. Ground water for industry. *Technol. Ireland* **8,** 12–16.
WYLLIE, P. J. and TUTTLE, O. F. 1961. Experimental investigation of silicate systems containing two volatile components; part II. *Am. J. Sci.* **259,** 128–143.
ZIEGLER, W. H. 1975. Outline of the Geological history of the North Sea. *In* Woodland, A. W. (Editor). *Petroleum and the Continental shelf of north-west Europe* **1,** *Geology* 165–187. Applied Science Publishers, London.

M. J. Russell,
Department of Applied Geology,
University of Strathclyde,
Glasgow G1 1XW, Scotland.

# Mesozoic vertical movements in Britain and the surrounding continental shelf

*1977 Celebrity lecture of the Geological Societies of Edinburgh and Glasgow*

***P. E. Kent***

---

Vertical movement constitutes an aspect of global tectonics which has been overshadowed in recent years by the spectacular evidence of lateral displacement of the continents. Nevertheless it is of critical importance in relation to sedimentation and hence to the distribution of economic minerals, as well as to concepts of global structure. On land in Britain the investigation of contemporaneous movements has a long history, and for the Jurassic period in particular our terrain is classic ground. Concepts were later extended and modified as deep borings in the basins supplemented data from outcrops; they have been further augmented by the large amount of data from adjoining offshore areas, particularly from the North Sea. Land and sea complement and explain each other's features. Interpretation of the tectonic development has been confused by analogies with regions of intra-cratonic rifting in other parts of the world, although there are fundamental differences between the different situations in secular and spatial terms. This paper emphasises the critical evidence as it is known from direct observation, and highlights unexplained problems.

## 1. Introduction

> "In the preparation of dishes for the connoisseurs
> of World Tectonics we seem to have become obsessed by
> plates and neglected basins"—A. Hallam 1975.

Concepts of the epeirogenic control of Mesozoic sedimentation in and around Britain have developed progressively through the last century as increasingly accurate and more widely extended information has become available. In the early stages some of the concepts were quite misleading: for example outcrop study of parts of the succession particularly sensitive to differential movement gave rise to an interpretation of the role of "Jurassic axes" out of all proportion to their total effect. Availability of long sections in the basins led to changes in these views, and

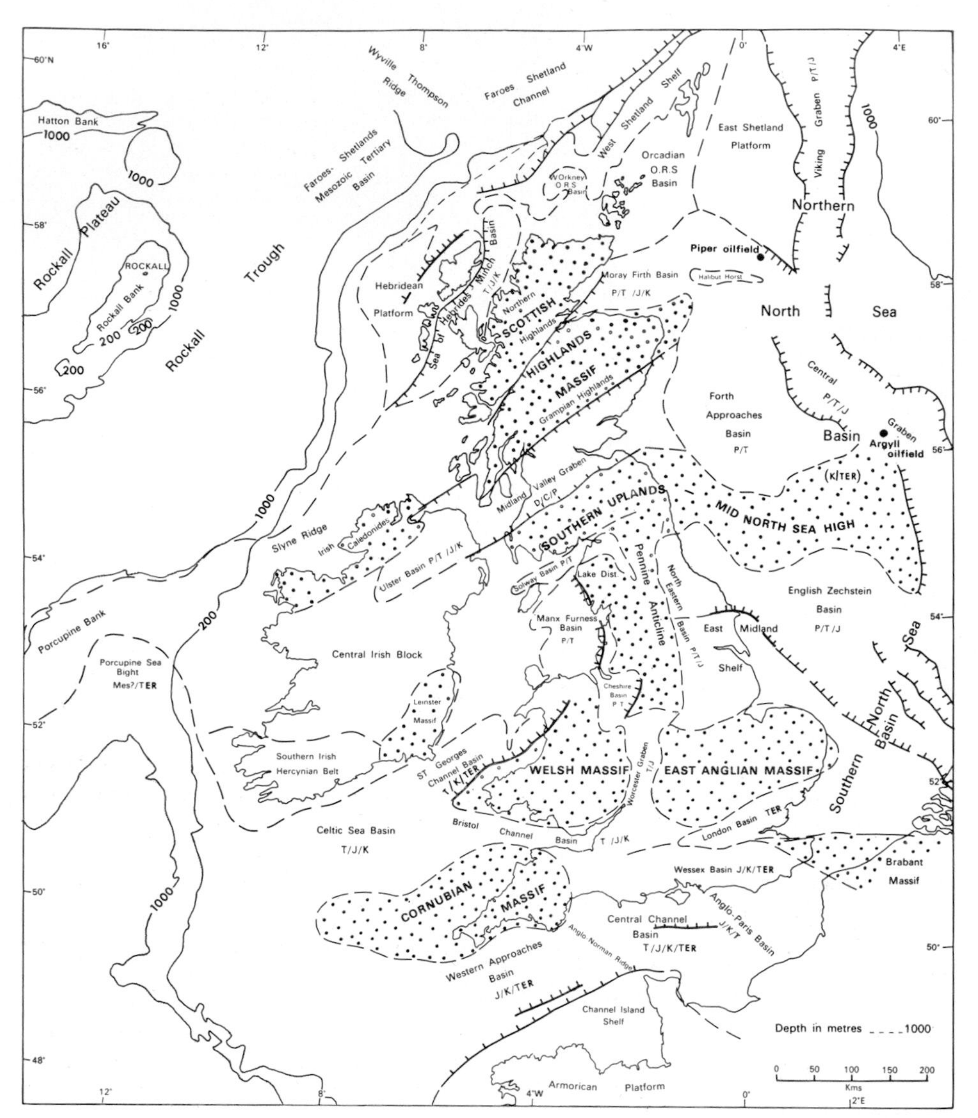


Fig. 1. Regional structure of Britain: massifs and basins. D or O.R.S. Devonian, C Carboniferous, P Permian, TR Triassic, J Jurassic, K Cretaceous, T Tertiary.

increased emphasis on the greater importance of broad blocks such as the East Anglian massif or London platform (the swells of some authors). More recently, data from offshore have led to recognition of taphrogenic control of basin development, initially offshore but also in landward areas, and realisation that the so-called "Alpine movements" in Britain were often very much earlier in date than the classic Tertiary orogeny of southern Europe. Linked with this, we are now beginning to relate the history of the Upper Palaeozoic basins to those of the Mesozoic in southern Britain, in Scotland and offshore.

## 1a. The nature of epeirogenic movement

Epeirogenic movements are the most widespread but one of the least-well understood tectonic agencies. In some degree they must relate to the new global tectonics, but as has been stressed elsewhere they tend to be far longer lasting, and have a tendency to show analogous features independently of relation to plate margins. The very large horizontal dimensions of some basins suggest that their subsidence is controlled by sub-crustal processes: the Congo basin might be quoted as an example of this.

Simple basin development on either a continental or country-wide scale is subject to modification due to discontinuities and variation within the crystalline basement. Ancient fractures in particular provide lines of weakness which persist for long periods in shelf (foreland) areas, in contrast to the dominant effect of the sedimentary load in miogeosynclinal basins. The distinction between some of these factors is becoming clearer in the British area.

During Mesozoic times the land area of Britain was almost entirely one of shelf-type of deposition with its local discontinuities, while in the deeper basins now occupied by the surrounding seas an approach to geosynclinal thicknesses is developed. On land the Cheshire basin might almost be included in this category, with 3600 m or more of sediments deposited during the Permo-Triassic period, but measurements of this order onshore and offshore are related to taphrogenic (rifted) subsidence.

## 1b. The post-Hercynian tectonic pattern

Modern Britain shows a scatter of mountains or hilly areas of old rocks, separated by basins of Permian and Mesozoic sediments. It is known from the evidence of shoreline deposits, unconformable relationships and thickness trends that this is for the most part a long-existing relationship: some areas—the Scottish Highlands, the Southern Uplands, the Pennines (with the Lake District), Wales and Cornubia—were marginal to the main basins from the Trias onwards (Kent 1949). Our highland areas are thus the relics of an ancient archipelago. To the list of exposed Palaeozoic massifs has to be added the concealed East Anglian or London–Brabant massif, originally part of a Carboniferous and older ridge extending from Wales into Brabant, and still retaining here a thin veneer of Mesozoic sediments (Fig. 1).

The history of the positive areas has been analysed by the present writer elsewhere (1975a). Most of these were probably near sea-level in Mesozoic times, suffering an alternation of shallow submergence with minor erosion, as in the case of East Anglia; probably only the Scottish Highlands provided a continuing large scale source of sediment. They are sometimes referred to as "Hercynian massifs", but with the exception of Cornubia all have histories as positive structures extending back at least into the Lower Carboniferous and are mainly blocks essentially Caledonian in structure.

In the modern seas we can trace continuity of the same pattern. In the northwest, the Outer Hebrides originally formed a positive area separated from the mainland by the Mesozoic basins of the Minches and Sea of the Hebrides. A positive area in Northern Ireland may have continued westwards in the submerged Slyne ridge, although nothing is yet known of its history. The Southern Uplands are the present emergent part of a Palaeozoic and Mesozoic ridge which extended across the North Sea into Denmark—the mid North Sea high. As shown below, the Mesozoic history of the emergent and submerged sections of this ridge were closely similar.

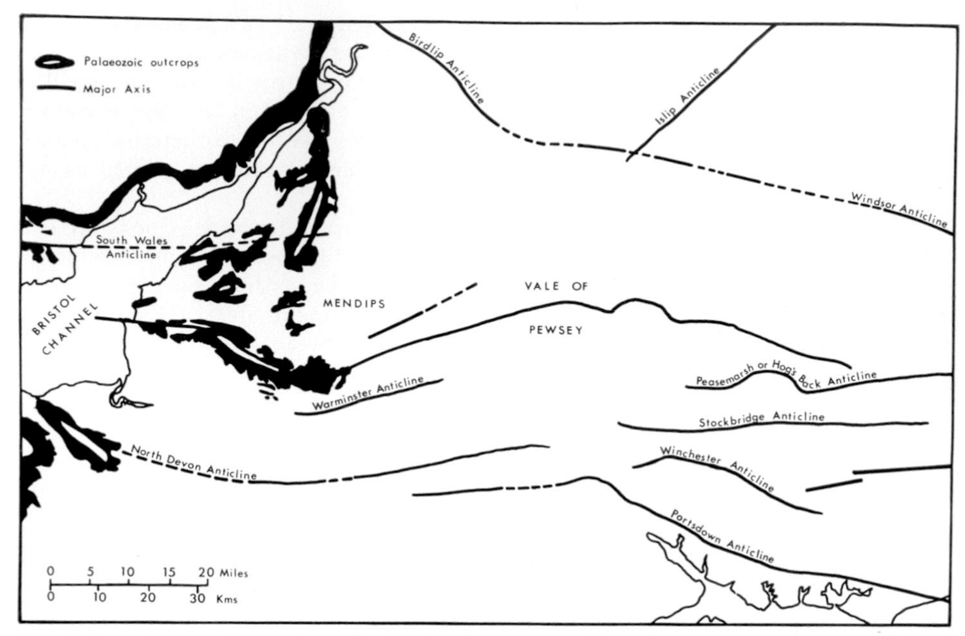


Fig. 2. Jurassic axes in southern England—the concept of persistent contemporary anticlinal uplifts controlling sediment thickness (after Arkell 1933).

Wales formed a discrete mass during the Mesozoic, bounded by sedimentary basins on all sides, but the Palaeozoic spur of the Cornubian peninsula probably continued for a hundred miles or more to the southwest between the basins of the Western Approaches.

The widespread continued emergence of these areas of Palaeozoic rocks (which themselves show a history of buoyancy throughout the Upper Palaeozoic) emphasises the basic fact that the British area was part of a broad Mesozoic shelf, and that the various basins are due to inter-cratonic subsidence. No true geosyncline is known in Britain or beneath the British seas after the Armorican geosynclinal subsidence, now largely concealed beneath southern England and the Channel.

## 2. Regional appraisal

### 2a. Mesozoic differential movement in southern England

Southern England has seen a bigger change in the understanding of deep structure than any other part of Britain, and in important respects has presented a model for other areas. Early views were dominated by the concepts of Godwin-Austen (1856) and Arkell (1933), that the surface folds, the "Jurassic axes", represented the lines of major long-term uplift affecting not only the whole of the Jurassic but in many cases the Palaeozoic rocks beneath (Fig. 2). The contrary view came from the petroleum industry, from G. M. Lees and his associates, who recognised that the mode of folding of such anticlines as Portsdown and Kingsclere represented free folding in a thick sedimentary column (Lees and Cox 1937). Drilling to the base of

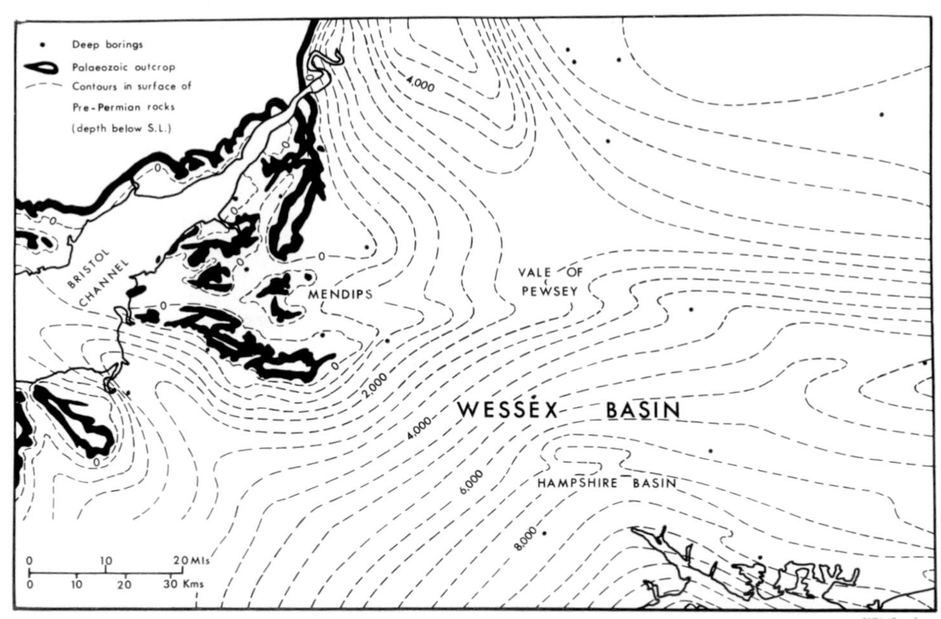


Fig. 3. Contours on the sub-Permian floor in southern England showing the minimal effect of local differential movement as compared with regional subsidence (after Kent 1949).

the Jurassic on a number of anticlines in southern England confirmed this concept; compilation of the results further showed that the Jurassic rocks of southern England occupy the broad "Wessex basin", which developed contemporaneously with Jurassic deposition (Kent 1949). The "Mendip axis", which had been regarded as a dominant contemporary anticlinal feature continuing eastwards at depth into Hampshire, was recognised as an unfolded shelf-like projection on the western margin of the Jurassic basin.

There was a subsequent complication in the history of basin development in northwest Europe which is only now becoming better understood. Lamplugh (1919) showed that the Wealden area is a broad anticline developed on the site of a Jurassic basin, a case of inversion which was then assumed to be essentially Alpine (mid-Tertiary) in age (Fig. 4). Tertiary history as compiled by Stamp (1934) could subsequently have cast doubts on this dating, for the Weald formed a rising area marginal to the subsiding London basin during the Eocene. In fact later data, stemming from continental Europe, but well documented also in the North Sea, shows that inversion phenomena are typically end-Cretaceous and earliest Tertiary (Ziegler 1975 fig. 17).

The phenomenon is also recognised further west in the Wessex basin. Boreholes in the western Hampshire basin show that the Tertiary syncline of East Dorset and southwest Hampshire coincides with a pre-Upper Cretaceous uplift in the Jurassic rocks, so that not only Wealden but Upper Jurassic formations are commonly absent beneath the Albian and later formations. The Purbeck anticline, in contrast, coincides with the belt of maximum thicknesses of Upper Jurassic and Lower Cretaceous. The same relationship continues into the Isle of Wight, with maximum thicknesses of Lower Greensand, Wealden and later Jurassic in the anticlinal

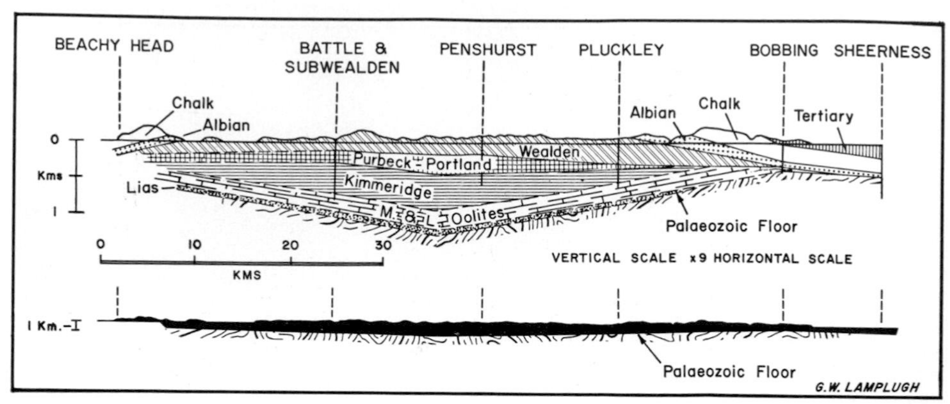


Fig. 4. Section through the Weald: a classical inversion structure. The lower figure is drawn at true scale (after Lamplugh 1919).

southern half of the island, but (on geophysical evidence) a markedly defective column beneath the Albian of the Tertiary basin to the north (Fig. 5).

In the English Channel to the south of the Isle of Wight comparable large structures have been mapped by sea floor sampling (Smith and Curry 1975). Their history is at present deduced from geophysical surveys without drilling control, but appears to be comparable with that of the land area—large monoclinal uplifts like that of Purbeck–Isle of Wight coinciding with areas formerly subject to major subsidence during the Jurassic (Fig. 6; Ziegler 1975).

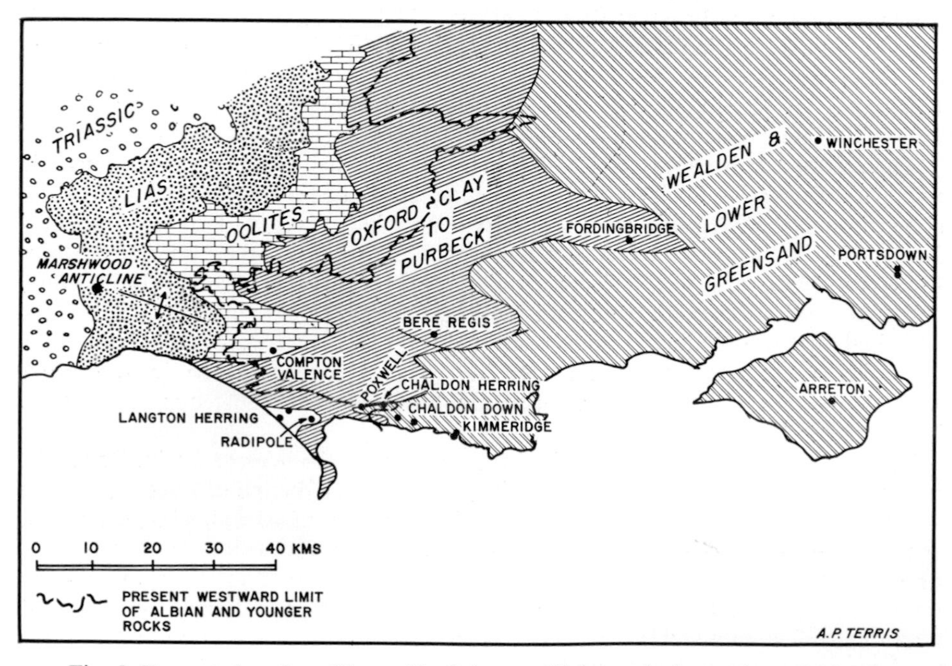


Fig. 5. Truncated surface ("incrop") of the pre-Albian rocks in southern England.

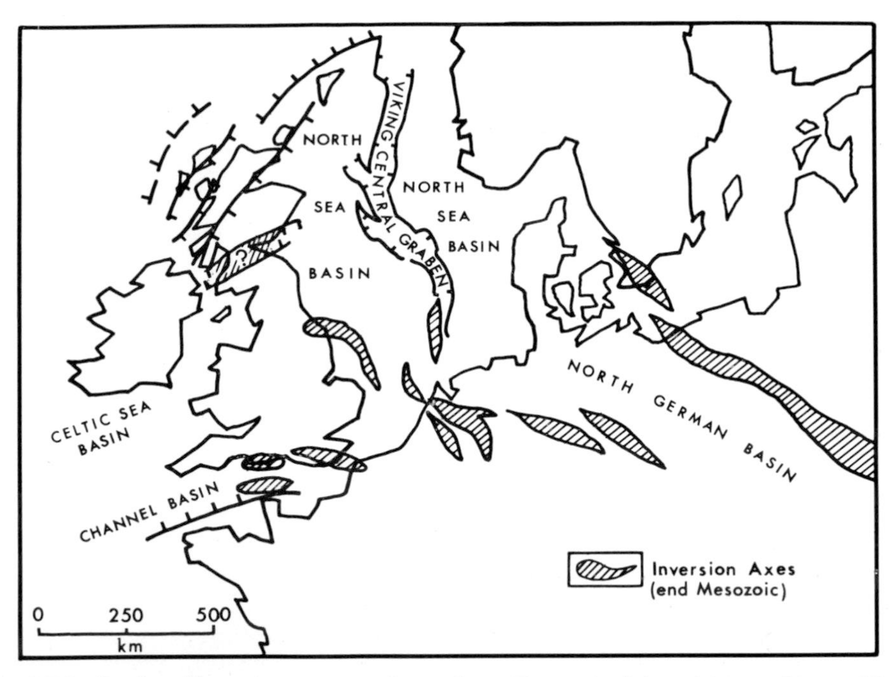


Fig. 6. Distribution of inversion structures in northwest Europe (mainly end-Mesozoic). Modified from Ziegler (1975).

## 2b. Eastern England and the southern North Sea

The post-Carboniferous basin of northeastern England developed from Permian times onwards as a subsiding area bounded on the south by the East Anglian massif (London platform) and on the west by the Pennines. Through the Permian and most of the Triassic this constituted a single simple area of subsidence, but at the end of the Trias differentiation into two linked basins began, with the slowly subsiding Market Weighton area separating two areas of more rapid subsidence, the Cleveland basin to the north and the Lincolnshire basin to the south (Kendall 1905; Arkell 1933). It became evident from later study that the belt of contemporary shallows was not a classic "anticline", as the earlier authors had assumed, but was essentially an unfolded buoyant block, in which the base of the Lias is closely parallel to the base of the unconformable Chalk over a distance of 20 km (Kent 1955, 1975a). This block had not been a major barrier between areas of contrasting sedimentation, but sedimentary facies changed progressively across much broader areas from mid-Lincolnshire into the Yorkshire basin.

Regional appraisal including the evidence from deep test wells in Lincolnshire and Yorkshire shows that the simple straight Jurassic outcrops extending from Grantham northwards to beyond Market Weighton are the surface stigmata of a shelf area which existed through the Carboniferous, Permian and Mesozoic—the East Midlands shelf—which has remained without any major dislocation through this long period (Kent 1966). The Market Weighton block is the northernmost tip of this unfolded area, joined to the shelf by a gentle monocline and bounded along its northern edge by a faulted hinge near Acklam (Fig. 7).

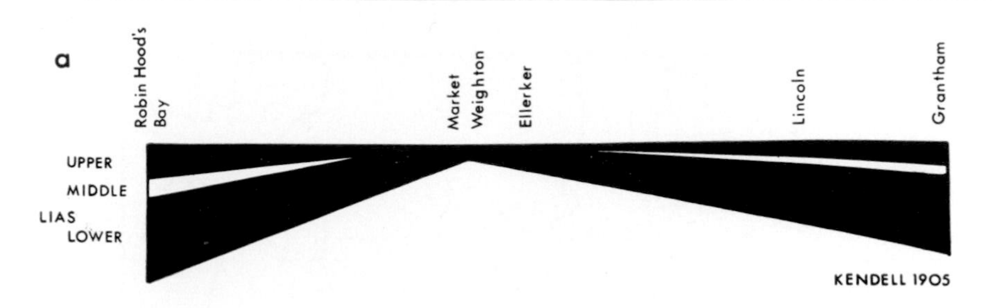


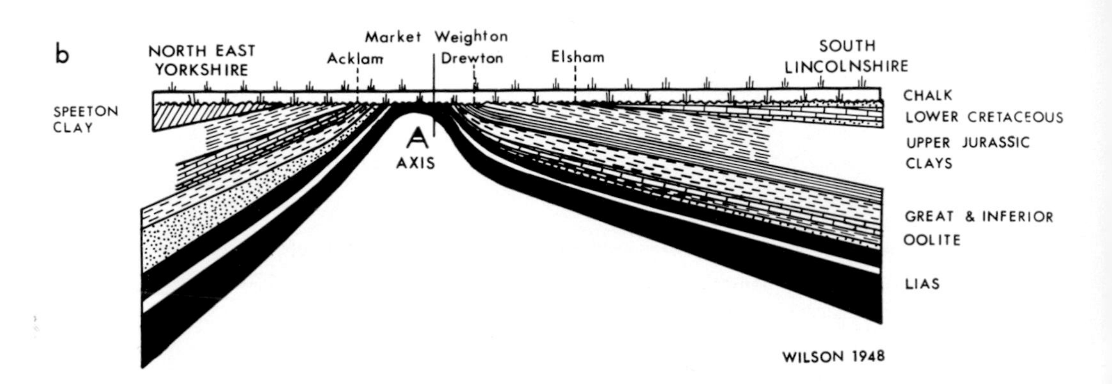


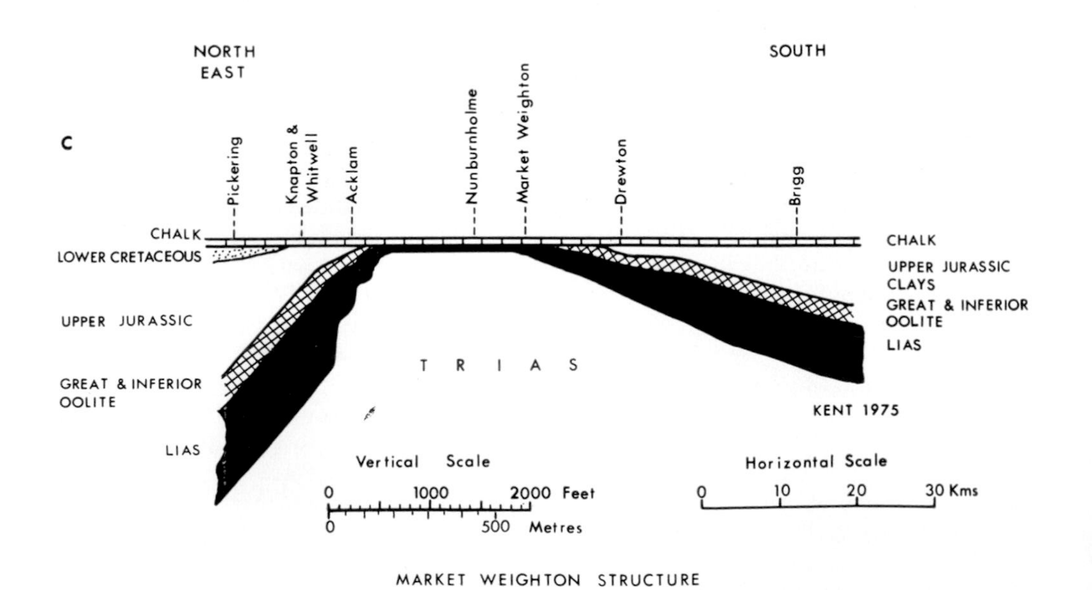


Fig. 7. The Market Weighton structure: sections illustrating developing concepts from contemporary anticlinal folding to a buoyant rigid block.

Offshore data from wells and seismic surveys show that the East Midlands shelf continues for *c*. 50 km offshore, and that its complex faulted edge continues east and then southeast to divide the shelf from the Sole Pit trough. Particular interest arises from this relationship, for the Sole Pit trough is an intra-Jurassic syncline which, like the Weald, has suffered inversion, the inversion being firmly dated as late late Cretaceous (Hancock and Scholle 1975). The landward continuation of the Sole Pit trough is provided by the Cleveland basin, with its axial Cleveland Hills anticline. There is aeromagnetic and palaeotemperature evidence that the deep structure beneath the anticline is a gentle syncline (compare the Weald) (Kent 1974 p. 16); regional compilation shows that the Jurassic of this area was deposited in a subsiding syncline, bounded by contemporary complex faults at its junction with the Market Weighton block (the pre-Cretaceous faults of the Howardian Hills). The amplitude of the folding of the Cleveland anticline measured in Jurassic rocks is much greater than the Pliocene uplift, and the evidence is compatible with a late Cretaceous–early Tertiary date for the inversion of the East Yorkshire Mesozoic basin, as will be shown in a forthcoming I.G.S. regional review (Kent 1978).

Eastern England, however, shows an additional feature, in that the Jurassic block-basin relationship inverted in the Cretaceous has an earlier parallel; the Cleveland area was basinal relative to the East Midland shelf also in the early Carboniferous, but before Permian deposition had similarly suffered inversion, so that the Upper Carboniferous was largely eroded over the basin, but survives on the former block beneath the Market Weighton area. There is no record of such a double phase of inversion in the North Sea, although there is some indication of a comparable history further south in the East Midlands, possibly in the area of the Carboniferous Widmerpool Gulf.

## 2c. The Southern Uplands, the Midland Valley of Scotland and the central North Sea

It has been shown elsewhere that the Southern Uplands with the Midland Valley to the north and the Solway basin to the south form a conjugate system which corresponds to the mid North Sea high flanked by basins to north and south (Kent 1975b). The Upper Palaeozoic history of the two positive areas is sufficiently close to indicate that they are essentially a single structural unit.

The Midland Valley of Scotland is a rift filled with Upper Palaeozoic strata (Lower Devonian to Permian) between the Highlands and the Southern Uplands. Recently deep crustal studies—gravity, later confirmed by the L.I.S.P.B. seismic line—have indicated that this rift coincides with a Lower Palaeozoic horst, a reversal of displacement on the boundaries possibly of 5–10 km. This is an inversion in the opposite sense to those in the Mesozoic, described above, and is perhaps unique in Britain.

The Midland Valley rift is not generally grouped with the Mesozoic basins, for the latest rocks within its landward limits are Permian volcanic rocks with a minor occurrence of Trias. Nevertheless it lies north of the Southern Uplands block which was positive relative to the Solway Permo-Trias and Jurassic basin, and it would be logical to assume that like the other flanking trough of the Southern Uplands, it originally contained Mesozoic sediments. Additionally, at its northeastern end in the North Sea it passes offshore into a Permo-Triassic downwarp containing halokinetic salt beds, implying a considerable thickness ($>2$ km) of clastic and evaporitic post-Carboniferous sediments. Following the Midland Valley to the southwest, moreover, its continuation in Northern Ireland is associated with the thickening of the

Lias and the Chalk; its boundary fault limits on the south the spread of Tertiary basalts and the rift includes the Oligocene Lough Neagh subsidence. Thus the Irish section, at least, had a history of post-Triassic movement, and the Scottish Midland Valley could have been similar.

The limited range of Mesozoic sediments in the Scottish Midland Valley itself may thus be a function of erosion, perhaps following a later Mesozoic inversion in its middle section, for it would be expected to have shared, at least in part, the Mesozoic subsidences recognised in its longitudinal continuations. Palaeotemperature studies could throw light on this problem, by indicating whether the existing rocks have formerly been more deeply buried.

## 2d. Eastern Scotland and the northern North Sea

Through most of the Mesozoic epoch, the northern mainland of Scotland appears to have been a source area for sediments accumulating in basins to east and west. The present rectilinear outline of eastern Scotland is in part related to fault lines, and Mesozoic faulting also provides the characteristic tectonic control within the North Sea basin, with slivers of Mesozoic rocks surviving on the Brora–Helmsdale coast to give some indication of the early tectonic history of the basin margin.

The dominant feature of the northern North Sea is the central Viking graben system. South of the Moray Firth this is flanked on the west by a broad shelf, thinly covered with Cretaceous and later sediments, which is partly the (downfaulted?) continuation of the Grampians and partly the shelved-out termination of the Central valley. The Moray Firth itself is a very strongly faulted basin of Permo-Triassic, Jurassic and Cretaceous rocks. North of it the East Shetland platform is a further area of shallow pre-Permian rocks which remained stable and unfolded through the Mesozoic and Tertiary.

Boreholes in the Moray Firth basin show a much higher proportion of sand in the Lower Cretaceous and Jurassic than is seen in the adjoining central North Sea areas, and there is evidence in the Brora–Helmsdale area that this is derived from the west, from the adjoining Highlands. Indications of contemporary intra-Jurassic movements of the boundary faults have long been known, and sharp thickness changes across the faults within the Moray Firth basin may also be due to contemporary movement. Alternatively they could be later tear faults, bringing different parts of the original Jurassic basin into juxtaposition.

The history of movement of offshore faults and fault blocks is becoming well documented as the oilfields are developed (Bowen 1975; Brennand and van Veen 1975; Pennington 1975; Williams *et al.* 1975; Howitt 1974). Distribution of the marine Permian shows that the rifting began at least as early as the Zechstein, and the expanded thicknesses of Trias, in rifts in the eastern North Sea and in the northerly Viking graben itself indicate contemporary fault movement measuring up to a thousand metres in some cases. Within the Jurassic sequences there is seismic evidence that faults die out upwards, and phases of fault movement are indicated at the end of the Lias, immediately pre-Callovian, late Kimmeridge and early Cretaceous (Halstead 1975). As Howitt (1974) and others have shown, only minor local faulting occurred after the early Cretaceous, even though the main period of opening of the adjoining North Atlantic was Tertiary.

Unconformities and stratigraphical breaks are best seen on the contemporary high fault blocks; between them, in the main graben, deposition was probably much more nearly continuous. Thus the Brent field on one of a series of fault scarps close to the Viking graben shows angular unconformities between Upper Jurassic and

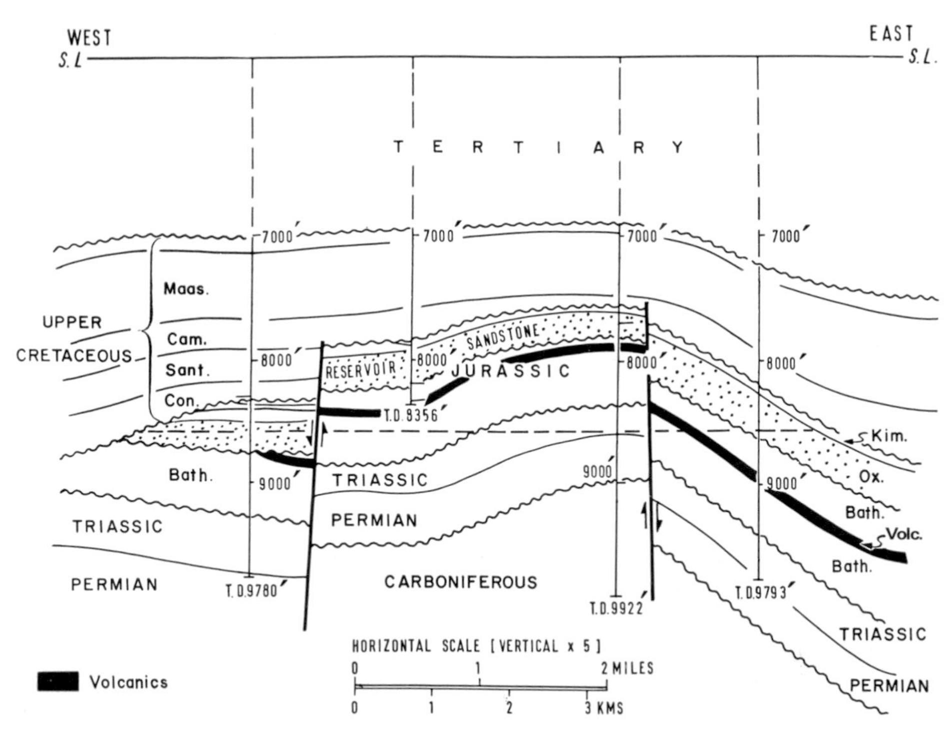


Fig. 8. Section through the Piper oilfield, northern North Sea, illustrating intermittent intra-Jurassic erosion on a slowly subsiding block (after Williams *et al.* 1975).

Aptian (Bowen 1975). In the Piper field, which occupies the crest of a gently warped fault-block near the edge of the Viking graben east of the Moray Firth, the oldest beds proved are Middle Carboniferous; these are unconformably overlain by Trias. On the Trias in turn rests non-marine Middle Jurassic (Bathonian) with angular unconformity. The Middle Jurassic is partly volcanogenic, and a basalt flow provides a useful structural marker, showing that the block was again tilted and eroded before deposition of the transgressive Upper Oxfordian marine sands. These grade upwards into dark Kimmeridgian shales. There is a time gap above the dark shales, but Lower Cretaceous (Barremian) follows without angular discordance. Aptian–Albian overlap these beds and were themselves faulted and eroded before deposition of the Turonian–Coniacian (Williams *et al.* 1975). Thus this block shows a long history of relative movement, extending from pre-Triassic to mid-Cretaceous. It was, nevertheless, uplift which did not result in any major erosion during Jurassic or Cretaceous times; as far as can be judged the sediments on this and other blocks were repeatedly lightly trimmed but never deeply eroded. In other documented cases, such as the Auk and Argyll oil fields, Jurassic rocks have been entirely removed from the high standing areas, and remain only in the graben, so that Lower Cretaceous transgresses onto Trias or Permian (Fig. 8).

It is evident that vertical movement of blocks and troughs in the North Sea long pre-dates the opening of the North Atlantic, and has to be ascribed to very long-term tension. Furthermore, there is no evidence known to the present writer to show that the Viking graben originated by collapse of a dome, as several authors have recently postulated on the basis of model studies. On the contrary, the middle of the North Sea can be shown to have been structurally depressed in the Devonian, Carboniferous, late Permian, Trias, Lias, late Jurassic and following periods. From at least the late Permian to middle Cretaceous this negative movement was fault related.

As contrasted with a rift such as the Red Sea, with which it is often compared, the North Sea thus represents progressive basinal development over a significant part of Phanerozoic time; its fault patterns are highly complicated, vulcanicity is of very minor importance and the time of its development is only very imperfectly related to that of the adjoining ocean.

## 2e. The West Shetland and Hebridean basins

Offshore northwest Scotland includes faulted basins of two different types, the external basin immediately west of the Orkneys and Shetlands, and the internal basin of the Sea of the Hebrides, both of which are within the limits of the continental basement. The two basins lie in line and control of both by the Minch fault system has been suspected, although the throws of the major faults in the two areas are in opposite directions.

An authoritative account of the West Shetland basin by Cashion (1975) shows that the 1000–2000 m boundary fault, the Shetland Spine fault, marks the inner edge of a block tilted away from the ocean, shown by formational thicknesses to have developed during the Permo-Triassic, Jurassic, early and early late Cretaceous. Later Cretaceous and Tertiary are draped over this structure, thickening oceanward without the complication of major fault control. Thinning onto the high edge of the tilted block is mostly piecemeal without major truncation below the Upper Cretaceous. Faults are variously truncated at the Cimmerian unconformity (pre-Lower Cretaceous), by the Upper Cretaceous and by the Tertiary (Fig. 9).

Southwards, subsidiary faults are developed within the main tilted block, and the resulting sub-blocks show independent movement of a similar kind, particularly in the Jurassic. The basement ridge which bounds the West Shetland basin on the oceanward side is itself the summit of an analogous west-facing fault scarp, on the ocean side of which 4000 m of Tertiary and Cretaceous rest on hypothetical Jurassic.

Mesozoic sedimentation in this area is hence directly fault controlled, and although they are located on the oceanic edge, origin of the fault controlled basins extends backwards in time long before the opening of the North Atlantic Ocean.

In the area further south, in the Sea of the Hebrides and its adjoining land areas, Mesozoic distribution is dominated by three major faults, the Minch fault complex flanking the Outer Hebrides, the Camasunary–Skerryvore fault extending SSW from southern Skye past Coll and Tiree, and the Great Glen fault which crosses Mull and continues offshore southwestwards towards Ulster (Binns *et al.* 1973, 1975). During the Mesozoic these fractures acted as "trapdoor" faults, downthrown on the southeastern sides, so that on the Minch and Camasunary faults, at least, the New Red and Jurassic sediments tend to be thickest near the western side of each fault basin. Outcrops of both the Upper Cretaceous and the Tertiary volcanic rocks are frequently limited by the major faults, indicating that the final movements were post-Eocene, but both locally transgress across major faults which limit the Jurassic and earlier formations. Thus this area also shows contemporary fault-controlled basins active during the Mesozoic, plus widely transgressive Cretaceous and Tertiary volcanic rocks followed by a lesser renewal of faulting (Figs 10, 11).

Hence in both the West Shetlands and the Sea of the Hebrides Mesozoic sedimen-

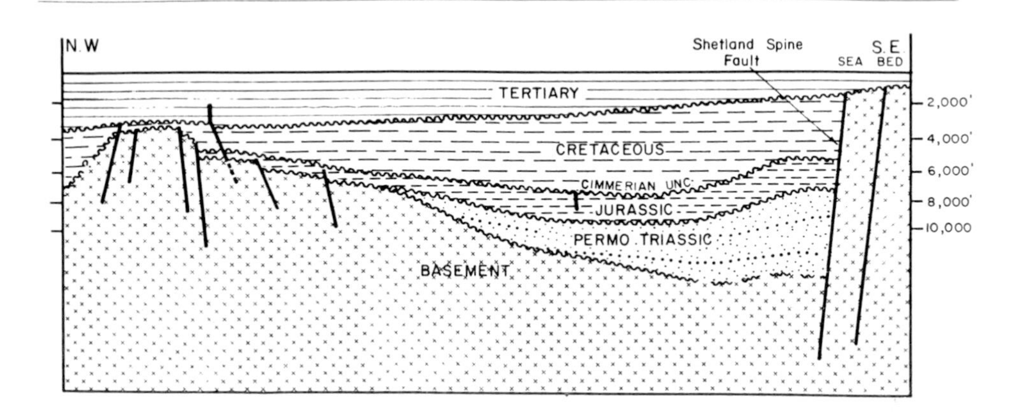


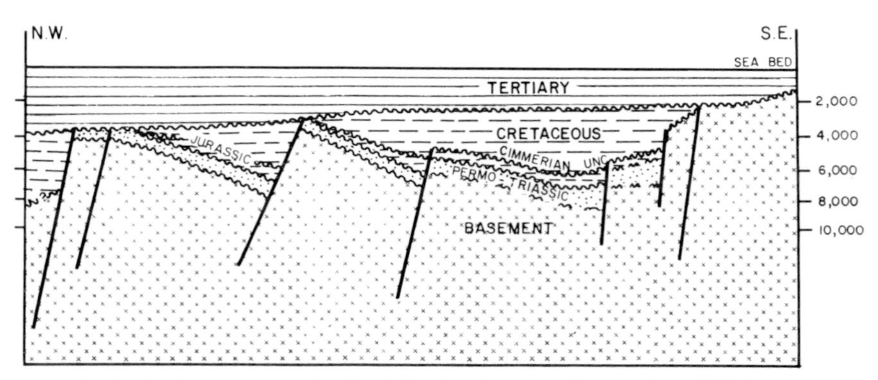


Fig. 9. Section through the West Shetland basin (after Cashion 1975).

tation was controlled by contemporary tilted fault-block basins, with the difference that the former were tilted eastwards, the latter (for no obvious reason) in the opposite direction. The Tertiary volcanic centres which later developed in the Sea of the Hebrides tend to be grouped near the large faults, but show no extension along the strike and their location appears to be fundamentally due to a different, no doubt much deeper, tectonic control.

## 2f. Western England and the Irish Sea basins

The inshore and offshore Mesozoic basins which surround the Welsh massif had a common history in at least the Permo-Triassic. On land the Worcester graben and Cheshire basin are major subsidences (2000–4000 m depth) with monoclinal or faulted margins; the latter area is also cut by large faults internally. In east Lancashire the onshore edge of the northern Irish Sea basin is also controlled by faults; both there and offshore the deposition during the Permo-Trias amounted to 3000 m plus (incomplete thickness) (Colter and Barr 1975).

These three basins have lost the later sediments which were no doubt deposited widely across them, for Jurassic remnants survive in the Worcester graben and the Cheshire basin, and there can be little doubt that at least the Upper Cretaceous was deposited over all but the highest ground (George 1974). From Cardigan Bay southwards, however, Jurassic. Cretaceous and Tertiary rocks survive and demons-

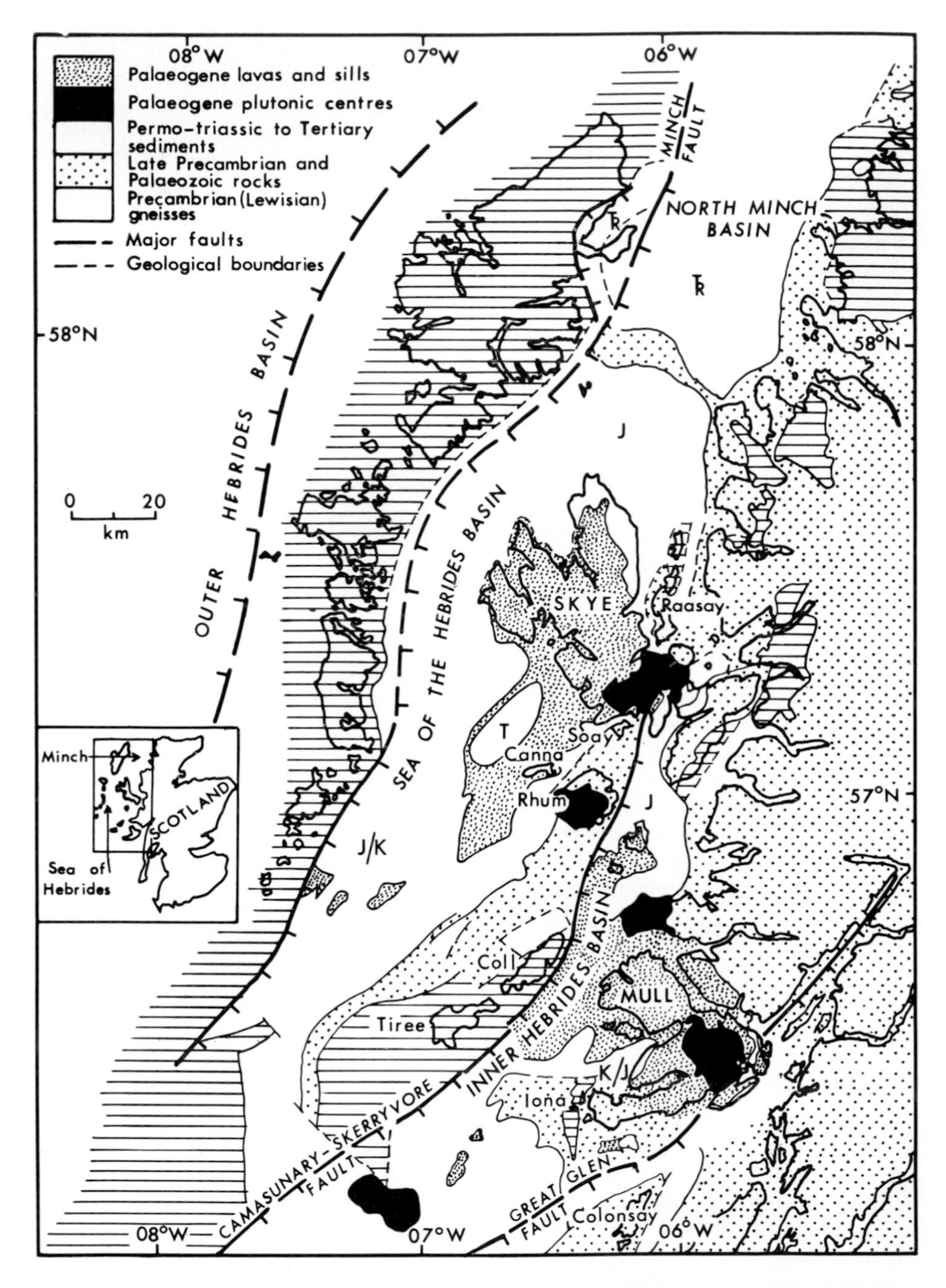


Fig. 10. Structural map of the Hebridean area, cut by large faults which control Mesozoic thicknesses (based on Binns *et al.* 1975).

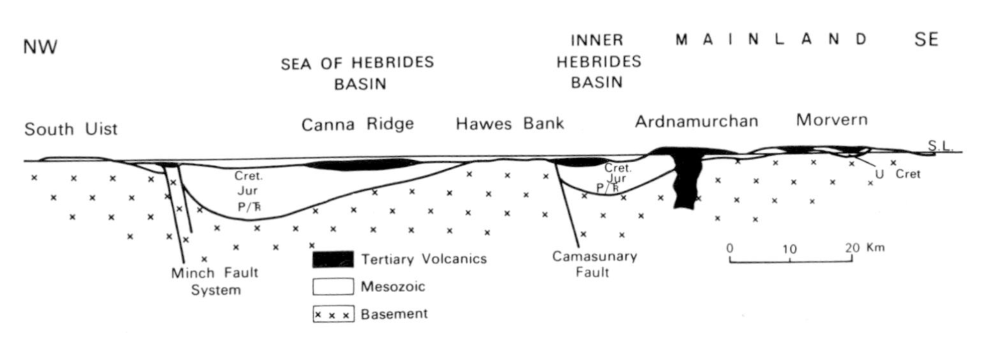


Fig. 11. Cross-section of the Sea of the Hebrides.

trate that the main period of graben formation had ended by the late Cretaceous. It is fair to assume that all these basins shared a similar tectonic history, although the chronology is not as precisely controlled as e.g. that in the northern North Sea.

## 3. Synthesis

The Mesozoic history of the British area has been essentially one of vertical tectonics in a cratonic environment, with widespread basin subsidence which was initially fault-controlled but later developed by simple downwarping aggregating 3–8 km or more in different areas.

The negative aspect of vertical movements has been dominant in the region with positive intermittent displacement of intervening blocks relative to sea level. Except for the Highlands, uplift of associated Caledonian and Hercynian massifs has been small, as shown for instance by the survival of Permo-Triassic relics in the Southern Uplands, Cretaceous on the Scottish mainland at Morvern, Triassic relics (redeposited in the Pliocene) in the southern Pennines, Eocene relics in southwest Wales, and Permian and Oligocene in rifts on the Cornubian peninsula. The geometry of the stratal relationships between basins and blocks both offshore and onshore is one of attenuation (often fault assisted) of successive formations; the blocks were almost continuously buoyant but, again except for the Highlands, none was deeply eroded from the Permian onwards. Many in fact show comparable stability from late Devonian times.

The fault-controlled phase of the vertical movements in the British seas pre-dated the opening of the North Atlantic, demonstrating that here (as fairly universally on the world's Atlantic-type coasts) the lateral opening was preceded by a much longer period of gentle tension than subsequently marked the development of new oceanic crust. The pre-opening tension ended late in the early Cretaceous—again world-wide—with a date much more uniform than the times of the opening of adjacent oceans (Kent 1977). In this region there is a further complication of inversion of many of the troughs at the end of the Mesozoic—in mainland Europe, in the English Channel and in the North Sea. The Midland Valley of Scotland may prove to be a further case, but other troughs in northern Britain behaved normally.

Thus, fundamental processes are more complicated in space and time than the simple plate tectonic concept would indicate, and continued investigation of deep-seated relationships on the margin of northwest Europe should make a major contribution to their understanding, and hence to interpretation of the structure of other continental margins.

*Acknowledgments.* Thanks are expressed to British Petroleum Company Ltd. for assistance with the figures and to Mr. A. P. Terris for Figure 5.

## References

ARKELL, W. M. 1933. *The Jurassic System in Great Britain.* Clarendon Press, Oxford.

BINNS, P. E., MCQUILLIN, R., FANNIN, N. G. T., KENOLTY, N. and ARDUS, D. A. *et al.* 1975. Structure and stratigraphy of sedimentary basins in the Sea of the Hebrides and the Minches. *In* Woodland, A. W., 93–102.

——, —— and KENOLTY, N. 1974. The geology of the Sea of the Hebrides. *Rep. Inst. geol. Sci.* **73/14.**

BOWEN, J. M. 1975. The Brent oil-field. *In* Woodland, A. W., 353–360.

BRENNAND, T. P. and VAN VEEN, F. R. 1975. The Auk oil-field. *In* Woodland, A. W., 275–281.

CASHION, W. W. 1975. The geology of the west Shetland Basin. In *Offshore Europe '75 Conference Papers.* Paper OE-75 216. *Offshore Services Magazine*, Spearhead Publications Ltd., Kingston-upon-Thames.

COLTER, V. S. and BARR, K. W. 1975. Recent developments in the geology of the Irish Sea and Cheshire Basins. *In* Woodland, A. W., 61–75.

GEORGE, T. N. 1974. Prologue to a Geomorphology of Britain. *Spec. Publ. Inst. Brit. Geogr.* No. 7, 113–125.

GODWIN-AUSTEN, R. 1856. On the possible extension of the Coal Measures beneath the south-eastern part of England. *Q. J. geol. Soc. Lond.* **12,** 38–46.

HALLAM, A. 1975. *Jurassic Environments.* Cambridge University Press.

HALSTEAD, P. H. 1975. Northern North Sea faulting. In *NPF–Jurassic Northern North Sea Symposium 1975 Proceedings.* Paper 10. Norwegian Petroleum Society (NPF).

HANCOCK, J. M. and SCHOLLE, P. A. 1975. Chalk of the North Sea. *In* Woodland, A. W., 413–425.

HOWITT, F. 1974. North Sea oil in a World Context. *Nature, Lond.* **249,** 700–703.

KENDALL, P. F. 1905. Subreport on the concealed portion of the coalfields of Yorkshire, Derbyshire and Nottinghamshire. *Final. Rep. Roy. Comm. on Coal Supplies* Pt **9** App. 3, 188–205.

KENT, P. E. 1949. A Structure Contour Map of the Surface of the Buried Pre-Permian Rocks of England and Wales. *Proc. Geol. Assoc.* **60,** 87–104.

——1955. The Market Weighton structure. *Proc. Yorks. geol. Soc.* **30,** 197–227.

——1966. The structure of the concealed Carboniferous rocks of north-eastern England. *Proc. Yorks. geol. Soc.* **35,** 323-352.

—— 1974. Structural History. *In* Rayner, D. H. and Hemingway, J. E. (Editors). *The geology and mineral resources of Yorkshire.* 13–28. Yorks. geol. Soc.

—— 1975a. The Tectonic Development of Great Britain and the surrounding seas. *In* Woodland, A. W., 3–28.

—— 1975b. Review of North Sea Basin development. *J. geol. Soc. Lond.* **131,** 435–468.

—— 1977. The Mesozoic development of aseismic continental shelves. *J. geol. Soc. Lond.* **134,** 1–18.

—— 1978. The Tees to the Wash. British Regional Geology. *Inst. geol. Sci.* H.M.S.O.

LAMPLUGH, G. W. 1919. The Structure of the Weald and analogous tracts. *Q. J. geol. Soc. Lond.* **75,** lxxiii-xcv.

LEES, G. M. and COX, P. T. 1937. The Geological Basis of the present search for oil in Great Britain by the D'Arcy Exporation Co. Ltd. *Q. J. geol. Soc. Lond.* **93,** 156–194.

PENNINGTON, J. J. 1975. The geology of the Argyll field. *In* Woodland, A. W., 285–291.

SMITH, A. J. and CURRY, D. 1975. The Structure and geological evolution of the English. Channel. *Phil. Trans. R. Soc. Lond.* Ser. A., **279** (1288), 3–20.

STAMP, L. D. 1934. *An Introduction to Stratigraphy* (*British Isles*). Thomas Murby & Co., London. 2nd Edit.

WILLIAMS, J. J. *et al.* 1975. The Piper oil-field, UK North Sea: a fault-block structure with Upper Jurassic beach-bar reservoir sands. *In* Woodland, A. W., 363–377.

WILSON, V. 1948. British Regional Geology *East Yorkshire and Lincolnshire.* H.M.S.O., London.

WOODLAND, A. W. (Editor) 1975. *Petroleum and the continental shelf of northwest Europe. 1. Geology.* Applied Science Publishers, Barking.

ZIEGLER, P. A. 1975. North Sea basin history in the tectonic framework of north-western Europe. *In* Woodland, A. W., 131–149.

Sir Peter Kent,
N.E.R.C. Alhambra House,
27–33 Charing Cross Road,
London WC2., England.

# Evolution of fault-controlled ensialic basins in northwestern Britain

*A. C. McLean*

Fault-bounded basins in northwestern Britain and its shelf, which contain New Red Sandstone and younger rocks, may be grouped into (i) a Marginal Belt, stretching inwards for 150 km from the shelf edge, within which basins trend NE–SW, parallel to the continental margin, and (ii) an inner, Clyde Belt, up to 150 km wide, which extends to the northwest, certainly as far as the Highland Boundary fault, and probably beyond that to Skye. To the southeast, it stretches to the Vale of Clwyd and Cheshire. Within it, faults are predominantly of NW–SE and NNW–SSE trend. It changes character as it crosses older NE–SW-trending crustal blocks. It came into existence in Westphalian or possibly Namurian times, and the largest subsidence took place in Permo-Triassic times.

The development of the Marginal Belt is explained by a mechanism which would relate it to the continental margin and to a pre-Triassic Rockall trough. The spatial distribution of the Clyde Belt makes any simple relationship to the formation of the continental margin unlikely. It is the product of a NE–SW tension, similar in orientation and age to stress which affected other areas of northwest Europe even further from the margin. The reactivation of part of the Clyde Belt in Tertiary times, suggests that the stress may be related to early stages of plate transport from a new spreading axis. The localization of strain within the Clyde Belt is not explicable by a single older structure at depth along its entire length, even though a scatter of older structures, including granite masses, control the positions of individual faults and troughs. It is suggested, on the basis of stratigraphical evidence in the Midland Valley, that flexuring of the crust as the North Sea basin formed, culminated in the Clyde region and that a crustal peculiarity of this type, if present along the entire belt, may be the structural control on its location.

## 1. Introduction

Exploration of the continental shelf to the west of Scotland was started in the mid-1960's by university-based geophysicists; shortly afterwards the Continental Shelf Unit North and the Marine Geophysics Unit of the Institute of Geological Sciences (I.G.S.) commenced a systematic reconnaissance. Gravity, magnetic and shallow seismic-profiling surveys, buttressed by some refraction lines and by deep

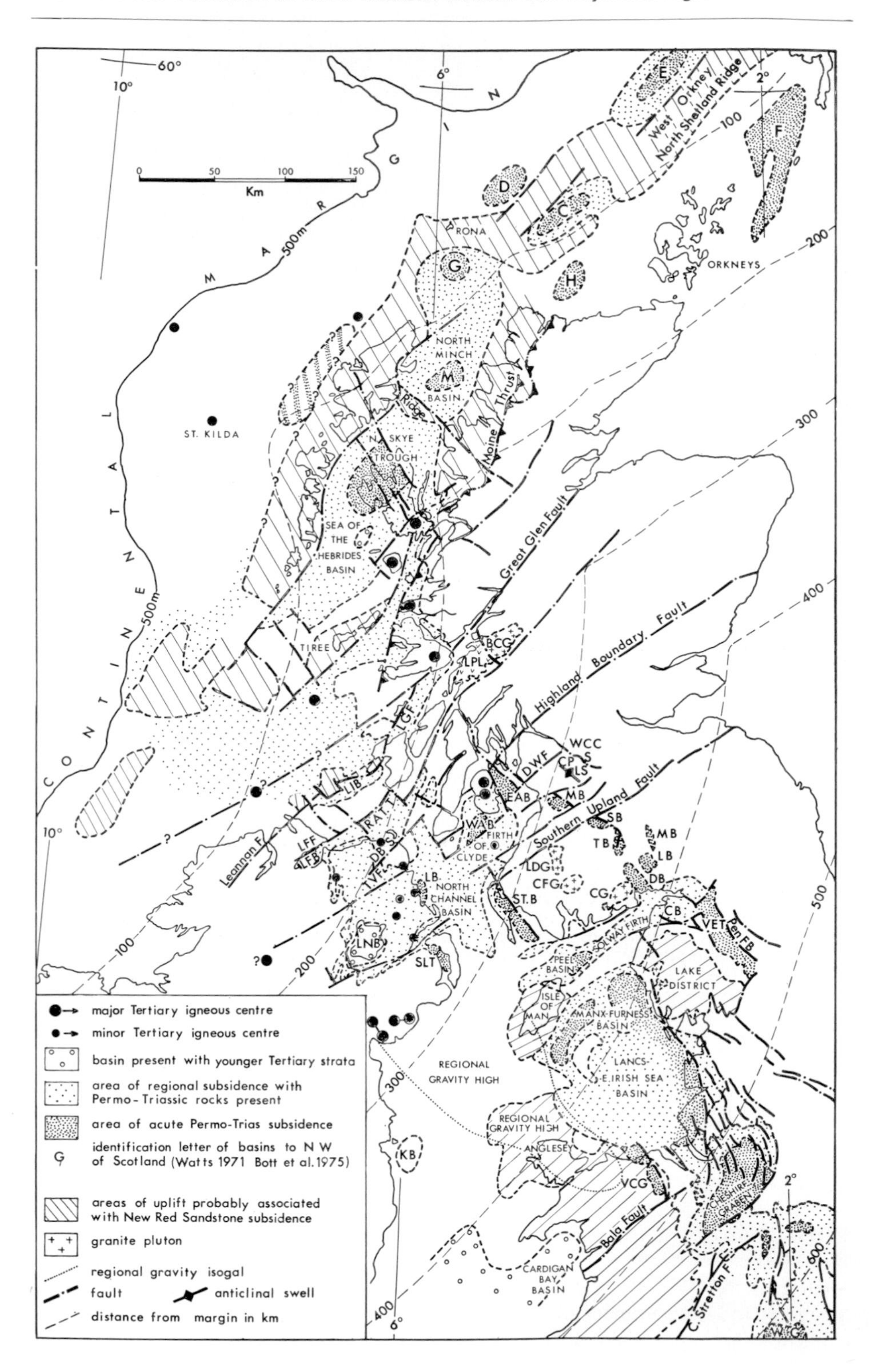
60°
10°
6°
2°
0 50 100 150
Km
CONTINENTAL MARGIN
500m
ST. KILDA
RONA
NORTH MINCH BASIN
N. SKYE TROUGH
SEA OF THE HEBRIDES BASIN
TIREE
ORKNEYS
West Orkney
North Shetland Ridge
Moine Thrust
Great Glen Fault
Highland Boundary Fault
Southern Upland Fault
Leannan F.
FIRTH OF CLYDE
NORTH CHANNEL BASIN
SOLWAY FIRTH
PEEL BASIN
ISLE OF MAN
MANX-FURNESS BASIN
LAKE DISTRICT
LANCS-E.IRISH SEA BASIN
REGIONAL GRAVITY HIGH
ANGLESEY
CHESHIRE GRABEN
Bala Fault
CARDIGAN BAY BASIN
C. Stretton F.
100
200
300
400
500
600
major Tertiary igneous centre
minor Tertiary igneous centre
basin present with younger Tertiary strata
area of regional subsidence with Permo-Triassic rocks present
area of acute Permo-Trias subsidence
identification letter of basins to N W of Scotland (Watts 1971 Bott et al. 1975)
areas of uplift probably associated with New Red Sandstone subsidence
granite pluton
regional gravity isogal
fault
anticlinal swell
distance from margin in km

seismic-reflection surveys (McQuillin and Bacon 1974), had been carried out over most of the western Scottish shelf by the mid-1970's and had defined the regional structure. Confirmatory shallow (up to about 50 m) drilling in drift and solid, and sampling (Chesher *et al.* 1972), followed on the geophysical surveys and gave a general picture of the main outcrops. The region still largely lacks information from a network of seismic-reflection lines and deep borings of the type associated with prospecting for oil: the one exception is the waters west of Shetland (Kent 1975; Whitbread 1975). In consequence, a substantial knowledge of the stratigraphy of the region is not yet available, and what exists is usually dependent upon observations on land and the use of densities and velocities for the identification of formations. As the literature relating to these reconnaissance investigations is extensive, only selected references are given here.

The structures discovered on the shelf include a series of deep basins (Fig. 1), the infill of which consists largely of Upper Carboniferous and Permian rocks together with Mesozoic rocks, mostly of New Red Sandstone facies. Many are bounded by faults, some of which have moved during Mesozoic times, indicating that at least some of these structures are basins of sedimentation whose development may be inferred from the stratigraphy of their infill. The nature of the infill does not, however, permit a refined stratigraphy, or define a sea-level datum for absolute subsidence or uplift. The proximity of deep basins to the continental margin is a common phenomenon along the Atlantic margins and suggests that the development of basins is part of its developmental cycle. Hence the basins to the west of Scotland may be a clue to the age of the adjacent continental edge and, in turn, to the age of the Rockall trough, which lies further west. This oceanic, or quasi-oceanic, region is interpreted by many, but not all, as the site of an early spreading axis between the micro-continent of Rockall plateau and western Europe, which ceased to function before the presently active Reykjanes ridge came into existence. There is disagreement about the age of the trough, with the majority of authors

Fig. 1. Areas of subsidence where New Red Sandstone is preserved in northwest Britain. The map also shows the important faults, the areas of ?New Red Sandstone uplift and granite plutons close to the subsidence, Tertiary igneous centres, and distances from the −500 m contour, which is used to define the continental margin.
Sources of data are acknowledged in the text. The following abbreviations are used:

| | | | |
|---|---|---|---|
| BCG | Ben Cruachan granite | LPL | Lorne Plateau Lavas |
| LGF | Loch Gruinart fault | LIB | Loch Indaal basin |
| RAT.T | Rathlin trough | LFF | Lough Foyle fault |
| LFB | Lough Foyle basin | DBS | Dungiven–Ballycastle syncline |
| TVF | Tow Valley fault | LB | Large basin |
| LNB | Lough Neagh basin | SLT | Strangford trough |
| DWF | Dusk Water fault | WCCS | West Central Coalfield syncline |
| CPLS | Clyde Plateau Lavas swell | EAB | East Arran basin |
| MB | Mauchline basin | WAB | West Arran basin |
| SB | Sanquhar basin | TB | Thornhill basin |
| MB | Moffat basin | LB | Lochmaben basin |
| DB | Dumfries basin | LDG | LochDoon granite |
| CFG | Cairnsmore of Fleet granite | CG | Criffel granite |
| ST.B | Stranraer basin | CB | Carlisle basin |
| VET | Vale of Eden trough | PenFB | Pennine Fault belt |
| VCG | Vale of Clwyd graben | WG | Worcester graben |
| KB | Kish basin | SFB | Solway Firth basin |

## Table 1. Mechanism of basin subsidence

| BOTT 1976 | THERMAL | LOAD | STRESS |
|---|---|---|---|
| PRIMARY AGENTS | Rise in geisotherms | Increased load | Regional tension |
| | a. lithospheric rupture - - - - - - - - - - - - - - - - - - - - - | - - - - - - - - - - - - - - - - | at a new ocean ridge. |
| | b. massive replacement of uppermost mantle by asthenosphere. | a. thick sedimentation especially on outer shelf. | b. local surface load, and thicker crust. |
| | c. emplacement of <u>huge</u> basic or ultrabasic bodies. | b. subcrustal thinning. | c. imperfect isostatic processes. |
| | d. change of lower crust to granulite or eclogite facies. | | d. increased curvature of lithosphere (doming, membrane tectonics). |
| | | | e. drag of convection currents on bottom of lithosphere. |
| | | | Secondary stress<br>f. major strike-slip movement along irregular fault-plane. |
| PRIMARY PROCESSES | 1. Thermal expansion (for a. and b.: distance from centre to flank 200-300 km; uplift of order 1.5 km over period <u>up to</u> 50 m.y. | Isostatic adjustment<br>a. by flow asthenosphere (for <u>c</u> 1.5 km over 25 m.y.; flexural parameter of lithosphere <u>c</u> 200 km). | 1. Creep of ductile lower lithosphere towards new shelf edge with brittle failure in upper crust. |
| | 2. Erosion of swell. | b. by flow of partial melt fluid. | 2. Flow towards a rising mountain range or granite pluton, with corresponding subsidence. |
| | 3. Contraction which decreases exponentially with time (for a. and b. over <u>c</u> 50 m.y. to allow <u>c</u> 2 km of sediment to accumulate). | flexure of elastic lithosphere. | 3. Normal faulting to base of elastic lithosphere with (i) compl. fault bounding a wedge, width determined by thickness of lithosphere. (30-60 km wide wedge formed). (ii) horst and graben developed by isostatic response to wedge shape. |
| | | | 4. (f. above) local compression and tension with "pull-apart" troughs. |
| SECONDARY EFFECTS IN LITHOSPHERE | 1. Initial decrease of thickness of 'elastic' lithosphere (to 30-60 km) affects<br>a. flexure under sediment load.<br>b. concentration of regional tension by "necking".<br>2. Load of denser igneous and "phase change" rocks in crust.<br>3. Load of sediments, compensated by regional isostasy.<br>4. Local stretching above forceful intrusion (<u>but</u> thermoelastic stress is negligible). | Load is a secondary effect <u>only</u>, unless partial melt fluid is displaced. | Uprise of hot fluids and heat into uppermost mantle and crust facilitated; igneous activity may be triggered. |

favouring an early Cretaceous date. The arguments are summarised by Russell (1976) who favours an opening in "earliest Permian" times, and extends an older argument that events on the outer shelf, when recorded by subsidence, are the best evidence yet available of the history of the margin.

General accounts of the western Scottish shelf basins appear in Woodland (1975) and the broad pattern of outcrops is shown on the I.G.S. (1972) map—The Sub-

Table 2. Intra-continental subsidence mechanisms criteria of primary cause

| | THERMAL EVENT | STRESS EVENT |
|---|---|---|
| STRATIGRAPHIC | 1. Igneous activity, if any, precedes subsidence.<br>2. Uplift precedes subsidence by up to 50 m.y. No complementary uplift other than isostatic adjustment at rims during late stage.<br>3. Subsidence affects a wide area. If the event is at the base of the crust --<br>400 to 600 km across basin<br>Uplift of order (max) 1.5 km<br>Subsidence of order (max) 3 km<br>4. Uplift and subsidence are affected only by thermal properties of rocks. Control by earlier structures is small. Little or none of the subsidence is along faults.<br>5. Gradual change of flexure of lithosphere, and shape of basin with time.<br>6. Subsidence decreases exponentially with time over 50 m.y. (but eustatic changes may confuse the record). | 1. Igneous activity, if any, is contemporaneous with faulting. High geothermal gradient assists flexuring and failure.<br>2. Complementary uplift is contemporaneous with subsidence (unless lithosphere reacts by attenuation).<br>3. Width of areas of subsidence/uplift are related to thickness of "elastic" lithosphere. If brittle upper layer is 10 to 20 km thick and is homogen. graben is 30 to 60 km wide. Subsidence up to 5 km, and amount is inversely proportioned to width of graben, and proportioned to ease of uplift of horsts.<br>4. Subsidence along normal faults or hinge lines over basement faults is usual. Location of faults and areas of subsidence are affected by inhomogeneities in crust (eg. granites, older faults).<br>5. Graben widens upwards.<br>6. Subsidence may be spasmodic and rapid. |
| CRUSTAL | 7. Crust below basinal sediment is thinner (c 2 km) than surroundings. Regional positive (after stripping sediment) Bouguer anomaly over basin, plus contributions from<br>(i) imperfect isostatic adjustment in centre<br>(ii) igneous masses in crust<br>(iii) changes to denser metamorphic facies in lower crust. | 7. Average crustal thickness not changed significantly (unless there has been crustal attenuation). |

Pleistocene Geology of the British Isles and the adjacent continental shelf, 1:2500000. A brief review by Hall and Smythe (1973) examines the distribution and structure of shelf basin of northwest Britain with respect to the continental margin, in response to other views published by Hallam (1972). They, and earlier authors recognised that not all Mesozoic basins in northwest Britain simply parallel the shelf margin, and that not all are explicable by one mechanism (cf. Bott 1976b).

In this account, the description of the basins is augmented by results from the Firth of Clyde–North Channel region and is extended to embrace similar New Red Sandstone basins on land. The latter information gives a picture of distribution, not arbitrarily limited by a coast-line, and provides in places a stratigraphy which is more detailed than is available for most of the marine areas. The distribution of basins infilled with upper Carboniferous or younger rocks indicates the presence of two groups, each of which forms a structural belt defined by its geometry relative to the margin and by the predominant trend of individual basins (Fig. 1; the principal sources of information for this map include the I.G.S. Continental Shelf Map (1972), the I.G.S. Tectonic Map (Dunning 1966), Hall and Smythe (1973) and other publications referred to elsewhere in the text). The Marginal Belt of basins lies within a zone stretching 150 km continentwards from the 500 m contour. It trends NE–SW, sub-parallel to the margin and stretches from west of Shetland to northwest of Ireland. The second zone is less clearly a unity as it suffers some change of character along its length. It trends NNW–SSE, making a distinct angle with the shelf edge, and it crosses successive NE–SW-oriented structural blocks which influence its development regionally. Individual faults and basins within it commonly trend NW–SE. It extends from the Cheshire graben certainly as far as the Highland Boundary fault and probably as far as the Little Minch, where it coincides with the Marginal Belt. It is well developed in the Firth of Clyde region, and as the name Clyde appears not only there, but in modified form in another part (Vale of Clwyd), it is referred to here as the Clyde Belt.

## 2. Models and mechanisms of subsidence

Before describing the zones of subsidence, it is appropriate to briefly review what is known about the mechanisms that may produce them, and the criteria by which mechanisms may be distinguished. An understanding of the origin and development of the New Red Sandstone basins in northwest Britain has been sought usually either (i) by trying to recognise a stratigraphic pattern (cf. Kent 1976) or a structural pattern (cf. Burke 1976) in the regional geology of the Atlantic margins or (ii) by the study of models based on possible mechanisms of subsidence. A few specific areas of subsidence, such as the Michigan basin (Haxby *et al.* 1976) have been analysed quantitatively in the latter way. Basins in NW Britain have been instanced as examples, but attempts to explain individual structures in quantitative detail are rare.

Present views and knowledge of mechanisms of subsidence have been summed up by Bott (1976a), and it is his grouping into mechanisms based on a thermal, or on a stress event as the prime agent that is followed and extended in Table 1. The criteria by which a primary cause of subsidence (thermal or stress) may be recognised, are listed in Table 2. Increased load with isostatic adjustment is normally a secondary effect only, since light sediments usually displace denser mantle. However, load may be more important if a partial melt fraction were displaced as might happen in special geological conditions with the lower density of the melt possibly allowing sinking of comparable magnitude to sediment accumulation (Scheidegger and O'Keefe 1967).

The parameters of the thermal models and processes are those derived by Sleep (1969, 1971), and Sleep and Snell (1976) from theoretical considerations and from the evidence of cooling at a mid-ocean ridge. The time constant of contraction, 50 m.y., may be in error by 10 m.y. (Sleep 1971 p. 326). The flexural parameter of the elastic upper layer of the lithospere which is important in defining its reaction to loads, such as thick sediment on the outer shelf (Watts and Ryan 1976) or denser rocks produced by a phase change (Haxby *et al.* 1976), is determined directly in a few areas (Sleep and Snell 1976 p. 127). The variation of thickness of the elastic lithosphere, as geisotherms rise and fall, is inferred to be from 32 to 60 km in the Michigan basin.

Stress as a prime agent of subsidence in different geological conditions has been investigated by Bott and co-workers, and the parameters of the stress-event models quoted in Table 1, come from their papers. These include studies of stress at a new continental margin (Bott 1971; Bott and Dean 1972, 1973) where creep oceanwards, induced by the gravitational energy of the new slope, takes place below the elastic lithosphere. The general features of this model are in accord with estimates of creep in the lithosphere made by Murrell (1976). The width of zone affected by such stress, about 150 km landwards from the continental margin, can be estimated directly from the structure of simpler marginal regions (Hall and Smythe 1973 pp. 54–55). Subsidence may also be produced within cratonic areas if regional tension produces normal faulting extending to a deeper ductile level in the lithosphere (Vening Meinesz 1950). The width of graben produced depends critically on the thickness of the brittle upper layer. The values in Table 1 are based on Bott (1976b) who presumes the layer to be 10 to 20 km thick. These are at the lower end of other estimates of its thickness (cf. Murrell 1976) and the 30 to 60 km width derived by Bott is probably a minimal figure, rather than a typical one. The flow of ductile, deeper rocks towards a rising mountain range (Bott 1964a), or stresses concentrated at the margin of a granite pluton (Bott 1965 pp. 196–200; 1967 p. 159) may also produce subsidence in adjacent areas. Local subsidence along major fault-lines may result from secondary effects of the main displacement (cf. Cloos 1931; Steel 1976).

Variation in the dimensions of these models of subsidence depends on local geological conditions. The precise figures are open to qualification, but the *order* of magnitude is certain, and may be used as a criterion. In attributing specific subsidence to heat or stress as the prime agent, it should, however, be borne in mind that in many cases the other will also be a secondary complicating effect.

## 3. The Marginal Belt

Within the area of Figure 1, the basins of the Marginal Belt extend from west of the Shetlands to northwest of Ireland, and are continued to the southwest as far as the northern end of the Porcupine seabight by the Slyne–Erris trough. The basins between the Minch and Shetland were discovered by geophysical studies (Bott and Watts 1970; Watts 1971; Browitt 1972; Bott and Browitt 1975). The seismic-refraction results in Basin E (West Shetland basin—Fig. 1) are consistent with the presence of a Mesozoic–Tertiary infill of 3 to 4 km resting on Palaeozoic or Torridonian rocks (cf. Bott 1975). Subsequent exploration for petroleum has given more structural detail, but not specific stratigraphic information (Kent 1975 pp. 23–24; Whitbread 1975 pp. 50–52). The West Shetland boundary fault is at the southeastern margin of the basin where the Mesozoic–Tertiary infill is thickest. It has a downthrow of more than 3·5 km. The Rona fault belt lies on the northwestern side of the basin, and has even greater displacement. Further south, the northern Minch conceals a single basin infilled with thin Jurassic sediments resting on over 3 km of Triassic rocks (Kent 1975 pp. 23–24) which in turn overlie upper Palaeozoic and Torridonian (upper Precambrian) rocks.

The crustal structure of this part of the shelf has been investigated as part of a seismic-refraction experiment (North Atlantic Seismic Project) and the first interpretation (Smith and Bott 1975) indicates that the crust thins by about 20% from the Caledonian fold-belt of the northern Highlands to the foreland area, west of the Moine thrust. The Moho lies at a uniform depth of 26 km under the Marginal Belt.

In the Little Minch–northern Skye area, gravity, deep seismic-reflection and seismic-refraction surveys have been made (Smythe *et al.* 1972). There a trough containing about 1·5 km of ?New Red Sandstone is bounded on the northeastern side by a NW–SE fault. This is overstepped by Jurassic strata and inferentially it was active during the deposition of the ?New Red Sandstone.

Outliers of New Red Sandstone on the coasts of the Minch (Steel *et al.* 1975) and of Jurassic rocks on Skye and Raasay (Morton 1965; Turner 1966; Sykes 1975) provide more stratigraphical information about this part of the Marginal Belt than elsewhere along it. The evidence is partial, but enough to show that subsidence had started by Permo-Triassic times (Steel and Wilson 1975) with associated movement of about 4 km (p. 198) along the Minch fault. A mid-Jurassic geography like the present one, with land in the area of the Outer Hebrides, is recognised by Hudson (1964). This supports other evidence that the present structural controls on morphology existed by early Mesozoic times with the basins of the Marginal Belt forming, or having formed, by then.

In the Sea of the Hebrides there are two NE–SW-trending basins bounded on the west by faults (Eden *et al.* 1973; Binns *et al.* 1974, 1975). The Sea of the Hebrides (southern Minch) basin extends south-eastwards from the North Minch basin, and the Inner Hebrides basin lies west of the Camasunary fault. Thick New Red Sandstone is present locally. One sample from west of the Minch fault has been shown to be Late Permian in age, and indirect evidence indicates that Carboniferous rocks are present in the same area (Binns *et al.* 1975 p. 98). Marine Jurassic sediments, up to 0·9 km thick, are covered by thin (20 m) Upper Cretaceous marine strata. Tertiary sediments of Eocene age are associated with the Tertiary lavas and south of Skye there is a small basin infilled with 1 km of younger (?Oligocene) strata (Smythe and Kenolty 1975).

The North Minch and Sea of the Hebrides basins are tilted towards the bounding faults and towards the continental margin (cf. Hall and Smythe 1973, p. 56). This generally westward tilt is not a consistent feature of this part of the continental shelf as, for example, the West Shetland basin tilts eastwards (Kent 1975 p. 24). To the south of Barra, shallow seismic-reflection results (Bailey *et al.* 1974) and magnetic results (Bailey *et al.* 1975) outline the continuation of the West Orkney–North Shetland–Outer Hebrides ridge more clearly than they define any sedimentary troughs present. However the Marginal Belt does cross this area as another deep basin, the Slyne–Erris trough, is present (Clarke *et al.* 1971; Max and Riddihough 1975; Kent 1975 p. 25) off the coast of northwest Ireland.

In the inner shelf area of the Malin sea, to the north of Ireland, geophysical surveys have been made, or reviewed, by Riddihough (1968), Riddihough and Young (1971), Max and Riddihough (1975), Dobson and Evans (1974) and Dobson *et al.* (1975). Over much of the area, knowledge of the deeper geology is less certain than in areas further north, but there do not appear to be deep basins present, comparable with those of the Marginal Belt. The deeper fault-bounded troughs at its eastern fringe are associated with the Clyde Belt and show evidence of persistent movement along major NE–SW faults into New Red Sandstone times.

The basins of the Marginal Belt are present at distances up to 150 km from the 500 m depth-contour while in the Minch–Sea of the Hebrides area the basins lie immediately west of the Moine thrust on the thinner, more vulnerable crust of the foreland (Fig. 1). The two factors of distance from the margin and crustal structure, influence their location elsewhere. Parallel to the zone of subsidence, there is a line of relative uplift of comparable dimensions, the North Shetland–West Orkney–Rona–Slyne ridge, which brings Lewisian rocks to crop-out in the Outer Hebrides. A gravity low to the west of Lewis has been interpreted as a sedimentary basin (I.G.S. Continental Shelf Map 1972) but more probably is produced by Lewisian granitic rocks, with only a thin cover of younger sediments present over much of the shelf to the west of the Outer Hebrides (Kent 1975 pp. 24–25).

The simple pattern of the belt is confused by Basin F (Fig. 1). It has some features of the marginal basins and its subsidence may be influenced by the mechanism which produced them. However, its siting and development are controlled principally by the sub-crop of the Walls boundary fault (a continuation of, or splay from, the Great Glen fault) which it covers (Bott and Browitt 1975 p. 370), and it probably has genetic affinities with the Moray Firth basin (Sunderland 1972; Chesher and Bacon 1975).

The general character of the Marginal Belt—the complementary uplift, its siting on thinner crust, and the subsidence along faults—is consistent with stress being the important primary agent that produced the basins. The stratigraphy indicates that these stresses were associated with subsidence in Permo-Triassic and Jurassic times. This subsidence took place within a zone parallel to the present continental margin and at less than 150 km from it. These relationships are readily understandable in terms of mechanisms of oceanward flow of ductile lower lithosphere towards a new continental slope, with basin formation in the upper brittle layer (Bott 1971) *if the margin, west of Scotland, existed by Permo-Triassic times.* Alternative hypotheses such as the view that this normal faulting was part of a regional extension *preceding* later continental rupture, leave these relationships less meaningful coincidences. For example, if the stresses that produced the Marginal Belt were a forerunner of the formation of the slope, it would be puzzling that the development of a later margin did not follow the line of deep troughs—an obvious line of weakness in the region—rather than run parallel to them at an unexplained distance of 150 km to the west. There are other considerations, none conclusive from which to infer an age for the continental margin of Scotland, and implicitly an age for the Rockall trough. However the development of the basins of the Marginal Belt, considered in isolation, favour the model of Russell (1976) of an opening in early Permian times.

## 4. The Clyde Belt

### 4a. General account

The geology of the Clyde Belt is shown in Figure 1. Sources of regional data additional to those listed previously, and Audley-Charles (1970a, b) who deals with the distribution of Triassic rocks, are:

(i) North Skye–Little Minch. Smythe *et al.* (1972) have investigated seismically a deep trough first recognised by Tuson (1959; cf. Bott and Tuson 1973)—the North Skye trough of Figure 2.

(ii) Firth of Clyde–North Channel. The results of geophysical and geological marine surveys are in course of publication (The solid geology of the Clyde sea area: a description of Sheet 55N/6W; McLean and Deegan 1978.)

(iii) Southwest Scotland and northwest England. New Red Sandstone basins surveyed by gravity include the Mauchline basin (McLean 1966), Sanquhar basin (McLean 1961), Stranraer basin (Mansfield and Kennett 1963), Dumfries and Lochmaben basins (Bott and Masson-Smith 1960), an unexposed basin near Wigtown (Parslow and Randall 1973), and the Vale of Eden trough (Bott 1974). These reveal thicknesses of between 1 to 2 km of New Red Sandstone in the deeper basins. The lower parts of the succession are concealed by progressive overlap. There is stratigraphic thickening towards the centres of basins with evidence of contemporaneous subsidence, often controlled by NW–SE faults. In addition, Deegan (1973 pp. 26–27) gives geological evidence of renewed movement along an older fault at the margin of a Carboniferous basin in Kirkcudbrightshire, in Jurassic times. Upper Carboniferous rocks are present in some Southern Uplands basins (Greig 1971), as well as under the Mauchline basin and Vale of Eden.

(iv) Northern Ireland. The results of land-based gravity surveys are given by Cook and Murphy (1952) and (Bullerwell 1961). A gravity low with a NNW–SSE trend, which reaches the coast at Larne, and another anomaly of lower amplitude at Strangford Lough, indicate the presence of New Red Sandstone basins. The increase of thickness of Triassic rocks near Larne to a total of more than 1·2 km is largely due to the presence of rock salt (Wilson 1972 pp. 44–45), and suggests that localised down-warping was occurring in Triassic times at the same time as broader subsidence was taking place over a wide area of northeast Ireland (Manning, in discussion Audley-Charles 1970b pp. 75–77). This subsidence was of the order of 0·6 km (Manning *et al.* 1970 pp. 35, 39) with some deepening under Lough Neagh (Cook and Murphy 1952 p. 24). Thick local developments of Triassic rocks (more than 1·5 km in the Port More borehole) are present in the Dungiven–Ballycastle syncline (cf. Cook and Murphy 1952 p. 21), and probably the Lough Foyle syncline and their continuations seawards at the margins of the Rathlin trough (Dobson and Evans 1974). Both folds lie close and parallel to the major NE–SW faults ofTow Valley and Lough Foyle. The Loch Indaal basin (Dobson *et al.* 1975) has a similar relationship to the Loch Gruinart fault in the Malin sea (Dobson and Evans 1974).

(v) Malin sea. Surveys are described by Riddihough (1968); Riddihough and Young (1971); Dobson and Evans (1974) and Dobson *et al.* (1975).

(vi) Northeastern Irish Sea. Gravity surveys by Bott (1964b, 1968); Bott and Young (1971) and Cornwell (1972) delineated an Irish Sea basin, separated from a smaller Peel–Solway–Carlisle basin to the north by a ridge connecting the Isle of Man to the Whitehaven area. Later work is described by Wright *et al.* (1971) and Bacon and McQuillin (1972). Exploratory drilling (Colter and Barr 1975) and seismic work has confirmed that a thick sequence of Permo-Triassic rocks are present (2·4+ km of Trias and 0·5 km of Permian) overlying upper Carboniferous rocks. NNW–SSE faults, initiated in pre-Permian times and reactivated later, are important, especially in the eastern part of the Irish Sea basin. The Cheshire basin has a comparable New Red Sandstone stratigraphy (Colter and Barr 1975 p. 65; see also Evans 1970; Thompson 1970), and an inherited NE–SW trend produced by subsidence along a major fault at its southeastern margin, which may be related to the Church Stretton fault system (p. 63). Faults of NW–SE trend are less prominent. Onlap and overstep occur at the western margin of both basins, against a structural high, which coincides with a regional gravity high trending 310° (Maroof 1974). The Precambrian rocks of Angelsey crop out at its southeastern end, but near Ireland a significant component of the gravity is produced by the Tertiary igneous centres. The Vale of Clwyd graben, at the southwestern margin of the Irish Sea basin is a NNW–SSE-trending, pre-Triassic faulted fold involving Carboniferous rocks, which has been reactivated by later down-faulting. Its structure is similar to that of the faulted New Red Sandstone basins which formed in the Southern Uplands.

The continuity of the Clyde Belt from northern Skye to Cheshire, and its structural unity, are less than obvious at first inspection. Its complex pattern of outcrops compared with the Marginal Belt, and the change of character along its length arise because it crosses major NE–SW-trending crustal units each of which shows a particular character of subsidence, and because the results of two principal types of subsidence can be recognized. In the case of the first type, narrow (*c.* 10 km) wide troughs trending NNW–SSE or NW–SE developed where acute localised subsidence took place along one, or occasionally two, bounding faults. Usually the single fault lies on the northeastern side of the trough. Between the East Arran basin and the Vale of Eden trough (Fig. 1) the troughs are strung along a NW–SE line. Similar troughs occur 80 km to the southwest, and a further 60 km to the southwest, the Larne Basin and Strangford Lough trough appear to lie at the southwesterly limit of their distribution. Small "rifts" trending NW–SE cross the northeasterly extension of the Southern Uplands massif (the mid North Sea high) (Armstrong 1972), but in northwest Britain subsidence in New Red Sandstone times is limited to the Clyde Belt, although there are also older traverse structures (cf. Horne 1975). The line of New Red Sandstone basins *crosses* the Southern Uplands rather than being spaced uniformly along it. In the case of the second type, there are broad (about 150 km wide) outcrops of New Red Sandstone along the Clyde Belt within which subsidence is evenly spread; contemporaneous faulting along NW–SE lines is not important. The thickness of New Red Sandstone preserved is of the order of 2 km at the centre of the West Arran basin and under Lough Neagh. This gentle subsidence is not continuous along the Clyde Belt but is best developed on major blocks, including the Midland Valley graben, which have had a previous history of relative subsidence in early Carboniferous times. In the Midland Valley the distribution of New Red Sandstone rocks is controlled by downwarping at the margins of the graben, rather than by movement of the marginal faults, and the regional outcrop shows a slight elongation in a NE–SW direction.

Areas of uplift (taking the base of the lower Carboniferous rocks at sea-level as a convenient datum, and criterion) are associated with the subsidence, for example, the Lake District block which separates the Vale of Eden trough from the Furness basin.

The broad sagging coincides roughly with the acute fault-bounded structures along most of the belt, but its axis of greatest New Red Sandstone thicknesses lies to the southwest of the main development of the troughs.

Another complicating feature of the belt is that some deep troughs containing New Red Sandstone trend NE–SW, i.e. transverse to the predominant alignment, but parallel to the older major faults. In northeast Ireland, the deep basins to the northwest of the Larne basin–Strangford Lough trough line, including the Dungiven–Ballycastle syncline, trend NE–SW and lie close to the downthrown sides of major faults. In structural development, they may be similar to, but younger than, the acute faulted synclines along the NE–SW-trending Kerse Loch fault in Ayrshire (McLean 1966). If so, the problem becomes why movements of normal type along the NE–SW faults persisted longer in certain parts of the graben than in others. As yet there is insufficient evidence to be certain whether the larger NE–SW basins in the Solway Firth, and in the northeastern fringes of the Irish Sea, are associated with major NE–SW faults.

The structural unit which lies furthest to the northwest, and which from its character may be part of the Clyde Belt, is the North Skye trough (Fig. 2). It developed in pre-Jurassic times by subsidence along a NW–SE fault. A block of the same dimensions, bounded by the Loch Maree fault is raised on its northeastern side. It is separated from similar elements of the Clyde Belt, such as the East Arran basin, by a distance of 200 km. If the stress or other agent which produced the belt affected the crust as far as Skye, then the development of a trough on the thin crust of the foreland (Smith and Bott 1975) and the absence of obvious subsidence on the thicker crust of the Highlands (Bamford *et al.* 1976) is readily understandable, provided the stresses were of a magnitude to cause failure only where

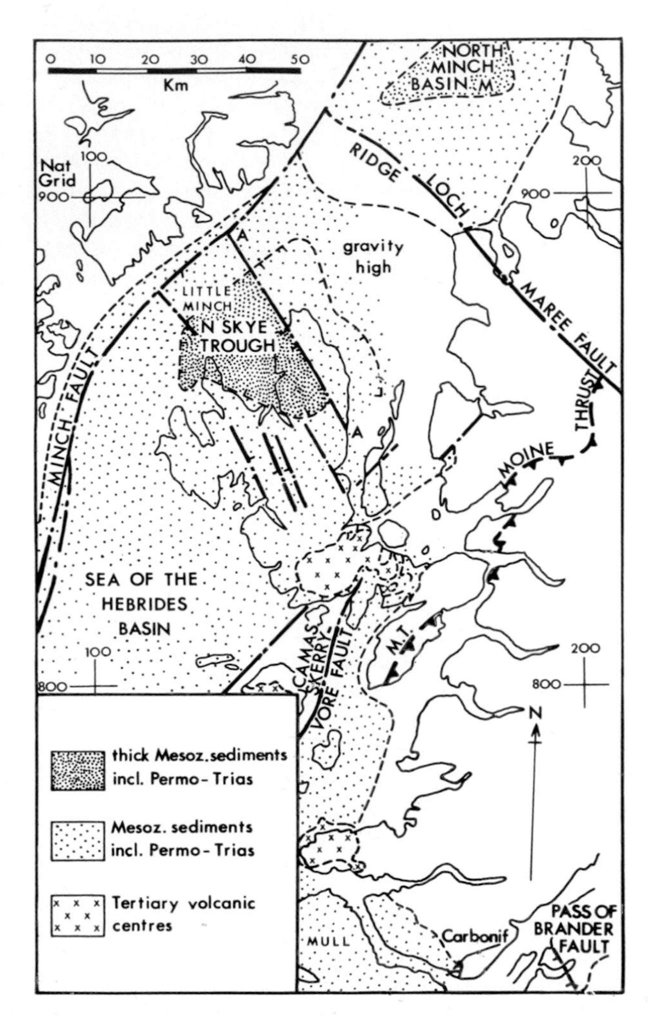


Fig. 2. The North Skye trough and the geology of its environs (after Smythe *et al.* 1972 and the publications of I.G.S). The outcrop of Carboniferous strata on the Sound of Mull, is similar to that further southwest at Bridge of Awe. MT is Moine thrust.

concentrated by local geological circumstances. This suggestion would explain the lack of clear continuity, but would still require a large scale correlation to link Skye to the Clyde Belt. Fortunately there is some evidence to support the hypothesis. The most likely place in the Highland massif to find any trace of an interpolated regional tension along the Clyde Belt, is the southwestern end of the Cruachan granite. Secondary stress at the margin, and isostatic stress would tend to make the granite rise and the adjacent country rock subside. A large fault trending NW–SE, the Pass of Brander fault, is present in the expected position, with a small outlier of Carboniferous sediments, the precise age of which is uncertain, on the southwestern, downthrown side (Geol. Surv. Scotland 1-inch Sheet 45; Kynaston and Hill 1908;

Johnstone 1973). The fragmentary evidence is consistent with a structure that is like one of the Southern Uplands basins, for example the Sanquhar basin which has been deeply eroded with only the bottom infilling preserved. It fits a model in which the crust of the central Highlands withstood regional tension except where the margin of the Etive granite masses provided a critical increment of stress.

At the southeastern end of the Clyde Belt (Fig. 1) the New Red Sandstone subsidence and associated NW–SE faulting are present as far as the Cheshire graben. Beyond there, subsidence, as on the Worcester graben, is along predominant N–S lines which appear to truncate the belt. There is major block faulting of Triassic rocks along NW–SE lines further to the southeast in the vicinity of Derby and Leicester (Geol. Surv. U.K. Gravity Overlay Map, Sheet 11 1956; see also Wills 1970). It is separated from the Cheshire graben by the N–S zone, but the Clyde Belt may persist in modified form into the English Midlands. There was a continuous zone of subsidence from Cheshire to the North Sea in Bunter times (Warrington 1970 p. 203). Comparable zones of NW–SE faults associated with Permo-Triassic and Jurassic subsidence exist in the southern North Sea (Kent 1975b p. 447) and in the southern Netherlands (De Sitter 1956 p. 149). In the Roer Valley rift movement along NW–SE fault—started at the end of the Permian, reached a maximum in the Jurassic, and was revived in younger Tertiary times. While these regions are clearly not part of the Clyde Belt, they show many corresponding features.

The features of the areal distribution of the Clyde Belt that seem to be important clues to its origin are (i) its trend is markedly discordant with the continental margin, and (ii) its development, in terms of amount of subsidence or other deformation, bears no simple correlation with distance from the margin, as might be expected if a flaw were propagated horizontally inwards from the margin. The subsidence is irregular, but is overall greatest at the southeastern end of the belt. The Clyde Belt does not appear to be related directly to the continental margin other than that subsidence is roughly contemporaneous in it and in the Marginal zone.

## 4b. In southwestern Scotland–Firth of Clyde–North Channel–northern Ireland

The Firth of Clyde and southwestern Scotland are suitable regions in which to study the stratigraphy and structure of the Clyde Belt in more detail (Figs 3, 4). Most of the area is covered by geophysical surveys, there is much subsurface evidence of the deeper geology of Ayrshire from coal-mining and data for the sea areas is available (McLean and Deegan 1978). The events in post-Devonian history of possible significance to the development of the Clyde Belt are summarised in Table 3. There are few, if any, reliable radiometric dates from lavas in the region but the time-boundaries given indicate the order of elapsed time (Francis and Woodland 1964; Smith 1964; Tozer 1964).

Differential subsidence along major NE–SW faults, which was the predominant tectonic activity in early Carboniferous and Namurian times (Francis 1965a, p. 323), persisted at some localities into early Permian times. In southern Ayrshire it gave way to subsidence along NW–SE trends and movement along normal faults in late Carboniferous times (Mykura 1967 p. 87), as the acute structures formed. The major faults, marginal to the Mauchline basin, are younger than the other NW–SE faults of Ayrshire. Most of the infill of the New Red Sandstone basins is of Bunter facies, of uncertain age (?Permian). The stratigraphy of the Upper Carboniferous strata in the Sanquhar basin (Geol. Surv. Scotland 1-inch Sheet 15; Simpson and Richey

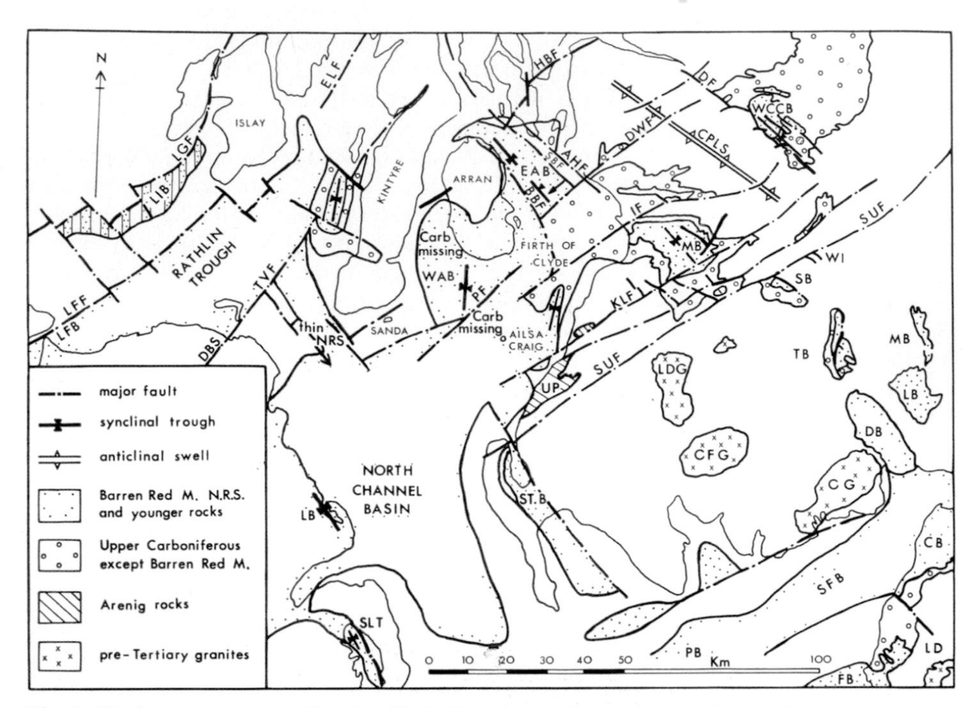


Fig. 3. Geological map showing the Clyde Belt in southwest Scotland, the Firth of Clyde, the North Channel and northern Ireland (after McLean and Deegan 1978) and the eastern Malin sea (Dobson *et al.* 1975); for abbreviations see Fig. 1, in addition, SUF is Southern Uplands fault, KLF is Kerse Loch fault, ELF is Ericht-Laidon fault, DF is Dechmont fault, BBF is Brodick Bay fault and WCCB is West Central Coalfield basin.

1936), where Coal Measures overlap the Passage Group at the southwestern margin, indicate that the basin had an identity by late Carboniferous times. Relative subsidence in late Carboniferous times is evident elsewhere. A sudden thickening of strata across the NW–SE trending Brodick Bay fault (Fig. 3) which bounds the East Arran basin, is inferred from seismic-refraction results and coastal geology, by Hall (1978).

In contrast to the evidence of the acute structures forming by late Carboniferous times, with the main subsidence later, the broad zone of subsidence coincides with a region of uplift in Carboniferous times. The New Red Sandstone of the West Arran basin rests directly on pre-Carboniferous rocks. On the western flank of the West Arran basin, the Old Red Sandstone is also attenuated, and ?Arenig rocks are brought close to the surface (Dobinson 1978). A detailed history of uplift of the area to the northeast of the basin is preserved on Arran. It occurred sporadically during most of Carboniferous times. The Carboniferous succession is thin on Arran, compared with its development to the northeast in the Midland Valley graben. It tapers from 0·4 km on the northeast coast to zero on the eastern side of the island, by a mixture of lateral change of thickness, intra-Carboniferous unconformities, and overstep by the basal Permian rocks (Geol. Surv. Scotland 1-inch Special Sheet, Arran; Tyrrell 1928 p. 44).

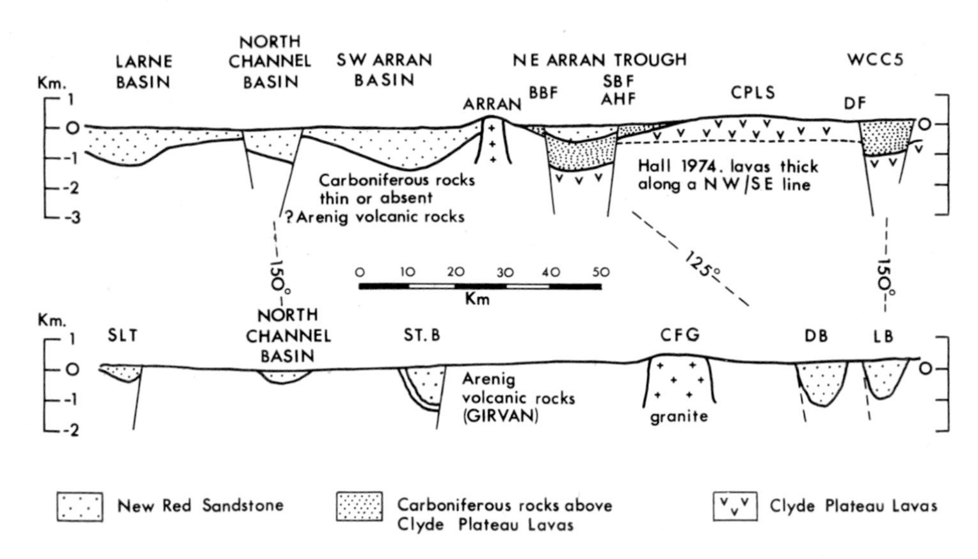


Fig. 4. Representative sections across the Clyde Belt in the Firth of Clyde–North Channel region; (i) from the Larne basin to the West Central Coalfield syncline and (ii) from Strangford Lough trough to the Lochmaben basin. Abbreviations used are BBF Brodick Bay fault, SBF Sound of Bute fault, AHF Ardrossan Harbour fault, CPLS Clyde plateau lava swell, DF Dechmont fault, SLT Strangford Lough trough, STB Stranraer basin, CFG Cairnsmore of Fleet granite, DB Dumfries basin, LB Lochmaben basin.

Volcanicity occurs sporadically throughout the Carboniferous succession within this part of the Clyde Belt *and also elsewhere in the Midland Valley graben*. The Clyde Plateau Lavas of north Ayrshire, Renfrewshire and west Lanarkshire form a pile thicker along a NW–SE line. The outcrops are controlled by this feature, rather than by a simple anticline (Hall 1974). A NW–SE trend has also been recognised in the late stage vulcanicity in the Campsies (Whyte and MacDonald 1974 p. 195). The distribution of the Passage Group Lavas (Richey *et al.* 1930 p. 200 and fig. 25) is centred on the Firth of Clyde while late Carboniferous intrusions are concentrated close to its shores in north Ayrshire (intermingled with Tertiary intrusions). Simpson and Richey (1936 p. 85) have pointed out that the distribution of late Carboniferous and early Permian vents may be correlated with the line of New Red Sandstone basins Mauchline–Sanquhar–Thornhill. These features suggest a possible causal relationship between igneous activity in Carboniferous times and the subsidence of the Clyde Belt. Certainly the crust in this region must have been heated and suffered *some* thermal swelling during much of the Carboniferous period. The correlation becomes less significant, however, when it is borne in mind that other areas of the Midland Valley were heated at the same time. The west Fife area had late Carboniferous igneous activity (Francis 1965b) as did central Ayrshire. but had no subsequent subsidence in New Red Sandstone times.

## 4c. Origin and formation of the belt

All criteria indicate that the fault-bounded basins are produced by tension. Their width of between 10 and 20 km means, however, that they are much too narrow to

Table 3. Chronology of volcanicity and subsidence in the Clyde Belt

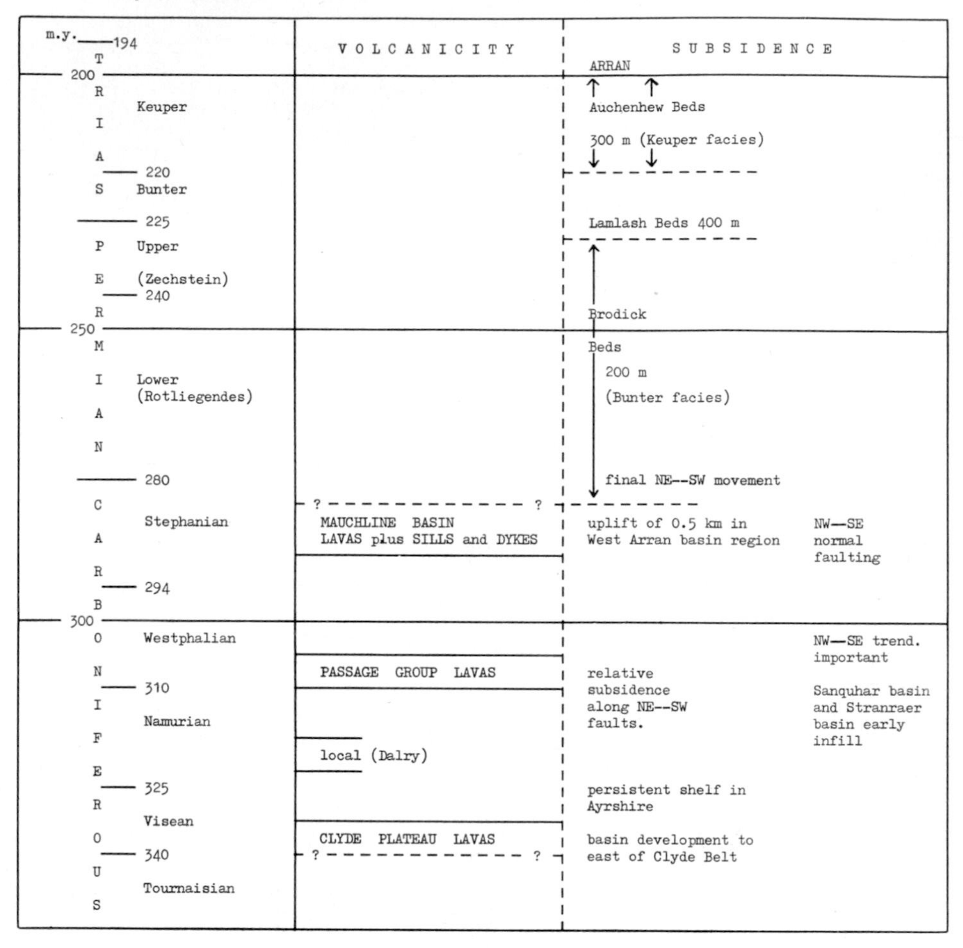

| m.y. | Period | Stage | VOLCANICITY | SUBSIDENCE (ARRAN) | SUBSIDENCE |
|---|---|---|---|---|---|
| 194–200 | TRIAS | | | | |
| 200–220 | TRIAS | Keuper | | Auchenhew Beds 300 m (Keuper facies) | |
| 220–225 | TRIAS | Bunter | | Lamlash Beds 400 m | |
| 225–240 | PERMIAN | Upper (Zechstein) | | Brodick | |
| 240–280 | PERMIAN | Lower (Rotliegendes) | | Beds 200 m (Bunter facies) final NE--SW movement | |
| 280–294 | CARBONIFEROUS | Stephanian | ? MAUCHLINE BASIN LAVAS plus SILLS and DYKES ? | uplift of 0.5 km in West Arran basin region | NW—SE normal faulting |
| 294–310 | CARBONIFEROUS | Westphalian | PASSAGE GROUP LAVAS | relative subsidence along NE--SW faults. | NW—SE trend. important; Sanquhar basin and Stranraer basin early infill |
| 310–325 | CARBONIFEROUS | Namurian | local (Dalry) | | |
| 325–340 | CARBONIFEROUS | Visean | CLYDE PLATEAU LAVAS ? ? | persistent shelf in Ayrshire; basin development to east of Clyde Belt | |
| 340– | CARBONIFEROUS | Tournaisian | | | |

have been produced by a mechanism involving the base of the elastic lithosphere (Bott 1976b), or any deeper interface. Where normal faults on both sides define the limits of the trough (e.g. East Arran basin), they converge at probably less than 15 km below the surface. Their general character—a synclinal sag bounded by one or two major faults and with no complementary narrow horst blocks—is similar to the rifts produced in clay models by Cloos (1931). In these, the location and development of a normal-faulted synclinal wedge in a succession of younger strata is controlled by a linear flaw in rigid basement rocks. To postulate what type of geological structure might serve as such a flaw under any one basin is not too difficult. To suggest the nature of structure that gives an alignment of basins from the Midland Valley into the Southern Uplands, and is implicitly continuous across the Southern Upland fault is difficult, and makes the mechanism in its simplest form less than probable.

The igneous activity and the uplift in the Firth of Clyde region are *qualitatively* consistent with a thermal event producing the broad regional subsidence, but this cannot be the *prime* cause. The area is too small, and any cooling and contraction would have taken too long. No adjustment of a thermal model—of the volume of intruded rock, of its temperature, or of the level of its emplacement—that involves reasonable assumptions can avoid the conclusion that the regional subsidence, as well as the faulted troughs, are the products of tension. The difficult problems that remain are firstly, to account for the source of the regional tension, and secondly to account for the concentration of strain along the NW–SE zone that became the Clyde Belt.

(i) *The source of regional tension.* For reasons given earlier, it is improbable that the regional tension was produced by a mechanism dependent on nearness to the continental margin, such as ductile flow to the slope (Bott 1971), or that the Clyde Belt is the failed arm of a triple junction. A NE–SW tension affects other regions of northwest Europe in early Mesozoic times and indicates that the stress was widespread though failure occurred only in certain areas. The tension may be related to the early stages of plate movement from a new spreading axis. This is suggested by the partial reactivation of the Clyde Belt when spreading from Reykjanes ridge began in early Tertiary times. Igneous centres (Roberts 1975, p. 6) and later Tertiary subsidence in the Sea of Hebrides and Lough Neagh occur in, or close to, the Clyde Belt between Skye and the northern Irish Sea (Fig. 1). Dyke swarms and the alignment of successive centres of activity within major centres, such as Arran (McLean and Wren 1978), indicate that a regional NW–SE tension reactivated the same area that was affected in late Carboniferous and New Red Sandstone times.

Simpler mechanisms involving plate movement, such as regionally variable coupling between plate and asthenosphere, with unevenness in the transmission of stress and in the rate of plate transport from region to region, seem to be inapplicable as plate movement from Reykjanes ridge was in a southeasterly direction; Mesozoic movement from a Rockall trough axis is likely to have been the same way, or to the east. The present regional stress in Scotland was inferred by Anderson (1951 pp. 143–144) to be an E–W relative tension.

(ii) *The localisation of strain in the Clyde Belt.* There is no obvious pre-existing line of weakness, crustal structure or other geological peculiarity that might explain the concentration of strain within the Clyde Belt. Indeed, the dominant trend of major pre-belt structures is NE–SW. There are older lines (cf. Horne 1975) such as Inverian and Laxfordian crush-belts near Skye, or granite margins in the Lake District, which may control the position of individual structural elements of the belt. But there is no single one which might explain *its* continuity from Skye to Cheshire. Any hypothesis about a deep line of weakness in the Precambrian basement meets the major problem that any such line is likely to be older than the Ordovician rocks seen at the surface, yet would straddle the Caledonian suture (cf. Phillips *et al.* 1976).

What is known of crustal structure in and around the Clyde Belt is only a basis for suggestions, rather than answers, about what may have caused its distinctive reaction to NE–SW regional tension in New Red Sandstone times. Changes in character of the belt in Scotland indicate that subsidence took place more readily where the crust is relatively thin. Theory also predicts an intensification of stress in such areas. Within the Midland Valley, a thickness of 6·6 km of Lower Old Red

Sandstone strata in Kincardineshire (Armstrong and Paterson 1970) thins regionally along the main outcrop in Strathmore and the Higland Border, until only 1·8 km is present in Arran (Friend *et al.* 1963). A marked reduction in total thickness of Carboniferous strata also occurs from Fife to the shelf region of Ayrshire (Francis 1965a) and continues across Arran. If a corresponding reduction in the crustal column occurs, then the crust is thinner under the Firth of Clyde and the Clyde Belt than elsewhere in the graben. The outcrops of Arenig rocks on the shores of the Firth, south of Girvan, are related to this same structural culmination which developed prior to the New Red Sandstone subsidence. The regional tilt to the east-northeast was accentuated during Tertiary times as the northern British area was tilted towards a subsiding North Sea basin. The principal evidence of a palaeoslope related to this Tertiary movement, extending across western Scotland, and culminating in roughly the region of the Firth of Clyde, is the river system. Eastward flowing rivers have their watershed near the west coast. There is no published seismic evidence to confirm or refute this hypothesis that the crust is thinner under the Clyde Belt than elsewhere on *each* NE–SW-trending crustal unit that it crosses, and it is offered only as a suggestion for considering future results.

## References


ANDERSON, E. M. 1951. *The dynamics of faulting*. (2nd Ed.) Oliver and Boyd, Edinburgh.

ARMSTRONG, G. 1972. Review of the geology of the British Continental Shelf. *The Min. Eng.* **131,** 463–477.

ARMSTRONG, M. and PATERSON, I. B. 1970. The Lower Old Red Sandstone of the Strathmore region. *Rep. Inst. geol. Sci.* No. 60/12, 23 pp.

AUDLEY-CHARLES, M. G. 1970a. Stratigraphical correlation of the Triassic rocks of the British Isles. *Q. J. geol. Soc. Lond.* **126,** 19–47.

—— 1970b. Triassic palaeogeography of the British Isles. *Q. J. geol. Soc. Lond.* **126,** 49–89.

BACON, M. and MCQUILLIN, R. 1972. Refraction seismic surveys in the north Irish Sea. *J. geol. Soc. Lond.* **128,** 613–621.

BAILEY, R. J., GRZYWACZ, J. M. and BUCKLEY, J. S. 1974. Seismic reflection profiles on the continental margin bordering the Rockall Trough. *J. geol. Soc. Lond.* **130,** 55–69.

——, BUCKLEY, J. S. and KIELMAS, M. M. 1975. Geomagnetic reconnaissance on the continental margin of the British Isles between 54° and 57°N. *J. geol. Soc. Lond.* **131,** 275–282.

BAMFORD, D. *et al.* 1976. A Lithosphere Seismic Profile in Britain—I. Preliminary Results. *Geophys. J. R. astron. Soc.* **44,** 145–160.

BINNS, P. E., MCQUILLIN, R. and KENOLTY, N. 1974. The geology of the Sea of the Hebrides. *Rep. Inst. geol. Sci.* No. 73/14, 43 pp.

——, FANNIN, N. G. T., KENOLTY, N. and ARDUS, D. A. 1975. Structure and stratigraphy of sedimentary basins in the Sea of the Hebrides and the Minches. *In* Woodland, A. W. (Editor). *Petroleum and the continental shelf of north-west Europe.* 1. *Geology*, 93–102. Applied Science Publishers, Barking.

BOTT, M. H. P. 1964a. Formation of sedimentary basins by ductile flow of isostatic origin in the upper mantle. *Nature, Lond.* **201,** 124–129.

—— 1964b. Gravity measurements in the north-eastern part of the Irish Sea. *Q. J. geol. Soc. Lond.* **120,** 369–396.

—— 1965. The deep structure of the northern Irish Sea—a problem of crustal dynamics. *Colston Pap.* **17,** 179–204.

—— 1967. Geophysical investigations of the northern Pennine basement rocks. *Proc. Yorks. geol. Soc.* **36,** 139–168.

—— 1968. The geological structure of the Irish Sea basin. *In* Donovan, D. T. (Editor). *Geology of Shelf Seas*, 93–115. Oliver and Boyd, Edinburgh.

—— 1971. Evolution of young continental margins and formation of shelf basins. *Tectonophysics* **11,** 319–327.

—— 1974. The geological interpretations of a gravity survey of the English Lake District and the Vale of Eden. *J. geol. Soc. Lond.* **130,** 309–331.

—— 1975. Structure and evolution of the North Scottish shelf, the Faeroe Block and the intervening region. *In* Woodland, A. W. (Editor). *Petroleum and the continental shelf of north-west Europe.* 1. *Geology*, 105–115. Applied Science Publishers, Barking.

—— 1976a. Mechanism of basin subsidence—an introductory review. *Tectonophysics* **36,** 1–4.

—— 1976b. Formation of sedimentary basins of graben type by extension of the continental crust. *Tectonophysics* **36,** 77–86.

—— and BROWITT, C. W. A. 1975. Interpretation of geophysical observations between the Orkney and Shetland Islands. *J. geol. Soc. Lond.* **131,** 353–371.

—— and DEAN, D. S. 1972. Stress systems at young continental margins. *Nature Phys. Sci., Lond.* **235,** 23–25.

—— 1973. Stress diffusion from plate boundaries. *Nature, Lond.* **243,** 339–341

—— and MASSON-SMITH, D. 1960. A gravity survey of the Criffel Granodiorite and the New Red Sandstone deposits near Dumfries. *Proc. Yorks. geol. Soc.* **32,** 317–332.

—— and TUSON, J. 1973. Deep structure beneath the Tertiary volcanic regions of Skye, Mull and Ardnamurchan, North-West Scotland. *Nature Phys. Sci., Lond.* **242,** 115–116.

—— and WATTS, A. B. 1970. Deep sedimentary basins proved in the Shetland-Hebridean Continental Shelf and Margin. *Nature, Lond.* **225,** 265–268.

—— and YOUNG, D. G. G. 1971. Gravity measurements in the north Irish Sea. *Q. J. geol. Soc. Lond.* **126,** 413–434.

BROWITT, C. W. A. 1972. Seismic refraction investigation of deep sedimentary basin in the continental shelf west of Shetland. *Nature, Lond.* **236,** 161–163.

BULLERWELL, W. 1961. The gravity map of Northern Ireland. *Irish Nat. J.* **13,** 254–257.

BURKE, K. 1976. Development of graben associated with the initial ruptures of the Atlantic Ocean. *Tectonophysics* **36,** 93–112.

CHESHER, J. A. *et al.* 1972. I.G.S. Marine drilling with m.v. *Whitethorn* in Scottish waters 1970–71. *Rep. Inst. geol. Sci.* No. 72/10, 25 pp.

—— and BACON, M. 1975. A deep seismic survey in the Moray Firth. *Rep. Inst. geol. Sci.* No. 75/11, 13 pp.

CLARKE, R. H., BAILEY, R. J. and TAYLOR-SMITH, D. 1971. Seismic reflection profiles of the continental margin to the west of Ireland. *Rep. Inst. geol. Sci.* No. 70/14, 67 pp.

CLOOS, H. 1931. Zur Experimentellen Tektonic. *Die Naturwissenschaften* **19** (11).

COLTER, V. S. and BARR, K. W. 1975. Recent developments in the geology of the Irish Sea and Cheshire Basins. *In* Woodland, A. W. (Editor). *Petroleum and the continental shelf of north-west Europe.* 1. *Geology*, 61–73. Applied Science Publishers, Barking.

COOK, A. H. and MURPHY, T. 1952. Measurements of gravity in Ireland. Gravity survey of Ireland north of the line Sligo–Dundalk. *Geophys. Mem. Dublin Inst. Adv. Stud.* **2,** pt. 4, 1–36.

CORNWELL, J. D. 1972. A gravity survey of the Isle of Man. *Proc. Yorks. geol. Soc.* **39,** 93–106.

DEEGAN, C. E. 1973. Tectonic control of sedimentation at the margin of a Carboniferous depositional basin in Kirkcudbrightshire. *Scott. J. Geol.* **9,** 1–28.

DE SITTER, L. U. 1956. *Structural geology.* McGraw-Hill, London.

DOBINSON, A. 1978. Geophysical studies south and west of Kintyre. *Rep. Inst. geol. Sci.* No. 78/9.

DOBSON, M. R. and EVANS, D. 1974. The geological structure of the Malin Sea. *J. geol. Soc. Lond.* **130,** 475–478.

——, EVANS, D. and WHITTINGTON, R. 1975. The offshore extension of the Loch Gruinart Fault, Islay. *Scott. J. Geol.* **11,** 23–35.

DUNNING, F. W. 1966. Tectonic map of Great Britain and Northern Ireland. *Geol. Surv. U.K.*

EDEN, R. A., DEEGAN, C. E., RHYS, G. H., WRIGHT, J. E. and DOBSON, M. R. 1973. Geological investigations with a manned submersible in the Irish Sea and off western Scotland 1971. *Rep. Inst. geol. Sci.* No. 73/2, 27 pp.

EVANS, W. B. 1970. The Triassic salt deposits of north-western England. *Q. J. geol. Soc. Lond.* **126,** 103–123.

FRANCIS, E. H. 1965a. Carboniferous. *In* Craig, G. Y. (Editor). *The Geology of Scotland,* 309–357. Oliver and Boyd, Edinburgh.

—— 1965b. Carboniferous–Permian igneous rocks. *In* Craig, G. Y. (Editor). *The Geology of Scotland,* 359–382. Oliver and Boyd, Edinburgh.

—— and WOODLAND, A. W. 1964. The Carboniferous period. *In* Harland, W. B. *et al.* (Editors). *The Phanerozoic Time-Scale,* 221–232. *Q. J. geol. Soc. Lond.* **120s,**

FRIEND, P. F., HARLAND, W. B. and HUDSON, J. D. 1963. The Old Red Sandstone and the Highland Boundary in Arran. *Trans. Edinb. geol. Soc.* **19,** 363–425.

GREIG, D. C. *et al.* 1971. The South of Scotland (3rd Ed.). *Brit. reg. Geol. Geol., Surv. and Mus., Lond.*

HALL, J. 1974. A seismic reflection survey of the Clyde Plateau Lavas in North Ayrshire and Renfrewshire. *Scott. J. Geol.* **9,** 253–279.

—— 1978. Seismic refraction studies in the Clyde area. *Rep. Inst. geol. Sci.* No. 78/9, 43–48.

—— and SMYTHE, D. K. 1973. Discussion of the relation of Palaeogene ridge and basin structures of Britain to the North Atlantic. *Earth Planet, Sci. Letters* **19,** 54–60.

HALLAM, A. 1972. Relation of Palaeogene ridge and basin structures and vulcanicity in the Hebrides and Irish Sea regions of the British Isles to the opening of the North Atlantic. *Earth Planet. Sci. Lett.* **16,** 171–177.

HAXBY, W. F., TURCOTTE, D. L. and BIRD, J. M. 1976. Thermal and mechanical evolution of the Michigan Basin. *Tectonophysics* **36,** 57–75.

HORNE, R. R. 1975. Possible transverse fault control of base metal mineralisation in Ireland and Britain. *Ir. Nat. J.* **18,** 140–144.

HUDSON, J. D. 1964. The petrology of the sandstones of the Great Estuarine Series, and the Jurassic palaeogeography of Scotland. *Proc. Geol. Assoc.* **75,** 499–528.

INSTITUTE OF GEOLOGICAL SCIENCES 1972. The Sub-Pleistocene Geology of the British Isles and the adjacent continental shelf. 1:2 500 000.

JOHNSTONE, G. S. 1973. The Grampian Highlands (3rd Ed. with amendments). *Brit. reg. Geol., Geol. Surv. and Mus., Lond.*

KENT, P. E. 1975a. The tectonic development of Great Britain and the surrounding seas, 3–28. *In* Woodland, A. W. (Editor). *Petroleum and the continental shelf of north-west Europe.* 1. *Geology,* 3–28. Applied Science Publishers, Barking.

—— 1975b. Review of North Sea Basin development. *J. geol. Soc. Lond.* **131,** 435–468.

—— 1976. Major synchronous events in continental shelves. *Tectonophysics* **36,** 87–91.

KYNASTON, H. and HILL, J. B. 1908. The geology of Oban and Dalmally. *Mem. geol. Surv. U.K.*

MCLEAN, A. C. 1961. A gravity survey of the Sanquhar Coalfield. *Proc. R. Soc. Edinb.* **B 68,** 112–127.

—— 1966. A gravity survey in Ayrshire and its geological interpretation. *Trans. R. Soc. Edinb.* **66,** 239–265.

—— and DEEGAN, C. 1978. A synthesis of the solid geology of the Clyde region. *Rep. Inst. geol. Sci.* No. 78/9, 93–114

—— and WREN, A. E. 1978. Gravity and magnetic studies in the Lower Firth of Clyde. *Rep. Inst. geol. Sci.* No. 78/9, 7–27.

MCQUILLIN, R. and BACON, M. 1974. Preliminary report on seismic reflection surveys in sea areas around Scotland, 1969–73. *Rep. Inst. geol. Sci.* No. 74/12, 7 pp.

MANNING, T. D., ROBBIE, J. A. and WILSON, H. E. 1970. Geology of Belfast and the Lagan Valley. *Mem. geol. Surv. North Irel.*

MANSEFIELD, J. and KENNETT, P. 1963. A gravity survey of the Stranraer sedimentary basin. *Proc. Yorks. geol. Soc.* **34,** 139–151.

MAROOF, S. I. 1974. A Bouguer anomaly map of southern Great Britain and the Irish Sea. *J. geol. Soc. Lond.* **130,** 471–474.

MAX, M. D. and RIDDIHOUGH, R. P. 1975. A geological framework for the continental margin to the west of Ireland. *Geol. J.* **11,** 109–118.

MORTON, N. 1965. The Bearreraig Sandstone Series (Middle Jurassic) of Skye and Raasay. *Scott. J. Geol.* **1,** 189–216.

MURRELL, S. A. F. 1976. Rheology of the Lithosphere—experimental indications. *Tectonophysics* **36,** 5–24.
MYKURA, W. 1967. The Upper Carboniferous rocks of south-west Ayrshire. *Bull. geol. Surv. G.B.* **26,** 23–98.
PARSLOW, G. R. and RANDALL, B. A. O. 1973. A gravity survey of the Cairnsmore of Fleet granite and its environs. *Scott. J. Geol.* **9,** 219–231.
PHILLIPS, W. E. A., STILLMAN, C. J. and MURPHY, T. 1976. A Caledonian plate tectonic model. *J. geol. Soc. Lond.* **132,** 579–609.
RICHEY, J. E., ANDERSON, E. M., MACGREGOR, A. G. 1930. The geology of North Ayrshire. *Mem. Geol. Surv. U.K.*
RIDDIHOUGH, R. P. 1968. Magnetic surveys off the north coast of Ireland. *Proc. Roy. Irish Acad.* **66B,** 27–41.
—— and YOUNG, D. G. G. 1971. Gravity and magnetic surveys of Inishowen and adjoining sea areas of the north coast of Ireland. *Proc. geol. Soc. Lond.* **1664,** 215.
ROBERTS, D. G. 1975. The solid geology of the Rockall Plateau. *In* Harrison, R. K. (Editor). *Expeditions to Rockall 1971–72*, 3–10. *Rep. Inst. geol. Sci.* No. 75/1.
RUSSELL, M. J. 1976. A possible Lower Permian age for the onset of ocean floor spreading in the northern North Atlantic. *Scott. J. Geol.* **12,** 315–323.
SCHEIDEGGER, A. E. and O'KEEFE, J. A. 1967. On the possibility of the origination of geosynclines by deposition. *J. geophys. Res.* **72,** 6275–6278.
SIMPSON, J. B. and RICHEY, J. E. 1936. The geology of the Sanquhar Coalfield and adjacent basin of Thornhill. *Mem. Geol. Surv. U.K.*
SLEEP, N. H. 1969. Sensitivity of heat flow and gravity to the mechanism of sea-floor spreading. *J. geophys. Res.* **74,** 542–549.
—— 1971. Thermal effects of the formation of Atlantic continental margins by continental break-up. *Geophys. J. R. astron. Soc.* **24,** 325–350.
—— and SNELL, N. S. 1976. Thermal contraction and flexure of mid-continent and Atlantic marginal basins. *Geophys. J. R. astron. Soc.* **45,** 125–154.
SMITH, D. B. 1964. The Permian period. *In* Harland, W. B. *et al.* (Editors). *The Phanerozoic Time-Scale,* 211–220. *Q. J. geol. Soc. Lond.* **120s.**
SMITH, P. J. and BOTT, M. H. P. 1975. Structure of the crust beneath the Caledonian foreland and Caledonian Belt of the North Scottish Shelf region. *Geophys. J. R. astron. Soc.* **40,** 187–205.
SMYTHE, D. K. and KENOLTY, N. 1975. Tertiary sediments in the Sea of the Hebrides. *J. geol. Soc. Lond.* **131,** 227–233.
——, SOWERBUTTS, W. T. C., BACON, M. and MCQUILLIN, R. 1972. Deep sedimentary basin below northern Skye and the Little Minch. *Nature Phy. Sci., Lond.* **236,** 87–89.
STEEL, R. J. Devonian basins of western Norway. *Tectonophysics* 207–224.
——, NICHOLSON, R. and KALANDER, L. 1975. Triassic sedimentation and palaeogeography in Central Skye. *Scott. J. Geol.* **11,** 1–13.
—— and WILSON, A. C. 1975. Sedimentation and tectonism (Permo–Triassic?) on the margin of the North Minch Basin. *J. geol. Soc. Lond.* **131,** 183–202.
SUNDERLAND, J. 1972. Deep sedimentary basin in the Moray Firth. *Nature, Lond.* **236,** 24–25.
SYKES, R. M. 1975. The stratigraphy of the Callovian and Oxfordian stages (Middle-Upper Jurassic) in northern Scotland. *Scott. J. Geol.* **11,** 51–78.
THOMPSON, D. B. 1970. The stratigraphy of the so-called Keuper Sandstone formation (Scythian–?Anisian) in the Permo-Triassic Cheshire Basin. *Q. J. geol. Soc. Lond.* **126,** 151–181.
TOZER, E. T. 1964. The Triassic period. *In* Harland, W. B. *et al.* (Editors). *The Phanerozoic Time-Scale, Q. J. geol. Soc. Lond.* **120s.**
TURNER, J. 1966. The Oxford Clay of Skye, Scalpay and Eigg. *Scott. J. Geol.* **2,** 243–252.
TUSON, J. 1959. *A geophysical investigation of the Tertiary volcanic districts of western Scotland.* Univ. Durham Ph.D. thesis (unpub.).
TYRRELL, G. W. 1928. The geology of Arran. *Mem. Geol. Surv. U.K.*
VENING MEINESZ, F. A. 1950. Les graben africains, résultat de compression on de tension dans la croûte terrestre? *Bull. Inst. R. Colon. Belg.* **21,** 539–552.
WARRINGTON, G. 1970. The stratigraphy and palaeontology of the "Keuper" Series of the central Midlands of England. *Q. J. geol. Soc. Lond.* **126,** 183–223.
WATTS, A. B. 1971. Geophysical investigations on the continental shelf and slope north of Scotland. *Scott. J. Geol.* **7,** 189–218.
—— and RYAN, W. B. F. 1976. Flexure of the lithosphere and continental margin basins. *Tectonophysics* **36,** 25–44.
WHITBREAD, D. R. 1975. Geology and petroleum possibilities west of the United Kingdom. *In* Woodland, A. W. (Editor). *Petroleum and the continental shelf of north-west Europe.* 1. *Geology*, 45–60. Applied Science Publishers, Barking.

WHYTE, F. and MACDONALD, J. G. 1974. Lower Carboniferous vulcanicity in the northern part of the Clyde Plateau. *Scott. J. Geol.* **10,** 187–198.
WILLS, L. J. 1970. The Triassic succession in the central Midlands. *Q. J. geol. Soc. Lond.* **126,** 225–283.
WILSON, H. E. 1972. *Regional geology of Northern Ireland.* H.M.S.O., Belfast.
WOODLAND, A. W. (Editor) 1975. *Petroleum and the continental shelf of north-west Europe.* 1. *Geology.* Applied Science Publishers, Barking.
WRIGHT, J. E. *et al.* 1971. Irish Sea investigations 1969–70. *Rep. Inst. geol. Sci.* No. 71/19, 55 pp.

A. C. McLean,
Department of Geology,
University of Glasgow,
Glasgow G12 8QQ, Scotland.

# Distribution of basement under the eastern North Atlantic Ocean and the Norwegian Sea

*M. Talwani*

Results of drilling in the North Atlantic and the Norwegian Sea combined with results of geophysical surveys have yielded information about the age and nature of basement in these areas. These data are used to obtain the history of the evolution of the Norwegian Sea and the early opening of the North Atlantic. The opening of the Norwegian Sea is complex, with a number of jumps of the ridge axis. Special problems in this area relate to the presence of continental fragments within the Norwegian Sea, the presence of areas of unusual basement elevation, the occurrence of an "opaque" layer on seismic reflection records, evidence for two simultaneous axes for sea-floor spreading within a small area, and the unusual tectonic features present in the ocean which are related to the start of sea-floor spreading. In the North Atlantic recent syntheses of geophysical data have refined the dates and geometry of the early opening.

## 1. Introduction

Basement has been reached by deep drilling at a number of locations on the mid-Atlantic ridge. However, as the sediment thickness increases away from the ridge crest, it becomes progressively more difficult to drill down to basement. In fact, nowhere has basement been reached south of 60°N in the deep ocean under the continental margin off western Europe. This makes it very difficult to determine the age and nature—continental or oceanic—of the basement under many parts of the continental margin. One appeals then to geophysical methods which permit determination of the past position of the continents. The first part of this paper reviews the information to reconstruct the relative positions of Europe and North America as far back as is reliably possible. Speculation is then possible about the age and nature of the basement close to the line of initial opening which now lies under the continental margins.

The second part of the paper deals principally with drilling results in the

Norwegian Sea where basement of various ages was reached in diverse locations. These results enable the testing of models of evolution of this complicated area.

## 2. Reconstruction

### 2a. Early opening of the North Atlantic

The position of Europe and North America prior to the time of the opening of the North Atlantic was first modelled by Bullard *et al.* (1965), by matching the 500 fathom bathymetric contours on either side. Although this reconstruction still serves as a good approximation to the position of these two continents prior to the time of opening, it suffers from uncertainties. These uncertainties principally arise from two kinds of factors. One is whether the 500 fathom curve indeed marks the boundary between ocean and continent and secondly, whether opening has taken place between areas such as Rockall bank and Porcupine bank and Europe on the one hand, and between areas such as Flemish Cap and Orphan Knoll and North America on the other hand (Fig. 1). These uncertainties still remain and will not be resolved except by drilling to basement. However by dating magnetic lineations a reliable reconstruction can be made for times somewhat later than the time of initial opening. The method of obtaining basement ages by dating of linear magnetic anomalies from sea-floor spreading has been used, for example, by Pitman and Talwani (1972), and Williams and McKenzie (1971) in the North Atlantic. By identifying magnetic anomalies on either side of the mid-Atlantic ridge, Pitman and Talwani (1972) were able to obtain the position of North America with respect to Europe at various ages by rotating North America in such a way that the anomaly lineations for a particular age on the west side of the North Atlantic coincided with the corresponding lineation on the east side after rotation. Pitman and Talwani (1972) were able to identify anomalies as old as anomaly 31 (nomenclature of Heirtzler *et al.* 1968) in the area between the Azores–Gibraltar ridge and the Charlie Gibbs fracture zone (Fig. 1). However, north of the Charlie Gibbs fracture zone, anomalies only as old as anomaly 24 were identified. This meant that reconstruction of the North Atlantic could only be made with certainty back to the time of anomaly 24 (late Palaeocene).

The magnetic time scale which relates magnetic anomalies to absolute ages has since gradually evolved from the one originally given by Heirtzler *et al.* (1968). There is, however, no agreement at the present time on the best time scale. The time-scale of Heirtzler is generally used here although ages in the early Tertiary are almost certainly too great by about 5–7 m.y. Some of the key lineations are assumed here to have the following ages:

| *Anomaly No.* | *Age* | *Reference* |
|---|---|---|
| 5D | 18 m.y. | Blakely (1974) |
| 6A | 24 m.y. | Blakely (1974) |
| 7 | 27 m.y. | Heirtzler *et al.* (1968) |
| 13 | 38 m.y. | Heirtzler *et al.* (1968) |
| 31 | 67 m.y. | LaBrecque *et al.* (1977) |
| 33 | 74 m.y. | LaBrecque *et al.* (1977) |
| 34 | 82 m.y. | LaBrecque *et al.* (1977) |

Kristoffersen (1977a, 1977b) has recently been able to make a better reconstruc-

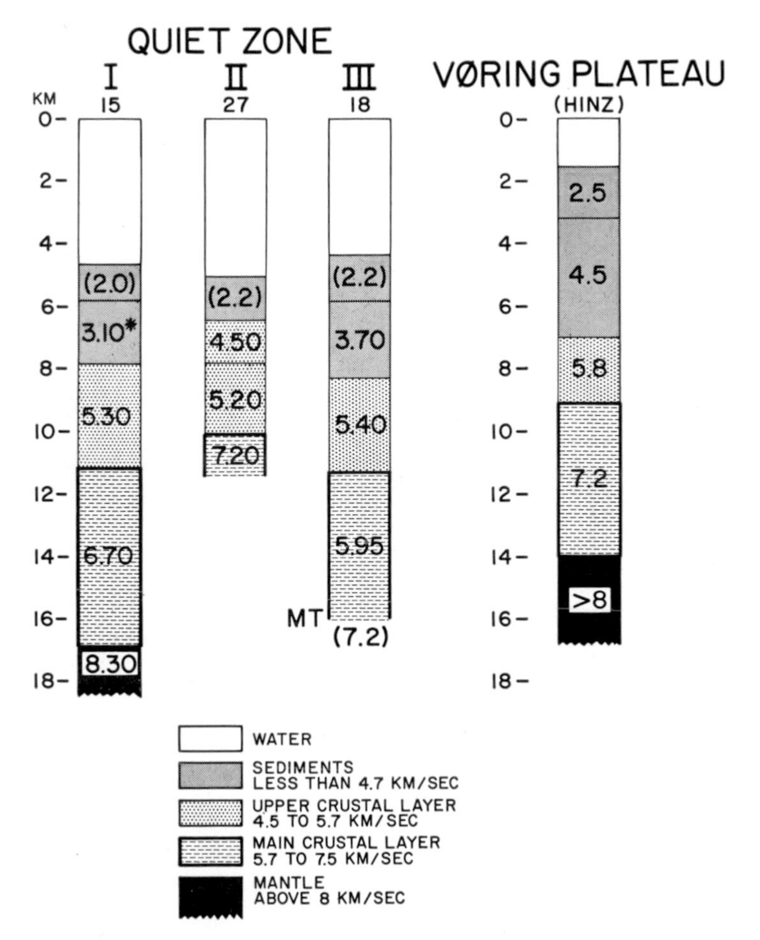


Fig. 4. Crustal section obtained by Hinz and Moe (1971) on the Vøring plateau compared with three sections obtained in the magnetic quiet zone south of Australia (Talwani *et al.* 1978). * denotes assumed velocity.

Kristoffersen has suggested that prior to anomaly 34 time the spreading axis between Ireland and Newfoundland continued into the Rockall basin. A prominent magnetic anomaly in the centre of the Rockall basin which appears to lie on a continuation of the reconstructed position of the ridge axis lends support to this hypothesis. By somewhat arbitrarily choosing a position for the pole of opening prior to anomaly 34 (Santonian) at 85·5°N, 136·2°E, in a reference frame fixed to Europe, Kristoffersen was further able to rotate the "continent-ocean boundaries" between Newfoundland and Ireland to coincide, thereby nearly closing Rockall basin. This rotation moved Greenland 165 km closer to Norway at a latitude of about 60° and about 85 km closer at a latitude of 75°. Objections to this model are the very narrow width of the Faeroe–Shetland channel and the large overlaps that are produced between Greenland and Svalbard when this rotation is made. It may be that this pole

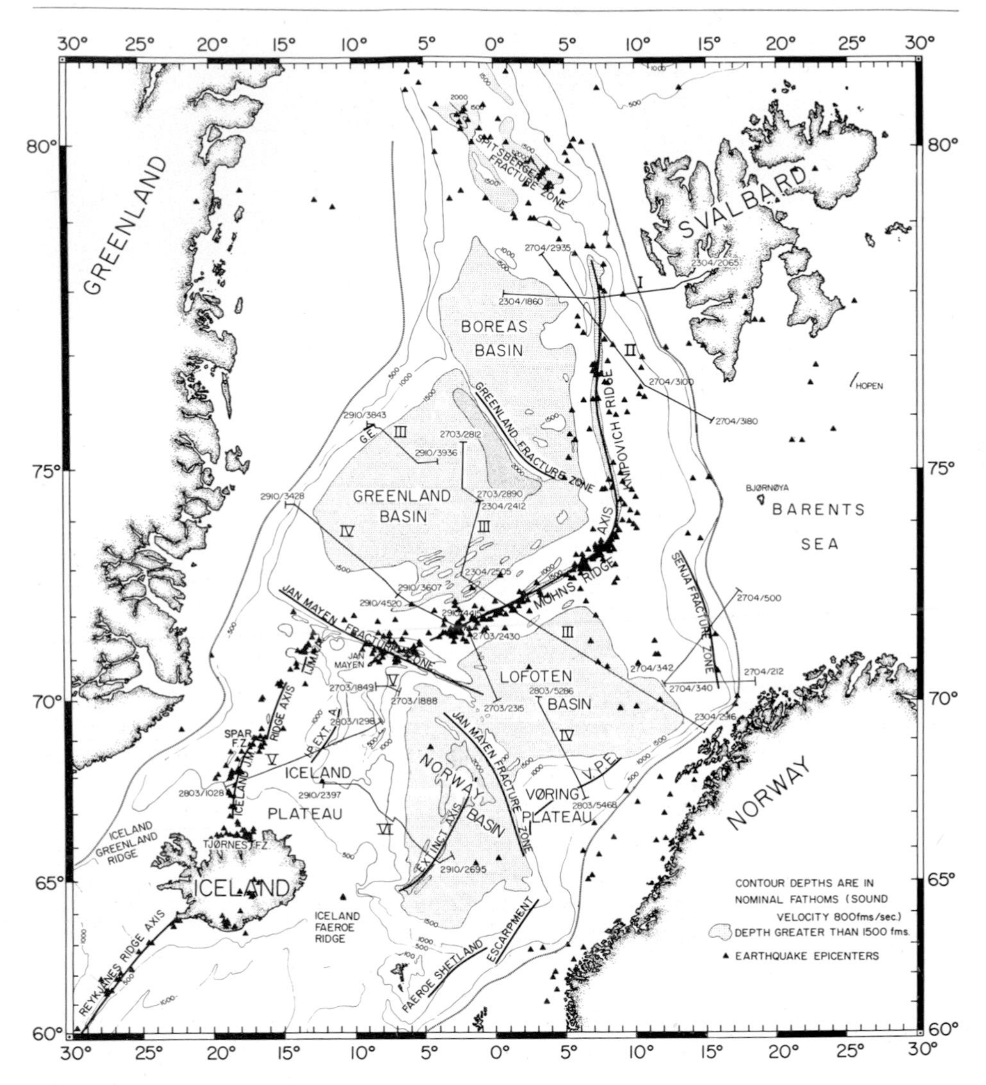


Fig. 5. Earthquake epicentres, bathymetry and major structural features in the Norwegian–Greenland Sea (after Talwani and Eldholm 1977).

of opening in the Santonian lay much farther south in which case the opening would be mainly confined to the southern part of the Rockall basin and the area between Ireland and Newfoundland and would cause no opening in the Norwegian Sea.

The second possibility that the opening from Rockall all the way up to the area of northern Norway occurred much earlier in the Mesozoic or late Palaeozoic, leaves unanswered the question of where the northern continuation of the opening between Newfoundland and Ireland might lie. Kristoffersen's location for anomaly 34 appears to preclude the presence of pre-anomaly 34 crust north of the Charlie Gibbs fracture zone between Hatton–Rockall and Labrador (see the position of anomaly 34 in Fig. 1).

When Talwani and Eldholm (1972) suggested that the area of the Vøring plateau was underlain by subsided continental crust, they appealed in part to the very quiet magnetic field over this area. It should also be pointed out that the basins in this area appear to be continuous (Fig. 3) with the grabens in the North Sea which have been caused by continental rifting. Also noted is the high gravity belt which crosses the Møre basin and which, according to Talwani and Eldholm (1972) is associated with rocks of Precambrian age, is not offset by any presumed opening of the basins by sea-floor spreading.

The seismic refraction data in this area are few and their results are not conclusive. A section by Hinz and Moe (1971) on the Vøring plateau is shown in Figure 4. The velocities and thicknesses are neither typically oceanic nor typically continental. This section is compared with seismic sections obtained in the area of the magnetic quiet zone south of Australia. That area is generally considered to have evolved by rifting and it was found that crustal sections determined seismically at a number of locations showed enormous variation in structure. These sections generally fell into three groups which are shown in Figure 4. The structure in group I resembles that of Hinz's section suggesting that Hinz's section for the Vøring basin is not atypical of rifted continental crust.

It is concluded that the manner of formation of the Mesozoic basins lying between Greenland and Norway is not yet clearly established but probably they formed by rifting in a manner similar to the Central and Viking grabens of the North Sea. The basement of the basins between Norway and Greenland is very difficult to drill because it lies at great depth but very detailed seismic refraction work might shed some further light on the problem. However, the southern part of Rockall basin which has magnetic anomalies, may have opened just prior to anomaly 34 time by sea-floor spreading, together with the sea-floor between Ireland and Newfoundland. The origin of the northern part of the Rockall basin and Faeroe–Shetland channel poses some of the same problems as do the Mesozoic basins of the Norwegian Sea.

## 3. Implications of drilling results

### 3a. Vøring plateau

Not only are there questions about the Vøring basin, which is the sedimentary basin underlying the inner part of the Vøring plateau, but opinions also differ about the origin of the outer Vøring plateau. The inner and the outer Vøring plateaus are separated by the Vøring plateau escarpment (Figs 5 and 7). This buried escarpment has been described by Talwani and Eldholm (1972). A single-channel seismic reflection profile which used a small airgun as a sound source is shown in Figure 6. This profile crosses the Vøring plateau escarpment at 67·6°N, 5·8°E, along an azimuth of about 175°N. Acoustic basement is clearly visible on the left side of the figure. It drops suddenly at the escarpment on the right side and the seismic record shows a number of reflectors which form the deep sedimentary basin—the Vøring basin. A field of diapiric bodies can also be seen near the right side of the figure. Drilling was carried out to acoustic basement at Sites 338 and 342 (*Glomar Challenger* Leg 38), on the diapiric bodies (Sites 339 and 340) and in the basin in between (Site 341). A tracing of the seismic reflection section obtained on the *Glomar Challenger* together with the summary of findings in the drill holes are shown in Figure 8. At Sites 338 and 342, the drill recovered basalt; the radiometric age of the rock recovered was 47 m.y. at Site 338 and 44 m.y. at Site 342 (Kharin *et al.*

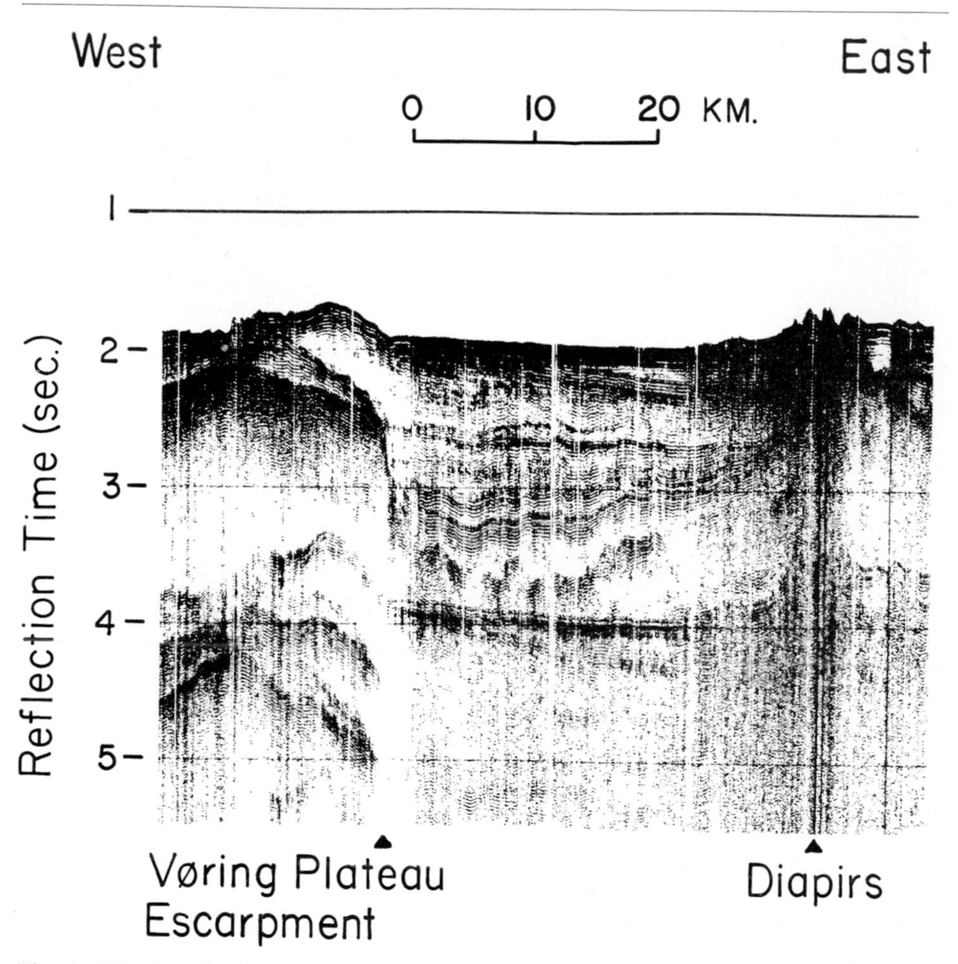


Fig. 6. Seismic reflection profiler section across the Vøring plateau escarpment obtained on *Vema* Cruise 30.

1976). The age of the overlying sediment at Site 338 was determined from faunal evidence as early Eocene. The oldest sediment recovered at Site 342 was early Miocene, but that discrepancy is attributed to the fact that the older sediments (Fig. 8) are absent at that site. Figure 7 shows Sites 338 and 342, located in an area where the basement age is expected to be older than anomaly 24, but younger than anomaly 25 which is not present anywhere in the Norwegian Sea. The palaeontological ages are not in conflict with the established ages from the magnetic lineations. The radiometric ages are somewhat younger, but explanations such as argon loss may account for the discrepancies. The hole at Site 341 in the Vøring basin only penetrated Miocene sediments but since a large thickness of sediment lies below the point drilled, the presence of much older sediments beneath the site cannot be disputed.

Recent seismic multichannel data obtained by the Bundesanstalt für Geowissenschaften und Rohstoffe and by the Institute Français du Petrole (K. Hinz and

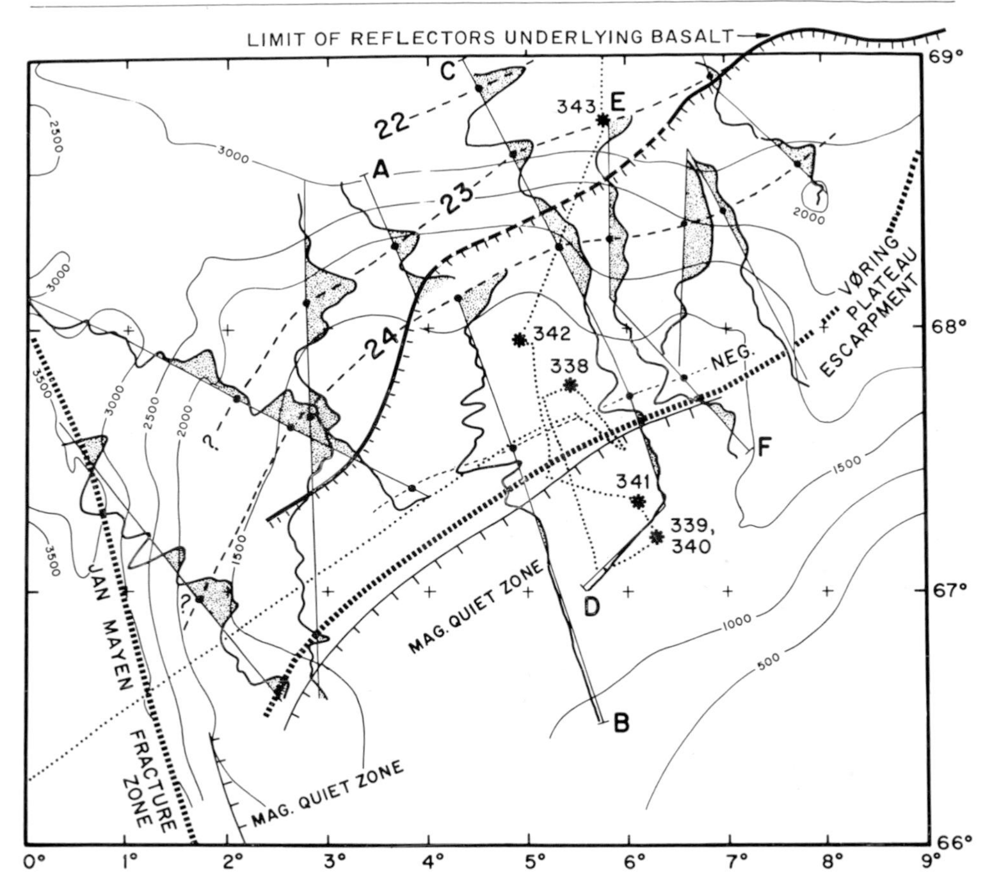


Fig. 7. Magnetic anomalies plotted perpendicular to selected ship tracks on the Vøring plateau. Bathymetric contours (light lines) are in metres. Drill sites on *Glomar Challenger* Leg 38 are shown. The outer limit of reflectors underlying the basalt layer is drawn on the basis of information supplied by K. Hinz (pers. comm.) and L. Montadert (pers. comm.).

L. Montadert, personal communication) show, however, that the situation is not as simple as the above findings would suggest. They found that beneath the basalt layer in the eastern part of the outer Vøring plateau there exist seismic reflectors which in some cases are flat-lying, but in other cases are seaward dipping. Below these reflectors lies another strong reflector which they termed "true basement". Since neither these reflectors nor "true basement" has been penetrated by the drill, uncertainties exist about the nature of the outer Vøring plateau. For example, it has been suggested that the eastern part of the Vøring plateau is not underlain by oceanic crust but by Mesozoic crust (continental or oceanic) that has been covered by Tertiary lavas produced at the time of opening of the Norwegian Sea. Hinz and Montadert's surveys extended northwards from the Vøring plateau to the Lofoten basin (Fig. 5) and reflectors have been shown to exist under the basalt layer in both regions. They were able to map the seaward limit of these reflectors. That limit is plotted in Figure 7. The region in which these reflectors exist extends not only east of magnetic anomaly lineation 24, as it does in the central Vøring plateau, but

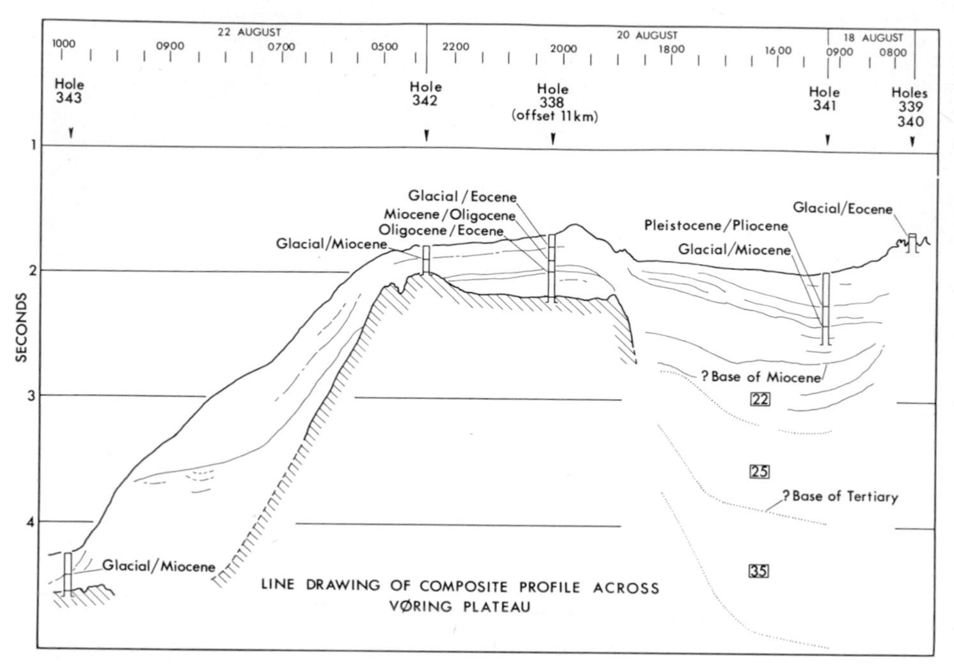


Fig. 8. Line drawing composite of seismic reflection profile across the Vøring plateau obtained aboard *Glomar Challenger* (track indicated by dotted line in Fig. 7), together with summary of drilling results at Sites 338–343. Acoustic basement as indicated by the *Challenger's* single channel seismic reflection profiling system is emphasized by oblique shading. After Talwani *et al.* (1976).

it crosses through anomaly 24 at 69°N, 07°E and it actually extends as far seaward as magnetic anomaly lineation 23. These circumstances can be explained if either anomaly 23 was misidentified, or the anomaly is not a sea-floor spreading type anomaly at all. However, these explanations are rather doubtful since the anomalies north of the Vøring plateau are among the best identified anomalies in the Norwegian Sea. The preferred explanation is that the entire area westward of the Vøring plateau escarpment is oceanic and has been generated since the time of Tertiary opening with the layered reflectors being associated with the very earliest time of opening. They were covered shortly after the time of formation by lava flows which extend from the location of the identified anomalies all the way east to the Vøring plateau escarpment. Proof of this can only be obtained by drilling through the reflectors down into "true basement".

Further investigation is needed for two associated problems. One is why is there a basement high, that is, why does the basement in the outer Vøring plateau, that seems to be associated with the initiation of spreading, lie higher than the oceanic basement in the Norway and Lofoten basins? The second concerns the smoothness of basement and the possibility that the occurrence of early lava flows, which blanket the very earliest basement relief, may be a more common phenomenon than at present suspected, and may explain the smooth basaltic layering in many parts of the world's oceans as, for example, in the Caribbean Sea.

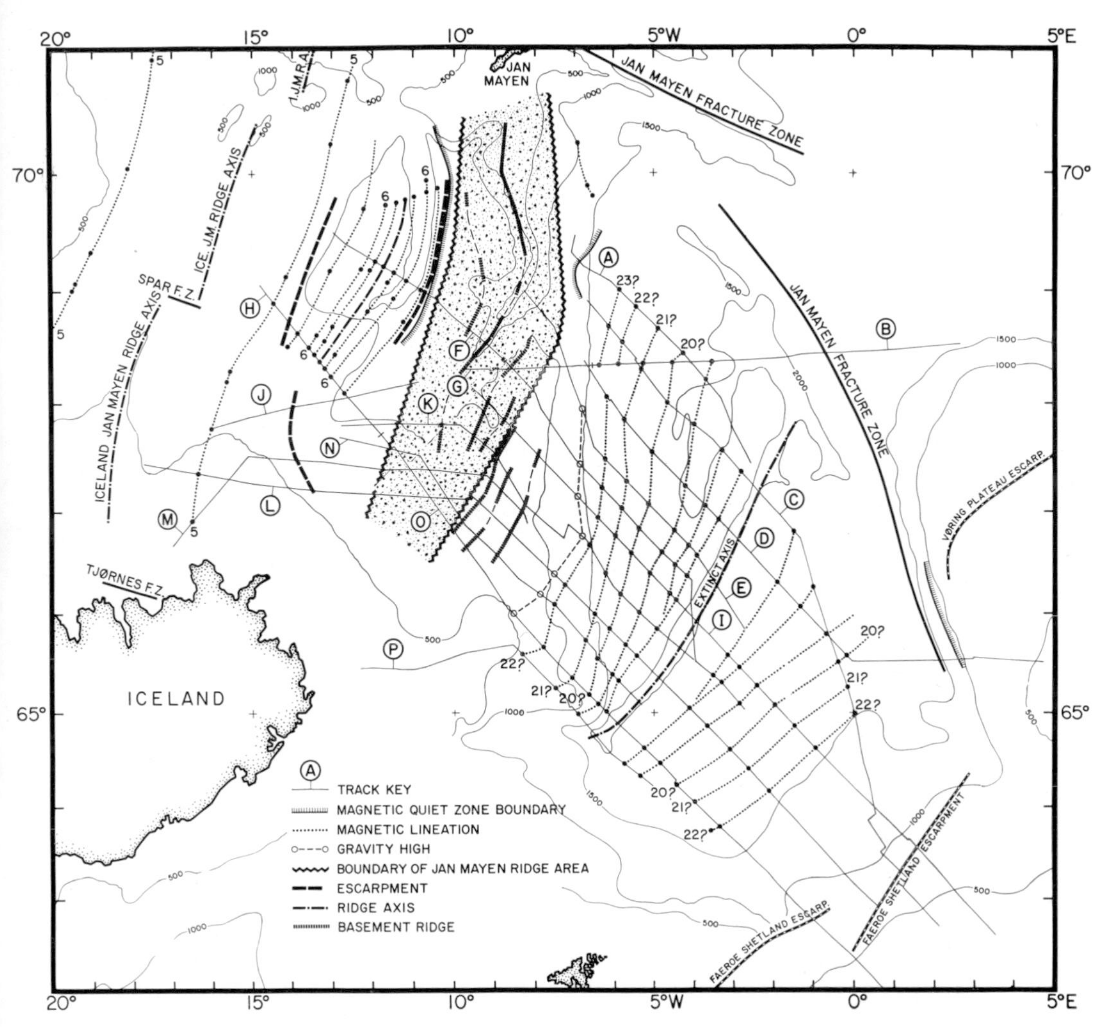


Fig. 9. Prominent structural features, ship tracks and magnetic lineations in the area lying between Jan Mayen fracture zone and the Iceland–Faeroe ridge. After Talwani and Eldholm (1977).

## 3b. Norway Basin, Icelandic plateau and adjacent areas

While the magnetic anomalies are symmetrically distributed about the Mohns ridge axis and about the Reykjanes ridge axis (Fig. 2) they are not symmetrically distributed in the intervening area. In particular, the evolution of the area lying between the Jan Mayen fracture zone to the north and Iceland and the Iceland–Faeroe ridge to the south has been complicated in part because the spreading axis has "jumped" several times. However, some magnetic anomalies can be identified in this area and a scheme of evolution can be proposed. Because of the young age of this area and consequent small thickness of sediments, it has been possible to test this scheme of evolution by drilling down to basement. This is the only area of the

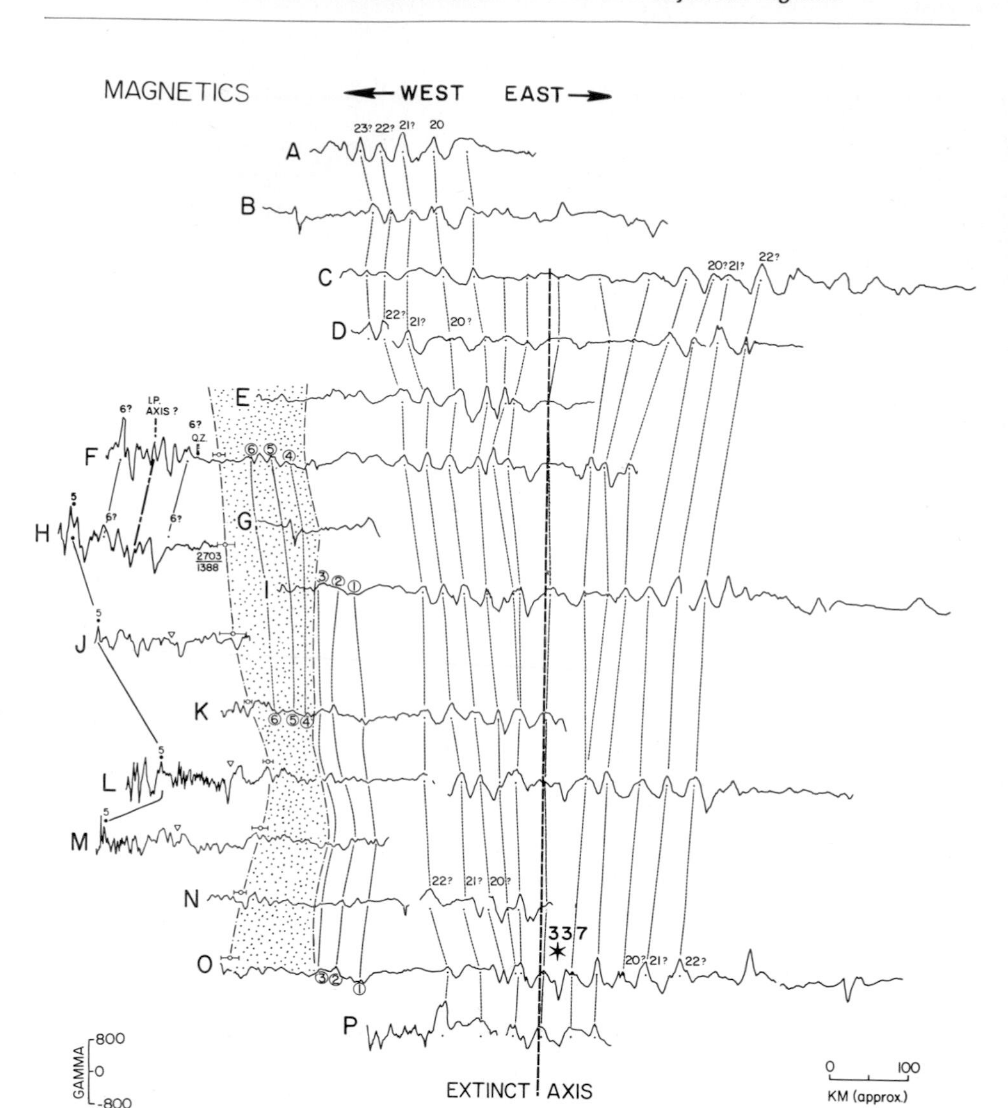


Fig. 10. Projected magnetic (total intensity, IGRF removed) profiles across the Norway basin and the southern part of the Iceland–Jan Mayen ridge area. Identification of anomalies younger than anomaly 20 in the Norway basin is uncertain. The location of the extinct axis, as well as basement ridges 1 (circled), to 6 (circled), are based on gravity data. The region of strong gravity gradients on the western side of the Jan Mayen ridge area is indicated by the bars with a circle in the centre; the ridge itself is indicated by stippled pattern. Triangles in the Icelandic plateau area indicate topographic escarpments (after Talwani and Eldholm 1977).

eastern North Atlantic Ocean in which such verification has been possible in regions away from the ridge crest. Drilling thus constitutes a very powerful tool for study here.

Magnetic lineations in the area between Jan Mayen fracture zone and Iceland and the Iceland–Faeroe ridge are shown in Figures 2, 9 and 12. Several investigators, including Johnson *et al.* 1972, have contributed to the unravelling of the evolutionary details here. In the following the detailed solutions suggested by Talwani and Eldholm (1977) are principally considered. The area which lies south of the Jan Mayen fracture zone and north of Iceland and the Iceland–Faeroe ridge can be divided into four parts (Fig. 9); (1) the Norway basin which roughly lies within the 1500 fathom contour and has an extinct axis; spreading around the extinct axis slowed sometime after anomaly 20 time and stopped around anomaly 7 time; D.S.D.P. Site 337 is located in this area, (2) the Jan Mayen ridge region, shaded in Figure 9, that is presumed to be continental in origin; D.S.D.P. Sites 346, 347 and 349 are located on this ridge, (3) the region produced by spreading about the extinct Icelandic plateau axis lying roughly between 10° and 14°W. In the northern part (between 69°N and 70°N) this region is bounded on the east and west by topographic ridges or escarpments denoted by heavy dashed lines. Spreading took place roughly between anomaly 6A time to anomaly 5D time in this area, (4) the Iceland–Jan Mayen ridge area, where spreading about roughly the present axis has gone on since anomaly 5 time.

Magnetic profiles which have been used to delineate these areas are shown in Figure 10. The location of the profiles is given in Figure 9. D.S.D.P. sites occupied during Leg 38 have been superimposed on a physiographic map of the area shown in Figure 11.

We now proceed to examine how the drilling results corroborate the suggested framework of the evolution of the area of the Norway basin and adjacent areas.

## 3c. Extinct axis in the Norway basin

Leg 38 of the Deep Sea Drilling Project tried to drill as close as possible to the topographic valley associated with the extinct axis in order to obtain samples of basement associated with the youngest oceanic crust in the Norway basin. However, as seen in Figure 13 (Profile EF) there is a considerable thickness of sediments at the extinct axis, and therefore D.S.D.P. Site 337 was located about 20 km east of the extinct axis. This site lies on the steep slope of the magnetic anomalies between the central maximum and the first minimum east of it (Fig. 10). Basement consisted of tholeiitic basalt typical of mid-ocean ridges. The presence of several layers of breccia believed to represent pillow lavas leaves little doubt that the basalt here was extruded. As far as the age of basement is concerned, there is some conflict between the radiometric ages which range from 17·5 to 25·5 m.y. and that determined from fossils in the overlying sediment. These sediments are poorly dated, but indicated a possible range of middle Oligocene to late Eocene. To reconcile the two kinds of data tentative assignment was of the youngest palaeontological data to this site, estimated to be about 29 m.y. This age can be used to estimate the age at which the spreading axis became extinct in the Norway basin. By assuming that the half-rate spreading rate was between 0·5 cm/y. and 0·8 cm/y. just before the time that this spreading axis became extinct and using the distance of 20 km of the site from the extinct axis, an age of 25–27 m.y. would be indicated for the extinct spreading axis. This age agrees well with the age of anomaly 7 (27 m.y.). Shortly after that time spreading ceased in the Norway basin and the spreading axis jumped west. It is pointed out, however, that because the range in the palaeontologically determined age for overlying sediments is large and because of discrepancies with the radiometrically determined age, the age at which the spreading axis in the Norway basin became extinct is correspondingly uncertain.

D.S.D.P. Site 350 lies in an area between regions (1) and (2) discussed above. This

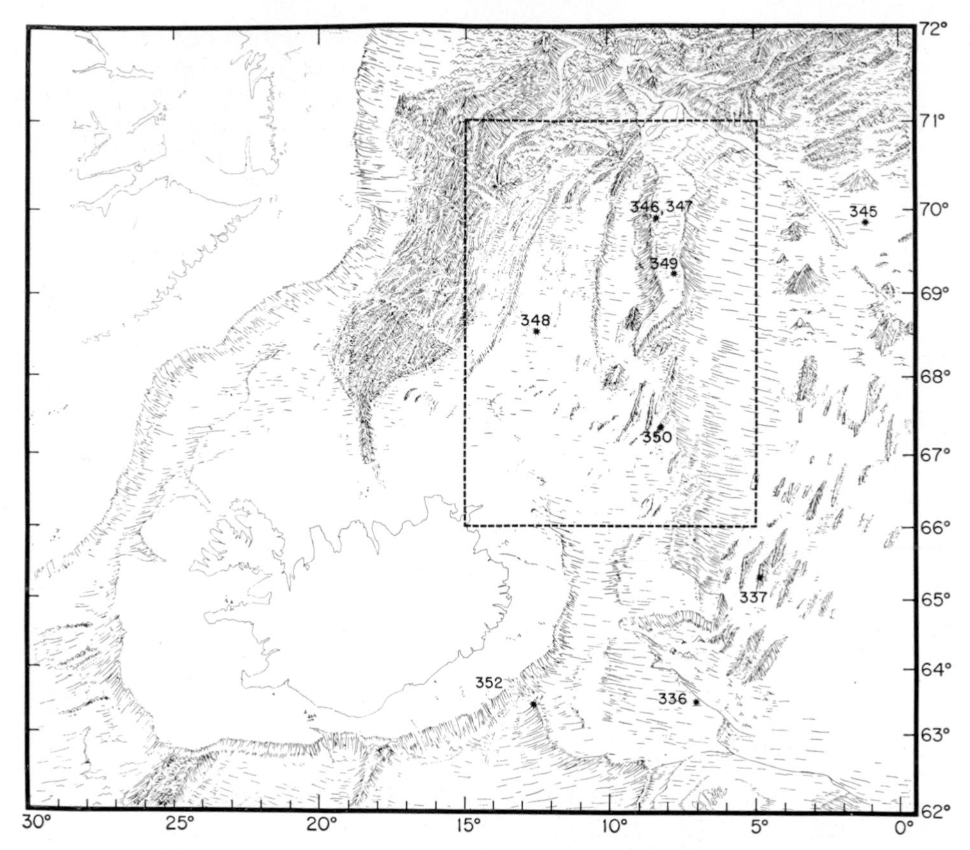


Fig. 11. Physiographic diagram of the area north of Iceland including the Jan Mayen ridge. The outlined area indicates the area of the map in Fig. 12. DSDP sites referred to in the text are located by number. This map was supplied by Gleb Udintsev. Site 352 should be approximately 1°W of the position shown.

area has been designated 3 (circled), in Figure 12. Morphologically this area has been considered a part of the Jan Mayen ridge because it lies at an elevation higher than the elevation of the Norway basin (Fig. 11). However, Talwani and Eldholm (1977) have suggested that this area was probably created by sea-floor spreading contemporaneously with the development of the Norway basin. The reason for this suggestion is as follows. The magnetic lineations in the Norway basin (Figs 9 and 10) show that while anomaly lineations older than 20 are parallel to each other, the younger anomalies tend to crowd southward in a fan-shaped pattern. From geometrical considerations, and comparison with the magnetic data on the Reykjanes and Mohns ridges, another area must be present where new oceanic crust was created between anomaly 20 time and the time that the Norwegian spreading axis became extinct (anomaly 7 time). It can be seen in Figure 9 that the northern segment of the Faeroe–Shetland escarpment which forms the eastern boundary of the Norway basin is not parallel to the 1000 fathom curve which formed the eastern boundary of the morphological Jan Mayen ridge, but rather it is parallel to the western boundary of area 3 (circled),(Fig. 12). In other words, if area 3 (circled), is

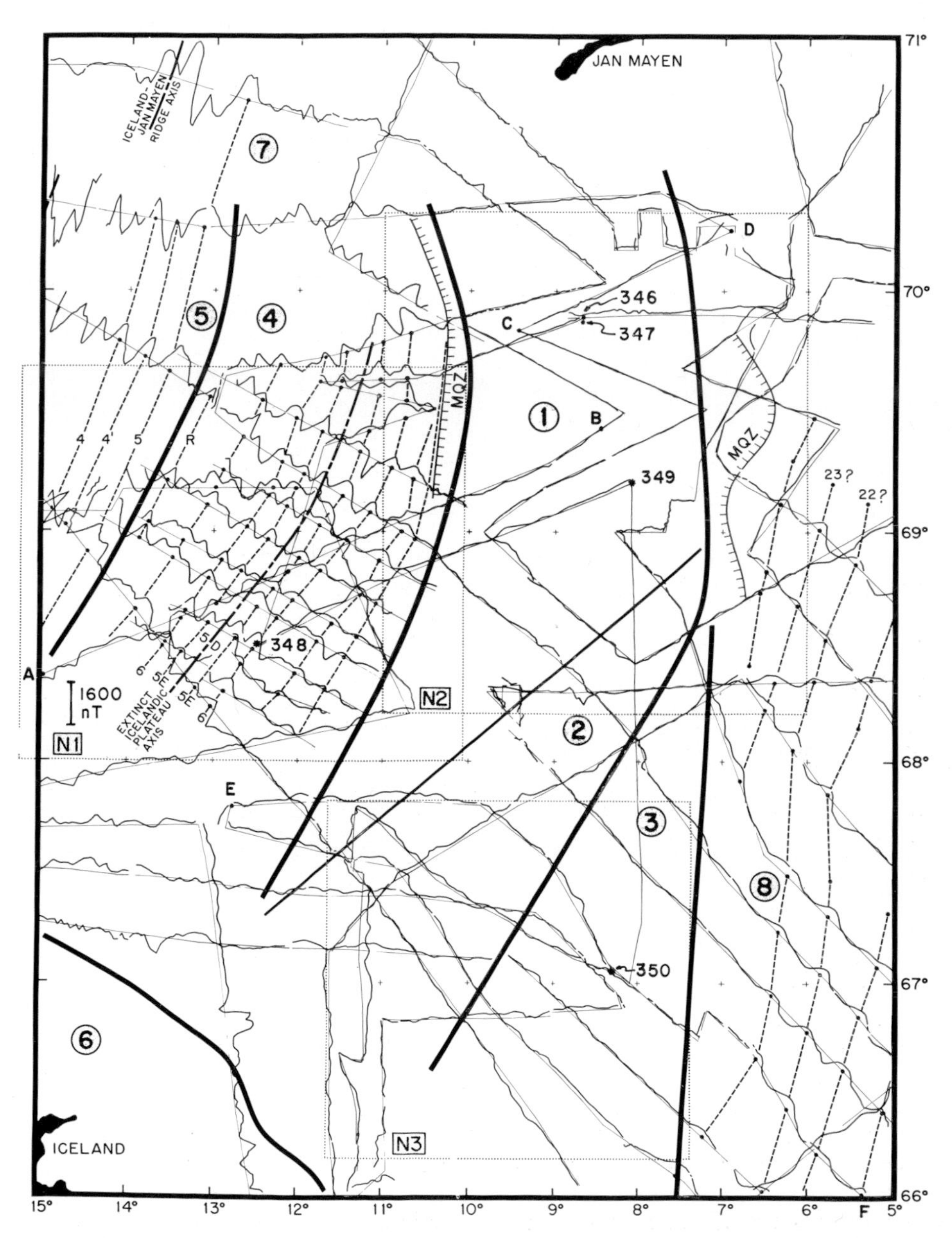


Fig. 12. Magnetic anomalies plotted along track in the area of the Jan Mayen ridge (after Grønlie and Talwani 1978). The anomaly identification in the area of the extinct Icelandic plateau axis is from Chapman and Talwani (1978). Drilling Sites 346–350 on *Glomar Challenger* Leg 38 are indicated. Areas 1 (circled), and 2 (circled), in this figure together correspond to area (2) in the text. Area 1 (circled), represents the blocky northern Jan Mayen ridge, 2 (circled), the irregular southern Jan Mayen ridge. Areas 8 (circled), 4 (circled), and 5 (circled), in the figure correspond to areas (1), (3) and (4) in the text. Area 3 (circled), lies between areas (2) and (3) and is discussed in the text.

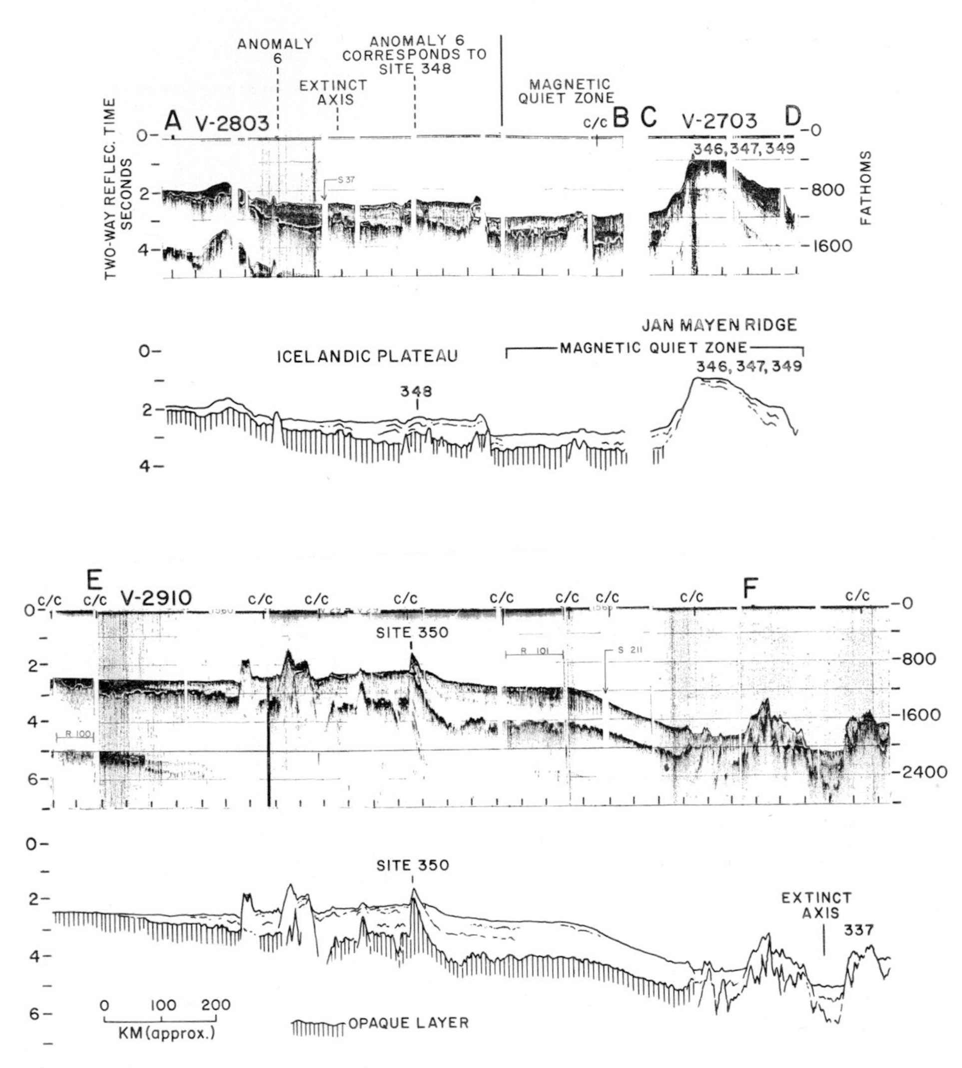


Fig. 13. Seismic reflection records along profiles AB, CD and EF; profiles are located in Fig. 12.

included within the Norway basin, then the Norway basin does not diminish in width from north to south, thereby satisfying a geometrical consideration which is imposed by the rigid motion of plates.

All of these considerations prompted Talwani and Eldholm (1977) to suggest that region 3 (circled), was created contemporaneously with the Norway basin by some form of sea-floor spreading. If this hypothesis is right, the age of basement in this area would lie between the ages of anomaly 20 and 7, that is between about 45 and 27 m.y. ago.

Seismic reflection records show that a seismically "opaque" layer underlies sediment in area 3 (circled), and continues eastward towards the Norway basin. Drilling

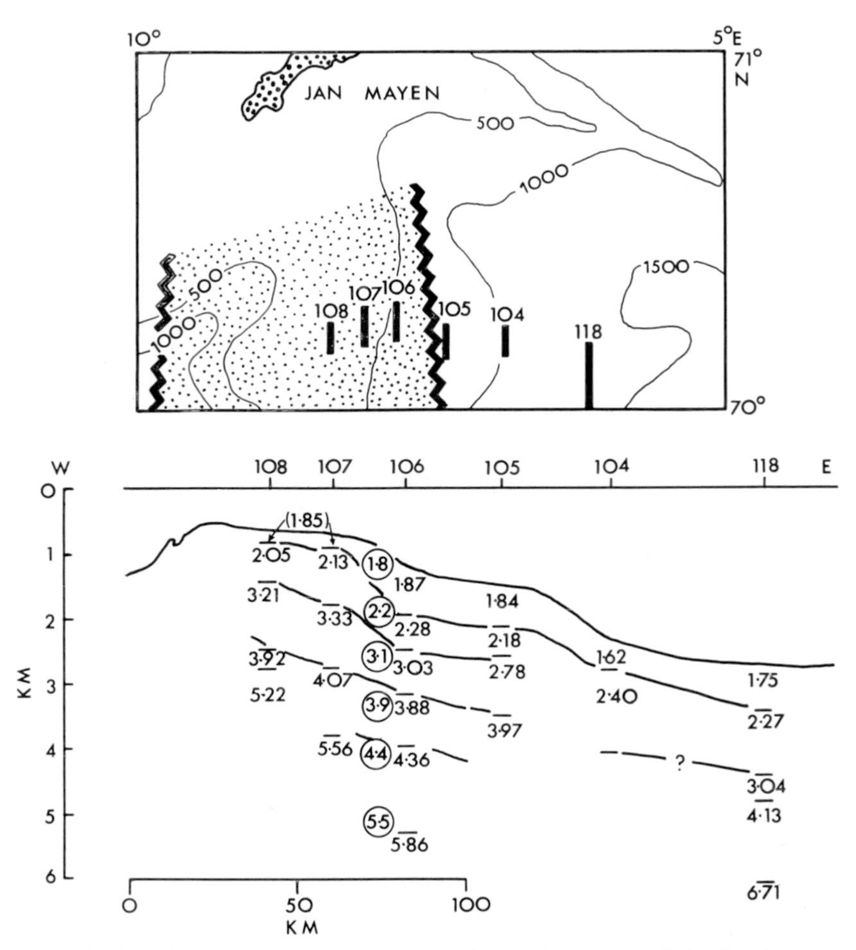


Fig. 14. Seismic refraction results obtained on the northern part of the Jan Mayen ridge. (Talwani and Eldholm 1977).

at D.S.D.P. Site 350 revealed that the opaque layer consists of basalt. There is considerable scatter in radiometric ages but values of 40–44 m.y. appear to be indicated (Kharin *et al.* 1976). These are not in conflict with the late Eocene age determined palaeontologically from the overlying sediments and fall within the range proposed by Talwani and Eldholm (1977) for the area, assuming that the opaque layer forms "true basement" or lies very close above it. It is particularly important that the drilling results at D.S.D.P. Site 350 are considerably different from the results elsewhere on the main topographic block of the Jan Mayen ridge at D.S.D.P. Sites 346, 347 and 349 where basement was not reached.

## 3d. Jan Mayen ridge and its southern continuation

The morphology of the northern block-like part of the Jan Mayen ridge is shown in Figure 11. Seismic reflection records of the area are shown in Figure 13 (Profile CD). A composite of the Jan Mayen ridge from the reflection data and from the

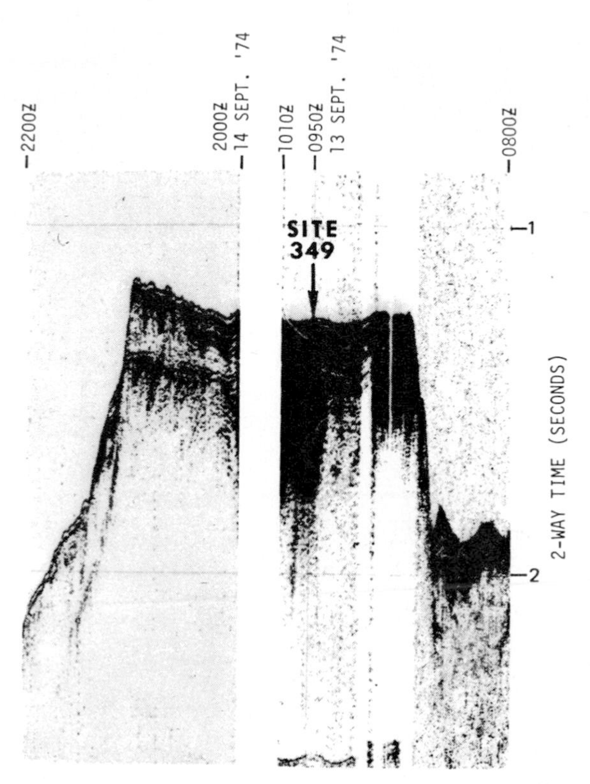


Fig. 15. Profiler record showing sub–bottom seismic reflectors at Site 349 on the Jan Mayen ridge (after Talwani *et al.* 1976).

refraction data (Fig. 14) is also shown in Figure 13. Figures 13 and 14 show that on the block-like Jan Mayen ridge, parallel horizontal reflectors exist just underneath the bottom (actually, because of the poor quality of the seismic records, these reflectors are seen best not at the Site of 346 and 347, but, as shown in Figure 15, at a point somewhat further south). These sediments are associated with a velocity of about 1·85 km/sec. The refraction results show that below the flat-lying top layers, lower layers exist which dip to the east. These are also seen in some reflection profiles. They are steeply dipping and are truncated by a prominent reflector which forms the base of flat-lying sediments on top. The layer with a velocity of greater than 5 km/sec. (Fig. 14) is supposed to represent basement. Neither the reflection records nor the refraction results are clear on where the basement lies near the western edge of the block. By extrapolating the apparent steep slope of the basement surface westwards, and noting the steep topographic western slope of the Jan Mayen ridge, it was considered, prior to drilling, that it was possible that basement cropped out on the topographic slope and in fact might form the western slope of the Jan Mayen ridge. Site 346 was located near the western edge of the topographic platform. The drill penetrated through the flat-lying sediments on the top, through the unconformity associated with the prominent reflector, and to the dipping reflectors below it. However, basement was not reached. Nor was it reached at Site 347 located even further to the southwest, closer to the steep slope. At Site 347, as at Site 349, the prominent reflector was penetrated and drilling progress was very slow below in a siliceous mudstone. Basement was not reached. From the drilling results it seems quite probable that this siliceous mudstone, and not a basement ridge, forms the western slope of the distinct topographic block of the Jan Mayen ridge.

At all three sites the prominent reflector was penetrated at about 120 m. This represents an unconformity which is especially clear lithologically at Site 349 where it is marked by a basal conglomerate. Above the unconformity are glacigene sediments as well as sediments extending from middle Miocene to middle Oligocene in age. Below the unconformity are sediments that range in age from early Oligocene to late (and middle?) Eocene. Because basement was not reached at any of the sites and because the oldest sediments found are not older than Eocene and therefore do not predate the Tertiary opening of the Norwegian Sea, one cannot directly deduce that the Jan Mayen ridge existed as a part of Greenland before the Norwegian Sea opened. However, several factors indicate that basement much older than Eocene underlies the Jan Mayen ridge. Figure 14 shows that a large thickness of sediments underlies the sediment with velocities 2·2 km/sec. that yielded Eocene fauna. Presumably the sediments at the base of the sedimentary column are much older. In the Norwegian Sea and the North Sea the boundary between layers with seismic velocity 2·2 km/sec. and 3·1 km/sec. is the boundary between the Tertiary and Cretaceous and the layers with velocity about 4·4 km/sec. there represent Mesozoic and perhaps even Palaeozoic sediments. A similar situation could exist here, that is, the lowest sediments underlying the Jan Mayen ridge could be Mesozoic or even Palaeozoic in age.

In its reconstructed position (Fig. 3) the Jan Mayen ridge lies adjacent to Vøring basin. It is interesting to note that the magnetic signature in the northern Jan Mayen ridge is very quiet (Fig. 12, area 1 (circled), as is the signature under the Vøring basin. The magnetic signature is somewhat noisier in the southern part of the Jan Mayen ridge corresponding to a similar situation under the Møre basin. These considerations suggest that the basement under the Jan Mayen ridge is similar to the basement under the Vøring and Møre basins.

Geophysical data have been used to trace the continuation of the structures underlying the Jan Mayen ridge to the southwest. Figures 9, 11 and 12 show that south of about 68·5°N the Jan Mayen ridge bends to the southwest. North of about 68·5°N the Jan Mayen ridge is characterized by a prominent topographic block-like feature. It is on this block-like feature that Sites 346, 347 and 349 are located. However, this block seems to be broken up into fragments and these fragments appear to lie progressively deeper as one proceeds to the southwest, forming topographic blocks or ridges, that are buried under sediments. These buried fragments have a characteristic look on the single-channel seismic profile records. They appear as interruptions of an acoustically opaque layer (Eldholm and Windisch 1974) and have been informally referred to as "holes". Figure 16, which is a compilation of seismic reflection records across the Jan Mayen ridge clearly shows how the block-like Jan Mayen ridge changes in its character to the southwest. The southern profiles shown on this figure have been interpreted to contain continental blocks buried deep under sediments and between adjacent pieces of ocean crust. The continental fragments appear as holes while the oceanic crust is represented by the opaque layer. It would be of great interest to verify the presence of continental crust by drilling at locations such as these.

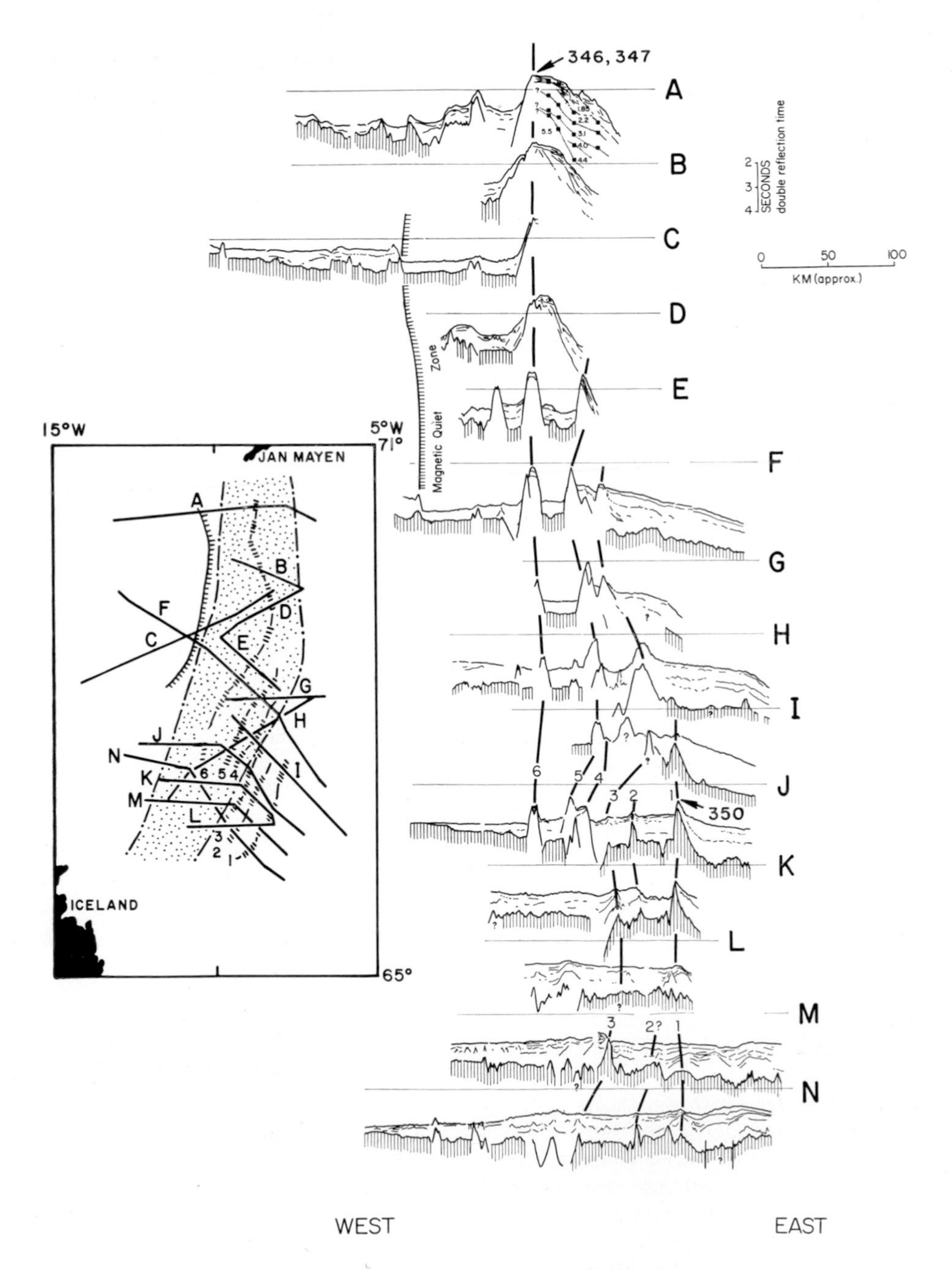


Fig. 16. Tracing of seismic reflection profile records across the Jan Mayen ridge. (After Talwani and Eldholm 1977).

### 3e. Extinct axis on the Iceland plateau

Between the now active Iceland–Jan Mayen ridge and the Jan Mayen ridge which is presumed to be a continental fragment, lies a zone which is characterized by the presence of magnetic anomalies. This zone is designated 4 (circled), in Figure 12. The magnetic lineations in this zone were first examined by Johnson *et al.* (1972), and identified more recently by Chapman and Talwani (1978). The magnetic lineations range from anomaly 5D to anomaly 6A. Anomaly 5D lies on the extinct axis, its age (according to the time scale of Blakely 1974) is about 18·3 m.y. Anomaly 6A has an age of 24·3 m.y., therefore spreading about the intermediate axis on the Iceland plateau roughly encompasses the period between 18 and 24 m.y. ago.

Site 348 was located on prominent positive magnetic anomaly which was identified by Chapman and Talwani (1978) as anomaly 6 (Fig. 12). The drill penetrated a seismically "opaque" layer. This layer can be seen in the seismic profiler records in Figure 13. The opaque layer here is basalt; the age determined radiometrically is 18·8±1·7 m.y. (Kharin *et al.* 1976) which corresponds rather well to the age of 21 m.y. for magnetic anomaly 6.

Although the magnetic anomalies clearly show an axis of symmetry in this intermediate zone of spreading on the Iceland plateau, no topographic axis corresponds with the magnetic axis. Also, if the opaque layer constitutes basement, it is not the rough basement one usually associates with mid-ocean ridges, but rather, is a smooth basement. Consideration of the acoustically opaque layer is given in section 3g.

### 3f. Iceland–Faeroe ridge

The Iceland–Faeroe ridge connects Iceland with Iceland–Faeroe block. This ridge is shallow and has a nearly flat top at a depth of about 400 m. It does not mark the position of an active or an extinct spreading centre such as the Reykjanes and Mohns ridge or the extinct axis in the Norway basin, but rather lies almost perpendicular to the direction of spreading. Most recent authors consider the Iceland–Faeroe ridge to be a "Proto-Iceland". Just as Iceland now stands higher than the Reykjanes ridge to the south, and the Iceland–Jan Mayen ridge to the north, the Iceland–Faeroe ridge stands higher than the Reykjanes basin to the south and the Norway basin to the north.

A sediment isopach map of the Iceland–Faeroe ridge is shown in Figure 17 and some profiles across it are shown in Figure 18. As drilling aboard the *Glomar Challenger* requires a minimum of about 100 m of sediments to spud in and the thickness of sediment on the top of this ridge is so small, drilling had to be carried out at two sites on the northern and southern flanks of the ridge in areas where the sediment is somewhat thicker.

The magnetic anomalies over this ridge are generally large in magnitude, but short in wavelength implying a shallow source such as basalt flows. They are not lineated in the fashion typical of sea-floor spreading type anomalies and for this reason the age of the ridge cannot be obtained from magnetic anomalies but has been deduced from geometrical considerations (Bott 1974; Talwani and Eldholm 1977). As the Iceland–Faeroe ridge lies between Iceland, which is presently active, and the continental margin, its age must be older than that of Iceland, but younger than the time of initiation of Tertiary spreading in the Norwegian Sea. Talwani and Eldholm (1977) believe that the spreading axis jumped from the Iceland–Faeroe ridge to Iceland at about anomaly 7 time. On this basis the age of the ridge must lie somewhere between the time of anomaly 7 and the time of anomaly 24. It has yet to be

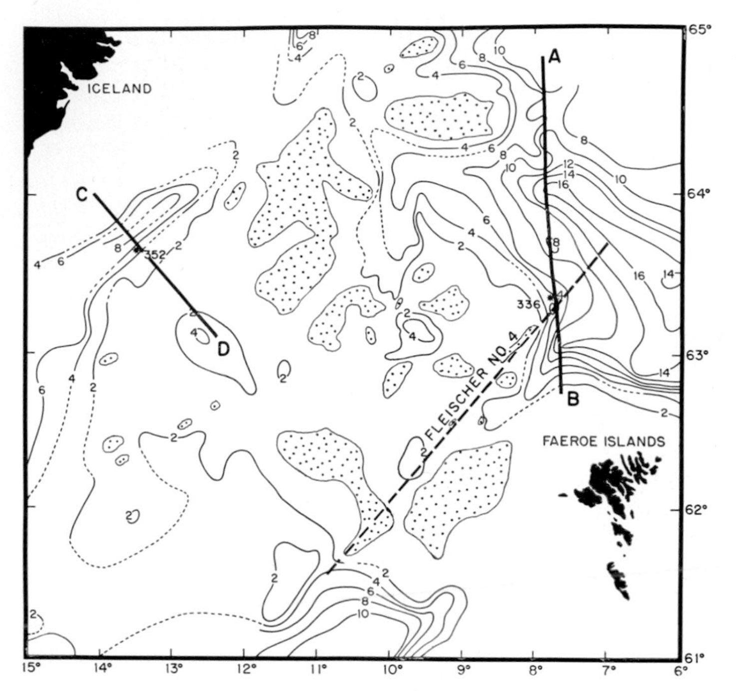


Fig. 17. Sediment isopach map of the Iceland–Faeroe ridge. Stippled areas on the ridge are barren of sediment; contours are in tenths of seconds double reflection time.

determined whether it was subaerial in the past and formed a land bridge between Eurasia and Greenland, and, if so, when it subsided below sea level.

Drilling at Site 336 revealed the presence of "rubble" immediately above the basalt which suggested subaerial weathering. The basalt, which is petrologically typical of mid-ocean ridge basalts, contains structures which indicate that it was extruded above sea level. It now lies at a depth of more 1300 m below sea level. Therefore the site has subsided by at least this amount. Although the actual situation may have been considerably more complicated, it is assumed here, for the sake of simplicity, that the isochrons on the Iceland–Faeroe ridge have a roughly NE strike, i.e. they lie perpendicular to the length of the ridge. Site 336 lies on the northern flank of the ridge. Isochrons passing through it also pass over the crest of the ridge where basement is now exposed at a depth of nearly 400 m. But the depth of bottom at Site 336 is 830 m, basement lies 485 m below the bottom, therefore basement at Site 336 lies at a depth of more than 900 m greater than the depth of basement of the same age along the crest of the ridge. Ignoring erosion and differential subsidence due to loading, the point on the crest of the Iceland–Faeroe ridge with the same age as basement at Site 336 was more than 900m above sea level when the basalt basement was extruded close to sea level at Site 336. Using subsidence curves available in the literature it can be assumed that it takes about 30 m.y. for the crest at 900 m to subside to sea level. The age of basement at Site 336 is approximately 45 m.y. Therefore the highest point on the ridge on the same isochron as Site 336 subsided to sea level 15 m.y. ago.

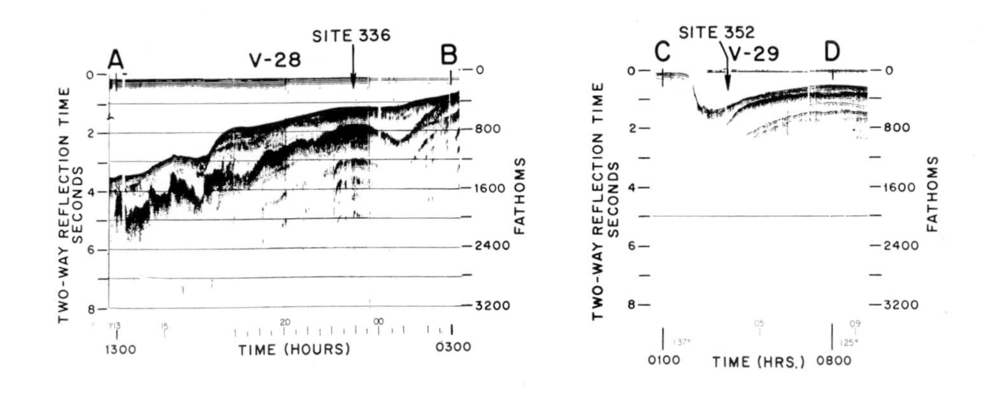


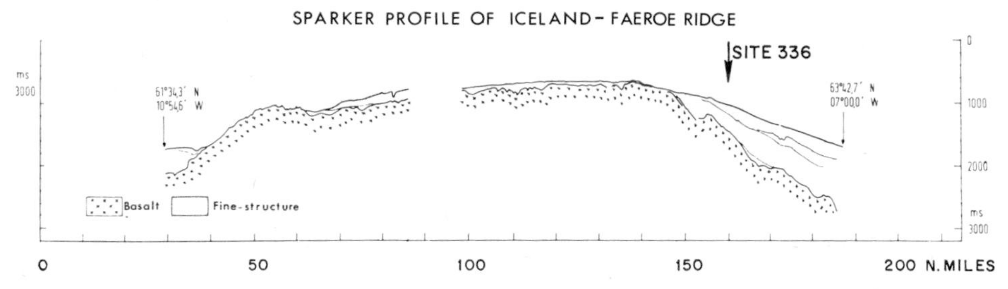


Fig. 18. The top seismic reflection profiles were obtained by *Vema* over the Iceland–Faeroe ridge. The bottom seismic profile tracing was obtained by Fleischer *et al.* (1974; Profile 4) aboard the *Meteor;* location of these profiles is given in Fig. 17.

At this time those points on the crest of the Iceland–Faeroe ridge for which the age of basement is younger than 45 m.y. were still above sea level; those with ages older than 45 m.y. had already subsided below sea level. The erosional history of the subaerial Iceland–Faeroe ridge is not known, nor is it certain that, when first formed, all points of the ridge stood at the same elevation. If it was formed between roughly 55 and 27 m.y. (anomaly 24 and 7 time), perhaps the oldest points on the ridge did not subside below sea level until about 25 m.y. (55 minus 30 m.y.). The youngest points subsided even later than 15 m.y. ago. Therefore 25 m.y. is probably the best estimate of the date at which the land bridge between Eurasia and Greenland was breached. Even after this time parts of the Iceland–Faeroe ridge were still presumably above sea level. It was not until some time in the mid to late Miocene that the entire ridge subsided below sea level.

At Site 352 on the southern flank of Iceland–Faeroe ridge, basement was not reached (drilling had to be stopped because of poor sea conditions), but the contrast in faunal evidence between Site 336 and 352 suggests that a barrier to the circulation between these two sites existed at least until late-middle Oligocene times. Faunal evidence is lacking on whether this barrier existed past late or middle Oligocene times because the Miocene and Pliocene sediments are absent at both sites.

The age of basement rock at Site 336 was determined between 40·4 and 43·4 m.y. (Kharin *et al.* 1976). This age lies between 55 m.y., the estimated time of opening of the Norwegian Sea, and 27 m.y., the time at which according to Talwani and Eldholm (1977) the ridge axis jumped from the Iceland–Faeroe ridge westward to form Iceland.

## 3g. The seismically "opaque" layer

It is perhaps not realized as often as it ought to be that the layering determined seismically depends on the nature of equipment used, particularly the frequency of the sound source, the nature of the initial pulse (sharp or reverberating), and the intensity of the sound source. An important feature of the seismic records in the area of the Icelandic plateau, obtained aboard the ship *Vema*, with single channel seismic equipment and a small airgun sound source, has been the near ubiquitous presence of an acoustically opaque layer (Eldholm and Windisch 1974) below which this seismic equipment could not penetrate. This opaque layer is generally quite smooth in appearance and therefore not typical of oceanic basement. One material suggested for this opaque layer has been that it is an ash layer that drapes basement rocks, thereby obliterating basement relief. Such an ash layer would present a strong target for reflection of sound waves and would appear as acoustic basement. If the opaque layer is an ash layer, it is difficult to imagine how the "holes" in this layer could exist, as shown for example in the southernmost profiles in Figure 16. These profiles show the opaque layer to be interrupted by blocks that rise above it. This geometry is very puzzling if the opaque layer is a sedimentary layer that was deposited rather recently and uniformly over a wide area.

Drilling at Sites 348 and 350 penetrated into the opaque layer and showed it to be a layer of basalt. The age, especially of the basaltic rocks at Site 348, is so close to that predicted from magnetic anomalies that it is probable that this layer of basalt was consolidated at a time very close to the time of formation of this area by sea-floor spreading. In a sense, the basalt penetrated at Sites 338 and 342 on the Vøring plateau may also be considered to appear as a seismically opaque layer for single channel reflection experiments, although multichannel seismic reflection has shown the presence of reflectors below it.

Perhaps a simple explanation for the opaque layer can be given as follows. The opening of a new axis of sea-floor spreading is associated with extensive areas of basalt flows. If these flows persist over large distances even after sea-floor spreading has started and an oceanic floor has been formed, a situation will develop where the younger, newly formed ocean-floor is covered by widespread basalt flows. It is possible that this is what has happened both in the Vøring plateau area and in the Icelandic plateau region. To test these ideas it is important to be able to drill through this basalt layer to determine what lies underneath. In both these areas the material underlying the basalt flow constitutes oceanic crust, since magnetic lineations indicating sea-floor spreading can be identified.

## 3h. Areas of unusual elevation

On a regional scale of hundreds or thousands of kilometres, several areas associated with the Norwegian Sea–Reykjanes ridge, Iceland–Jan Mayen ridge, and the Iceland–Icelandic plateau, are unusually high with respect to other mid-ocean ridges as judged by the application of age versus depth curves (Sclater *et al.* 1971). For sea-floor older than Miocene two areas have unusually high elevation—the Iceland–Faeroe ridge and the Reykjanes ridge. Depths to basement are not

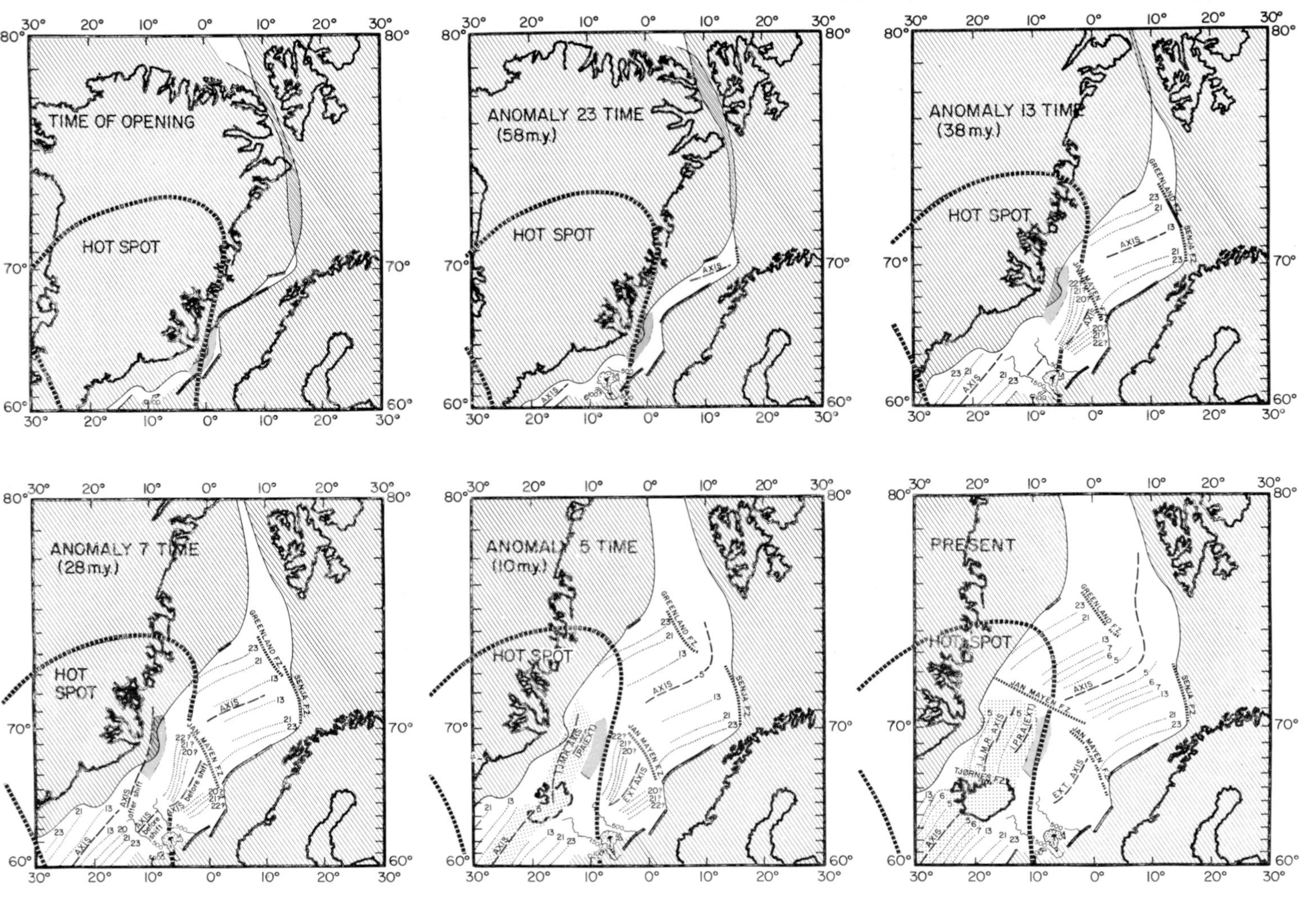


Fig. 19. Norwegian–Greenland Sea shown at various stages of evolution (after Talwani and Eldholm 1977).

unusually shallow in the Norway basin or on the flanks of the Mohns ridge. However, after the shift of the ridge axis at anomaly 7 time, not only did areas of unusually high elevation persist in the Reykjanes ridge and Iceland, but in addition several areas generated since that time; the area north of Iceland consisting of the Iceland plateau, the Iceland–Jan Mayen ridge and the southern end of the Mohns ridge, are also unusually shallow. Thus on a regional scale the areas of large elevation appear to persist over a long time. They have persisted over the Iceland–Iceland Faeroe ridge system as well as the Reykjanes ridge since the initiation of spreading in the Norwegian Sea, and over Iceland, the Iceland–Jan Mayen ridge, and the Icelandic plateau since anomaly 7 time. The existence of areas of unusual elevation has been attributed to hot spots (Wilson 1965). Vogt and Avery (1974) and Haigh (1973) have also associated hot spots with such areas. If the hot spot hypothesis is valid in the area of the Norwegian Sea, then the hot spots must have remained under the Iceland–Faeroe ridge, Iceland and the Reykjanes ridge since the initiation of spreading in the Norwegian Sea, but must not have initially underlain the Norway basin. On the other hand, after the shift of the spreading axis to the Iceland–Jan Mayen ridge area the axis must lie over a hot spot. With these constraints, Talwani and Eldholm (1977) constructed Figure 19. This figure shows the various stages in the opening of the Norwegian Sea from the early Tertiary opening to the present. It also shows that by postulating a large hot spot that is nearly stationary with respect to the Reykjanes ridge, areas of high elevation are generated when the axis of spreading lies within the hot spot as it does for the Reykjanes ridge and Iceland; areas of low elevation are generated where the spreading axis is outside the hot spot as for the Norway basin and the Mohns ridge, but that after the shift of the axis after anomaly 7 time west of the Jan Mayen ridge, the axis of spreading again lies over the hot spot thereby generating unusually shallow oceanic crust.

Areas of unusually high elevation are also present locally at a scale of tens of kilometres. These areas appear to be associated with the initiation of sea-floor spreading or with shifts in the spreading axes. The high elevations associated with the basement on the seaward side of the Vøring plateau and Faeroe–Shetland escarpment are examples of this. Figure 13 shows the Icelandic plateau where spreading was active between about 18 to 24 m.y. and the boundaries of this area are outlined by topographic highs or escarpments. The one on the west side lies beyond anomaly 6, under the lettering V–2803 in this figure. On the east side there is an escarpment just west of the boundary with the Magnetic Quiet Zone. These highs are probably associated with the initiation of spreading in this region.

The unusual elevation of the basement at the time of the opening is not a phenomenon limited to the Norwegian Sea, but is most easily seen here because of the young age of opening. Gravity highs associated with basement highs that coincide with the oldest magnetic anomalies west of South Africa (Talwani and Eldholm 1973) and south of Australia (König and Talwani 1977) point to the same phenomenon; Rabinowitz and LaBrecque (1977) have discussed this on an ocean-wide basis.

Where a new spreading axis opens on top of a hot spot, the volume of basalt extruded appears to be even further increased. The unusually high elevation of the basement seaward of the Vøring plateau escarpment may be a reflection of this.

*Acknowledgments.* This work was supported by contract N00014–75–C–0210 with the Office of Naval Research, U.S. Navy. I thank those numerous individuals who have supplied unpublished information as cited in the text.

## References

BLAKELY, R. J. 1974. Geomagnetic reversals and crustal spreading rates during the Miocene. *J. geophys. Res.* **79,** 2979–2986.

BOTT, M. H. P. 1974. Deep structure, evolution and origin of the Icelandic Transverse Ridge. *In* Kristjansson, L. (Editor). *Geodynamics of Iceland and the North Atlantic Area* **11,** 33–47. D. Reidel, Dordrecht.

—— 1975. Structure and evolution of the Atlantic floor between northern Scotland and Iceland. *Norges Geol. unders.* **316,** 195–199.

BULLARD, E., EVERETT, J. E. and SMITH, A. G. 1965. The fit of the continents around the Atlantic. In *A Symposium on Continental Drift. Phil. Trans. R. Soc. Lond.* A **258,** 41–51.

CHAPMAN, M. and TALWANI, M. 1978. An extinct spreading center on the Icelandic Plateau.

ELDHOLM, O. and WINDISCH, C. 1974. The sediment distribution in the Norwegian–Greenland Sea. *Geol. Soc. Am. Bull.* **85,** 1661–1676.

FENWICK, D. K. B., KEEN, M. J., KEEN, C. E. and LAMBERT, A. 1968. Geophysical studies of the continental margin northeast of Newfoundland. *Can. J. Earth Sci.* **5,** 483–500.

FLEISCHER, U., HOLZKAMM, F., VOLLBRECHT, K. and VOPPEL, D. 1974. Die struktur des Island–Faroer–Ruckens aus geophysikalischen Messungen. *Sonderdruck aus der Deutschen Hydrograph. Zeitsch.* **27,** 97–113.

GRØNLIE, G and TALWANI, M. 1978. A geophysical study of the Jan Mayen Ridge.

HAIGH, B. I. R. 1973. North Atlantic oceanic topography and lateral variations in the upper mantle. *Geophys. J. R. astron. Soc.* **33,** 405–420.

HEIRTZLER, J. R., DICKSON, G. O., HERRON, E. M., PITMAN, W. C. III, and LE PICHON, X. 1968. Marine magnetic anomalies, geomagnetic field reversals, and motions of the ocean floor and continents. *J. geophys. Res.* **73,** 2119–2136.

HINZ, K. and MOE, A. 1971. Crustal structure in the Norwegian Sea. *Nature, Lond.* **232,** 187–190.

JOHNSON, G. L., SOUTHALL, I. R., YOUNG, D. W. and VOGT, P. R. 1972. Origin and structure of the Iceland Plateau and Kolbeinsey Ridge. *J. geophys. Res.* **77,** 5688–5696.

JONES, E. J. W., EWING, M., EWING, J. and EITTREIM, S. L. 1970. Influences of Norwegian Sea overflow water on sedimentation in the northern North Atlantic and Labrador Sea. *J. geophys. Res.* **75,** 1655–1680.

KHARIN, G. N., UDINTSEV, G. B., BOGATIKOV, O. A., DIMITRIEV, J. I., RASCHKA, H., KRUEZER, H., MOHR, M., HARRE, W. and ECKHARDT, F. J. 1976. K/Ar age of the basalts of Norwegian–Greenland Sea Glomar Challenger Leg 38, Deep Sea Drilling Project. *In* Talwani, M. *et al.* (Editors). *Initial Reports of the Deep Sea Drilling Project* **38,** 755–760. U.S. Government Printing Office, Washington.

KÖNIG, M. and TALWANI, M. 1977. A geophysical study of the southern continental margin of Australia, Great Australian Bight and western sections. *Geol. Soc. Am. Bull.* **88,** 1000–1014.

KRISTOFFERSEN, Y. 1977a. *Labrador Sea: A geophysical study.* Columbia Univ. Ph.D. thesis (unpubl.).

——1977b. Sea floor spreading and the early opening of the North Atlantic. *Earth Planet. Sci. Lett.* **36.**

LABRECQUE, J., KENT, D. E. and CANDE, S. C. 1977. Revised magnetic polarity time scale for Late Cretaceous and Cenozoic time. *Geology* **5,** 330–335.

LAUGHTON, A. S. 1972. The southern Labrador Sea—A Key to the Mesozoic and Early Tertiary evolution of the North Atlantic. *In* Laughton, A. S. *et al.* (Editors). *Initial Reports of the Deep Sea Drilling Project* **12,** 1155–1179. U.S. Government Printing Office, Washington.

PITMAN, W. C. III and TALWANI, M. 1972. Sea-floor spreading in the North Atlantic. *Geol. Soc. Am. Bull.* **83,** 619–646.

RABINOWITZ, P. and LABRECQUE, J. 1977. The isostatic gravity anomaly: key to the evolution of the ocean-continent boundary at passive continental margins. *Earth Planet. Sci. Lett.* **35,** 145–150.

ROBERTS, D. G. 1974. Structural development of the British Isles, the continental margin, and the Rockall Plateau, *In* Burk, C. A. and Drake, C. L. (Editors). *The Geology of Continental Margins*, 340–360. Springer-Verlag, New York.

—— 1974. Marine geology of the Rockall Plateau and trough. *Phil. Trans. R. Soc. Lond.* A. **278,** 447–507.

SCLATER, J. G., ANDERSON, R. N. and LEEBELL, M. 1971. Elevation of ridges and evolution of the central eastern Pacific. *J. geophys. Res.* **76,** 1888–1915.

SCRUTTON, R. A. 1972. The crustal structure of Rockall Plateau microcontinent. *Geophys. J. R. astron. Soc.* **27,** 259–275.

TALWANI, M. and ELDHOLM, O. 1972. The continental margin off Norway: A geophysical study. *Geol. Soc. Am. Bull.* **83,** 3575–3608.

TALWANI, M. and ELDHOLM, O. 1973. The boundary between continental and oceanic crust at the margin of rifted continents. *Nature, Lond.* **241,** 325–330.
—— and ——1977. Evolution of the Norwegian–Greenland Sea. *Geol. Soc. Am. Bull.* **88,** 969–999.
——, MUTTER, J., HOUTZ, R. and KÖNIG, G. 1978. Crustal structure and evolution of the area underlying the magnetic quiet zone on the margin south of Australia. *Bull. Am. Assoc. Petrol. Geol. (in press).*
——, UDINTSEV, G., *et al.* 1976. *Initial Reports of the Deep Sea Drilling Project* **38.** U.S. Government Printing Office, Washington. 1256 pp.
VAN DER LINDEN, W. J. H. and SRIVASTAVA, S. P. 1975. The crustal structure of the continental margin off central Labrador. Offshore Geology of Eastern Canada. *Geol. Surv. Can. Pap.* **74–30,** 233–245.
VOGT, P. R. and AVERY, O. E. 1974. Detailed magnetic surveys in the northeast Atlantic and Labrador Sea. *J. geophys. Res.* **79,** 363–389.
WILLIAMS, C. A. and MCKENZIE, D. 1971. The evolution of the north-east Atlantic. *Nature, Lond.* **232,** 168–173.
WILSON, J. T. 1965. Submarine fracture zones, aseismic ridges and the International Council of Scientific Union line proposed western margin of the East Pacific Ridge. *Nature, Lond.* **207,** 907–910.

Manik Talwani,
Lamont-Doherty Geological Observatory
of Columbia University,
Palisades, New York,
NY10964, U.S.A.

# The origin and development of the continental margins between the British Isles and southeastern Greenland

*M. H. P. Bott*

The presence of the Rockall–Faeroe microcontinent complicates the continental margins between Britain and southeastern Greenland. Two older complementary margins occur to the east of the microcontinent and two younger complementary margins occur to the west of it. Prior to continental splitting, Triassic and Jurassic graben formation affected the region from the North Sea to East Greenland. This phase was of longer duration and more extensive than along the more southern Atlantic margins. It may be attributed to a crustal stretching mechanism possibly associated with the northern pre-split period and with splitting further south.

The margins east of the Rockall microcontinent probably formed during the early Cretaceous, although definitive magnetic anomalies are lacking. The margins adjacent to Ireland, the Outer Hebrides and east of the Rockall plateau are steep with a narrow transition between presumed oceanic and continental crust. Sediments post-dating the split are unusually thin for an inactive margin. Subsidence appears to have been minimal, except locally such as in the Porcupine bight. Further north, location of the margins is more problematical although the Faeroe–Shetland channel is commonly interpreted as a northward extension of the presumed oceanic Rockall trough. Rockall trough and the Faeroe–Shetland channel both shallow towards the intervening Wyville-Thomson ridge, as a result of gentle increase in crustal thickness towards the ridge.

The west Rockall and southeast Greenland margins formed in the Palaeocene as a result of the last phase of Atlantic splitting. Their formation has been influenced by the nearby Icelandic transverse ridge and the postulated hot spot beneath it. These margins display greater subsidence than those east of the microcontinent, but the west Rockall margin is sediment starved giving rise to a gentle slope. The anomalous southeast Greenland margin displays pre-split sediments at depth beneath the rise, overlying subsided continental crust. An angular unconformity marking onset of contour current activity occurs within the overlying post-split Tertiary sediments of the rise. North of 62°N, the crustal thickening towards the continent occurs to the west of the slope.

## 1. Introduction

This paper discusses the influence of the developing North Atlantic Ocean on the continental borderlands between Britain and southeastern Greenland south of the

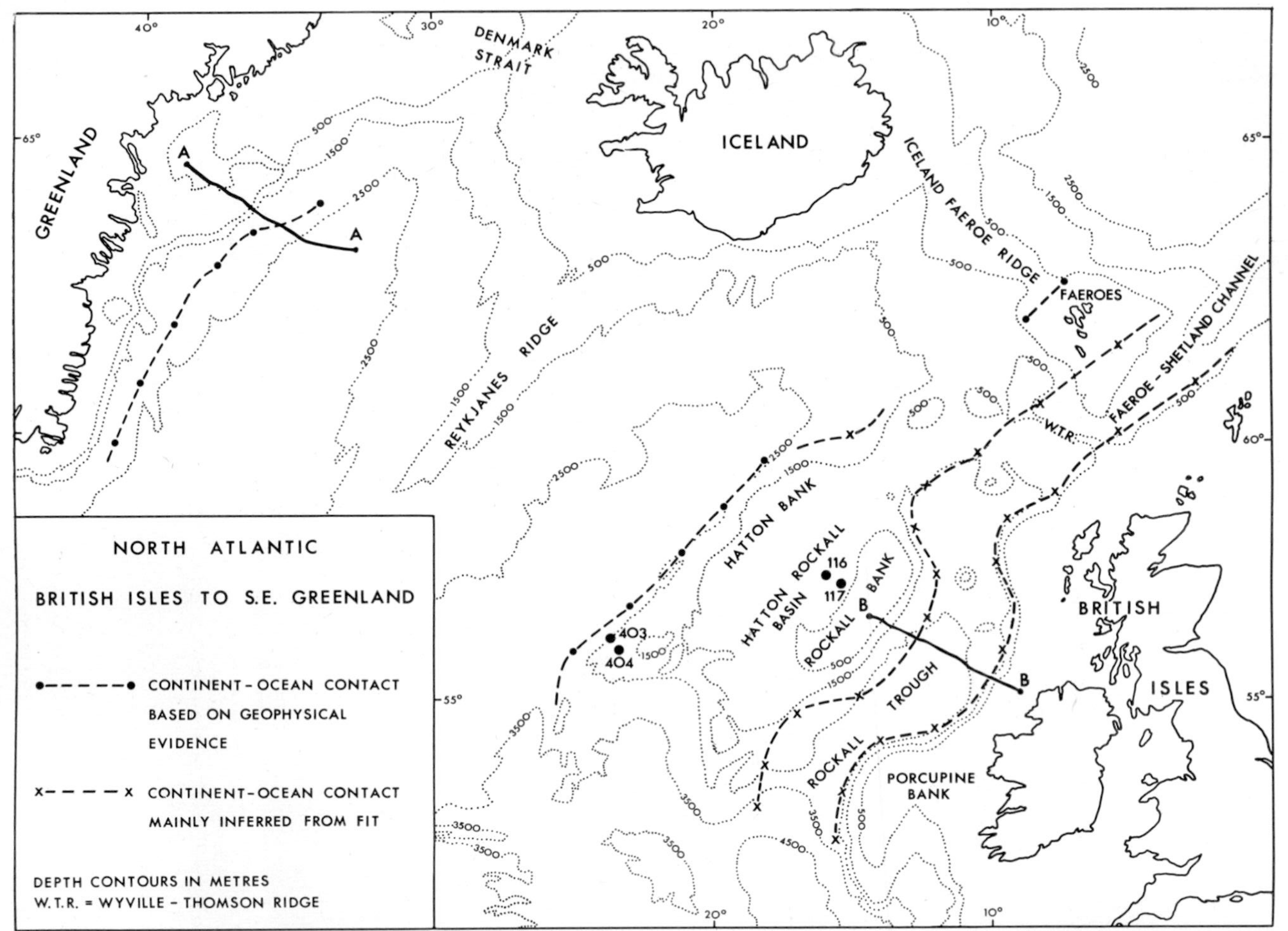


Fig. 1. Bathymetric map of the Atlantic Ocean between the British Isles and Greenland, showing interpreted continent–ocean crustal contacts and D.S.D.P. holes 116, 117, 403 and 404. Depth contours based on unpublished data kindly provided by Dr A. S. Laughton.

Icelandic transverse ridge. This part of the Atlantic (Fig. 1) has been complicated by persistent migrations of the spreading axis in contrast to the much simpler South Atlantic development, giving rise to microcontinents and additional continental margins. Thus four continental margins apparently lie between Britain and Greenland. Two younger complementary margins lie to the west of the Rockall–Faeroe microcontinent, and two older but more controversial complementary margins lie to the east of it. These margins have also been affected by the development of the nearby highly anomalous Icelandic transverse ridge, with its associated early Tertiary continental volcanism and its anomalously thick oceanic crust. Further margins of intermediate age lie south of the microcontinent but these are outside the scope of the paper.

## 2. Development of the North Atlantic

The Atlantic south of the Azores has evolved by relatively simple separation of opposite margins, starting about 180 m.y. ago between North America and North Africa and about 130 m.y. ago between South America and Africa. North of the Azores, however, the evolution has been more complicated because of repeated changes in the position of the spreading axis producing four main branches over different time intervals—the Labrador Sea, the Reykjanes basin and its northward extension, the Rockall trough, and the Bay of Biscay. Present knowledge of development stems from the study of oceanic magnetic anomalies, deep sea drilling and plate tectonic reconstructions.

Greenland and the Rockall-Faeroe microcontinent started to separate about 60 m.y. ago on the Heirtzler *et al.* (1968) time scale* as dated by magnetic anomaly 24 (Vogt and Avery 1974; Featherstone *et al.* 1977). The Labrador Sea opened between about 80 and 45 m.y. ago spanning the period between magnetic anomalies 32 and 19 (Laughton 1972), with a triple junction with Reykjanes ridge active between 60 and 45 m.y. ago. Grand Banks and Iberia probably started to separate about 115 m.y. ago, but an earlier age of 137 m.y. is possible as dating of the J-anomaly is still uncertain (Keen *et al.* 1977). The end of the opening of the Bay of Biscay can be dated at about 73 m.y. by anomaly 31–32 at the fossil triple junction west of the Bay (Williams 1975), and it presumably started opening 115 m.y. ago or later. As it opened, there was anticlockwise rotation of Iberia relative to France and left-lateral displacement related to oblique spreading about an E–W axis. An earlier period of left-lateral transform faulting between Iberia and France may have preceded the opening, a possibility which causes difficulty in assessing the origin of Rockall trough.

How did Rockall trough and its northward extensions form? Distinctive magnetic anomalies are lacking but a pre-73 m.y. ago age is indicated because magnetic anomaly 31–32** cuts across the mouth of the trough (Williams and McKenzie 1971; Roberts 1975) without evidence for a triple junction. It is possible that Rockall trough and Faeroe–Shetland channel formed by subsidence of continental crust (Talwani and Eldholm 1972) but a seafloor spreading origin is favoured here

* Recent work suggests the Heirtzler *et al.* (1968) time scale may slightly overestimate ages.

** According to Talwani's contribution to this symposium, this anomaly has been re-identified as 34 of about 80 m.y. age because of (i) the relatively good fit of opposite margins (Fig. 1), (ii) the indications of underlying oceanic crust (see below), and (iii) the dissimilarity of the long 200 km wide graben-like trough with known features of continental subsidence. Assuming an oceanic origin is correct, the age of formation remains problematical. Bott and Watts (1971) suggested a Permo-Triassic age and Russell (1976) has recently advocated an early Permian age. An early Mesozoic age would involve opening of the trough and its northward extensions deep within a continental interior with transform faults linking the ends to other plate boundaries. Other workers, including Hallam (1971); Bailey *et al.* (1971); Laughton (1971) and Roberts (1971, 1975) have favoured a late Jurassic or Cretaceous age. A late opening subsequent to the split of Grand Banks and Iberia is also favoured in this paper because of the above mentioned problem of intra-continental formation and the lack of any evidence of Permo-Triassic or Jurassic subsidence along the margins (see below). The lack of distinctive magnetic anomalies is consistent with the Rockall trough forming during part or all of the Cretaceous normal polarity period, 111 to 85 m.y. ago. The Rockall trough may have been connected to the spreading axis further south by a transform fault (e.g. Laughton 1972) or by an oblique spreading axis.

Thus the margins between the Rockall–Faeroe microcontinent and Greenland formed about 60 m.y. ago in the Palaeocene. The margins bordering Rockall trough may have been split apart between about 110 and 95 m.y. ago during the Albian or Cenomanian. As a background for discussing the development of the margins, the relevant stages of Atlantic opening can be summarized as follows:

(i) About 230–180 m.y. ago (Triassic): early rift stage pre-dates split between North America and North Africa.

(ii) About 180–115 m.y. ago (Jurassic and early Cretaceous): separation of North America and North Africa, with left lateral shear movement between Europe and Africa.

(iii) About 115 m.y. ago to present: separation of Grand Banks and Iberia.

(iv) About 115–75 m.y. ago: opening of the Bay of Biscay and possibly the Rockall trough–Faeroe–Shetland channel during this interval, details still obscure.

(v) About 80–45 m.y. ago: opening of Labrador trough.

(vi) About 60 m.y. ago to present: separation of the Rockall–Faeroe microcontinent from Greenland, with contemporaneous opening of the Labrador trough for the first 15 m.y. of this interval.

## 3. Permo-Triassic and Jurassic basin formation

An early pre-split graben stage of about 50 m.y. duration appears to be typical of rifted inactive margins although it is not ubiquitous. This may accompany thermal uplift of the lithosphere prior to splitting which can give rise to the tensional stress system in the upper crust (Kusznir and Bott 1977). In the British region fault-controlled basin subsidence occurs extensively in the Permo-Triassic and continues through the Jurassic. The relationship to Atlantic development has given rise to recent controversy (Hallam 1972; Hall and Smythe 1973). Such basins, however, have not been observed on the Rockall–Faeroe microcontinent and are not conspicuous along the immediate margins of the Rockall trough. In the British and North Sea region, this period of fault-controlled basin formation covers a much more extensive area and lasts much longer than at a typical inactive margin. Dingle (1976) has referred to it as an example of a "tension failed-arm margin".

It is not yet clear whether this Permo-Triassic and Jurassic basin formation can be related to the development of the Atlantic Ocean as Dingle suggested. If the Rockall trough formed in the early Permian (Russell 1976) then the basins might be related to a pre-existing margin, but the absence of significant graben formation along the Rockall margins is against this hypothesis. If Rockall trough formed in the Cretaceous, then the Permo-Triassic and Jurassic basins must have formed by crustal stretching within a continental interior. A mechanism for subsidence has recently been suggested (Bott 1976) assuming tensile stresses affected the crust. The origin of this stress system is obscure. In the Permo-Triassic the tension may perhaps result from doming associated with a very early phase related to Atlantic formation, and in the Jurassic it may result from the stress system caused by opening further south.

## 4. The margins of Rockall trough

The Rockall trough (Scrutton and Roberts 1971; Roberts 1975) is flanked by relatively steep continental margins along the whole east side and along the west side south of 58°. Further north on the west side there is a series of shallow banks with depressions between them, and the margin is poorly developed. The trough shallows towards its north end, and the height of the slope consequently decreases from above 3 km at the south end to less than 1 km in the north. The continental rise is only locally developed on the east side, particularly near the Hebrides Terrace seamount where there are fans. The more well developed rise on the west side, associated with the Feni ridge, imparts an asymmetry to the trough. Three conspicuous seamounts occur within the trough.

The south Rockall trough contains up to 5 km of nearly flat-lying sediments on an assumed early to mid-Cretaceous oceanic basement (Fig. 2). A prominent reflector R4 separates Eocene and earlier oozes from Oligocene cherty beds (Roberts 1975). Miocene and later deposits forming the Feni ridge dominate the west part of the overlying sediments whereas terrigenous deposits including fans predominate in the east. The sediment thickness is rather less in the northern part of the trough. Sediments are very thin or even absent from the Wyville–Thomson ridge.

The sedimentary structure of the margins (Stride *et al.* 1969; Himsworth 1973; Bailey *et al.* 1974; Roberts 1975; Riddihough and Max 1976) is highly atypical of a passive margin. The acoustic basement occurs at relatively shallow depth being typically shallower than half a kilometre. There is no evidence for pre-split graben type sediments on either side of the Rockall trough or for significant sediments of Permo-Triassic or Jurassic age. The thin marginal sedimentary succession records two main phases of subsidence and sedimentation of presumed late Cretaceous and Tertiary age separated by a period of uplift and erosion of suggested Palaeocene age (Stride *et al.* 1969; Bailey *et al.* 1974). The only thick sequence appears in the Porcupine seabight. Lack of thick sediments on the west margins of the trough can be attributed to lack of sediment supply, but on the east side the major factor appears to be lack of adequate subsidence. The sediment story is consistent with an early to mid Cretaceous origin for the trough under unusually cool thermal conditions and the Palaeocene uplift may be associated with the initiation of the Icelandic transverse ridge structure and the associated early Tertiary continental volcanism.

The continental crust is about 28 km thick beneath Porcupine bank (Whitmarsh *et al.* 1974) and about 31 km beneath Rockall bank (Scrutton 1972). Gravity

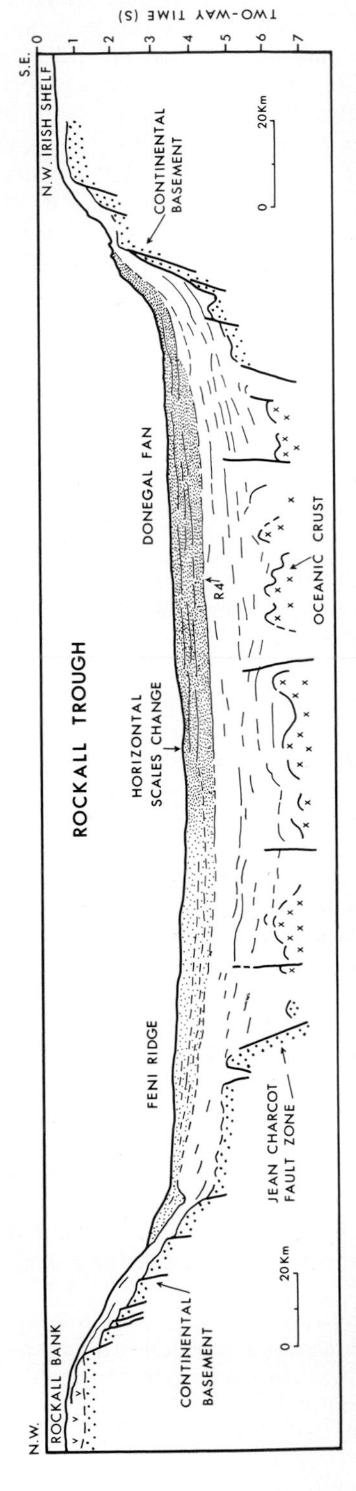


Fig. 2. Seismic reflection section across the Rockall trough along line BB (Fig. 1), redrawn from Roberts (1975). Note the slight change in horizontal scale near the middle of the profile. Post R4 sediments are predominantly contourites beneath Feni ridge and fan deposits towards the east.

interpretations (Gray and Stacey 1970; Scrutton 1972) indicate about 7 km of sub-sedimentary crust beneath the southern Rockall trough, supported by a refraction line re-interpretation by Jones *et al.* (1970) and by preliminary results from the Hebridean Margin Seismic Project (Armour and Bott 1977). A relatively short crustal transition within 10–20 km is indicated by gravity interpretations across the Porcupine bank (Gray and Stacey 1970) and the Hebridean margins (Himsworth 1973), consistent with the steep slopes. Gravity interpretations also indicate that the crust beneath the trough thickens towards the north end, explaining the northward shallowing and reduction in height of the slope. The Wyville–Thomson ridge is itself underlain by a root of locally thickened crust (Himsworth 1973).

The geometry of the opposite margins of the Rockall trough show general similarity but do not fit exactly. In particular, there is an apparent overlap of Porcupine bank (Bullard *et al.* 1965). Roberts (1975) has suggested that the continent-ocean crustal contact lies near the Jean Charcot fracture zone beneath the western rise and generally lies near the foot of the slope on the east side. With these assumptions, a reasonable fit between the opposite margins can be obtained, without need for stretching or rotation of Porcupine bank, by a rotation of 2·7° about a pole at 76°N, 90°E as shown in Figure 1. According to this interpretation, the direction of spreading is 116° in the trough, which is parallel to the Wyville–Thomson and the Ymir ridges and gives good agreement with prominent geometrical features on opposing margins. The fit does cause some problems north of the Wyville–Thomson ridge where much of the eastern shelf of the Faeroe Islands appears to be oceanic, or alternatively one can assume that part of the opening here occurred to the west of the Faeroes as in the recent reconstruction of Smythe *et al.* (1977). An implication of the fit presented here is that very considerable subsidence has affected a narrow belt of continental crust beneath the slopes and locally beneath parts of the rise, in contrast to the relatively minor subsidence of the outer shelf.

The cooling oceanic lithosphere subsides nearly exponentially after initial formation (Sclater and Francheteau 1970). The Rockall trough has probably subsided by 2–3 km by this mechanism and by a further 1–2·5 km as a result of sediment loading. There has thus been differential vertical movement of about 3–5 km between the trough and the relatively stable shelves adjacent to it. This has probably taken place by normal faulting beneath the margins (Fig. 3), from the evidence of the seismic profiles (Roberts 1975). The faulting was probably most vigorous during the Cretaceous, dying off during the Tertiary.

If the continent-ocean crustal contact occurs at or near the foot of the slope, then continental crust beneath the slope and inner rise has subsided by up to 4 km without much associated uplift of the adjacent oceanic crust. This implies early thinning of the continental crust beneath the slope and rise (Fig. 3) to satisfy the transverse gravity profiles. The amount of subsidence appears to be rather greater than predicted by the thermal hypotheses of Sleep (1971) or Falvey (1974), and is probably not produced by "necking" (Kinsman 1975) because of absence of a pre-split graben stage. It is therefore suggested that the thinning occured by creep of the low density continental crust towards the seaward side of the margin (Bott 1971) possibly supplemented by thermal subsidence. The displaced continental material may have become widely dispersed in the topmost mantle beneath the trough. In contrast, the relatively minor subsidence of the shelf may be attributed to either thermal or creep mechanisms or both.

The highly unusual lack of major shelf subsidence along much of the Rockall

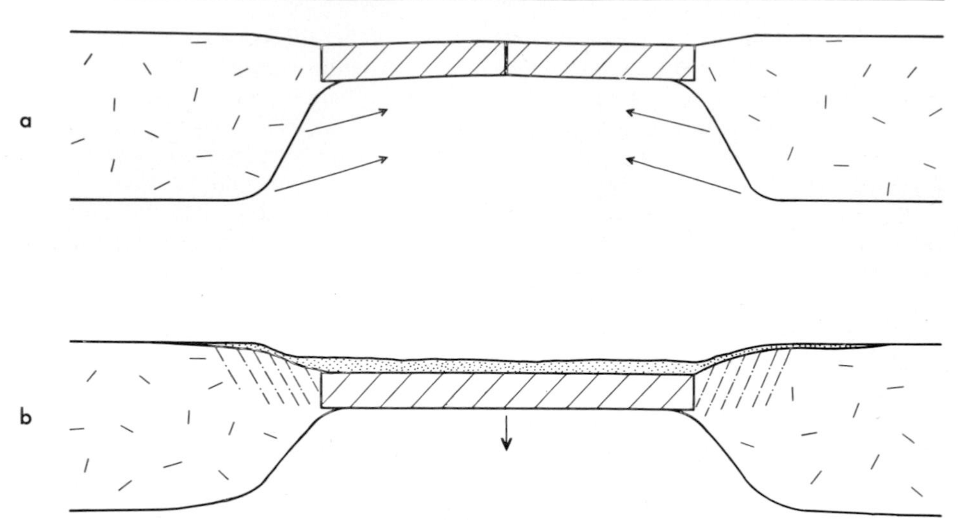


Fig. 3. Suggested geodynamic development of Rockall trough and its margins:
(a) Formation of Rockall trough by Cretaceous seafloor spreading, with approximately contemporaneous thinning of adjacent continental crust (by creep) to form the slope.
(b) Subsequent sinking of oceanic lithosphere by cooling and sediment loading with differential vertical movement across the slope taken up by normal faulting at shallow depth and creep beneath.

trough margins can be explained by absence of a pre-split graben stage and by a relatively short period of active spreading. The adjacent continental borderlands thus remained relatively cool, and subsidence by either thermal or creep mechanisms was inhibited except immediately adjacent to the crustal contact.

There are exceptions to this general stability in the Porcupine bank region. The bank has subsided slightly as a marginal plateau, but the basement is up to 7 km deep beneath the seabight (Roberts 1975). In the writer's opinion, this deep basement can be better explained by subsidence of continental crust (Scrutton *et al.* 1971), rather than by an oceanic origin with clockwise rotation of the Porcupine bank (Stride *et al.* 1969), as associated transform faulting and compression to the north are difficult to identify. The enhanced subsidence in this region can be explained by the continuing presence of a spreading centre from anomaly 32 onwards when the Rockall spreading centre had become extinct. This subsidence may be caused by creep (Bott 1971) and thermal subsidence (Sleep 1971). Other exceptions to stability are the bathymetric troughs which have developed between the shallow banks between the Rockall plateau and the Faeroe Islands.

## 5. The northwest Rockall and southeast Greenland margins

The northwest Rockall and southeast Greenland margins were split apart just before the time of magnetic anomaly 24. A slight misfit of the margins (Laughton 1972) can best be explained in terms of subsided continental crust rather than pre-anomaly 24 spreading (Featherstone *et al.* 1977). North of about 63°N on the Greenland side anomalies 23 and 24 cut out against an eastward swing of the margin (Johnson *et al.* 1975; Featherstone *et al.* 1977), whereas on the Rockall side there is a complementary widening of the ocean floor of this age just north of the Hatton bank (Fig. 1). This may be interpreted as a local westward migration of spreading axis northward of this feature shortly after the split, the result being a straightening out of an original sinuous spreading axis (Featherstone *et al.* 1977). An implication is that the hot spreading axis was near the Greenland margin north of 63° for longer than normal.

The physiography of the two complementary margins shows strong contrasts (Fig. 1). The margin northwest of the Rockall microcontinent lacks a shelf break and has a relatively gentle slope spread out over 100 km. North of the Hatton bank, the margin is broken by depressions between the shallow banks. The southeast Greenland margin between 60° and 63·5°N has a narrow shelf and steep slope paralleling the coastline, but north of 63·5° the slope swings eastwards and the shelf widens considerably. The continental rise is well developed.

These contrasts in physiography arise more from differences in history of sedimentation than from geodynamic factors. The margin northwest of the Rockall plateau is a "starved" margin in that post-split sediments are exceptionally thin, being generally less than 0·5 km thick (Roberts 1975; Montadert *et al.* 1976). This is attributed to lack of sediment supply rather than to absence of subsidence. In contrast, much thicker post-split sediments (up to about 2 km) overlie a gently seaward sloping basement at the southeast Greenland margin (Featherstone *et al.* 1977). These sediments are divisible into two groups separated by an erosional unconformity visible on reflection records north of 62°N (Figs 4 and 5). The underlying slightly seaward dipping group of transparent sediments forms the slope and part of the inner rise, and is interpreted as lithified oozes of about Eocene age formed prior to strong contour current activity. The younger more strongly stratified group overlie these Eocene oozes with angular unconformity, the plane of the unconformity dipping towards the east (Fig. 4) and cropping out on the rise north of about 62·5°N (Fig. 5). The younger sediments are interpreted as contour current deposits of Miocene or younger age, contemporaneous with other such deposits in this region (Jones *et al.* 1970; Roberts 1975) which formed when the Icelandic transverse ridge first subsided below sea level (Vogt 1972). The contour currents evidently cut the plane of unconformity and have continued to cut the slope and keep parts of the inner rise free of post Eocene sediments.

A series of seaward dipping older reflectors has been detected unconformably underlying the Tertiary sediments of the southeast Greenland margin south of 63°N between anomaly 24 and the continental slope as seen in Figure 4 (Featherstone *et al.* 1977). A narrow strip of comparable sediments has been detected on the west margin of Hatton bank. These presumed older sediments appear to have formed an extensive pre-split sedimentary basin of possible Mesozoic age (Fig. 5). The continental split appears to have cut through this basin, leaving most of it on the Greenland side where it now overlies attenuated and subsided continental crust.

The continent-ocean contact can be tentatively identified as lying just to the landward side of magnetic anomaly 24 off southeast Greenland south of 63°N and off Hatton bank (Fig. 1). Just north of this, the contact on both sides is offset slightly towards the east. Off southeast Greenland the crustal contact appears to show a magnetic signature and is marked on the seismic reflection records by the change from oceanic basement to the older reflectors (Fig. 4). Further north on the Greenland side, the position of the contact becomes uncertain except by continental reconstruction, but on the east side it may follow the feet of the Lousy, Bailey and Faeroe banks, running into the Iceland–Faeroe ridge where its location on the crest

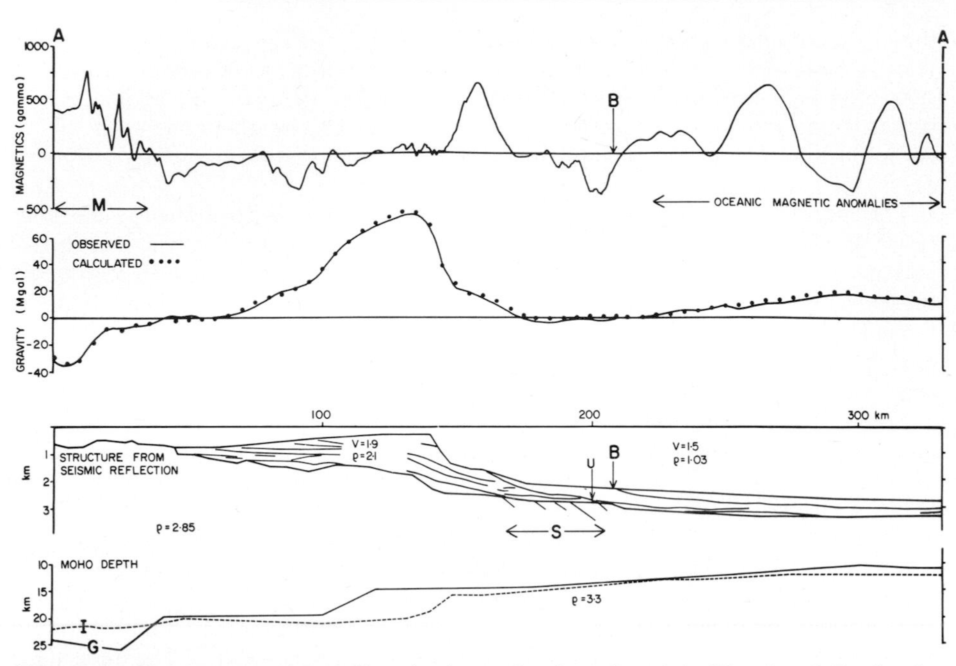


Fig. 4. Profile across southeast Greenland margin along line AA (Figs 1 and 5), showing magnetic anomaly, free air gravity anomaly, sediment structure from reflection profiling, and depth to Moho interpreted by gravity modelling (G) and assuming Airy isostatic equilibrium (I). $\varrho$ is the rock density used for gravity model in g/cm³ and V is the seismic velocity in km/s. B is the interpreted position of the continent–ocean crustal contact, U is the intra-Tertiary unconformity, and S marks the region of observed pre-Tertiary basement reflectors assumed to overlie continental crust. After Featherstone *et al.* (1977).

occurs just to the east of the bathymetric step between the northwest Faeroes shelf and the ridge crest (Bott *et al.* 1976).

The deep crustal structure across the northwest Rockall margin has not been studied in detail, but gravity interpretation across the southeast Greenland margin (Featherstone *et al.* 1977) indicates a gradational transition in crustal thickness corresponding approximately to the slope south of 62°N but occurring landward of it further north (Fig. 4). It is clear that the crustal transition beneath the west Rockall and southeast Greenland margins is much more spread out than beneath the margins east of the microcontinent. This can be attributed to the much hotter conditions of formation of the later pair of margins, probably brought about by igneous penetration into the lower continental crust in the Palaeocene.

D.S.D.P. holes 403 and 404 drilled about 30 km east of anomaly 24 above presumed continental crust reveal the subsidence history of the northwest Rockall margin (Montadert *et al.* 1976). The deepest rocks penetrated were clastic sediments of late late Palaeocene to early Eocene age of littoral or inner shelf aspect, deposited near sea level during and shortly after rifting. The thin overlying sequence of unfaulted rocks starts with Middle Eocene siliceous chalks deposited in middle bathyal depths. Subsidence of over 2·5 km has occurred since the split, and the most rapid subsidence probably occurred prior to the Middle Eocene. This suggests that the gentle continental slope west of Hatton bank was caused by post-split subsi-

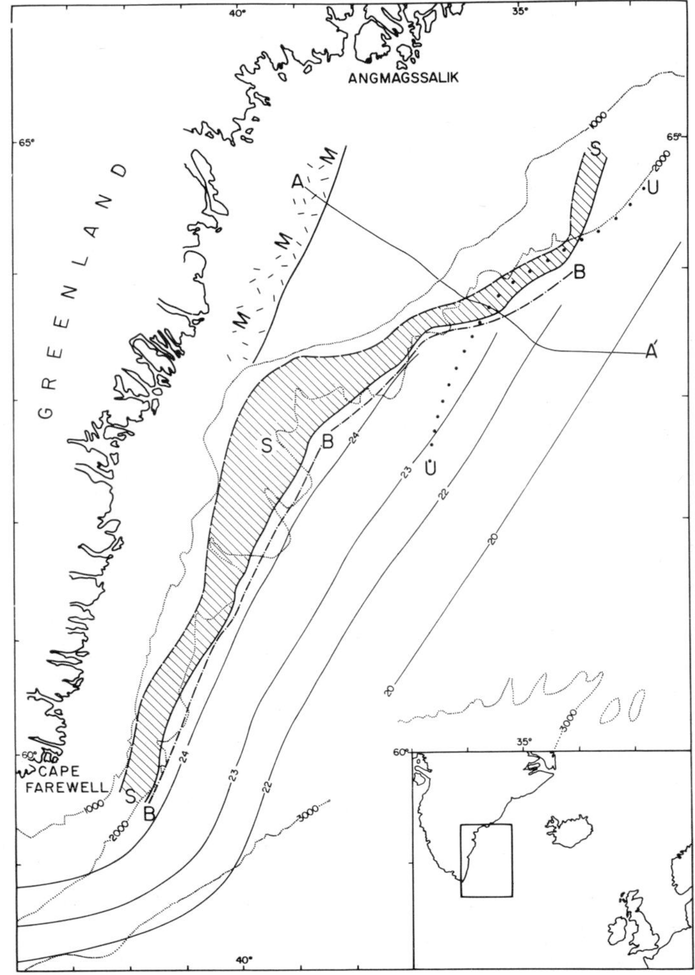


Fig. 5. Structure of the southeast Greenland margin showing the interpreted continent-ocean crustal contact B, the contact between the intra-Tertiary unconformity and the basement U, the region of pre-Tertiary reflectors assumed to overlie subsided continental crust S, and the region of short wavelength magnetic anomalies on the shelf M. Bathymetric contours at 1000 m interval after Johnson *et al.* (1975). After Featherstone *et al.* (1977).

dence of continental crust adjacent to the margin, without conspicuous faulting, in a sediment-starved environment.

The southeast Greenland margin apparently underwent a similar history of subsidence of up to about 3 km, as indicated by over 1 km thickness of older Tertiary sediments beneath the slope and the inner continental part of the rise (Fig. 4). Apparently much of this subsidence occurred prior to onset of contour current activity. The Greenland margin, however, differs from that of Rockall plateau in the greater areal extent of continental subsidence affecting the region occupied by the postulated pre-split sedimentary basin (Fig. 5).

Another subsidence problem is posed by the Rockall–Hatton basin (Roberts *et al.* 1970; Matthews and Smith 1971; Bott 1972; Laughton *et al.* 1972). This basin connects to the margin south of Rockall plateau, but the differential subsidence of over 2 km apparently started about the time of the Rockall–Greenland split. D.S.D.P. holes 116 and 117 (Laughton *et al.* 1972) place the main subsidence in the Palaeocene–Eocene, with further differential subsidence in the Miocene. The basin overlies thinner crust than beneath the Rockall bank (Scrutton 1972) suggesting isostatic subsidence related to thinning crust. Lack of evidence of sufficient faulting precludes origin by crustal extension or necking (Matthews and Smith 1971; Bott 1972).

Evidence of extensive igneous activity in the Palaeocene suggests that the subsidence of the Rockall and Greenland complementary margins and of the Rockall–Hatton basin may have been encouraged by heating of the crust by igneous intrusion. The mechanism of Sleep (1971) involving thermal uplift and crustal thinning by subaerial erosion meets difficulties for the following reasons: (i) subsidence of about 2·5 km without thick sediments requires a crustal thinning by erosion of over 10 km; (ii) the hypothesis requires a gap of about 50 m.y. for uplift and erosion prior to marine sedimentation, which is not evident in this region; and (iii) it is difficult to account for the strongly differential subsidence of Rockall–Hatton basin. It is tentatively suggested that the hypothesis of Bott (1971) may account for the subsidence being an isostatic response to crustal thinning by creep of lower crustal material towards the margin, with strong early Tertiary subsidence favoured by the high temperature.

## 6. Influence of Icelandic transverse ridge

A ridge of elevated crust of oceanic origin extends from the East Greenland shelf to the continental Faeroes block, Iceland itself being situated where it crosses the mid-ocean ridge. This ridge can be divided into two main segments, the northwest trending Iceland–Faeroe ridge which is underlain by oceanic crust about 30 km thick and probably formed contemporaneously with the last phase of Labrador Sea opening between anomalies 24 and 19, and the Iceland block more recently formed. The Wyville–Thomson ridge may possibly represent an early embryonic stage in the development of the transverse ridge.

The transverse ridge formed by a much larger than normal thickness of crust being differentiated from the underlying mantle at the spreading centre, and this has normally been attributed since Wilson (1963) to an underlying hot mantle, caused either by a plume (Morgan 1971; Vogt 1971) or by some sort of convective overturn (Bott 1973). The hot underlying mantle has also given rise to a more gentle shallowing of the ocean depths towards Iceland, attributed by Haigh (1973) to higher than normal temperatures rising towards a focus beneath Iceland.

The Icelandic hot spot and the associated transverse ridge have influenced the margins of our region in several ways. *Firstly*, an anomalous margin occurs where the transverse ridge abuts against continental crust at the junction of the Iceland–Faeroe ridge with the Faeroe microcontinent, where oceanic and continental crusts do not differ greatly in thickness. This margin appears to have been tectonically stable, with little evidence of much shelf subsidence and some indication of slight downfaulting on the oceanic side. The crustal contact gives rise to a gravity anomaly because of the slightly thicker and less dense crust of the Faeroe block (Bott *et al.* 1971). It also gives rise to converted seismic phases at the truncation of the oceanic layering (Bott *et al.* 1976). Similar margins may occur in the Denmark Strait and at the ends of the Wyville–Thomson ridge but these have not yet been investigated.

*Secondly,* extensive volcanism affected the continental borderlands during the early Tertiary, dying off shortly after spreading started. Lithospheric temperatures were probably raised by magma penetration causing slight doming of some parts of the region. The Palaeocene uplift and erosion of the British margin postulated by Stride *et al.* (1969) and Bailey *et al.* (1974) may be attributed to this. Also the widespread crustal thinning and subsidence affecting the margins between Greenland and the microcontinent and forming the Rockall–Hatton basin may be attributed to raised lower crustal temperatures caused by early Tertiary magma penetration. *Thirdly*, the change in sedimentary regime following the onset of contour currents may be attributed to the subsidence of the Icelandic transverse ridge below sea level (Vogt 1972).

## 7. Conclusions

1. The southeast Greenland and northwest Rockall margins formed in the Palaeocene when Greenland and Rockall split apart. It is suggested that the more controversial margins of the Rockall trough formed by seafloor spreading during the early or middle Cretaceous.

2. The origin of the Permo-Triassic and Jurassic basins prevalent in the British and North Sea region is obscure. They appear to be graben-type basins, formed under crustal tension which may be related to the earliest phases of Atlantic development.

3. The margins of the Rockall trough are unusual in the apparent absence of a pre-split graben stage and in the relatively slight amount of shelf subsidence which is attributed to a cool origin. The continent-ocean contact is interpreted as lying near the foot of the slope and locally beneath the rise, suggesting strong subsidence of a narrow strip of continental crust beneath the slope. It is suggested that the strong early subsidence resulted from crustal thinning during and just after the split, and that the margins have subsequently developed by normal faulting beneath the slope and inner rise as the oceanic lithosphere subsided relative to the adjacent continents as it cooled.

4. The complementary margins of southeast Greenland and northwest Rockall differ in physiography mainly because of differing sediment histories. The Rockall margin is sediment-starved but the Greenland margin displays two groups of Tertiary sediments of moderate thickness separated by an erosional unconformity attributed to contour currents. Continental crust beneath both margins has subsided by about 3 km or more since the split, mostly during the early stages. The Tertiary sediments on the southeast Greenland rise are underlain by an extensive area of older pre-split sediments on subsided continental crust. This subsidence, and that of the Rockall–Hatton basin, is associated with crustal thinning which probably resulted from marginward creep of lower continental crustal material rather than from uplift and erosion. The crustal thinning was probably stimulated by raised lower crustal temperatures which may have been caused by magma penetration at the time of the split.

5. The Icelandic transverse ridge, a post-split product of the Icelandic hot spot, is associated with an anomalous margin at its Faeroe end. The contour current activity which strongly influenced the marginal sediments to the south of it started when the ridge subsided below sea level. The magma penetration into the continental

lithosphere and crust during the Palaeocene, which may have caused slight doming in some regions and strong marginal subsidence elsewhere, is attributed to early activity of the Icelandic hot spot.

*Acknowledgments.* I am grateful to Mr. D. G. Roberts for discussions and permission to use Figure 2, Dr A. S. Laughton for permission to use unpublished bathymetry, and Dr C. E. Keen for allowing reference to unpublished evidence on the Iberia-Grand Banks date of splitting. The manuscript has been typed by Mrs. Hilda Winn.

## References

Armour, A. R. and Bott, M. H. P. 1977. The Hebridean margin seismic project of 1975. (Abstract.) *Geophys, J. R. astron. Soc.* **49,** 284.

Bailey, R. J., Buckley, J. S. and Clarke, R. H. 1971. A model for the early evolution of the Irish continental margin. *Earth Planet. Sci. Lett.* **13,** 79–84.

——, Grzywacz, J. M. and Buckley, J. S. 1974. Seismic reflection profiles of the continental margin bordering the Rockall Trough. *J. geol. Soc. Lond.* **130,** 55–69.

Bott, M. H. P. 1971. Evolution of young continental margins and formation of shelf basins. *Tectonophysics* **11,** 319–327.

—— 1972. Subsidence of Rockall Plateau and of the continental shelf. *Geophys. J. R. astron. Soc.* **27,** 235–236.

—— 1973. The evolution of the Atlantic north of the Faeroe Islands. *In* Tarling, D. H. and Runcorn, S. K. (Editors). *Implications of continental drift to the earth sciences* **1,** 175–189. Academic Press, London and New York.

—— 1976. Formation of sedimentary basins of graben type by extension of the continental crust. *Tectonophysics* **36,** 77–86.

——, Browitt, C. W. A. and Stacey, A. P. 1971. The deep structure of the Iceland–Faeroe Ridge. *Marine geophys. Res.* **1,** 328–351.

——, Nielsen, P. H. and Sunderland, J. 1976. Converted *P*-waves originating at the continental margin between the Iceland–Faeroe Ridge and the Faeroe Block. *Geophys. J. R. astron. Soc.* **44,** 229–238.

—— and Watts, A. B. 1971. Deep structure of the continental margin adjacent to the British Isles. *In* Delany, F. M. (Editor). *The geology of the East Atlantic continental margin* **2,** *Inst. geol. Sci. Rep.* No. 70/14, 89–109.

Bullard, E., Everett, J. E. and Smith, A. G. 1965. The fit of the continents around the Atlantic. *Phil. Trans. R. Soc. Lond. A* **258,** 41–51.

Dingle, R. V. 1976. A review of the sedimentary history of some post-Permian continental margins of Atlantic-type. *An. Acad. bras. Ciênc.* **48** (Suplemento), 67–80.

Falvey, D. A. 1974. The development of continental margins in plate tectonic theory. *Aust. Pet. Explor. Assoc. J.* **14,** 95–106.

Featherstone, P. S., Bott, M. H. P. and Peacock, J. H. 1977. Structure of the continental margin of south-eastern Greenland. *Geophys. J. R. astron. Soc.* **48,** 15–27.

GRAY, F. and STACEY, A. P. 1970. Gravity and magnetic interpretation of Porcupine Bank and Porcupine Bight. *Deep-sea Res.* **17,** 467–475.
HAIGH, B. I. R. 1973. North Atlantic oceanic topography and lateral variations in the upper mantle. *Geophys. J. R. astron. Soc.* **33,** 405–420.
HALL, J. and SMYTHE, D. K. 1973. Discussion of the relation of Palaeogene ridge and basin structures of Britain to the North Atlantic. *Earth Planet. Sci. Lett.* **19,** 54–60.
HALLAM, A. 1971. Mesozoic geology and the opening of the North Atlantic. *J. Geol.* **79,** 129–157.
—— 1972. Relation of Palaeogene ridge and basin structures and vulcanicity in the Hebrides and Irish Sea regions of the British Isles to the opening of the North Atlantic. *Earth Planet. Sci. Lett.* **16,** 171–177.
HEIRTZLER, J. R., DICKSON, G. O., HERRON, E. M., PITMAN, W. C. and LE PICHON, X. 1968. Marine magnetic anomalies, geomagnetic field reversals, and motions of the ocean floor and continents. *J. geophys. Res.* **73,** 2119–2136
HIMSWORTH, E. M. 1973. *Marine geophysical studies between northwest Scotland and the Faeroe Plateau.* Ph. D. thesis, University of Durham.
JOHNSON, G. L., SOMMERHOFF, G. and EGLOFF, J. 1975. Structure and morphology of the west Reykjanes Basin and the southeast Greenland continental margin. *Marine Geology* **18,** 175–196.
JONES, E. J. W., EWING, M., EWING, J. I. and EITTREIM, S. L. 1970. Influence of Norwegian Sea overflow water on sedimentation in the northern North Atlantic and Labrador Sea. *J. geophys. Res.* **75,** 1655–1680.
KEEN, C. E., HALL, B. R. and SULLIVAN, K. D. 1977. Mesozoic evolution of the Newfoundland basin. *Earth Planet. Sci. Lett.* **37,** 307–320.
KINSMAN, D. J. J. 1975. Rift valley basins and sedimentary history of trailing continental margins. *In* Fischer, A. G. and Judson, S. (Editors). *Petroleum and global tectonics*, 83–126. Princeton University Press.
KUSZNIR, N. J. and BOTT, M. H. P. 1977. Stress concentration in the upper lithosphere caused by underlying visco-elastic creep. *Tectonophysics* **43,** 247–256.
LAUGHTON, A. S. 1971. South Labrador Sea and the evolution of the North Atlantic. *Nature, Lond.* **232,** 612–617.
—— 1972. The southern Labrador Sea—a key to the Mesozoic and early Tertiary evolution of the North Atlantic. *In* Laughton, A. S., Berggren, W. A. *et al. Initial reports of the deep sea drilling project* **12,** 1155–1179. U.S. Government Printing Office, Washington.
—— *et al.* 1972. Sites 116 and 117. *In* Laughton, A. S., Berggren, W. A. *et al. Initial reports of the deep sea drilling project* **12,** 395, U.S. Government Printing Office, Washington.
MATTHEWS, D. H. and SMITH, S. G. 1971. The sinking of Rockall Plateau. *Geophys. J. R. astron. Soc.* **23,** 491–497.
MONTADERT, L. *et al.* 1976. Glomar Challenger sails on leg 48. *Geotimes* **21,** No. 12, 19–23.
MORGAN, W. J. 1971. Convection plumes in the lower mantle. *Nature, Lond.* **230,** 42–43.
RIDDIHOUGH, R. P. and MAX, M. D. 1976. A geological framework for the continental margin to the west of Ireland. *Geol. J.* **11,** 109–120.
ROBERTS, D. G. 1971. New geophysical evidence on the origins of the Rockall Plateau and Trough. *Deep-sea Res.* **18,** 353–359.
—— 1975. Marine geology of the Rockall Plateau and Trough. *Phil. Trans. R. Soc. Lond. A* **278,** 447–509.
——, BISHOP, D. G., LAUGHTON, A. S., ZIOLKOWSKI, A. M., SCRUTTON, R. A. and MATTHEWS, D. H. 1970. New sedimentary basin on Rockall Plateau. *Nature, Lond.* **225,** 170–172.
RUSSELL, M. J. 1976. A possible Lower Permian age for the onset of ocean floor spreading in the northern North Atlantic. *Scott J. Geol.* **12,** 315–323.
SCLATER, J. G. and FRANCHETEAU, J. 1970. The implications of terrestrial heat flow observations on current tectonic and geochemical models of the crust and upper mantle of the Earth. *Geophys. J. R. astron. Soc.* **20,** 509–542.
SCRUTTON, R. A. 1972. The crustal structure of Rockall Plateau microcontinent. *Geophys. J. R. astron. Soc.* **27,** 259–275.
—— and ROBERTS, D. G. 1971. Structure of Rockall Plateau and Trough, north-east Atlantic. *In* Delany, F. M. (Editor). *The geology of the East Atlantic continental margin* **2,** *Inst. geol. Sci. Rep.* no. 70/14, 77–87.
——, STACEY, A. P. and GRAY, F. 1971. Evidence for the mode of formation of Porcupine Seabight. *Earth Planet. Sci. Lett.* **11,** 140–146.
SLEEP, N. H. 1971. Thermal effects of the formation of Atlantic continental margins by continental break up. *Geophys. J. R. astron. Soc.* **24,** 325–350.
SMYTHE, D. K., KENOLTY, N., RUSSELL, M. J. and SCRUTTON, R. A. 1977. Re-interpretation of the early history of the Rockall Trough and the northern North Atlantic, and its economic implications. *Unpublished mss.*
STRIDE, A. H., CURRAY, J. R., MOORE, D. G. and BELDERSON, R. H. 1969. Marine geology of the Atlantic continental margin of Europe. *Phil. Trans. R. Soc. Lond. A* **264,** 31–75.
TALWANI, M. and ELDHOLM, O. 1972. Continental margin off Norway: a geophysical study. *Geol. Soc. Am. Bull.* **83,** 3575–3606.

VOGT, P. R. 1971. Asthenosphere motion recorded by the ocean floor south of Iceland. *Earth Planet. Sci. Lett.* **13,** 153–160.
—— 1972. The Faeroe–Iceland–Greenland aseismic ridge and the western boundary undercurrent. *Nature, Lond.* **239,** 79–81.
—— and AVERY, O. E. 1974. Detailed magnetic surveys in the northeast Atlantic and Labrador Sea. *J. geophys. Res.* **79,** 363–389.
WHITMARSH, R. B., LANGFORD, J. J., BUCKLEY, J. S., BAILEY, R. J. and BLUNDELL, D. J. 1974. The crustal structure beneath Porcupine Ridge as determined by explosion seismology. *Earth Planet. Sci. Lett.* **22,** 197–204.
WILLIAMS, C. A. 1975. Sea-floor spreading in the Bay of Biscay and its relationship to the North Atlantic. *Earth Planet. Sci. Lett.* **24,** 440–456.
—— and MCKENZIE, D. 1971. The evolution of the north-east Atlantic. *Nature, Lond.* **232,** 168–173.
WILSON, J. T. 1963. Evidence from islands on the spreading of ocean floors. *Nature, Lond.* **197,** 536–538.

M. H. P. Bott,
Department of Geological Sciences,
University of Durham,
Durham DH1 3LE,
England.

# The siting of Tertiary vulcanicity

*I. R. Vann*

The Tertiary volcanic province of northwest Britain forms part of the larger Thulean province evidence of which extends over a large part of the Scottish continental shelf to the Faeroes and Greenland. A review of the extensive available data allows estimates to be made of the total volume of basalt produced during this event and its rate of production. Such estimates, when combined with petrological and geological data, provide an internally consistent thermal model of Mantle conditions at this time. Processes within the Mantle controlled the gross distribution of igneous activity. The localisation of igneous activity to intrusive centres remains an unexplained phenomenon. The intrusion of large volumes of basaltic magma to form cylindrical masses extending through much of the Crust appears contrary to the expected behaviour of a low viscosity liquid. The hypothesis that this localisation is induced by the generation and diapiric uprise of granitic magma, although lacking a sound mechanical basis, appears tenable. The intrusion of dykes and dyke swarms in the Tertiary igneous province of Scotland has been extensively studied using a variety of mathematical models. Although the intrusion of individual dykes is controlled largely by the structure of the host rock, on a large scale the dyke swarms can be mainly attributed to the nature of the stress field at the time of intrusion. Computer models employing Finite Element Analysis techniques predict with some accuracy many of the features observed in the dyke swarms. Hypotheses that the distribution of the swarms is controlled by major en-echelon shear fractures or other crustal inhomogeneities can be further examined in the light of the information produced by these modelling techniques.

## 1. Introduction

The Tertiary igneous province of Britain forms part of the larger Thulean igneous province which includes the early Tertiary igneous rocks of Greenland, the Faeroe islands and the submarine intrusions and extrusions of the Rockall plateau and the European continental shelf (Fig. 1). Throughout more than a century of investigation a vast amount of geological data has been gathered about all aspects of these rocks. These works have been reviewed by Stewart (1965); Richey *et al.* (1961) and Noe-Nygaard (1974).

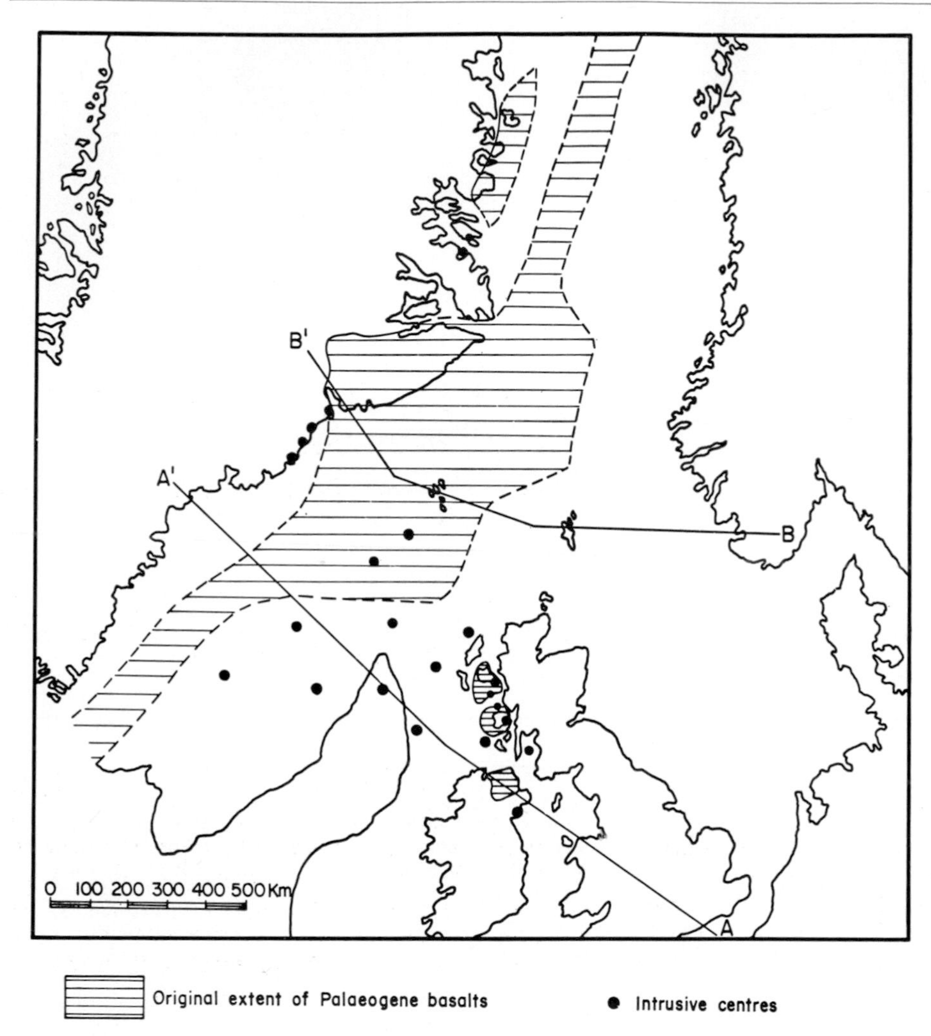


Fig. 1. Map of the distribution of Thulean vulcanicity. Some of the igneous centres, particularly in and around Rockall trough may be late Cretaceous. The submarine distribution of basalts is largely speculative.

In western Britain the igneous activity consists of a series of roughly concentric intrusive complexes, usually interpreted as the eroded remnants of central volcanoes. Associated with some centres are thick successions of basaltic lavas, in general older than the intrusive complexes. Dykes were intruded throughout the history of magmatic activity, arranged in parallel swarms with a general NW–SE trend.

The country rocks into which the magma was intruded vary in age and type from Archean granulites, through Moine and Dalradian schists, Palaeozoic and Mesozoic sediments to isolated pockets of Tertiary sediments. Despite this wide range, however, certain generalisations may be made. These are that (a) all the Tertiary centres lie within areas of marked Mesozoic subsidence and (b) most of the centres lie on or close to major faults with complex pre-Tertiary histories. The emplacement of the

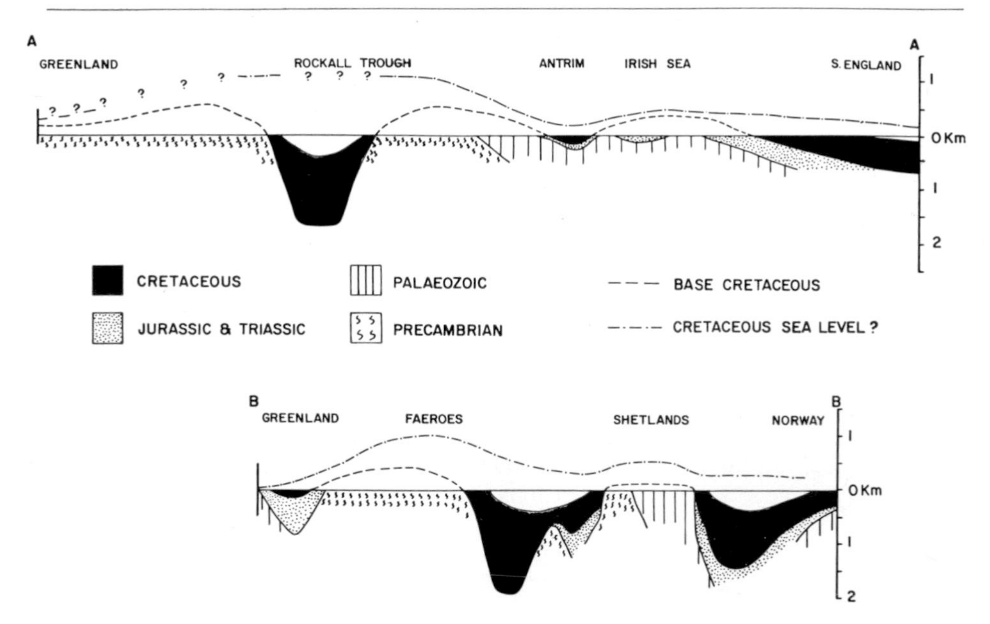


Fig. 2. Cross sections from Europe to East Greenland in early Tertiary times. (See Fig. 1 for locations.) The zero datum is Palaeogene sea level. The amount of uplift in the early Tertiary is shown by the elevation of the Cretaceous sea-level.

dyke swarms, on the other hand, does not appear to bear any simple relationship to either the structure or the stratigraphy of the areas they traverse.

The vast amount of data that has been gathered about these rocks have formed the basis of a number of important theoretical studies, e.g. the mechanics of intrusion (Anderson 1938), and has inspired fundamental geological concepts, e.g. the magma series and magma types of Harker (1904) and Bailey (1924). Despite these studies, however, no coherent picture of the genesis of the Province has emerged.

## 2. A thermal model of the origin of the Thulean province

### 2a. Early Tertiary uplift

At the end of the Cretaceous a marked period of uplift began in northern Europe and Greenland and sedimentation ceased over a wide area, continuing only in the deeper parts of the North Sea, Rockall trough and the continental margins bordering the already formed North Atlantic, south and west of the European–Greenland continental mass.

Bott (1975) has suggested that the basal Tertiary unconformity apparent over much of the area of the Thulean province may result from uplift associated with early Tertiary igneous activity. The evidence of late Cretaceous to early Tertiary uplift in northwest Europe and Greenland comes from the detailed stratigraphy of both land and sea areas.

The precise amplitude of this uplift can only be assessed in areas where the stratigraphical succession is complete enough to allow separation of the various phases of Mesozoic subsidence and uplift that had occurred prior to the early

Tertiary uplift. In the Hebridean region for instance, Palaeogene lavas rest directly on rocks ranging in age from Precambrian to late Cretaceous and it is impossible to confidently separate the effect of end Cretaceous uplift from many other preceding events.

According to George (1975 p. 114) the Palaeogene "rocks of southern England may be regarded as fluctuating offlap from an emergent dome of Chalk". These lowermost Tertiary rocks are dominantly terrigeneous except in the Paris basin where chalk sedimentation continued throughout the Danian without unconformity. However the tectonism that produced the unconformity was not simple, the uplift being complicated by subsidence in fault bounded basins in the western parts of the British shelf and in the North Sea.

In southern Britain the amplitude of post Cretaceous uplift can be estimated with some accuracy. The Chalk was deposited in a water depth of approximately 300 m (Hancock and Scholle 1975). The Landenian deposits of southern England (the Thanet Sands, Woolwich and Reading Beds) were deposited on an erosional surface which had degraded the underlying Chalk to a depth of 200 m in the region (Curry 1965). Further west in Devon the basal Tertiary beds descend on to Albian Greensands implying the erosion of perhaps 500 m of upper Cretaceous deposits. Further north evidence on land of the extent of this erosional surface is scant but "In notional extension of structures north and east it (the amplitude of erosion) is likely to have been greater in Oldland Wales and northern England, where the surface in Palaeozoic rocks was presumably exposed to prolonged subaerial denudation throughout Palaeogene times" (George 1975). In the Irish Sea, Cretaceous rocks are largely absent from the Mesozoic sequence beneath the basal Tertiary unconformity and only as the Irish Sea merges southwards into the Celtic Sea does Chalk re-appear in any thickness (Dobson *et al.* 1973).

In the North Sea the upper Cretaceous is developed mainly as Chalk but in the Viking graben it passes laterally into marl and shale (Hancock and Scholle 1975). The Chalk was deposited in water depths of a few hundred metres but the shales of the Viking graben were a deeper water facies, being restricted to the zone of maximum subsidence (Ziegler 1975). In the Viking and Central grabens of the North Sea, chalk deposition continued into the lower Tertiary with the deposition of the Danian Chalk of the Ekofisk area. Elsewhere in the North Sea basin, Palaeogene clastic sediments rest unconformably on pre-Tertiary rocks. The influx of clastic sediments into the northern North Sea basin marked the onset of uplift of Oldland Britain. Deltaic deposits prograded eastward from the region of the present day Moray Firth and the Shetland platform (Kent 1975; Parker 1975).

Cross-sections of the northern North Sea (e.g. Kent 1975) show a large thickness of upper Cretaceous rocks to the east of the bounding fault of the East Shetland platform. As these rocks contain little or no clastic material, it may be assumed that much of the movement of this fault is post-Cretaceous and upper Cretaceous deposits of some thickness extended westwards over the Shetland Platform in pre-Tertiary times. Traces of such deposits have been largely removed and early Tertiary clastic sediments now rest directly on a Caledonian basement implying an uplift of the Platform of at least several hundred metres during the early Tertiary.

On the other side of the Atlantic in East Greenland the stratigraphical record is less complete and little is known of either the structure or stratigraphy of the Greenland continental shelf. However, Koch and Haller (1971) state that "The end of the Mesozoic, perhaps, marks the most drastic change in the morphotectonic environment of East Greenland since the Precambrian. At the turn of the era the

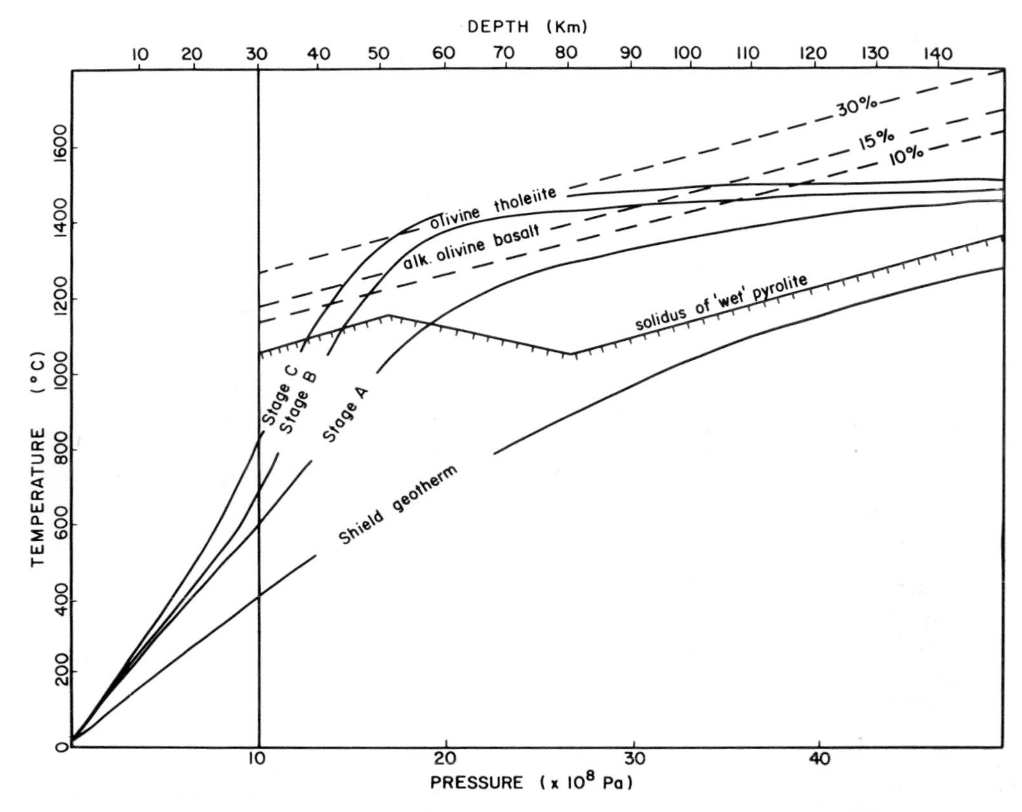


Fig. 3. A pressure-temperature diagram for the partial melting of 'wet' pyrolite after Green (1971). The percentages of partial melt formed are shown by the dashed lines. Geothermal gradients (solid lines) for a number of stages of evolution of the Thulean province are shown (see text) as well as a typical Shield geotherm.

whole central East Greenland region was epeirogenically uplifted and began to erode, thus exhuming what had formally been the Mesozoic shelf".

Throughout the Thulean province then, such data as are preserved appear to indicate the presence of a basal Tertiary unconformity that increases in amplitude towards the centre of the province. This unconformity attains a maximum value well in excess of 1500 m and may be as great as 2 km in the region of the Faeroes (Fig. 2).

## 2b. The thermal anomaly

Petrogenetic models for the rocks of the Thulean province suggest that the basalt originated from high degrees of partial melting of the mantle at depths between 50 km and 100 km (Thompson *et al.* 1972; Thompson 1974). The igneous activity in the province began with the extrusion of large quantities of alkali olivine basalt. This phase was generally followed by a period of tholeiitic magmatism although these events often overlapped. Accepting the petrogenetic models of Green (1971) for the melting of wet pyrolite and the contention of Thompson (1974) that many of the basalts of the province are in their primary state, the temperatures at depths between 50 km and 100 km must have lain between 1200°C and 1400°C (Fig. 3).

Prior to the rise in temperature which produced the Tertiary basalts the geothermal gradient in the Thulean province was probably rather greater than that of the Precambrian shield areas shown in Figure 3. This area lay close to both the Rockall trough, an area of probable Mesozoic sea-floor spreading, and the North Sea where there had been active rifting throughout most of the Mesozoic. During most of the Mesozoic then, the area of the Thulean province may have had a geothermal gradient similar to that shown as Stage A in Figure 3. During the late Cretaceous the thermal high that had persisted in the area for a long period probably declined somewhat, accounting for the extensive subsidence at that time. At the end of the Cretaceous it is probable that temperatures in the depth range 50 km–100 km were between 1000°C and 1200°C. It follows that the early Tertiary thermal event probably increased the temperature of the asthenosphere by approximately 200°C.

The expansion of the asthenosphere and the uplift of the overlying lithosphere due to a temperature rise of 200°C can be simply estimated employing an isostatic model and ignoring the effect of erosion. An isostatic model for the response of the lithosphere is probably valid since the wavelength of the thermal anomaly must have been approximately 2000 km if surface vulcanicity is restricted to the area overlying the partial melt zone. This is significantly greater than the normally assumed complete compensation area of $3° \times 3°$. The effect of erosion on such a model is to increase the amount of uplift.

The isostatic uplift due to the low density pillow underlying the area is given by

$$h_L = h_P \left( \frac{\rho_M - \rho_P}{\rho_L} \right) \qquad \text{(Eq. 1)}$$

where $h_L$ is the amount of uplift

$h_P$ is the thickness of the anomalously hot layer (50 km)

$\rho_M$ is the density of the normal Mantle (3400 kg/m$^{-3}$)

$\rho_L$ is the density of the Upper Crust (2500 kg/m$^{-3}$)

and $\rho_P$ is the density of the Mantle within the anomalously hot layer.

The density of the anomalous mantle $\rho_P$ can be estimated using a volume thermal expansion $\alpha_V$ in the range $3{\cdot}5 \pm 0{\cdot}5 \times 10^{-5}$ °C$^{-1}$ (Skinner 1966) and a 20% partial melt of liquid $\rho_B$ with density 2900 kg/m$^{-3}$ (Vogt 1974).

$$\rho_P = \rho_M(1 - \alpha_V \,.\, \Delta T).\ 0{\cdot}8 + \rho_B \,.\, 0{\cdot}2 \simeq 3300 \text{ kg/m}^{-3}$$

Substituting this value in Equation 1 yields $h_L \simeq 2$ km. This value is remarkably similar to the maximum uplift recorded in the area and hence this model appears to provide a consistent model for both magmatic and tectonic activity in the area.

## 2c. The causes of the thermal anomaly

That a thermal anomaly in the Upper Mantle existed in early Tertiary times is evident from both the extensive vulcanism and the stratigraphical record. The origin of this anomaly is open to speculation, but any model must invoke some form of convective mass transfer, whether it be cellular convection, single convective overturn (Bott 1975), a mantle plume (Vogt and Johnson 1975) or two phase convection (McKenzie and Weiss 1975). Such mechanisms must be invoked because of the low thermal conductivity of the mantle and the consequent long time scale, of the order of several tens of millions of years, required for the build up and decline of a conductively transferred thermal event.

A minimum value for the heat flow into the base of the zone of magma generation can be calculated if the average rate of basalt production in the zone is known. This average rate can be estimated for the province. The total volume of basalt produced during this volcanic event may have been in excess of $5 \times 10^6$ km$^3$ or $1{\cdot}5 \times 10^{19}$ kg.

The event lasted less than $10^7$ years or $3 \times 10^{15}$ s. Hence the total heat transferred to the surface, assuming a latent heat of fusion of 420 J.kg$^{-3}$ and a specific heat of 1300 J.kg$^{-3}$ °C$^{-1}$, was $6 \times 10^{25}$ J and the rate of heat transfer was $2 \times 10^{10}$ W. The total area covered by the Tertiary province was approximately $4 \times 10^{12}$ m$^2$. Hence the equivalent heat output of the production zone was $5 \times 10^{-3}$ W.m$^{-2}$. The model of Richardson (1975) suggests that the conductive heat flow at a depth of 100 km is approximately $2 \times 10^{-2}$ W.m$^{-2}$ in continental areas. A similar value has been calculated by Roy *et al.* (1972) for heat flow at this depth below the Canadian Shield. Hence it is apparent that an adequate heat flow was available even by conduction to support the magma production. This heat flow, however, is not great enough to support asthensopheric temperatures in excess of 1200°C required to produce tholeiitic basalt. In order to achieve temperatures of this magnitude the heat at the base of the partial melt zone must have been in the order of 0·1 W.m$^{-2}$, a five fold increase over the normal heat flow at such a depth.

McKenzie and Weiss (1975) demonstrate that a large increase in heat flow through the mantle promotes convection in the Upper Mantle. Assuming the convecting layer to be 700 km thick, the thermal time constant for a ten fold increase in heat flow is approximately 30 m.y. In this model a temperature increase of approximately 200°C in the asthenosphere is achieved in approximately 20 m.y. Although this period is short compared with the time constant of a conductive system, it appears to be too long to adequately explain the rapid onset of events in the Tertiary province. Both the speed with which the uplift was accomplished and the short time range of the intense volcanic event, appear to favour a time constant of less than 10 m.y. McKenzie and Weiss (1975) also point out that this time constant is probably directly dependent on the thickness of the convecting layer and that considerations of the motion of the major plates demand two scales of convection within the mantle. Large scale convection demands the movement of the lithospheric plates, whilst on a smaller scale there may exist longitudinal convective rolls (Fig. 4). Where these two types of convection intersect at plate boundaries, models predict a variety of transient effects which may account for the short thermal time constant required by the geological evidence in the Thulean province.

During late Cretaceous times, the temperature distribution beneath the province was approximately that shown in Stage A in Figure 3. Towards the end of this era a thermal anomaly developed in the outer core producing perhaps a ten fold increase in heat flow into the base of the Upper Mantle. This increased heat flow produced active convection in the Upper Mantle commencing at approximately 65 m.y. ago. Within a few million years (Stage B) the convective system had increased the temperature in the asthenosphere by 200°C. Partial melting of the mantle in this zone was around 15%, producing significant quantities of alkali olivine basalt and a marked upwarp of the overlying lithosphere. By this time the Greenland and European plates were beginning to separate and large volumes of basalt were being extruded and intruded into the crust. Increased heat flow and the passage of magma through the crust promoted partial melting within the lower crust and granitic magmas were forming. The temperature in the asthenosphere continued to increase rapidly and within a short space of time partial melting locally reached 30%. At this stage (C) magmatic activity was dominated by tholeiitic magmas forming large intrusive masses and extensive dyke swarms. This stage lasted for perhaps 5 m.y. over the whole province. After this period the partial melt zone, originally a broad linear feature orientated along a NNW–SSE axis, became concentrated into a smaller circular zone of high activity on the NE–SW linear partial melt zone under-

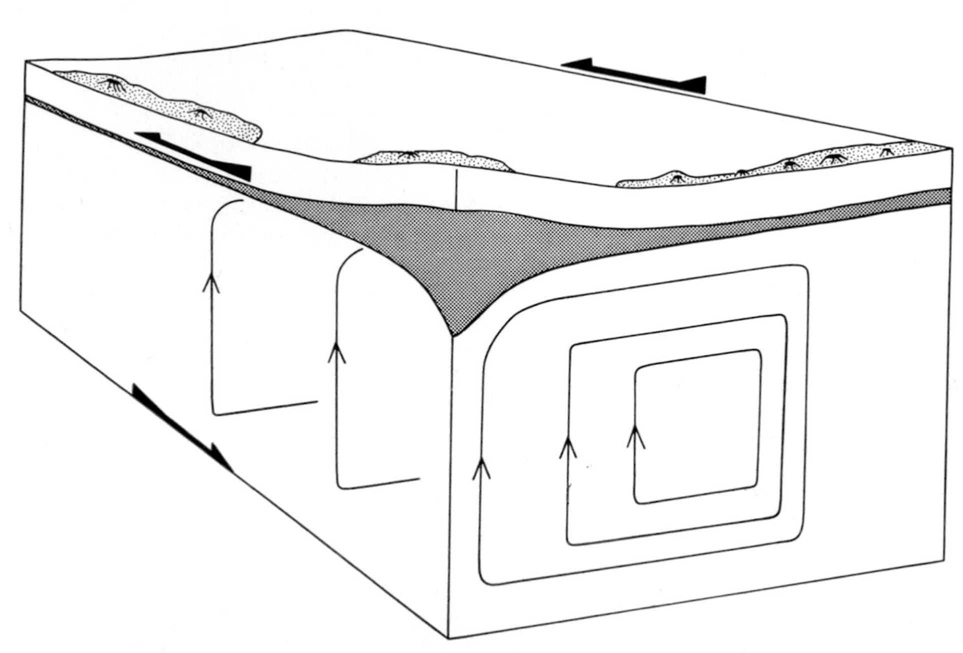

Fig. 4. A block diagram of the two scale convective system adapted from McKenzie and Weiss (1975). The large arrows indicate the direction of plate motion and the return flow above the Lower Mantle. Smaller arrows and flow lines indicate the motion of the longitudinal convective rolls. Extensive partial melting takes place in two zones (black) approximately at right angles.

lying the now well developed Reykjanes ridge (Stage D). This circular feature has continued to be active until the present day producing a local excess of magma in the formation of the oceanic lithosphere responsible for the production of the Iceland–Faeroe ridge, the Davis Strait ridge and the anomalous structure of Iceland.

## 3. The emplacement of the igneous centres

The major igneous centres of the British Tertiary province have recently been investigated by geophysical methods (Bott and Tuson 1973; McQuillin *et al.* 1975). These studies have shown that the centres are underlain by large masses of basic and ultrabasic rocks. The Skye centre, for instance, has been shown to expand rapidly downwards from a surface diameter of 10 km to 25 km at a depth of 15 km.

The mechanism of intrusion of these large bodies of basaltic magma is poorly understood and as yet no sound mechanical theory for this phenomenon exists. Both theoretical studies and experimental data (Pollard 1973; Ramberg 1967) indicate that the intrusion of liquids should depend on the viscosity contrast between the intruding liquid and the host medium. Any intrusive liquid with a viscosity less than that of a granitic magma should intrude in relatively narrow fissures forming various types of sheets. The orientation of these types of intrusion are controlled by the stress acting in the host rock. Hence it appears that the huge masses of the centres may consist of large numbers of stacked sheet intrusions. This

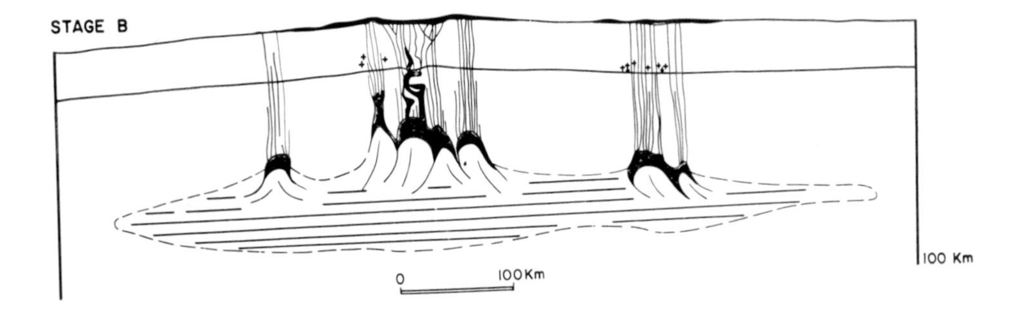


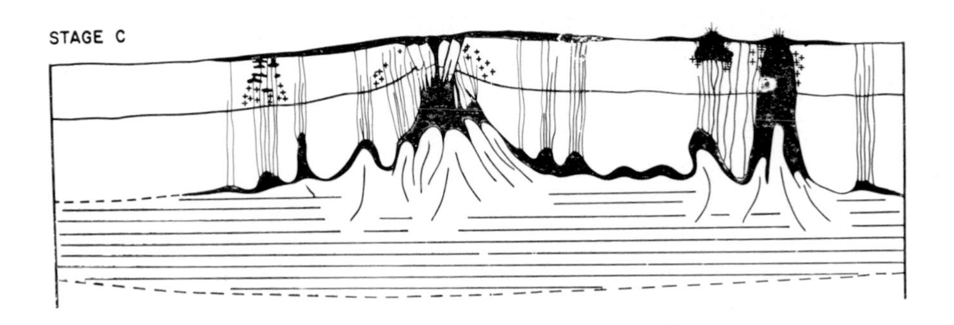


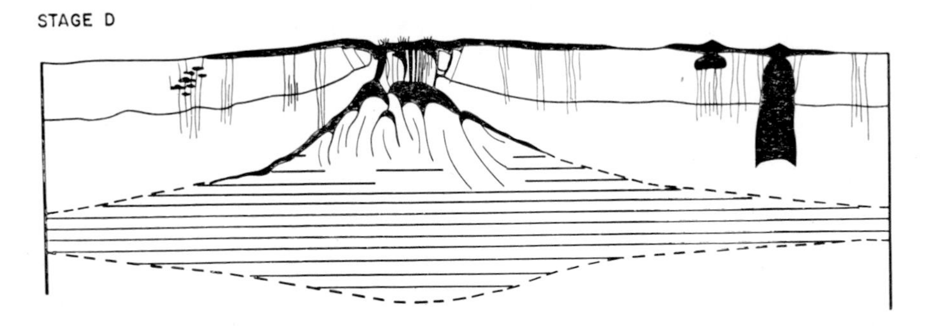


Fig. 5. NW–SE cross sections from Britain to Greenland showing the active stages of development of the Thulean province. During Stage B a zone of extensive partial melting developed (ruled), promoting the rise of diapirs. This rise concentrated magma into pockets (black) from which dykes were intruded through the lithosphere to feed extensive lava fields. The passage of magma through the crust induced local partial melting (crosses). During stage C the partial melt zone was more extensive and large volumes of tholeiite were being produced. Where there had been previous partial melting of the crust the central complexes were formed, in some cases stretching through the crust and possibly into the mantle below, whilst in others only forming high level magma chambers. In stage D times activity became concentrated below the forming Reykjanes ridge and true oceanic lithosphere was emplaced. This stage has continued up to the present.

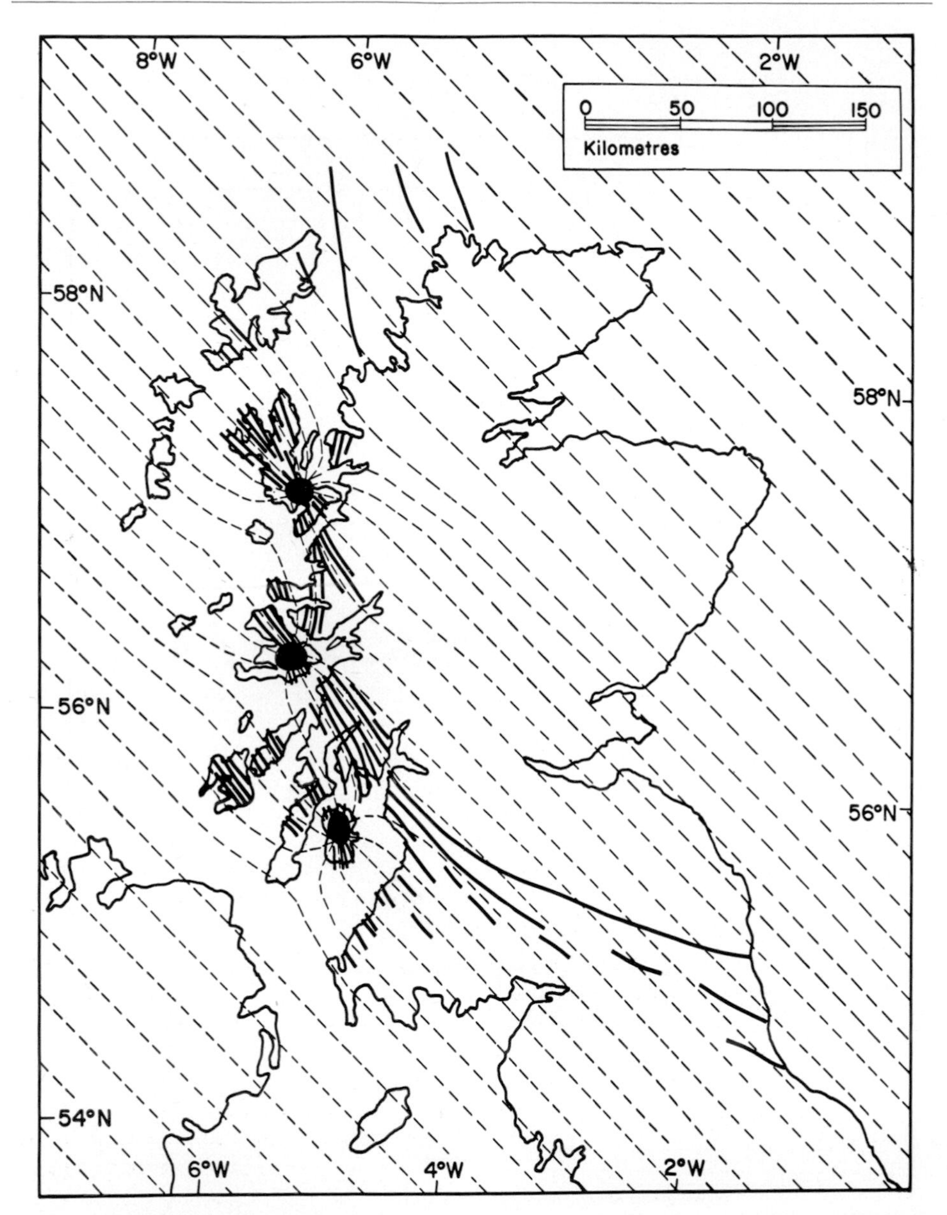


Fig. 6. Comparison of the distribution of Tertiary dykes in northern Britain (solid lines) and the stress trajectories produced by the Finite Element model described in the text.

contention is supported by the limited data available from surface exposures of the centres (Walker 1975).

Even if the hypothesis of stacked sheet intrusions is accepted, the explanation of the phenomenon remains obscure. In a non-hydrostatic stress field basalt should ascend by dyke intrusion. The commonly observed mode of basalt eruption is from fissures and present day examples of central volcanoes erupting basalt are rare.

Many of the centres lie on or close to major faults trending NE–SW and within areas of extensive alkali basalts lava fields predating the central volcanoes. These lavas were fed from early dyke swarms (Mattey *et al.* 1977) with a NW–SE trend and the central intrusions occur at the intersection of these early dyke swarms with the major faults.

Walker (1975) has suggested that the emplacement of the basaltic central intrusions follows the diapiric rise of granitic magmas formed by partial melting of crustal rocks. Whilst the model is not wholly corroborated either by field observation or the mechanical relationship between the diapiric rise of granite and the emplacement of a complex of sheet intrusions, the association is qualitatively reasonable. It is probable that the generation of granitic magma was promoted at these localities by the concentration of dyke intrusion in the pre-existing fault zone and of hot circulating fluids in this more permeable region.

## 4. The emplacement of the dyke swarms

### 4a. The Hebridean model

The geometry of the dyke swarms of northwest Britain has recently been the subject of intensive studies (Sloan 1971; Spieght 1972; Knaap 1973). These studies contain voluminous data relevant to the problem outlined by Richey (1932), that is, what controls the distribution and orientation of the dyke swarms?

The general NW–SE trend of the swarms has usually been taken to reflect the orientation of the maximum principal stress in the region during the emplacement of these intrusions. The deviations from this trend, particularly the swings of the Skye and Mull swarms into more N–S orientations north of Mull and Arran have been more contentious issues, however (Fig. 6). Richey (1939) suggested that these swings might be due to the influence of a crustal "zone of weakness" of approximately N–S trend. He further suggested that the plutonic centres of Skye, Mull, Ardnamurchan, Rhum and Arran lie at the intersection of this zone with the major NE–SW faults. This view has been supported by Tyrrell (1949), Auden (1954), Knaap (1973) and McIntyre *et al.* (1975). Contrary to this is the view of Harker (1904), that the orientation of the dyke swarms shows a poorly defined radial pattern around each of the central intrusions and that deviations from the NW–SE trend are due to the influence of the centres.

An investigation of the distribution of the Tertiary dyke swarms of this area employing Finite Element Analysis (Zienkiewicz and Cheung 1967) has been undertaken by the author. In this study the Hebridean region is modelled by a plate in which the plutonic centres of Skye, Mull and Arran were represented by pressurised holes. Both the boundary conditions, that is the forces acting on the boundaries of, and within, the plate and the elastic properties of the whole plate or small sub-areas within it were varied to obtain a stress distribution which best matched the statistical data on the observed distribution of dykes (Speight 1972; Sloan 1971; Knaap 1973).

The simplest type of model employed in the study was that of homogeneous, isotropic plates subjected to a NW–SE orientated compressive stress and equal magma pressures at each of the centres. In this type of model the resultant stress distribution is approximately dependent on the ratio

$$k = (P_m - S_1)/(S_1 - S_2)$$

where $P_m$ is the magma pressure applied at the plutonic centres, $S_1$ is the maximum

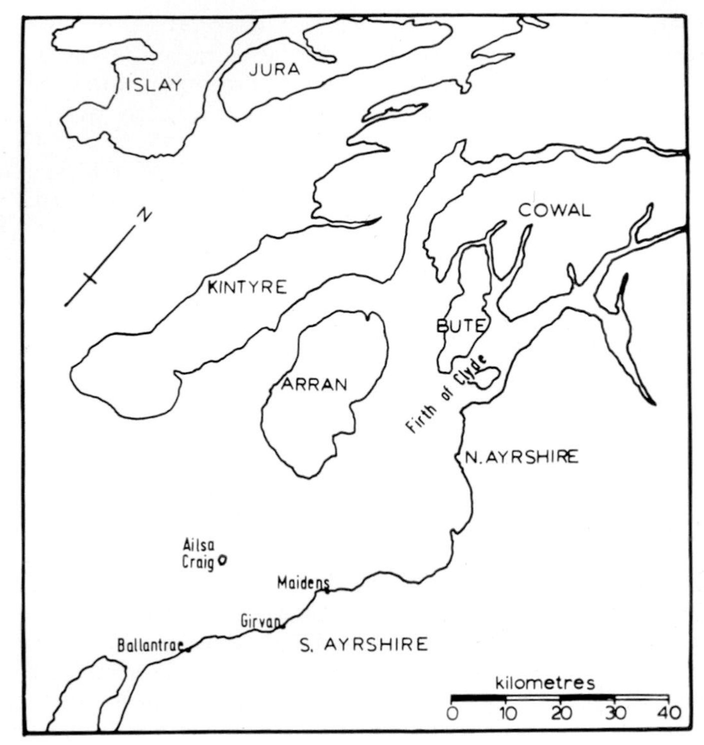


Fig. 7. Locality map of the area around the Arran igneous centre.

principal regional stress, and $S_2$ is the minimum principal regional stress. The best fit of the model to the data was obtained with a value of $k = 2{\cdot}0$. The results of this model are shown diagrammatically in Figure 6.

## 4b. The Arran igneous centre model

The area around the Arran centre (Fig. 7) was studied in greater detail employing a sub-area mesh (Fig. 8) In this type of study it is necessary to use the nodal displacements derived from the overall model as boundary conditions on the sub-mesh. The internal boundary conditions are forces acting radially outwards on the nodes surrounding the internal hole which simulates the intrusive centre. These forces are calculated such that the magma pressure is the same as in the original model.

The $S_1$ stress trajectories produced by this model are compared with the pattern of Tertiary dykes taken from the 1:633,600 geological map of Great Britain (Stubblefield 1964) and the mean trend data of Knaap (1973); (Fig. 9). In general the agreement between the predicted and observed trends is good but much of the area in which the most critical comparisons could be made lies beneath the waters of the Firth of Clyde.

The changes in orientation of the Mull dyke swarm in Cowal and of the few members of the swarm which continue on the southeastern side of the Firth of Clyde are very similar to those predicted by the stress trajectories. A close corres-

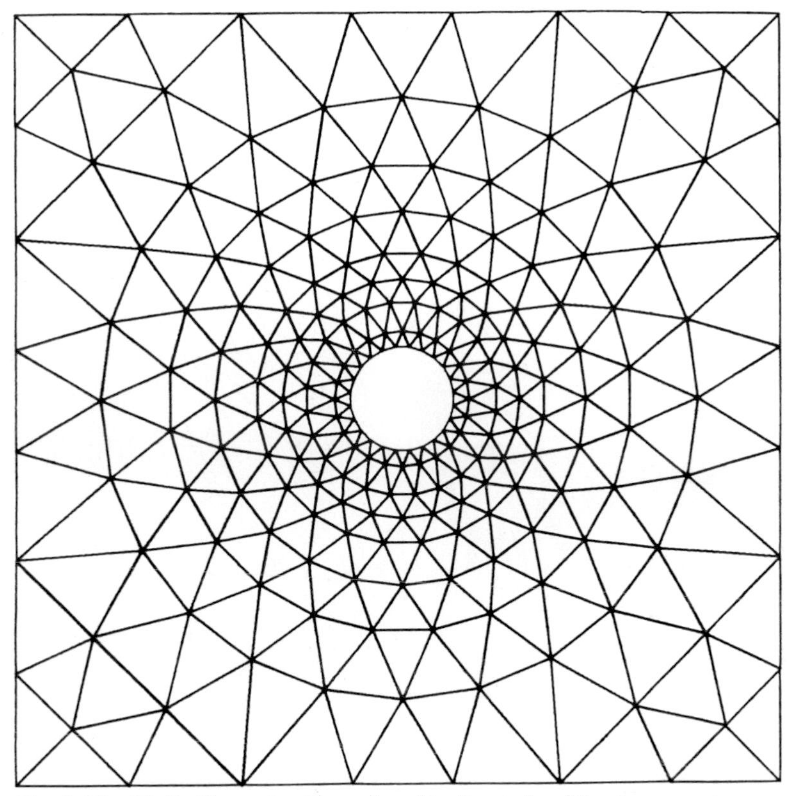

Fig. 8. The Finite Element grid employed in the study of the Arran centre.

pondence of actual and predicted trends also occurs in areas both northwest and southeast of the centre but these examples are less important since the presence of the igneous centre has little effect on the orientation of the stress trajectories in these areas.

Notable misfits occur to the east and west of the igneous centre in north Ayrshire and Kintyre. In the latter area the dykes are dominantly undersaturated alkali olivine basalts which appear from cross-cutting relationships to be earlier than the tholeiites which form the vast majority of the dykes of the Mull and Arran swarms (Knaap 1972; Lamacraft 1977). This relationship has been observed in other parts of the igneous province (Bailey *et al.* 1924 p. 359). It has been suggested (Mattey *et al.* 1977) that this reflects the fact that many of the alkali olivine basalt dykes were intruded contemporaneously with the extrusion of the plateau lavas and in general predate the formation of the plutonic centres and the associated intrusion of tholeiitic dykes. If this suggestion holds true for the dykes of Kintyre then the trend of the dykes of alkali olivine basalt should not be related to the stress distribution around the intrusive centre and should maintain a relatively constant NW–SE orientation throughout the area. In Ayrshire too, undersaturated basaltic intrusions are common and in north Ayrshire and the Cumbraes these greatly outnumber tholeiites, perhaps accounting for the dominance of the NW–SE trend in this area.

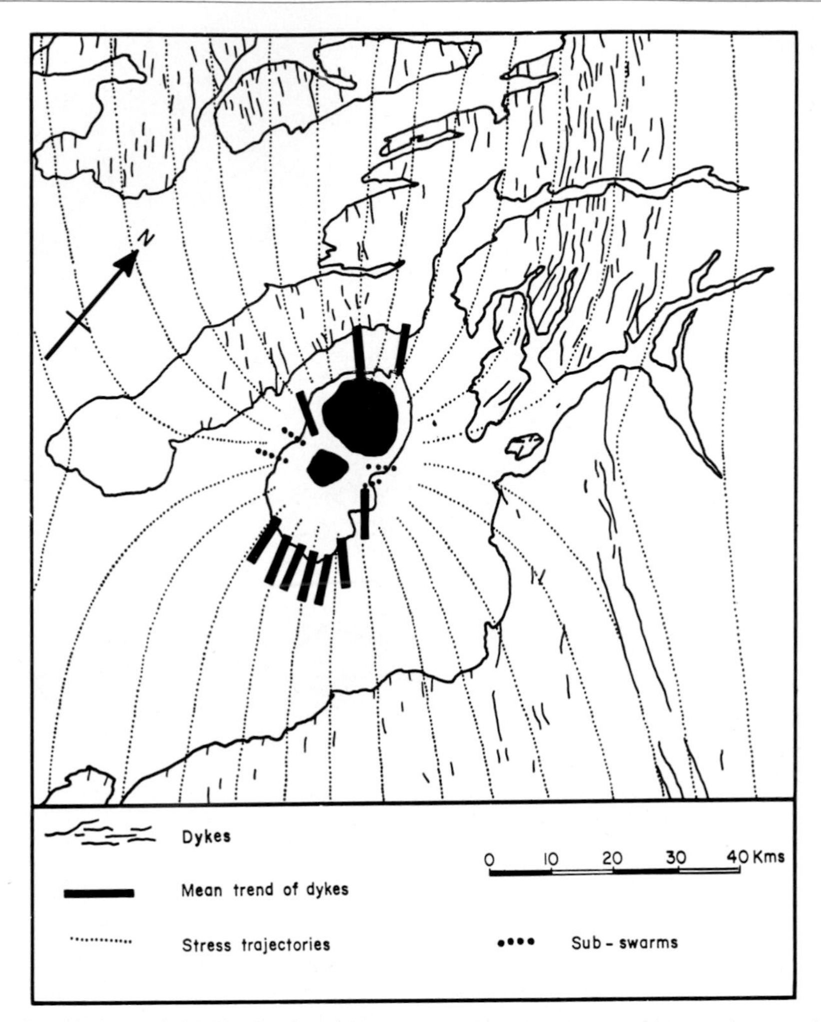


Fig. 9. Comparison of the distribution dykes around the Arran centre with maximum principal stress trajectories produced by the model. The mean trend and sub-swarm data are from Knaap (1973).

(i) *An anisotropic model.* Although this model adequately describes the re-orientation of the Mull dyke swarm in Cowal, north of the Arran centre, the suggestion that the swing is due to a N–S zone of weakness cannot be discounted on this evidence alone and is worthy of further investigation. Within this region there are a number of structural peculiarities of which the apparent southerly deflection of the Highland Boundary Fault resulting in a narrowing of the Midland Valley graben from 85 km in Scotland to 45 km in Ireland is the most obvious regional anomaly (Fig. 10).

The NNE–SSW trending faults of the Loch Tay set change direction from a mean of 030° to a mean of 005° in the region of the Upper Firth of Clyde. The Loch Tay set of faults have been correlated with the N–S rucks of north Ayrshire (Paterson 1954). A further anomaly is the NNW–SSE elongation of the magma chamber

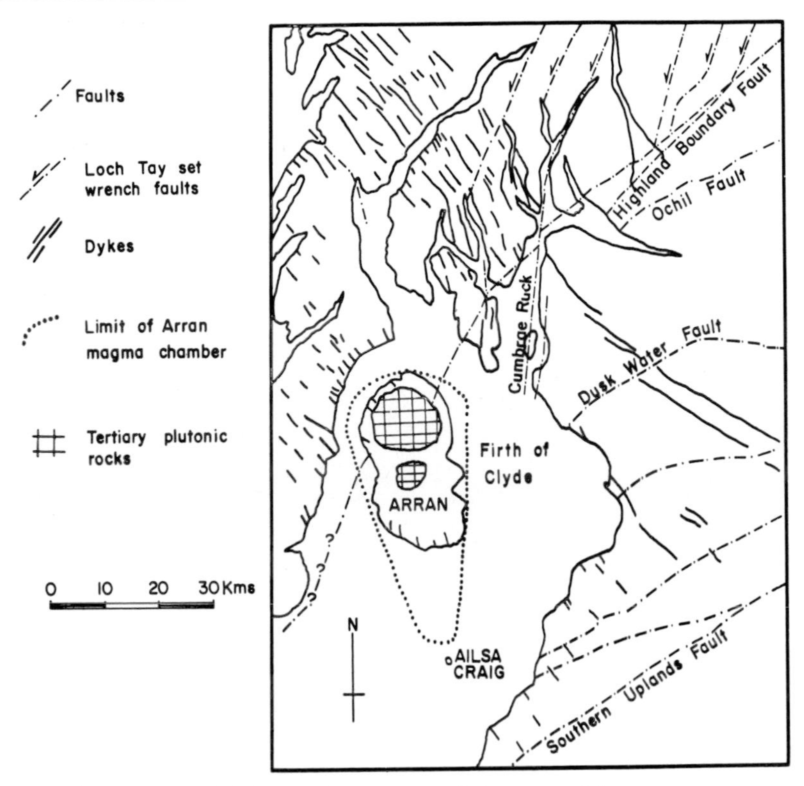


Fig. 10. Map of dykes and major tectonic features of the Arran area.

which underlies the Arran centre, geophysical evidence suggesting that it stretches from below the Northern Granite of Arran to Ailsa Craig (Fig. 10).

Zones of weakness with an approximate N–S trend in Scotland and Ireland have been postulated by Russell (1972). Basing his arguments largely on the distribution of base metal mineralisation Russell suggests that these "geofractures" are regularly spaced lineaments owing their origin to the separation of Rockall from Europe. Although Russell does not postulate the existence of a geofracture beneath the Firth of Clyde such a feature would fill a gap in an otherwise regular pattern (McIntyre *et al.* 1975).

In an attempt to model the re-orientation of stress which might be produced by such a structure, a 30 km wide N–S anisotropic zone was introduced into the Finite Element model in the Firth of Clyde region. A shear modulus, G, of 10 G Pa was assigned to the elements of this zone. This value of G is one-quarter of that assigned to the other elements of the model. This low value could be realistic only if the rocks of this zone contain a relatively high proportion of mica orientated in vertical N–S planes. This is because the micas have the lowest G values of any of the common minerals, with less than 6 G Pa in biotite and phlogopite (Birch 1966). The presence of micas in such zones in the lower crust is not unlikely. If analogy can be made with shear zones in high grade gneisses the two most obvious changes which occur are a general decrease in grain size and hydration of the mafic mineral phases (Watterson 1975).

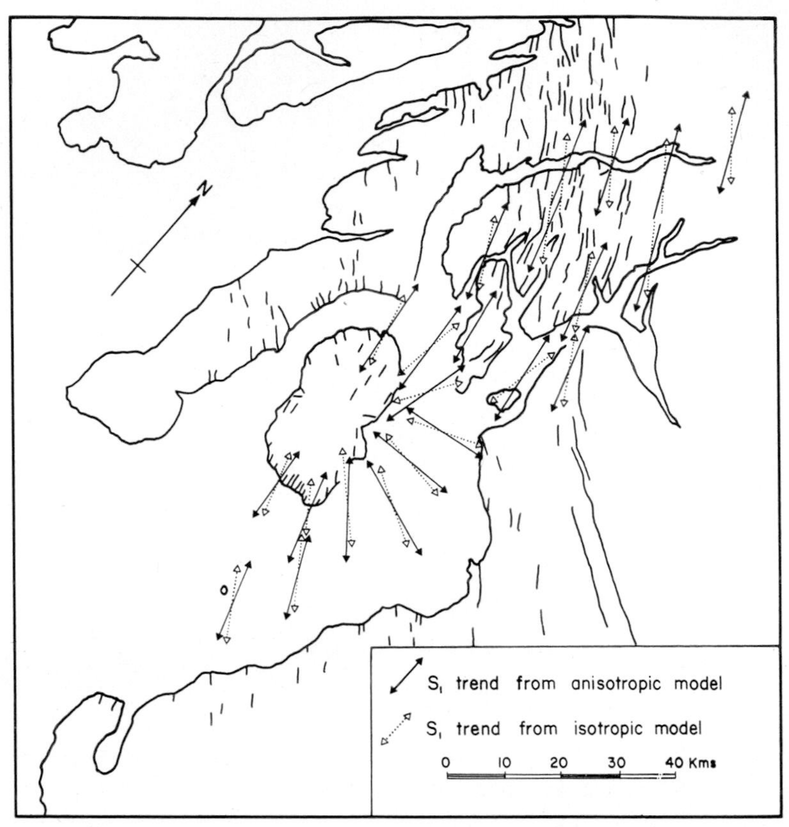


Fig. 11. Comparison of maximum principal stress trends produced by isotropic and anisotropic models. The distribution of arrows is limited to the supposed N–S anisotropic zone. Elsewhere both models produce trends as in Fig. 9.

The effect of the introduction of this zone on the stress distribution is shown in Figure 11. The maximum principal stress is re-orientated into a direction more nearly parallel with the longitudinal direction of the anisotropic elements whilst the stress orientations outside the anisotropic zone show only minor changes of less than 5° in the elements bordering the zone and less than 2° at greater distances.

The introduction of the anisotropic zone tends to produce a poorer correlation with the dyke trends of northern and western Cowal than does the original model. In Bute, Arran and eastern Cowal it is not possible to discriminate between the two models whilst elsewhere the data for critical comparison is not available. A feature of the anisotropic zone which tends to discriminate against it is the fact that it produces a relatively sharp bend in the stress trajectories at the boundaries of the zone. Although this deviation is less than 15°, there does not appear to be any similar abrupt change of direction in the dykes of the region.

Despite these qualifications the misfit of the anisotropic model is not so great as to exclude the possibility that such a zone exists in this area. This is more particularly true since this zone, if it exists, lies at some depth within the crust and hence the stress field at the present depth of erosion would have been modified to an unknown extent by the elastic response of the overlying rocks.

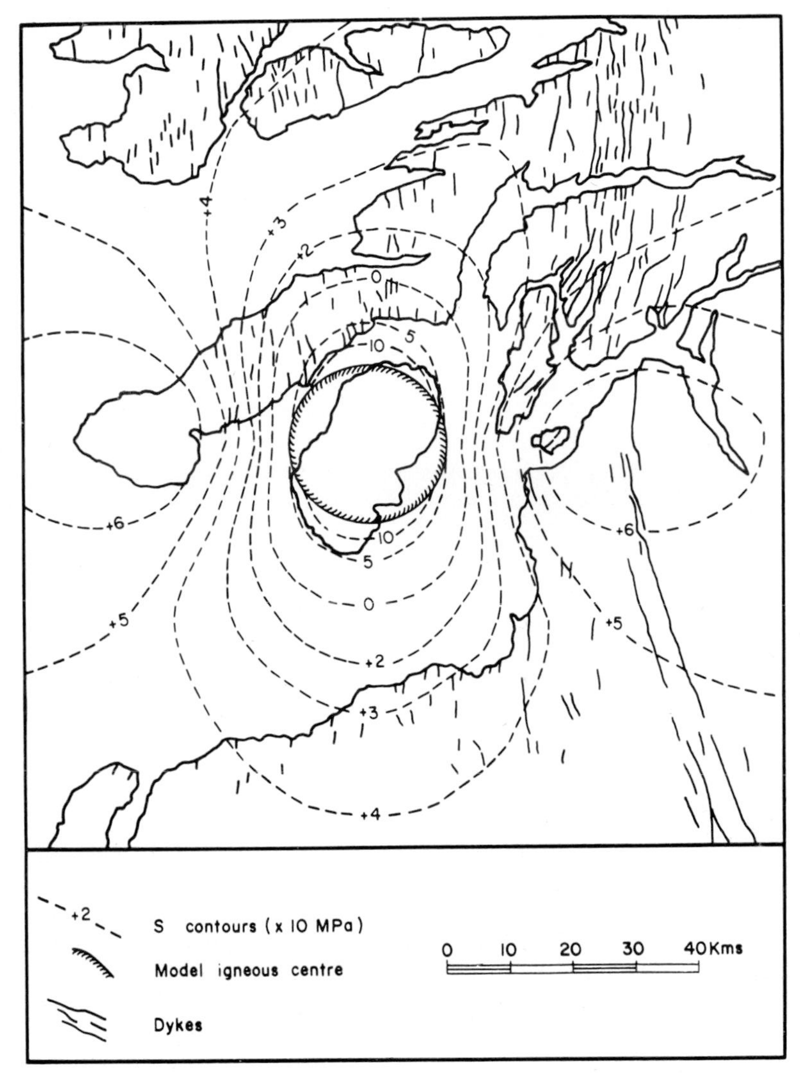


Fig. 12. Contoured values of minimum principal stress around the Arran centre assuming a magma pressure of 225 M Pa and a maximum regional applied stress of 100 M Pa.

(ii) *The distribution of dykes.* The restriction of dykes of the Arran swarm largely to quadrants lying northwest and southeast of Arran may be due to the pattern of variation of the minimum principal stress associated with the igneous centre (Fig. 12). The negative values of minimum principal stress which surround the centre are tensions whereas at greater distances this quantity becomes compressive and shows two almost symmetrically situated maxima. These maxima correspond approximately to points of hydrostatic stress associated with the presence of the pressurised cylinder. It seems reasonable to assume that dykes are most likely to form in areas where the minimum principal stress is small and particularly in zones in which it is tensile since these should contain an abundance of open cracks. These zones also lie close to the igneous centre and hence an adequate magma supply.

The zones of high compressive minimum principal stress tend to inhibit dyke intrusion. Dykes which do intrude these zones might be expected to have a more random distribution of orientation than those of other areas because of the near hydrostatic nature of the stress field in these zones. If the dykes of Kintyre are ignored because of their different petrological nature, the vast majority of the dykes of both the Mull and Arran swarms lie within an area bounded by the 40 M Pa contour in Figure 12. The only major exceptions to this are the few members of the Mull swarm which traverse the Firth of Clyde and continue southeastwards into northern England. Although these dykes are tholeiitic and petrologically similar to the rest of the Mull swarm, they are distinctive in both their thickness and their length. It is perhaps unsurprising that dykes of such exceptional dimensions should continue uninterrupted by the zone of high stress.

The distribution of stress does not accurately reflect the intensity of the dyke swarms. The values of the minimum principal stress in Figure 12 predicts a maximum dyke intensity in southeast Arran whilst the measured maximum of both dilation and the number of dykes occurs along the south coast of the island (Tyrrell 1928; Knaap 1973). The maximum dyke intensity on the Ayrshire coast occurs south of Girvan whilst the stress minimum occurs 20 km further north. This disagreement may be due to the inappropriateness of the circular model of the igneous centre. The actual form of the igneous mass is oval in plan with its major axis orientated N–S (Fig. 10). The minimum principal stress around a pressurised elliptical cylinder attains a minimum value where the curvature of the cylinder is greatest. Hence in this case the minimum value probably lay close to the southern termination of the oval magma chamber.

In an attempt to confirm the relationship between the distribution of dykes and the minimum principal stress an iterative Finite Element model, employing the "stress transfer method" (Zienkiewicz *et al.* 1968) was used. In the model each element is tested against an arbitrary failure criterion and if the element fails the nodal forces acting on it are altered so that the stress perpendicular to the direction of failure or intrusion becomes equal to a prescribed value of magma pressure. The criterion, $S_3 \leqslant 0$ was used in this model and the magma pressure equal to that within the igneous centre was applied to the intruded elements.

The results of the progressive intrusion model are shown in Figure 13. On the first iteration intrusion occurs only in those elements which lie within the zone of tension produced by the centre. On the second iteration the zone of intrusion grows rapidly in the direction of maximum principal stress reaching the edge of the model. This is an artificial boundary and the true extent of the intrusive zone cannot be estimated.

During the third iteration the zone widens slightly and if the model were larger would no doubt lengthen considerably in the direction of maximum principal stress. In the fourth iteration few elements fail and the stable position is attained with no further failure on subsequent iterations. This applies only to the area of the model and it is probable that the length of the intrusive zone would continue to increase. This process probably continues so long as adequate magma supplies are available.

Comparison of the intrusive zone with the contoured values of the minimum principal stress shows that the limit of the zone close to the centre lies approximately midway between the 40 M Pa and 50 M Pa contours (Fig. 13). This limit agrees well with the distribution of dykes around the Arran centre.

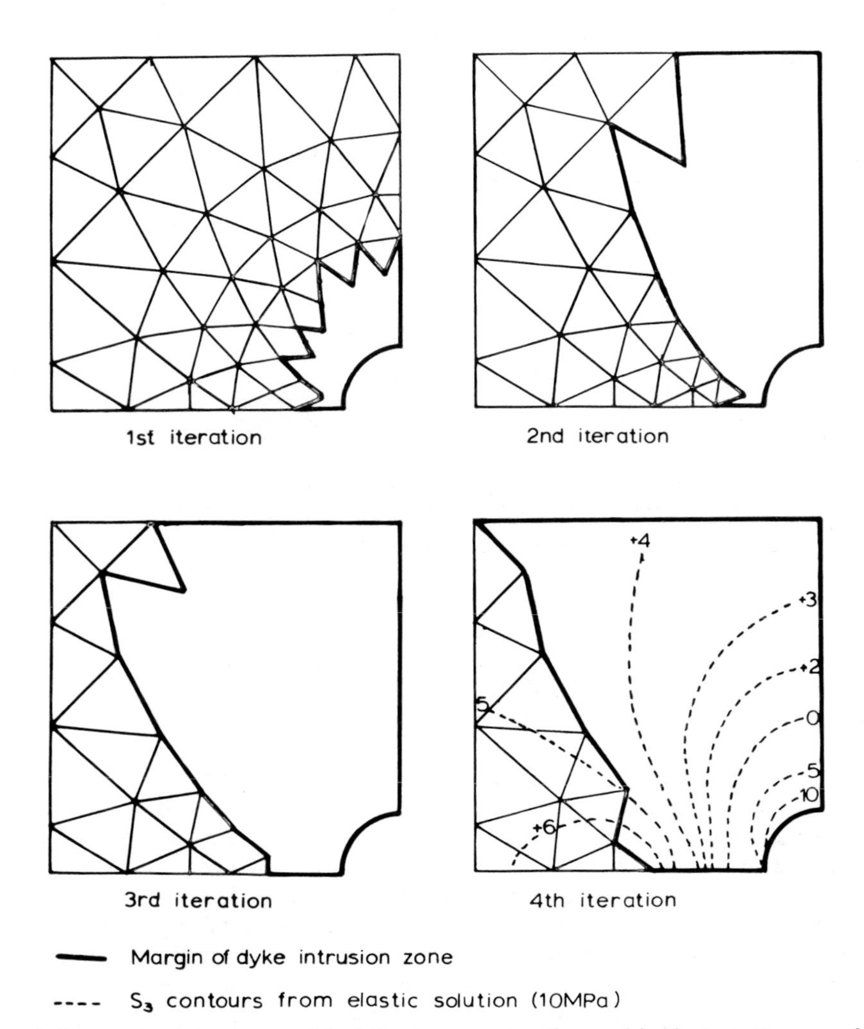


Fig. 13. Progressive intrusion model of the Arran centre discussed in the text. Because of the approximate nature of the boundary conditions imposed on this model the solution is symmetrical and only the top left hand quadrant of the model is shown here.

The model of the progressive development of the intrusive zone also predicts the amount of dilation produced by the intrusion of the dyke swarms (Fig. 14). The dilation recorded along a 25 km section of the south coast of Arran is 7% (Tyrrell 1928), although, by splitting this into hundred metre sections Knaap (1973) has shown that peak values of 12% are attained. Along the north coast the average dilation is approximately 2% with a peak value of 5%. In both these areas the model predicts approximately 2% dilation and, although this is in reasonable agreement with the data from northern Arran it does not approach the value attained in the southern part of the island.

Two factors may contribute to the discrepancy. Firstly, it has already been suggested that the magma chamber underlying Arran may stretch southwards beneath the south coast as a ridge of magma radically altering the local stress condi-

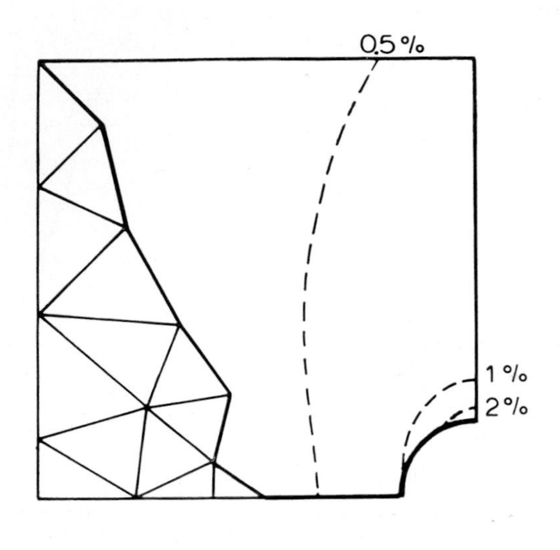


Fig. 14. Dilation due to intrusion predicted by the progressive intrusion model (dashed lines).

tions. Secondly, the model represents only a single intrusive event. If this area was subjected to several periods of intrusion separated by time intervals sufficiently long to allow relaxation of the stresses set up by the previous intrusive phase, much larger dilations could occur.

Along the Ayrshire coast Knaap (1973) records dilations in excess of 0·5 % with a maximum value of 2 % whilst the model predicts values of approximately 0·5 % over a section of coastline bounded by the 40 M Pa contour in Figure 12. Along the east coast of Kintyre opposite northwest Arran the average dilation is 1 % with a maximum of 4 %. These values decrease rapidly northwestwards and along the west coast of Kintyre the values only locally exceed 0·5 %. In the model this area lies largely in the zone between 0·5 % and 1 % dilation.

The correlation between predicted and observed values in all these areas is good and this study appears to confirm the distinct nature of the dykes of Islay and Jura where Knaap (1973) records dilations locally greater than 5 %.

## 5. Conclusions

1. The widespread unconformity at the base of the Tertiary was produced by an uplift of perhaps 2 km centred on the Faeroe region. This uplift can be accounted for by a temperature rise of 200°C in the Upper Mantle. This temperature rise is the same as that required by petrogenetic models of Thulean basalts.

2. The Upper Mantle thermal anomaly can be explained by a two scale convection model with a time constant of approximately 10 m.y.

3. The major igneous centres occur at the intersection of early dyke swarms with

major NE–SW faults. The emplacement of these large bodies of basic material may have been initiated by the formation of granitic partial melts of crustal materials in the areas of high heat flow at these intersections.

4. The orientation of the Tertiary dyke swarms of the Hebridean area can be modelled with reasonable accuracy by simple models. The centres of Skye, Mull and Arran dominate the patterns of dyke distribution exposed on land in the area.

5. A more detailed model of the Arran area corroborates the fit of the model with observed dyke trends and in particular with the change in trend across the Cowal Peninsula. The areal distribution of maxima and minima of the minimum principal stress around the Arran centre provide an explanation for the variation in observed dyke intensity throughout this region.

6. The hypothesis that the region of the Upper Firth of Clyde is underlain by a N–S trending zone of weakness causing the re-orientation of the Mull dyke swarm in the Cowal Peninsula was tested by introducing an anisotropic zone into the models. Although this model represents an extreme case even here resolution is not sufficient to discriminate between the different hypotheses.

7. A progressive model of intrusion has been developed. When applied to a model of the Arran area this technique accurately predicts the distribution of dykes.

8. Dilations predicted by the progressive intrusion model were generally lower than observed values especially in areas close to the igneous centre. Much of this discrepancy may be due to multiple phases of intrusion around the igneous centre.

*Acknowledgments.* This work is part of the Ph.D. study of the author at the University of Glasgow. Thanks are due to Drs A. McLean, J. Hall and D. Powell for fruitful discussion.

## References

Anderson, E. M. 1938. The dynamics of sheet intrusion. *Proc. R. Soc. Edinb.* B, **58,** 242–251.
Auden, J. B. 1954. Drainage and fracture patterns in northwest Scotland. *Geol. Mag.* **91,** 337–351.
Bailey, E. B., Clough, C. T., Wright, W. B., Richey, J. E. and Wilson, G. V. 1924. The Tertiary and post-Tertiary geology of Mull, Loch Aline and Oban. *Mem. geol. surv. U.K.*
Birch, F. 1966. *In* Clark, S. P. (Editor). Handbook of physical constants. *Mem. geol. Soc. Am.* **97,** 97–121.
Bott, M. H. P. 1975. The structure and evolution of the North Sea Basin, the Faeroe Block and the intervening region. *In* Woodland, A. W. (Editor). *Petroleum and the Continental Shelf of north-west Europe.* 105–116. Applied Science Publishers, Barking.
—— and Tuson, J. 1973. Deep structure beneath the Tertiary volcanic regions of Skye, Mull and Ardnamurchan, north-west Scotland. *Nature phys. Sci., Lond.* **242,** 114–116.
Curry, D. 1965. The Palaeogene beds of south-east England. *Proc. Geol. Ass.* **76,** 151–174.
Dobson, M. R., Evans, W. E. and Whittington, R. E. 1973. The geology of the southern Irish Sea. *Rep. Inst. geol. Sci.* No. 73/11.
George, T. N. 1974. Prologue to the Geomorphology of Britain. *Inst. Br. Geogr. Spec. Pub.* **7,** 113–125.
Green, D. H. 1971. Composition of basaltic magmas as indicators of conditions of origin; applications to oceanic volcanism. *Phil. Trans. R. Soc. Lond.* A. **268,** 707–725.
Hancock, J. M. and Scholle, P. A. 1975. Chalk of the North Sea. *In* Woodland, A. W. (Editor). *Petroleum and the Continental Shelf of north-west Europe.* 413–427. Applied Science Publishers, Barking.
Harker, A. 1904. The Tertiary igneous rocks of Skye. *Mem. geol. Surv. U.K.*
Kent, P. E. 1975. The tectonic development of Great Britain and the surrounding seas. *In* Woodland, A. W. (Editor). *Petroleum and the Continental Shelf of north-west Europe.* 3–28. Applied Science Publishers, Barking.
Knaap, R. J. 1973. *The form and structure of the Tertiary dyke swarms of Arran.* Ph.D. thesis, University of London.
Koch, L. and Haller, J. 1971. Geological map of East Greenland. *Medd. Grønland.* 183.

LAMACRAFT, H. 1977. *The geochemistry of the Mull, Islay and Jura dyke swarms.* Ph.D. thesis, University of London.
MACINTYRE, R. M., MCMENAMIN, T. and PRESTON, J. 1975. K-Ar results from western Ireland and their bearing on the timing and siting of Thulean magmatism. *Scott. J. Geol.* **11,** 227–249.
MCKENZIE, D. and WEISS, N. 1975. Speculations on the thermal and tectonic history of the Earth. *Geophys. J. R. astron. Soc.* **42,** 131–174.
MCQUILLIN, R., BACON, M. and BINNS, P. E. 1975. The Blackstones Tertiary igneous complex. *Scott. J. Geol.* **11,** 179–192.
MATTEY, D. P., GIBSON, I. L., MARRINER, G. F. and THOMPSON, R. N. 1977. The spatial geochemistry of Lower Tertiary low-alkali tholeiite dykes from the regional swarms of the Isle of Skye, NW Scotland. *Mineral. Mag.* **41,** 273–286.
NOE-NYGAARD, A. 1974. Cenozoic to recent volcanism in and around the North Atlantic basin. *In* Nairn, A.E.M. and Stehli, F.G. (Editors). *The ocean basins and margins 2 The North Atlantic.* 391-443. Plenum Press, New York.
PARKER, J. R. 1975. Lower Tertiary sand development in the Central North Sea. *In* Woodland, A. W. (Editor). *Petroleum and the Continental Shelf of north-west Europe.* 447–453. Applied Science Publishers, Barking.
PATERSON, E. M. 1954. Notes on the Tectonics of Greenock-Largs Uplands and Cumbraes. *Trans. geol. Soc. Glasg.* **21,** 430–435.
POLLARD, D. D. 1973. Derivation and evaluation of a mechanical model for sheet intrusion. *Tectonophysics* **19,** 233–269.
RAMBERG, H. 1967. *Gravity, deformation and the Earth's Crust studied by centrifugal models.* Academic Press. New York.
RICHARDSON, S. W. 1975. Heat Flow Processes. In *Geodynamics Today.* R. Soc. Lond. spec. pub., 123–131.
RICHEY, J. E. 1932. Tertiary ring structures in Great Britain. *Trans. geol. Soc. Glasgow* **19,** 42–140.
—— 1939. The dykes of Scotland. *Trans. geol. Soc. Edinb.* **13,** 393–435.
——, MACGREGOR, A. G. and ANDERSON, F. W. 1961. *British Regional Geology, Scotland; the Tertiary volcanic districts.* (3rd ed.) H.M. Geol. Surv.
ROY, F. R., BLACKWELL, D. D. and DECKER, E. R. 1972. Continental heat flow. *In* Robertson, E. C. (Editor). *The nature of the solid Earth.* 504–543. McGraw-Hill, New York.
RUSSELL, M. J. 1972. North-south geofractures in Scotland and Ireland. *Scott. J. Geol.* **8,** 75–84.
SKINNER, B. J. 1966. *In* Clark, S. P. (Editor). Handbook of Physical constants. *Geol. Soc. Am. Mem.* **97,** 75–96.
SLOAN, T. 1971. *The structure of the Mull Tertiary dyke swarm.* Ph.D. thesis. University of London.
SPEIGHT, J. M. 1972. *The form and structure of the Tertiary dyke swarms of Skye and Ardnamurchan.* Ph.D. thesis, University of London.
STEWART, F. H. 1965. Tertiary igneous activity. *In* Craig, G. Y. (Editor). *The Geology of Scotland.* 417–465. Oliver & Boyd, Edinburgh and London.
STUBBLEFIELD, C. J. 1964. *1:633,600 geological map of Great Britain.* Geological Survey of Great Britain, London.
THOMPSON, R. N. 1974. Primary basalts and magma genesis. *Contr. Mineral. Petrol.* **45,** 317–341.
——, ESSON, J. and DUNHAM, A. C. 1972. Major chemical variations in the Eocene lavas of the Isle of Skye, Scotland. *J. Petrol.* **13,** 219–353.
TYRRELL, G. W. 1928. The Geology of Arran. *Mem. geol. Surv. U.K.*
—— 1949. The Tertiary igneous geology of Scotland in relation to Iceland and Greenland. *Meddr. dansk. geol. Foren.* **2,** 413–440.
VOGT, P. R. 1974. Volcano height and plate thickness. *Earth Planet. Sci. Lett.* **23,** 337–348.
—— and JOHNSON, G. L. 1975. Transform faults and longitudinal flow below the midoceanic ridge. *J. geophys. Res.* **80,** 1399–1428.
WALKER, G. P. L. 1975. A new concept in the evolution of the British Tertiary centres. *J. geol. Soc. Lond.* **131,** 477–498.
WATTERSON, J. 1975. Mechanism for the persistence of tectonic lineaments. *Nature, Lond.* **253,** 520–522.
ZIEGLER, P. A. 1975. North Sea Basin History in the Tectonic Framework of North-Western Europe. *In* Woodland, A. W. (Editor). *Petroleum and the Continental Shelf of north-west Europe.* 131-149. Applied Science Publishers, Barking.
ZIENKIEWICZ, O. C. and CHEUNG, Y. K. 1967. *The Finite Element Method in structural and continuum mechanics.* McGraw-Hill, London.
——, VALLIPAN, S. and KING, I. R. 1968. Stress analysis of rock as a 'non-tension' material. *Geotecthnique* **18,** 56–66.

I. R. Vann,
B.P. Petroleum Development Ltd.,
Dyce,
Aberdeen, AB2 ABP, Scotland.
(Formerly University of Glasgow.)

# *Part 5*

# Present day processes

# Fissure swarms and Central volcanoes of the neovolcanic zones of Iceland

*K. Saemundsson*

The large scale structure of the neovolcanic zones in Iceland and the nearest offshore segments joining them to the mid-ocean ridges is dominated by fissure swarms having variable trends. They are characteristically about 10 km wide, between 50–100 km long and occur in both dextral and sinistral en echelon arrays. The structural diversity of the swarms is related to their position with regard to the locus of highest lava production in the central part of Iceland and also with regard to the zone of active plate accretion (axial rifting zones). Cone shaped stratovolcanoes are located in so-called lateral rift zones well outside the zone of active plate accretion. Within the axial rift zones shield-like Central volcanoes form with siliceous rocks, calderas and high temperature geothermal fields. The Central volcanoes are an integral part of the Fissure swarms especially towards the middle of Iceland where lava production is greatest. In the extreme southwest and northeast of Iceland nearest to the submerged mid-ocean ridges, siliceous rocks vanish and Central volcanoes become less conspicuous as lava production diminishes. The axial rift zones are shown to be under tensional stress acting parallel with the direction of spreading, whereas the lateral rift zones are characterized by horizontal shear stress. The episodic character of rifting and volcanism is discussed in the light of a current rifting episode in northern Iceland. Here an interplay between magmatic processes in a Central volcano and rifting in the associated Fissure swarm has been observed. Some large scale features of the neovolcanic zones are explained as a result of mantle plume activity.

## 1. Introduction

This account deals with some of the main features of the geology of Iceland particularly the neovolcanic (or young volcanic) zones. The structure of the volcanic pile is reviewed. It was generated by the same processes as are operating in the neovolcanic zones of today (Bödvarsson and Walker 1964). Erosion has dissected the older parts of the lava pile so that it is possible there to view the structures three dimensionally and trace the development of the pile with time. The neovolcanic zones of Iceland are made up of discrete Fissure swarms and

Central volcanoes (Saemundsson 1971; Tryggvason 1973; Saemundsson 1974; Walker 1975; Jakobsson *et al.* 1978). The Fissure swarms are regarded as the surface expression of dyke swarms many of which have been mapped in the older parts of the lava pile (Walker 1974). I propose that a distinction should be made between axial and lateral rift zones. The former mark the boundary of active plate accretion continuing the trace of the mid-Atlantic ridge axis through Iceland, whereas the latter are located well off this axis in areas underlain by older crust. As yet little is known about the mechanism of rifting in Iceland, but recent events in the axial rift zone of northern Iceland are beginning to throw some light on this question (Björnsson *et al.* 1977). Crustal stresses deduced from fracture patterns indicate different stress regimes in the lateral and axial rift zones.

## 2. The volcanic pile of Iceland

The exposed volcanic pile of Iceland reveals up to 1500 m of vertical sections, that is about a third of the average depth to seismic layer 3 which possibly constitutes the base of the extrusives (Palmason 1971; Gibson and Piper 1972; Fridleifsson 1977). Layer 3 in Iceland is considered as equivalent to layer 3 of the oceanic crust which implies that the Icelandic crust is oceanic in character rather than continental. Much of our basic knowledge of the structure of the volcanic pile of Iceland is due to the researches of G.P.L. Walker in eastern Iceland carried out in the years 1955–1965, summarized in Walker (1974). The exposed volcanic pile is built predominantly of tholeiitic lavas and subordinate sediment of volcanic origin. Alternating sequences of olivine tholeiite and tholeiitic lavas occur which are traceable as stratigraphical units over long distances (Walker 1957, 1974). The same rock types occur in the neovolcanic zones as fairly distinct stratigraphical successions indicating a cause of deep origin which is not yet satisfactorily explained. Local concentrations of acid and intermediate lavas mark the sites of Central volcanoes (Walker 1963) which are also characterized by copious basaltic volcanism.

Dykes cutting nearly at right angles through the lava pile tend to occur in swarms (Walker 1960). A systematic relationship exists where dyke swarms pass through the Central volcanoes (Walker 1963), but dyke swarms also exist where this relationship is unclear. The dyke swarms are up to 40 km long and commonly about 5–10 km broad and thus approach closely the dimensions of Fissure swarms of the neovolcanic zones to be described later.

In the Central volcanoes irregular sheets and sometimes minor gabbro or granophyre intrusions add to the vertical dykes causing a much greater proportion of intrusive rocks in their roots (Walker 1963). The raised geothermal gradient caused by the intrusions gave rise to temporary hydrothermal convection cells which altered the core of the Central volcanoes to propylitized rock far above the zeolite facies metamorphism of the deeper levels of the volcanic pile elsewhere.

The dyke trend changes from dominantly NE–SW in the southern part of Iceland to nearly N–S in the northern part (Fig. 7), just as do the volcanic and nonvolcanic fissures of the neovolcanic zones of the present day. This has not yet been interpreted satisfactorily but obviously some general principle is involved.

The lavas of the pile dip gently on a regional scale. It has been shown that the dips increase gradually from near zero at the highest exposed levels of the pile to about 5–10° at sea level (Walker 1974). Also the stratigraphical units of the pile generally thicken when traced down the direction of dip. This implies that the pile tilted gradually as it grew thicker. The tilt of the lavas has been related to the growth of the lava pile in a stationary volcanic zone where lateral migration and subsidence occurred as a result of crustal extension and loading due to the growing thickness of the pile (Bödvarsson and Walker 1963; Palmason 1973; Daiginiere *et al.* 1975). The law of superposition predicts that the oldest exposed rocks in Iceland should occur in the furthest northwest, north and east. Radiometric dating of the lowest exposed levels in the east indicate that the oldest rocks in that area are just over 13 m.y. (Ross and Musset 1976). As yet ages from the northwest and north are very fragmentary with a cluster around 16 m.y. for a deep stratigraphical level in the northwest (Moorbath *et al.* 1968). Ages from the north of Iceland indicate that the oldest part of the lava pile there may be well over 10 m.y. (Aronson and Saemundsson 1975). The ages appear unexpectedly low if compared to the age of the sea bottom from magnetic anomaly identification (Fleischer 1974). The explanation may lie in the very significant overlap of subaerial lavas, the ages being derived from the uppermost 1000 m of the pile below which another 3–4 km of lavas at least must be expected. Mean lava extrusion rates have been estimated for many kilometre thick lava sequences in eastern and western Iceland (McDougall *et al.* 1976, 1977), as one lava per 11,000–13,000 years. This corresponds to an average accumulation rate of about 700 m/m.y. The sections upon which these estimates are established avoid Central volcanoes where higher extrusion rates occur during their active periods. The local thickening around the Central volcanoes is compensated by breaks at the ends of their active periods as shown by lavas from new sources in the rift zone banking up against them.

Structural relationships indicate that the lava pile grew as lenticular units along dyke swarms (Gibson and Piper 1972), the major ones passing through Central volcanoes, where these units attain their greatest thickness. The Central volcanoes are preserved as entities in the volcanic pile, indicating that they grew, moved off towards the margin of the current volcanic zone together with the dyke or Fissures swarms and then became extinct. New ones replace them over the more or less stationary deep-seated zone of magma generation. Despite their height of several hundred metres above their surroundings, the Central volcanoes normally become buried under younger lavas which are seen to pile up against them from new sources situated in the direction of dip. Sometimes a period of erosion and considerable sediment accumulation seems to have followed the extinction of the Central volcanoes and Fissure swarms. Most of the plant-bearing sediment horizons in western Iceland that are traceable over dozens of kilometres along strike appear to have formed in this way. The intervals of sediment accumulation were of the order of perhaps 300,000 years (McDougall *et al.* 1977) compared with a lifetime of 300,000–500,000 years for the Central volcanoes themselves (Saemundsson and Noll 1974; Fridleifsson 1973).

The basalt plateau is heavily denuded and dissected by glacial erosion which becomes negligible on approaching the neovolcanic zones where the pile is still in the process of growth. Headward valley erosion towards the neovolcanic zones is retarded because lavas from there are channelled into the valleys (Saemundsson and Noll 1974), thus the dissection seldom reveals stratigraphically continuous rock formations that date from the uppermost Matuyama and Brunhes polarity epochs.

The detailed history of growth of the lava pile including possibly major shifts of the rift axes (Piper 1973; Saemundsson 1974) depends heavily on radiometric dating

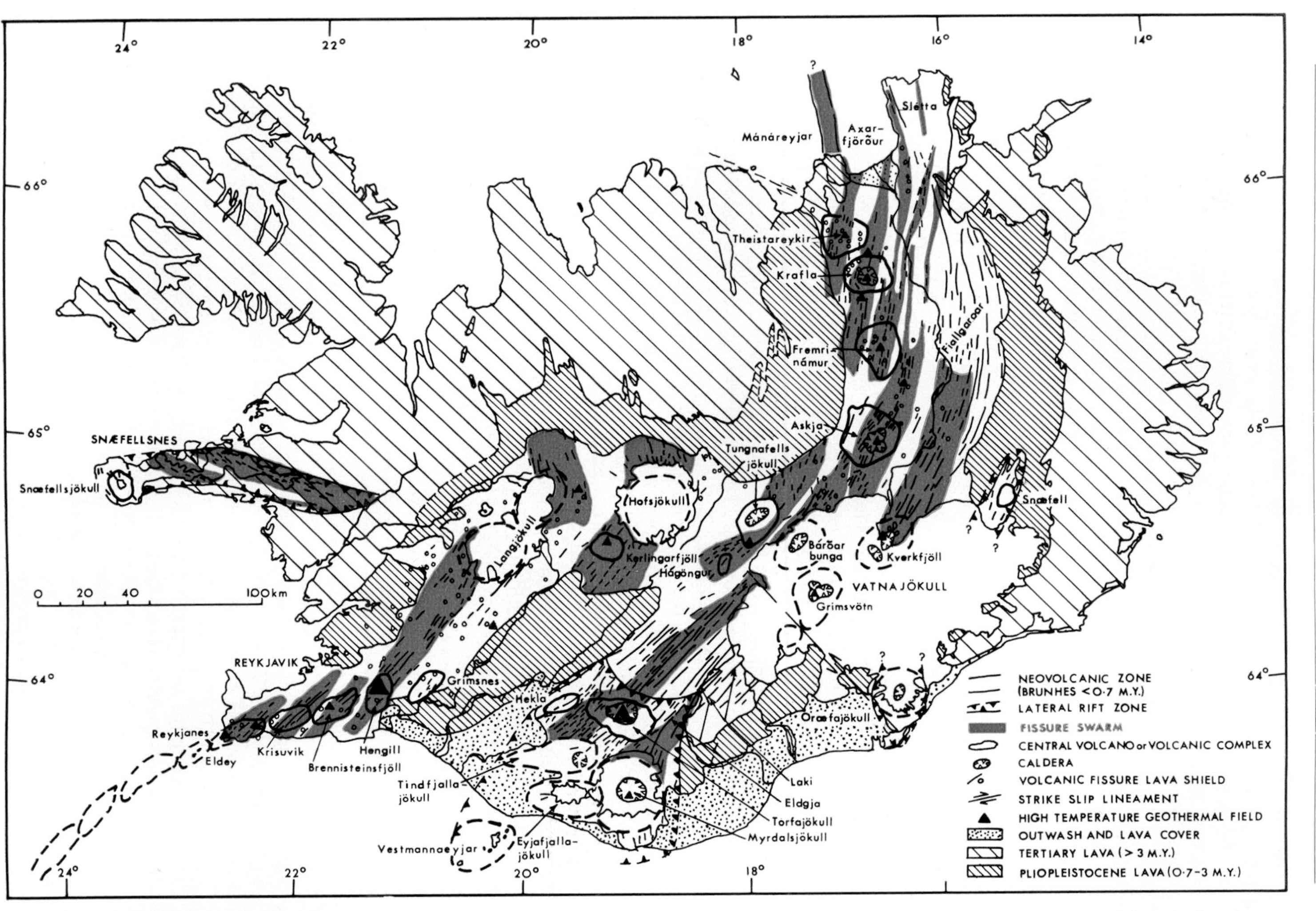


Fig. 1. Structural map of the neovolcanic zones in Iceland. The neovolcanic zones are here defined as embracing rocks erupted during the later stages of the Brunhes polarity epoch. Two dormant or extinct units (Snaefell and Fjallgardar) which have not erupted in post-glacial time are included.

of the lavas. Work of this kind is in progress but the results (Ross and Musset 1976; McDougall *et al.* 1976, 1977; Aronson and Saemundsson 1975) are still too fragmentary to allow any firm conclusions to be made. Ultimately an age map of the lava pile will show how the whole volcanic edifice grew.

## 3. The neovolcanic zones

### 3a. Axial and lateral neovolcanic rift zones

The neovolcanic zones cross Iceland in a rather complicated way with several branches. In the southwest one branch runs into the Reykjanes ridge. In the north one branch connects with the Kolbeinsey ridge along the Tjörnes fracture zone. The large scale structure of the neovolcanic zones is dominated by Fissure swarms and Central volcanoes. The trend of the Fissure swarms is variable but rather uniform within each branch of the neovolcanic zones. They typically form en echelon arrays which may be dextral or sinistral, the controlling factor being the direction of maximum tensional stress which is parallel to the direction of plate movement. The Fissure swarms are commonly about 10 km wide and their length varies from 30 to over 100 km. Their limits are most easily defined in those branches of the neovolcanic zones that strike obliquely to the direction of spreading. In those branches that strike NE–SW the limits are difficult to define and the swarms coalesce. Figure 1 shows the Fissure swarms in the neovolcanic zones. Most of them pass through Central volcanoes which are the loci of highest lava production and are also defined by the presence of acid rocks and high temperature geothermal fields. Of about 20 well defined Central volcanoes in the neovolcanic zones ten have developed calderas, a rather different picture from a widely held view that there exists only one (Williams and McBirney 1968).

Jakobsson (1972) found that the neovolcanic zones in Iceland emitted during postglacial time either tholeiitic or transitional to alkalic lavas, the former being erupted along the ridge crest where spreading is active and the latter erupted on the flanks of the ridge where spreading is negligible. Sigurdsson (1970) came to similar conclusions about the Snaefellsnes volcanic zone which he viewed as a transcurrent fault zone contrasting with the central tholeiitic zones of rifting and volcanism.

The neovolcanic zones can be divided into two main types. One type has branches that mark the trace of the plate boundary where active plate growth is taking place. These are *the axial rift zones* (Saemundsson 1974; Schäfer 1975) and are defined by very pronounced linear tensional tectonics. They are flanked by volcanic piles which dip and become progressively younger towards them, without major time gaps in the strata. The depth to seismic layer 3 is in the range 2·5–4·5 km (Palmason 1971). It should be noted that the axial rift zone of Walker (1975) refers to the smaller scale feature of a Fissure swarm. The flank zones of Jakobsson are the second type—*the lateral rift zones*. They include branches of the neovolcanic zones which are distinctly unconformable upon older piles of volcanics that have sometimes suffered erosion before the younger volcanism set in. They have poorly developed features of tensional tectonics. The depth to seismic layer 3 is greater than normal in southern Iceland (Palmason 1971) which is an area where large volcanoes occur. The heat flow is lower in the lateral rift zones than in the axial rift zones (Palmason 1973). A further important feature is the position of the lateral rift zones within regions under horizontal shear stresses as discussed later.

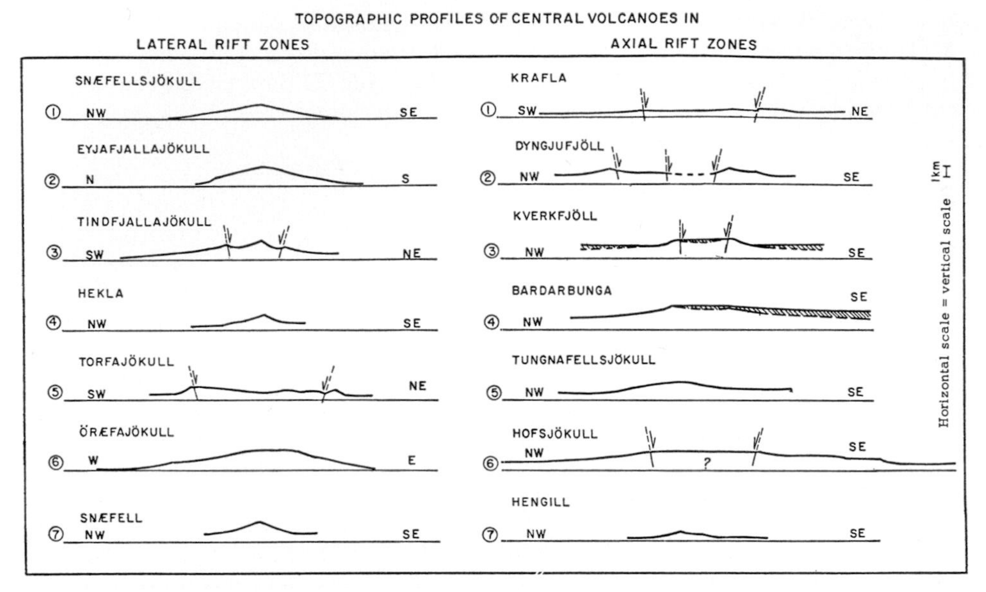


Fig. 2. Topographic profiles of Icelandic central volcanoes. The base line is in each case sea level. Ice is indicated by hachuring.

In Figure 1 the Snaefellsnes volcanic zone and the southern part of the eastern volcanic zone represent the lateral rift zones. The stratovolcanoes, Öraefajökull and Snaefell, also fall into this second type mainly because they appear superimposed on an older eroded infrastructure.

The Central volcanoes of the axial rift zones commonly differ in shape and height above their surroundings from those of the lateral rift zones (Fig. 2) which tend to be cone shaped when viewed perpendicular to the fissure trend (the cone often is deformed to a ridge parallel to the fissure trend). And they often have a large summit crater (Fig. 4). The Central volcanoes of the axial rift zone have lower relief and a well defined central crater is not present (Fig. 3). This difference may be obscured where subglacial volcanism has predominated such as in the Vatnajökull region. The main reason for the difference is probably tectonic, with large scale subsidence of elongated strips of country being general along Fissure swarms of the axial rift zones, but being negligible along the swarms of the lateral rift zones. This may well cause the subdued topographic expression of the Central volcanoes of the axial rift zones. However, it is likely that the mode of eruption and physical properties of the magma erupted also plays a role. The Central volcanoes of the lateral rift zones have a much higher proportion of siliceous rocks (Jakobsson 1972) and hence also of pyroclastics among their eruption products; in fact almost all the pyroclastic siliceous tephra produced in postglacial time in Iceland originated there. Our knowledge of the history and construction of these volcanoes is poor except for Hekla (Thorarinsson 1967) but they are stratovolcanoes.

The Central volcanoes of the axial rift zones mainly erupt highly fluid tholeiite and sometimes also olivine tholeiite lavas that form the gently sloping flanks of a broad, low dome. The two types of rock appear to correlate well with the eruptive fissures and lava shields which are distinctive in the morphology and lithology of

Fig. 3. View towards south along the Krafla fissure swarm. Note the undisturbed borders on either side of the 6–7 km broad Gjástykki graben zone in the foreground. The Krafla composite shield volcano is conspicuous as a broad elevation enclosing the flat caldera bottom (marked) and traversed by the fissure swarm. The branched and sinuous character of the faults is clear from the picture and also the tendency to split into en echelon segments. (Photo by Oddur Sigurdsson, Feb. 1977).

Fig. 4. View of the Eyjafjallajökull stratovolcano (1617 m, rising from sea level) from the north-west. The E–W volcanic axis and the large summit crater, $2\frac{1}{2}$ km wide, are prominent. A fissure swarm displaying normal faults and tension cracks similar to the Krafla swarm of Fig. 3 is not present. (Photo by Bessi Adalsteinsson, Jan. 1977).

their lavas (Walker 1971, 1974). Repeated eruptions of the two types together with occasional more siliceous rocks give rise to a composite volcano that is properly referred to as a shield volcano and should not be mixed up with a monogenic lava shield as pointed out by Walker (1971). Abundant subglacial volcanism often has caused a distortion of this shape by producing piles of hyaloclastite at the eruptive sites instead of lava flows that would add to the flanks.

The total volume of lava erupted within the neovolcanic zones during postglacial time has been estimated as 400–500 km$^3$ (Thorarinsson 1965; Jakobsson 1972). Of this 85% was erupted within the axial rift zones and only 15% in the lateral rift zones (Jakobsson 1972).

## 3b. Structure of the lateral rift zones

Sigurdsson (1970) recognised three WNW–ESE trending lines of Pleistocene to Recent volcanism within the E–W oriented Snaefellsnes volcanic zone. The westernmost of those includes the Snaefellsjökull stratovolcano. Stratovolcanoes have not developed on the two eastern lines although acidic rocks are present very locally where magma eruption is greatest on the Fissure swarms. The southernmost part of the eastern volcanic zone which is considered here as a lateral rift zone is characterised by a dense cluster of large stratovolcanoes. Hekla, Tindfjallajökull and Eyjafjallajökull grew along ENE–WSW and east–west oriented volcanic axes whereas Myrdalsjökull to the east of them has N–S oriented tectonic features. The island group of Vestmannaeyjar off the south coast forms a separate volcanic field focussed about Heimaey which last erupted in 1973. Hekla and Torfajökull nearest to the intersection with the axial rift zone have the most prominent Fissure swarms among these volcanoes. The main difference in the tectonics as compared to the axial rift zone to the north is that large scale tensional faulting and graben formation is inconspicuous until well north of these volcanoes.

The rate of volcanic production is variable in the lateral rift zones, being lowest in the eastern part of Snaefellsnes. It is highest in Mýrdalsjökull which is estimated to have produced 30–35 km$^3$ of lava during postglacial time (Thorarinsson 1975). The volcanic production is not balanced by subsidence along the Fissure swarms as in the case of the axial rift zones, hence the marked topographic expression of the stratovolcanoes. Instead of subsidence and lateral drift within elongated fissure swarms, a regional thickening of seismic layer 2 indicates subsidence on a regional scale to compensate the extra load. The life span of stratovolcanoes in the lateral rift zones is little known but they are a fairly recent feature of Icelandic geology, being absent from the volcanic pile except for one example in Snaefellsnes (Sigurdsson 1970), which is of lower Pleistocene age (Moorbath *et al.* 1968). Possibly the development of the lateral rift zones is related to the large scale shifts of the axial rift zones proposed to have taken place geologically recently (Saemundsson 1974; Aronson and Saemundsson 1975).

## 3c. Structure of the axial rift zones

Volcanism along the axial rift zones varies with respect to volcanic production and volcanic structures along the general SW–NE trend across Iceland. There is a pronounced culmination of volcanic activity in a transverse E–W zone in central Iceland which has produced clusters of Central volcanoes forming the socle of the great ice caps of that region. Towards the southwest and northeast a gradual lowering of the topographic relief correlates with a decline in volcanic production and the disappearance of siliceous rocks and Central volcanoes. Walker (1975) has

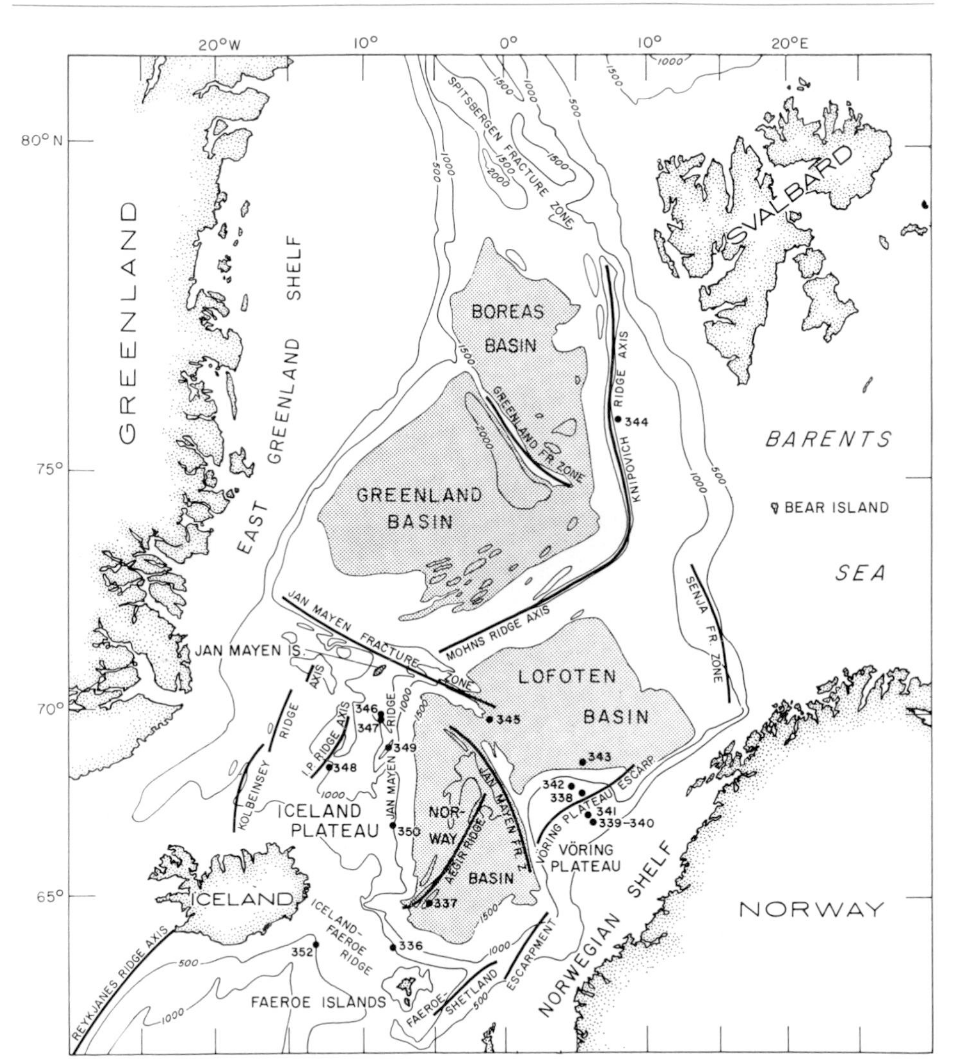


Fig. 2. Index map of Norwegian–Greenland Sea showing main geographical features and location of D.S.D.P. sites (modified from Talwani *et al.* 1976). Bottom contours in nominal fathoms (sound velocity 800 fms/sec.).

northwest Europe (Figs 1, 2). The western boundary is formed by active mid-ocean spreading ridges; so that in the North Atlantic Ocean north of the Charlie Gibbs fracture zone and south of Iceland, the Reykjanes ridge forms the boundary. In the Norwegian–Greenland Sea, the Kolbeinsey ridge forms the boundary between Iceland and the Jan Mayen fracture zone, and Mohns ridge and Knipovich ridge the boundary north of the Jan Mayen fracture zone (Figs 1, 2).

Coverage of oceanic sedimentation in this paper extends as far south as the Charlie Gibbs fracture zone, and as far north as the southwest-facing continental slope that forms the continental margin of the Barents Sea (Figs 1, 2). The oceanic region is divided into separate deep marine basins by NW-trending transverse structures such as the fracture zones and the Icelandic transverse ridge, which consists of the Iceland–Faeroe ridge and Iceland. Small continental plates such as the Jan Mayen ridge and Rockall plateau, which were detached from the surrounding continents by early phases of rifting and subsequently isolated by westward shifting of the locus of active spreading, also divide the oceanic region into separate deep marine basins.

The E—W-trending Charlie Gibbs fracture zone is a transform fault separating two parts of the North Atlantic Ocean that underwent different histories of opening (Fig. 1). The escarpment marking the fracture zone forms a partial barrier to southward flowing sediment gravity flows.

The NW-trending aseismic Icelandic transverse ridge (Bott 1974, 1975b) extends from the Faeroe Islands to the southeastern continental margin of Greenland and forms a prominent barrier to the movement of marine water and sediment (Fig. 1). It separates ocean basin areas of the northern North Atlantic Ocean from those of the Norwegian–Greenland Sea and probably developed in the early Cenzoic, concurrently with sea-floor spreading, as an elongate volcanic plateau originally underlain by a mantle diapir or hot spot (Schilling 1976). Iceland is thought to have developed over a mantle plume or hot spot (Vogt 1972; Schilling 1973) during the past 20 m.y.; its oldest exposed rocks have been dated as $16{\cdot}0 \pm 0{\cdot}3$ m.y. (Moorbath *et al.* 1968).

The NW-trending Jan Mayen fracture zone is a transform fault marked by a prominent escarpment that extends from the eastern continental margin of Greenland to the southern edge of the Vøring plateau (Fig. 2). The escarpment forms a partial barrier to movement of water and sediment, and separates two ocean basins, the Lofoten basin to the north and Norway basin to the south. The prominent active volcanic island of Jan Mayen is located directly south of the fracture zone, near its intersection with Jan Mayen ridge.

The NE-trending continental fragment underlying Rockall plateau divides the Ocean south of the Icelandic transverse ridge into two separate deep marine basins (Fig. 1). The elongate Rockall trough to the east is underlain by a long-inactive spreading ridge and has received thick sediments derived mainly from continental source areas to the east and northeast. The elongate Iceland basin west of the Rockall plateau contains considerably less sediment cover because the Rockall plateau apparently never formed a major source terrane. Both the Rockall trough and the Iceland basin have been partly filled by sediment transported southward and longitudinally down the axes of the two basins by cold Norwegian Sea bottom water (Fig. 1).

In the southern part of the Norwegian–Greenland Sea, the elongate Jan Mayen ridge continental microplate separates the area east of the active Kolbeinsey ridge into two separate ocean basins (Fig. 2). The easternmost basin, the Norway basin, is underlain by the now inactive Aegir spreading ridge and contains a relatively thin accumulation of sediments. The western basin also is underlain by an inactive spreading ridge, the Icelandic plateau ridge (Fig. 2). The plateau does not contain a thick accumulation of sediments because of its elevation and because land source areas are distant.

The Lofoten basin is relatively large and is unbroken by fracture zones or small continental plates (Fig. 2). It contains a thick accumulation of sediment derived mainly from eastern continental source areas.

The general history of sea-floor spreading and opening of the North Atlantic and

Norwegian–Greenland Sea regions have been summarized by Le Pichon and Fox (1971); Pitman and Talwani (1972); Bott (1973, 1975a, b); Pitman and Herron (1974); Vogt and Avery (1974); Roberts (1975a, b); Bailey (1975); Laughton (1975); Hallam and Sellwood (1976) and Talwani and Udintsev (1976). The geology and history of sedimentation on the continental shelf adjacent to Britain and Norway have been summarized in Woodland (1975); Pegrum *et al.* (1975) and Whiteman *et al.* (1975). Early studies of deep marine sedimentation from analyses of piston cores are summarized by Ericson *et al.* (1961, 1964). The palaeo-oceanographic and palaeobiogeographic evolution of the North Atlantic and Norwegian–Greenland Sea areas were respectively summarized by Berggren and Hollister (1974) and Schrader *et al.* (1976). Leg 12 of the D.S.D.P. drilled holes in the Northeast Atlantic (Laughton *et al.* 1972) and Leg 38 in the Norwegian–Greenland Sea (Talwani *et al.* 1976). The locations of these drilling sites, plus more recently drilled sites on Leg 48 (Montadert *et al.* 1976), are shown on Figures 1 and 2. The characteristics of sediments in cores from the Norwegian–Greenland Sea are summarized by White (1978) and Nilsen and Kerr (1978a), and those from the Northeast Atlantic by Davies and Laughton (1972).

## 2. General characteristics of deep sea sedimentation

Deep sea sediments are deposited by a variety of processes, including (1) sediment gravity flows derived from basin margins, (2) vertical settling of biogenic pelagic detritus, (3) introduction of detritus by volcanic eruptions, ice-rafting, cosmic dust, and winds and (4) transportation, resuspension, and redeposition of sediment by bottom currents. The sedimentary record of the region considered contains sediments deposited by most of these processes. Deep sea sediments reported in this study have been examined through cores, especially those obtained by the D.S.D.P., shallow piston cores and box cores and seismic reflection profiles.

The ocean basins have been largely filled by hemipelagic and gravity-flow deposits derived from the Norwegian, Svalbardian, British and Irish continental margins and from the margins of Iceland. Major westward-prograding deep sea fans are present west of the continental margins of Britain, Norway, and the Barents Sea. Deposition of biogenic pelagic sediments was widespread at various intervals, particularly where the sea temperatures were warmer and where the continental margins were submerged. Such submergence of shelves by transgression commonly results in trapping of coarse grained sediment on the shelves, inhibiting transportation and deposition of terrigenous detritus in the deeper ocean basins.

Throughout the history of the region, abundant volcanic material has been contributed to the deep marine basins from volcanic activity associated with the development of the early Tertiary Brito-Arctic volcanic province, spreading ridge volcanism, and later subaerial volcanism on Iceland and Jan Mayen. In the northern part extensive ice-rafted glaciomarine sediments were deposited during the Pliocene and Pleistocene on the ocean floors.

Southward movement of cold bottom water from the Norwegian Sea through Faeroe Bank channel (Crease 1965; Worthington 1969; Worthington and Volkmann 1969; Jones *et al.* 1970) to the Iceland basin (Steele *et al.* 1962; Dietrich 1967) and across the Wyville–Thomson ridge to the Rockall trough (Ellett and Roberts 1973) has resulted in extensive deposition of sediment drifts in both areas. Deposition of volcanogenic turbidites via the Maury channel (Cherkis *et al.* 1973), a major deep sea channel not associated with a deep sea fan complex, has taken place west of the Rockall plateau.

In rifted-margin ocean basins such as the northern North Atlantic Ocean and Norwegian–Greenland Sea, sedimentation is generally most strongly influenced by the effects of continued subsidence of older ridge-formed ocean floor away from the active spreading ridge, and by the progressive increase in distance of the active spreading ridge from the dominant continental margin sediment source areas, as new oceanic crust forming at the spreading ridge adds to the overall width of the ocean basin. Consequently stratigraphical sequences in such ocean basins should, in general, record a progressive deepening of ridge-formed depositional sites through time, particularly those sites having thick sedimentary covers deposited over a long period of time on older oceanic crust. The stratigraphical sequences deposited on younger oceanic crust are generally thinner and include sediments deposited over shorter periods of time. These sediments are typically finer grained and less terrigenous in character, because at the time of their deposition, the continental margins or sediment source areas were already quite distant.

However, the present region is not formed of simple rifted-margin ocean basins (Figs 1, 2). It is complicated by historical and geometrical aspects including (1) dominance of longitudinal transport and deposition in the Iceland basin and Rockall trough of sediments derived from the Norwegian Sea overflow water rather than sediments derived laterally from continental margins; (2) westward shifting of the main spreading ridge in the Atlantic from an extinct ridge beneath the Rockall trough to the presently active Reykjanes ridge, and in the southern part of the Norwegian–Greenland Sea from the extinct Aegir ridge to the extinct Icelandic plateau ridge north of Iceland and west of the Jan Mayen ridge, and finally to the presently active Kolbeinsey ridge; (3) large volcanic plateaus, probably developed over mantle diapirs or hot spots strongly influenced water circulation and sedimentation patterns, such as the Iceland–Faeroe ridge, Iceland, and the Vøring plateau; (4) continental fragments such as the Jan Mayen ridge and Rockall plateau, detached by westward shifting of the location of the active spreading ridge, subdivide the ocean basins into several sub-basins; (5) extensive fracture zones such as those of Charlie Gibbs and Jan Mayen that form prominent escarpments trending more or less perpendicular to the continental margins and which have probably strongly influenced sedimentation patterns, and (6) the direction of plate motion and location of the pole of rotation have changed during the course of the rifting apart first of Greenland–Europe from North America and then of Greenland–North America from Europe, causing a more complicated juxtaposition of oceanic geographical and tectonic elements.

D.S.D.P. sites located on abyssal plains and continental rises of sedimentary basins generally provide the most complete records of terrigenous sediment gravity-flow sedimentation. Ridges in ocean basins may also preserve complete sedimentary records, but are more commonly thinner and contain dominantly pelagic or hemipelagic sediments. In addition, ridges commonly contain unconformities in their sedimentary records. In the Norwegian Sea, only Site 345 in the southwest Lofoten basin provides a fairly complete and continuous record of upward changes from non-marine to turbidite to hemipelagic to biogenic to glaciomarine sedimentation. In the Northeast Atlantic Ocean, Site 115 was drilled into volcanogenic turbidites associated with the Maury channel, and Site 114 into sediment drifts of the Gardar ridge on the east flank of the Reykjanes ridge.

Above a probable non-marine basalt at Site 345 is a reddish weathered basaltic

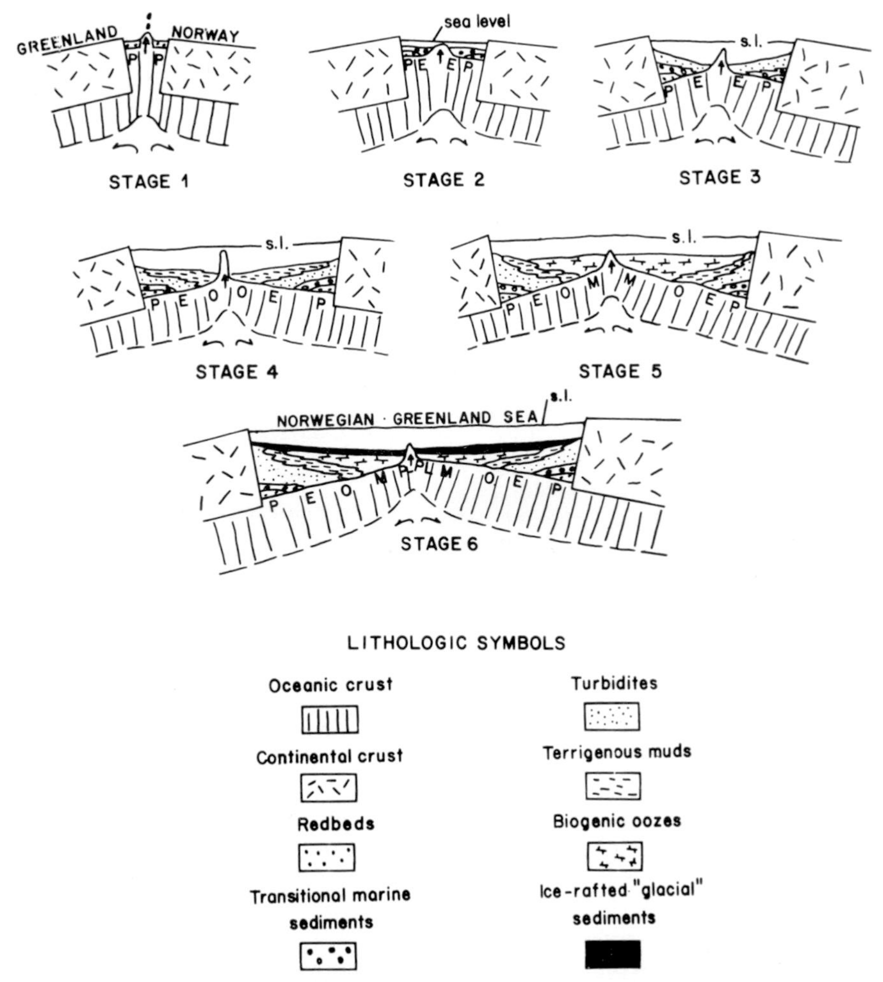


Fig. 3. Diagrammatic sketches showing sequential development of the Norwegian–Greenland Sea area and stratigraphical relationships. Stage 1, Palaeocene, initial rifting, deposition of redbeds on subaerially weathered basalts. Stage 2, early Eocene, incursion of oceanic waters, subsidence, deposition of transitional marine sediments. Stage 3, late Eocene, deposition of turbidites. Stage 4, Oligocene, deposition of terrigenous mudstone in centre of basin. Stage 5, Miocene, deposition of biogenic ooze in centre of basin. Stage 6, Pliocene and Pleistocene, deposition of glaciomarine sediments ice-rafted blanket. Symbols in oceanic crust indicate age of crust: P, Palaeocene; E, Eocene; O, Oligocene; M, Miocene; P–PL, Pliocene and Pleistocene.

breccia and a thin sequence of shallow marine sandstone, conglomerate, and sandy mudstone of early Oligocene age that records a rapid transition from non-marine to deep marine sedimentation. A thick turbidite sequence that, in general, thins and fines upward overlies these transitional sediments and is in turn overlain by Oligocene terrigenous mudstone. Stratigraphically overlying the mudstone are Miocene biogenic sediments and ice-rafted Pliocene and Pleistocene deposits. Thus, the vertical sequence at Site 345 fits the general model of rifted-margin abyssal-plain oceanic sedimentation: sediments become less terrigenous and more biogenic

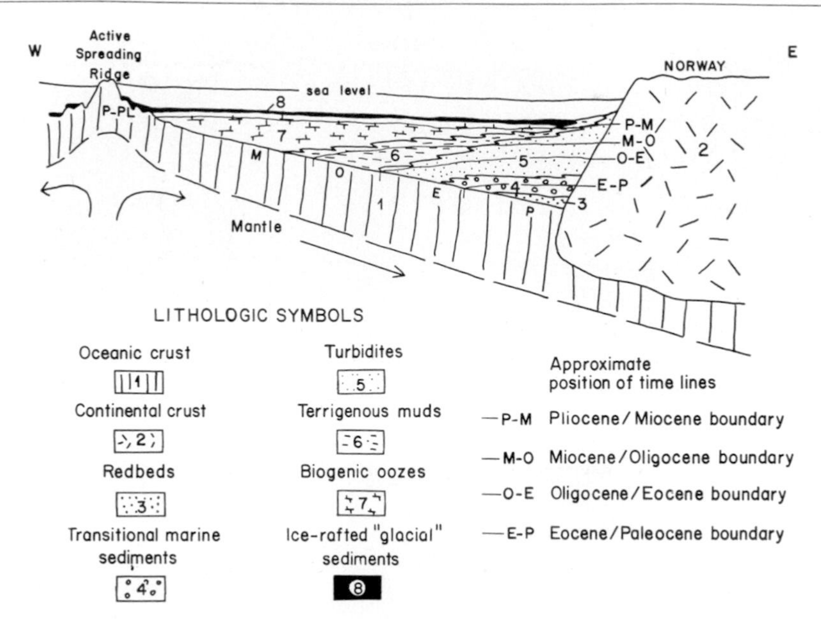


Fig. 4. Diagrammatic cross-section of Lofoten basin, showing stratigraphical relations. Symbols same as Fig. 3.

upward through time, and sediments record a progressive deepening at individual sites through time.

This model is shown diagrammatically for the northern Norwegian–Greenland Sea in Figure 3. Initial sedimentation in the rift system, which developed on and within the originally joined Greenland and Norway–Spitsbergen continental blocks, consists of subaerially extruded and weathered Palaeocene or Eocene basalt, possibly overlain locally by lateritic palaeosols. As spreading continued, the rift area widened and subsided rapidly, with incursion of marine waters. Deposition of shallow marine Eocene sediments was followed by deposition of thick proximal turbidites. Pebbly mudstone and deformed mudstone deposited by debris flows, slumps, sediment gravity flows and characterized by syndepositional folding and preconsolidation deformation, mark this initial interval of rapid opening and subsidence of the still narrow rifted basin.

With continued spreading and widening of the ocean basin, newly formed Oligocene oceanic crust received more distal fine grained turbidites, followed by terrigenous to hemipelagic mudstone derived mainly from the increasingly distant continental margins. By Miocene time, the continental margins were too far away to influence sedimentation on newly formed ocean crust, and biogenic sedimentation, mainly of diatomaceous, spiculitic, and nannofossil-rich oozes, became dominant. Sedimentation rates on new crust were increasingly slower, as the continental margin source areas were displaced farther from the newly created and subsiding sites of ocean floor sedimentation. Finally, in the Pliocene and Pleistocene, extensive ice cover resulted in the deposition of ice-rafted material over the entire ocean floor and directly on to basalt near the active spreading ridges (Fig. 4). Near the continental margin, terrigenous sediments are continuously deposited as deep sea fans and continental rise accumulations, and do not typically show the vertical changes characteristic of more distant oceanic basin sites. These deposits are derived from the nearby continent and transported mainly by bottom flowing turbidity currents and floodwater overflows during interglacial periods. However, marginal sites in areas such as the eastern Lofoten basin were not drilled, so we have little record of this continental margin sedimentation.

Intermixed sporadically through all of the sequences are volcanic ash layers derived from various sources such as the adjacent continents (e.g. volcanic centres in North Scotland and East Greenland), individual oceanic volcanic islands standing above sea level (e.g. Jan Mayen Island), and the volcanic plateaus built up by voluminous outpourings of lava (e.g. Iceland, Iceland–Faeroe ridge, and probably the Vøring plateau). The ash layers vary in composition, depending on the source and are persistent but irregularly distributed components of the Norwegian–Greenland Sea sedimentary record.

## 3. The Rockall trough and adjacent deep sea areas

The Rockall trough is a NE-trending 3000 m deep basin thought by Roberts (1975b) and Laughton (1975) to have been generated by Early to Late Cretaceous sea-floor spreading (Fig. 1). The spreading was contemporaneous with the opening of Bay of Biscay during the earliest phase of opening of the northern North Atlantic Ocean when the Greenland–North America–Rockall plate split away from Europe (Roberts 1971). It probably contains more than 2500 m of sediment (Ewing *et al.* 1973) and opens southwestward into the Northeast Atlantic Ocean. The trough is bounded on the west by a continental fragment, the Rockall plateau, on the east by the continental slope of Britain and Ireland, on the north by the transverse Wyville–Thomson ridge, and on the south by Porcupine abyssal plain east of the Charlie Gibbs fracture zone. The Porcupine Seabight, a shallower basin to the southeast of the Porcupine bank, trends approximately parallel to Rockall trough and is probably developed on thinned continental crust. To the northwest, the shallow NE-trending Hatton–Rockall basin is developed on the continental crust of the Rockall plateau between Hatton bank and Rockall bank. Roberts (1971, 1975a, b) has provided the most comprehensive summaries of the geology of these areas, based mainly on seismic reflection profiles and D.S.D.P. data from Sites 116 and 117 in the Hatton–Rockall basin and Sites 114 and 115 southwest of the Iceland basin; most of the summary presented in this and the following section is taken directly from Roberts (1971a).

Some cold Norwegian Sea water overflows a 1722 m deep gap in the Wyville–Thomson ridge and flows southward to the Rockall trough. Much transported sediment has been deposited both within the gap and in the Rockall trough. Holocene sediments are thin to absent on the Wyville–Thomson ridge but are abundant southward in the Rockall trough, where fine grained Norwegian Sea-derived sediments are mixed with coarser quartzose sediments derived from the British and Irish continental margins (La Touche and Parra 1976). Feni ridge (Fig. 5) is a variably oriented sediment drift best developed along the western margin of southern Rockall trough (Fig. 1); to the north, Feni ridge is not morphologically distinguishable, even though extensive sediment drifts are present. The ridge and its morphologically unexpressed northward extension are characterized by large scale undulations that trend parallel to the ridge axis, small scale bed roughness, non-depositional and, or, erosional channels, and an unconformable basal contact with underlying basement. Two seamounts in the deepest axial portion of Rockall trough along the eastern margin of the Feni ridge, Anton Dohrn seamount and Rosemary bank, are surrounded by well developed moats that are flanked by outer ridges (Roberts *et al.* 1973). A third non-moated seamount, that of the Hebrides terrace, is present at the base of the Scottish continental slope to the east.

The lack of canyons cut into the eastern margin of the Rockall plateau and the presence of the Feni ridge sediment drift indicates that little sediment is or has been supplied to the Rockall trough from the Rockall plateau. This contrasts markedly with the abundant canyons developed on the west-facing Scottish continental slope; submarine canyons are most prominent on the slope west of northwest Ireland, through which sediments are transported westward. Two large deep sea fans, Barra (2500 km$^2$) and Donegal (9000 km$^2$), have developed along the eastern margin of the Rockall trough. The fan-channel systems transport sediment westward and then southward along the axis of Rockall trough to the east of the Feni ridge; the fans grade laterally outward and southward to smooth-bottomed basin-plain deposits. One prominent channel follows the base of the slope before joining the axis of the Rockall trough north of the Porcupine bank.

The Porcupine Seabight contains terrigenous slope sediments that are cut by several canyons. It appears similar in origin to the Hatton–Rockall basin, and reflection profiles indicate onlapping stratigraphical relations and progressive subsidence of the Seabight (Fig. 5).

R4 is a prominent reflector present in the Atlantic west of the Rockall pleateau, in the Hatton-Rockall basin, and also in the Rockall trough. It is thought to be 37·5 m.y. old and provides useful control on the history of sedimentation in the Rockall trough (Fig. 5). It was penetrated by drilling at Sites 116 and 117. West of the Rockall plateau it pinches out on oceanic crust dated magnetically as 37 m.y. old. Sediments above the R4 reflector are cherty oozes that were probably deposited 27 m.y. ago. In the Rockall trough, the western sediment drift and eastern fan and basin-plain turbidite deposits overlie the oozes and probably indicate relatively continuous terrigeneous sedimentation in the trough from 27 m.y. ago to the present.

The pre-R4 sequence in the Rockall trough is more difficult to describe. On the Rockall plateau the pre-R4 sequence is much thinner than in the Rockall trough, and correlations with reflectors in the Rockall–Hatton basin are less certain. Reflectors X, Y, and Z in the Rockall trough have been recognized and correlated over large areas (Fig. 5a). Strata between Reflector Z and underlying oceanic basement are thought to consist of pelagic sediments draped over oceanic crust in the centre of the trough and laterally infilling deep sea fans along the eastern and western margins of the trough. Reflector Z is thought to be 100 m.y. old and to have developed after spreading had ceased, coevally with a major transgression beginning about 100 m.y. ago. Reflector Y is thought to be 76 m.y. old. The intra-Z–Y sequence is acoustically transparent and conformable on the older sediments; this sequence overlaps the Jean Charcout fault zone, indicating that faulting on the west margin of the trough had ceased prior to 76 m.y. (Fig. 5). Reflector X in the Rockall trough is thought to be about 60 m.y. old. It is unconformable on older sequences in many areas, possibly reflecting shifting in active spreading to the west of the Rockall plateau and development of the Charlie Gibbs fracture zone. The intra-R4–X sequence commonly thickens at the base of the slope along the eastern margin of the Rockall trough, suggesting that sedimentation rates were high between 60 and 37 m.y. ago.

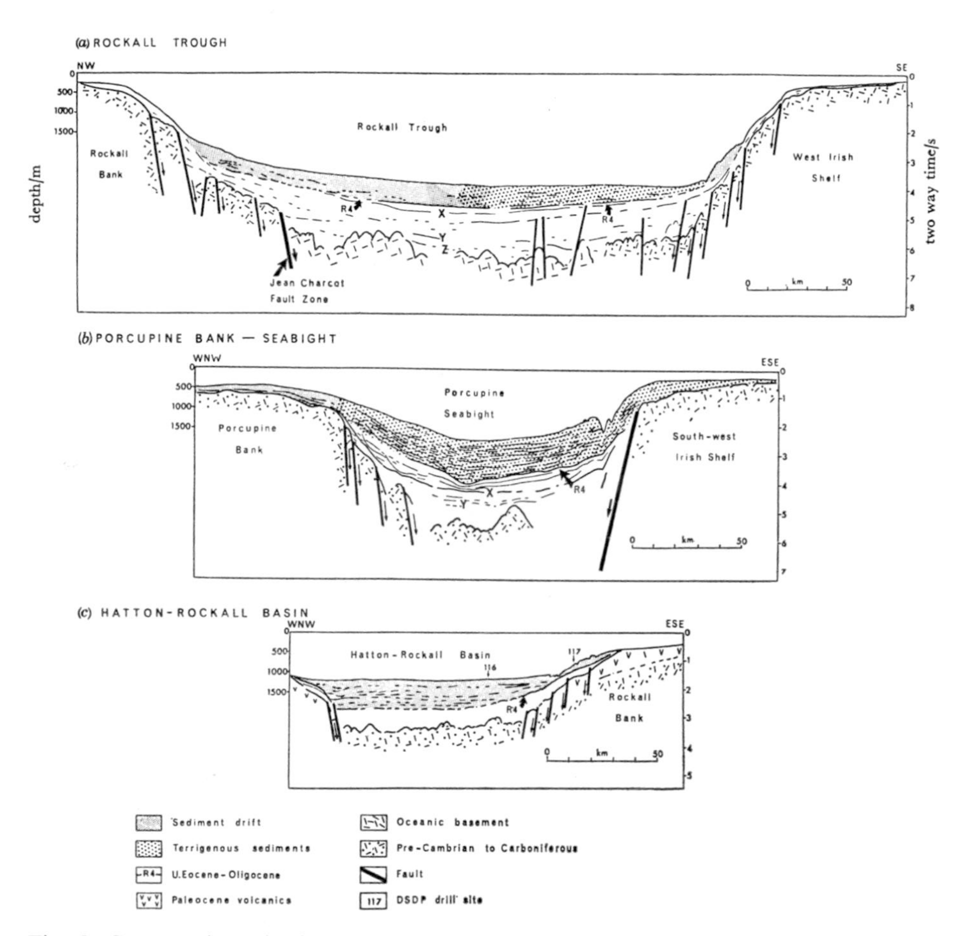


Fig. 5. Comparative seismic reflection sections across the Rockall trough, Hatton-Rockall basin, and Porcupine Seabight (from Roberts 1975a).

The R4 reflector at the base of Oligocene cherty oozes is generally conformable with older sequences. It is most prominently developed in the axial and western parts of the Rockall trough and in the Hatton–Rockall basin. In the eastern part of the trough and in Porcupine abyssal plain, the overlying cherty oozes are thin to absent, indicating continued introduction of terrigenous sediments from eastern source areas and diminished terrigenous supply from the Rockall plateau. The cherty oozes may represent deposition during an interval of somewhat closed circulation in the North Atlantic. Subsidence of the Iceland–Faeroe ridge or establishment of the Faeroe–Shetland channel and the Faeroe Bank channel overflow systems for Norwegian Sea water may have ended the interval of closed circulation and initiated deposition of sediment drifts in late Oligocene time. Deposition of sediment drifts on the topographically higher Hatton–Rockall basin may result from northeastward flow of sediment-laden intermediate-level Labrador Sea water.

## 4. The Iceland basin and the east flank of the Reykjanes ridge

This oceanic area lies between the Reykjanes ridge and the west-facing western slope of the Rockall plateau (Fig. 1). More than 1400 m of sediment, based on an assumed sediment velocity of 2 km/sec., is shown in this basin by Ewing *et al.* (1973). The ocean crust here is divided into two different segments by an old E–W-trending fracture zone located west of the south end of Hatton Bank; ocean floor to the north is thought to be younger than 60 m.y., and ocean floor to the south, as far as Charlie Gibbs fracture zone, is thought to be as old as 76 m.y. The western continental margin of the Rockall plateau gradually merges with a continental rise and is characterized by poorly developed shelf-like banks.

The ocean floor south of the Rockall plateau has an undulating relief characteristic of deposition of sediment drifts. Well developed moats encircle several seamounts, and several low sediment ridges extend southward from the Charlie Gibbs fracture zone. The pre-R4 sequence is draped over oceanic crust and was apparently partly derived from erosion of the Rockall plateau (Roberts 1975a). The post-R4 sequence consists of sediment drifts of variable thickness. Surface sediments on the north Reykjanes ridge are enriched in Fe, Mn, Cu, Cr and Zn, similar to sediments of other active spreading ridges (Horowitz 1974).

The east flank of the Reykjanes ridge is topographically relatively smooth and dominated by the broad Gardar ridge, a major sediment drift deposited by southward flowing Norwegian Sea bottom water (Johnson and Schneider 1969; Davies and Laughton 1972; Ruddiman 1972). The Gardar ridge is about 1000 km in length and contains many features, such as moating around seamounts and undulations, characteristic of bottom current deposition and erosion. The pre-R4 sequence thins and pinches out west of the Rockall plateau on oceanic crust that is 37 m.y. old. The post-R4 sediment drift of Gardar ridge is 1·8 km thick and the drift sediments grade laterally eastward into volcanogenic turbidites of the Maury channel system. Thinner drift sediments (1·0 km thick) cover the southwest flank of Rockall plateau. Roberts (1975a) suggested that sedimentation rates had in general increased into Pleistocene time.

The Maury channel system separates the two areas of sediment drift deposition (Fig. 1). It extends southwestward from the Faeroe Bank channel possibly as far as the Biscay abyssal plain, and follows the axis of greatest basin depth. It contains highly reflective turbidites 0·3 km thick, but is not bounded by prominent levees (Ruddiman 1972; Cherkis *et al.* 1973). The source of the volcanogenic sediments in the Maury channel system penetrated at D.S.D.P. Site 115 may be primarily the south slope of Iceland, where several submarine canyons are apparently tributary to Maury channel (Davies and Laughton 1972).

## 5. The Faeroe–Shetland channel

The Faeroe–Shetland channel is about 1200 m deep and is underlain by thinner crust than that beneath both the adjacent Scottish continental shelf and the Faeroe Islands block (Bott and Watts 1971). More than 2500 m of sediment, based on an assumed sediment velocity of 2 km/sec., has accumulated in the channel (Ewing *et al.* 1973). The floor of the channel contains flat-lying sediments that thicken northeastward toward the Norwegian Sea (Bott 1975b). The age and character of these sediments is not known, but because the channel probably opened by rifting at the same time as both the Rockall trough to the south and the deep sedimentary basin on the Vøring plateau east of the Vøring plateau escarpment to the north (Bott 1975b), the oldest sediments filling the channel may be of Mesozoic age.

Southward movement of cold Norwegian Sea bottom water through the channel and across the Wyville–Thomson and Ymir ridges to the Rockall trough (Ellett and Roberts 1973) and westward to the Northeast Atlantic Ocean (Crease 1965; Jones *et al.* 1970; Ellett and Martin 1973) has probably inhibited deposition in the channel, especially during postglacial time. This contention is supported by the thinness or absence of Holocene sediments both in the channel and on Wyville–Thomson ridge and the northern derivation of sediments in the channel (La Touche and Parra 1976).

## 6. The Iceland–Faeroe ridge

Sediment cover is very thin on the flat and smooth topped Iceland–Faeroe ridge, whose crest lies about 400 m below sea level; bedrock is exposed at the surface of the ridge in places. The ridge is separated from the topographically higher Iceland and Faeroe blocks by steep submarine scarps. D.S.D.P. Sites 336 and 352 were drilled on the northeastern and southwestern flanks of the ridge, respectively (Talwani *et al.* 1976). Basalt was reached at Site 336. The ridge is aseismic and its plateau basalts are thought by Schilling (1976) to have been extruded over a mantle plume or diapir.

Grab samples of surface sediments from the ridge crest contain large amounts of ice-rafted and volcanogenic material, increasing amounts of biogenic carbonate and silicate detritus with depth and distance from Iceland, and rare volcaniclastic turbidites (Horowitz 1974). The grain size of ridge-crest sediments increases toward Iceland. The sediments are also very poorly sorted (bimodal or trimodal), indicating a probable mixture of variably sized volcanic, glacial, and biogenic detritus. Metal concentrations in sediments near the ridge crest are not enriched, as they are on Reykjanes ridge (Horowitz 1974).

LaTouche and Parra (1976) concluded that during cold glacial intervals, Norwegian Sea water was prevented from moving south, resulting in deposition of locally derived sediment on the ridge and in the Northeast Atlantic Ocean. In contrast, during warm interglacial intervals and at the present time, fine grained sediment is deposited on and south of the ridge by southward-flowing sediment-laden Norwegian Sea water.

On the southwest side of the Iceland–Faeroe ridge at Site 352, at least 54 m of foraminiferal-rich glaciomarine sediment overlies middle Oligocene nannofossil ooze at least 68 m thick. On the northeast side of the ridge at Site 336, 159 m of glaciomarine sediment overlies a thick ( $>300$ m) sequence of Eocene to Oligocene nannofossil-poor mud and clay that in turn overlies a lateritic palaeosol developed on basalt of Eocene age. The lateritic palaeosol is about 21 m thick, and upward increases in Al, ferric iron, and Ti coupled with upward decreases in Si, ferrous iron, Mg, Ca, Na, and K, plus the mineralogical changes shown on Figure 6, prove the subaerial origin of the palaeosol (Nilsen and Kerr 1978b). The contrast in Oligocene microfossils on opposite sides of the ridge indicates that the ridge formed a barrier to southward migration of Norwegian Sea water in Oligocene time (Talwani *et al.* 1976).

The sea floor at Site 336 is situated about 400 m lower than the ridge crest, and

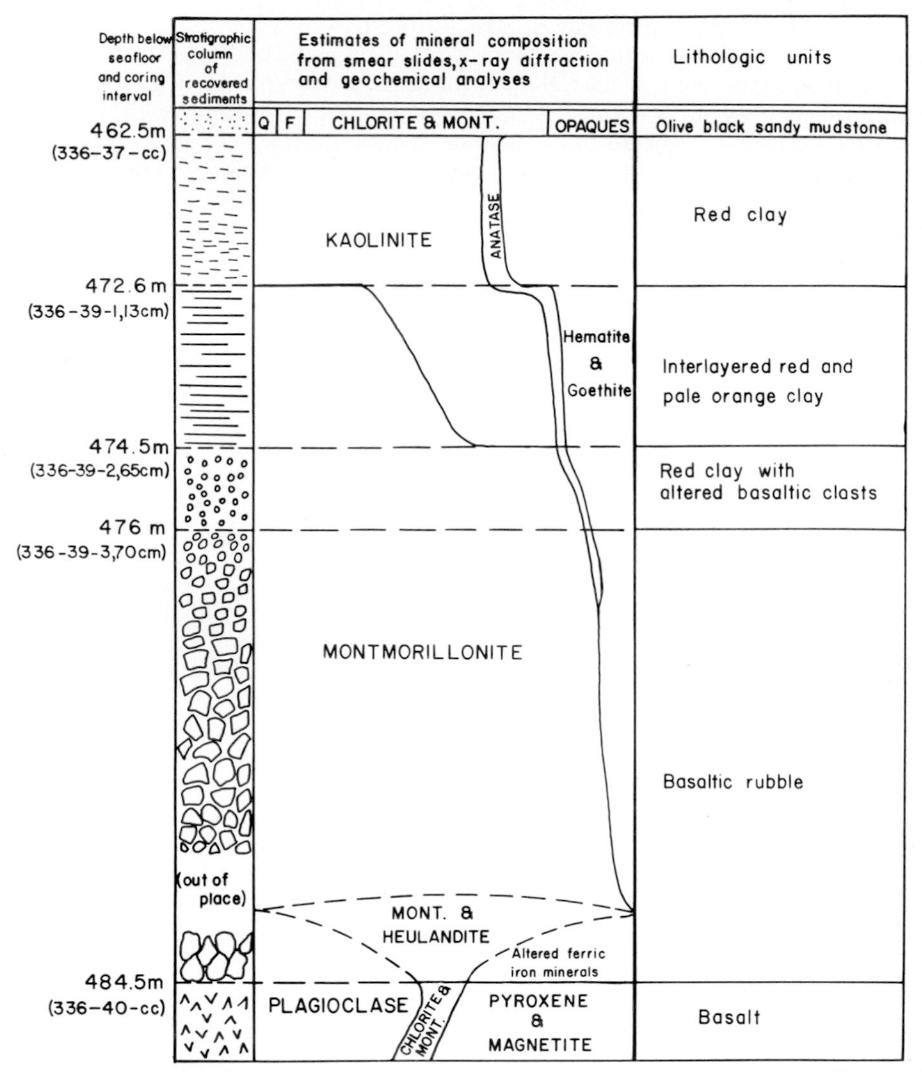


Fig. 6. General mineralogy of lateritic palaeosol from D.S.D.P. Site 336 (from Nilsen and Kerr 1978a).

the Eocene lateritic palaeosol is about 900 m lower than the ridge crest. Thus, at the time the subsiding ridge flank was at sea level, the ridge crest stood high above sea level. From 45 m.y. ago to 30 m.y. ago, the ridge flank subsided at a rate of 40 m/m.y., suggesting that the ridge crest did not subside below sea level until at least Oligocene time. Thus, the ridge probably formed a land bridge between Greenland and Europe prior to that time, permitting migration and movement of land mammals (Nilsen and Kerr 1978b). The ridge gradually subsided below sea level as active plateau volcanism shifted westward and Iceland developed astride the now extinct Icelandic plateau spreading ridge and the presently active Kolbeinsey ridge.

## 7. The Norway basin and the Icelandic plateau

This area contains two extinct spreading ridges, the Aegir ridge in the Norway basin and the Icelandic plateau ridge, and a continental fragment, the Jan Mayen ridge. The average thickness of sediments in the Norway basin is 0·84 km, whereas that on the higher Icelandic plateau to the west is 0·27 km (Eldholm and Windisch 1974). Most of the sediment cover in the Norway basin is of terrigenous origin, and the Aegir ridge is defined by a continuous sediment-filled rift valley (Eldholm and Windisch 1974). Palaeocene volcanogenic lutite was recovered in one piston core from the eastern Norway basin south of the Vøring plateau (Saito *et al.* 1967), an area of probable sea-floor erosion. The eastern flank of the Icelandic plateau contains a continuous sequence of sediment that acoustically is relatively transparent; other portions of the plateau are underlain by smooth acoustic basement that may consist of volcanic debris (Eldholm and Windisch 1974). Jan Mayen ridge is underlain by a thick sequence of turbidites that apparently was originally deposited at the East Greenland continental margin and subsequently rifted eastward. A thin cover of glaciomarine sediments unconformably overlies the turbidite sequence.

D.S.D.P. Site 337, drilled about 20 km east of the inferred extinct Aegir ridge axis, penetrated 113 m of sediment over 18–24 m.y. old basalt (Talwani *et al.* 1976). The sedimentary sequence consists of an upper 32 m of probable glaciomarine sediment and a lower, poorly dated 81 m of pelagic clay and interlayered nannofossil ooze of Oligocene and possibly younger age. The older pelagic sequence represents ridge-top deposition where bottom-seeking turbidity currents could not deposit sediment.

D.S.D.P. Site 348 was drilled on the Icelandic plateau between the active Kolbeinsey ridge and the continental Jan Mayen ridge, over the axis of the extinct Icelandic plateau spreading ridge. The sequence penetrated consists of 64 m of Pleistocene glaciomarine sediment, 138 m of Miocene to Pleistocene biogenic siliceous sediment with some terrigenous mud and clay, and 257 m of Oligocene(?) to lower Miocene terrigenous mudstone with some thin layers of nannofossil ooze. Scattered basalt pebbles are present near the base of the sedimentary sequence, which rests on basalt thought to be about 21 m.y. old. The thickness of the terrigenous part of the sedimentary sequence probably reflects relatively rapid sedimentation during the time that the ridge axis was close to the East Greenland continental margin; some sediment may also have come from erosion of the Jan Mayen ridge.

D.S.D.P. Sites 346, 347, 349 and 350 were drilled on the Jan Mayen ridge and on a smaller ridge extending southwards. Basement was penetrated only at Site 350, where 41 m.y. old basalt was recovered. The glaciomarine sequence is as thick as 91 m at these sites and overlies Oligocene and Miocene pelagic and hemipelagic sediment as thick as several hundred metres. Well-lithified Eocene turbidites and pebbly mudstones at least 200 m thick are below the pelagic sediments. These turbidites were deposited on the eastern continental margin of early Tertiary Greenland before westward shifting of the spreading ridge.

## 8. The Vøring plateau

The Vøring plateau lies west of the central Norwegian continental shelf at an average depth of 1450 m below sea level. The southwest boundary of the plateau is formed

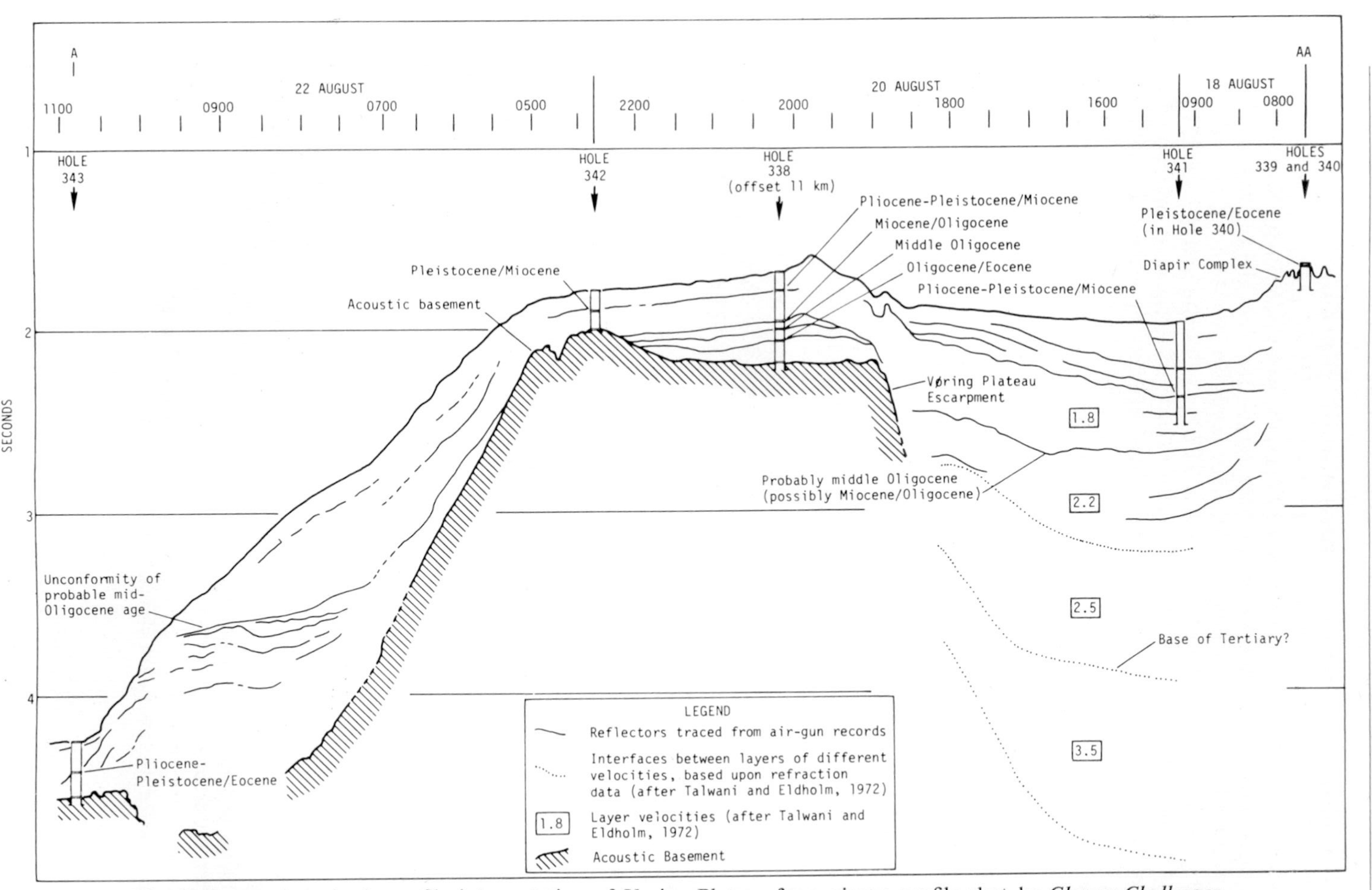


Fig. 7. Composite seismic profile interpretation of Vøring Plateau from airgun profile shot by *Glomar Challenger* between Sites 339/340 and 343 showing D.S.D.P. site positions, significant reflectors, and basement surface, which is emphasized by oblique shading (from Talwani *et al.* 1976). Refraction data (layers shown by dotted lines, velocities given in boxes) taken from Talwani and Eldholm (1972) and plotted on the assumption that the base of the 1·8 km/sec. layer is equivalent to the mid-Oligocene reflector.

by the Jan Mayen fracture zone. A buried escarpment (Vøring plateau escarpment) divides the plateau into an outer plateau underlain by basalt at a depth of less than 1 km, and an inner plateau underlain by as much as 8 km of sedimentary strata (Talwani and Eldholm 1972). The basement ridge west of the Vøring plateau escarpment has been interpreted as oceanic crust (Talwani and Eldholm 1972), a continental fragment (Hinz 1972), and as a volcanic plateau constructed over a mantle diapir (Vogt and Avery 1974). D.S.D.P. Sites 338 to 343 were drilled on the plateau, and the sedimentary geology of the plateau was summarized by Caston (1976).

Eocene terrigenous sediments, derived in part from erosion of basalt highs on the plateau, or alternatively from Greenland when it was closer to the Vøring plateau, were deposited at the western base of the plateau (Site 343) and on the plateau (Site 338). These sediments grade upward into pelagic siliceous oozes of middle Eocene age at Site 343 and of late Eocene age at Site 338 (Fig. 7). Oligocene, Miocene, and Pliocene pelagic sediments grade upward into poorly sorted Pliocene and Pleistocene muds and pebbly muds of glaciomarine origin. The topographically higher parts of the plateau west of the escarpment did not sink to a level at which sediment could accumulate until early Miocene time (Fig. 7). Evidence of volcanic activity is most prevalent in lower Eocene to middle Oligocene sediments, although ash layers are also common in the lower Miocene and Plio-Pleistocene.

The Vøring plateau escarpment was apparently formed by a major fault that has been active throughout Tertiary and probably Quaternary time (Caston 1976). The thick sedimentary sequence east of the escarpment at Site 341 contains a Pliocene and Pleistocene sequence of interlayered biogenic and terrigenous sediments and a middle and upper Miocene sequence of siliceous biogenic oozes. Diapirs of Eocene siliceous oozes containing montomorillonitic clay penetrated at Sites 339 and 340 along the eastern margin of the plateau probably developed as a result of density imbalances created by deposition of Pliocene and Pleistocene terrigenous sediments over older biogenic oozes.

## 9. The Lofoten basin

The average thickness of sediment in the Lofoten basin is 1·3 km (Eldholm and Windisch 1974). The sedimentary fill is divisible into two parts, a lower sequence of acoustically relatively transparent sediment that is generally conformable on oceanic basement, and an upper opaque turbidite sequence that is interlayered locally with transparent material (Eldholm and Windisch 1974). The sediment thickness increases northeastward toward the continental rise adjacent to the Barents Sea; basement is deeply buried beneath this part of the rise and cannot be seen in seismic profiles beneath the sedimentary sequence (Eldholm and Windisch 1974). Large deep sea fans have prograded westward and southward into Lofoten basin from this margin, and prograding sedimentary sequences have also been noted on the outer shelf and upper slope of the Barents Sea (Eldholm and Ewing 1971).

At the northwestern edge of the Lofoten basin, turbidites thin and are dammed against Mohns ridge, forming a narrow abyssal plain beyond the margins of the prograded deep sea fans. To the north, near the transition from the Mohns to the Knipovich ridge, northward-derived turbidites and continental slope sediments fill low areas in the crest of the spreading ridge, and some sediments have overflowed into the rift valley itself.

D.S.D.P. Site 345 was drilled at the southwestern corner of the Lofoten basin adjacent to the Jan Mayen fracture zone (Talwani *et al.* 1976). Above subaerially weathered basalt at this site is 431 m of upper Eocene and Oligocene turbidites that include pebbly mudstone and some chaotic deposits, 285 m of Oligocene and lower Miocene clay, mud, transitional silicic sediment and foraminiferal ooze, and 46 m of Pliocene(?) and Pleistocene glaciomarine mud and sandy mud with some interlayered foraminiferal ooze. The implications of this sequence for understanding the history of sedimentation in laterally infilled rifted-margin ocean basins is discussed in the introduction to this paper (Fig. 4). Biogenic sediments are not prominent constituents of the sequence until Oligocene and early Miocene time, when free movement of water between the North Atlantic and the Norwegian–Greenland Sea had been established. The glaciomarine sequence is relatively thin, reflecting the large distances between the site and adjacent continental margins.

D.S.D.P. Site 344 was drilled on a ridge along the east flank of the Knipovich ridge (Talwani *et al.* 1976). Drilling penetrated a 377·5 m thick sequence of Miocene or lower Pliocene to Pleistocene terrigenous mud, sandy mud, pebbly mud, and clay above a basalt dyke or sill dated at 3 m.y. Only the lowest 6 m of the sedimentary sequence, which contains turbidites, may not be of glaciomarine origin. The great thickness of glaciomarine sediments probably reflects the northern position of the site and abundance of sediment supplied from the nearby shelf margin west of Svalbard and the Barents Sea.

## 10. Summary and conclusions

The complex history of sea floor spreading in the Northeast Atlantic Ocean and Norwegian–Greenland Sea areas, combined with a shifting history of volcanic activity, intermittent southward movement of cold Norwegian–Greenland Sea water into the North Atlantic, and extensive glacial effects, have yielded a diverse and complex pattern of late Mesozoic and Cenozoic sedimentation. Virtually all of the major sedimentary facies characteristic of deep marine sedimentation are present in the two marine areas.

Although the sedimentary sequences in the two areas are probably basically similar to late Mesozoic rifted-margin basin deposits of the North and South Atlantic Oceans, there are several major differences: (1) evaporites, which characterize deposits in the early stages of rifting in the Atlantic further to the south, apparently did not form, possibly because climatic conditions were different, the newly formed ocean basins subsided too rapidly to permit deposition of evaporites in marginal marine environments, or the basins were not rimmed by sills controlling inflow and outflow of water; (2) the ocean basin northward is much narrower, shallower, and the bordering continental areas are much higher than to the south in the Atlantic, so that terrigenous rather than biogenic sedimentation is most prominent; (3) the location of the areas in more northern latitudes resulted in more widespread deposition of ice-rafted glaciomarine sediments in Pliocene and Pleistocene time; (4) southward flow of sediment-laden Norwegian Sea bottom water and northeastward flow of sediment-laden Labrador Sea water at intermediate depths have yielded more extensive deposition of longitudinal sediment drifts than to the south; (5) mid ocean channel or canyon systems, in this area represented by Maury channel, may be more extensive in the northern areas, and (6) deep sea fans adjacent to continental margins are larger and more prominently developed in the northern areas, whereas abyssal plains are larger and more extensive further to the south where the ocean floor is wider and older. The shallowness of ocean basins in the North Atlantic Ocean north of the Charlie Gibbs fracture zone and in the Norwegian–Greenland Sea compared to ocean basins in other parts of the world, and the greater relief of the surrounding land masses compared to land masses surrounding other ocean basins, may possibly be caused by extensive mantle plume or hot spot activity (Vogt and Avery 1974). Thus the unusual type and history of sea floor spreading in this area have contributed importantly to the development of a somewhat unique record of oceanic sedimentation.

Initial opening of the Rockall trough in Late Jurassic(?) or Early Cretaceous time resulted in deposition of terrigenous turbidites at the margins of the trough and transparent pelagic sediments over topographically higher crust in the central part of the trough. Later subsidence and warping of the Rockall plateau diminished the supply of terrigenous sediment to the western margin of the trough, but yielded deposition of acoustically transparent sediments in the Hatton–Rockall basin on the plateau (Fig. 5). The Hatton–Rockall basin gradually subsided through Cenozoic time, accompanied by Eocene volcanism (Roberts 1969). During an interval of possibly closed circulation in the North Atlantic, Eocene and Oligocene cherty oozes were deposited in the trough and on the plateau. Following an episode of regional warping, subsidence, uplift, and faulting in Oligocene time, the modern depositional systems in the trough and adjacent areas developed. Terrigenous turbidites, including large deep sea fans fed by canyons cut into the Irish and Scottish continental slopes, filled the eastern part of Rockall trough, and sediment drifts fed by southward-flowing Norwegian Sea bottom water developed into the Feni ridge and filled the western part of the trough. Sediment drifts, probably derived from northeastward-flowing Labrador Sea water at intermediate depths, were deposited in the Hatton–Rockall basin above the Eocene and Oligocene cherty oozes (Roberts 1975a).

Maury channel developed west of the Rockall plateau in the deepest part of the Iceland basin. The channel extends southward for more than 1000 km and may cross the Charlie Gibbs fracture zone, terminating in Biscay abyssal plain. Volcanogenic detritus is transported southeastward along the channel system from the south edge of Iceland and from the Faeroe Islands block (Davies and Laughton 1972; Ruddiman and Glover 1972; Roberts 1975a). To the west, the 1000 km along Gardar ridge sediment drift infills the mid ocean ridge topography of the east flank of the Reykjanes ridge (Johnson and Schneider 1969; Davies and Laughton 1972; Ruddiman 1972). The Gardar ridge developed on post-40 m.y. ocean crust by deposition of fine grained sediments transported through the Faeroe–Shetland channel and the Faeroe Bank channel to the Iceland basin by southeastward flowing Norwegian Sea bottom water.

The Iceland–Faeroe ridge formed a barrier to southward flow of Norwegian Sea water in early Tertiary time. The flat top of the ridge is almost sediment-free, reflecting rapid movement of water across the ridge crest in modern times. Following subsidence of the ridge and establishment of southward flow of Norwegian Sea water around the Faeroe Islands block, extensive sediment drifts were deposited to the south on the Feni and Gardar ridges. Lateritic palaeosols of probable Eocene age on the northeastern flank of the ridge indicate that the ridge stood above sea level during early Tertiary time. The ridge probably formed a land bridge connecting western Europe to Greenland–North America at this time (Nilsen and Kerr 1978b).

A major unconformity separating Palaeogene sediments from glaciomarine sediments probably resulted from non-deposition or erosion on the ridge during the time that it subsided below sea level and the flow of water between the Norwegian Sea and North Atlantic commenced.

In the Norwegian Sea, sediment fill is relatively thin because of the number of ridges and the elevation of the Icelandic plateau. Pelagic sediments were draped over over the extinct Aegir and Icelandic plateau spreading ridges after the ridges subsided. Eventually the ridges were completely buried by sediment, major components of which are terrigenous clay and silt derived from the Greenland or Norwegian continental margins when the Norwegian–Greenland Sea was much narrower. The continental fragment that forms the Jan Mayen ridge is underlain by well-lithified lower Tertiary turbidites probably deposited originally on the East Greenland continental margin. These turbidites are overlain unconformably by Oligocene and Miocene pelagic and hemipelagic sediments and Pliocene and Pleistocene glaciomarine sediments.

The Vøring plateau contains dominantly pelagic sediments on top of basalts on its western part and thicker pelagic sediments in the deep inner basin on the eastern part of the plateau (Caston 1976). The inner part of the plateau may contain as much as 8 km of sediment fill, including sediments of Mesozoic age; it may have developed at the same time as Rockall trough by similar rifting processes.

The Lofoten basin contains thick prograding deep sea fan deposits and basin-plain turbidites derived largely from the Barents Sea marginal slope. The sediment fill overlies subaerially weathered basalt and consists of, in ascending order, pebbly mudstone, turbidites, terrigenous mud, biogenic pelagic clay, and terrigenous glaciomarine deposits. The sedimentary sequence in Lofoten basin most typically reflects the sequence of events associated with basin-floor filling of rifted-margin ocean basins that are dominated by lateral infilling of terrigenous sediments derived from marginal continental source areas.

*Acknowledgments.* I thank my shipboard colleagues on D.S.D.P. Leg 38, particularly the sedimentologists, for assistance in many aspects of this research and for valuable discussions. D. R. Kerr, formerly of the U.S. Geological Survey, completed extensive work on the lateritic palaeosols recovered at D.S.D.P. Hole 336. I thank J. Gardner and J. Yount of the U.S. Geological Survey for providing helpful reviews of this paper.

## References

Bailey, R. J. 1975. Sub-Cenozoic geology of the British continental margin (lat. 50°N to 57°N) and the reassembly of the North Atlantic late Paleozoic supercontinent. *Geology* **3,** 591–594.

Berggren, W. A. and Hollister, C. D. 1974. Paleogeography, paleobiogeography and the history of circulation in the Atlantic Ocean. *In* Hay, W. W. (Editor). *Studies in paleo-oceanography,* 126–186. *Soc. Econ. Paleontologists and Mineralogists Spec. Publ.* **20.**

Bott, M. H. P. 1973. The evolution of the Atlantic north of the Faeroe Islands. *In* Tarling, D. H. and Runcorn, S. N. (Editors). *Implications of continental drift to the earth sciences* **1,** 175–189. Academic Press, London and New York.

—— 1974. Deep structure, evolution and origin of the Icelandic transverse ridge. *In* Kristjansson, L. (Editor). *Geodynamics of Iceland and the North Atlantic area,* 33–47. Reidel Publ. Co., Dordrecht and Boston.

Bott, M. H. P. 1975a. Structure and evolution of the North Scottish Shelf, the Faeroe Block and *of* the intervening region. *In* Woodland, A. W. (Editor). *Petroleum and the continental shelf North-West Europe,* **1,** *Geology,* 105–113. Applied Science Publisher, Barking.

—— 1975b. Structure and evolution of the Atlantic floor between northern Scotland and Iceland. *In* Whiteman, A., Roberts, D. and Sellevoll, M. A. (Editors). *Petroleum geology and geology of the North Sea and northeast Atlantic continental margin,* 195–199. *Norges Geol. Unders.* **316.**

—— and Watts, A. B. 1970. Deep sedimentary basins proved in the Shetland–Hebridian continental shelf and margin. *Nature, Lond.* **225,** 265–268.

Caston, V. N. D. 1976. Tertiary sediments of the Vøring Plateau, Norwegian Sea, recovered by Leg 38 of the Deep Sea Drilling Project. *In* Talwani, M. *et al. Initial Reports of the Deep Sea Drilling Project* **38,** 761–782. U.S. Govt. Printing Off., Washington.

Cherkis, N. Z., Fleming, H. S. and Feden, R. H. 1973. Morphology and structure of Maury Channel, North East Atlantic Ocean. *Geol. Soc. Am. Bull.* **84,** 1601–1606.

Crease, J. 1965. The flow of Norwegian Sea Water through the Faeroe Bank Channel. *Deep Sea Res.* **12,** 143–150.

Davies, T. A. and Laughton, A. S. 1972. Sedimentary processes in the North Atlantic Ocean. *In* Laughton, A. S. *et al. Initial Reports of the Deep Sea Drilling Project* **12,** 905–934. U.S. Govt. Printing Off., Washington.

Dietrich, G. 1967. The International "Overflow" Expedition (ICES) of the Iceland–Faeroe Ridge, May–June 1960—a review. *Rapp. Proces-Verb. Reunions Cons. Perm. Int. Explor. Mer.* **157,** 268–275.

Eldholm, O. and Ewing, M. 1971. Marine geophysical survey in the southwestern Barents Sea. *J. geophys. Res.* **76,** 3832–3841.

—— and Windisch, C. C. 1974. Sediment distribution in the Norwegian–Greenland Sea. *Geol. Soc. Am. Bull.* **85,** 1661–1676.

Ellett, D. J. and Martin, J. H. A. 1973. The physical and chemical oceanography of the Rockall Channel. *Deep-Sea Res.* **20,** 585–625.

—— and Roberts, D. G. 1973. The overflow of Norwegian Sea Deep Water across the Wyville–Thomson Ridge. *Deep-Sea Res.* **20,** 819–835.

Ericson, D. B., Ewing, M. and Wollin, G. 1964. Sediment cores from the Arctic and subarctic. *Science* **144,** 1183–1192.

——, Ewing, M., Wollin, G. and Heezen, B. C. 1961. Atlantic deep-sea cores. *Geol. Soc. Am. Bull.* **72,** 193–285.

Ewing, M., Carpenter, G., Windisch, C. C. and Ewing, J. 1973. Sediment distribution in the oceans—the Atlantic. *Geol. Soc. Am. Bull.* **84,** 71–87.

Hallam, A. and Sellwood, B. W. 1976. Middle Mesozoic sedimentation in relation to tectonics in the British area. *J. Geol.* **84,** 301–321.

Hinz, K. 1972. Der Krustenaufbau des Norwegischen Kontinentalrandes (Vøring Plateau) under der Norwegischen Tiefsee zwischen 66° and 68°N nach seismichen Untersuchungen. "*Meteor*" *Forschung.* **10,** 1–16.

Horowitz, A. 1974. The geochemistry of sediments from the northern Reykjanes Ridge and the Iceland–Faeroes Ridge. *Mar. Geol.* **17,** 103–122.

Johnson, G. L. and Schneider, E. D. 1969. Depositional ridges in the North Atlantic. *Earth Planet. Sci. Lett.* **6,** 416–422.

Jones, E. J. W., Ewing, J. I. and Eittreim, S. L. 1970. Influences of Norwegian Sea overflow water on sedimentation in the northern North Atlantic and Labrador Sea. *J. geophys. Res.* **75,** 1655–1680.

Kent, P. E. 1978. Mesozoic vertical movements in Britain and the surrounding continental shelf. *In* Bowes, D. R. and Leake, B. E. (Editors). *Crustal evolution in northwestern Britain and adjacent regions*, 309-324. *Geol. J. Spec. Issue* No. 10.

LaTouche, C. and Parra, M. 1976. Mineralogie et geochimie des sediments Quaternaires de l'Ocean Atlantique nord–oriental (Mer de Norvege–Golfe de Gascogne)—essais d'interpretations sedimentologiques. *Mar. Geol.* **22,** 33–69.

Laughton, A. S. 1975. Tectonic evolution of the northeast Atlantic Ocean—a review. *In* Whiteman, A., Roberts, D. and Sellevoll, M.A. (Editors). *Petroleum geology and geology of the North Sea and northeast Atlantic continental margin,* 169–193. *Norges Geol. Unders.* **316.**

—— *et al.* 1972. *Initial Reports of the Deep Sea Drilling Project,* **12.** U.S. Govt. Printing Off., Washington.

LePichon, X. and Fox, P. J. 1971. Marginal offsets, fracture zones, and the early opening of the North Atlantic. *J. geophys. Res.* **76,** 6294–6308.

McLean, A. C. 1978. Fault-controlled ensialic basins in northwestern Britain. *In* Bowes, D. R. and Leake, B. E. (Editors). *Crustal evolution in northwestern Britain and adjacent regions*, 325–346. *Geol. J. Spec. Issue* No. 10.

Montadert, L. *et al.* 1976. Glomar Challenger sails on Leg 48. *Geotimes* **21,** 19–23.

Moorbath, S., Sigurdsson, H. and Goodwin, R. 1968. K–Ar ages of the oldest exposed rocks in Iceland. *Earth Planet. Sci. Lett.* **4,** 197–205.

NILSEN, T. H. and KERR, D. R. 1978a. Turbidites, redbeds, sedimentary structures, and trace fossils observed in DSDP Leg 38 cores and the sedimentary history of the Norwegian–Greenland Sea. *In* Talwani, M. *et al. Initial Reports of the Deep Sea Drilling Project,* Supplement **38.** U.S. Govt. Printing Off., Washington.
—— 1978b. Paleoclimatic and paleogeographic implications of a lower Tertiary laterite (latosol) on the Iceland–Faeroe Ridge, North Atlantic region. *Geol. Mag.* **115.**
PEGRUM, R. M., REES, G. and NAYLOR, D. 1975. *Geology of the North-West European continental Shelf,* **2,** *The North Sea.* Graham Trotman Dudley Ltd., London.
PITMAN, W. C. III and HERRON, E. M. 1974. Continental drift in the Atlantic and the Arctic. *In* Kristjansson, L. (Editor). *Geodynamics of Iceland and the North Atlantic area,* 1–15. Reidel Publ. Co., Dordrecht and Boston.
—— and TALWANI, M. 1972. Sea-floor spreading in the North Atlantic. *Geol. Soc. Am. Bull.* **83,** 619–646.
ROBERTS, D. G. 1969. New Tertiary volcanic centre on the Rockall Bank, eastern North Atlantic Ocean. *Nature, Lond.* **223,** 819–820.
—— 1971. New geophysical evidence on the origin of the Rockall Plateau and Trough. *Deep-Sea Res.* **18,** 353–360.
—— 1975a. Marine geology of the Rockall Plateau and Trough. *Phil. Trans. R. Soc. Lond.* (A) **278,** 447–509.
—— 1975b. Tectonic and Stratigraphic Evolution of the Rockall Plateau and Trough. *In* Woodland, A. W. (Editor). *Petroleum and the continental shelf of North–West Europe* **1,** *Geology,* 77–91. Applied Science Publishers, Barking.
—— HOGG, N. H., BISHOP, D. G. and BINNS, P. 1973. Sediment distribution around moated seamounts in the Rockall Trough. *Deep-Sea Res.* **21,** 175–184.
RUDDIMAN, W. B. 1972. Sediment redistribution on the Reykjanes Ridge—seismic evidence. *Geol. Soc. Am. Bull.* **83,** 2039–2062.
—— and GLOVER, L. K. 1972. Vertical mixing of ice-rafted volcanic ash in North Atlantic sediments. *Geol. Soc. Am. Bull.* **83,** 2817–2835.
SAITO, T., BURCKLE, L. H. and HORN, D. R. 1967. Paleocene core from the Norwegian Basin. *Nature, Lond.* **216,** 357–359.
SCHILLING, J.-G. 1973. Iceland mantle plume. *Nature, Lond.* **246,** 141–143.
—— 1976. Rare-earth, Sc, Cr, Fe, Co, and Na abundances in DSDP Leg 38 basement basalts—some additional evidence on the evolution of the Thulean Volcanic Province. *In* Talwani, M. *et al. Initial Reports of the Deep Sea Drilling Project* **38,** 741–750. U.S. Govt. Printing Off., Washington.
SCRADER, H. J., BJORKLUND, K., MANUM, S., MARTINI, E. and VAN HINTE, J. 1976. Cenozoic biostratigraphy, physical stratigraphy and paleo-oceanography in the Norwegian–Greenland Sea, DSDP Leg 38 paleontological synthesis. *In* Talwani, M. *et al. Initial Reports of the Deep Sea Drilling Project,* **38,** 1197–1211. U.S. Govt. Printing Off., Washington.
STEELE, J. H., BARRETT, J. R. and WORTHINGTON, L. V. 1962. Deep currents south of Iceland. *Deep-Sea Res.* **9,** 465–474.
TALWANI, M. and ELDHOLM, O. 1972. Continental margin off Norway—a geophysical study. *Geol. Soc. Am. Bull.* **83,** 3575–3606.
—— and UDINTSEV, G. 1976. Tectonic synthesis. *In* Talwani, M. *et al. Initial Reports of the Deep Sea Drilling Project* **38,** 1213–1242. U.S. Govt. Printing Off., Washington.
—— *et al.* 1976. *Initial Reports of the Deep-Sea Drilling Project,* **38.** U.S. Govt. Printing off., Washington.
VOGT. P. R. 1972. Evidence for global synchronism in mantle plume convection, and possible significance for geology. *Nature, Lond.* **240,** 338–342.
—— and AVERY, O. E. 1974. Tectonic history of the Arctic basins—partial solutions and unsolved mysteries. *In* Herman, Y. (Editor). *Marine Geology and Oceanography of the Arctic seas,* 83–117. Springer-Verlag, New York.
WHITE, S. M. 1978. Sediments of the Norwegian–Greenland Sea, DSDP Leg 38. *In* Talwani, M. *et al. Initial Reports of the Deep Sea Drilling Project,* Supplement **38.** U.S. Govt. Printing Off., Washington.
WHITEMAN, A., ROBERTS, D. and SELLEVOLL, M. A. (Editors) 1975. Petroleum geology and geology of the North Sea and northeast Atlantic continental margin. *Norges Geol. Unders.* **316,** 367 pp.
WOODLAND, A. W. (Editor) 1975. *Petroleum and the continental shelf of North-West Europe* **1,** *Geology.* Applied Science Publishers, Barking.
WORTHINGTON, L. V. 1970. The Norwegian Sea as a Mediterranean basin. *Deep-Sea Res.* **17,** 77–84.
—— and VOLKMANN, G. H. 1969. The volume transport of Norwegian Sea overflow water in the North Atlantic. *Deep-Sea Res.* **12,** 667–676.

Tor H. Nilsen,
U.S. Geological Survey,
Menlo Park,
California 94025, U.S.A.

# Author index

Numbers in bold type refer to pages on which references are listed and those followed by an asterisk refer to text figures.

# Place index

# Subject index